Innovations in Landscape Research

Series Editor

Lothar Mueller, Leibniz Centre for Agricultural Landscape Research (ZALF), Müncheberg, Brandenburg, Germany

Aims & Scope

The Springer series "Innovations in Landscape Research" presents novel methodologies and technologies to understand, monitor and manage landscapes of the Anthropocene. The aim is to achieve landscape sustainability at high productivity. This includes halting degradation of landscapes and their compartments, developing cultural landscapes, and preserving semi-natural landscapes. Clean water and air, fertile and healthy soils for food and other ecosystem services, and a green and bio-diverse environment are attributes of landscapes for the survival and well-being of humans who inhabit them.

How do landscapes function? How do future landscapes look like? How can we sustainably develop intensively used and stressed kinds of landscapes? Scientific innovations and decision tools are key to answer and solve those challenging questions. The series will inform about advanced methods and results of disciplinary, interdisciplinary and transdisciplinary work in landscape research. It presents a broad array of methods to measure, assess, forecast, utilize and control landscapes and their compartments. These include field and laboratory measurement methods, methods of resource evaluation, functional mapping and risk assessment, and sensing methods for landscape monitoring, advanced methods for data analysis and ecosystem modeling, methods and technologies for optimizing the use of multi-functional landscapes, for the bioremediation of soil and water, and basics and procedures of landscape planning. The series provides a new view on landscapes with some focus on scientific and technological innovations, on soils and problems of optimizing agricultural landscapes under conditions of progressive urbanization. Landscape research in a globalized world of the Anthropocene is based on gathering big data and scenario modeling. International long-term experiments and agri-environmental monitoring systems will deliver data for ecosystem models and decision support systems.

Edited volumes of this series will address the following topics at high priority: Status and Trends of Landscape Research; Understanding Key Landscape Processes; Landscape Services, Functions and Biodiversity; Assessing Soil Resources and Quality; Water Resource and Quality Monitoring; Landscape Monitoring Concepts and Studies; Landscape Sensor and Monitoring Technologies; Landscape Modeling and Decision Support; Agricultural Soil and Plant Management; Basics and Tools for Landscape Planning; Tools for Water and Wetland Management; Forest Management and Agroforestry; Rehabilitation of Degraded Landscapes.

The books of this series are a source of information for researchers, teachers, students, and stakeholders interested in the topics of landscape science and related disciplines. They present status analyses, methodical chapters and case studies showing the practical relevance and feasibility of novel decision tools and technologies. Thus, the books of this series will be a particular valuable information basis for managers and decision makers at various levels, from local up to international decision bodies.

An author/editor questionnaire, instructions for authors and a book proposal form can be obtained by contacting the Publisher.

More information about this series at http://www.springer.com/series/16118

Lothar Mueller · Frank Eulenstein
Editors

Current Trends in Landscape Research

Springer

Editors
Lothar Mueller
Guest Scientist
Leibniz Centre for Agricultural Landscape Research (ZALF)
Müncheberg, Brandenburg, Germany

Frank Eulenstein
Leibniz Centre for Agricultural Landscape Research (ZALF)
Müncheberg, Brandenburg, Germany

ISSN 2524-5155 ISSN 2524-5163 (electronic)
Innovations in Landscape Research
ISBN 978-3-030-30068-5 ISBN 978-3-030-30069-2 (eBook)
https://doi.org/10.1007/978-3-030-30069-2

This Springer imprint is published by the registered company Springer Nature Switzerland AG
The registered company address is: Gewerbestrasse 11, 6330 Cham, Switzerland

Editors' Preface

Dear Reader,

We are pleased that the title of this book and the contents of some of its chapters have caught your interest. The subject is, of course, landscapes and their exploration. If you are about to click straight on to the next page, or put the book back down, you should know that our titles and topics are more than just scientific talk about regional and local problems caused by tensions between humans and nature. They examine existential questions of mankind from different perspectives. These questions and challenges are always related to specific regions and landscapes; to their functions and values for us.

Landscapes emerged as a result of cosmic and geological processes. This is the basis on which soils, water, flora, fauna, diverse ecosystems and we ourselves (as the dominant species) could develop. Landscapes are crucial for our existence, our plans and our actions. We have used and designed them. They are a living archive that bears witness to our cultural history. A variety of landscapes shape the present face of our planet and secure our existence and well-being. Landscapes will be crucial factors in our survival as a species. Their condition can be used as an indicator of how good the chances are for our descendants.

What are the Framework Conditions and Motivations for This Book?

The grave problems and conditions facing humans and nature in the twenty-first century are basically well known. All processes of nature are subject to permanent change, hence the constant changes to our planet's landscapes. However, the existential problems of today are man-made. These problems consist in a crowded world with a deficient world order, with extreme gradients in terms of ensuring basic human needs and rights, with unresolved social, economic, ethno-cultural, and religious issues and conflicts. This affects humanity and nature. Effects include the increasing degradation of our resources of fresh water, healthy food, fertile soil, clean air, energy and raw materials. These entail the irreversible disruption of food webs and natural cycles, the extinction of species and the man-made aspects of climate change.

People's fundamental rights have been addressed in the United Nations Sustainable Development Goals, which are considered a compass for all responsible citizens and decision-makers. In terms of landscapes, the European Landscape Convention already provides good guidelines for responsible action. The implementation of such goals and conventions requires wise decisions at all levels. At the regional level, the problems and conflicts arising from tension between humans and nature are very specific. Today's landscapes are multifunctional and networked in a globalized world. Depending on a landscape's natural features and the desired development goals, the functional mechanisms and gaps in our knowledge can be addressed.

The state and development of landscapes can and must be explored. Knowledge has to be disseminated and understood. Only then can improvements be made for the benefit of man and nature. As editors, we see it as our mission to support this process of sharing knowledge about solutions for more sustainable landscapes.

What Can Landscape Research Do?

Landscape research can significantly assist the process of knowledge generation. This includes coming up with scientific and technological innovations to better understand, monitor and manage landscapes, and making responsible decisions about them that must be based on ethical and humanistic principles. Landscape research creates insights and solutions at the regional, national and international level. Many of the decision-making algorithms, models, technical innovations and environmental standards that have been developed are transferable and only need to be refined locally.

We would like to go into more detail. What general criteria and indicators can be used to explore the characteristics of phenomena and processes in landscapes and identify landscape functions? How should reliable, efficient monitoring systems for essential landscape features be designed? How do we perceive landscapes and how can we shape or preserve them as sustainably as possible? How can decision-making processes at the landscape level be made professionally, involving dialogue between science, citizens, decision-makers and other stakeholders? Our aim is to address these problems and communicate current knowledge.

Many scientific book projects end by addressing and analyzing problems. Many offer impressive proof of how mankind can destroy nature and eventually the foundations of our own existence. That is right. However, such projects often give the impression that there are no alternatives or solutions.

Why the New Springer Book Series "Innovations in Landscape Research" and Why This Book?

Within our series of books, and in this particular book, we would like to go one step further: to reveal concepts and solutions. That is not possible without reviewing existing knowledge about landscapes and their exploration, analyzing basic conditions and providing status analyses and definitions. This book, "Current Trends in Landscape Research", is dedicated to that task. Other books in this series will then deal more with specific sub-problems of landscape research and highlight the latest developments. Scientific and technological innovations and information technologies are needed more than ever to keep a finger on the pulse of the landscape and to initiate sustainable solutions.

Landscape research—and this book—can help to overcome boundaries and barriers, especially linguistic ones. One main aim is for knowledge to be exchanged within the international community of scientists. Language barriers have prevented the broad exchange of novel ideas and technologies across the globe. Russia has great traditions in landscape research. Numerous contributions by Russian authors reveal the advantages to be gained by improving the East-West dialogue in science. The book will benefit many authors by encouraging them to improve the quality of their disciplinary work and launch inter- and trans-disciplinary projects for landscape sustainability.

What Is the Basic Content of This Book and What Key Message Do We Want to Convey?

The book consists in 4 parts:

1 The Essence and Mission of Landscape Research
2 Concepts for Landscape Assessment and Evolvement
3 Tools for Landscape Planning
4 Landscape Characterization: International Case Studies

Every part contains 5–9 chapters written by invited authors.

All in all, we would like to inform you about the interesting and important contributions that landscape research can make to leave us and our descendants with a planet where pleasant, meaningful life is possible.

This information is not just for decision-makers and responsible citizens. It also applies to scientists who, following the mainstream, only publish within their specialty and no longer have time to read. As a result, they cannot understand ecosystems and barely communicate with colleagues from other disciplines.

Landscape research implies a focus on regional studies. However, there are increasing numbers of international joint projects and well-networked scientists from international teams who can develop and implement innovative, sustainable solutions in interdisciplinary and transdisciplinary work.

Comments and Thanks to Authors and Supporters

To develop new solutions and projects, different approaches, perspectives and opinions are required. This is ensured in the book by its internationally experienced, innovative authors and research teams from different disciplines and from different regions of the world. They are responsible for the content of their chapter, and are free to point out aspects of their study from their individual perspective. It is desirable and natural for their approaches and opinions to diverge. Consequently, as editors, we also accept approaches and conclusions that are not shared by us in every technical detail.

In some chapters, trade names are used to provide specific information. Mentioning a trade name does not constitute a guarantee of the product by the authors or editors. Neither does it imply an endorsement by the authors or editors of comparable products that are not named.

The editors would like to thank all the book's authors, contributors and supporters for their work and engagement. It was our pleasure to serve as editors by coordinating and reviewing the manuscripts of highly motivated scientists drawing on considerable expertise.

Ms. Anne Koth (Dresden) provided proofreading for some chapters with care and scientific expertise. We would like to thank Springer International for picking up the topic and publishing this book. Special thanks also to Michael Leuchner (Aachen). While working for Springer, he was the publishing editor responsible for launching the book series "Innovations in Landscape Research". The cooperation with Robert Doe and his team as the publishing editor for this book was also constructive and smooth.

Finally, we are aware that the publication of this book can only offer a snapshot of a broad and diverse topic. We are hopeful that you will gain some information and inspiration for your own work. You are encouraged to find an individual, optimum approach for drawing conclusions and acting imaginatively.

Müncheberg and Letschin/Sydowswiese, Germany — Lothar Mueller
Müncheberg and Paulinenaue, Germany — Frank Eulenstein
April 2019

Contents

Editors and Contributors

About the Editors

Lothar Mueller is a retired scientist from the Leibniz Centre for Agricultural Landscape Research (ZALF) in Muencheberg, Germany.

Lothar Mueller studied Agriculture at the Humboldt University in Berlin and graduated from the Academy of Agricultural Science of the GDR (Dr. agr.) in 1974. He has worked at the Research Centre of Soil Fertility (1974–1991), and the ZALF Institutes of Soil Landscape Research and Landscape Hydrology (1991–2014).

His strongest background is in soil science and agriculture. Jointly with an international team he developed the Muencheberg Soil Quality Rating, an acknowledged approach for assessing the quality of global soils for cropping and grazing. Over the past 5 years his scientific interest and international network was broadened on the field of landscape monitoring and management.

Lothar Mueller has published more than 350 papers, amongst them more than 100 peer-reviewed publications and book chapters. Since 2013 he has been an Honorary Professor at the State Agricultural University of the Kuban in Krasnodar (Russian Federation). He was appointed as an Honorary Doctor of the Russian Research Institute of Rice in Krasnodar, of the Pryanishnikov Institute for Agrochemistry in Moscow and of the Institute for Soil Science and Agrochemistry in Almaty, Kazakhstan.

Frank Eulenstein is a scientist at the Leibniz Centre for Agricultural Landscape Research (ZALF) in Muencheberg, Germany. He studied Agriculture at the Georg August University of Göttingen and graduated there at the Institute for Soil Science in 1992 (Dr. sc. agr.).

Since 1993 he has worked as scientist at the ZALF Institute of Land Use Systems. His fields of interest are water, matter and energy fluxes in agricultural systems and the practical questions of cropping and fertilisation. Currently, he works in the Research Area "Land Use and Governance", workgroup "Sustainable Grassland Systems" of ZALF.

Frank Eulenstein has published more than 300 scientific papers. He has more than 20 years experience as a manager (CEO) in the segments of the development of organic fertilizers and classical agricultural trade.

Since 2013 he has been an Honorary Professor at the State Agricultural University of the Kuban in Krasnodar (Russian Federation). He was appointed as an Honorary Doctor of the Russian Research Institute of Rice in Krasnodar and of the Institute for Soil Science and Agrochemistry in Almaty, Kazakhstan.

Contributors

Dana Ali Department of Landscape Design and Ecosystem Management, Faculty of Agricultural and Food Sciences, American University of Beirut, Beirut, Lebanon

Per Angelstam School for Forest Management, Faculty of Forest Sciences, Swedish University of Agricultural Sciences (SLU), Skinnskatteberg, Sweden

Marc Antrop Landscape Research—Department of Geography, Ghent University, Ghent, Belgium

Debbie Bartlett Faculty of Engineering and Science, University of Greenwich, Gillingham, Kent, UK

Nicole Bauer Swiss Federal Research Institute for Forest, Snow and Landscape WSL, Birmensdorf, Switzerland

Viktoria R. Bityukova Faculty of Geography, Department of Economic and Social Geography of Russia, Lomonosov Moscow State University, Leninskie Gory, Moscow, Russia

Winfried Blum Institute of Soil Research, University of Natural Resources and Life Sciences, Vienna, Austria

Vladimir Bocharnikov Far-Eastern Branch, Pacific Geographical Institute, Russian Academy of Sciences, Vladivostok, Russia

Janine Bolliger Swiss Federal Research Institute for Forest, Snow and Landscape WSL, Birmensdorf, Switzerland

Lluís Brotons Centre Tecnològic Forestal de Catalunya (CEMFOR-CTFC), Solsona, Spain;
CREAF, Cerdanyola del Vallés, Spain;
CSIC, Cerdanyola del Vallés, Spain

Subhan Arif Budiman Department of Soil Science, University of Jember, Jember, Indonesia

Elena Bukvareva Biodiversity Conservation Center, Moscow, Russia

Robert M. Campellone National Wildlife Refuge System, U.S. Fish and Wildlife Service, Falls Church, VA, USA

Kristina M. Chouinard Migratory Bird Program/National Wildlife Refuge System, U.S. Fish and Wildlife Service, Jackson, TN, USA

Denis Couvet UMR MNHN-CNRS-SU 'CESCO', Paris, France

Fatima Cruz Sustainable Forest Management Research Institute, University of Valladolid, Palencia, Spain

Ralf Dannowski Leibniz Centre for Agricultural Landscape Research (ZALF), Muencheberg, Germany

Nikolai M. Dronin Faculty of Geography, Lomonosov Moscow State University, Moscow, Russia

Veerle Van Eetvelde Landscape Research—Department of Geography, Ghent University, Ghent, Belgium

Ursula Eisendle University of Salzburg, Salzburg, Austria

Frank Eulenstein Leibniz Centre for Agricultural Landscape Research (ZALF), Muencheberg, Germany;
Kuban State Agrarian University, Krasnodar, Russia;
Mitscherlich Academy for Soil Fertility, Paulinenaue, Germany

Alexander Evseev Faculty of Geography, Lomonosov Moscow State University, Moscow, Russia

Mariia Fedoriak Department of Ecology and Biomonitoring, Chernivtsi National University, Chernivtsi, Ukraine

Nicholas A. Fisichelli Climate Change Response Program, National Park Service, Fort Collins, CO, USA;
Schoodic Institute at Acadia National Park, Winter Harbor, ME, USA

Thomas Foken Bayreuth Center of Ecology and Environmental Research (BayCEER), University of Bayreuth, Bayreuth, Germany

John M. Francis JMF Research Associates, Philadelphia, USA

Amy E. Frazier School of Geographical Sciences and Urban Planning, Arizona State University, Tempe, AZ, USA

Monika Frielinghaus Kuratorium Boden des Jahres, Geldern, Germany

Tanja Fugiel Prof. Schaller UmweltConsult GmbH, Munich, Germany

Abror Gafurov Helmholtz-Zentrum Potsdam Deutsches GeoForschungsZentrum GFZ, Telegrafenberg, Potsdam, Germany

John A. Gallo Conservation Biology Institute, Corvallis, OR, USA

Greta Gaudig Institute of Botany and Landscape Ecology, Greifswald University, Partner in the Greifswald Mire Centre, Greifswald, Germany

Maria Gerasimova Faculty of Geography, Lomonosov Moscow State University, Moscow, Russia;
V.V. Dokuchaev Soil Science Institute, Moscow, Russia

Lars Gerlitz Helmholtz-Zentrum Potsdam Deutsches GeoForschungsZentrum GFZ, Telegrafenberg, Potsdam, Germany

Leah V. Gibbons School of Sustainability, Arizona State University, Tempe, AZ, USA

Frank Glante German Environment Agency, Dessau-Rosslau, Germany

Johannes Gnaedinger Prof. Schaller UmweltConsult GmbH, Munich, Germany

Mathias Göckede Max-Planck-Institute for Biogeochemistry, Jena, Germany

Karsten Grunewald Leibniz Institute of Ecological Urban and Regional Development, Dresden, Germany

Mahmud H. Haghighi Helmholtz-Zentrum Potsdam Deutsches GeoForschungs Zentrum GFZ, Telegrafenberg, Potsdam, Germany

Vivi Trisna Handini Department of Agribusiness, University of Jember, Jember, Indonesia

Michael Haubold-Rosar Research Institute for Post-Mining Landscapes (FIB), Finsterwalde, Germany

Virgilio Hermoso Centre Tecnològic Forestal de Catalunya (CEMFOR-CTFC), Solsona, Spain;
CREAF, Cerdanyola del Vallés, Spain

Lionel Hertzog Forest and Nature Lab, Ghent University, Melle-Gontrode, Belgium

Jörg Hoffmann Julius Kühn-Institut (JKI) Bundesforschungsinstitut Für Kulturpflanzen, Institut Für Strategien Und Folgenabschätzung, Nachhaltige Landwirtschaft Und Biodiversität, Kleinmachnow, Germany

Guang Hu School of Civil Engineering and Architecture, Zhejiang Sci-Tech University, Hangzhou, China

Julia Illigner Helmholtz-Zentrum Potsdam Deutsches GeoForschungsZentrum GFZ, Telegrafenberg, Potsdam, Germany

Zita Izakovičová Institute of Landscape Ecology, Slovak Academy of Sciences, Bratislava, Slovakia

D. John Pierce Washington Department of Fish and Wildlife, Wildlife Program, Olympia, WA, USA

Michael Jones Department of Geography, Norwegian University of Science and Technology, Trondheim, Norway

Tomasz Joniak Department of Water Protection, Faculty of Biology, Institute of Environmental Biology, Adam Mickiewicz University Poznań, Poznań, Poland

Lev K. Kazakov Faculty of Geography, Lomonosov Moscow State University, Moscow, Russia

Peter J. Kedron School of Geographical Sciences and Urban Planning, Arizona State University, Tempe, AZ, USA

Alexander V. Khoroshev Faculty of Geography, Lomonosov Moscow State University, Moscow, Russia

Felix Kienast Swiss Federal Institute for Forest, Snow and Landscape Research WSL, Birmensdorf, Switzerland

Dirk Knoche Research Institute for Post-Mining Landscapes (FIB), Finsterwalde, Germany

Michael Köhl Universität Hamburg, Center for Earth System Research and Sustainability World Forestry, Hamburg, Germany

Elena M. Koptseva Department of Geobotany and Plant Ecology, Faculty of Biology, St. Petersburg State University, St. Petersburg, Russia

Tatiana Krasovskaya Faculty of Geography, Lomonosov Moscow State University, Moscow, Russia

Valery V. Kravchenko V.B. Sochava Institute of Geography of the SB RAS, Irkutsk, Russia

Jörn Lauterjung Helmholtz-Zentrum Potsdam Deutsches GeoForschungs Zentrum GFZ, Telegrafenberg, Potsdam, Germany;
Central Asian Institute for Applied Geosciences (CAIAG), Bishkek, Kyrgyzstan

Jiajia Liu Forest and Nature Lab, Ghent University, Melle-Gontrode, Belgium; College of Life Sciences, Zhejiang University, Hangzhou, Zhejiang, China

Joseph R. Lujan U.S. Fish and Wildlife Service, National Wildlife Refuge System, Albuquerque, NM, USA

William A. Mackaness School of Geosciences, Edinburgh, UK

Dmitry Mandychev Central Asian Institute for Applied Geosciences (CAIAG), Bishkek, Kyrgyzstan

Michael Manton Faculty of Forest Science and Ecology, Vytautas Magnus University, Akademija, Kaunas, Lithuania

Cristina Mattos Esri Deutschland Group GmbH, Kranzberg, Germany

Ronald J. McCormick Environmental Quality and Protection, Bureau of Land Management, Washington, DC, USA

Blair M. McKenzie The James Hutton Institute, Dundee, UK

Thomas A. Miewald U.S. Fish and Wildlife Service, North Pacific Landscape Conservation Cooperative/National Wildlife Refuge System, Portland, OR, USA

László Miklós Institute of Landscape Ecology, Slovak Academy of Sciences, Bratislava, Slovakia

Gerhard Milbert Kuratorium Boden des Jahres, Geldern, Germany

Sarah Milliken Department of Landscape, School of Design, University of Greenwich, Gillingham, UK

Wilfried Mirschel Leibniz Centre for Agricultural Landscape Research (ZALF), Muencheberg, Germany

Bolot Moldobekov Central Asian Institute for Applied Geosciences (CAIAG), Bishkek, Kyrgyzstan

Alejandra Morán-Ordóñez Centre Tecnològic Forestal de Catalunya (CEMFOR-CTFC), Solsona, Spain;
CREAF, Cerdanyola del Vallés, Spain

Lothar Mueller Leibniz Centre for Agricultural Landscape Research (ZALF), Muencheberg, Germany;
Kuban State Agrarian University, Krasnodar, Russia

Brent A. Murry U.S. Fish and Wildlife Service, Caribbean Landscape Conservation Cooperative, San Juan, PR, USA

Yurij Pautov Silver Taiga Foundation for Sustainable Development, Syktyvkar, Russia

Pravitasari Anjar Pratiwi Department of Agribusiness, University of Jember, Jember, Indonesia

Guy M. Robinson Department of Geography, School of Social Sciences, University of Adelaide, Adelaide, SA, Australia;
Department of Land Economy, University of Cambridge, Cambridge, UK

Jörg Römbke ECT Oekotoxikologie GmbH, Flörsheim, Germany

Mohammad Rondhi Department of Agribusiness, University of Jember, Jember, Indonesia

Olga Ruhovich Pryanishnikov-Institute for Agrochemistry, Moscow, Russia

Elmira Salnjikov Soil Science Institute, Belgrade, Serbia

Abdulla Saparov Kazakh Research Institute of Soil Science and Agrochemistry Named After U. U. Uspanov, Almaty, Kazakhstan;
Research Centre for Ecology and Environment of Central Asia (Almaty), Almaty, Kazakhstan

Joerg Schaller Prof. Schaller UmweltConsult GmbH, Munich, Germany;
Esri Deutschland Group GmbH, Kranzberg, Germany

Uwe Schindler Kuban State Agrarian University, Krasnodar, Russia;
Mitscherlich Academy for Soil Fertility, Paulinenaue, Germany

Tilo Schöne Helmholtz-Zentrum Potsdam Deutsches GeoForschungsZentrum GFZ, Telegrafenberg, Potsdam, Germany

Abdysamat Shakirov Central Asian Institute for Applied Geosciences (CAIAG), Bishkek, Kyrgyzstan

Azamat Sharshebaev Central Asian Institute for Applied Geosciences (CAIAG), Bishkek, Kyrgyzstan

Askhad K. Sheudzhen Kuban State Agrarian University, Krasnodar, Russia

Daniel R. Shively Watershed, Fish, Wildlife, Air & Rare Plants Program, U.S. Forest Service, Washington, DC, USA

Jana Špulerová Institute of Landscape Ecology, Slovak Academy of Sciences, Bratislava, Slovakia

Nico Stolarczuk Helmholtz-Zentrum Potsdam Deutsches GeoForschungsZentrum GFZ, Telegrafenberg, Potsdam, Germany

Mikhail Strunin Federal State Budgetary Institute "Central Aerological Observatory" of ROSHYDROMET (FSBI "CAO"), Dolgoprudny, Moscow Region, Russia

Olga I. Sumina Department of Geobotany and Plant Ecology, Faculty of Biology, St. Petersburg State University, St. Petersburg, Russia

Aryo Fajar Sunartomo Department of Agricultural Extension, University of Jember, Jember, Indonesia

Viktor G. Sychev Pryanishnikov-Institute for Agrochemistry, Moscow, Russia

Franziska Tanneberger Institute of Botany and Landscape Ecology, Greifswald University, Partner in the Greifswald Mire Centre, Greifswald, Germany

Vitaly V. Terleev Peter the Great Saint-Petersburg Polytechnic University, St. Petersburg, Russia

Alex G. Topaj Agrophysical Research Institute, St. Petersburg, Russia

Maria Gabriella Trovato Department of Landscape Design and Ecosystem Management, Faculty of Agricultural and Food Sciences, American University of Beirut, Beirut, Lebanon

Marcel van Oijen Centre for Ecology & Hydrology, Penicuik, UK

Kris Verheyen Forest and Nature Lab, Ghent University, Melle-Gontrode, Belgium

Flurina M. Wartmann Swiss Federal Research Institute for Forest, Snow and Landscape WSL, Birmensdorf, Switzerland

Mingjian Yu College of Life Sciences, Zhejiang University, Hangzhou, Zhejiang, China

Dmitry Zamolodchikov Biological Faculty, Lomonosov Moscow State University, Moscow, Russia;
Centre for Ecology and Productivity of Forests of Russian Academy of Sciences, Moscow, Russia

Cornelia Zech Helmholtz-Zentrum Potsdam Deutsches GeoForschungsZentrum GFZ, Telegrafenberg, Potsdam, Germany

Alexander Zubovich Central Asian Institute for Applied Geosciences (CAIAG), Bishkek, Kyrgyzstan

Part I
The Essence and Mission of Landscape Research

Chapter 1
Territory and/or Scenery: Concepts and Prospects of Western Landscape Research

Marc Antrop and Veerle Van Eetvelde

Abstract Landscape has etymologically multiple meanings, hence a wide variety of interpretations and approaches to study it. The word originated in Western Europe during the thirteenth century, denoting both a tract of land organized by people and its visual appearance. The way the land was organized and shaped created a distinct scenery. Therefore, the landscape is essentially a holistic concept. As a spatial unit, it characterizes the identity of the land of a community and defines a territory where custom rights apply. Both territory and scenery are manifestations of local, regional or national relationships between a community and the way it is using the environment. The landscape also gives the basis of cultural identity and defines value systems. During history, this dynamical relationship changed and so did the meanings and ways of seeing the landscape. This resulted in different approaches in the study of the landscape and its visualization. We summarize the development of landscape research along these two dimensions, territory and scenery. We discuss how landscape research differentiated and diffused with the Western culture all over the world. Many of the shifts were caused by cultural differences, in particular, linguistic and semiotic, resulting in a growing divergence of interpretations and applications. Also, environmental, economic and technological driving forces steered the development of landscape research, adding new related concepts, new methods to analyze and new perspectives. In a globalizing world, the polarization and compression of the geographical time–space resulted in faster urbanization and land abandonment simultaneously, so all landscapes are affected. The demand grew for more proactive and applied research and transdisciplinary approaches to landscape management. Formal definitions, in particular, the European Landscape Convention, offer platforms for integrating all fragmented efforts.

M. Antrop (✉) · V. Van Eetvelde
Landscape Research—Department of Geography, Ghent University,
Krijgslaan 281, S8, 9000 Ghent, Belgium
e-mail: Marc.antrop@ugent.be

V. Van Eetvelde
e-mail: veerle.vaneetvelde@ugent.be

L. Müller and F. Eulenstein (eds.), *Current Trends in Landscape Research*,
Innovations in Landscape Research, https://doi.org/10.1007/978-3-030-30069-2_1

Keywords Western culture · Landscape disciplines · Landscape science · History · Holism · Approaches · Ways of seeing

1.1 Introduction

Landscape research developed along different paths following conceptual changes of the meaning of the word landscape and changing perspectives of seeing the landscape. Landscape research is a recent generic term for all studies 'dealing with the landscape,' ranging from pure academic research, landscape design, artistic expression and visualization, and landscape policy. However, since the time human lifeways started transforming the environment, people have been 'dealing with the landscape' but called it very differently. Antrop and Van Eetvelde (2017) and Antrop (2019) summarize the history of landscape research and the development of different disciplines involved, as well as the important moments of change caused by new paradigms and emerging new technologies. The general picture is one of a tree growing and stretching out divergent branches. Landscape research today is not one unified discipline in a specific academic research domain, but an attempt to integrate all activities 'dealing with the landscape' embracing the whole of the alpha, beta and gamma domains of science, purely academic and applied. Earlier attempts for such an integrated approaches were found in the concept of nature (von Humboldt 1849–1858; Wulf 2015), in geography as 'landscape science' (*Landschaftskunde*, *Landschaftslehre*) (Kant 1764; Sauer 1925; Granö 1929; Passarge 88, 89; Neef 1967; Paffen 1973; Granö and Paasi 1997), in holism (Smuts 1926) and in landscape ecology (Forman and Godron 1981; Naveh and Lieberman 1994; Zonneveld 1995; Moss 1999).

We will first focus on the linguistic and semiotics of the concept landscape to formulate the fundamental holistic nature of the concept. The two basic pillars of the meaning of the word landscape are territory and scenery. We will use these to explain important new landscape-related concepts and the important consequences for research and policy. We summarize the possible approaches as different ways of seeing or perspectives on the landscape. A key set of new concepts that are essential for reading and understanding the landscape is presented. Landscapes are dynamic, and change is their nature. The driving forces help understanding how transformation occurs and how to make scenarios for future landscape development.

1.2 The Development of Landscape Research

1.2.1 From the Description of Nature to a Proactive Transdisciplinary Science

From the perspective of Western European culture, where the word landscape originated, Antrop and Van Eetvelde (2017) recognized six consecutive phases in the development of landscape research:

1. The proto-scientific phase of describing, depicting, imagining and designing landscapes (before the nineteenth century)
2. The emerging scientific research: the landscape as object of study of geography (nineteenth–early twentieth century)
3. The landscape seen from the air: aerial photography with its holistic perspective and the emergence of historical geography (approximately 1920–1960s)
4. The loss of the holistic synthesis due to specialization and quantification in a world driven by economic and urban expansion (approximately 1960–1980s)
5. The humanistic approach and the revival of landscape ecology for dealing with increasing economic and environmental crises (approximately 1980–1990s)
6. The 'landscape crisis' caused by increasingly faster changes and growing uncertainty, and the shift toward applied and transdisciplinary landscape studies (since approximately 1990).

With the global spreading of the Western European culture, landscape and related concepts were introduced in many parts of the world and adapted to fit in the local culture. Antrop (2019) gives an overview of the divergence of the landscape concept and key research domain in three core regions of the world, i.e., North America, Europe–Middle East and Southeast Asia. As an example, Fig. 1.1 illustrates the filiation of the concept of landscape ecology and its diffusion of all over the world.

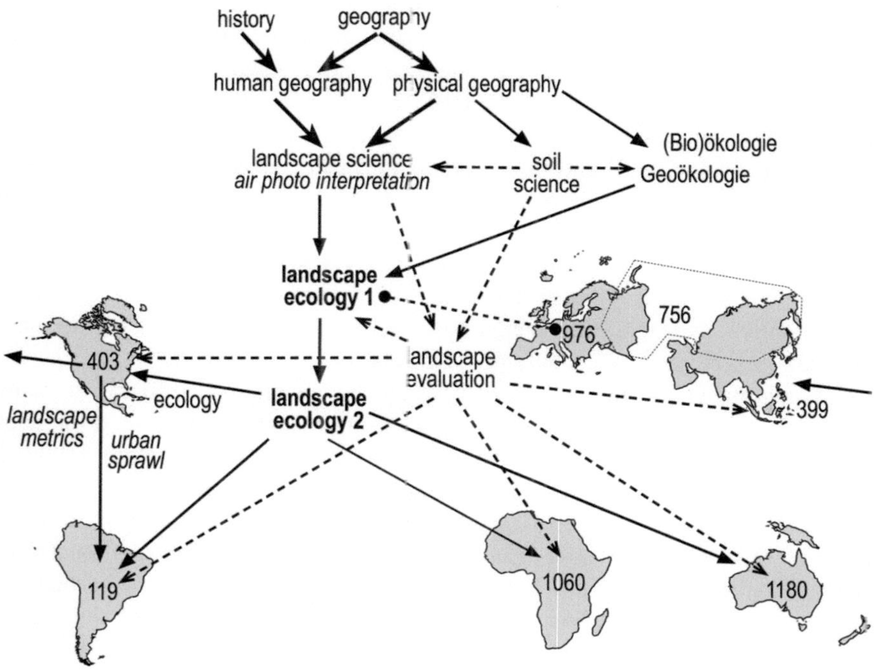

Fig. 1.1 Filiation and diffusion of landscape ecology over the world. The numbers are the Google hits of 'landscape ecology' in thousands (situation 2018)

1.2.2 Filling the Gap: Landscape as an Integrating Concept

Failing to make the landscape a 'universal' core concept in scientific research (Paffen 1973) reduced landscape studies after the Second World War to marginal descriptions and monographs, being considered the ultimate synthesis mainly in regional and historical geography. This approach became obsolete with the growing specialization and the introduction of quantitative methods for analyzing and modeling. The regional synthesis, and with it the study of the landscape, became lost when the 'new orientation' of geography expanded. In many countries, interdisciplinary geography split in social/human geography and physical geography, the first belonging to the alpha and the second to the beta sciences.

Jan Zonneveld called it 'the gap in geography' (Zonneveld 1980), deplored the loss of the holistic synthesis and predicted that this gap would rapidly become filled by new disciplines focusing on the importance of holism and interdisciplinary research.

Indeed, the renewed focus on landscape studies developed from different sources. In the applied natural sciences emerged land evaluation and the need for landscape surveying and mapping, in particular, in developing countries where detailed maps were missing. Aerial photography became the most important tool in surveying and methods developed for making landscape classification maps (Zonneveld 1995).

Noteworthy is the development in England where the fast-growing interest for the local landscape by the general public was more important than the academic interest. This ultimately stimulated new research, in particular in landscape history and archaeology (Taylor 2006). The interest for the landscape in cultural geography, initiated by Sauer mainly based on the territorial concept (Sauer 1925), but was heavily criticized by Hartshorne (1939) and the developing 'regional sciences.' However, the landscape found a new élan in humanistic geography where the focus shifted more toward the scenery concept and landscape as a social construct driven by culture (Tuan 1974; Lowenthal 1975; Cosgrove and Daniels 1988; Olwig 1996).

All these approaches had different ways of reading the landscape and used different methods. One common shared landscape concept was still missing. Forced by continuous and aggravating environmental deterioration, an interdisciplinary cooperation became progressively necessary.

An important momentum was the Veldhoven meeting in 1981 when many researchers 'dealing with the landscape' from the most diverse disciplines (Tsjallingii and de Veer 1982) initiated the revival of landscape ecology followed by the founding of the International Association of Landscape Ecology in 1982. This new 'umbrella of thinking' of landscape ecology opened a transdisciplinary approach to landscape research (Zonneveld 1995). As Moss (1999) formulated it: "To me, landscape ecology is simply about the study of landscapes and of the need to derive understanding about landscapes in order to enhance our abilities to manage them more effectively. Landscape ecology is not the only field to focus on the landscape, but it has emerged in the last few decades because, quite clearly, existing approaches that sought to address a whole range of landscape scale environmental issues were proving to be inadequate" (p. 138).

1.3 Etymology and Semiotics in Action

1.3.1 The Word 'Landscape' in Common Language

One of the oldest references to the word landscape is found in the Dutch language around the year 1200 AD. The word '*lantscap*' (also spelled as '*lantscep*' and '*landscip*') denotes a tract of land or region. It is composed of the word 'land,' in the sense of a bordered territory, and the suffix-*scep*, which refers to reclaiming or creating, as in '*scheppen*' (to shuffle) and in the German '*Landschaft*'—'*schaffen*' = to make (Antrop and Van Eetvelde 2017). In Old English, this became '*landscip*,' the suffix having the meaning of -*ship* as in kinship and friendship (Olwig 1996; Cosgrove 2002). Thus, it focuses on land cleared for settlement in a rural community. Consequently, it also made the distinction between the rural and the urban lands (town, nowadays also referred to as townscape). In this sense, landscape is intimately connected to the terrain, the soil and the territory, i.e., the area created and managed by the community who lives there. In the Western culture, land also means property and wealth and many customs and laws were created to regulate the use of rights for all. When land clearing was defined by the physical conditions and the available technology, the organization and shaping of the land reflected the cultural identity of the local community or landowner. Thus, landscape became the visual manifestation of the interaction between land and people, and the identity of a society. In Western culture, the representation of the landscape in prints and paintings only became a distinct genre in pictorial arts in the fifteenth century. In particular, in the Low Countries, it became a commercially successful genre, making the word landscape synonymous with a painting of a landscape scenery. With the Dutch Golden Age of the seventeenth century, the word became introduced in English language in the meaning of a painting or scene of the countryside.

The landscape does not only refer to a tangible and complex reality 'out there' that can be described and analyzed using objective scientific methods, but also refer to the subjective experience of its perception generating a 'mindscape' as well (Lowenthal 1975; Cosgrove and Daniels 1988). Unsurprisingly, the approaches to the landscape are very diverse and not always clearly defined. Most disciplines and interest groups are dealing with the same tract of land but conceive different landscapes. There are different 'ways of seeing the landscape,' and its meaning shifts with the context, with the background of the observers and users, and with the methods and technology to study it (Cosgrove 2002).

1.3.2 Overcoming the Babylonian Confusion

In common language, the word landscape has multiple meanings and, according to the focus one makes, different perspectives, readings and interpretations are possible. Different linguistic interpretations and translations are added to the complexity. The beginnings of landscape research were dominated by a search of *the* scientific definition of the new object of study in geography, but failed to resolve in 'universal' accepted definition (Zonneveld 1995; Olwig 1996; Claval 2004; Antrop 2005). To clarify the exact meaning one is using, adjectives were added, such as natural or cultural landscape, rural or urban landscape or designed landscape, and even in a metaphorical sense as in political landscape. Nevertheless, such precisions raised new problems and did not solve the loss of holistic thinking about the landscape as shown in the following examples.

Natural and cultural landscape.

The German geographers from the end of the nineteenth and first half of the twentieth century debated a lot on the definition of landscape (Paffen 1973). They introduced the difference between natural and cultural landscapes as the main paradigm. The distinction was characteristic of an environmental determinism of the cultural landscape. Geographers elsewhere opposed to this deterministic approach (Sauer 1925; Vidal de la Blache 1922; Claval 2004), which caused the development of environmental 'possibilism' for the development of the cultural landscape. Jones and Daugstad (1997) discussed the many problems in using this distinction as it causes a fundamental split in the holistic nature of the concept landscape. Nowadays, there is the awareness that the environment is globally affected by human activities. So, it is more relevant to consider the landscapes in a continuum on an intensity scale of human impact. As Sauer formulated it: "culture is the agent, the natural area is the medium, the cultural landscape is the result" (Sauer 1925). It placed humans as important actors of the changing environment, which was recognized in the first important symposium on *Man's Role in the Changing Face of the Earth* in June 1955 (Thomas 1956).

When concepts as natural and cultural landscapes have become obsolete in scientific research, it is not the case in policy as can be seen, for example, in the definition of 'cultural landscape' by UNESCO (see further). In many countries, the protection of landscape and nature depends on different administrations and follows different rules and legislation.

Rural and urban landscapes.

The distinction between rural and urban landscapes is based on two different lifestyles and visions about the environment. This distinction was already used in Roman times and clearly does not correspond to the distinction of built versus unbuilt land. Rural areas often refer to agricultural (agrarian) land that contains villages and small towns. Forests, wasteland and wilderness are not included. In the British tradition, the rural landscape is called countryside and the French speak of *la campagne*. The difference between rural and urban is more defined by customs and law than the landscape composition and scenery. However, the distinction is still used in policy, such as in the Rural Development Programs 2007–2013 of the EU (European Commission 2008).

Geographers studied the rural landscape from a structural, functional and historical perspective. They considered it as a specific form of cultural landscape, characterized by settlement patterns, land use zoning and field systems. Rural landscapes reflect a long history of land reclamation and management that can be studied as a palimpsest to reveal successive layers of time depth (Claval 2004).

In the past, the distinction between rural and urban was clear, spatially, legally and socially. With the Industrial Revolution and the general urban sprawl, stimulated by the increasing mobility by railway and cars, the border between rural and urban has become fuzzy. Also, the settlement types form a continuous gradient according to size, area and power (Antrop 2000) and 'the urban' needed to be redefined as well. Highly dynamical urban areas affect more than their morphological extent. New concepts such as functional urban regions (Cheshire 1995), functional urban areas (Antikainen 2005; OECD 2012) and urban morphological zones (European Environmental Agency 2002) have been introduced. Many of these contain fragments of the former natural or rural landscape.

Sublime and ordinary landscapes.

This pair of extreme landscape types is based on a scale of aesthetic appreciation and preference. It basically refers to the scenery, but not merely to the visual landscape, but to the experience with all senses when standing in or crossing the landscape. Sublime and spectacular landscapes refer to the nineteenth-century romantic travel experiences and in a lesser extent to the beautiful and the picturesque landscapes depicted in arts (Schama 1995; Bell 1999). The ordinary or quotidian landscapes are the ones experienced daily by urban commuters (Groth and Bressi 1997).

1.3.3 Formal Definitions of Landscape

Formal definitions are based on an agreement between different parties and are consolidated in a convention that all parties agree to apply in their policy and legislation. Two formal definitions of landscape have an international scope and stimulate integrated research of the landscape.

Cultural landscapes in the UNESCO World Heritage Convention.

In 1992, the UNESCO World Heritage Convention introduced a new category for the list of protected properties, i.e., cultural landscapes (UNESCO 1992). It contains the first formal definition of landscape. Cultural landscapes are to "represent the 'combined' works of nature and man."

Three main categories are recognized:

(1) *Designed landscapes* have been created intentionally by men, such as gardens and parks. They are constructed for aesthetic (and sometimes political) reasons and are often associated with monumental ensembles.
(2) *Organically evolved landscapes* are the result of and have developed from the interactive process between a specific culture and in response to its natural environment. They fall into two sub-categories:

    (a) *Relict (or fossil) landscapes* are the ones that still show characteristic material features resulting from the processes that made them but came to an end.
    (b) *Continuing landscapes* are the ones that are sustained by a still active traditional way of life in the contemporary society.

(3) *Associative cultural landscapes* refer symbolically to powerful religious, artistic or cultural associations of the natural element rather than material cultural evidence.

The UNESCO Convention only applies on 'properties' that are selected for their 'outstanding universal value.' Categories (1) and (3) are clearly 'special' landscapes which are often considered spectacular or sublime. Category (2) deals with traditional agrarian and pastoral landscapes, which constitute the main part of characteristic landscape diversity in the countryside. Important problems related to this category are already recognized in the two sub-categories. When lifestyle changes and does not sustain anymore the landscape it created over centuries, the living landscape dies and becomes a relic. Properties that lose their qualities are put on a list of endangered world heritage.

The European Landscape Convention.

In 2000, the Council of Europe presented the European Landscape Convention (ELC) for signature, which entered already into force in 2004. Article 1 formulates clearly a consistent landscape definition and several other related definitions:

1. "Landscape is defined as an area, as perceived by people, whose character is the result of the action and interaction of natural and/or human factors.
2. Landscape policy means an expression by the competent public authorities of general principles, strategies and guidelines that permit the taking of specific measures aimed at the protection, management and planning of landscapes.
3. Landscape quality objective means, for a specific landscape, the formulation by the competent public authorities of the aspirations of the public with regard to the landscape features of their surroundings.
4. Landscape protection means actions to conserve and maintain the significant or characteristic features of a landscape, justified by its heritage value derived from its natural configuration and/or from human activity.
5. Landscape management means action, from a perspective of sustainable development, to ensure the regular upkeep of a landscape, so as to guide and

harmonize changes which are brought about by social, economic and environmental processes.

6. Landscape planning means strong forward-looking action to enhance, restore or create landscapes."

(http://conventions.coe.int/Treaty/en/Treaties/Html/176.htm)

The definition of landscape is a consensus between policy-makers and scientific experts from the committees on diversity and landscape (CO-DBP, representing IUCN) and cultural heritage (CC-PAT, representing ICOMOS). The definition fits closely the etymological meanings of the original word. It refers to a territory or land unit and to the perception people have of it, but broader than the scenic aspects alone. The character refers to its holistic quality and as well as the identity and uniqueness. Finally, the interactions between natural processes and human activities make the landscape a dynamic, always changing phenomenon. However, the perspective is clearly human-centered, which makes it more specific than the general organism-centered focus as used in landscape ecology (Fahrig 2005; Mac Nally 2005).

Most important is the scope of the ELC, which encompasses the entire territory of the member states and "covers natural, rural, urban and peri-urban areas. It includes land, inland water and marine areas" (ELC, Article 2). There are no non-landscapes. This makes it completely different from the UNESCO definition of cultural landscapes.

The great merit of this convention is that it offered a broadly accepted and operational definition of landscape including its inherent complexity and holistic character. This allowed initiating many programs for studying the landscape as never before and stimulating interdisciplinary and international cooperation. This is remarkable as the convention has not been a legal basis such as an EU directive and no financial means are provided.

1.4 Perspectives on the Landscape

1.4.1 Territory and Scenery

The two pillars of the landscape are territory and scenery. Territory refers to a delimited tract of land owned by someone has rights over its use. Land is a property and means also value, mostly expressed in monetary terms. This makes that the concept of land is fundamentally different from that of the landscape. The landscape can be experienced by everyone, at least as an observer. The landscape is a common good, and its quality is expressed by its scenery and some rights of using it.

A territory is marked in space by borders and landmarks, separating it from neighboring territories. Often, a settlement is the control center from which the land is organized and the landscape is shaped. Characteristic is a land zoning of intensively used and valuable *infields* near the settlements, surrounded by more extensively used *outfields* and further away vast areas of wilderness (forest and marshland). For each zone, special regulations are applied. Outfields and wilderness remained common land until the important land clearings were needed to absorb the growing population. Growing population caused growing territories, which ultimately connected, and competition for the land could result in territorial disputes.

Antrop and Van Eetvelde (2017) showed that the settlement–territory model can be used as a holistic building block (a *holon*) to understand the spatial configuration of the landscape, allowing landscape typology and classification. Also, scenic elements have been proven significant for the visual control of the territories.

1.4.2 Ways of Seeing

Cosgrove (2003) distinguished two basic approaches or discourses to the landscape, which he called ecological and semiotic. The ecological landscape discourse focuses on the complex interactions of natural processes that shape land areas and the interaction with human activities. The semiotic landscape discourse puts emphasis on the context and processes through which cultural meanings are shaping our world. Thus, the landscape becomes a social construct full of symbolic meaning. The first discourse is typical in natural sciences such as geography and landscape ecology, while the second is found in social sciences, such as humanistic and cultural geography.

Luginbühl (2012) discussed the relationship between society and landscape, and between the landscape as a tangible reality and its representation. He recognized that two groups of societies exist: one without the notion of landscape and one with the notion of landscape. The first group has only a utilitarian or symbolic relation with the land and does not contemplate the landscape. The second group contemplates the landscape and recognizes different types and regions, which receive proper names that reflect their identity or character. Also, various artistic expressions and representations are characteristics in this approach.

Antrop and Van Eetvelde (2017) defined four perspectives for the study of landscape according to the (virtual) positions of the observer: standing in the landscape or the 'interior' perspective, the bird's-eye perspective from above the landscape, the inner or mental perspective and the transcendent abstract perspective. Thus, the landscape can be comprehended or modeled using different paradigms and theories in three different ways:

- As a complex spatial system of objects (also referred to as elements) and continuous phenomena varying continuously in space. The elements of that spatial system are in dynamical interaction. The following concepts are used in this approach: structure, pattern, functions, ecosystem, change and dynamics. System theory is the basic paradigm of this approach.
- As a scene or image that can be described using concepts and rules of perception, which are also related to preference and experience. Theories of environmental perception and Gestalt psychology (also based on holism) apply here as well as design principles.

- As an existential phenomenon with strong symbolic meanings and values. This fits in Cosgrove's semiotic discourse and is the approach followed in arts, philosophy and sociology. Basic concepts here are: homeland, heritage, history, genius loci, character and identity.

1.4.3 Multiple Scales

The way the landscape can be studied depends primarily on the scale it is observed. The spatial scale defines the extent of the landscape seen as well as the degree of detail (or *grain*) that can be observed. Scale is a function of grain and extent: scale = *f*(grain, extent). Scale, grain and extent define the outcomes of spatial analysis using landscape metrics (Li and Wu 2004; Cushman et al. 2008).

In geography and cartography, the scale is expressed as the ratio between standard lengths represented on the map to its corresponding length on the terrain. Hence, a 'small scale' means a long view distance so objects are represented very small and a broad coverage (extent) is seen. In this case, small objects cannot be represented on maps in their real size and shape as they are not resolved with the given grain size (i.e., pixel resolution or minimum mappable unit—MMU). They have to be generalized and represented by symbols. Similarly, 'large scale' means that a small extent is seen in great detail (fine grain or high resolution).

The confusion arises when the reverse is meant, which is often the case in applications of landscape research in landscape architecture and planning. Here, a 'large project' refers to a large area (extent), and a 'small project' to a small area, which has to be represented, respectively, on a small-scale and large-scale map. To avoid this confusion, scale levels are often denoted according to the policy levels: micro, meso and macro, or better local (municipal), regional (provinces, district, etc.), national and international.

Besides the spatial scale, landscape research uses also temporal or timescales, and organization and planning scales. Table 1.1 gives and overview of the dimensions of the scale concept is different application domains in landscape research.

Essential in landscape research is alternatively zooming in and out on the observed landscape, i.e., changing scales, and thus revealing the overall landscape structure and the detailed context of each element and changing perspectives (Forman and Godron 1986).

Table 1.1 Dimensions of the scale concept and application domains in landscape research (after Antrop and Van Eetvelde 2017)

Application domain	Degree of detail	Coverage size
Spatial or geometrical scale	Resolution, grain, MMU	Extent
Time or temporal scale	Time interval	Duration
Organization of planning scale	Agent (actor)	Competence domain

1.4.4 Classification: Typology and Chorology

Classification is the phase in scientific research where all the individual and unique objects that have been described are systematically ordered. This results in a general framework that helps understand, formulate theories and make extrapolations. Zonneveld (1989, 1994) defined two approaches to the classification of land units: (1) by agglomeration, leading to typification, similar to taxonomy, and (2) by subdivision of larger spatial units into smaller and more detailed ones. These are also called, respectively, typology and a chorology and result in hierarchical classifications according to the scale of organization. The methods for the two approaches differ, but combinations are possible (Antrop and Van Eetvelde 2017).

Landscape classification is not only important in pure academic research, but also in applications in land evaluation (Zonneveld 1995) and various forms of landscape character assessment (Swanwick 2002; Aldred and Fairclough 2002) and in visual landscape assessment and design (Fry and Sarlov Herlin 1995). The results are also presented in different forms. Early landscape classifications consisted of atlases (Aalen et al. 1997), i.e., a collection of thematic maps. More recent forms consist of digital map databases that can be consulted interactively. Some of these 'atlases' are also heavily illustrated with iconographic and photograph documents and contain essays about different themes. When the approach chronologically tell the history and genesis of the landscapes in a specific region, they are called landscape biographies (Kolen et al. 2015).

Landscape classification gained new significance with the European Landscape Convention, in particular with the special measures that were proposed: "each Party undertakes: (a.i) to identify its own landscapes throughout its territory, (a.ii) to analyze their characteristics and the forces and pressures transforming them, (a.iii) to take note of changes, and (b) to assess the landscapes thus identified, taking into account the particular values assigned to them by the interested parties and the population concerned" (Council of Europe 2000, Art. 6 Special measures, C Identification and assessment).

1.5 Landscape-Related Key Concepts

1.5.1 Holism and Landscape as a Multi-scaled Hierarchical System

Clearly, landscape contains elements from the natural, physical world and from the cultural world and human society. The concept bridges the alpha and beta domains of science in search of a unified integration. This was already the goal of the early naturalists, clearly expressed by Alexander von Humboldt, when setting off his voyage to the Americas in June 1799: "I must find out about the unity of nature" (Nicolson 1995). He was not speaking of the landscape, which he mainly saw as a visual representation of nature in painting, but his unifying concept was nature. He did not use either the word *Gestalt* or holism, which both would be introduced much later. The German concept Gestalt was introduced by psychologists between the two World Wars in relation to the way human perception and learning work. As perception is an essential part of experiencing the landscape, Gestalt theory is significant in landscape research as well (Antrop and Van Eetvelde 2017). Gestalt theory also supports the concept of *holism*, which was first introduced by Jan Christiaan Smuts in 1926. In his book 'Holism and Evolution,' he based his theory on Darwin's theory of evolution and Einstein's theory of general relativity to conclude that holism expresses "the tendency in nature to form wholes that are greater than the sum of the parts through creative evolution" (Smuts 1926).

Naveh and Lieberman (1994) used later term the *Total Human Ecosystem* to denote the holistic hierarchy and the organization of the eco-sciences, in which landscape ecology forms a significant scale level. Important is also the concept of a '*holon*,' defined as an open subsystem in this hierarchy that functions partially autonomous depending on its situation in the whole context. Ies Zonneveld, a Dutch landscape ecologist active in land evaluation and one of the founders of landscape ecology, considered land units at different scale levels as holons that he called 'black boxes' (Zonneveld 1989, 2005).

1.5.2 Models and Layers: Reducing Landscape Complexity

To help understanding landscapes, the reduction of the complexity is necessary as well as searching relationships between the composing parts. Several models have been developed for different purposes. Antrop and Van Eetvelde (2017) describe different groups of models related to the different perspectives discussed above:

- Model 1: elements (discrete objects), components (continuous phenomena), structures (ensembles of elements and components that are spatially and functionally related).
- Model 2: points, lines, polygons and surfaces are basic primitives for geocoding landscape features into a GIS (Burrough and McDonnell 1998).
- Model 3: patch, corridor, matrix and mosaic are the basic primitives used in landscape ecology (Forman and Godron 1986).
- Model 4: mass, screen, space, and open and solid volumes are examples of primitives used in the analysis of the visual landscape (Bell 2004).
- Model 5: landmark, district, path, node and edge are the primitives introduced by Lynch (1960) in the mental mapping of places.

Often, several themes are described and mapped in separate layers. Map overlaying of selected themes allows to study relationships and reveal patterns of coherence. Decomposing a landscape in layers is a common research technique, which existed already long before GIS made it easy.

1.5.3 Some Meta-Realities of Landscape

Dealing the 'more' than the sum of the components.

In the holistic sense, landscape as 'a whole is more than the sum of its composing parts.' Principally this means that a reductionist approach analyzing all components separately never can be added together to describe this holistic 'more.' This was one of the arguments to dismiss the holistic approach as not operational and useless.

However, holism is not claiming such a reductionist analytic approach. Instead, it advocates to define the real meaning of each composing element in its contextual relationship to the whole. Also, it supports methods to describe meta-properties of the holistic ensemble. Examples of such meta-properties of the landscape are diversity, heterogeneity, coherence, information entropy and order. The use of landscape metrics has been an essential innovation from landscape ecology in landscape research (McGarigal and Marks 1995; Botequilha Leitão and Ahern 2002; Li and Wu 2004, 2007; Ode et al. 2008) (Fig. 1.2).

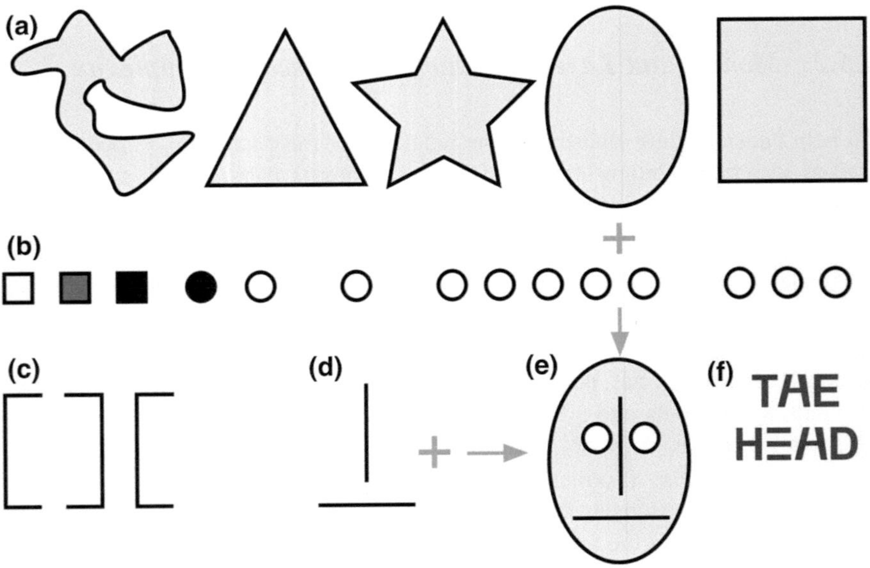

Fig. 1.2 Holism at work—pattern analysis, Gestalt rules and context: (**a**) regular geometrical shapes form a hard Gestalt with a pronounced (symbolic) meaning; (**b**) series of similar element group together according to their spatial pattern (discrete elements forming lines), shape and color, and continuity is restored when elements are missing; (**c**) linear shapes can form borders of (invisible) spaces (a room and corridor); (**d**) the vertical dimension dominates: both lines have equal length, but we cannot help see the vertical one longer; tall objects in the landscape form landmarks; (**e**) the power of context: the whole is more than the sum of the components and (**f**) each element gets its meaning from its place in the whole: the information ('message') of this pattern is transferred by signs we attribute a meaning, so we can decode the message. Some signs are confusing and do not fit in our decoding table, the alphabet in this case. Nevertheless, we can read and understand the message as these signs get the full meaning from the context, even when elements are missing

Composition and configuration.

The landscape is structurally defined by its contents and by the spatial arrangement of the items that make the content. The first property is called *composition* and the second property *configuration*. The distinction between both is important in pattern analysis, in particular when using landscape metrics for the quantitative description and expression of meta-properties such as landscape diversity and heterogeneity (McGarigal and Marks 1995) (Fig. 1.3).

Coherence, fragmentation and order.

Coherence is defined as how the parts of something fit (logically) together or are internally consistent. Consequently, it also refers to a degree of order. In the context of landscape research, it refers to the degree in which landscape elements or features are consistently associated with each other in space or time. In landscape research, coherence is defined in various ways. Van Mansvelt (1997) classified the concepts of the ecological coherence of the rural landscape into three groups: the vertical (on site), the horizontal (landscape-level) and the cyclical (temporal) coherence. Phipps (1984) called to the first two groups spatial–ecological and topological organization. In spatial analysis, coherence is described by spatial autocorrelation, statistically represented by correlograms and often measured by Moran's I metric (Mander et al. 2010). Spatial autocorrelation is scale dependent, in particular, dependent on the grain or spatial resolution of the data (Burel and Baudry 2003). Thus, it also relates to spatial heterogeneity and fragmentation.

Zones and fields–borders and fuzziness–gradients and ecotones.

The territorial concept implies a mosaic of areal 'tiles' or patches that are bordered in some way. The borders can be crisp or fuzzy and materialized (as with stonewalls or hedgerows) or not (as administrative borders which are just abstract lines on a map). Basically, the tiles forming the mosaic are considered internally homogeneous and different from the adjacent ones. In choropleth maps, they are represented by zones, which demand special precautions when making interpretations. In reality, most of the zones representing real landscape features on maps are not homogeneous in reality. Their internal heterogeneity can be revealed when zooming in on direct landscape observation. This is not the case of zoning plans, where each zone means one homogenous status of the land within the zone.

Many forces and processes in landscape are not bound by bordered zones, but show gradients and variation in space. These are in fact (power) 'fields' where the intensity of a characteristic variable changes from place A to place B (Longley et al. 2001). Conceptually, these are *geographical surfaces* that are represented on maps by isolines (Unwin 1981). This is the case for physical components such as slope degree, water table depth, temperature and rainfall. Map overlaying of different landscape themes of result in gradients that according to their width and scale are considered transition zones or slivers (Granö 1929). In landscape ecology, borders and edges are conceived as *ecotone* transition zones, and gradients represent *eco-clines* and *eco-fields* (Farina and Belgrano 2006; Farina 2009).

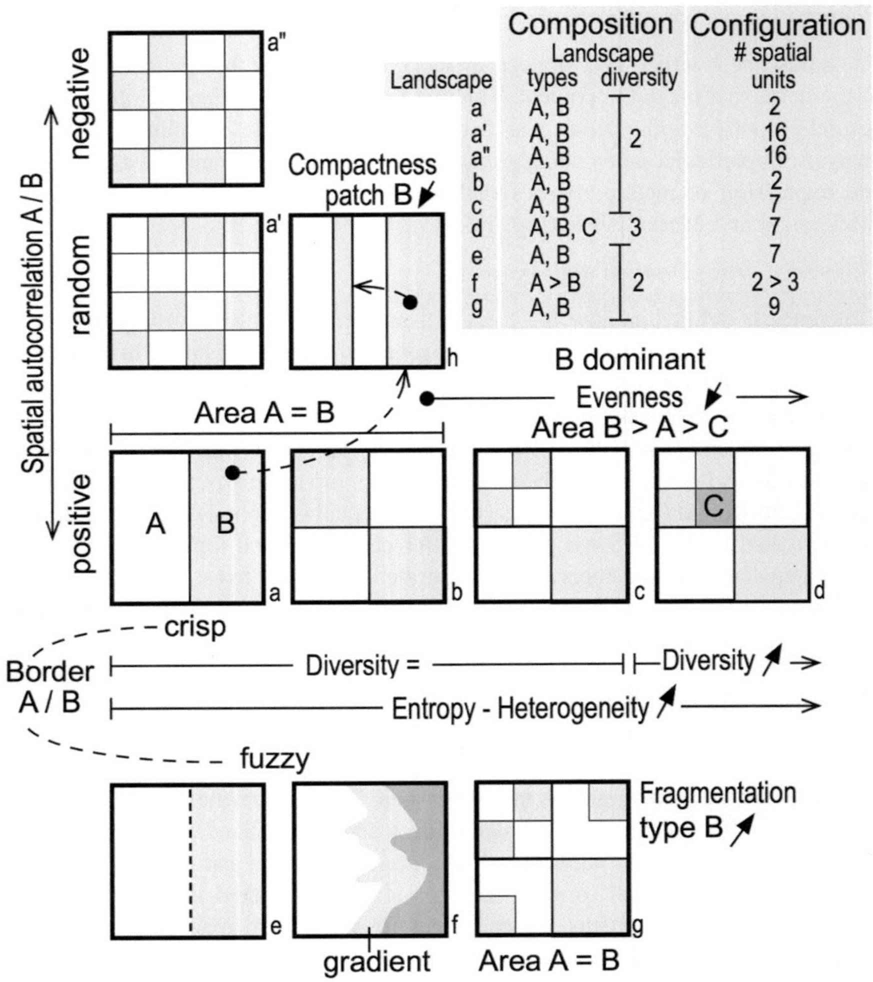

Fig. 1.3 Composition, configuration, landscape diversity, heterogeneity, entropy, fragmentation and autocorrelation. The shades show the different categories of landscape types A, B and C that compose the landscape, and their number gives the landscape diversity or richness. The spatial arrangement of the landscape units ('patches') defines the configuration of the landscape. The complexity of the pattern increases with the number of spatial units. Complexity combined with diversity defines the landscape heterogeneity which can be described by information entropy measures such as Shannon's diversity and evenness. The way units of the same type are spatially distributed defines the fragmentation, contagion, interspersion and autocorrelation between the patch types. The area/edge length ratio gives the compactness of the patches. Borders can be crisp or fuzzy and eventually form transition zones or gradients

In human geography, political and administrative units are normally represented as zonal features, as is also land use and land cover. However, in urban studies, many features show gradients away from the city centers. Urban places act as centers of gravity for many features and activities. A classic example is the population density and the provision of services, like transportation. Gravity models are used to analyze the patterns. Examples are the land use zoning model by von Thünen, the central place theory by Christaller, which gives a basis for the hexagonal model settlement of settlement patterns (Antrop and Van Eetvelde 2017). In settlement geography and historical geography, gravity models are also significant. They explain the distribution of the *infield–outfield* zoning or the distribution of *hortus-ager-saltus-silva* (Antrop and Van Eetvelde 2017).

Also, scenic aspects of the landscape can be represented by zones and fields and show gradients and fuzziness. Frequently used are viewsheds and geographical surfaces (Nijhuis et al. 2011).

1.6 Drivers of Landscape Transformation

1.6.1 Landscapes Change Naturally

Landscapes are dynamical systems and are continuously transforming. The basic paradigm is balancing the spatial, morphological structure with its functioning or the continuous interaction between patterns and processes as described by Forman and Godron (1986):

> An endless feedback loop:
>
> Past functioning has produced today's structure;
>
> Today's structure produces today's functioning;
>
> Today's functioning will produce future structure.

This is not only one of the basic paradigms of landscape ecology, but is also significant in landscape management when it comes to controlling landscape changes. Basically, two ways of acting are possible: (1) acting upon the spatial structure, e.g., making material changes in spatial planning, and (2) steering upon the processes, e.g., by-laws and regulations (Antrop and Van Eetvelde 2017).

According to the time considered, the distinction is made between landscape genesis and landscape history. The first encompasses the geological timescale. The second roughly corresponds to the period where the human impact on the landscape is obvious and hence stretches beyond the written record of history into prehistory. The first is studied by geology, palaeogeography and climatology and ecology, and the second by historical ecology, landscape archaeology and historical geography. However, landscapes can have very deep time depths and most of the research demands an interdisciplinary approach between all these disciplines.

1.6.2 *From Human in the Land to the Great Anthropocene Acceleration*

Early humans wandered through the landscape almost without leaving any trace. The agricultural revolution in the Neolithic meant also the creation of landscapes and settlements that changed gradually the face of the earth. During history, cultural and technological changes correspond to successive landscape transformations, some of which wiped away the previous ones. The frequency and magnitude of landscape transformation varied in time. The speed, frequency and magnitude of the landscape changes showed an exponential increase since the eighteenth century, corresponding to a series of linked revolutions: the agricultural innovations and 'invasion' of new foreign crops, the Industrial Revolution and political–social revolutions (e.g., the French and American revolutions) and World Wars. All these changes made the transition from an agrarian–rural to an urban–industrial way of life with corresponding landscape changes (Antrop and Van Eetvelde 2017). Antrop (1997) called the landscapes before these devastating changes 'traditional' in the sense they were sustainable for at least three generations and hence constituted a common heritage. Nowadays, landscape changes so quickly that everyone will live in several completely transformed environments during one's lifetime. According to the philosopher Lemaire (2002), this causes a growing discontent and discomfort and creates a mentality that landscapes and the environment are makeable and hence have no inherent value.

These landscape transformations are coupled with important environmental transformations that characterize the trajectory of the Anthropocene, which Steffen et al. (2015) called the Great Acceleration. There are indications that since 1800 AD 'techno-stratigraphic' deposits globally mark the landscapes and it makes sense to place the beginning of the Anthropocene here, with the Industrial Revolution (Saul and Waterton 2019). Steffen et al. (2007) recognized two stages in the Anthropocene: the industrial phase until 1945/1950 and the Great Acceleration since then (Fig. 1.4).

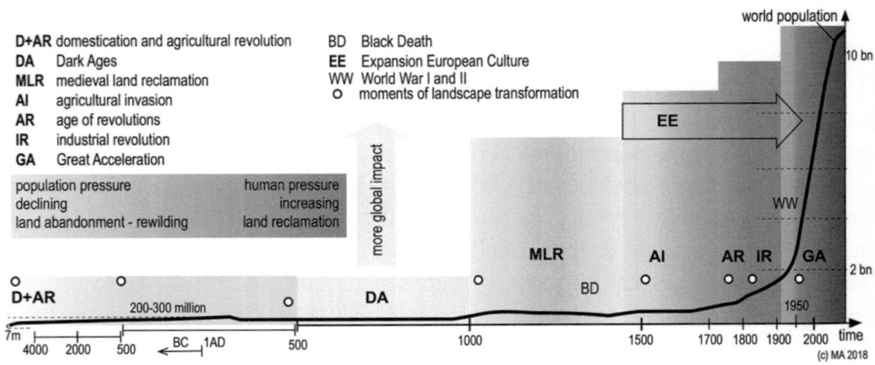

Fig. 1.4 Periods and scale of human impact related to the evolution of world population and moments of landscape transformation in Europe

1.6.3 The Landscape: A Highly Dynamical System Far from Equilibrium

In 1955, the Nobel Prize Winner Ilya Prigogine wrote "the majority of the phenomena studied in biology, meteorology, astrophysics and other subjects are irreversible processes which take place outside the equilibrium state" (Prigogine 1955). Prigogine (1987) spoke of dissipative systems, which could suddenly change when their characteristic factor became unstable. Landscape certainly belongs to the 'other subjects' and can be seen a manifestation of the actual state of a complex system far from equilibrium. Pinto-Correia et al. (2018) speak of a landscape in disequilibrium and illustrate this with examples of the current landscape transitions in rural Europe.

Following the ideas of Laszlo (1994), Naveh (2007) saw the development of different landscape types as successive bifurcations in evolution caused by unstable conditions related to cultural advancement. He recognized following revolutionary cultural steps resulting in new landscape levels: fire making > natural–pristine, toolmaking > sub-natural, agriculture > seminatural, Industrial Revolution > agro-silvo-pastoral, Industrial Revolution > industrial–agro-urban and information revolution > postindustrial. The new states are only temporary semi-stable until new disrupting factors occur (Fig. 1.5).

1.6.4 The General Trends and Driving Forces

The general trends of contemporary environmental changes and landscape transformations are well known as are the main, global driving forces (Vos and Klijn 2000; Klijn 2004; Antrop and Van Eetvelde 2017). At the global scale, two discourses drive competing agendas: an economical discourse of the free market and trade (the WTO agenda) and an ecological discourse focusing on (the UN agenda

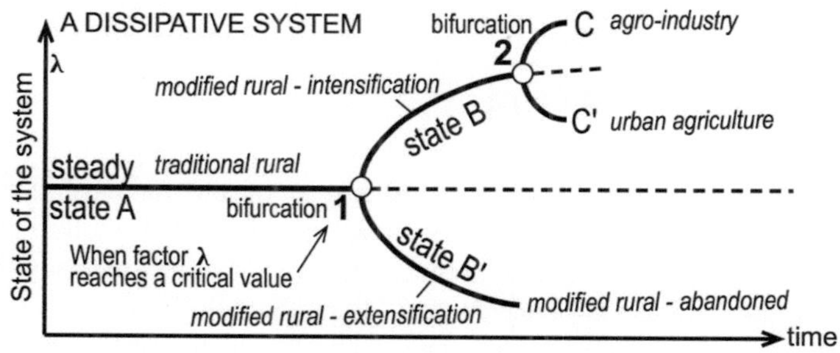

Fig. 1.5 Landscape as a dissipative system. When all factors that sustain a traditional rural landscape become unstable, sudden disruptive changes may occur, creating new landscape types such as the ones indicated here

for sustainable development) (Primdahl and Swaffield 2010). However, the practical outcomes on the local level and the processes involved remain poorly understood and very unpredictable (Pinto-Correia et al. 2003, 2018).

Plieninger et al. (2016) made a review of the 144 studies on the driving forces of landscape change in Europe. Following Geist and Lambin (2002), they made the distinction between proximate and underlying drivers. The first group consists of human activities at the local level, such as agricultural expansion or extension of settlements. The underlying drivers comprise social and natural processes, such as population dynamics, agricultural policies, markets and cultural factors, which have an indirect impact on the national or global level. They found how driving forces have been addressed according to medium-term timescales and local spatial scales. Their review showed that land abandonment/extensification is the most prominent (62% of cases) proximate driver of landscape change. They also found that distinct combinations of mainly political/institutional, cultural and natural/spatial underlying drivers are determining landscape change.

Jepsen et al. (2015) made a survey of the evolution of the land management regimes in 28 European countries over the past 200 years and related these to changing technological, institutional and economic drivers. They recognized seven successive land management regimes with different effects on the management. They denoted these the era of peasantry, innovations and rights, intensification, industrialization, collectivization, de-intensification and commercialization and environmental awareness. They concluded that land system changes are not unidirectional developments following predefined trajectories, but path-dependent processes that are affected by various drivers and sudden events.

These complex interactions of very different driving forces act at scales ranging from global to local. The combined effect can be summarized as a polarization of the geographical space between relatively small core areas of intensification and vast areas of extensification (Antrop and Van Eetvelde 2017). The core areas are characterized by a concentration of people and activities in urban and industrial areas, congested by traffic and fragmented by infrastructure. The peripherical areas suffer from land abandonment, lack of essential services and rewilding. The core areas are often located in along coastal zones and rivers in areas with high risk for calamities.

1.6.5 The Uncertain Future

The uncertainty about our future grows because of the increasing speed and magnitude of the contemporary environmental changes and the numerous agents making changes often acting in a non-concerted way. The uncertainty refers to the predictability (probability) of some future—as predicted by scenarios—becoming realized. The uncertainty of a scenario depends on the quality of the data it is based on, the time horizon in the future, the feasibility of the goals set, the means invested and the consistency of the willingness to achieve this by all stakeholders involved (Fig. 1.6).

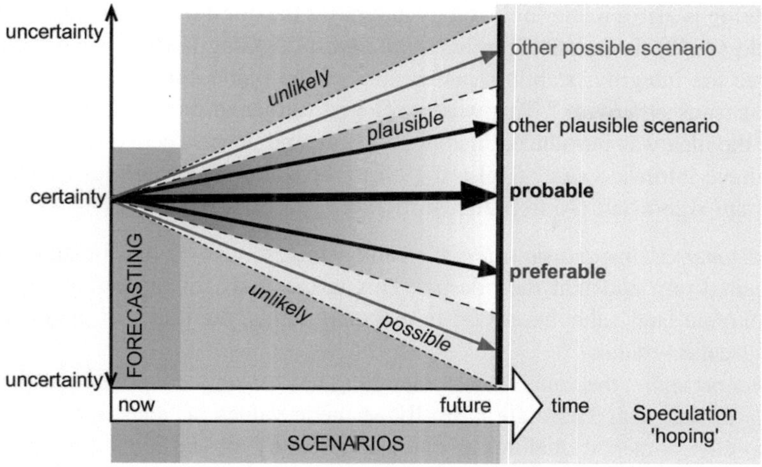

Fig. 1.6 Uncertainty and predicting the future and the probabilities of the outcome. On the short term, forecasting gives the most certain outcomes. Predictions based on facts and scenarios range from probable over preferable, plausible to possible outcomes, and some are unlikely. Even the most probable scenario has no absolute certainty to become realized as planned. When data are lacking or unreliable, which becomes often the case for very long-term speculations, or the scenario models are wrong, no probability of any outcome can be guaranteed

1.7 The Value of Landscape

Both approaches to landscape, territory and scenery, are valuated, but differently. The value of a tract of land depends primarily on its natural resources and the productivity it may yield. Utility defines value, which is often expressed in monetary terms as the price of the land. The potential for different uses is essential. Classic examples are changing wasteland and forest or agricultural land to urban development. Such changes are essential to increase the benefits for the landowners, but might be perceived by others as a value loss of the landscape qualities. In this case, the value is based on other qualities, such as ecological ones, and most often on scenic qualities. While the territory of a landowner is a private good, the scenery of the landscape is regarded as a common good.

It all comes to an ethical matter, and an assessment of right or wrong forms a specific perspective of the observer. The distinction between the concepts' quality and value is essential. Qualities describe properties and characterize the landscape area or a place, making it distinct from another area or place. Qualities can be measured principally on four different scales: Nominal measurement is qualitative description, or ranked on an ordinal scale, and measured quantitatively using interval and ratio scales. In a second step, qualities can be assigned values according to predefined criteria, depending on the perspectives and goal of the valuation. For example, the assessment of the 'outstanding universal value' of a property to be listed on the UNESCO World Heritage is an example of a qualitative evaluation by experts following the criteria defined in the World Heritage Convention.

Valuing is also a matter of axiology and raises the question of right and wrong as Leopold (1949) formulated it in his *land ethics*: "A thing is right when it tends to preserve the integrity, stability, and beauty of the biotic community. It is wrong when it tends otherwise." This was developed further in the *environmental ethics* where the idea was introduced that "wild nature and healthy ecosystems as holistic entity have intrinsic value" (Rolston 1988). The following categories of landscape values are significant (Antrop 2012):

- *Instrumental or extrinsic value*: the value which something has for someone as a means to an end that they desire. This is the basis for the discussion of the economic land value expressed in monetary terms, the utility of landscape and landscape services.
- *Inherent value*: the value which something has for someone, but not as a means to a further end. These are typically aesthetic values of landscape scenery but also environmental qualities as quietness, healthy air and feeling of space.
- *Intrinsic value*: the value something has. No appeal needs to be made for those for whom it has value. It is simply valuable and is so independently of anyone finding it valuable. For landscape qualities, this is the most difficult to make operational and rather similar to the intrinsic value of nature.

1.7.1 Essential Landscape Qualities

There are numerous landscape qualities, and each can have assigned values in very different ways all depending on the perspective and goals of the assessment. Many qualities are abstract concepts with multiple facets. Hence, they are described by multiple attributes, variables or criteria, which can be used for subsequent assessment, qualitatively or quantitatively, or monetary or not. Thresholds can be defined as well, e.g., to delineate spatial units of defined critical values.

Several lists of criteria to assess landscape qualities are available. An example is found in the UNESCO World Heritage criteria for selection (https://whc.unesco.org/en/criteria/). However, no heuristic is provided on how to use these. Very popular, in particular in environmental impact assessment, is the use of comparative tables, such as the popular Leopold matrix (Leopold et al. 1971). Using GIS, even interactive and participatory assessment became possible (Eastman et al. 1993).

Antrop and Van Eetvelde (2017) defined three main groups of criteria for landscape assessment, depending on the content of the landscape (i.e., territorial perspective), the perception and experience (i.e., the scenery perspective) and the use of the landscape (i.e., the right of use and access). For example, the first group consists of criteria as rarity, uniqueness, authenticity, coherence, diversity and heterogeneity. The second group contains criteria as legibility, identity, character, order, naturalness, variation and atmosphere. The third group has accessibility, utility and stewardship.

From these, landscape diversity, character and identity are the essential ones as concisely formulated in the First Assessment of Europe's Environment (the Dobříš Assessment):

> The richness and diversity of rural landscapes in Europe is a distinctive feature of the continent. There is probably nowhere else where the signs of human interaction with nature in landscape are so varied, contrasting and localised.
>
> Despite the immense scale of socio-economic changes that have accompanied this century's wave of industrialisation and urbanisation in many parts of Europe, much of this diversity remains, giving distinctive character to countries, regions and local areas (European Environmental Agency 1995).

1.7.2 A Common Heritage and a Human Right

The landscape is regarded as both a natural and a cultural heritage, meaning it witnesses from our past and still has actual significance in our society (Antrop 2005). Subtle linguistic differences are important here. The Cambridge Dictionary defines heritage as all "features belonging to the culture of a particular society, such as traditions, languages, or buildings, that were created in the past and still have historical importance." The Oxford Dictionary adds "things of special architectural, historical or natural value that are preserved for the nation," essentially including 'nature' as well. On the contrary, patrimony, or inheritance, is a property that someone gets from an heir and is often private. Heritage refers more to common values for the public and society. The UNESCO World Heritage considers 'properties' to be listed. These differences indicate that heritage can be seen as (i) immovable, real estate or movable, material artifacts, and (ii) as intangible. The first group refers to objects and areas, and is territorial, and the second group refers to perception, thus more scenic aspects, and to traditions, hence practices in using the land. This makes a difference when it comes to protection and conservation. In many countries, landscape protection is situated in administrations of nature conservation (for natural properties) or spatial planning (for real estate, cultural landscapes and monuments). Consequently, landscape protection, conservation and management are fragmented over several policy domains, which do not always pursue the same goals and consider the same values.

The European Landscape Convention (Council of Europe 2006) "recognises a human right to landscape" (p. 18), in particular in the context of public participation where landscape is seen "both as a human right and a social responsibility" (p. 120). Linking landscape and human rights is a complex matter and varies again with a territorial or scenic perspective (Egoz et al. 2011).

1.7.3 The Economics of Landscape and Landscape Services

Certain landscape qualities have economic values, of which some can be expressed in monetary terms. This became the recent research field of the economics of landscape (Price 2019). It covers both the territorial aspects, such as the price of the land in relation to its potential utility or location, and scenic aspects as the access price to certain viewpoints and the price of lodging depending on the view.

Processes in the landscape result in functions that have value as well, which introduced the concept of *landscape services* as a special and extended form of ecosystem services (de Groot et al. 2010). Tremorshuizen and Opdam (2009) see landscape services as a bridge between landscape ecology and sustainable development.

1.7.4 Landscape as an Indicator

Most policy domains demand easily obtainable, reliable, up-to-date and transparent indicators to monitor and assess their actions (Parris 2003). Examples are sustainable development (Agenda 21), multi-functional agriculture (Common Agricultural Policy), environmental reporting (European Environmental Agency), economy and development (OECD), trade agreements (WTO), ecosystem and nature conservation (Ramsar Convention), landscape policy and protection (European Landscape Convention, UNESCO World Heritage Convention). These policy domains have direct or indirect impact on the landscape, both on the territorial, spatial and structural characteristics and its scenery. For example, OECD defined agricultural landscapes as "the visible outcomes from the interaction between agriculture, natural resources and the environment, and encompass amenity, cultural, and other societal values," and hence the visual landscape can be used as an indicator (Wascher 2003).

As a result of these demands, several kinds of methods for measuring the landscape have been developed (Botequilha Leitão et al. 2006). Uuemaa et al. (2009) gave an overview of the available landscape metrics and indices. They are used mostly in analyzing changes in land use/cover patterns, but proved to be still less powerful in assessing functions and values (Uuemaa et al. 2012).

1.8 Prospects on Future Landscapes

1.8.1 Paradoxes—Sustainable Landscapes?

A paradox arises when a positive process—desired and well intentioned—initiates parallel negative side effects or feedbacks. A classic example is described in the

Tragedy of the Commons by Hardin (1968) and related to the concept of carrying capacity. Carrying capacity has been defined "the average maximum number of individuals of a given species that can occupy a particular habitat without permanently impairing the productive capacity of that habitat" (Rees 2001). Carrying capacity depends on the potential of natural and cultural resources of a tract of land. It is described by the logistic growth model that strives toward a 'sustainable' level of a population after a stage of almost exponential growth. The concept was used to define the ecological footprint to express the impact of a population on the environment. Different types of carrying capacity can be recognized (Mitchell 1979). For example, biophysical carrying capacity is used to define the number of animals on grazing land without disrupting the ecosystem. Behavioral carrying capacity is used in social sciences and refers to human pressure, population density, congestion and crowding.

Carrying capacity and the related concepts are variables that depend on lifeways and technological innovations (Antrop 1991). Most landscape transformations in history were made to increase the carrying capacity of the land people lived on. The most important ones were the introduction of agriculture and sedentary lifeways and the different forms of urbanization. Among the many other innovations to increase the carrying capacity are manuring, crop rotation, irrigation and multi-functional land use.

As landscapes change continuously, the idea of sustainable landscapes might seem to be a paradox (Antrop 2006). He formulated two possible perspectives on landscape sustainability. The first perspective focuses upon the preservation of inherent landscape qualities and values. These consist of both natural resources, such as biodiversity, and cultural capital, such as heritage. Sustainable preservation demands maintaining the practices and functions that created and managed these qualities. The second perspective focuses upon sustaining rural economies in areas suffering land abandonment. Hence, landscape sustainability is in this case not the target, but landscape is a means to sustain and develop a lifeway. A classic example that also results in a paradox is the development of tourism. Tourism often proposed a solution to stimulate new development in declining regions. The tourism paradox describes the inconsistent nature of tourism's relationship with the environment. The tourism carrying capacity could be defined as the possible maximum number of visitors or accessibility of a site without deterioration of the physical environment and landscape scenery, so that the qualities for tourist attraction and satisfaction become lost (Salerno et al. 2013). Often, valuable landscape sites can become lost completely.

Sustainability also depends on the scale, and time horizon one is aiming. The first perspective clearly demands a long-term and mostly unmonetized value. The time horizon is here 'as long as possible.' On the contrary, in economy, the scale, the time horizon and planning horizons are a priori defined and they depend on the investment engaged. Landscapes are seen as makeable products that have to be depreciated.

Essentially, many global processes, both the natural ones and certainly also the demographic and economic ones, are not sustainable, but steer a unidirectional evolution. Economical processes are characterized by increasing productivity, reducing the time/cost distance (time–space compression), networking via hubs (world cities) and replacing or displacement to sustain market profit. At the local scale, the responses to these global drivers attempt sustainability, for the simple reason that for local people, displacement and migration are always the worst options (Antrop and Van Eetvelde 2017). Options at the local scale are multi-functional land use (specialization, intensification, upscaling), restructuring and innovation and maintenance. Sustaining landscapes at the local and regional scale is important for protection of the natural and cultural heritage and lifeways (Buttimer 2001), the natural and social capital (Potschin and Haines-Young 2006), landscape and ecosystem services (Wu 2013) and aesthetic qualities (Nohl 2001).

1.8.2 Climate Change, Migration and Energy Transition

During the geological history of the earth, climate change, tectonics and geo-morphogenesis have been the main factors of landscape formation and transformation. As long no people were witnessing this, concepts such as landscape, territory and scenery were irrelevant. It is the awareness of the risk for the existing lifeways that makes these natural processes—'as perceived by people'—significant. The details of the spatial impact of climate change are still uncertain. Nevertheless, we can be certain that it will affect all types of landscapes and ecological processes and system services and will affect the behavior of people (Opdam et al. 2009).

During history, climatic changes in some regions threatened subsistence, so whole populations moved on to more comfortable areas. Migration is primarily linked to climate changes and natural calamities combined with a loss of vital natural resources, often leading to social crises and conflicts. Climate changes and migration changed civilizations and landscapes several times (Zhang et al. 2007, 2011). Population movements are now associated with terms such as displacement, relocation and refugees. Concerning the landscape, new challenging questions arise: How do landscape and place attachment affect the newcomers' experience? How can decisions of resettlement also be based on the existing landscape qualities and identity? How can the principle of 'a right to landscape' from the European Landscape Convention be implemented? (Egoz et al. 2011, 2018; Egoz and De Nardi 2017).

The actual lifeways are characterized by increasing demand for energy consumption. During history, all energy transitions marked the landscape profoundly, as both territories and scenery. Pasqualetti and Stremke (2018) proposed a conceptual framework for a typology of what is now called *energy landscapes*. They recognized thirteen new landscape types based on substantive (the type of energy source), spatial (scale) and temporal (duration) qualifications. Five are based on renewable energy sources: wind, biomass, solar energy, geothermal energy and hydropower. The others are based on fossil fuels (peat, coal, oil, natural gas, unconventional fossil fuels, e.g., tar sands) and nuclear energy, and on collated energy (using two energy sources) and complex energy (using several technologies together in one landscape).

1.8.3 Dealing with the Landscape

Stakeholders, competencies and participation.

Selman (2006) showed the important differences that exist between the valuing of *insiders* and *outsiders* when planning at a landscape scale. Both can be 'expert' or 'layman.' Insiders are often also 'locals' or residents, while outsiders can be public authorities or recreationists and tourists, just passing by.

Studying environmental politics, Hägerstrand (2001) identified two main types of competencies stakeholders may hold: the territorial and spatial competence. The territorial competence is the one (private and public) landowners possess to make material changes in the land and is by far the most important one. The spatial competence refers to the power of the politicians, civil services, administrations and agencies have to regulate, stimulate and prohibit actions in land-management, which Hägerstrand called 'symbolic transactions.' These are organized in two ways: by functional specialization, i.e., sector authorities, and geographically by a specific territorial authority, such as a nature reserve.

Naveh (2007) argued that a transdisciplinary approach and participation are mandatory to solve all divergent issues in landscape planning. Since the Aarhus Convention, public participation in decision making is required (United Nations 1998) and included in the European Landscape Convention (Jones 2007). According to the ELC, participation includes "the general public, local and regional authorities, and other parties with an interest in the definition and implementation of landscape policies" (Council of Europe 2000, art. 5c).

Different types and levels of participation are recognized, ordered according to the engagement of the citizens: informing, consultation, cooperation, communication, advisory committees, management boards, partnership and community control (Jones and Stenseke 2011). Often, only the lower levels of participation are applied in practice, such as informing, consulting and communicating. Participation in landscape transformation decisions cannot be expected to result in stable solutions forever. In the present rapidly changing society, participatory processes need to be creative and flexible and have to include risk and uncertainty for long-term outcomes (Roe 2019).

Landscape as the scene of action.

The landscape scenery shows the result of actions. Olwig (2002) demonstrated how big landowners consciously used landscaping of their large estates to exhibit their power and wealth and to promote their ideology or political vision. However, most of the highly valued traditional landscapes today are not made by planning and design, but resulted from the labor of peasants and local communities trying to survive (Muir 2000).

Local landscape management becomes increasingly driven by centrally defined policies and decisions made by outsiders and at scales far above the local level. Often, this demands a long chain of communication with an important time–lag and a lot of noise. Intentions and objectives at the top level ensuing indirectly in distorted side-effects at the bottom. It is a challenge to understand how the information transfer works, evaluate the outcomes for both the policy intentions and the local development and find new balances and priorities beneficial for locals and outsiders (Pinto-Correia et al. 2006).

Landscape as an integrating concept.

As a holistic phenomenon, the landscape combines very different perspectives in one place or area, which makes it a common integrative concept for all stakeholders involved. 'Planning at a landscape scale' allows to agree on different visions and goals (Selman 2006). Nassauer (2012) proposed the local landscape to serve as a flexible boundary concept for landscape ecology, offering a common ground to scientists and practitioners with different backgrounds, values and interests. However, Pinto-Correia et al. (2018) defined five inherent tensions when it comes using the landscape as an integrating concept:

1. Landscape has no boundaries, but management arrangements depend on well-defined ones. Essentially, this means that while landscapes are perceived as scenery, policy and management exercise a territorial authority.
2. Many policies have significant landscape impacts, but landscape competence itself is not addressed. A classic example is the impact of the EU CAP on landscapes, while the European Commission has no direct competence for direct actions in the landscape and does not assume real responsibility in the matter.
3. The rural landscape is increasingly multi-functional, but in most regions, it is still structured as if agriculture were the only function. This monofunctional thinking is found in all sectors aiming at full control of their own singular interests. Multi-functionality should be a basic principle in all landscape planning.
4. Landscape heritage should be preserved, but a living landscape is a developing landscape. The general principle is that what loses its function and becomes obsolete degraded and finally disappears. Protection by 'freezing' as a museum is a paradox. This is the problem, for example, trying to preserve cultural landscapes as the 'organically evolved landscapes' in the UNESCO World Heritage.
5. Landscape is a common good, but it managed individual tract of land. Basically, this is the dichotomy between landscapes as scenery and territory.

Effects of introducing formal definitions of the landscape.

When early landscape research was mainly descriptive and academic, it gradually broadened its scope and field of application. After the Second World War, landscape research made a leap with aerial photograph survey and applications in land evaluation. A new step forward in landscape research came with the introduction of formal definitions of the landscape and the goals and measures formulated in international conventions relevant to the landscape. As an example, Table 1.2 summarizes the approaches and knowledge before and after the introduction of the European Landscape Convention (ELC) in 2000.

Table 1.2 Difference between landscape concepts, research and applications before and after the European Landscape Convention (ELC)

Ante-ELC	Post-ELC
Definition of landscape	
Many different definitions and concepts – Territorial vision in regional, historical geography – Scenic vision in landscape architecture and gardening and pictorial arts – Natural and cultural landscape	One formal definition – Both territorial and scenic vision – Character and identity – Interaction of natural processes and human activities – All is landscape, and there are no non-landscapes
Significance of landscape	
Landscape as heritage – Only special landscapes and places: the sublime, the spectacular, the iconic and the symbolic – Landscape as environment for nature and human living	**Landscape as common heritage and a human right** – All natural, rural, urban and peri-urban areas, including land, inland water and marine areas
Landscape research	
From classical geography – Regional monographs – Chorology of national and European landscapes (atlases) **From historical geography** – Landscape genesis and history – Typology of settlements, field systems and rural landscapes **From landscape ecology** – Scale, context, holons and hierarchy – Dynamics of change, processes and spatial structure – Composition and configuration patterns and landscape metrics **From archaeology** – Geographical and territorial context of sites as explaining factor	**Academic research: integrated landscape studies** – Multi- and interdisciplinary – Using common methods and technology (GIS, (spatial) statistics, remote sensing, etc.) **Applied research** – For the people, by the people – Transdisciplinary – Participation of the public and civil authorities **Perspective of sustainable development** – Conserve, protect, maintain and create landscapes

1.9 Prospects on Landscape Research

Considering the measures proposed in the landscape significant conventions, the actual landscape dynamics and the potential effects of the major driving forces, the following list of research priorities for the future can be formulated (after Antrop 2008):

1. There is no such thing as a non-landscape. So, all landscapes should be considered in all policy domains. This makes the landscape a central theme in all strategic, environmental and policy assessments. It means also that appropriate data and methods need to be widely available.

2. The contribution and impact of different landscape disciplines in policy are often very unequal and biased. To contribute equally, all landscape disciplines need to develop, besides pure academic research, also practical tools for applications in policy-making.
3. Much landscape research is still descriptive (the actual situation and the historical development) and mainly contributes to the understanding and knowledge of landscapes. Landscape research should be also proactive and focus more on prognosis and scenarios for the future.
4. Although inter- and transdisciplinary approaches are hype, rare are really integrated landscape studies. All disciplines need to work together from the beginning of a research project to the end. They have to understand each other's language (jargon) and formulate the problems and solutions together.
5. Participation is too often restricted to 'informing the public' or 'making inquiries.' Real and fair stakeholder participation should start at the beginning of a project. This is often the key to a successful implementation afterward.
6. Indicators of landscape change that are significant for policy-makers should be developed and tested (Parris 2004). Complex, abstract and 'black box' indicators are occasionally not transparent enough to be understood easily and, therefore, are not trusted. This is often the case with landscape metrics, which demand cautious interpretation and understanding the effects of factors such as scale, time and data quality. Also, thresholds need to be defined.
7. Expressing landscape values and services in monetary terms could be only helpful for some aspects. Most landscape values, however, are immaterial and cannot be monetized meaningfully. Economical assessments already applied in environmental impact studies could be adapted. This can be a strategic research option, as most of the driving forces that cause landscape change are economical. Possible examples are linking heritage to recreational and touristic services.
8. Available digital data with full spatial (and transborder) coverage are dealing with environmental themes such as topography, soils, land cover, vegetation zones and climate variables. Most landscape data are obtained indirectly from these environmental themes and inferred from census data collected from statistical sample units that are seldom representative for the actual landscape variation. There is a bias in favor of themes related to the material content (terrain, land cover, composition and configuration) of the landscape, while information about processes, functions and cultural themes is often lacking or fragmentary.
9. There is an urgent need for efficient monitoring of landscape changes to keep the date reliable. Landscape changes more rapidly than data are collected.
10. Administrative and political borders cause policies to increase landscape fragmentation. These territories are often discordant to landscape structures, and processes in the landscape are rarely controlled by such borders. The subsidiary principle stimulates bottom-up participatory planning and policy, promoting decision at the local level, which is often adapted to the landscape scale. This causes landscape fragmentation and may deteriorate identity and character of the landscape as a whole. The practical implementation of applying the 'aspirations' of the (local) public may not be favorable for a holistic view on the landscape.

1.10 Conclusions

The landscape is a complex, multi-scale and dynamical system, which essential nature is holistic. The etymology of the word combines intimately two essential properties, territory and scenery. Hence, the word landscape has multiple meanings and there are several ways of seeing the landscape. Humans are dealing with the landscape since they began organizing the land to fit their lifeways and shaped the landscape in a characteristic way reflecting their identity, community and culture. Scientific research with specific landscape as object began in the eighteenth century in geography, after landscape in arts was already a successful genre in painting and the literature, which stimulated landscaping as custom practice and the precursor of landscape architecture and planning.

In the early beginnings, landscape research integrated all aspects of the meaning and aimed at a holistic synthesis. The specialization of the sciences after the Second World War caused a temporary loss of the holistic approach and interest in the landscape. The fast environmental and societal changes in the late twentieth century resulted in the search of a new interdisciplinary synthesis, which was found in a renewed interest in landscape research. When in the past, landscape research was mainly descriptive, the present development focuses more on proactive and trans-disciplinary applications in policy, planning and management.

References

Aalen FHA, Whelan K, Stout M (eds) (1997) Atlas of the Irish rural landscape. Cork University Press, Cork, p 352

Aldred O, Fairclough G (2002) Historic landscape characterization—taking stock of the method. The National HLC Method Review 2002, English Heritage & Somerset County Council

Antikainen J (2005) The concept of Functional Urban Area—findings of the ESPON project 1.1.1. Informationen Raumentwicklung Heft 7:447–452

Antrop M (1997) The concept of traditional landscapes as a base for landscape evaluation and planning. The example of Flanders Region. Landscape Urban Plann 38:105–117

Antrop M (2000) Changing patterns in the urbanized countryside of Western Europe. Landscape Ecol 15(3):257–270

Antrop M (2005) Why landscapes of the past are important for the future. Landscape Urban Plann 1–2:21–34

Antrop M (2006) Sustainable landscapes: contradiction, fiction or utopia? Landscape Urban Plann 75:187–197

Antrop M (2012) Intrinsic values of landscapes. In: Papayannis T, Howard P (eds) 2012. Reclaiming the Greek Landscape, Athens, Med-INA, pp 31–42

Antrop M (2019) A brief history of landscape research. In: Howard P, Thompson I, Waterton E, Atha M (eds) The Routledge companion to landscape studies (2nd edn). Routledge, London and New York, pp. 1–15

Antrop M, Van Eetvelde V (2017) Landscape perspectives—the holistic nature of landscape. Landscape series, vol 23. Springer, p 436

Antrop M (1991) Rethinking carrying capacity. In: Brandt J, Agger P (eds) Proceedings of the European of the international association for landscape ecology (IALE) on practical landscape ecology. Roskilde University Centre: Roskilde, vol III, pp 55–64
Antrop M (2008) Landscapes at risk: about change in the European landscapes. In: Dostal P (ed) Evolution of geographical systems and risk processes in the global context. Charles University Prague, pp 57–79
Bell S (2004) Elements of visual design in the landscape. Spon Press, London, p 196
Bell S (1999) Landscape. Pattern, perception and process. Taylor & Francis Group, 344p
Botequilha Leitão AB, Ahern J (2002) Applying landscape ecological concepts and metrics in sustainable landscape planning. Landscape Urban Plann 59(2):65–93
Botequilha Leitão A, Miller J, Ahern J, McGarigal K (2006) Measuring landscapes. A planner's handbook. Island Press, London, 245 pp
Burel F, Baudry J (2003) Landscape ecology: concepts, methods and applications. Science Publishers, Inc., Enfield, New Hampshire, USA, 362 pp
Burrough PA, McDonnell RA (1998) Principles of geographical information systems. Oxford University Press, Oxford, 333 pp
Buttimer A (ed) (2001) Sustainable landscapes and lifeways: scale and appropriateness. Cork University Press, Cork, 398 pp
Cheshire P (1995) A new phase of urban development in Western Europe? The evidence for the 1980s'. Urban Stud 32(7):1045–1063
Claval PL (2004) The languages of rural landscapes. In: Palang H, Sooväli H, Antrop M, Setten S (eds) European rural landscapes: persistence and change in a globalising environment. Kluwer Academic Publishers, Dordrecht, pp 11–40
Cosgrove D, Daniels S (1988) The iconography of landscape. Cambridge University Press, Cambridge, p 318
Cosgrove D (2002) Landscape and the European sense of sight—eyeing nature. In: Anderson K, Domosh M, Pile S, Thrift N (eds) Handbook of cultural geography. Chapter 12. SAGE Publications, London, pp 249–268
Cosgrove D (2003) Landscape: ecology and semiosis. In: Palang H, Fry G (2003) Landscape interfaces. Cultural heritage in changing landscapes. Kluwer Academic Publishers, Dordrecht, 15–20 pp
Council of Europe (2000) European landscape convention. Firenze, 20.10.2000. http://conventions.coe.int/Treaty/en/Treaties/Html/176.htm. Accessed 15 Mar 2019
Council of Europe (2006) Landscape and sustainable development—challenges of the European landscape convention. Council of Europe Publishing, Strasbourg, p 213
Cushman SA, McGarigal K, Neel MC (2008) Parsimony in landscape metrics: strength, universality, and consistency. Ecol Ind 8(5):691–703
de la Blache PV (1922) Principes de géographie humaine. Armand Colin, Paris (trans: Principles of human geography), de Martonne E (ed). Holt, New York
Eastman JR, Kyem PAK, Toledano J, Jin W (1993) GIS and decision making. UNITAR Explorations in GIS Technology, vol 4. UNITAR, Geneva, p 112
Egoz S, Jørgensen K, Ruggeri D (eds) (2018) Defining landscape democracy: a path to spatial justice. Edward Elgar Publishing, Cheltenham, UK. https://doi.org/10.4337/9781786438348
Egoz S, Makhzoumi J, Pungetti G (eds) (2011) The right to landscape: contesting landscape and human rights. Routledge, London and New York, p 337
Egoz S, De Nardi A (2017) Defining landscape justice: the role of landscape in supporting wellbeing of migrants, a literature review. Landscape Res. https://doi.org/10.1080/01426397.2017.1363880
European Environmental Agency (1995) Europe's environment—the Dobríš assessment. European Environment Agency, Copenhagen, Denmark, p 680
European Environmental Agency (2002) Urban morphological zones 2000 (UMZ2000)
European Commission (2008) http://ec.europa.eu/agriculture/rurdev/index_en.htm. Accessed 15 Mar 2019

Fahrig L (2005) When is a landscape perspective important? In: Wiens JA, Moss MR (eds) Issues in landscape ecology. Studies in landscape ecology. Cambridge University Press, Cambridge, pp 3–10
Farina A, Belgrano A (2006) The eco-field hypothesis: toward a cognitive landscape. Landscape Ecol 21(1):5–17
Farina A (2009) Ecology, cognition and landscape: linking natural and social systems. Landscape series, vol 11. Springer Science & Business Media, 161 pp
Forman RTT, Godron M (1981) Patches and structural components for a landscape ecology. Bioscience 31(10):733–740. In: Wiens JH, Moss MR, Turner MG, Mladenoff DJ (eds.) Foundation Papers in Landscape Ecology. Columbia University Press, New York, pp 246–253
Forman R, Godron M (1986) Landscape ecology. Wiley, New York, p 619
Fry GL, Sarlov Herlin I (1995) Landscape design: how do we incorporate ecological, cultural and aesthetic values in landscape assessment and design principles? In: Griffiths GH (ed) Landscape ecology: theory and application. Proceedings IALE-UK conference, Reading, 51–60 pp
Geist HJ, Lambin EF (2002) Proximate causes and underlying driving forces of tropical deforestation. Bioscience 52(2):143–150
Granö JG (1929) Reine Geographie. Eine methodologische Studie beleuchtet mit Beispielen aus Finnland und Estland. Helsinki, Helsingfors. Acta Geogr 2(2):202 pp
Granö O, Paasi A (eds) (1997) Pure geography. In: Granö JG (trans: Hicks M). The John Hopkins University Press, Baltimore and London, 191 pp
de Groot RS, Alkemade R, Braat L, Hein L, Willemen L (2010) Challenges in integrating the concept of ecosystem services and values in landscape planning, management and decision-making. Ecol Complex 7:260–272
Groth P, Bressi TW (eds) (1997) Understanding ordinary landscapes. Yale University Press, New Haven, 272 pp
Hardin G (1968) The tragedy of the commons. Science 162(3859):1243–1248
Hartshorne R (1939) The nature of geography. A critical survey of current thought in the light of the past. Ann Assoc Am Geogr 29(3–4):173–658
Hägerstrand T (2001) A look at the political geography of environmental management. In: Buttimer A (ed) Sustainable landscapes and lifeways: scale and appropriateness. Cork University Press, Cork, pp 35–58
Jepsen MR, Kuemmerle T, Müller D, Erbe K, Verburg PH, Haberl H, Vesterager JP, Andric M, Antrop M, Austrheim G, Björn I, Bondeau A, Bürgi M, Bryson J, Caspar G, Cassaro JF, Conrado E, Chromy P, Daugirdas V, Van Eetvelde V, Elena-Rosselló R, Gimmi U, Izakovicova Z, Jancák V, Jansson U, Kladnik D, Kozak J, Konkoly-Gyurów E, Krausmann F, Mander Ü, McDonagh J, Pärn J, Niedertscheider M, Nikodemus O, Ostapowicz K, Pérez-Soba M, Pinto-Correia T, Ribokas G, Rounsevell M, Schistou D, Schmit C, Terkenli TS, Tretvik AM, Trzepacz P, Vadineanu A, Walz A, Zhllima E, Reenberg A (2015) Transitions in European land-management regimes between 1800 and 2010. Land Use Policy 49:53–64
Jones M (2007) The European landscape convention and the question of public participation. Landscape Res 32(5):613–633
Jones M, Daugstad K (1997) Usages of the "cultural landscape" concept in Norwegian and Nordic landscape administration'. Landscape Res 22(3):267–281
Jones M, Stenseke M (eds) (2011) The European landscape convention. Challenges of participation. Landscape series, vol 13. Springer, Dordrecht, 326 pp
Kant I (1764) Observations on the feeling of the beautiful and sublime [Beobachtungen über das Gefühl des Schönen und Erhabenen]. (eds and trans by P. Frierson, P. Guyer). Cambridge University Press, Cambridge, p 396 (2011)
Klijn JA (2004) Driving forces behind landscape transformation in Europe, from a conceptual approach to policy options. In: Jongman RHG (ed) The new dimensions of the European landscapes Alterra Green World Research, Landscape Europe. Wageningen UR frontis series, vol 4. Springer, pp 201–218

Kolen J, Renes H, Hermans R (eds) (2015) Landscape biographies. Geographical, historical and archaeological perspectives on the production and transmission of landscapes. Landscape and heritage studies. Amsterdam University Press, p. 437
Laszlo E (1994) The choice: evolution or extinction. A thinking person's guide to global issues. C. P. Putnam & Sons, New York, p 215
Lemaire T (2002) Met open zinnen. Natuur, landschap, aarde. Amsterdam, Ambo, p 358
Leopold A (1949) The land ethic. A Sand County Almanac. Oxford University Press, New York, p 182
Leopold LB, Clarke FE, Hanshaw BB, Balsley JR (1971) A procedure for evaluating environmental impact, vol 645. US Geological Survey Circular, Washington DC, p 13
Li H, Wu J (2004) Use and misuse of landscape metrics. Landscape Ecol 19(4):389–399
Li H, Wu J (2007) Landscape pattern analysis: key issues and challenges. In: Wu J, Hobbs RJ (eds) Key topics in landscape ecology. Studies in landscape ecology. Cambridge University Press, pp 39–61
Longley PA, Goodchild MF, Maguire DJ, Rind DW (2001) Geographic information systems and science. Wiley, New York, p 454
Lowenthal D (1975) Past time present place: landscape and memory. Geogr Rev 65(1):1–36
Luginbühl Y (2012) La mise en scène du monde. Construction du paysage européen. Paris: CRNS Editions. 432 pp
Lynch K (1960) The image of the city. The MIT Press, Cambridge, MA, London, 194 pp
Mac Nally R (2005) Scale and an organism-centric focus for studying interspecific interactions in landscapes. In: Wiens JA, Moss MR (eds) Issues in landscape ecology. Studies in landscape ecology. Cambridge University Press, Cambridge, pp 52–69
Mander Ü, Uuemaa E, Roosaare J, Aunap R, Antrop M (2010) Coherence and fragmentation of landscape patterns as characterized by correlograms: a case study of Estonia. Landscape Urban Plann 94(1):31–37
Van Mansvelt JD (1997) An interdisciplinary approach to integrate a range of agro-landscape values as proposed by representatives of various disciplines. Agric Ecosyst Environ 63:233–250
McGarigal K, Marks BJ (1995) FRAGSTATS: spatial pattern analysis program for quantifying landscape structure. USDA Forest Service General Technical Report PNW-351
Mitchell B (1979) Geography and resources analysis. Longman, London, p 399
Moss MR (1999) Fostering academic and institutional activities in landscape ecology. In: Wiens JA, Moss MR (eds) Issues in landscape ecology. International association for landscape ecology, fifth world congress, Snowmass Village, CO, USA, 138–144
Muir R (2000) The new reading the landscape. Fieldwork in landscape history. University of Exeter Press, Exeter, p 256
Nassauer JI (2012) Landscape as medium and method for synthesis in urban ecological design. Landscape Urban Plann 106:221–229
Naveh Z, Lieberman A (1994) Landscape ecology: theory and application. Springer, New York, p 360
Naveh Z (2007) Transdisciplinary challenges in landscape ecology and restoration ecology—an anthology. Landscape series, vol 6, XVII. Springer, Dordrecht, 423 pp
Neef E (1967) The theoretical foundations of landscape study (Die theoretischen Grundlagen der Landschaftslehre) (trans) Geographisch-Kartographische, Anstalt, ed., 18–38. VEB Hermann Haack, Gotha/Leipzig. In: Wiens JH, Moss MR, Turner MG, Mladenoff DJ (eds) Foundation papers in landscape ecology. Columbia University Press, New York, pp 225–245
Nicolson (1995) A historical introduction. In: von Humboldt A (ed) Personal narrative of a journey to the equinoctial regions of the new continent. Abridged and translated by Jason Wilson and a historical introduction by Malcolm Nicholson. Penguin Books, London, p 311
Nijhuis S, van Lammeren R, van der Hoeven F (2011) Exploring the visual landscape, research in urbanism series, vol 2. IOS Press, Amsterdam, p 334
Nohl W (2001) Sustainable landscape use and aesthetic perception—preliminary reflections on future landscape aesthetics. Landscape Urban Plann 54:223–237

Ode Å, Tveit MS, Fry G (2008) Capturing landscape visual character using indicators: touching base with landscape aesthetic theory. Landscape Res 33(1):89–117
OECD (2012) Defining Functional Urban Areas (FUA) for the OECD metropolitan database. OECD, 9 pp
Olwig KR (1996) Recovering the substantive nature of landscape. Ann Assoc Am Geogr 86 (4):630–653
Olwig KR (2002) Landscape, nature and the body politic: from Britain's renaissance to America's new world. University of Wisconsin Press, Madison, p 299
Opdam P, Luque S, Jones KB (2009) Changing landscapes to accommodate for climate change impacts: a call for landscape ecology. Landscape Ecol 24(6):715–721. https://doi.org/10.1007/s10980-009-9377-1
Paffen K (ed) (1973) Das Wesen der Landschaft. Wege der Forschung, Band XXXIX, Wissenschaftliche Buchgesellschaft, Darmstadt, 514 pp
Parris K (2003) Agricultural landscape indicators in the context of the OECD work on Agri-10 environmental indicators. In: Dramstad W, Sogge C (eds) Agricultural impacts on landscapes; developing indicators for policy analysis. Proceedings from the NIJOS/OECD expert meeting on agricultural landscape indicators in Oslo, Norway, 7–9 Oct 2002, pp 10–18
Parris K (2004) Measuring changes in agricultural landscapes as a tool for policy makers. In: Brandt J, Vejre H (eds) Multifunctional landscapes. Theory, values and history, vol I. WIT Press, Southampton, 193–218 pp
Pasqualetti M, Stremke S (2018) Energy landscapes in a crowded world: a first typology of origins and expressions. Energy Res Soc Sci 36:94–105
Passarge S (1919–1921) Grundlage der Landschaftskunde, 3 vol. Friedrichsen, Hamburg
Passarge S (1921–1930) Vergleichende Landschaftskunde, 5 vol. Reimer, Berlin
Phipps M (1984) Rural landscape dynamics: the illustration of some key concepts. In: Brandt J, Agger P (eds) Proceedings of the first international seminar of IALE on the methodology in landscape ecological research and planning, vol I, Theme I. Roskilde University Centre, Roskilde, pp 47–54
Pinto-Correia T, Gustavsson R, Pirnat J (2006) Bridging the gap between centrally defined policies and local decisions—towards more sensitive and creative rural landscape management. Landscape Ecol 21(3):333–346
Pinto-Correia T, Cancela d'Abreu A, Oliveira R (2003) Landscape areas in Portugal—can they be a support for applying indicators? NIJOS/OECD expert meeting on agricultural landscape 7–9 Oct 2002, Oslo, Norway, NIJOS rapport 7/2003, pp 309–316
Pinto-Correia T, Primdahl J, Pedroli B (2018) European landscapes in transition—implications for policy and practice. Cambridge University Press, 330 pp
Plieninger T, Draux H, Fagerholm N, Bieling C, Bürgi M, Kizos T, Kuemmerle T, Primdahl J, Verburg PH (2016) The driving forces of landscape change in Europe: a systematic review of the evidence. Land Use Policy 57:204–214. https://doi.org/10.1016/j.landusepol.2016.04.040
Potschin MB, Haines-Young RH (2006) Landscapes and sustainability. Landscape Urban Plann 75(3–4):155–161
Price C (2019) Researching the economics of landscape. In: Howard P, Thompson I, Waterton E, Atha M (eds) The Routledge companion to landscape studies, 2nd edn. Routledge, London, pp 387–401
Prigogine I (1955) Introduction to Thermodynamics of Irreversible Processes. Charles C. Thomas Publisher, Springfield, IL, p 115
Prigogine I (1987) Exploring complexity. Eur J Oper Res 30(2):97–103
Primdahl J, Swaffield S (Eds.) (2010) Globalization and agricultural landscapes—change patterns and policy trends in developed countries. Cambridge University Press, Cambridge, 292 pp
Rees W (2001) Ecological footprint, concept of. In: Encyclopedia of biodiversity, vol 4. Academic Press, Waltham, MA, pp 229–244
Roe M (2019) Landscape and participation. In: Howard P, Thompson I, Waterton E, Atha M (eds) The Routledge companion to landscape studies, 2nd edn. Routledge, London, pp 402–417

Rolston HI (1988) Environmental ethics. Temple University Press, Philadelphia, p 373
Salerno F, Viviano G, Manfredi EC, Caroli P, Thakuri S, Tartari G (2013) Multiple carrying capacities from a management-oriented perspective to operationalize sustainable tourism in protected areas. J Environ Manage 128:116–125
Sauer CO (1925) The morphology of landscape. University of California Publications in Geography, 2; reprinted in Leighly J (ed) (1974) Land and life. University of California Press, Berkeley, CA, pp 315–350
Saul H, Waterton E (2019) Anthropocene landscapes. In: Howard P, Thompson I, Waterton E, Atha M (eds) The Routledge companion to landscape studies, 2nd edn. Routledge, London, pp 139–151
Schama S (1995) Landscape and memory. A. Knopf, New York, p 652
Selman P (2006) Planning at the landscape scale. Routledge, London, p 213
Smuts JC (1926) Holism and evolution. The Macmillan Company, New York, p 376
Steffen W, Broadgate W, Deutsche L, Gaffney O, Ludwig C (2015) The trajectory of the Anthropocene: the great acceleration. Anthropocene Rev 2(1):81–98. https://doi.org/10.1177/2053019614564785
Steffen W, Crutzen J, McNeil JC (2007) The Anthropocene: are humans now overwhelming the great forces of nature? Ambio 36:614–621
Swanwick C (2002) Landscape character assessment. Guidance for England and Scotland. The Countryside Agency, Scottish Natural Heritage. https://www.nature.scot/sites/default/files/2018-02/Publication%202002%20-%20Landscape%20Character%20Assessment%20guidance%20for%20England%20and%20Scotland.pdf. Accessed 15 Mar 2019
Taylor C (2006) The making of the English landscape and beyond: inspiration and dissemination. Landscapes 2:96–104
Thomas WL (ed) (1956) Man's role in the changing face of the earth. The University of Chicago Press, Chicago, p 1193
Tremorshuizen JW, Opdam P (2009) Landscape services as a bridge between landscape ecology and sustainable development. Landscape Ecol 24(8):1037–1052
Tsjallingii SP, de Veer AA (eds) (1982) Perspectives in landscape ecology. Centre for Agricultural Publishing and Documentation, Wageningen
Tuan Y-F (1974) Topophilia: a study of environmental perception, attitudes and values. Prentice Hall, Englewood Cliffs, NJ, 260 pp
UNESCO (1992) Cultural landscapes. http://whc.unesco.org/en/culturallandscape/#1. Accessed 15 Mar 2019
United Nations (1998) Aarhus convention on access to information, public participation in decision-making and access to justice in environmental matters. United Nations, Aarhus. http://www.unece.org/fileadmin/DAM/env/pp/Publications/Aarhus_Implementation_Guide_interactive_eng.pdf. Accessed 15 Mar 2019
Unwin D (1981) Introductory spatial analysis. Methuen, London, p 212
Uuemaa E, Mander Ü, Marja R (2012) Trends in the use of landscape spatial metrics as landscape indicators: a review. Ecol Ind 28:100–106. https://doi.org/10.1016/j.ecolind.2012.07.018
Uuemaa E, Antrop M, Roosaare J, Marja R, Mander Ü (2009) Landscape metrics and indices: an overview of their use in landscape research. Living Rev Landscape Res 3:1–28. http://lrlr.landscapeonline.de/Articles/lrlr-2009-1/
von Humboldt A (1849) COSMOS: Sketch of a physical description of the universe, Vol. II, Translated by Sabine E, 7th edn, Longman, Brown, Green and Longmans, London
von Humboldt A (1858) Cosmos – a sketch of the physical description of the universe, Vol. I, Translation Otte, E.C., John Hopkins University Press, Baltimore and London
Vos W, Klijn J (2000) Trends in European landscape development: prospects for a sustainable future. In: Klijn J, Vos W (eds) From landscape ecology to landscape science. WLO, Wageningen, Kluwer Academic Publishers, pp 13–30
Wascher D (2003) Overview on agricultural landscape indicators across OECD countries. In: Dramstad W, Sogge C (eds) Agricultural impacts on landscapes; developing indicators for

policy analysis. Proceedings from the NIJOS/OECD expert meeting on agricultural landscape indicators in Oslo, Norway, 7–9 Oct 2002, pp 19–36
Wu J (2013) Landscape sustainability science: ecosystem services and human well-being in changing landscapes. Landscape Ecol 28:999–1023
Wulf A (2015) The Invention of Nature – Alexander von Humboldt's New World. A.A.Knopf, New York, 473 pp
Zhang DD, Brecke P, Lee HF, He Yuan-Qing, Zhang J (2007) Global climate change, war, and population decline in recent human history. PNAS 104(49):19214–19219
Zhang DD, Lee HF, Wang Cong, Li Baosheng, Pei Qing, Zhang J, An Ylun (2011) Causality analysis of climate change and large-scale human crisis. PNAS 108(42):17296–17301
Zonneveld JIS (1980) Het gat in de geografie. K.N.A.G. Tijdschrift, XIV 2:93–98
Zonneveld IS (1989) The land unit—a fundamental concept in landscape ecology and its applications. Landscape Ecol 3(2):67–89
Zonneveld IS (1995) Land ecology. SPB Academic Publishing bv, Amsterdam, 199 pp
Zonneveld IS (2005) The land unit as a black box: a Pandora's box? In: Wiens JA, Moss MR (eds) Issues in landscape ecology. Studies in landscape ecology. Cambridge University Press, Cambridge, pp 331–345

Chapter 2
Landscape Concept in History of Russian (Soviet) Geography

Nikolai M. Dronin and John M. Francis

Abstract Our historical review aims to clarify the modern meaning of the term "landscape" in the context of historical evolution of the notion in the twentieth century. A remarkable feature of the history of Russian geography was the coexistence of multiple readings of the term "landscape." In the 1910s, landscape notion had an important human dimension, whereas Soviet geographers interpreted landscape as a principally natural object. Nor was there ever a monopoly on reading the term "landscape" in the Soviet period. For some Soviet geographers landscapes had definite scale (Solntsev's landscape) while for many others, the term was applicable for a complex of any scale and rank. For some scientists the terms "landscape" and "geosystem" were synonymous but for others—not. All too often collisions of different interpretations of landscape were a matter of definition. Imperfect argumentation resulted in marginalization of the study of landscape morphology, the most original concept in Soviet geography, and blurring the term. Most modern Russian geographers use the term "landscape" as a synonym of natural complex applicable to units at any spatial scale. The term "geosystem" also is used widely by modern geographers referring in a generalizing and non-selective way to all natural complexes regardless of their spatial scale and rank. For the bulk of modern Russian geographers, the terms "landscape," "natural complex," and "geosystem" are loosely interchangeable.

Keywords Soviet geography · Landscape · Anthropogeography · Physical-geographical regionalization · Geosystem

---

N. M. Dronin (✉)
Faculty of Geography, Lomonosov Moscow State University, Leninskie Gory, 119991 Moscow, Russia
e-mail: ndronin@gmail.com

J. M. Francis
JMF Research Associates, Philadelphia, USA

L. Müller and F. Eulenstein (eds.), *Current Trends in Landscape Research*, Innovations in Landscape Research, https://doi.org/10.1007/978-3-030-30069-2_2

2.1 Introduction

"Landscape" as an organizing concept in Russian geography appeared at the very beginning of the twentieth century and still widely used by modern Russian geographers. At its core, landscape theory presents the earth's surface as composed of landscapes, natural units or cells of highly interrelated natural components. From the very beginning the idea in Russian geography was characterized by extreme confusion in interpreting the very notion "landscapes." It could refer to most anything—scenery, individual territories, and homogeneous territories—and was applied at very different spatial scales—from local to regional. In the work of a single author, one readily finds quite diverse uses of the term "landscape." This ambiguity was somehow sustained during the Soviet period. Clearly, most geographers were little concerned about the multitude of possible readings of the key term in their discipline or with resolution of the ensuing ambiguities. This flexibility, on the one hand, guaranteed long-standing of the notion but, on the other hand, any science should avoid ambiguity of its key term.

There is a rich literature on the historical evolution of the landscape concept in the first half of the twentieth century (Zabelin 1959, 1969; Isachenko 1971; Sukhova 1981; Preobrazhenskij and Makarov 1988; Dronin 1999). Nor does the most recent work on the history of Russian geography go beyond 1960 (Shaw and Oldfield 2016). Our review covers the same period from the early 1900s to the late 1960s, as we see these years as the most important for shaping contemporary Russian physical geography. Many historians noticed unsettled coexistence of different interpretations [*traktovki*] of the term landscape, especially, in early years. The diverse interpretations used to be regarded as difference in personal views of Russian geographers. At the same time, different readings of the term landscape found in the works of a single author were considered by some authors as an example of a logical flaw (Isachenko 1971; Sukhova 1981; Dronin 1999). In general the phenomenon of coexistence of different interpretations of the term has never been a distinct subject for historical and epistemological investigation. We show that this historical phenomenon is important for understanding the modern state of landscape geography in Russia.

2.2 Diffusion of Landscape Notion in Russian Geography (1900–1910s)

Many authors have agreed that the appearance of the landscape concept was associated with the crisis of geographical science in the late nineteenth century. Semenov-Tyan-Shanskij (1928) stated that the crisis manifested itself as a comprehensive uncertainty as to the subject of geography as a fundamental science. According to Zabelin (1959, 1969), the crisis was associated with the "exhaustion of descriptive methodology of geographical science." Nobody would perceive geography as a real science if it only presented nature and folk portrayals of different countries. Isachenko (1971) saw the explanation in the appearance of discrete disciplines (hydrology, meteorology, geomorphology, and others) with the concomitant fracturing of geography as holistic science.

This identity crisis of geographical science should be regarded in the broader context of the so-called Big Science project of the late of nineteenth century (Shaw and Oldfield 2016). New universities and scientific institutions were appearing in industrialized states.

In Russia, geography was beginning to establish itself within higher education system after a special decree issued by Ministry of Education in 1884. Russian geographers welcomed the appearance of their science at universities in Moscow, Saint Petersburg, Kazan, Kiev, Kharkiv, and Novorossijsk, but they had no answers to the tough questions about their subject, its structure, and methodology. In 1875, a special survey among professors of Russian universities revealed uncertainty and skepticism as to the location of the subject within the system of higher education. Most thought of geography as a supplementary discipline somewhere within the historical-philological faculty (Esakov 1959). In 1887, the Russian Geographical Society (RGO) proposed to establish a separate geographical division within the faculty of physics and mathematics. The new division would comprise a chair of general geography and several additional ones on astronomy, geodesics, physics, chemistry, comparative linguistics, world literature, world history, and history of Russia (Petri 1892a). While some geographers found the project "beautiful" (Krotov 1892), others described Dokuchaev's characterization of geography as "diffluent in all directions" or Krasnov's description as a "vinaigrette of sciences" (Krasnov 1912).

This uncertainty around geography had been overcome with spreading ideas of a German geographer Alfred Hettner. In 1905, he published a methodological article "Substance and Methods of Geography" in a German journal *Geographische Zeitschrift,* specializing in human geography. Hettner proposed a classification of sciences into three groups of disciplines: systematic, chronological, and chorological. The goal of the systematic sciences (physics, chemistry, biology) would be to discover the universal natural laws while chronological (history), and chorological sciences would study, respectively, phenomena in relation to time (history and philology) and space (geography and astronomy). Implementation of this classification with respect to geography implied a principled rejection of the search for general laws for Earth as a whole as such research would fall in the turf of physics as systematic science. It occasioned doubt as to the scientific status of General Physical Geography which some geographers regarded as a variety of natural philosophy but not positivistic science.[1] Following Hettner, the task of geographical science would be to investigate the spatial differentiation of natural

[1]Petri (1892b), the head of the first chair of geography at Saint Petersburg University, wrote that geography could be a successor to the Hegelian philosophical system, which, in his words, had collapsed some time earlier.

and human phenomenon on the earth's surface as interrelated following natural boundaries or landscapes, but not, for example, administrative or political ones.[2]

Thus, Hettner's ideas offered fertile soil for the bulk of Russian geographers at sea between the extremes of geography as the highest synthesis of geology, geophysics, hydrology, meteorology, and other disciplines (Petri 1892b) and a heterogeneous collection of the spatial aspects of the naturalist and humanitarian disciplines (Anuchin 1949). Geography as a chorological science studies quite different phenomena in their interaction with Earth's surface. This approach resolved another thorny issue, namely how to combine natural and anthropogenic elements in geographical studies. A German geographer, Ed. Banze, who had done yeoman's work in promoting Hettner's ideas in Russia,[3] claimed that any phenomenon could be investigated by "chorological" methods—fancy ladies' hats, for example, if they were interconnected with other elements of landscape (Yarilov 1913).

In the 1910s, Hettner's chorological model gained great popularity among Russian geographers. On October 11, 1913, no less an authority on questions of geography than Lev Semenovich Berg delivered a programmatic speech *The subject and tasks of geography* at the RGO in which he vigorously affirmed Hettner's view of geography as the chorological science (Berg 1915). Following Hettner, Berg decisively rejected General Physical Geography. This antagonistic attitude toward global geographical research would become an index of the principal commitment of those Russian geographers (e.g., Borzov 1912) who identified as firm supporters of the chorological vision.

This first diffusion of the chorological reading in Russian geography was characterized by extreme confusion in interpreting the very notion "natural complex" or "landscapes." It is safe to say that nobody knew what the term meant. It could refer to most anything and could be applied at very different spatial scales—from local to regional. In the work of a single author, one readily finds quite diverse uses of term. In one and the same publication, "landscape" could be synonymous with "scenery," with depictions of smells and sound at sunset, while used at the same time to denote the macro-regions of European Russia (e.g., Central Agricultural Region) (Nechaev 1905; Borzov 1908).[4] Clearly, most geographers

[2]Systematic sciences also have geographical aspects. For example, as a systematic science economics could investigate spatial aspects of economy. But these two disciplines "geographical economy" and "economical geography" would be, in principle, distinct. The former studies peculiarities of spatial distribution of economic phenomenon, while the latter investigates how the economic phenomena interact with all other natural and human elements in landscapes.

[3]Russian geographers first encountered the chorological concept through the publications of Edward Banze (Dronin 1999). Some of the geographers (e.g., Berg) criticized Banze for his "vulgarian–feuilleton" language.

[4]Understanding landscape as a scenery *[peizazh]* was one of the earliest interpretations of landscape notion. As early as in 1905, A. P. Nechaev published a small book *Pictures of the Motherland*. According to Sukhova (1981), the book was the first publication in Russian geography in which landscape was called as the major subject of geographical science. Nechaev defined landscape as formed by "all objects located on the Earth surface—stones, streams, plants, animals,

were little concerned about the multitude of possible readings of the new key term in their discipline or with resolution of the ensuing ambiguities. Sometimes, they mentioned competing views of colleagues but did little to take issue with the disparities. On occasion, they referred to views of their colleagues but did so wrongly. Berg in one publication (Berg 1915) rebuked Hettner for considering administrative states as landscapes, although Hettner had never advanced such a claim. This evident failure to endow the key notion underlying their endeavor with precise definition of scope and structure can be only explained by the observation that the major aim of Russian geographers during this period was to justify the place of geography as a fundamental discipline in the family of other university sciences. Chorology offered this justification.

An alternative version of origin of Russian landscape geography was proposed by Soviet historians in the course of ideological reconstruction of Soviet science in the 1930s. According to this version not Hettner, but a nineteenth-century Russian soil scientist, V. V. Dokuchaev, was proclaimed true founder of landscape theory. However, there is considerable historic evidence against this version of history. In the beginning of the twentieth century, Soviet historians were prone to proclaim Dokuchaev the master of thought [*vlastitel' dum*] of all Russian geographers (Preobrazhenskij et al. 1997). This is not true. When Dokuchaev died in 1903, geographers did not react to the passing of this renowned soil scientist. There was only A. A. Kruber's brief obituary which did not mention any of Dokuchaev's contributions to geography (Kruber 1904). Some Soviet geographers have even identified Berg as Dokuchaev's student. However, there is direct evidence from Berg himself that he learned of Dokuchaev's work on natural zones only in 1927 from an article written by the soil scientist K. D. Glinka (Berg 1962). These discussions on natural zones were of a popular character and appeared in a local newspaper Caucasus in 1889 and were republished only in 1949 (Dokuchaev 1949).

---

and humans with diverse results of their activities." The leading role in formation of scenery was given to geomorphological structure of a site, while vegetation and cultural objects were essential but secondary. In 1908, A. A. Borzov published a book under the similar title *Pictures on Geography of Russia*. In the preface he stressed that "landscape" could be synonymous with "scenery," with depictions of smells and sound at sunset (Borzov 1908). The book presented descriptions of 12 regions of European Russia [borrowed from the scheme of Kruber (1898)] with reference to their scenery characteristics. For example, the Central Agricultural Region was described with such language as: "…all contours and colors here are so elegant, so soft, all transitions are so gentle, that only here could have appeared the melancholic and heartfelt poetry of Turgenev and the graceful miniatures of Fet. The month of May is captioned: greeneries are fresh and brighter being washed by recent rain." (p. 9).

There was no reflection of this exotic interpretation in Russian geography. For example, in German geography this interpretation of term "landscape" was justified as a positivistic attempt to restrict geographical research to visual subjects thus excluding mental ones like ideas, faiths, morality, etc.

2.3 First Interpretations of "Landscape" (1920s)

Despite of all political and social turmoil, the 1920s witnessed an impressive expansion of the education system in Soviet Russia. All universities were open to young people without regard to entrance examinations. Petrograd (as the newly Russified Sankt-Peterburg was briefly known), where the first Institute of Geography was established in 1918, emerged as the center of education in the field. The institute employed, among others, such famous scientists as Berg himself (a biologist and geographer), the statistician V. P. Semenov-Tyan-Shanskij, the botanist V. N. Sukachev, the geologist A. E. Fersman, the meteorologist P. I. Brounov, the economist S. V. Bernshtein-Kogan, the linguist and historian N. Y. Marr, and the physiologist L. A. Orbeli. By 1919, the institute had enrolled 100 students, by 1920, 340, and by 1923, 641, when an additional 889 were registered as auditors. However, postgraduate statistics were less impressive: In 1921, just five people were enrolled in postgraduate studies, a number that reached 50 by 1925. In 1926, the Institute of Geography was united with the Geography Faculty of Leningrad University. This was accompanied by a reduction in the number of students.

In general, publication by Russian geographers was modest, especially, in the first half of the 1920s. In 1925, a collection of methodological articles by Russian geographers under the title *Questions of Country Studies* (*Voprosy Stranovedeniya*) was published by the Institute of Geography. The collection opened with a translation of Hettner's famous paper of 1905 *The State of the Art and Methods of Geography* (Hettner 1925). The contributors to the collection all paid tribute to Hettner and the chorological methodology. Berg, in his article "Geography and its place among other disciplines", repeated the principal arguments in favor of the chorological method and the understanding of geography as a science of landscapes. He accepted characterization of General Physical Study as embracing only landscapes on earth's surface. The collection is interesting for the presence of papers written by economic geographers who argued that that the subject of economic geography was "the economic region" which spatially either somehow "correlated with" (S. V. Bernshtein-Kogan) or essentially "matched" (A. A. Grigor'ev) landscapes. Their participation in this collection made sense in respect to the ongoing discussion between regional and branch approaches in economical geography. Both scientists found purchase in Hettner's distinction between systematic and chorological disciplines. The chorological interpretation gained important support as well from secondary school methodologists who found the chorological approach appealing as a way to present geography to younger students (Sukhova 2018).

At the same time, lacking crisp definition in practice, the term "landscape" itself continued to present a confused profile in the work of leading geographers. Some continued to employ the term as synonymous with "scenery," as a visually fixed harmony whole of different natural and cultural elements (*peizazh*). Semenov-Tyan-Shanskij, in particular, promoted this interpretation in his very original work *Region and Country* (1928). He believed that Berg shared this understanding of "landscape." Indeed in his article of 1925, Berg defined "landscape" abstractly as "an organism" and "harmonic unit" and added the term "scenery" in brackets to illustrate his thought. At the same time Berg, insisting on homogeneity as criterial for landscape delineation, criticized Hettner for his reading of landscape as administrative or historical territory such as Crimea. Berg argued that Crimea was not sufficiently homogeneous to be taken as a landscape and consisted rather of several landscapes. At the same time, Berg offered as his own example of "landscape" the entire West Siberian plain. Thus, in the 1920s, three major and divergent interpretations of the term "landscape" coexisted in Russian geography: landscape as scenery, as individual region, and as homogeneous territory. Geographers provided no arguments to substantiate their choice in favor of one or another of these interpretations. Nor did they offer counterarguments challenging alternative definitions. Sometimes they provided contradicted examples of landscapes. In his 1925 collection, Berg again offered unsorted examples of landscape: sand deserts, river valleys, and west Siberian plain. Hettner's own position was more consistent: In his major theoretical work *Geography, its history, meaning, and methods*, published in Germany in 1925 and translated into Russian in 1930, he reflected on different interpretations of landscapes: the individual (Crimea), the homogeneous (steppe), and the functional (river basin). He concluded that, depending on the purpose at hand, any of these three variants was appropriate.

The same confusion can be seen with respect to the spatial scale of "landscape." Most geographers, including Hettner, considered landscape as one taxon in the hierarchy of natural complexes: the most general being "parts of the world" (continents), followed by macro-region (oblast' or zone), then landscape and site (*mestnost*) as the most specific. Thus, most geographers thought of landscape as a unit located somewhere between regional and local. They provided no arguments, however, for this hierarchy or for the presence of one privileged taxon—landscape. It was mere intuition that dictated that landscape should be more important than any other taxon (site is too small, zone is too large). Berg (1925) proposed a hierarchy of natural complexes on the analogy with the classical biological taxonomy: sites correspond to species, landscape to genus, and zone to family. However, there is a key difference between geographical and biological taxonomies. There is no requirement that plants or animals of a particular taxon (species, genus, or family) be distributed contiguously. According to Berg, sites constitute a homogeneous landscape and several neighboring landscapes constitute a zone, also a relatively homogeneous unit. Armand (1975) said of the logical problem for any such regionalization: We can't demand of nature to locate similar subjects together.

However, a number of publications in the 1920s presented pioneer landscape maps with concrete geographical descriptions of landscapes of different regions of Russia. In some works, landscapes were interpreted as individual units or regions (*rajony*) (Dobrynin 1924; Gunbina 1928). The first graduates of the Institute of Geography—they had been able to attend Lev Berg's "Introduction to the study of geographical landscapes"—arrived in the field ready to map landscapes as types of territories. They discovered that if homogeneous territories existed at all, they were miniscule. They dubbed them "elementary landscapes" or "micro-landscapes" (Markus 1922; Larin 1926). Other geographers raised questions about the methodological warrant of combining "elementary landscapes" into "landscapes" at regional scale. The problem was that "elementary" landscapes could be quite diverse in their natural characteristics (relief, geology, soil, and vegetation) although located in close proximity (Gozhev 1930a, b; Polynov 1926). The famous soil scientist Polynov (1925) proposed that different elementary landscapes could be united into more general ones by a hidden factor. His suggestion was that within one landscape all "elementary" landscapes (micro-landscapes) had one dominant type of spatial interrelation. Although he provided some anecdotal evidence of such spatial interrelations of different complexes (mostly involving runoff and deposition of soil materials), his idea was not particularly instructive and not supported by Soviet geographers either at the time or later.

2.4 "Dehumanization" of Landscape (Late 1920s)

In the 1920s, the geographical faculty of Russian universities was organized along three principal lines: physical geography (i.e., hydrology, geomorphology, etc.), country studies (i.e., landscape study), and human geography (i.e., anthropogeography). Human geography included a number of courses such as the systematic geography of human races, demographic statistics, colonial geography, ethnography of USSR, geography of resettlement regions, history of culture, and economic geography. In 1929, leftist students initiated protests against instruction in human geography as "bourgeois relics" (Bykov 1980). According to historical materialism, the main driver of human history was the class struggle while geographical conditions played a minor role in the fortunes of nations. Human geography was denounced as an example of vulgar geographical determinism. As a result of the student protests, in 1929, all disciplines of human geography except economical geography were terminated. Despite endless and stormy criticism by external and internal opponents, economic geography as an area of instruction survived here and there. In 1931, the RGO ethnography division was closed. Human geography disappeared from Soviet geography forever.[5]

The denunciation of human geography concerned the landscape geography. Interactions between humanity and nature had been a favorite theme for many chorologists. Hettner first came to the attention of Russian geographers with the appearance of his *European Russia: Anthropogeographical etudes* translated and published in Russia in 1907 (Hettner 1907).[6] V. P. Semenov-Tyan-Shansky in his *Region and Country* (1928) presented many original observations documenting

[5]In the Moscow State University Geography Department, this structure has changed little since 1929. There are more than dozen chairs in physical geography disciplines (physical geography and landscape study, hydrology, geomorphology, etc.) and a sole chair in economical geography.

[6]It is worth to remind that his most important methodological publication of 1905 became acknowledged by Russian geographers only in the early 1910s.

how natural environments influence people. Berg in his first textbook *Landscape-geographical zones of the USSR* (1931) systematically described a close correspondence between landscape features and folk culture in boreal forests, forest-steppe, steppe zones, etc. The art of anthropogeography was to discern the very fine influences of nature on ordinary human life. There was no ambition to explore the geographical drivers of human history as did, for example, Lev Gumilev in his theory of ethnogenesis. Another target of landscape theory was the impact of human presence on the landscapes themselves, also important for geographers. Semenov-Tyan-Shanskij (1928) was the first to propose a classification of landscapes according to the extent of their anthropogenic transformation. A similar system and the very term "cultural landscape" were proposed by Berg (1931). As a result of the suppression of anthropogeography from Soviet science, landscape studies focused on exploration of natural landscapes and their transformation by human activities. This understanding of the scope of landscape study, inherited from the ideological discussions of the 1930s, is still characteristic of modern Russian geography.

2.5 Landscape as Homogenpous Territory (1930s)

In the very beginning of the 1930s, Marxist philosophers initiated the philosophical reconstruction of Soviet sciences. Each academic discipline would be required to prove its compliance with the fundamental "laws" and other principals of historical and dialectical materialism. Reconstruction got its start a bit earlier in economic geography than in physical-geographical disciplines. Publication of the Russian translation of the main theoretical work of A. Hettner's *Geography: its history, nature, and methods* in 1930 (Hettner 1930) was the best fuel for ideological battles that time.

In landscape geography, the most important "reconstruction" publication was that of M. A. Pervukhin's *On subject and method of modern physical geography* (1932) in the most prominent philosophical journal *Under the Flag of Marxism*. In his article, Pervukhin declared the existence of a serious crisis in the chorological approach in Soviet geography. To his eye, this approach did not resemble the truly scientific methods of hydrology, meteorology, and other disciplines of physical geography. The latter used precise methods and analyzed primary data, while chorological geography utilized descriptive methods and compilation of secondary data. He referenced works of Hettner and Berg. Berg, he claimed for example, in his *Landscape-geographical zones of the USSR* (1931), crammed all natural and human phenomena including relief features and ethnographic composition into "the Procrustean bed" of natural zones. Pervukhin called for the further development of landscape studies at a new level. He regarded the history of the landscape concept as passing through three stages: In the first stage, landscape was understood as scenery, in the second, as individual region, and in the new stage, landscape should be investigated as a type of territory. Semenov-Tyan-Shanskij represented the first stage, Hettner was "stuck" somewhere between the first and second stages, while Berg's work belonged entirely to the second stage. Pervukhin was thus the first to clearly distinguish different interpretations of landscapes within in Soviet geography, distinctions which most other Russian geographers had poorly noted.[7]

During the 1930s, Pervukhin had sought to be consistent in promoting his view of landscape as a type of territory. Some other geographers joined him in this view. There were already in use a number of terms for small, homogeneous natural territories: elementary landscapes (Polynov), micro-landscapes (Larin), natural landscapes of first order (Gozhev), landscapes in the narrow sense (Berg), and types of area (Pervukhin). Most such terms had been proposed in the 1920s. But the key difference between landscape works in the 1920s and those of the 1930s was that earlier researchers had regarded these units only as parts of a genuine, larger landscape while the later ones regarded them as genuine instances of landscapes. At the first geographical congress in 1934, Pervukhin (1934) made it clear by proposing that field landscape study should include two major steps—identification of elementary landscapes and their classification as types of area. He stressed that elementary landscapes could belong to an identical type even if they were remote from each other. These methodological recommendations were advanced in several works in the late 1930s. For example, Ponomarev (1937) mapped elementary landscapes of the steppe of Priural'e. He mapped every small depression occupied by birch thicket and denoted each as a "concrete elementary landscape." These concrete "elementary landscapes" differed from each other by size of depressions, their depth, and details of species composition, but all could be classified as a single "elementary landscape" (similar to species in a biological taxonomy). Then, this type could be combined with a second type of "elementary landscape" represented, for example, by small depressions covered by aspen thickets. This would be a "geographical landscape" (similar to a genus). Larger region or natural zones (steppe, forest-steppe, and other) consist of very different types of geographical landscapes. It is why zones were called in this work as subjectively drawn individual territories in contrast with Berg's view zones as classification units (types of territories).

Despite its compelling logic and consistent analysis of landscapes as types, Pervukhin's interpretation was not shared by the bulk of geographers. Many (e.g., Gozhev 1934; Berg 1936) continued to imagine landscapes as fundamental natural units of regional scale despite the clear logical defects of their work. "Elementary landscape" could not provide a secure anchorage for landscape study. Works similar to Ponomarev's produced a sense of gridlock rather than a sense of new horizons opening before geographers. Polynov (1925) was the first geographer who proposed that different elementary landscapes could be united into more general ones by a hidden factor. His suggestion was that within one landscape all "elementary" landscapes (micro-landscapes) had one dominant type of spatial

[7]It is difficult to understand why Berg (1936) even in the second edition of his textbook *Landscape-geographical zones of the USSR* repeated obviously inconsistent instances of landscapes: on the one hand, a pine standing with lichen flora as one type of ecosystem, on the other, Valdaj Upland, which was an individual unit (region), very diverse in terms of types of ecosystems.

interrelation. Although he provided some anecdotal evidence of such spatial interrelations of different complexes (mostly involving runoff and deposition of soil materials), his idea was not particularly instructive and not supported by Soviet geographers either at the time or later. However, his publication was important due to definite formulation of the problem which seemed to be solved by the study of landscape morphology elaborated by N. A. Solntsev in 1947.

2.6 Landscape as Individual Territories (1940s)

The postwar period was extremely difficult in many regards and no actual field research had been conducted for some time. Even where geographers were able to undertake actual fieldwork, they were able to find here and there only elementary landscapes, areas small in size and relatively homogeneous in soil-vegetation cover. They wanted something different—to discover large scale yet somehow indistinguishable areas corresponding more closely to the image of the main cells of nature packed onto Earth's surface.

Sometimes geographers expressed this aim quite simply. S. Kalesnik, a geographer at Leningrad University, who preferred the term "geographical envelope" to "unified process" but was, at the same time, sympathetic to the landscape approach, devoted several paragraphs of his textbook *Fundamentals of General Earth Study* (*Osnovy Obshchego Zemlevedeniya*) to discourse on the size of the "true landscape" (Kalesnik 1947). He offered his own definition of landscape as a part of the "geographical envelope" with individual structure but typically presented the surface of the Earth as having considerable spatial extension. Elementary landscapes could be only parts of "true landscapes." At the same time, Kalesnik ran into a wall in his efforts to define the size of a true landscape. "It is impossible to decide whether a marsh or lake is a landscape or not. A very small bog in the forest, a pond in the steppe are probably not landscapes but only components of landscapes. But boggy Poles'e or Vasjugan'e[8] are already landscapes, and the Caspian Sea captures a group of landscapes." (p. 474) In another place, he again warns geographers not to consider every hassock or pond or slope a landscape. One additional argument of a practical nature was that identification of Gozhev's small units as landscapes would, as they number in the millions or tens of millions, the present geographers with an insuperable descriptive challenge. These citations spectacularly demonstrate that by the end of 1940s, 25 years after landscapes had been declared the central focus of geographical science, no Soviet geographer had a clear vision of what they were.

In 1946–1947, a geographer at Moscow University, Nikolai Solntsev, seemed to succeed in addressing the challenge of identifying the true landscape during fieldwork in the Moscow region. Solntsev called his discovery "the study of

[8]Poles'e is a very large lowland in Belarus, Vasugan'e is an even larger boggy plain in Western Siberia (spreading across three administrative provinces: Tomsk, Novosibirsk, and Omsk).

landscape morphology." The first presentation of his approach was delivered at the Second Geographical Congress in 1947. According to Solntsev, a landscape could be identified in the field thanks to a morphology consisting of multiple small units (micro-landscapes) whose combination remained generally constant within a single landscape. A change in the bundle of micro-landscapes would signal to the researcher that he had entered a new landscape. Thus, each landscape was characterized by a specific combination of micro-landscapes.[9] Solntsev offered a simple justification for his approach. The most important element of landscape was its geological-geomorphological foundation. It was this factor that determined the combination of smaller and simpler units. Solntsev (1949) described the importance of his discovery this way: Geographers following Berg recognized landscape as the main subject of geographical study "but for long it was not clear what this territory was—large or small, how it could be identified in the field, which parameters one needed to know in order to distinguish it from other landscapes" (p. 6). His simple approach and explanation were well received by Soviet geographers. Many believed that the defining character of this fundamental notion had at last been established, 35 years after its first appearance in Russian geography and 20 years after Polynov's insight on landscape as "homogenous inhomogeneous" [in words Isachenko (1951)].

2.7 Landscape as Physical-Geographical Unit (1950s)

In this period, research and instruction in geography became a massive enterprise in the Soviet Union. By the end of the 1950s, geography was being taught in 31 of the country's 39 universities and in 62 pedagogical institutes. Two the most important centers of geographical education remained Moscow University and the Leningrad University. Moscow State University still developed the concept of landscape as the key subject of physical geography. The position of Leningrad University was in favor of General Earth Study. To bridge the divide between general (global) physical geography and (regional) landscape geography, Leningrad geographers proposed the notion of "the geographical envelope" as the major focus of Soviet geography. The "geographical envelope" was considered by them as a natural complex of the highest rank. The major trend in Soviet physical geography became the unification of the majority of Soviet geographers under a new scheme of physical geography as a science of natural complexes of different rank—from landscape to the geographical envelope.

According to the new concept, Soviet physical geography combined three disciplines in one coherent system: General Physical Geography, the investigation of

[9]Solntsev renamed micro-landscapes in *fatsii* and *urochishch* (as combination of several fatsii) in Russian, denoting a separate terrain with particular features, a word difficult to render literally in English.

the geographical envelope, the study of landscapes as landscape or regional geography, and physical-geographical regionalization as the exploration of natural complexes on scales intermediate between landscape and the geographical envelope. Though, the notional core of physical geography—the geographical envelope—was a rare subject for these discussions which remained restricted to secondary questions such as its lowermost and its uppermost borders (in km).

The main focus of discussion was physical-geographical regionalization, its methodology, and logical basis. Disagreement marked every question, including the most basic ones. Reflecting general disagreement as to the logical basis of regionalization, some geographers held that all units of all ranks of physical-geographical regionalization were homogeneous territories, while others considered them individual (unique) territories. Armand (1952, 1955) insisted that regionalization was the same procedure as classification: "classification of territories via mapping is regionalization" (Armand 1975, p. 161). Kalesnik (1955) held the view that regionalization was a particular approach to scientific analysis alongside typology and taxonomy. Isachenko (1965) held that regionalization and classifications were contrasting and distinctive procedures: The former individualize objects, while the latter seek common features in different objects. Some geographers (Gvozdetskij 1957; Milkov 1954) proposed that there were two kinds of regionalization—typological and individual—and that their differences were caused by the existence of two kinds of natural complexes—internally homogeneous ones and internally diverse. For example, some complexes although formed by different processes show a similarity in soil and vegetation cover. Such complexes would belong to "typological" units. Some territories of the same genesis could be quite diverse internally and were subjects for individual regionalization. Rodoman (1956) suggested a very original idea that individual regionalization was a special case of typological regionalization when one type was presented by only one region. Finally, most geographers reached agreement that the goal of regionalization was to identify initially individual regions which could then be classified by type of territory (Mikhajlov 1956; Isachenko 1965). This understanding continues to dominate Russian geographies today.

Anyway, the view of landscape as a basic unit of analysis in geography was questioned by the majority of Soviet geographers. In the course of several discussions about the privileged position of landscape, geographers came to the conclusion that all complexes of all ranks differ from each other but not in a principled way (Armand 1952). Isachenko (1953) and others made attempts to elaborate schemes of physical-geographical regionalization in which landscapes (in Solntsev's terms) occupied a basic (privilege) position in the hierarchy of complexes. But their schemes had been the target of quite reasonable criticism by other geographers. Kalesnik (1955) finally recognized the futility of such attempts and the use of "landscape" to refer to any complex of any rank came under renewed scrutiny (Milkov 1954; Obsuzhdenie 1956).

This was a serious blow to landscape theory. Landscape geographers held that landscapes were basic entities which constitute Earth's entire surface. Landscapes were analogous to cells in biology or atoms in chemistry. Elements of lower rank figured only as components of landscapes. Any component of higher rank (region, province, oblast, or whatever) belonged to a different taxonomy, determined with reference to classificatory systems beyond the landscape. The difference was one of principle—landscapes were units that could be identified only in the field, all other ranks were the result of armchair analysis. The determination of landscapes and their boundaries required actual fieldwork. According to Solntsev's instruction, landscape could be identified with reference to typical combination of repeated, smaller, and sometimes contrasting, complexes. However, this instruction did not cover all possible field situations. Landscapes could be quite diverse internally making it impossible to partition them into, say, regular cells with the expectation that all cells would be similar in terms of a set of smaller complexes. Landscapes could be very patchy. Moreover, a landscape could comprise units located in serial order (e.g., a macro-slope or landscape surrounded by a large lake) but without spatial repetition.[10] Geographers had to rely on intuition for identification of landscapes in the field, a quality that endowed landscape, in contrast with complexes of higher rank, with a deep reality. As it was directly open to visual scrutiny landscape carried for geographers, an absoluteness, they could observe it, take pictures of it, dig in it, and cross it on foot. Some included direct observability as a defining property (Lidov1949; Saushkin 1951).

Increasing misunderstanding and even mistrust of landscape geographers led to their some kind of isolation. Still, they continued fieldwork, mapping landscapes in different regions of the country and did so with great enthusiasm. Field landscape mapping began in many universities across the country (Kazakova 1955). Solntsev's methodology of landscape mapping spread through the country. When in 1955 the RGO excluded landscape geography from the agenda of the All-Union Geographical Congress, landscape geographers gathered in Leningrad for an alternative assembly. To settle the scandal, Kalesnik, President of the RGO, sent an official greeting to the first All-Union seminar of landscape geographers. Thus, was launched a tradition of annual seminars on landscape mapping. Other seminars were held in L'vov (1956), Tbilisi (1958), Riga (1959), Moscow (1960), and Alma-Ata (1963). The seminars came with a principled condition: All presentations were required to offer results of concrete fieldwork along with new landscape maps. An important part of each seminar was an extended field excursion. Tellingly, even at the fourth seminar in Riga (1959), of 70 presentations on landscape research, only two were the work of geographers from the Institute of Geography of the Academy of Sciences.

[10]These examples were given by D. Armand, one of the critics of Solntsev's "study on morphology of landscape."

2.8 Landscape as Geosystem (1960s)

In the 1960s, geographical science continued its expansion in Soviet universities and institutes of the USSR. There were already several new geographical scientific centers, for example, the Geography Department of Voronezh University led by professor F. N. Milkov and the Institute of Geography of the Siberian division of the Academy of Science in Irkutsk headed by academician V. B. Sochava. However, the centers that made the weather in theoretical matters in Soviet geography were still the Moscow University and Leningrad University.

In the 1960s, many questions that were first raised in the 1950s still provoked confusion. These concerned the number of ranks, concrete indications for each step in regionalization, the character of natural borders (fuzzy or sharp?), objectivity[11] and subjectivity of regionalization, and others. All these questions served rather as the reasons for divergence of different (especially provincial) scientific schools. This centrifugal trend was clear and explained the prolonged character of many discussions which had never been resolved in Soviet geography.

Some of the discussions were no doubt especially painful for landscape geographers. For example, during the 1960s and peaking in the 1970s there was prolonged discussion of the so-called most powerful component of landscape. According to Solntsev's study on landscape morphology, the strongest component of landscape was its geological-geomorphological foundation (Solntsev 1961). Solntsev maintained this position throughout his entire academic career (see Solntsev 1984) because it derived logically from his definition of landscape. The geological-geomorphological component determined the set of smaller complexes which regularly intersected within the borders of the landscape. This point was self-evident for scientists who had field experience in landscape mapping and who confirmed this thesis many times while delineating landscape borders. But for other geographers who knew about Solntsev's landscapes only theoretically, it was easy to present elementary evidence of how different components impact other components in a natural complex. Some evidence favored climate (Armand 1975), some—biota (Sochava 1978; Isachenko 1991), some, both (Sochava 1978), and some, three of them (lithology, biota, and climate) (Krauklis 1979). Milkov (1975, 1977), the leading geographer at Voronezh University, went further by proposing a classifications of landscapes according to the most salient component, a factor that varied with the situation. According to the most salient component (climatogenic, lithogenic, biogenic, anthropogenic, etc.), he distinguished 12 categories of landscapes. Isachenko (1974) disagreed only with the category "anthropogenic" as, in his view, humans could generate anthropogenic complexes only at levels lower than landscape. One can only imagine how daunting such discussions were for landscape geographers! Their landscapes were directly observable, simple, natural, and the

[11]Despite recognition that different schemes of regionalization were very different even for the same territory, all authors insisted on the objective character of regionalization and that outlined "regions" exist objectively.

most important feature in the hierarchy of natural complexes, and certainly formed on the basis of geological-geomorphological foundations rather than climate or biota or some combination thereof.

In the 1960s, meanwhile, a new generation of geographers was coming to Soviet geography, one imbued with a modern way of thinking and relatively free of ideological stereotypes. They wanted nothing to do with discussions of physical-geographical regionalization, geographical envelope, and the hierarchy of natural complexes. They wanted geographical science to be a modern one with quantitative methods, modeling, and more sophisticated theory.[12] Many younger geographers thought that systems analysis would modernize Soviet geography. However, the first was academician Sochava (1963), who suggested a new term "geosystem" to apply to all natural complexes regardless of rank—from elementary landscape to geographical envelope. Many older geographers welcomed this move as they traditionally thought of geographical units as a manifestation of close interrelations of different components of landscape. Only a few (Milkov 1967) criticized the new term as excessive. However, the younger generation of Soviet geographers went further, questioning whether physical-geographical units including landscape possessed features of real geosystems. There were two positions on this question. Solntsev (1977) suggested that analysis of system properties was the correct path rather than the identification of the individual properties of particular components. In this case, any object could be explored in relation to its structure, elements, environment, links, and integrity. He regarded systems analysis as a new "paradigm" of physical geography in Kuhn's terms. But others considered systems were only objects if their integrity, spatial, and temporary borders were determined by energy flows and matter (Armand 1966; Retejum 1971). According to the second view not all complexes could be automatically categorized as "geosystems." According to some geographers, among traditional complexes only the lowest rank—Gozhev's elementary landscapes and Solntsev's*urochishche* (several associated elementary landscapes, for example, located in the upper and lower parts of a slope)—could be considered geosystems but not landscapes and natural complexes of higher ranks. Retejum (1977) referred to his research of landscapes submerged by a water reservoir, claiming that such large units as landscape do not demonstrate functional integrity. As might be expected, many geographers showcased river basins as examples of geosystem (D'yakonov 1974). Armand (1975) joined the discussion suggesting a distinction between territorial natural complexes (all ranks) and non-territorial geosystems. He presented the first

---

[12]This was, to a great extent, a matter of taste. V. N. Solntsev (son N. A. Solntsev), who actively promoted the systems approach, characterized the language of theoretical discussions of older geographers as "tongue-tied" (Solntsev 1977). To some extent, it was a generational conflict (literally, a father/son thing). A. D. Armand (son of D. L. Armand) also considered as an essential property of geosystems their capacity for self-regulation. He also stated that information flow could be taken into account along with energy and matter flow in "geosystems." However, D. L. Armand criticized renaming of well-known processes in terms of information theory and called on his son personally to call a cat a cat.

classification of geosystems, dividing them into hydro-meteorological (systems of atmospheric circulation), hydro-geomorphological (a ravine), and ecosystems (a lake). This move reached its logical conclusion in a scheme of classification of different geosystems of Earth's surface as elaborated by Retejum (1977). Among more than a dozen types of geosystems, Solntsev's landscape occupied only the modest position of a so-called geoit—a special type of geosystem formed by the presence of a large geological body that clearly determines (transforms) both direction and intensity of energy and matter flows in the natural environment.

For landscape geographers, such sorts of geosystem classification would mean yet once again scrapping landscape as the linchpin in the basic nature complex and relegating it to the status of a single item in a panoply of peripheral features stipulated by geographical geosystem theory (after dissolving Solntsev's landscape in the hierarchy of physical-geographical units). Fortunately, (for landscape geographers, at least), such a radical view failed to gain support among Soviet geographers and remained a separate intellectual exercise. Most Soviet geographers were content with formally renaming natural complexes regardless of their rank as "geosystems," thus effectively neutralizing the efforts of modernizers.

2.9 Conclusion

The landscape notion has survived the turbulence of quite different historical moments. Paradoxically, however, landscape theory has never been close to dominant. It has never been unreservedly supported by Russian (or Soviet) geographers. On the contrary—the concept has been questioned by many leading Soviet geographers. Most curious is the fact that until the late 1940s, i.e., after three decades of development, none of its advocates were able to identify landscapes in the field or to map them! The vitality of the landscape notion can only be explained by strong adherence to the idea of the existence of definite geographical regions (there was no alternative mode for conceptualizing geographical reality)—a variety of archetype in geographical thinking. In other words, the failure of geographers to map geographical regions for decades did not imply that regions or landscapes were non-existent. It is especially true for the first two decades when the highest priority was to demonstrate the authority of geography to stand as a fundamental discipline alongside others at Russian universities.

The multitude of interpretations of the term "landscape" was a remarkable feature of Russian geography in the twentieth century. Reading the term "landscape geography" in the 1910s was a very different from that in the 1930s and later. Some geographers held an exotic view of landscape as "scenery" or a visually fixed harmony whole of different natural and cultural elements. The human dimension of landscape was very important for Russian geographers in the first two decades of the century. After 1930 Soviet geographers interpreted landscape principally as a natural object. Nor was there ever a monopoly on reading the term "landscape" in the Soviet period. For some Soviet geographers landscapes had definite scale (Solntsev's landscape) while for many others the term was applicable for complexes of any physical-geographical scale and rank. In the 1960s this diverse picture was supplemented by notion of "geosystem". For some Soviet scientists the terms "landscape" and "geosystem" became synonymous but for others—not.

Reflection on different interpretations of landscape had been always weak in Russian geography. Different readings of the term could be easily found in the work of a single author. Either most geographers hardly realized that collision of different interpretations of landscape was a matter of definition. When Berg rejected examples of landscape suggested by Hettner (the Crimea), he did it so wrongly as Hettner did not regard "homogenous" as a principal criterion of landscape. Berg's argument against Hettner was obviously tautological: Landscape must only denote homogeneous territory; Crimea is not homogeneous. Therefore, Crimea is not a landscape. Berg contended that landscape was "harmony" and "organism" but failed to argue why homogeneous territory should be regarded as "harmony" or "organism." In the 1950s most Soviet geographers denounced Solntsev's landscape by questioning its privileged position in the system of physical-geographical regionalization. They did so wrongly as this criterion—status as a basic physical-geographical unit—was absent in the study of landscape morphology. For N. Solntsev units of any higher rank possessed much less actuality than landscapes identified "in field." In his logic other physical-geographical units could only result of classification of "true" landscapes. In other words, the difference of landscape and other physical-geographical units was non-issue in his study of landscape morphology. In the 1960s a new generation of Soviet geographers rejected Solntsev's approach on the ground that landscape boundaries were not formed by distinct matter/energy flow (as supposed to take place in a river basin). Either they did it wrongly. By definition landscape was a combination of repeated, smaller, and sometimes contrasting, natural complexes. That is it. It was a logical flaw to question the very existence of these combinations because of dispersed character of matter and energy flow within the territories. The historical fact is that such imperfect argumentation resulted in marginalization of the study of landscape morphology and blurring the term "landscape". Most modern Russian geographers use the term "landscape" as a synonym of a natural complex applicable to units at any spatial scale. The term "geosystem" also is used widely by modern geographers referring in a generalizing and non-selective way to all natural complexes regardless of their spatial scale and rank. For the bulk of Russian geographers the terms "landscape," "natural complex" and "geosystem" are loosely interchangeable. Finally, the study of landscape morphology, the most original concept in Soviet geography, has been further marginalized. Few modern geographers are trained to make actual landscape mappings in the field. Solntsev's aspiration to cover the entire USSR with landscape mappings and to give a unique name to each has not been realized. A handful of regions have been provided with detailed landscape maps, but even these lack clear practical applications. In regional and local environmental studies, Russian geographers prefer to use electronic land use maps (e.g., CORINA) which synch readily with satellite images and other global data sets.

References

Anuchin DN (1949) Geografiya (Geography). Izbrannye geograficheskie raboty (Selected geographical essays) OGIZ, Moscow: 29-39

Armand DL (1952) Printsipy fiziko-geograficheskogo rajonirovaniya (Principals of physical-geographical regionalization). Izvestiya AN SSSR. #1: 68–82

Armand DL (1955) Proiskhozhdenie i tipy prirodnykh granits (Origin and types of natural borders). Izvestiya VGO #3:266–278

Armand AD (1966) Prirodnye kompleksy kak samoregulirujushie informatsionnye sistemy (Natural complexes as self-regulated information systems) Izvestiya AN SSSR. #2: 15-21

Armand DL (1975) Nauka o landshafte (Science on landscape). Mysl', Moscow

Berg LS (1915) Predmet i zadachi geografii (Subject and tasks of geography). Izvestiya RGO. 51:463–475

Berg LS (1925) Geografiya i ee polozhenie v ryadu drugikh nauk (Geography and its place among other sciences). Voprosy stranovedeniya (Questions of country studies). Moskovskoe aktsionernoe izdatel'skoe obshchestvo, Moscow-Leningrad, pp 3–17

Berg LS (1931) Landshaftno-geograficheskie zony SSSR (Landscape-geographical zones of the USSR). Selkhozgiz, Moscow-Leningrad

Berg LS (1936) Fiziko-geograficheskie (landshaftnye) zony SSSR (Physical-geographical (landscape) zones of the USSR). Izdatel'stvo Leningradskogo universiteta, Moskva-Leningrad

Berg LS (1962) V.V.Dokuchaev i uchenie o goegraficheskix zonakh (V.V.Dokuchaev and geographical zones study). Istoriya russkix geograficheskix otkrytij (History of Russian geographical discoveries). Izdatel'stvo AN SSSR, Moscow

Borzov AA (1908) Kartiny po geografii Rossii (Pictures on geography of Russia) Izdatel'stvo Grosman i Knebel', Moscow. Issue 1

Borzov AA (1912) Geografiya (Geography) Prakticheskaya shkol'naya jenctsiklopedia pod redaktsiej N.V. Tulupova i P.M.Shestakova, Moscow, pp 148–163

Bykov VI (1980) Idejnye istoki "levizny" v jekonomicheskoj geografii nachala 30-kh godov (Ideological sources of leftish movement in economical geography in the beginning of the 1930s) Voprosy Geografii 115:44–52

D'yakonov KN (1974) Printsipy kompleksnosti fiziko-geograficheskogo rajonirovaniya i integral'nyj pokazatel' geograficheskoj provintsii (Principals of complex physical-geographical regionalization and integrative indicator of geographical province) Prirodnoe i selskoxozyajstvennoe rajonirovanie SSSR (Natural and agricultural regionalization of the USSR). Izdatel'stvo Moskovskogo universiteta, Moscow, pp 185–190

Dobrynin B (1924) Landshafty Dagestana (Landscapes of Dagestan). Zemlevedenie, pp 92–113

Dokuchaev VV (1949) K izucheniyu o zonakh prirody (Towards the study of the zones of nature) Izbrannye sochineniya (Selected essays) Gosudarstvennoe izdatel'stvo Sel'skokhozyaistvennoi Literatury, Moscow, vol III, pp 317–322

Dronin NM (1999) Jevolutsiya landshaftnoj kontseptsii v russkoj i sovetskoj fizicheskoj geografii (Evolution of landscape concept in Russian and Soviet physical geography). GEOS, Moscow

Esakov VA (1959) Geografiya v universitetakh Rossii v 60-80-e gody XIX veka. Bor'ba za sozdanie kafedr geografii (Geography in Russian universities in the 60s–80s of XIX century. Fighting for formation of geographical departments) Trudy instituta istorii estestvoznaniya i tekhniki. vol 27, pp 44–66

Gozhev AD (1930a) O nekotorykh predstavitelyakh zhivotnogo mira, vstrechennykh v predelakh zapadnoj chasti Tersko-Dagestansnskogo peschanogo massiva (About some representatives of animal kingdom met in western part of Tersk-Dagestan sand massive). Isvetisya GGO v.52, 3:269–286
Gozhev AD (1930b) Tipy peskov zapadnoj chasti Tersko-Dagestanskogo massiva i ikh khozyajstvennoe znachenie (Types of sand of western part of Tersk-Dagestan massive and their economic importance). Isvetisya GGO.v.52, 4:463–526
Gozhev AD (1934) K metodologii fizicheskoj geografii (To methodology of physical geography). Izvestiya GGO. V. LXYI, 4:487–512
Gunbina TN (1928) Landshaftnye rajony Arzamasskogo uezda Nizhegorodskoj gubernii (Landscape regions of Arzamas district of Nizhegorod province) Zemlevedenie: 45–71
Gvozdetskij NA (1957) Retsenziya na knigu A.G.Isachenko "Osnovnye problemy fizicheskoj geografii" (Review of the book A.G. Isachenko "Main problems of physical geography) Zemlevedenie 4:300–302
Hettner A (1907) Evropejskaya Rossia. Antropogeographicheskij jetud. (European Russia. Anthropogeographical Etude). Zemlevedenie, Moscow, 174 p
Hettner A (1925) Sushchnost' i metody geografii (Essence and methods of geography)Voprosy stranoveedeniya (Questions of country studies) Moskovskoe aktsionernoe izdatel'skoe obshchestvo, Moscow-Leningrad: 33–85
Hettner A (1930) Geografiya: ee istoriya, sushchnost' i methody (Geography: its history, nature and methods) Gosudarstvennoe izdatel'stvo, Moscow-Leningrad, 416 p
Isachenko AG (1953) Osnovnye voprosy fizicheskoj geografii (Main questions of physical geography). Izdatel'stvo Leningradskogo universiteta, Leningrad
Isachenko AG (1965) Osnovy landshaftovedeniya i fiziko-geograficheskogo rajonirovaniya (Principals of landscape study and physical-geographical regionalization) Vysshaya shkola, Moscow
Isachenko AG (1971) Razvitie geograficheeskikh idej (Development of geographical ideas). Nauka, Moscow
Isachenko AG (1974) O tak nazyvaemykh antropogennykh landshaftakh (About so called anthropogenic landscapes). Izvestia VGO 106, #1:70–75
Isachenko AG (1991) Landshaftovedenie i fiziko-geograficheskoe rajonirovanie (Landscape study and physical-geographical regionalization). Vysshaya shkola, Moscow
Kalesnik SV (1947) Osnovy obshchego zemlevedeniya (Fundamentals of general Earth study). Uchpedgiz, Moscow-Leningrad
Kalesnik SV (1955) Osnovy obshchego zemlevedeniya (Fundamentals of general Earth study) 2-oe izdanie, 2nd edn. Uchpedgiz, Moscow-Leningrad
Kazakova ON (1955) Iz opyta polevykh landshaftnykh issledovanij (From experience of field landscape exploration). Izvestiya VGO.#2:147–161
Krasnov AN (1912) Sovremennaya geografiya i ee novye techeniya (Modern geography and its new movements). Priroda, pp 793–828 (June)
Krauklis AA (1979) Problemy jeksperimental'nogo landshaftovedeniya (Problems of experimental landscape study) Nauka, Novosibirsk
Krotov PI (1892) Zadachi nauchnoj geografii i postanovka ee prepodavaniya v Imperatorskom Kazanskom Universitete (Tasks of scientific geography and system of its teaching in Imperial Kazan University) Ezhegodnik Imperatorskogo RGO. #2:341–368
Kruber AA (1898) Opyt razdeleniya Evropejskoj Rossii na rajony (Experience of division of the European Russia into regions). Zemlevedenie 3-4:175–184
Kruber AA (1904) V.V.Dokuchaev. Zemlevedenie 10, #4:65–72
Larin IV (1926) Opyt opredeleniya po rastitel'nomu pokrovu pochv, materinskikh porod, rel'efa, sel'skokhozyajstvennykh ugodij i drugikh jelementov landshafta v srednej chasti Ural'skoj gubernii (Experience of indication of soils, deposits, relief, agricultural lands and other elements of landscapes by vegetation in central part of Ural province). Trudy obshchestva izucheniya Kazakhstana (Proceedings of society on exploration of Kazakhstan) Kyzl-Orda. 7 (2):1–44

Lidov VP (1949) Iz opyta raboty po landshaftnomu kartirovaniju Prioksko-Terrasnogo gosudarstvennogo zapovednika (On experience of landscape mapping of the Prioksko-Terrasnyj state reserve). Voprosy geografii (16):179–190
Markus EA (1922) Podzolisto-bolotnye pochvy srednej chasti Kol'skogo poluostrova (Podzol and foggy soils in the central part of Kolsky peninsula) Petrograd
Mikhajlov NI (1956) O tipologicheskom fiziko-geograficheskom rajonirovanii (About typological physical-geographical regionalization). Voprosy geografii (39):8–27
Milkov FN (1954) Retsenziya. A.G.Isachenko "Osnovnye voprosy fizicheskoj geografii" (Review Iaschenko AG. "Main questions of physical geography") Izvestiya VGO (4):370–375
Milkov FN (1967) Osnovnye problem fizicheskoj geografii (Main problems of physical geography). Vysshaya shkola, Moscow
Milkov FN (1975) V zashhitu antropgennyx landshavtov (To defense anthropogenic landscapes). Izvestiya VGO 107
Milkov FN (1977) Antropogennoe landshaftovedenie, predmet izuchesniya i sovremennoe sostoyanie (Anthropogenic landscape studies, subject of investigation and modern state of art). Voprosy Geografii. Issue 106: 11-27
Nechaev AP (1905) Kartiny prirody. Tipichnye landshafty Rossii v svyazi s ee geologicheskim proshlym (Pictures of nature. Types of landscapes of Russia in connection with geological past). Sankt-Petersburg, 105 p
Obsuzhdenie (1956) knigi A.G.Isachenko "Osnovnye voprosy fizicheskoj geografii" (Discussion on the book Isachenko AG. "Main questions of physical geography"). Voprosy Geografii (39):200–205
Pervukhin MA (1932) O predmete i metode sovremennoj fizicheskoj geografii (On subject and method of modern physical geography) Za Marksistko - Leninskoe estestvoznanie. #1:83–126
Pervukhin MA (1934) Kartografirovanie tipov mestnosti (Mapping of types of sites). Trudy Pervogo Vsesojuznogo geograficheskogo s'ezda (Proccedings of the first all-union geographical congress) (4):3–5
Petri EJu (1892a) Geografiya kak predmet universitetskogo obrazovaniya (Geography as a subject of university education) Ezhegonik Imperatotskogo RGO. Sankt Petersburg
Petri EJu (1892b) Metody i printsipy geografii (Principals and methods of geography) Kartograficheskoe zavedenie Il'ina, Sankt Petersburg
Polynov BB (1925) Landshaft i pochva (Landscape and soil) Priroda. #1:74–84
Polynov BB (1926) Peski Donskoj oblasti, ikh pochvy i landshafty (Sands of Don region, their soils and landscapes) Trudy pochvennogo instituta im V.V.Dokuchaeva. Issue 1: 3-199
Ponomarev AN (1937) Jelementarnye landshafty abrazionno-jerozionnoj platform Urala v podzone raznotravno-kovyl'nykh steppej (Elementary landscapes of abrasion-erosion platforms of Urals in subzone of mixed herb-feather grass steppes) Zemlevedenie 39(1):58–67
Preobrazhenskij VS, Makarov VZ (1988) Razvitie landshaftovedeniya v SSSR (Development of landscape study in the USSR). VINITI, 200 p.
Preobrazhenskij VS, Aleksandrova TD, Maksomova LV (1997) Geografiya v izmenyajushchemsya mire (Geography in the changing world). IGRAN, Moscow, 273 p
Retejum AJu (1971) Sistemnyj podxod k problem fiziko-geograficheeskogo rajonirovaniya (System approach to physical-geographical regionalization). Prirodnoe i sel'skokhozyajstvennoe rajonirovanie SSSR (Natural and agricultural regionalization of the USSR). Izdatel'stvo Moskovskogo universiteta, Moscow, pp 191–196
Retejum AJu (1977) O faktorax i formakh uporyadochennosti prostranstva obolochki Zemli (About factors and forms of assembling space of the Earth envelope). Voprosy Geografii (104):84–104
Rodoman BB (1956) Sposoby individual'nogo i tipologicheskogo rajonirovaniya i ikh izborazhenie na karte (Ways of individual and typological regionalization and their presentation on maps). Voprosy Geografii 39:28–69
Saushkin JuG (1951) K izucheniju landshaftov SSSR, izmenennykh v processe proizvodstva (About exploration of landscapes, transformed by industrial development). Voprosy Geografii 24:276–299

Semenov-Tyan-Shanskij VP (1928) Rajon i strana (Region and Country) Gosudarstvennoe izdatel'stvo. Moscow-Leningrad

Shaw DJB, Oldfield JB (2016) The development of Russian Environmental Thought. Routledge, London and New York

Sochava VB (1963) Opredelenie nekotoryx ponyatij i terminov fizicheskoj geografii (Defining some notions and terms of physical geography). Doklady Instituta Geografii Sibiri i Dal'nego Vostoka. #3:50–59

Sochava VB (1978) Vvedenie v uchenie o geosistemakh (Introduction into study of geosystems). Nauka, Novosibirsk

Solntsev NA (1949) O morfologii prirodnogo geograficheskogo landshafta (On morphology of natural geographical landscape) Voprosy Geografii (16):61–86

Solntsev NA (1961) Nekotorye dopolneniya i utochneniya v voprose o morfologii landshafta (Some additions and corrections for question on morphology of landscape). Vestnik Moskovskogo Universiteta. Seriya 5. Geografiya. #3:53–57

Solntsev VN (1977) O trudnostyakh vnedreniya sistemnogo podxoda v fizicheskuju geografiju (About difficulties of introduction of system approach into physical geography). Voprosy geografii (104):20–36

Solntsev NA (1984) Problema ustojchivosti landshafta (Problem of sustainability of landscape). Vestnik Moskovskogo Universiteta. Seriya 5. Geografiya. #1:14–20

Sukhova NG (1981) Razvitie predstavlenij o prirodnom perritorial'nom komplekse v russkoj geografii (Development of notions on natural territorial complex in Russian geography) Nauka. Leningrad.

Sukhova NG (2018) Khorologicheskaya kontseptsiya v Rossii (pervaya tret' XX stoletiya) (Chorological conception in Russia (the first third of the XX century)). Izvestiya RGO 150, #3:58–78

Yarilov AA (1913) Geografiya i paleogeografiya (Geography and paleogeography). Zemlevedenie 1–2:1–20

Zabelin NM (1959) Teoriya fizicheskoj geografii (Theory of physical geography). Geografgiz, Moscow

Zabelin NM (1969) Jevolutsiya geograficheskoj nauki (Evolution of geographical science) Ocherki istorii i teoriya razvitiya nauki (Essays on history and theory of science development). Nauka, Mocsow, pp 325–360

Chapter 3
Landscape Ecology Culture and Some Principles of Sustainable Nature Use

Lev K. Kazakov

Abstract Economic activity, as a rule, is accompanied by negative changes to landscapes and ecological conditions, and by a depletion of natural resources. The main causes of these processes are analyzed. Two possible aspects of sustainable nature management are highlighted: ideas from the natural sciences about the formation of a culture of nature management, and the emergence of natural and human-made landscapes, cultural landscapes included. These ideas are based on the paradigm of evolutionary synergy. This serves as a methodological basis for developing principles and approaches to sustainable nature management. An important component of this paradigm is the concept of a culture of nature management. This concept roughly describes ecological culture, its components and the reasons why it lags behind the material and production culture of nature management, as well as describing landscapes and their resiliency. Definitions of basic concepts are a subject of theoretical and applied research. The consideration and development of various aspects of an ecological culture, including a culture of landscape ecology and nature management, is a key element of sustainable nature management. The principles underlying the landscape ecology planning of economic activity and the ecological and technological culture of production are mutually complementary. They make it possible to create a natural-economic system of cultural landscapes and sustainable nature management. Some mutually complementary approaches are presented. These can be used to optimize nature management in terms of landscape ecology, to preserve the most important elements of landscape diversity, and to create favorable environmental conditions.

Keywords Ecological culture of nature management · Ethics of nature management · Material and industrial culture · Landscape · Cultural landscapes · Natural and economic systems · Resiliency · Landscape ecology planning · Universal evolutionism · Ecological imperative

---

L. K. Kazakov (✉)
Faculty of Geography, Lomonosov Moscow State University, Leninskie Gory, 119991 Moscow, Russia
e-mail: kazakovlk@yandex.ru

L. Müller and F. Eulenstein (eds.), *Current Trends in Landscape Research*, Innovations in Landscape Research, https://doi.org/10.1007/978-3-030-30069-2_3

3.1 Introduction

The population of the earth is growing rapidly. Accordingly, economic activity has intensified, and more and more natural resources and territories are required. Based on material and technological forms of nature management, economic activity and consumer ideology have led to a rapid depletion of natural resources and to negative changes in landscapes and the ecological state of the environment (Millennium Ecosystem Assessment 2005). All this, accompanied by resource emergencies and environmental crises, is beginning to slow down the further development of mankind. Therefore, in the 1970–1990s, scientists and progressive public figures began to pay more attention to the creation of various models of sustainable development and the development of principles for sustainable nature management. At the same time, considerable attention is now paid to the generation of ideas about an ecological culture. It has become a fundamental theoretical and practical category, which is intended to underlie the paradigm of the "sustainable development" of a gradually ecologized technocratic, anthropocentric civilization.

Sustainable nature management. Two aspects of sustainable nature management can be considered:

(i) sustainable nature management without development: relatively efficient management without increasing damage to the ecological state of the environment, including the natural environment,
(ii) sustainable nature management with development: the development of economic activity and an increase in the efficiency of production, without increasing environmental damage to landscapes and the state of the environment.

The first aspect, nature management without development, is realizable only if renewable natural resources are used. In the field of peasant economics, it is of limited use only for small farms or closed communities living according to the traditions of subsistence farming in remote areas, such as the "Old Believers" in Siberia.

The second aspect, nature management with development, is an element of the sustainable development of mankind while maintaining a favorable ecological situation, natural landscapes and biodiversity. This aspect of sustainable nature management includes the optimization of nature management and economic activities in general in terms of technology, resources and environmental, economic and landscape planning, based on the accelerated development of an ***ecological culture of nature management*** (ECNM). In an ECNM, in turn, a difference is made between its natural science and socioeconomic components.

Natural science aspects of sustainable nature management. The natural science fundamentals of nature management are based on modern ideas about the landscape envelope and its evolution under the influence of humanity into the anthroposphere and the noosphere. According to the concept of V.I. Vernadsky and P. Teilhard de

Chardin, the anthroposphere is increasingly moving into the noospheric stage of its development (Vernadsky 1967, 1988, 1994). Accordingly, the development of new scientific directions, patterns and concepts connected with this is accelerating. In particular, it has become obvious that the ecological culture of nature management lags far behind the material and industrial culture, which is focused on meeting people's exaggerated, growing needs. Because of this, as N. N. Moiseyev stated, the noosphere cannot arise by itself: it must be created by ecologizing thinking and developing an ecological culture of nature management (Moiseyev 1995; Kazakov 2003a, 2012; Kazakov and Chizhova 2001).

3.2 The Lagging Ecological Culture of the Use of Nature

Historical aspects. What are the reasons for the lag in ecological culture and when did it start? In the Paleolithic, humans were almost completely dependent on nature and in order to survive they had to know it well and adapt to it. In the Neolithic period, when a type of economic activity was born (the "Neolithic revolution") and technological mechanisms and methods (material and production culture) of nature management were actively developed, human dependence on various natural phenomena decreased. With the transition to a productive type of vital activity, in accordance with the landscape features of the territories in which they lived, there was a change in ethnic groups' adaptive instincts, mechanisms and skills. The culture of farming (cultivation of the land) and animal husbandry was born, the tools and technology of cultivation were improved, and irrigated agriculture was created. In other words, humans had learned to use nature to gain more material life benefits. As a result, the adaptive instinct began to become dulled, and the customs, collective consciousness, world view and behavior of people in ethnic communities changed. Humans began to stand apart and be alienated from nature. This is reflected in the Judean and Christian anthropocentric elements of cultures. In these creeds, humans appear as the highest divine creation and rule over all the living and inanimate nature on earth. As a result, the development of the environmental and humanitarian aspects of the cultural and environmental ethics of nature management began to lag significantly. As a result, an anthropocentric, material and pragmatic, production-based culture of nature management was formed. In it, all nature is viewed as an inexhaustible material and production resource, the value of which is determined by its usefulness to humans. That is, the pragmatic imperative was made the basis of the collective ideological culture of nature management—everything is right that is useful to humans, and natural resources are inexhaustible. The hypertrophied development of one of the components of culture, namely the technological culture of nature management, which is oriented toward meeting humans' increasing material demands and needs—often at the expense of nature—is leading to the degradation of nature and the depletion of its resources. Under the influence of the technological culture of nature management, a second, "humanized nature" was formed. Karl Marx wrote in 1868 that "… culture, if it develops spontaneously, and is not directed consciously … leaves a desert behind it …" (Marx 1964).

Some meanings of culture. Culture as a complex, fundamental concept has many definitions. From the point of view of the natural sciences, it can be defined as follows:

1. Culture is a way of life and the fruits of labor of a particular society, based on its collective mind and captured in anthropogenic-natural landscapes. In the broader philosophical understanding, culture is any superstructure of nature associated with human activity.

Other meanings are

2. The formation factor of a second, humanized nature.
3. The form or method of organizing the vital activity of mankind and its results in nature and society, focused on the best way of humans' adaptation to the environment.
4. The set of means and forms of a person's social adaptation to their environment: the techniques, technologies and rules for a person's purposeful material, industrial and spiritual life, aimed at optimizing their relationship with the environment.

The aggravation of environmental problems of nature management and the study of their causes determined the relevance of the accelerated development of an ***ecological culture of nature management***, or environmental management as a special theoretical and applied direction. When the anthroposphere enters the noospheric stage of development, an ecological culture based on science, education, high technologies and the ethics of nature management and life activity in general begins to play a leading role. According to V. I. Vernadsky, scientists, including geo-ecologists, are faced with the tasks of consciously organizing natural-anthropogenic landscapes through the co-adaptation of economic activity and nature. They cannot move away from this, as they are guided by the spontaneous growth of scientific knowledge, and are still being pushed today by growing environmental and economic situations (Vernadsky 1988).

3.3 Basics and Components of the Ecological Culture of Nature Management

Ecological culture is one of the theoretical and practical aspects of the concept of "culture." It underlies the paradigm of escaping the systemic ecological crisis and creating sustainable nature management within the development of a modern technocratic, anthropocentric civilization.

In the ecosystem model of the anthroposphere, the culture of nature management proceeds from the understanding that nature is a habitat, an arena for the economic activity of mankind and a source of the resources people need. In turn, humanity, with its economic activity, in evolving has already become a habitat for nature. That is, in this natural and economic ecosystem, nature and humans are fully fledged subjects of coexistence and interaction. The most active, reasoning factor in their joint long-term development, co-adaptation and co-evolution is humanity, with its collective mind and environmental ethics. In order to survive, it is obliged and compelled to maintain the auspicious conditions of the coexistence of nature and society. To do this, when cultural landscapes are designed as elements of the noosphere, natural ecosystems should become analog models for the ecologized technological culture of nature management. An important component of ecological culture is its humanitarian, spiritual part, represented by the ethics of nature management. It is based on education, traditions and a way of thinking that form an ecological outlook (Moiseyev 1995; Kazakov and Chizhova 2001; Kazakov 2013).

Nature management ethics can be defined as the voluntary restriction of freedom of action in order to preserve the material, spiritual and environmental benefits of nature for a long time. These restrictions are imposed by natural science concepts, education and the environmental imperative formed in the public consciousness.

The evolutionary synergetic concepts and paradigms of V. I. Vernadsky, P. Teilhard de Chardin, I. R. Prigogine, G. Hagen developed and concretized in the works of other scientists, including N. N. Moiseyev, serve as the natural science basis for the development of an ECNM (Kazakov and Chizhova 2001). They are the basis of methodological developments connecting universal evolutionism and the self-organization of nature. According to the synergistic paradigm which is of enormous ideological significance, the processes of the creation, development and evolution of open systems, regardless of their nature, are subject to a single algorithm and are characterized by an increase in the complexity and orderliness of their organizational structure. This paradigm has been tested in relation to explanatory models of the evolutionary transition of the biosphere into the noosphere, as well as the transformation of natural landscapes into natural-anthropogenic and cultural landscapes. They are special cases of the synergetic model of development (Vernadsky 1967, 1994, 1988; Kazakov 1999, 2008; Kazakov and Chizhova 2001). Table 3.1 shows the main components of ecological culture which define the principles of sustainable nature management that have to some extent developed in theoretical, methodical and practical plans.

For geography and landscape studies, aspects and areas of environmental culture are of great and direct importance, such as the creation of general schemes for introducing different types of environmental management in large regions, as well as planning schemes at the regional level and at the level of construction projects. The choice of the types of environmental management to be introduced in large

Table 3.1 Components of the culture of nature management

Ecological culture of nature management (ECNM)				
Material and production culture		World view (ways of thinking and understanding)		
Technological		Natural science	Ideological (humanitarian)	
Technologies production, ecological and technological features of the resources used, protective structures, filters, circulating systems, closure of technological systems, emission standards in the operating system, etc.	Accommodation, planning and construction aspects of ecological culture	Natural science models of the universe, the laws of organization and evolutionary development, universal evolutionism, synergy and self-development, landscape, landscape ecology and ecological/ geographical concept of nature management, ideas about the sustainability of landscapes	Beliefs, religions, cults, customs, traditions, ideological ecological flows and associations (alarmism, biocentrism, environmentalism, universal ethics, extending the scope of human ethics to include nature)	Philosophical, political, economic theorized constructions, models, legislative systems, to substantiate the interests of countries, peoples, ethnic groups, communities and their leaders in resources, the environment, power and other issues
Ecotechnologies of nature management, conservation and melioration of nature		Ethics of nature management, environmental imperatives (e.g., something is correct and allowed if it does not violate natural laws and the ecological balance)		

regions determines the environmentally safest and most cost-effective way to develop certain types of production, recreation or settlement in each natural area or regional landscape, and which landscapes should be left as specially protected or reserves. This is the level of the master plan for the development of production and settlement in a country or a large region thereof. The planning aspect suggests whether it is environmentally safer to place certain types of environmental management in the selected region or industrial zone, or in the agricultural or natural landscape. At this level, landscape planning is determined by the structure of a territory's landscape and ecological framework. Construction aspects determine the environmental safety of production or settlement, as well as their design within industrial zones and individual settlements and parts thereof. It is important to create or maintain all the elements in a territory's landscape-ecological framework, including sanitary protection zones and strips.

The natural science foundations of the ecological culture of nature management are the basis for its technological and humanitarian (ideological) components.

3.4 Landscape Ecology Approaches and Principles of Sustainable Nature Management

Scientific approaches, principles and methods behind sustainable, inexhaustible environmental management, especially of landscape-based farming, have been actively developed since the end of the 19th century and the early twentieth century.

Definitions of terms and basic concepts. Landscape. The term "landscape" (German *Landschaft*, Russian *Ландшáфт*, French *paysage*)—a type of terrain, a limited relatively homogeneous stretch of a terrain, region or country) is entrenched in geography and soil science and has acquired a deep scientific meaning in Russia. The concept of landscape is now widely used both in natural science and in the humanitarian sphere. In natural history, landscape has become a fundamental concept.

The definition of concepts is an important component of any science. In geography, landscape is a complex, multifaceted scientific concept. Therefore, it can have many definitions reflecting its different aspects. One thing they share in common is the idea of an interconnected set of landscape components (natural and other unity) and their binding to a certain territory. The following definitions of the scientific concept of "landscape" are the result of generalizations of many definitions available in Russian-language reference books, dictionaries, encyclopedias, textbooks and manuals on physical geography, landscape ecology and nature protection. In addition, the definitions take into account the views of the author and other scientists working in the field of theoretical and applied landscape science, engineering geography and geo-ecology (Berg 1947, 1958; Milkov 1970, 1973, 1986; Nikolaev 2000; Solntsev 2001; Sochava 1978; Kazakov and Chizhova 2001; Kazakov 2004, 2005, 2008, 2013).

1. As a natural territorial complex (NTC), a landscape is a morphologically (structurally) and functionally expressed part of the earth's mantle, formed in a narrow contact zone of the abiotic and biotic environments. Examples include the landscape of the earth, a continent or its parts.
2. L. S. Berg (one of the first disciples of V. V. Dokuchaev) defined a landscape as an area in which the nature of the elevation, climate, vegetation and soil cover, wildlife, population and human culture "merge into a single harmonious whole, typically repeated throughout the known zone of the Earth."
3. A natural geographical landscape is a natural territorial complex of any dimension. It stands out morphologically from its surroundings and is a genetically relatively homogeneous area of the earth's surface where a geographically stable set of naturally related and interacting natural components, functioning as a whole, producing a specific new substance, energy and information. The landscape is characterized by natural combinations of properties of surface sediments, mesoforms, climates, soils, hygrotopes, vegetation and animal populations. Under economic development, the landscape performs the functions of natural ecological and technological conditions of human

activity and also provides the basic resources for production, turning into a natural-anthropogenic landscape (NAL). Some synonyms of the term "natural geographical landscape" are "natural landscape," "landscape geosystem," "landscape complex" or "natural territorial complex" (NTC).

4. A natural geographical landscape is a NTC of a certain rank—the lowest regional level of the landscape hierarchy, relatively homogeneous in origin, with zonal and azonal features, forming on a genetically individual lithogenic morphostructure on the macro-level and characterized by naturally repeated interrelated combinations of natural complexes of local levels (facies, tracts and localities), known as its morphological parts, as well as its local climate. It acts as a link between local and regional landscape geosystems. The horizontal dimension is n 10–n 100 km^2.
5. As a natural territorial complex (NTC), a landscape is a relatively small, specifically homogeneous section of the earth's surface, delimited by natural boundaries, within which natural components are closely interconnected and mutually depend upon one another and human beings (with elements of their culture) and are historically adapted to one other.
6. As a typological or generic concept, the term "landscape" is used with an adjective denoting a species or other generalizing classification, reflecting its specificity and relative genetic and other homogeneity in terms of a certain attribute. Examples include taiga, steppe, marsh or mountain landscape(s), cultural, domestic or marginal landscape(s), natural or natural-anthropogenic landscape(s), ecotone, geochemical, eluvial or elemental landscape(s), industrial or agricultural landscape(s), spiritual and ideological landscape(s): political, ethno-cultural, folklore, sacral or criminal landscape(s).
7. The natural-anthropogenic landscape is a landscape transformed to some extent by economic activity (positive or negative), often saturated with various elements of material culture.
8. A cultural landscape (from Latin *cultura*—cultivation, processing) is purposefully transformed and regularly used by man for the sustainable production of environmental, material and spiritual benefits, a landscape complex that includes interrelated elements of culture and nature, functioning as a whole.
9. A geographical landscape is an environment-forming and resource-replicating geo-ecosystem that serves as a habitat and arena for the economic activities of socio-ethnic groups and communities. This interpretation of the term "landscape," supplementing its former classical definitions, offers a broader picture of its modern use.
10. A landscape can also be a visually limited part of the earth, its external appearance, perceived through the senses, the appearance, image and generally visible part of the terrain. In French literature, landscape architecture and landscape design, the terms "landscape" and "paysage" are often used as synonyms. Examples of a landscape include forest, forest-field, steppe, mountain, field, rural, urban, open, visually shielded, deep, multi-composition, etc.

Other definitions of the landscape, with their humanitarian interpretations, are also presented in textbooks, manuals and articles by the author. The concept that a landscape is fundamental is now applied in different spheres of human activity and fields of knowledge. The main feature of the extended concept of a landscape is that it is linked to a certain territory, terrain or surface, and that interrelated, interdependent elements perform certain functions in the natural, natural-anthropogenic or anthropogenic environment.

Examples of landscape modifications in the twentieth century. In the second half of the nineteenth and the early twentieth century in Russia, the questions and principles of optimizing nature management and inexhaustible, sustainable farming on a landscape basis were actively developed by Dokuchaev (Dokuchaev 1994), his students and followers. The practical implementation of these developments in the Soviet Union (Russia) was reflected in the State Plan for the electrification of the country (1920–1930) and, especially, in the State Plan for the transformation of nature (1948–1954). On the scientific basis of that time, the implementation of this plan allowed a system of interconnected forest belts of various ranks (state, district, kolkhoz, etc.) to be created in the steppe, forest-steppe and treeless southern regions of the non-black-earth zone of the European territory of the Soviet Union. Most of them were tied to river valleys and other erosional forms, following the thalwegs. Therefore, the forest belts fulfilled a complex role in anti-erosion, water protection, snow retention, anti-deflationary and other amelioration, with a favorable effect on crop yields and the local climate. However, in the years 1960–1990, funding for forest reclamation and fire prevention in the system of state and collective forest belts was gradually reduced, and in 1993, it stopped. Therefore, by 2015, many forest belts were degraded, or had been destroyed by fires and excessive economic activity.

In the countries of Western and Central Europe, such theoretical and practical developments in landscape ecology at the end of the twentieth century can be associated with attempts to create ecological networks to preserve the most valuable elements of the landscape and biodiversity.

Geo-ecological studies in developed regions have shown that the negative consequences of economic activity there are usually associated with a poorly developed culture of ecological technology and landscape planning. The negative effects, as well as economic losses, can be eliminated or reduced by using the landscape ecology approach to designing business activities. An important part of such an approach to optimizing nature management is the concept of landscape resiliency.

3.5 Landscape Resiliency

Resiliency (stability, resistance) is one of the most important properties of any natural, natural-economic and economic system. It determines the existence of landscapes and other geo-ecosystems, their development, efficiency and the favorableness of their economic use. Resiliency often determines the choice of the type and intensity of landscape use in economic activity and nature conservation. The natural resiliency of landscapes is one of the main prerequisites for efficient production. At the same time, the persistency of the negative properties of landscapes (waterlogging, salinization, etc.) complicates their amelioration, increases costs and reduces production efficiency.

In Russia, ideas about the resiliency of landscapes began to actively develop in the 1980–1990s. Research has established that landscape stability is a complex phenomenon, and therefore, it has many definitions (Moiseyev 1987, Kazakov 1999, 2008).

In a generalized form, the stability of landscape geosystems is their ability to remain relatively unchanged or vary within a structural and functional range, or return to it during the period of their life cycle or the cycle of external influence. Like any complex phenomenon, stability has many aspects. It is possible to assess the stability of landscape geosystems or NTCs in terms of the amplitude of natural fluctuations of their parameters within the invariant range or deviations from it, or the deviations of these parameters under anthropogenic loads. When the stability of geosystems is measured and evaluated, it acts simultaneously as a relative value and as a very specific concept. For example, a clear definition is needed of the type of resistance that is being tested (to mechanical, chemical effects, etc.), and a point of reference is required when measuring and evaluating—an invariant of a specific NTC or changes in similar parameters in adjacent geocomplexes of other types. The indicator used also needs to be specified. Even relying on the concept of an invariant, one should take into account the phase characteristics of geosystems changing in the course of their functioning or development, since many parameters of geosystems change the speeds and directions of their "drift" and how informative the data is in different phases of functioning and development. For example, in winter, the photosynthetic activity of plants and the erosion activity of sloping NTCs in Russia are significantly lower than in the spring and summer periods.

Examples of different kinds of landscape resiliency. The differences in the natural stability of landscape geosystems and their resistance to anthropogenic influences can be shown by the following examples. Thus, natural zonal tundra and forest-steppe landscapes, mudflow or avalanche geocomplexes in the mountains and valley geosystems on the plains, in modern environmental conditions, are very stable both in space and in time. However, they differ greatly in the dynamics (variability) of their states. It has been established that there are landscape geo-ecosystems with strongly and weakly fluctuating organizational structures. For example, geocomplexes of floodplains and gentle watersheds differ sharply in the dynamics of their structure and state. In NTC watersheds, fluctuations of their parameters relative to the mean are less than in floodplain geosystems. However, these are their stable norms or invariants under natural environmental conditions. That is, floodplain NTCs are stable in terms of their increased natural variability or dynamism. At the same time, their resistance to various specific anthropogenic influences varies. In particular, naturally occurring tundra and north-taiga geo-ecosystems react very unstably to acid pollution, and forest-steppe and dry-steppe landscapes react to this type of impact very poorly. Moreover, even the reaction to acid pollution in different landscapes can take different directions. In taiga landscapes—especially those composed of outwash sand, with poor plant nutrients in podzolic soils—zonal coniferous forests and moss-lichen communities are actively dying out under the influence of acid emissions. In the steppe zone, acid emissions are easily neutralized by chestnut and chernozem soils with a saturated base-absorbing complex. At the same time, geosystems with wormwood plant communities can even occur on alkaline soil varieties. That is, under the influence of the same pollutant in taiga and tundra landscapes, the effect of one of the factors limiting the biodiversity of geosystems, lack of nutrition, is increased. The effect of ash emissions on the ecological situation in the same geosystems will have the opposite effect: a positive one in the taiga and a negative one in the dry steppe.

The resistance of sloped and flat geosystems to mechanical loads caused by recreational use, motor vehicles and grazing varies significantly. For example, for dry whitewood bogs on poor, highly podzolic sandy soils, the permissible recreational load which does not lead to the development of landscape ecology crises, is 1–2 persons/ha. For natural territorial complexes (NTCs) with fresh grass birch forests on weakly podzolic light loamy soils, it increases to 15–20 people/ha. In the examples given, different properties of landscape NTCs are shown: the factors that influence landscapes' passive or static (buffer) stability to various types of anthropogenic loads (Kazakov 1999, 2008).

Factors, mechanisms and types of landscape resiliency. The factors and mechanisms behind the stability of landscape geosystems are divided into passive/static and dynamic. Passive/static factors are usually determined by the mass, capacity, stiffness or strength characteristics of a substance, or the power of the energy flow. The passive (static) stability of landscape complexes is manifested in their invariance with respect to their structural and organizational invariant within the "characteristic time cycle" of their development. In contrast, dynamic factors are related to the plasticity of landscape geosystems—their adaptive capabilities, and elasticity—and the ability of landscapes to quickly return to a state of relative dynamic equilibrium after the load is removed (IGAN USSR 1989). A common property or factor combining the passive and dynamic stability of landscapes is their hierarchical organization.

Studies in areas of anthropogenic influence show that the properties of the natural components of different landscapes have very different effects on their stability. Therefore, hard scales for assessing the resiliency of landscapes can produce errors when used in large areas with different landscapes. However, studies have revealed some patterns to how the stability of landscape geosystems depends on the properties of their individual components (lithogenic basis, moisture, climate, biota, soils).

Landscape components and their resiliency. Other things being equal, the following relationships were revealed between the properties of natural components and landscapes' stability when exposed to anthropogenic loads (Kazakov 1999, 2008).

(1) The gravitational or denudation potential of a territory—the larger it is, the less geo-resistant it is to denudation, erosion, mechanical stress and even to toxicants,
(2) The slope of the surface—the greater it is, the lower the stability, but with slopes less than 10° stability may fall due to possible waterlogging and low self-purification from pollutants,
(3) The length of the slope—the longer it is, the lower the stability,
(4) The mechanical composition of the soil is usually more resistant to stress when the NTC is composed of light loam and sandy loam, but the maximum can be shifted somewhat depending on the type of impact (when exposed to acid precipitation, the NTC's stability distribution graph is sharply asymmetric),
(5) Soil thickness—if loamy soils have a thickness of less than 1.0–1.2 m, then as it decreases, the stability of the NTC decreases,
(6) Hygrotopes (or moisture)—maximum resistance to stress in geo-ecosystems of fresh habitats, dry and wet, stability decreases,
(7) Climatic characteristics—NTCs with the optimum ratio of heat and moisture have the highest resistance (hydrothermal coefficient and coefficient of moistening are close to 1), while NTCs with pronounced limiting factors for heat and moistening and a wide range of oscillations have the lowest resistance and moderate winds of 2.5–4 m/s also contribute to the stability of landscapes,
(8) Soils—the greater the thickness of the humus horizon, the humus content, capacity and saturation with the bases of the absorbing complex of the soil, the more resistant the NTC,
(9) Biota—the more capacious and intensive biogeochemical circulation (BIC), the denser the projective surface coverage, the higher the stability of the NTC, conifers and forests are on average less resistant to impacts than hardwood, meadow-steppe grass species are more resistant than forest ones, and roadside grasses and other synanthropes have the most resistance, species with a deep and dense root system are more resistant than those with a superficial and loose one, modified plant communities in the middle of a highly productive stage are the most resistant to anthropogenic impacts (e.g., forests at 50–70 years of age)
(10) The following landscape geo-ecosystems are potentially more resilient: (a) those with increased diversity and repeatability (duplication) of structures, (b) those in the central range of typicality for their zone and region, (c) trans-accumulative landscapes are more stable compared to trans-eluvial ones, (d) those which are more ambitious in size and substance, higher hierarchical ranks (landscape zone > landscape > tract > facies).

The stability of dissipative landscapes of hills, which mainly dissipate matter and energy in the environment, is lowered. It is also reduced in the NTCs of the extreme accumulative units of landscape catenas characterized by maximum entropy.

In the Soviet Union (Russia), several maps have been published to assess the potential passive stability of landscapes of the territory of the USSR/Russia and its individual regions with regard to various types of pollution and erosion hazard. Examples of these include maps with an analysis of the geochemical prerequisites of landscape resistance to pollutants (Glazovskaya 1988), or a map of landscape resistance to acid emissions from thermal power plants (Kazakov 1999).

Due to the different stability of natural complexes, the same processes or environmental factors can, with varying likelihood, cause environmental crises in some geosystems and hardly affect others. Thus, in the areas of influence of acid emissions from thermal power plants and steel mills, damage and shrinkage of coniferous taiga forests in eluvial habitats are common. In the trans-accumulative units of the same landscape catenas, as well as in landscapes of deciduous forests and forest-steppe, there is less visible damage to vegetation in zones of influence of acid emissions is less. This is explained by the different stability or buffering of NTC data with respect to acid emissions. They can differ in their stability levels by as much as 50–200 times.

Hierarchical level of landscape geosystems and their resiliency. An important factor determining landscape geosystems' passive stability and other types of stability in natural and anthropogenic conditions is their hierarchical organization (Kazakov 2003b). The increased stability of geosystems at higher hierarchical levels is primarily based on their greater mass and area, and therefore on inertia. The stability of large regional geosystems, which include significant masses of matter and energy, can only be disturbed by the impact of a more powerful natural or anthropogenic factor than those required to change the state of small local geosystems. This is most clearly manifested differences in the passive stability of geosystems of different ranks. A similar pattern occurs in ecology as applied to living organisms: The individual is less stable than the population or species. Accordingly, landscape dominants are usually more stable with respect to subdominants, etc.

However, as natural geo-ecosystems evolved, other mechanisms were developed in addition to passive stability: dynamic mechanisms for overcoming crises, aimed at stabilizing environmental protection systems in the environment and their further development. The key mechanism is the various types of adaptive variability found in the structures and functions of geo-ecosystems that are in crisis situations. This mechanism determines the adaptive type of landscape resiliency.

Often, adverse factors that cause crises and even disasters in some organisms and landscape geo-ecosystems are favorable factors for the development and prosperity of others. As a result, the latter begin to flourish, functionally replacing the former and stabilizing the changed landscape as a whole during the environmental changes (ECs). It is not without reason that the image of the concept of "crisis" in Chinese consists of two hieroglyphs denoting "danger" and "opportunity." For example, river valleys or mudflow-avalanche complexes, though generally stable in natural environmental conditions, can easily change some elements of their planned structure. In floodplain landscapes, depending on the nature of the floods, some old landscape elements disappear, but new ones appear, new channels appear and the river banks are formed and disappear. Accordingly, the vegetation and soil are rebuilt. That is, depending on the specific states of the environmental parameters, landscape geosystems can change their structure somewhat and even sacrifice part of the NTC on smaller, local levels.

Adaptive capabilities of landscape geosystems (elasticity and plasticity). The greater stability of the landscapes at the highest hierarchical levels is determined not only by their greater inertia in terms of mass and size but also by their great adaptive capabilities. More complex geosystems of higher ranks are more diverse in their constituent structural elements than geosystems of lower ranks. Due to their greater diversity, the range of possible and permissible adaptive changes in the states of complex geosystems is wider, without any loss to their stability. Different landscape complexes included in complex geosystems react differently to interannual or even seasonal changes in weather conditions. In some, biological productivity increases, while in others, it decreases with the same changes to the hydrothermal environmental factors. As a result, the bioproductivity of the landscapes that contain plant associations, on average, changes less than the bioproductivity of each of the separate associations. A similar picture is observed in the landscapes of large river systems with diverse watersheds. There, too, the average changes in the water level in the main river artery change to a lesser extent as compared with the river geosystems, which have smaller and simpler catchments in the landscape plan.

A moderate agricultural development of the moraine-glacial plain geosystem as a whole will not lead to a loss of stability and complete degradation. At the same time, the same moderate loads on its slope elements or subsystems can lead to a loss of stability and a radical restructuring of some local landscape geosystems of lower rank when erosion is activated. As a result of such local adjustments to the landscape of the plain, it will retain its stability in general.

In the cases considered, the stability of geosystems is supported on the one hand by the ability of more diverse geosystems to better absorb external impacts, variously mediating them, and on the other hand, by the fact that geosystems which are more complex and diverse in structure are easier to rearrange in accordance with environmental changes (ECs). Such properties and mechanisms for maintaining the stability of geosystems can be called adaptive plasticity or elasticity.

Studies have shown that NTCs of the following types have greater adaptive stability due to the plasticity of geosystems: ecotone landscapes, due to the greater species diversity of elements and their ability to easily replace each other, NTCs with highly fluctuating modes of operation and structures, NTCs with a high variety of elements, actively developing NTCs at the secondary bioproductive stages of succession. Geosystems with pronounced limiting factors, with reduced diversity, have low plasticity and adaptive stability.

Ability of landscape geosystems to self-repair. Another of the mechanisms that support the stability of geosystems is their ability to self-repair after disruption. This ability is described as a landscape's elastic stability.

Examples include the rapid recovery of destroyed vegetation or intensive self-cleaning of pollutants. The stability of geosystems can therefore be evaluated by the speed of their self-healing. Thus, tundra landscapes are less stable measured by the criterion of self-restoration in comparison with floodplain geosystems, which can restore not only disturbed meadow-shrub vegetation, but even the lithogenic basis in 2–6 years. Landscapes of tropical rainforests, characterized by high-capacity, intensive biogeochemical circulation (IBC), also have a great restorative ability. However, this mechanism for maintaining the stability of landscapes works mainly with periodic and occasional impacts. If the disturbed geo-ecosystem is restored during the time between impacts, it is assessed as resistant to them. Comparing the stability of different NTCs, elastic stability is assessed by the speed of their self-healing, and passive resistance by the degree of degradation or alteration.

Analysis of landscape geosystems shows that the mechanism of sustaining stability due to self-healing works better in geosystems with powerful real energy flows. An example of this is the landscape geosystems of river valleys, with such a powerful system-forming factor as a water flow, possessing high-capacity and intensive IBC. Other examples are delta-type landscape geo-ecosystems with a powerful stream of nutrient and biophilic nutrient elements and landscapes of humid subtropical, tropical and equatorial forests. These geosystems are characterized by a powerful stream of solar radiation and a significant amount of precipitation supporting active and high-capacity IBC.

An analysis of the general mechanisms and processes that generally determine the stability of geosystems shows that the geosystems least resistant to anthropogenic influences are the following:

- relict and young geosystems whose structure and functioning are not fully consistent with the modern conditions of their natural environment,
- geosystems with increased or, conversely, reduced reserves of potential dissipation energy (dissipation), but with increased concentration potential of a substance (mountains, hills or lowlands),
- geosystems with pronounced limiting hydrothermal factors (tundra—lack of heat, desert—lack of moisture, swamps—excessive moisture) or trophic factors (geosystems on well-washed fluvioglacial or alluvial sands),
- stability decreases with a decrease in the hierarchical rank or level of geosystems, as well as geosystems decreasing from dominants to subdominants and rare NTCs.

The most stable landscape geosystems are those which are located at the penultimate, long-term, highly productive stages of restorative successions. They are characterized by relatively high passive stability, including natural fluctuations of the environment, high potential for directional development, enhanced bioproductivity and a variety of structures. These properties also determine the wide possibilities of their adaptive variability, which helps to preserve the stability of the geosystem as a whole. That is, a small artificial rejuvenation of climax geo-ecosystems and their maintenance at highly productive stages of successions is one of the important geo-ecological directions for maintaining geo-ecosystems in a steady state, even under conditions of increasing anthropogenic influence and development.

Economic activity destabilizes and violates the stability of landscapes in the environment both by itself and through the intensification of destructive natural processes. As a result, natural-economic systems (NESs) function less efficiently and dangerous crisis situations develop in nature and society. Therefore, one of the ways to optimize NESs and nature management is to stabilize them in the natural environment, by creating sustainable cultural landscapes and preserving natural diversity.

In landscape science and geo-ecology, ideas about the elements of ecological culture of nature management have been developed: cultural landscapes as natural-economic systems, ecological and landscape ecology frameworks of territories, ecological planning and design of economic activities (Vernadsky 1988; Dokuchaev 1994; Kazakov 1999, 2008, 2013).

3.6 Landscape Ecology Principles and Approaches to the Optimization of Nature Management and Natural-Economic Systems

Landscape ecology principles. The landscape ecology culture of nature management and environmental technology allows us to gently overcome multi-scale environmental crises by mutually adjusting and adapting natural landscapes and technologies of economic activity. This is the essence of co-adaptation and the joint, sustainable evolutionary development of nature and society.

Important components of the ecological culture of nature management include the rationalizing of life, tied to the landscape, the development of environmentally friendly high technologies, focused on minimizing the consumption of natural resources and waste products discharged to the environmental.

The types of nature management, specialization and technical levels of production (agriculture, mining or processing industries, recreation) form a different organizational structure of cultural and marginal landscapes. Studies in industrial regions with localized powerful pollutant emissions have allowed us to identify patterns in the formation of unfavorable environmental conditions and combat the degradation of landscapes (Kazakov and Chizhova 2001; Kazakov 2008).

The environmental problem of the degradation of nature can to some extent be solved, and crisis situations mitigated, with the help of technological methods and technical means, for example, the integrated use of extracted raw materials and circulating water supply systems. In analog ecosystem models and natural-economic systems implemented in practice with cultural landscapes, the waste of some industries is partly used in others and partly incorporated into natural processes, assimilating elements of nature. This allows the economic subsystems to function more efficiently, preserving the valuable properties and elements of natural complexes, biodiversity and a favorable environmental situation. However, with poorly organized, diffuse emissions and impacts on the landscapes of the regions, it is not possible to solve the problem of environmental degradation by technical methods alone.

A greater effect is achieved if, to solve this problem, a location is planned in terms of landscape ecology and the territorial structure of economic activity is organized. To prevent the development of acute environmental crises in industrialized regions, it is necessary to use all available scientific and methodological, administrative and educational principles and techniques, including those developed in the culture of landscape ecology and nature management.

Approaches to the optimization of natural-economic systems. Studying the interaction of different economic systems with landscapes shows that the following complementary approaches can be used to optimize nature management in terms of landscape ecology.

1. An ecological and geographical territorial approach associated with the planning and design of economic activities in terms of landscape ecology. For example, in large areas of Russia, from the points of view of ecology, economics, landscape ecology and hygiene, it is more expedient to design and place thermal power plants (TPPs) operating on solid fuel in the landscapes of the forest zone. TPPs using fuel oil are ecologically safer and more effective in steppe landscapes. If it is necessary to locate them in the forest zone, it is better to choose landscapes formed on carbonate rocks or cover loams for this purpose. The stability and productivity of landscapes composed of sandy soils with coniferous and mixed forests in the areas of influence of thermal power plants operating on fuel oil can be greatly reduced (Kazakov 1999, 2008).

A well-known example is the optimization planning of agricultural activities depending on the slope and slope exposure. Slopes with a gradient of less than 2–4° are optimal for arable farming. As the gradient increases, the intensity of erosion processes sharply increases and the stability and efficiency of the functioning of agro-landscapes of this type decreases.

2. Another approach to the geo-ecological optimization of natural-economic systems is an adaptational one. This is associated with adaptive adjustments of small, less stable natural complexes and their components to new environmental conditions. Examples of such adjustments are the application of fertilizers or neutralizing agents to contaminated geosystems, the creation of new geosystem elements or the replacement of the least stable ones (replacement of coniferous trees with other trees that are more resistant to pollutants). When agro-landscaping NESs, such new or specially conserved elements include a variety of anti-erosion flux-scattering and windbreak forest belts and grass strips on the slopes near the top of ravines and on convex watersheds.

Another example of adaptive adjustments of recreational areas to improve the resiliency of landscapes is phytomelioration and various technical measures.

Phytomelioration can be carried out by grafting and planting species of trees, shrubs and herbaceous plants which are resistant to trampling, or creating artificial multi-species lawns. Technical means of improving the resiliency of recreational landscapes include the creation of an artificial road and footpath network, small architectural forms, etc. These activities increase a landscape's ability to withstand excessive loads, while maintaining a high level of biodiversity and attractiveness. The stability of NTCs increases, and hence, the maximum permissible recreational loads on them can significantly increase, sometimes tenfold. Standards have been developed for the density of the road and path network, depending on the recreational loads in parks and forest parks.

An important geo-ecological trend in maintaining landscapes in a steady state under the conditions of various anthropogenic impacts is the artificial rejuvenation of climax communities and their maintenance at highly productive stages of succession. These activities and new elements of the natural-economic landscape dramatically increase its stability and efficient functioning.

3. The third approach is technological. It is associated with the geo-ecological optimization of the technology used in economic activity (production) and environmental protection measures. In particular, coal-fired thermal power plants located in arid and sub-arid landscapes, in order to comply with EC hygienic standards, must keep all their equipment running well in the long term, and adjacent geosystems in a steady state require more efficient ash-collecting filters and more powerful and efficient cooling systems.

Using these principles of co-adaptation and techniques aimed at ecological modifications of low-level landscape geosystems, in accordance with new environmental conditions, allows us to prevent environmental crises of nature management, or shift them to micro-levels, reducing possible damage. As a result, restructuring in nature comes with less negative consequences for both the landscape and humankind.

The landscape ecology planning of cultural landscapes and technologies of economic activity is one means of actively promoting the co-adaptation of human beings and their economic activity in the natural environment. At the same time, an important role in the conservation of nature and biodiversity is played by the idea of the territories' landscape ecology complexes (LECs), which determines the favorableness of the ecological situation. LECs are an important element of landscape ecology planning and environmental stability (Kazakov 2008).

3.7 Conclusions

1. The lagging ecological culture from the material and production culture of nature management and consumer ideology are the main causes of environmental and resource crises, hampering the development of humanity.

2. The ecologization of the technological culture of nature management, the development of landscape ecology regulation and the planning of economic activities, as well as environmental ethics, are the basis of sustainable nature management.
3. The further development of natural science and other fundamentals of the ecological culture of nature management and the definition of its basic concepts should contribute to the formation of a new ecological world view in the society.
4. The use of landscape ecology principles and approaches to optimize nature management in practice will allow many environmental problems of nature sustainability to be solved or mitigated.

References

Berg LS (1947) Geographical zones of the Soviet Union. vol 1. Moscow, 401 p (Берг Л.С. Географические зоны Советского Союза. Т. 1 М. 1947, 401 с.)

Berg LS (1958) Subject and tasks of geography M.: Publishing house of USSR Academy of Sciences, vol 2. Physical geography, pp 112–119 (Берг Л.С. Предмет и задачи географии М.: Издательство АН СССР, 1958. том 2. Физическаягеография. с. 112–119)

Dokuchaev VV (1994) More expensive than gold: Russian black soil. Moscow University Publishing House, 1994, 544 c. (Докучаев В.В. Дороже золота русский чернозем. М.: Изд-во Московского ун-та, 1994. 544 с.)

Glazovskaya MA (1988) Geochemistry of natural and technogenic landscapes of the USSR. High School, Moscow, 328 p (ГлазовскаяМ.А. (1988) Геохимияприродныхитехногенных хландшафтовСССР. М.: Высшаяшкола, 1988. 328 с.)

IGAN USSR (1989) Factors and mechanisms for the stability of geosystems. Moscow IGAN USSR, 333 p (Факторы и механизмы устойчивости геосистем. М.: ИГАН СССР, 1989. 333 с.)

Kazakov LK (1999) Stability of natural complexes. In: Scientific works of the MNEPU, ser. Ecology, issue 1. M., pp 26–36. Устойчивость природных комплексов // Научные труды МНЭПУ, сер. Экология. Выпуск 1. М. 1999. с. 26–36.)

Kazakov LK (2003a) Landscape Stability and ecological framework of the SB territory. Scientific Papers. "Landscape ecology" issue 3, RITS—alpha, Moscow, pp 26–38 (Казаков Л.К. Устойчивостьландшафтов и экологическийкаркастерриторииСб. научныхтрудов. «Ландшафтнаяэкология» вып.3, М.: РИЦ - Альфа, 2003, с. 26–38)

Kazakov LK (2003b) Geoecological aspects of stable development. Geography and environment. Science, St. Petersburg, pp 364–375 (КазаковЛ.К. Геоэкологическиеаспектыстаб ильногоразвития.ГеографияиокружающаясредаСПБ.:Наука, 2003, с. 364–375)

Kazakov LK (2004) Landscape science (natural and natural-anthropogenic landscapes). Publishing House MNEPU, Moscow, 264 p (КазаковЛ.К. (2004) Ландшафтоведение (природныеиприродно-антропогенныеландшафты). - М.:ИздательствоМНЭПУ, 2004, 264 с.)

Kazakov LK (2005) Theoretical and methodological interrelations in natural sciences and humanitarian interpretations of the term and the concept of "landscape". Probl Reg Ecol 3:6–12 (Казаков Л.К. Теоретико-методологические взаимосвязи в естественнонаучныхи гуманитарных трактовках термина и понятия «ландшафт». Проблемырег иональнойэкологии.№ 3, 2005.с. 6–12)

Kazakov LK (2008) Landscape science with the basics of landscape planning. Publishing Center "Academy", Moscow, 336 p (Казаков Л.К. Ландшафтоведение с основами ландшафтного планирования. М.: Изд. Центр «Академия», 2008. 336 с.)
Kazakov LK (2012) Stability and dynamics of landscapes as factors of nature management. Environmental management: theory, practice, education-M.: Geographical f-t MSU, pp 40–49 (Казаков Л.К. Устойчивость и динамика ландшафтов как факторы природопользования. / Рациональное природопользование: теория, практика, образование. М.: Географический ф-т МГУ, 2012. с. 40–49)
Kazakov LK (2013) Landscape studies: textbook for students of higher education institutions. M.:. Publishing Center "Academy", 336 c. (Казаков Л.К. (2013) Ландшафтоведение: учебник для студентов учреждений высшего профессионального образования. – М.: Изд. Центр «Академия», 2013. 336 с.)
Kazakov LK, Chizhova VP (2001) Engineering geography. Moscow Landros, 268p (Казаков Л.К., Чижова ВПИнженерная география. М.: Лэндрос, 2001. 268с.)
Marx K (1964) Letter to F. Engels, March 23, 1868-Marx K., Engels F. Works, vol. 32, M.: Publishing house of political literature, p 45.(Маркс К. Письмо Ф.Энгельсу, 23 марта 1868 г. – Маркс К., Энгельс Ф.Соч., т. 32, М.: Издательство политической литературы, 1964. с. 45)
Milkov FN (1970) Landscape sphere of the earth. - M.: "Thought", 207 p (МильковФ.Н. Ландшафтнаясферваземли. -М.: «Мысль», 1970. 207 с.)
Milkov FN (1973) Man and landscapes. M.: "Thought". 234 p (Мильков Ф.Н. Человек и ландшафты. М.: «Мысль». 1973. 234 с.)
Milkov FN (1986) Physical geography: the doctrine of landscape and geographical zoning. VSU Publishing House, Voronezh, 328 p. (МильковФ.Н. Физическаягеография: учениеоландшафтеигеографическаязональность. Воронеж: изд-воВГУ, 1986. 328 с)
Millennium Ecosystem Assessment (2005). Ecosystems and human well-being: synthesis. Island Press, Washington, DC, 137 p. https://www.millenniumassessment.org/documents/document.356.aspx.pdf. Accessed 16 Dec 2018
Moiseyev NN (1987) Algorithms of development. Moscow Science, 304 p (Моисеев Н.Н. Алгоритмы развития. – М.: Наука, 1987. – 304 с.)
Moiseyev NN (1995) Modern rationalism. Moscow MGVP KOKS, 376 p (Моисеев Н.Н. Современный рационализм. М.: МГВП КОКС, 1995. 376 с.)
Nikolaev VA (2000) Landscape studies. Seminar practical classes. M.: Published. Moscow University, 94 p.(НиколаевВ.А. Ландшафтоведение. Семинарские практические занятия. М.: Издат. Московскогоуниверситета. 2000. 94с.)
Sochava VB (1978) Introduction to the doctrine of geosystems. Science, Novosibirsk, 318 p (Сочава В.Б. Введение в учение о геосистемах. Новосибирск: Наука. 1978. 318с.)
Solntsev NA (2001) The doctrine of the landscape. M.: Publishing House of Moscow University, 384 p (СолнцевН.А. Учениеоландшафте. М.: Издательство Московского университета, 2001. 384 с.)
Vernadsky VI (1967) Biosphere. Moscow, Mir, 183 p (Вернадский В.И. (1967) Биосфера. М.: Мир, 1967. 183 с)
Vernadsky VI (1988) Philosophical thoughts of a naturalist. Moscow, Science, 520 p (ВернадскийВ.И. Философскиемыслинатуралиста. М.: Наука, 1988. 520 с)
Vernadsky VI (1994) Living matter and biosphere. Science, Moscow, 669 p (Вернадский В.И. Живое вещество и биосфера. М.: Наука, 1994. 669 с.)

Chapter 4
Landscape Science for Natural Resource Management in a Globalised World

Guy M. Robinson

Abstract This chapter addresses key issues tackled by landscape scientists in the study of natural resource management (NRM). It outlines the nature of landscape science, emphasising that its practitioners originate from several disciplines, but shares a focus on problems occurring at the intersection of the natural and human-created environments. It stresses the role of the modelling of land system dynamics in studies of NRM, but recognises that, increasingly, landscape science has sought to utilise information and knowledge that sits outside traditional scientific data. This is addressed in one of three sections dealing with prominent aspects of landscape science. The first of these addresses the issue of the need to engage with 'stakeholders': the land managers and interested parties engaged in or directly affected by NRM. The attitudes, values, behaviour, and knowledge of stakeholders plays a vital part in determining the nature and consequences of NRM, so understanding how these elements can be incorporated into schemes and policies has been a vital facet of landscape science. In Europe, the term NRM is largely synonymous with agri-environment as applied to policy (AEP) or in specific schemes (AES). Again, key stakeholders are land managers, with one aspect of their input addressed here, namely their stewardship of the land. The nature of stewardship is discussed, with different types of stewardship recognised and their significance for the management of cultural landscapes. The next section considers the growing importance of studies focused on ecosystem services (ESS) and their measurement. It is argued that the analysis of changing ESS can be used as a means of assessing the nature of agricultural transformations associated with the impact of both global and local forces. Finally, the need for landscape science to engage both with the general public and policy makers is discussed.

G. M. Robinson (✉)
Department of Geography, School of Social Sciences, University of Adelaide, Adelaide, SA 5005, Australia
e-mail: guy.robinson@adelaide.edu.au

G. M. Robinson
Department of Land Economy, University of Cambridge, 16-21 Silver Street, Cambridge CB3 9EP, UK

L. Müller and F. Eulenstein (eds.), *Current Trends in Landscape Research*, Innovations in Landscape Research, https://doi.org/10.1007/978-3-030-30069-2_4

Keywords Landscape science · Natural resource management · Agri-environment · Stakeholders · Ecosystem services · Stewardship · Engagement

4.1 Introduction

Research in landscape science focuses on the development of new methods and integrated assessment to help conceptualise and represent the land use system (Lovett et al. 2015). Recent advances include work on vulnerability assessments for different scenarios, examining landholder attitudes towards land use change, and the policy development implications of various changes in land management. This work engages directly with societal interests through the identification of processes and actions to assist communities adapt to both growing population numbers and limited resource availability at different spatial scales, set against the context of changing societal expectations and globalisation. This presents a challenging interplay between biophysical, societal, and policy elements, increasingly being brought into sharper focus by public acknowledgement of new pressures associated with global climate change and the effects of global markets on communities worldwide.

This chapter examines the development of research on landscape change in regional social–ecological systems. The research is centred on the desire to obtain an improved and more renewable balance between commercial land-based production and conservation of natural and semi-natural ecosystems. So, the chapter deals with tensions between production and conservation, especially (though not exclusively) in regions where land use is dominated by agricultural production. The research it reviews focuses on interconnections between the environment, the social setting and the principal issues arising when considering how changes to land use might deliver multiple benefits. The latter involves interrogating the interface between management, policy, and societal issues, which 'must be understood to enable different land users, locales, and regions to adapt to a myriad of future challenges' (Robinson and Carson 2013, p. 1).

The primary focus herein is on the application of landscape science with respect to natural resources management (NRM), a term that generally refers to a particular set of natural resources rather than the entire resources spectrum. NRM has been applied primarily to management associated with agriculture and protected areas, rather than other industries such as mining. The use of this term has its roots in late nineteenth-century North America when there was interest in understanding the ecological nature of the rangelands, and the emergence of a national conservation movement in the USA (Mitchell 1981). Its use has been popularised in Australia whilst in most other developed countries, a narrower terminology, 'agri-environment programmes/policies (AEP)', which incorporates conservation measures, has been employed more frequently, though the two terms are often used synonymously.

The chapter starts by providing a brief overview of the nature of landscape science before focusing on four key issues within the contemporary development of the discipline. The first of these is the need to incorporate the knowledge and views of ‘stakeholders’ (especially land managers and community groups) within models and scenarios created within landscape science. This presents landscape scientists with the challenge of dealing with qualitative and non-scientific information in conjunction with ‘hard’ data they may have collected. The next section deals with the incorporation of ecosystem services within landscape science, noting that much of the data assembled for ecosystems exists at a landscape scale and increasingly is being used to drive NRM policy. Finally, there is consideration of the need for landscape scientists to develop greater engagement with the general public and policy makers to extend the dissemination and impact of their work.

4.2 Landscape Science

‘Landscape’ itself is a chaotic concept as it has several components but is difficult to define. However, as virtually all landscapes are created or modified to varying degrees by people, then it is possible to regard them all as comprising ‘cultural’ landscape, though with gradients of human impact. Within these cultural landscapes, there are also different balances between nature and culture, the latter including the historical and cultural values attached to the landscape. For natural resource management, the focus is primarily on the natural and cultural values in the ‘everyday’ landscapes of farming. These possess a ‘concrete’ physical reality, but they are also experienced by individuals and groups who have different understandings of these landscapes. These understandings influence people’s behaviours and decision making, leading to different management decisions and consequent actions.

The term ‘landscape science’ refers to research seeking to understand the relationship between people and their environment, focusing on land use change and data referring to land resources at the landscape scale. It originates in different disciplines, including the natural sciences (notably landscape ecology) (Wu 2017), the human sciences (via human geography and human ecology) (Steiner 2008), and humanistic and symbolic approaches to landscape (Antrop 2000; Antrop and Van Eetvelde 2017). It deals with understanding of drivers of change, structures, processes, and their varying states in terms of anthropogenic uses of landscapes. Landscape science is a meta-discipline, evolving from synergistic theories, methods, and knowledge of different scientific disciplines applied to study landscapes. The disciplines include agricultural science, forestry science, conservation science, geoscience, biology, the social sciences (especially geography), engineering sciences, and mathematics. Thus, landscape science has fostered interdisciplinary and solution-oriented approaches that foster collaboration between scientists and stakeholders in society, in which landscapes are conceptualised as social–ecological systems resulting from the interaction between societal and natural processes (Norberg et al. 2008).

Table 4.1 Content of landscape science

• Environmental, economic, and social processes induced by land use or land use changes
• Spatial interactions of processes and driving forces in anthropogenic landscapes
• Studying multiscale processes and cross-scale dynamics
• Spatial data analysis and modelling
• Landscape systems analysis and landscape modelling (e.g. scaling approaches, scenario techniques, indicator identification, complexity reduction)
• Discrete and integrated impact assessment of land use
• Methods and understanding of participatory approaches for land use development and decision making in land management
• Fundamentals and tools for decision support systems as well as providing scientific methods in decision implementation of landscape management

Based on Antrop (2000)

Some of the key aspects and concerns of landscape science are summarised in Table 4.1. These overlap with the content of the closely related 'land system science': 'an interdisciplinary field [that] seeks to understand the dynamics of land cover and land use as a coupled human-environment system to address theory, concepts, models, and applications relevant to environmental and societal problems, including the intersection of the two' (Turner et al. 2007, p. 20, see also Verburg et al. 2015).

Both landscape science and land system science are multi- and interdisciplinary and operate at the meeting point of nature and society (Bloemers et al. 2010). So, researchers working in these areas consider the behaviour of people and communities, multi-level decision making that affects multifaceted land units, and the contexts within which decisions about landscape and land issues are made and in which the land itself exists (Lambin et al. 2006). Much of this research has used empirical analysis to tackle problems associated with human–environment interactions, including changing land use practices and associated changes in land cover (e.g. deforestation), responses to climate change and impacts of diverse environmental changes on agriculture, forestry, biomass production, and ecosystem functioning (e.g. Brose and Hillebrand 2016; Plieninger et al. 2015b).

There are some compelling reasons for research on NRM to use landscape science. Perhaps, the most overriding one is the growing realisation that successful NRM requires integration of the management of different types of resource, with the development of system understanding that can be best provided by landscape science. It is also the case that much management of NRM occurs at the landscape scale which is also the main scalar focus of landscape science.

Modelling of land system dynamics often accompanies studies of NRM, frequently involving structural analysis of complex interactions within the land system between people and the environment, generating information that assists evaluation of policy outcomes or develops new policy prescriptions. These models usually utilise physical, mathematical, or numerical approaches to help develop better understanding of existing processes and interactions and to tackle 'what if?' scenarios, e.g. the possible outcomes of a new policy or land management activity. The models make use of environmental data in various forms, with decision makers requiring information systems capable of integrating large and diverse datasets and also providing tools for analysing and visualising key information.

Models have been used to explore systems behaviour, for ex-ante assessments of policies, inputs to the planning process, and scenario creation (e.g. Zagaria et al. 2017). Increasingly, landscape science is developing new models, acknowledging that these will need to progress beyond single-sector foci to capture the complexity of human–environment interactions across different scales (e.g. Vaz 2016). Modelling strategies have been adapted to meet this challenge, including economic models (e.g. Munroe et al. 2014), models integrating socio-economic and environmental processes (Leimbach et al. 2012) and agent-based modelling for simulating complex decision making (Millington and Wainwright 2016). However, landscape science has explicitly attempted to develop methods combining 'hard' scientific data with input of a qualitative nature, in the form of knowledge from landholders, land managers, and other community stakeholders. These methods have often used models both as an aid in understanding how landscapes can be managed and in developing management scenarios that can incorporate mitigation of and adaptations to environmental change (Turner et al. 2016). A major concern in model formulation and operationalisation has been how to incorporate the views and actions of different human 'stakeholders'.

The starting point for much work in landscape science is the assembly of data, both quantitative and qualitative, with which to understand the complex, often nonlinear processes that frequently produce systems operating at or understandable at a landscape scale. Yet, an ongoing problem, also highlighted frequently in studies on climate change, is 'the lack of long-term data capable of showing the state of the environment and the direction in which it is heading' (Lefroy et al. 2012, p. 3). So, despite the frequent availability of data in various forms, interpretation and analysis must necessarily develop innovative ways of combining the different forms, including surrogates for historic and prehistoric changes and employing modelling that can both provide estimates for missing data and generate management scenarios.

The growth of studies incorporating inputs from different disciplines, groups, and individuals as well as comprising both quantitative and qualitative data has produced a growing realisation that policy development should acknowledge this multiplicity of approaches, sources, and inputs. To develop policy capable of producing future landscapes that can meet multiple goals and satisfy demands from various stakeholders, it is necessary to incorporate inputs from a broad spectrum of disciplines. This multiplicity must extend beyond academe to inputs from land managers and communities affected by policy, enabling participation in the research process and in formulating policy to embrace a wide variety of individuals and groups. It is this that forms the next aspect of landscape science addressed in this chapter.

4.3 Engaging Stakeholders

A central characteristic of recent research in landscape science has been its use of inputs from the individuals and groups who manage the land: individual land managers, rural communities, groups (including indigenous peoples), and other individuals who play a role in landscape management. In the past, the knowledge possessed by these managers (usually termed 'stakeholders') has often been neglected by researchers and policy makers. A key reason for this, as discussed by Raymond et al. (2010b) in an examination of various environmental management projects in different countries which were attempting to integrate knowledge from different sources, is that integration of stakeholder knowledge is inherently complex, and it is frequently dealt with in arbitrary fashion employing various knowledge integration perspectives by different researchers/policy makers. Western paradigms and scientific systems of knowledge do not readily address the full complexity of environmental management, nor are they sufficiently able to integrate perspectives of local stakeholders into the development of environmental management strategies (Eden 1996).

New approaches to environmental management have recognised the problem, developing approaches to landscape science that explicitly include local knowledge. This has been given various terms: community-based NRM (Dyer et al. 2014), sustainability science (Folke et al. 2016), and adaptive co-management (Wilson et al. 2018), which are all approaches recognising the need to integrate knowledge possessed by academic researchers (frequently extending across traditional disciplinary boundaries) and non-academic participants. They emphasise the need to build on different (and sometimes disparate) bodies of knowledge to tackle a research or applied question by formulating shared theory, methods, and new knowledge to promote common understanding of environmental management problems.

Previous reliance on 'hard' scientific data and numerically based models is giving way to the inclusion of different types of input to models and landscape analysis. This has generated scope for innovative approaches, such as participatory action research and the incorporation of qualitative data as a routine component of analysis. Furthermore, they frequently utilise participatory research methods, aiming to facilitate participatory, multi-level governance processes to both enhance the validity of knowledge elicited in research and to increase stakeholder inclusivity in decision making. This may follow iterative processes of knowledge creation, application, reflection, learning, and feedback to science or decision making. A common aim is to integrate knowledge across different spatial and temporal scales. These characteristics have become integral components of contemporary landscape science.

There has been a growing emphasis on direct engagement with policy makers, often explicitly embedding them within the research process rather than treating them as recipients of research findings or as commentators on modelled scenarios. Research designs incorporating knowledge exchange between different stakeholders are now more prominent, with landscape scientists taking responsibility for engaging with policy makers and politicians to remove traditional barriers and entrenched mindsets. Given the speed with which landscape science has evolved in the past two decades, there can be much hope that such shifts in approach and innovative frameworks within which the research develops can enable new successes to be realised.

A good example of innovative approaches being introduced into landscape science, and which laid a solid foundation for subsequent research, is the Landscape Logic project in Australia (Lefroy et al. 2012). This focused on knowledge discovery, integration, and 'broking' (maintaining a two-way flow of information between researchers, end-user partners, and stakeholders). The latter was intended to generate desirable management outcomes, explicitly incorporating the knowledge possessed by landholders and managers as well as that from scientists and stakeholders in government and industry. There were five basic components within the project's system thinking, which are typical of the landscape science approach:

- Research involved clearly defined, measurable, and widely accepted end points, including both environmental and socio-economic goals.
- The system boundaries were large enough to incorporate the major factors whilst not attempting to encompass all external devices.
- Research articulated each of the major influences acting on the variables of interest.
- It identified major factors providing appropriate or sufficient measurement of the effect of one major variable on another.
- Data collection was at a spatial and temporal resolution sufficient for testing of relationships.

As illustrated by Landscape Logic, a heightened role within landscape science for local knowledge has been strongly supported in many quarters in the past two decades (see Girard 2015). Such arguments recognise that knowledge can take many forms and can be derived from different sources, with government and science not having a monopoly on information that is of practical value in NRM. However, power relations may restrict locally held knowledge from being valued or utilised effectively by decision makers in government, especially the knowledge possessed by farmers, which 'represents various degrees of localised, expert, tacit, and implicit knowledge which may have been derived through formalised or informal processes' (Raymond et al. 2010b, p. 1767). Formalised processes do not usually generate experiential and local knowledge. Yet, this type of 'hybrid' knowledge which is obtained from various sources can be produced through social learning processes that combine scientific knowledge with that developed through personal experience, interpretation, and interaction (Stringer et al. 2017). Hybridity may then be explicitly applied in NRM, such as in agricultural development programmes, e.g. in Mexico's Yaqui Valley (McCullough and Matson 2016), agroforestry in India (Singh and Dhayani 2014), and European viticulture (Krzywoszynska 2016). Yet, in developing these programmes there are epistemological challenges to be overcome regarding fundamental differences in how people perceive the nature of knowledge or come to know something. Moreover, 'any knowledge integration process needs to be sufficiently flexible to take into account changes in perceptions emerging during the project and to deal with new information arising after application' (Raymond et al. 2010b, p. 1770).

Fortunately, there are now increasing numbers of good examples demonstrating successful knowledge integration between scientific and local, and top-down and bottom-up (e.g. Girard et al. 2015; Moschitz et al. 2015). These include projects facilitating adaptation to climate change (e.g. Xu and Grumbine 2014; Bocco and Napoletano 2017). For example, in south-east Australia, Cross and Ampt (2017) report on successful adoption of a rotational grazing scheme using few purchased inputs to develop a more sustainable agroecological farming system. They describe the adoption of the grazing system as occurring through the work of a community of practice (comprising farmers and their advisors) centred on farmers and their knowledge, but also with some input of scientific knowledge and principles. However, they acknowledge that there is often limited support from formal institutions (such as government departments) in the actual implementation of such schemes, especially for the integration of different types of knowledge or for a 'bottom-up' farmer-centred approach.

In discussing this issue, Bjerkes (2010) presents the argument that effective management of resources requires devolution of rights over resources and property to local user groups and organisations. This is not merely decentralisation from one government agency to another. Shared understandings are extremely important in the context of climate change, especially given the diversity of views about causes and their likely consequences (Raymond and Spoehr 2013). Robinson (2018) argues that if longer-term adaptive strategies are to be implemented with respect to climate change, then it is vital for government to develop more effective dialogue with communities of practice (such as farm systems' groups) and individual farmers (Fig. 4.1). Based on Bjerkes (2010), he presents a three-step approach to promote greater adaptation to climate change by farmers and the wider rural community:

1. Encourage groups of actors/stakeholders to reach a shared understanding of issues and identify their vision for the future.
2. Translate the vision of each group into action plans.
3. Support multiple cycles of joint and collaborative action.

Input to NRM from community stakeholders has been regarded as increasingly important, in part because this input can ensure that a variety of views is heard. There remains, though, the question of what constitutes successful engagement. In general, research has shown that understanding on both sides may be enhanced using qualitative methods, with deliberate attempts to link the values of individuals and groups to places. For example, when a particular ecosystem service directly positively impacts the viability or enjoyment of the community the more likely the community advisors are to support the management and preservation of that service. However, community consultation that deals with policies falling outside a

Fig. 4.1 A farm systems group inspecting livestock in the Yorke Peninsula, South Australia. This group formed one of a series of focus groups for research reported in Robinson (2018). Photograph by C. M. Raymond

group's or individual's immediate ambit tends to be regarded as ones in which they do not have a direct stake, and this can affect their advice and input (MacDonald et al. 2013).

4.4 Agri-Environment Schemes and Stewardship

Agri-environmental schemes (AES) are mechanisms enabling landowners and other individuals and bodies who manage land to be incentivised to manage their land in an 'environmentally friendly' fashion. The European Union (EU) has pioneered many such schemes and between 2007 and 2013 spent €23 billion on AES that affected 46.9 million ha as part of the Common Agricultural Policy (CAP). For 2014–2020, this spending is scheduled to grow to €25 billion, representing 30% of total spend compared with 25% in the previous period (European Commission 2013). The rise reflects the adoption of broader objectives, including climate change mitigation, promotion of organic farming and so-called climate and environment investment measures. In addressing these objectives, landscape scientists have adopted participatory techniques to assist in integrating multiple objectives into landscape planning. These have included collective visioning for the future, mapping of local stakeholder values to inform priorities for conservation, and segmenting land managers according to different types of values and/or farming motivations to improve targeting of policy (Forrester et al. 2015).

Overtime AESs have evolved and taken on the following three characteristics:

- increased complexity of scheme design and evaluation,
- a range of innovative collaborative governance models has been developed to support the design and evaluation of the next wave of AES, and
- encouraging collaboration across institutional levels.

Key issues in recent AES policy development are how to manage trade-offs between competing objectives, and how to engage local stakeholders in setting priorities for land management, as per the considerations raised above.

AESs have been a component of policy in the EU since 1986, with the designation of environmentally sensitive areas (Robinson 1994). However, AES and their associated payments for environmental/ecological services (PES) have a much longer history in North America. Probably, the world's largest and longest running PES programme is the Conservation Reserve Programme in the USA. This has operated in various guises since the 1930s (Hellerstein 2017) and currently provides around $1.8 billion p.a. through 766,000 contracts with farmers and landowners who place a total of 140,000 km^2 of what are defined as 'environmentally sensitive lands' into the programme. The farmers agree to plant 'long-term, resource-conserving covers to improve water quality, control soil erosion and enhance habitats for waterfowl and wildlife' (FSA 2014).

Another scheme on a vast scale is the much-celebrated Chinese PES scheme, the $43 billion Grain for Green programme, known also as the Conversion of Cropland to Forest and Grassland Programme or the Sloping Land Conversion Programme, established in 1999. This has removed marginal land from agricultural production to deliver ecological gains (Wang et al. 2017). It is China's largest ecological restoration programme and rewards farmers for not clearing forested slopes for farming, thereby reducing erosion and reducing the flow of sediments from farmland into rivers. Satellite remote sensing has revealed that across China's Loess Plateau, Grain for Green has increased tree cover by about 41%, with major increases in the enhanced vegetation index (EVI), leaf area index, and the fraction of photo-synthetically active radiation absorbed by canopies (Xiao 2014). By the end of 2012, Grain for Green had converted 9.1 million ha of cropland to forest and 0.64 million ha to grassland (Song et al. 2014).

In the EU, AES has included various farming intensity reduction measures, notably management of low-intensity pasture systems, integrated farm management, promotion of organic farming, conservation of high-value habitats, and conservation of target flagship species. There have been numerous different schemes, which are usually voluntary in terms of farmer participation, possessing different aims and levels of support, but many of which have attracted researchers' attention in terms of understanding farmer motivations. Financial incentives are usually provided to encourage participation, especially through promoting greater security of income, a commitment to the scheme's environmental conservation aims, the fit with farm management plans and lack of alteration required to existing management arrangements (Lastra-Bravo et al. 2015; Reed et al. 2017; Riley 2016). Typologies of farmers have been formulated based on the characteristics of participants and non-participants or on adopted pro-environmental behaviours (Burton 2014; McGuire et al. 2015; Riley 2011).

Many AES measures are aimed at securing a better balance between agriculture's environmental impacts and the demand for food. Yet, they have generally not focused on specified sustainability targets, though promotion of organic farming has been an option in several programmes. Through AES farmers have been encouraged to move away from the industrial, productivist model of farming as part of what has been termed 'the Naturalization Food and Agricultural (NFA) paradigm' (Rausser et al. 2015). The paradigm subsumes diverse arguments relating to food safety, obesity, the role of agriculture in climate change, the demise of family farming, and the concentration of agribusiness. It has grown in popularity, especially in the West, through increased interest in organic and environmentally friendly farming and concerns over negative externalities from industrialised agriculture. It is exerting growing political influence, despite champions of alternatives to the industrial–productivist system often struggling to produce economically rational policy proscriptions. In part, this is because the paradigm is an amalgam of various socio-economic and biophysical conditions and their attendant diverse potential 'solutions', which themselves may be problematic. For example, a wholesale change from an industrial model to farming systems favoured by NFA could seriously reduce food security if output is compromised (Leifield 2016). As shown in several studies (e.g. Servon and Pink 2015; Oostindie et al. 2016), the reality is that there is increasing coexistence of both globalised and local systems of production, serving different markets. This implies that policies which are applied on a 'one size fits all' across every producer may need to be reconsidered. More analysis of alternative environmental and resource policies is needed that addresses the effects of agri-industrial concentration, food safety, and food security, especially by reformulating welfare and political–economic arguments (Anderson et al. 2014; Rausser and Farrell 2016).

Better land stewardship is a central aim of all AES schemes targeted at farmers, usually as expressed in terms of delivering environmental gains. The premise is that farming involves the production of landscape, which frequently generates positive agri-ecological outcomes in addition to the main purpose of most farming, namely production of food and fibre. Yet, research by Raymond et al. (2016a, b) reveals that among farming communities, there are many different understandings of 'stewardship', with interpretations falling into four differentiated categories: environmental, production, holistic, and instrumental, as shown in Table 4.2.

Raymond et al's (2016b) study of 40 farmers in south-west England found the sample was almost evenly split between the first, third, and fourth of these categories, with a smaller proportion from the production category. This differentiation reflected tensions in the wider farming community regarding future agri-environment policy. Production respondents contended that AES neglects the prime need to encourage food production, whereas environmental and holistic

Table 4.2 Typology of farm stewardship

Environmental	Managing environmental features to enable protection for future generations This often involves implementing measures to encourage wildlife
Production	Maintaining productivity of land for future generations This may involve use of traditional farming techniques These farmers possess different conservation goals to those of the Environmental category
Holistic	Recognizing interactions and interdependencies between production and the agri-ecological system This can involve maintaining or enhancing landscape diversity through a mosaic of different land uses, including field boundaries, such as hedgerows
Instrumental	Defining stewardship in the context of government policy or an incentive Stewardship means acting in accordance with a given scheme in which land managers are paid for pro-environmental actions

Based on Raymond et al. (2016a, b)

farmers regarded AES as not supporting the development of a local 'green' food culture and its associated social infrastructure.

Bieling and Plieninger (2017) champion the role of landscape stewardship within landscape science. Their conceptualisation involves a proactive landscape science that does not only observe or respond to landscape change but also 'explores and supports innovative and passionate stewardship of ecosystems and landscapes, thus reinforcing feedback loops between humans and nature' (p. xiii). They view stewardship as both a management approach and an ethic which emphasises responsibility, collaboration, participation, and communication in the planning and management of land resources (see Gundersen and Mäkinen 2009). These sentiments are linked to the ideas of Aldo Leopold in his *Sand Country Almanack* (1949), which laid the foundations for re-engaging and inspiring people to take care of the land. Stewardship has been employed in various guises with respect to NRM, but especially to imply responsible use of natural resources, including conservation, with respect to agriculture and forestry that balances societal interests and needs, e.g. for a reliable food supply. 'Responsible use' by farmers generally reinforces their strong attachment to the land they farm (Raymond et al. 2010a).

Bieling and Plieninger (2017) were writing in part on the basis of their experiences in HERCULES (HERitage in CULtural LandscapES), a three-year research project funded by the European Union. The project's focus was on cultural landscapes in which the concepts of landscape and nature are interlinked (Fig. 4.2). Often, the general public and also some policy makers increasingly idealise the European 'countryside' as 'natural' and not the product of long-term and ongoing human interaction with the physical environment. However, it is also a view in which traditional human landscapes become valued cultural landscapes that need to be protected from 'insensitive' developments through conservation and preservation measures. Concepts such as nature, landscape, and cultural landscape can only

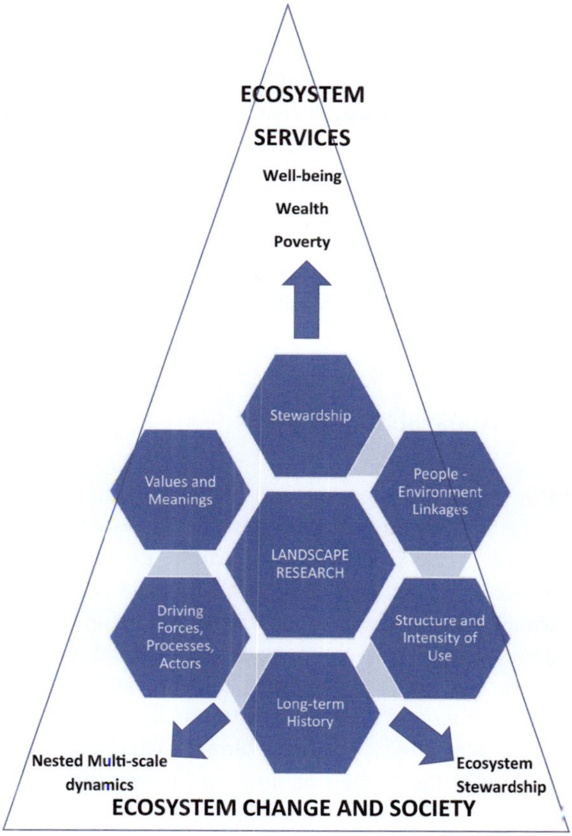

Fig. 4.2 HERCULES project: Exploring ecosystem change and society through a cultural landscape lens (based on T. Plieninger, http://www.hercules-landscapes.eu/blog.php)

be understood and interpreted in their historically specific social and cultural contexts, but the meanings attached to the landscape will tend to be produced by dominant groups and maintained by institutions representing their interests. Such ideas are reproduced, as well as contested, in the relations between these institutions and users of the landscape. It can be recognised that different sub-cultures also produce their own varieties of landscape, which reflects both physically and conceptually social relationships and constellations of power in society.

In managing cultural landscapes, there is always the issue of what change should be permitted or encouraged versus what should be conserved. In dealing with this latter question (in effect, 'what gets conserved?'), there are two distinct components. One is the designed elements in the landscape, such as buildings, monuments, and ownership structures, which typically are preserved or restored to a former 'optimal state'. The other is the evolved or living agricultural landscape for which it is generally acknowledged that many types of traditional agrarian landscape can ultimately only be preserved if the traditional forms of farming that led to their occurrence are maintained. Hence to preserve such landscapes, succession in the agri-ecosystem must be prevented. Of course, conservation creates its own landscapes, which raises issues about which landscapes are valued and protected through legislative instruments. In some cases, valued landscapes are restored or re-created (e.g. Robinson and Robinson 2016), adding an extra level of complexity into land management decisions. For landscape scientists, the issue is how best to present analysis and models that can provide clear information on contentious issues being considered by managers and policy makers. Some of this information may take the form of revealing the nature of historic and recent landscape changes, which can take many guises, but for quantitative changes measurement of ecosystem services has become increasingly popular (Di Minin et al. 2017; Naidoo et al. 2008; Plieninger et al. 2015a).

4.5 Ecosystem Services

In the past two decades, one of the popular ways in which interactions between people and the environment have been conceptualised has been through ecosystem services (ESS), enabling research to focus on how ecosystems (natural, semi-natural, and agri-ecological) provide benefits to people. These benefits were formally categorised in 2005 by the United Nations Millennium Ecosystem Assessment, which recognised four broad categories: provisioning (food and water), regulating (controlling climate and disease), supporting (nutrient cycles, crop pollination), and cultural (recreational and spiritual) (Millennium Ecosystem Assessment 2005). A huge volume of research has been developed on these services, in terms of their ecology, measurement, evaluation, and economic implications (Everard 2017). Accordingly, work on ecosystem services has become a key component of landscape science.

From a scientific perspective, Kremen (2005) suggests the following research agenda for the study of ecosystem services:

(a) identification of ecosystem service providers (ESPs), that is the species or populations that provide specific ecosystem services, and characterisation of their functional roles and relationships,
(b) determination of community structure aspects that influence how ESPs function in the landscape, e.g. compensatory responses that stabilise functions and non-random extinction sequences which can erode them,
(c) assessment of key environmental (abiotic) factors influencing the provision of services, and
(d) measurement of the spatial and temporal scales on which ESPs and their services operate.

Provision of ESS may be stabilised or increased with increasing biodiversity, which also benefits the variety of ESS available to society. Hence, understanding the relationship between biodiversity and the stability of an ecosystem is essential to the management of natural resources and their services. In this context, estimating the functional structure of an ecosystem and then combining it with information about individual species' traits can help understanding of the resilience of an ecosystem as environmental change occurs, e.g. through explicit human intervention or as a by-product of human actions, e.g. climate change.

Perhaps, the most dramatic changes to ESS in recent years have been associated with the impacts of globalisation, especially where global forces have been incorporated in processes of agricultural transformation. Nowhere has this been more evident than in China where some of the greatest changes to ESS worldwide have occurred, driven by massive socio-economic changes and innovation following adoption of economic reforms from December 1978 onwards (Anderson 2018; Zhou 2016). The transformation of Chinese agriculture has corresponded with dramatic economic growth, leading to substantial modifications to land use and land cover (Li et al. 2018). These have significantly affected biodiversity and ecosystem productivity, leading inevitably to modifications to the structure of ecosystems and so affecting ESS both positively and negatively (Liu et al. 2015).

Traditional farming methods, in China as elsewhere in the world, promoted soil fertility, irrigation, and crop/livestock production, which represented positive ESS in the form of soil and water conservation, and the balancing of the provisioning functions for grains with inputs from animal manures. Agriculture maintained healthy agri-ecosystems and could contribute to biodiversity conservation. However, through rapid urbanisation in recent decades, often involving substantial losses of farmland and the growth of intensive, industrial-style production methods, modern agriculture has been associated with major land use modifications and many changes to ESS. Outcomes include adverse effects on biodiversity; e.g. in China, several native species are endangered which is reducing important functions of ESS, including pest control and pollination (Wang and Zhou 2014).

Monitoring of ESS can occur at many scales, from broad global estimates to micro-scale analyses for individual villages (Robinson and Song 2018; Song et al. 2017) (Fig. 4.3). However, how to use knowledge gained about ESS in making decisions that in turn will impact on ESS is a concern still to be fully addressed by landscape scientists. For example, Martinez-Harms et al. (2015) refer to the potential offered by five core steps leading from ESS measurement to a management decision:

- identification of the problem and its social–ecological context,
- specification of objectives and associated performance measures,
- defining alternative management actions and evaluating the consequences of these actions,
- assessment of trade-offs and prioritisation of alternative management actions, and
- making management decisions.

To date ESS, assessments have largely focused on quantification and mapping, and hence, they have highlighted important areas for the supply and delivery of ESS, but closer links to management decisions need to be developed, especially through analyses of trade-offs between ESS (Castro et al. 2014) and using ESS to evaluate policy scenarios (Costanza et al. 2017). However, to use ESS more effectively in

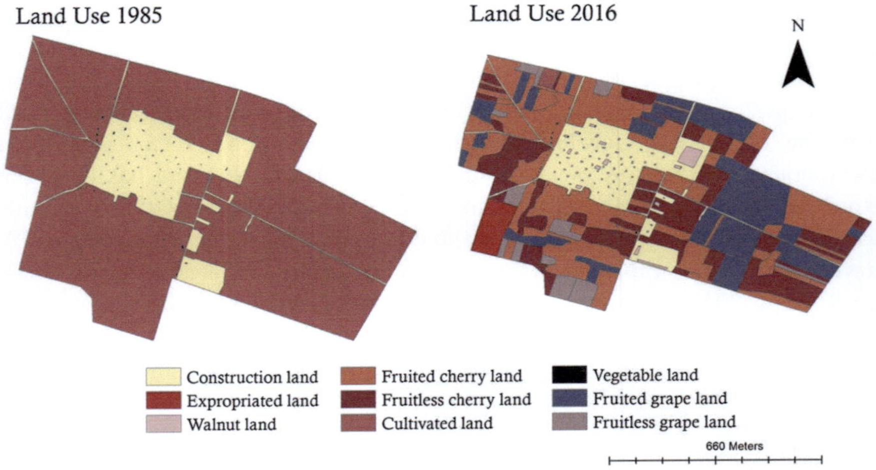

Fig. 4.3 Land use changes in Duling Village, Shaanxi Province, China, 1985–2016. These changes, reconstructed using participatory mapping, were used as the basis for calculating changes to ecosystem services by Song et al. (2017). Maps drawn by B. Song

decision making will require greater engagement with stakeholders, as discussed above with respect to landscape science in general, using deliberative and participatory methods. This also raises issues about how knowledge gleaned from the application of landscape science, including the use of ESS, can best be exchanged with those working outside academe.

4.6 Conclusion: Knowledge Exchange

One of the future challenges for landscape scientists is to find better ways of communicating their results to non-academic users if there is to be effective exchange of knowledge (Shuttleworth and Palang 2017, p. 809). Indeed, the concept of knowledge exchange has shifted. So, it is not merely scientists informing other scientists and policy makers about their findings, but it is part of a four-way process: from science to science, from science to citizens, from citizens to science, and from citizens to citizens (Shaw et al. 2017). This contributes to the development of 'solutions' that can bridge the gap between top-down expert views and bottom-up knowledge and perceptions. The growth of citizen science may have a role to play in this, though it is not a new phenomenon, having been popularised in North America in the 1990s, e.g. by the Annapolis River Guardians (Cliche and Freeman 2017; Robinson 1997). But it 'offers new approaches and opportunities for engagement between researchers, practitioners and communities' (Shuttleworth and

Palang 2017, p. 814). There remains a gap between researchers and practitioners, partly because of the lack of involvement of the latter in research and the perception that the impacts of landscape science research are often long-term, indirect, and nonlinear.

HERCULES, the EU-funded project on cultural landscapes referred to above, explicitly addressed the issue of knowledge exchange by considering basic differences in perspective between academic researchers and practitioners (see Tieskens et al. 2017). Its solution was to engage directly with local communities in case study areas, transmitting information to individual communities about themselves and other communities to show that local issues may have similarities to those experienced elsewhere. Similar to practice in several other landscape science projects, HERCULES used a 'knowledge hub' (for landscape practices) to store information that could be accessed by all involved in the project as well as providing an opportunity for citizen participants to upload information. The hub acted as an open platform to which all landscape actors, their networks, and organisations could contribute. It enabled visualisation of the interface between nature and society, using tools from Geographical Information Systems (GIS). Geospatial data were combined with layers of information relating to values and meaning, history, structure, and land use intensity, as well as driving forces, processes, and actors in landscape change. The hub provided the means for collecting both the contribution of experience-based local insights and scientific knowledge.

The hub could also be used as a crowdsourcing tool, collecting feedback and input from citizens as well as landscape professionals. The crowdsourced data offered the possibility to improve knowledge of the collective image of landscapes and to understand visually which landscape features most attract recreationists, visitors, and other user groups. However, uncertainty about the social and ecological representativeness of publicly shared photographs means that caution is required set against the increasing attractiveness of crowdsourced social media for landscape science research. In addition, multiplicity of outlets for communicating research results is another increasingly popular strategy that was also adopted in the HERCULES project, which utilised various ad hoc communication tools.

The nature of knowledge exchange has also become of increasing concern because of scientists' awareness of the need to communicate their findings about global environmental change to policy makers and citizens who need to be actively involved in taking urgent action to limit or prevent deleterious changes to socio-ecological systems, e.g. to limit global warming. The exchanges are essentially a social process but have been limited by lack of adequate conceptual development and appropriate methods to evaluate what are both complex and multifaceted exchanges. Fazey et al.'s (2014) analysis of 135 projects involving peer-reviewed evaluations of knowledge exchange revealed great variations between disciplines (seven types of exchange were recognised) and often a quite limited attempt at exchanging knowledge that merely emphasised one-way information flow from scientists to interested parties. This led to the conclusion that a clear methodology for knowledge exchange needed to be developed, involving a means, whereby end-users could evaluate knowledge exchange, think carefully about how knowledge exchange is conceptualised and formulate assumptions as to why it is expected to deliver its outcomes, and then taking steps to evaluate outcomes of the process.

In taking knowledge exchange seriously, landscape scientists are ensuring that their work gains greater impact and credibility. In recognising its importance, it is clear that knowledge exchange should be designed into research, with 'the needs of likely research users and other stakeholders … systematically represented in the research where possible, and long-term relationships must be built on trust and two-way dialogue between researchers and stakeholders in order to ensure effective co-generation of new knowledge' (Reed et al. 2014, p. 337). Only by involving a wide range of stakeholders will the work of landscape scientists ensure the necessary collaborations that will co-produce new knowledge and the more effective sharing and application of existing knowledge to manage change under global pressures in socio-ecological systems at the landscape scale.

References

Anderson K (2018) From taxing to subsidizing farmers in China post-1978. China Agric Econ Rev 10(1):36–47

Anderson K, Rausser G, Swinnen J (2014) Agricultural policy: a global view. pp 179–194. In: Van Alfen NK (ed) Encyclopedia of agriculture and food systems. Elsevier, Amsterdam, 464 p

Antrop M (2000) Geography and landscape science. Belgian Journal of Geography, Belgeo, special issue. In: 29th International Geographical Congress, vol ¼, pp 9–36

Antrop M, Van Eetvelde V (2017) Approaches in landscape research, pp. 61–80. In: Antrop M, Van Eetvelde V (eds) Landscape perspectives. Springer, Dordrecht, ISBN 978-94-024-1183-6

Bieling C, Plieninger T (eds) (2017) The science and practice of landscape stewardship. Cambridge University Press, Cambridge, UK, 402 pp, ISBN 978-1-107-14226-8

Bjerkes F (2010) Devolution of environment and resources governance: trends and future. Environ Conserv 37:489–500

Bloemers T, Daniels S, Fairclough G, Pedroli B, Stiles R (2010) Landscape in a changing world: bridging divides, integrating disciplines, serving society. Science Policy Briefing 41(October). European Science Foundation, Strasbourg, France. http://edepot.wur.nl/161371

Bocco G, Napoletano BM (2017) The prospects of terrace agriculture as an adaptation to climate change in Latin America. Geogr Compass 11(10)

Brose U, Hillebrand H (2016) Biodiversity and ecosystem functioning in dynamic landscapes. Philos Trans R Soc Lond B: Biol Sci May 19, 371 (1694). pii: 20150267. https://doi.org/10.1098/rstb.2015.0267

Burton RJF (2014) The influence of farmer demographic characteristics on environmental behaviour: a review. J Environ Manage 135:19–26

Castro AJ, Verburg PH, Martín-López B, Garcia-Llorente M, Cabello J, Vaughn CC, López E (2014) Ecosystem service trade-offs from supply to social demand: a landscape-scale spatial analysis. Landsc Urban Plan 132:102–110

Cliche L, Freeman L (2017) Applying integrated watershed management in Nova Scotia: a community-based perspective from the Clean Annapolis River Project. Int J Water Resour Dev 33(3):441–457

Costanza R, de Groot R, Braat L, Kubiszewski I, Fioramonti L, Sutton P, Farber S, Grasso M (2017) Twenty years of ecosystem services: how far have we come and how far do we still need to go? Ecosyst Serv 28:1–16

Cross R, Ampt P (2017) Exploring agroecological sustainability: unearthing innovators and documenting a community of practice in Southeast Australia. Soc & Nat Resour 30(5):585–600

Di Minin E, Soutullo A, Bartesaghi L, Rios M, Szephegyi MN, Moilanen A (2017) Integrating biodiversity, ecosystem services and socio-economic data to identify priority areas and landowners for conservation actions at the national scale. Biol Cons 206:56–64

Dyer J, Stringer LC, Dougill AJ, Leventon J, Nshimbi M, Chama F, Kafwifwi A, Muledi JI, Kaumbu JM, Falcao M, Muhorro S (2014) Assessing participatory practices in community-based natural resource management: experiences in community engagement from southern Africa. J Environ Manage 137:137–145

Eden S (1996) Public participation in environmental policy: considering scientific, counter-scientific and non-scientific contributions. Public Underst Sci 5(3):183–204

European Commission (2013) Overview of CAP reform 2014-2020. Agricultural Policy Perspectives Brief No. 5, December 2013. https://ec.europa.eu/agriculture/sites/agriculture/files/policy-perspectives/policy-briefs/05_en.pdf

Everard M (2017) Ecosystem services: key issues. Oxon, Routledge, Abingdon, p 188

Farm Service Agency (FSA) (2014) Conservation Reserve Program — State Acres for Wildlife Enhancement (SAFE) Approved Projects. FSA, United States Department of Agriculture, Washington DC, Conservation Fact Sheet

Fazey I, Bunse L, Msika J, Pinke M, Preedy K, Evely AC, Lambert E, Hastings E, Morris S, Reed MS (2014) Evaluating knowledge exchange in interdisciplinary and multi-stakeholder research. Glob Environ Change 25:204–220

Folke C, Biggs R, Norström AV, Reyers B, Rockström J (2016) Social-ecological resilience and biosphere-based sustainability science. Ecol Soc 21(3)

Forrester J, Cook B, Bracken L, Cinderby S, Donaldson, A (2015) Combining participatory mapping with Q-methodology to map stakeholder perceptions of complex environmental problems. Appl Geogr 56:199–208

Girard N (2015) Knowledge at the boundary between science and society: a review of the use of farmers' knowledge in agricultural development. J Knowl Manag 19(5):949–967

Girard C, Pulido-Velazquez M, Rinaudo JD, Pagé C, Caballero Y (2015) Integrating top–down and bottom–up approaches to design global change adaptation at the river basin scale. Glob Environ Change 34:132–146

Gundersen V, Mäkinen K (2009) Aldo Leopold and stewardship: Lessons for forest planning and management in the Nordic countries? Norsk Geogr Tidsskr 63(4):225–232

Hellerstein DM (2017) The US conservation reserve program: the evolution of an enrollment mechanism. Land Use Policy 63:601–610

Kremen C (2005) Managing ecosystem services: what do we need to know about their ecology? Ecol Lett 8:468–479

Krzywoszynska A (2016) What farmers know: experiential knowledge and care in vine growing. Sociologia Ruralis 56(2):289–310

Lambin EF, Geist HJ, Rindfuss RR (2006) Introduction: local processes and global impacts. In Lambin EF and Geist HJ (eds) Land-use and landcover change: local processes and global impacts. Springer, Berlin, Germany, pp 1–8. https://doi.org/10.1007/3-540-32202-7 1

Lastra-Bravo XB, Hubbard C, Garrod G, Tolón-Becerra A (2015) What drives farmers' participation in EU agri-environmental schemes? Results from a qualitative meta-analysis. Environ Sci Policy 54:1–9

Lefroy T, Curtis A, Jakeman A, McKee J (eds) (2012) Landscape logic: integrating science for landscape management. CSIRO Publishing, Collingwood, Victoria, Australia

Leifield J (2016) Current approaches neglect possible agricultural cutback under large-scale organic farming. A comment to Ponisio et al. Proc R Soc B—Biol Sci 282(1824), 20151623

Leimbach M, Popp A, Lotze-Campen H, Bauer N, Dietrich JD, Klein D (2012) Integrated assessment models: the interplay of climate change, agriculture and land use in a policy tool. In: Dinar A, Mendelssohn R (eds) Handbook on climate change and agriculture. Edward Elgar Publishing, Cheltenham, pp 204–229

Li J, Wang Z, Lai C, Wu X, Zeng Z, Chen X, Lian Y (2018) Response of net primary production to land use and land cover change in mainland China since the late 1980s. Sci Total Environ 639:237–247

Liu Y, Liao L, Long H, Qin J (2015) Effects of land use transitions on ecosystem services value—a case study of Hunan province. Geogr Res 34(4):691–700

Lovett A, Appleton K, Warren-Kretzschmar B, Von Haaren C (2015) Using 3D visualization methods in landscape planning: an evaluation of options and practical issues. Landsc Urban Plan 142:85–94

MacDonald DH, Bark R, MacRae A, Kalivas T, Grandgirard A, Strathearn S (2013) An interview methodology for exploring the values that community leaders assign to multiple-use landscapes. Ecol Soc 18(1)

Martinez-Harms MJ, Bryan BA, Balvanera P, Law EA, Rhodes JR, Possingham HP, Wilson KA (2015) Making decisions for managing ecosystem services. Biol Conserv 184:229–238

McCullough EB, Matson PA (2016) Evolution of the knowledge system for agricultural development in the Yaqui Valley, Sonora, Mexico. Proc Natl Acad Sci 113(17):4609–4614

McGuire JM, Morton LW, Arbuckle JG Jr, Cast AD (2015) Farmer identities and responses to the social–biophysical environment. J Rural Stud 39:145–155

Millennium Ecosystem Assessment (2005) Ecosystems and human well-being: synthesis. Island Press, Washington D.C.

Millington JD, Wainwright J (2016) Comparative approaches for innovation in agent-based modelling of landscape change. Land 5(2):13. https://doi.org/10.3390/land/502001

Mitchell LC (1981) Witnesses to a vanishing America: The nineteenth-century response, vol 770. Princeton University Press, Princeton, NJ

Moschitz H, Roep D, Brunori G, Tisenkopfs T (2015) Learning and innovation networks for sustainable agriculture: processes of co-evolution, joint reflection and facilitation. J Agric Educ Ext 21(1):1–11

Munroe DK, McSweeney K, Olson J, Mansfield B (2014) Using economic geography to reinvigorate land-change science. Geoforum 52:12–21

Naidoo R, Balmford A, Costanza R, Fisher B, Green RE, Lehner B, Malcolm TR, Ricketts TH (2008) Global mapping of ecosystem services and conservation priorities. Proc Natl Acad Sci 105(28):9495–9500. https://doi.org/10.1073/pnas.0707823105

Norberg J, Wilson J, Walker B, Ostrom E (2008) Diversity and resilience of social-ecological systems. In: Norberg J, Cumming GS (eds) Complexity theory for a sustainable future. Columbia University Press, New York, pp 46–79

Oostindie H, Van Broekhuizen R, De Roest K, Belletti G, Arfini F, Menozzi D, Hees E (2016) Sense and non-sense of local–global food chain comparison, empirical evidence from Dutch and Italian pork case studies. Sustainability 8(4):319. https://doi.org/10.3390/su8040319

Plieninger T, Bieling C, Fagerholm N, Byg A, Hartel T, Hurley P, López-Santiago CA, Nagabhatla N, Oteros-Rozas E, Raymond CM, Van Der Horst D (2015a) The role of cultural ecosystem services in landscape management and planning. Curr Opin Environ Sustain 14:28–33

Plieninger T, Kizos T, Bielin C, Le Dû-Blayo L, Budniok MA, Bürgi M, Crumley CL, Girod G, Howard P, Kolen J, Kuemmerle T (2015b) Exploring ecosystem-change and society through a landscape lens: recent progress in European landscape research. Ecol Soc 20(2):5. https://doi.org/10.5751/ES-07443-200205

Rausser GC, Farrell K (eds) (2016) Alternative agricultural and food policies and the Farm Bill. Giannini Foundation of Agricultural Economics, University of California, Berkeley

Rausser GC, Zilberman D, Kahn G (2015) An alternative paradigm for food production, distribution and consumption: a non-economist's perspective. Ann Rev Resour Econ 7:309–331

Raymond CM, Spoehr J (2013) The acceptability of climate change in agricultural communities: comparing responses across variability and change. J Environ Manage 115:69–77

Raymond CM, Brown G, Weber D (2010a) The measurement of place attachment: personal, community, and environmental connections. J Environ Psychol 30(4):422–434

Raymond CM, Bieling C, Fagerholm N, Martin-Lopez B, Plieninger T, (2016a) The farmer as a landscape steward: Comparing local understandings of landscape stewardship, landscape values, and land management actions. Ambio 45(2):173–184

Raymond CM, Reed M, Bieling C, Robinson GM, Plieninger T, (2016b) Integrating different understandings of landscape stewardship into the design of agri-environmental schemes. Environ Conserv 43(4):350–358

Raymond CM, Fazey I, Reed MS, Stringer LC, Robinson GM, Evely AC (2010b) Integrating local and scientific knowledge for environmental management. J Environ Manage 91(8):1766–1777. https://doi.org/10.1016/j.jenvman.2010.03.023

Reed MS, Stringer LC, Fazey I, Evely AC, Kruijsen JHJ (2014) Five principles for the practice of knowledge exchange in environmental management. J Environ Manage 146:337–345

Reed MS, Allen K, Attlee A, Dougill AJ, Evans KL, Kenter JO, Ho J, McNab D, Stead SM, Twyman C, Scott AS (2017) A place-based approach to payments for ecosystem services. Glob Environ Change 43:92–106

Riley M (2011) Turning farmers into conservationists? Progress and prospects. Geogr Compass 5 (6):369–389

Riley M (2016) How does longer term participation in agri-environment schemes [re]shape farmers' environmental dispositions and identities? Land Use Policy 52:62–75. https://doi.org/10.1016/j.landusepol.2015.12.010

Robinson GM (1994) The greening of agricultural policy: Scotland's Environmentally Sensitive Areas (ESAs). J Environ Planning Manage 37:215–225

Robinson GM (1997) Community-based planning: Canada's Atlantic Coastal Action Program (ACAP). Geogr J 172(3):25–37

Robinson GM (2018) Adaptation to climate change in farming communities: Observations from two research projects in South Australia. South Australian Geogr J 114(1). https://doi.org/10.21307/sagj-2018-006

Robinson GM, Carson DA (2013) Applying landscape science to natural resource management. Ecol Soc 18(1):32. https://doi.org/10.5751/ES-05639-180132

Robinson GM, Robinson SR (2016) 'Improving landscape or improving the Picturesque? Historic rural landscapes in the English Midlands'. Belgeo (4). https://doi.org/10.4000/belgeo.19664

Robinson GM, Song B (2018) Transforming the peri-urban fringe in China: the example of Xi'an-Xianyang. Sustainability 10:3932. https://doi.org/10.3390/su10113932

Servon LJ, Pink S (2015) Cittaslow: going glocal in Spain. J Urban Aff 37(3):327–340

Shaw BJ, Draux H, GarciaMartin M, Martin J, Bieling C (2017) Contributions of citizen science to landscape democracy: potentials and challenges of current approaches. Landsc Res 42:831–844

Shuttleworth S, Palang H (2017) Landscape research and knowledge exchange: learning from the HERCULES research project. Landsc Res 42(8):809–818. https://doi.org/10.1080/01426397.2017.1396305

Singh AK, Dhayani SK (2014) Agroforestry policy issues and challenges. In: Dagar JC, Singh AK, Arunachalam A (eds) Agroforestry systems in India: livelihood security and ecosystem services. Springer India, New Delhi, pp 367–372

Song X, Peng C, Zhou G, Jiang H, Wang W (2014) Chinese Grain for Green Program led to highly increased soil organic carbon levels: a meta-analysis. Sci Rep: 4460. https://www.nature.com/articles/srep04460

Song B, Robinson GM, Zhou Z (2017) Agricultural transformation and ecosystem services: a case study from Shaanxi Province, China. Habitat Int 69:114–125

Steiner F (2008) The living landscape: an ecological approach to landscape planning, 2nd edn. Island Press, Washington D.C., USA. ISBN-13: 978-1597263962, ISBN-10: 1597263966

Stringer LC, Reed MS, Fleskens L, Thomas RJ, Le QB, Lala-Pritchard T (2017) A new dryland development paradigm grounded in empirical analysis of dryland systems science. Land Degrad Dev 28(7):1952–1961

Tieskens KF, Schulp CJ, Levers C, Lieskovský J, Kuemmerle T, Plieninger T, Verburg PH (2017) Characterizing European cultural landscapes: accounting for structure, management intensity and value of agricultural and forest landscapes. Land Use Policy 62:29–39

Turner BL II, Lambin EF, Reenberg A (2007) The emergence of land change science for global environmental change and sustainability. Proc Natl Acad Sci USA 104(52):20666–20671

Turner KG, Anderson S, Gonzales-Chang M, Costanza R, Courville S, Dalgaard T, Dominati E, Kubiszewski I, Ogilvy S, Porfirio L, Ratna N (2016) A review of methods, data, and models to assess changes in the value of ecosystem services from land degradation and restoration. Ecol Model 319:190–207

Vaz E (2016) The future of landscapes and habitats: the regional science contribution to the understanding of geographical space. Habitat Int 51:70–78

Verburg PH, Crossman N, Ellis EC, Heinimann A, Hostert P, Mertz O, Nagendra H, Sikor T, Erb KH, Golubiewski N, Grau R (2015) Land system science and sustainable development of the earth system: a global land project perspective. Anthropocene 12:29–41

Wang Y, Zhou ZX (2014) Evaluation of urban agricultural land use efficiency based on urban agricultural multi-functionality—a case study of Xi'an City. Econ Geogr 34(7):129–134

Wang J, Peng J, Zhao M, Liu Y, Chen Y (2017) Significant trade-off for the impact of grain-for-green programme on ecosystem services in North-western Yunnan, China. Sci Total Environ 574:57–64

Wilson JR, Lomonico S, Bradley D, Sievanen L, Dempsey T, Bell M, McAfee S, Costello C, Szuwalski C, McGonigal H, Fitzgerald S (2018) Adaptive co-management to achieve climate-ready fisheries. Conserv Lett. https://doi.org/10.1111/conl.12452

Wu J (2017) Thirty years of Landscape Ecology (1987–2017): retrospects and prospects. Landscape Ecol 32:2225–2239. https://doi.org/10.1007/s10980-017-0594-8

Xiao J (2014) Satellite evidence for significant biophysical consequences of the "Grain for Green" Program on the Loess Plateau in China. J Geophys Res: Biogeosci 119(12):2261–2275

Xu J, Grumbine RE (2014) Integrating local hybrid knowledge and state support for climate change adaptation in the Asian highlands. Clim Change 124(1–2):93–104

Zagaria C, Schulp CJ, Kizos T, Gounaridis D, Verburg PH (2017) Cultural landscapes and behavioral transformations: an agent-based model for the simulation and discussion of alternative landscape futures in East Lesvos, Greece. Land Use Policy 65:26–44

Zhou Y (ed) (2016) Seeking changes: economic development in contemporary China. World Scientific Publishing, Singapore, p 288

Chapter 5
Landscapes, Their Exploration and Utilisation: Status and Trends of Landscape Research

Lothar Mueller, Frank Eulenstein, Wilfried Mirschel, Marc Antrop, Michael Jones, Blair M. McKenzie, Nikolai M. Dronin, Lev K. Kazakov, Valery V. Kravchenko, Alexander V. Khoroshev, Maria Gerasimova, Ralf Dannowski, Uwe Schindler, Olga Ruhovich, Viktor G. Sychev, Askhad K. Sheudzhen, Denis Couvet, Guy M. Robinson, Winfried Blum, Tomasz Joniak, Ursula Eisendle, Maria Gabriella Trovato, Elmira Salnjikov, Michael Haubold-Rosar, Dirk Knoche, Michael Köhl, Debbie Bartlett, Jörg Hoffmann, Jörg Römbke, Frank Glante, Olga I. Sumina, Abdulla Saparov, Elena Bukvareva, Vitaly V. Terleev, Alex G. Topaj and Felix Kienast

Abstract A new geological epoch has begun—the Anthropocene. Huge anthropogenic transformations of terrestrial landscapes over the past five decades have forced its declaration. Exploring of interaction of humans with nature in general, and with landscapes in particular, can be characterised properly by the terms 'landscape research' and 'landscape science'. Landscape science has been a traditional scientific discipline of geography. This is the case in Russia, whilst the terms geo-ecology and landscape ecology have become established in the

L. Mueller (✉) · F. Eulenstein · W. Mirschel · R. Dannowski
Leibniz Centre for Agricultural Landscape Research (ZALF),
Eberswalder Str. 84, 15374 Muencheberg, Germany
e-mail: agrarlandschaft@gmail.com

L. Mueller · F. Eulenstein · U. Schindler · A. K. Sheudzhen
Kuban State Agrarian University, 13 Kalinin Str., 350044 Krasnodar, Russia

M. Antrop
Landscape Research—Department of Geography, Ghent University, Ghent, Belgium

M. Jones
Department of Geography, Norwegian University of Science and Technology,
7491 Trondheim, Norway

B. M. McKenzie
The James Hutton Institute, Invergowrie, Dundee DD2 5DA, UK

N. M. Dronin · L. K. Kazakov · A. Khoroshev · M. Gerasimova
Faculty of Geography, Lomonosov Moscow State University,
Leninskie Gory 1, Moscow 119991, Russia

L. Müller and F. Eulenstein (eds.), *Current Trends in Landscape Research*,
Innovations in Landscape Research, https://doi.org/10.1007/978-3-030-30069-2_5

English-speaking scientific community. As landscapes are multifunctional, highly complex systems, landscape research is a platform for disciplinary, interdisciplinary and transdisciplinary research. Landscape research in the Anthropocene must aim to combine landscape sustainability with high quality and productivity. This mission is in accord with the Sustainable Development Goals of the United Nations and the provisions of the Landscape Convention of the European Council. It includes halting landscape degradation, developing cultural landscapes and maintaining semi-natural landscapes. Clean water and air, fertile and healthy soils for food and other ecosystem services and a green and biodiverse environment are attributes of landscapes for the survival and well-being of humans in coexistence with nature. Landscape research must generate knowledge, innovations and responsible decision rules for achieving these aims. Big data gathering and scenario modelling are

V. V. Kravchenko
Sochava Institute for Geography of the Russian Academy of Sciences,
Ulan Batorskaya 1, Irkutsk 664033, Russia

M. Gerasimova
V.V. Dokuchaev Soil Science Institute, Pyzhevskiy pereulok 7, Moscow 117047, Russia

F. Eulenstein · U. Schindler
Mitscherlich Academy for Soil Fertility, Prof.-Mitscherlich-Allee 1, 14641 Paulinenaue, Germany

O. Ruhovich · V. G. Sychev
Pryanishnikov-Institute for Agrochemistry, Pryanishnikova St., 31a, 127550 Moscow, Russia

D. Couvet
UMR MNHN-CNRS-SU 'CESCO', 36, Rue Geoffroy Saint-hilaire (Cp 41), 75231 Paris Cedex 05, France

G. M. Robinson
Department of Geography, School of Social Sciences, University of Adelaide, Adelaide, SA 5005, Australia

G. M. Robinson
Department of Land Economy, University of Cambridge, 16-21 Silver Street, Cambridge CB3 9EP, UK

W. Blum
Institute of Soil Research, University of Natural Resources and Life Sciences,
Peter-Jordan-Str. 82, 1190 Vienna, Austria

T. Joniak
Department of Water Protection, Faculty of Biology, Institute of Environmental Biology, Adam Mickiewicz University Poznań, Umultowska Str. 89, 61-614 Poznań, Poland

U. Eisendle
University of Salzburg, Hellbrunnerstr. 34, 5020 Salzburg, Austria

M. G. Trovato
Department of Landscape Design and Ecosystem Management, Faculty of Agricultural and Food Sciences, American University of Beirut, Beirut 1107 2020, Lebanon

E. Salnjikov
Soil Science Institute, Teodora Drajzera 7, 11000 Belgrade, Serbia

important for knowledge generation in a globalised world. International long-term experiments, observatories and monitoring systems will deliver data for comprehensive ecosystem models and decision support systems. Technical innovations must be imbedded in cultural solutions for the evolvement of landscapes. Springer International's new book series 'Innovations in Landscape Research' aims to support better understanding, monitoring and managing landscapes. It contains a multitude of approaches and data. Some focus is on technical innovations for agri-environmental monitoring, on land and water management and its implications

M. Haubold-Rosar · D. Knoche
Research Institute for Post-Mining Landscapes (FIB), Brauhausweg 2, 03238 Finsterwalde, Germany

M. Köhl
Universität Hamburg, Center for Earth System Research and Sustainability World Forestry, Leuschnerstrasse 91, 21031 Hamburg, Germany

D. Bartlett
Faculty of Engineering and Science, University of Greenwich, Central Avenue, Chatham Maritime, Kent ME4 4TB, UK

J. Hoffmann
Julius Kühn-Institut (JKI) Bundesforschungsinstitut Für Kulturpflanzen, Institut Für Strategien Und Folgenabschätzung, Nachhaltige Landwirtschaft Und Biodiversität, Stahnsdorfer Damm 81, 14532 Kleinmachnow, Germany

J. Römbke
ECT Oekotoxikologie GmbH, Böttgerstr. 2-14, 65439 Flörsheim, Germany

F. Glante
German Environment Agency, Woerlitzer Platz 1, 06844 Dessau-Rosslau, Germany

O. I. Sumina
Department of Geobotany and Plant Ecology, Faculty of Biology, St. Petersburg State University, Universitetskaya Naberezhnaya, 7/9, 199034 St. Petersburg, Russia

A. Saparov
Kazakh Research Institute of Soil Science and Agrochemistry Named After U. U. Uspanov, Al Faraby Ave 75b, Almaty 050060, Kazakhstan

A. Saparov
Research Centre for Ecology and Environment of Central Asia (Almaty), Almaty 050060, Kazakhstan

E. Bukvareva
Biodiversity Conservation Center, Vavilova Street 41, Office 2, 117312 Moscow, Russia

V. V. Terleev
Peter the Great Saint-Petersburg Polytechnic University, Polytechnicheskaya ul. 29, 195251 St. Petersburg, Russia

A. G. Topaj
Agrophysical Research Institute, Grazhdansky Pr. 14, 195220 St. Petersburg, Russia

F. Kienast
Swiss Federal Institute for Forest, Snow and Landscape Research WSL, Birmensdorf, Switzerland

for landscape sustainability. Authors present novel tools for ecosystem modelling and forecasting of landscape processes, and on creating knowledge, rules and approaches for handling the multifunctionality of landscapes. The coming book series may serve as a knowledge, data and communication basis for informed decisions regarding the development of landscapes. It will enlarge our horizon and field of action by building bridges between scientific communities, scientific disciplines, and researchers and citizens.

Keywords Landscape · Research · Status · Trends · Monitoring · Anthropocene · Ecosystems · Modelling · Data · Land use · Sustainability · Innovations · Cultural landscapes

5.1 Introduction: What Should We Know About Landscapes?

The effects of humans on the global environment continue to escalate. Scientists have debated and declared that the geological epoch of the Holocene has ended and a new epoch, the Anthropocene, has begun (Crutzen 2002; Zalasiewicz et al. 2011; Lewis and Maslin 2015). The Great Acceleration of the Anthropocene began in the second half of the twentieth century (Steffen et al. 2015). Evidence for this human-induced break in the history of our planet comes from rapidly increased levels of climate-warming CO_2 in the atmosphere. Other irreversible signals which can be found in sediments now and later are radioactive elements from nuclear bomb tests, plastic material and residues from abruptly extinct animals and plants (Zalasiewicz et al. 2011). Modern agricultural land use, necessary to feed an increasing population, producing high concentrations of nitrogen and phosphorus in soils, derived from artificial fertilisers, pig and chicken factories and other industry-style farming, is also significant for the declaration of the Anthropocene (Crutzen 2002; Zalasiewicz et al. 2011; Lewis and Maslin 2015). These accelerating interactions of humans with their environment have implications for all spheres, regions and landscapes on earth (Millennium Ecosystem Assessment 2005; United Nations 2015). Further growth in world population, climate change and other factors will exacerbate the challenges to maintain the capacity of ecosystems and livelihoods for coming generations.

Landscape research comprises all activities of knowledge generation about landscapes. Can improved knowledge help stop the degeneration of landscapes and initiate sustainable development? Our hypothesis is that it can, at least for many landscapes of the globe.

In this chapter, we aim to review and propose:

- the role of landscapes in the generation of knowledge about human–nature interactions with some focus on Europe and Northern Eurasia

- landscape transformations and their drivers with a focus on Central Europe
- challenges and opportunities of landscape research in the Anthropocene and
- innovations and knowledge transfer for improving landscape sustainability.

5.2 Landscape: A Subject of Identification for Citizens and Scientists

According to the European Landscape Convention (ELC), 'Landscape means an area, as perceived by people, the character of which is the result of the action and interaction of natural and/or human factors' (COUNCIL OF EUROPE 2000). Landscapes are territories of our planet characterised by discrete assemblies and numerous underlying functions. The concept of the landscape is a solid part of our language of operation. We are familiar with major landscape types, such as deserts, wetlands, mountains, steppes and forests. We distinguish between urban, peri-urban and rural or agricultural landscapes, and between natural or semi-natural and cultural landscapes. Landscapes are key elements of individual and social well-being (Fig. 5.1). They contribute to the formation of diverse cultures and local identities, constituting a basic component of the European natural and cultural heritage (COUNCIL OF EUROPE 2000). Management rules for World Heritage Cultural Landscapes apply to selected landscapes (Mitchell et al. 2009; Schenk and Weizenegger 2006). The European Landscape Convention (COUNCIL

Fig. 5.1 Landscapes are a source of individual and social well-being and spirituality of people. The photograph shows a sunrise scene in a European river and meadow landscape, full of local wildlife and migrating birds

OF EUROPE 2000) covers all types of landscapes and aims at their protection, management and planning with an emphasis on broad public participation (Jones and Stenseke 2011).

The typical *landscape scale* (landscape level), though not clearly defined (Ahern and Cole 2012), seems to be a common area of perception and identification of people. It comes close to a regional scale, a territory of about some tens to hundreds of square kilometres, the typical horizon of our ancestors, the hunters and gatherers. The landscape level has particular importance for addressing regional planning and conservation issues (Selman 2006; Sayer et al. 2013; Stenoien et al. 2018; Doyle-Capitman et al. 2018).

Any issue of long-term importance for human societies has its reflections in arts and sciences. In Germany, Russia and some other countries, landscape science (German terms: Landschaftskunde, Landschaftslehre, Russian term: ландшафтоведение (landshaftovedenie)) has developed as a discipline of geography. It reached its climax in the twentieth century. Now, the term landscape ecology is more common. Soil science and particularly pedology, the understanding of soils in the landscape, established by the Russian Geographer and Soil Scientist Dokuchaev (1883) and further developed by Glinka (1927), Mitscherlich (1931) and others has provided many inputs for soil and landscape science in Europe, Asia and America (Arend 2017; Moon and Landa 2017). Prominent initiators of landscape science were Berg (1915, 1930) in Russia, and Passarge (1919), Hettner (1905, 1927), and Neef (1967) in Germany. Zoning of landscapes according to their bioclimatic conditions has been a work of high practical relevance. L. S. Berg identified landscape zones of the Soviet Union (Berg 1930), and E. Meynen and J. Schmithüsen edited a handbook of natural landscape units of Germany (Meynen and Schmithüsen 1953-1962).

Later, in Europe and in the English-speaking sphere, the terms *landscape ecology* (German: Landschaftsökologie) and *geo-ecology* (German: Geoökologie), became more common (Troll 1939, 1971; Zonneveld 1995; Antrop 2000). Geo-ecology has become a discipline of higher education in European countries, whilst *landscape science* [ландшафтоведение (landshaftovedenie)] has remained a scientific and educational discipline in Russia as well (Kazakov and Chizhova 2001; Shaw and Oldfield 2007).

Research into the transformation of the physical environment by human activity has been undertaken in the German- and English-speaking worlds from the middle of the nineteenth century (Rowntree 1996; Jones 2003b; Potthoff 2013). G. P Marsh's seminal work on 'Man and Nature' (Marsh 1864) inspired the conservation movement in the USA. The term *cultural landscape* (German term: Kulturlandschaft) was coined by the first professor of geography at the University of Berlin, C. Ritter, in his Asian study to describe the cultural imprint left by humans through physically changing the land (Ritter 1832). The term was adopted by C. Vogel, J. Wimmer and most notably F. Ratzel and O. Schlüter to describe an area as modified by human activity (Vogel 1851; Wimmer 1885; Ratzel 1893, 1895-96; Schlüter 1903). N. Krebs defined the cultural landscape as settled regions transformed by human activity as opposed to non-settled regions where human life was subordinate to nature (Krebs 1922). Carl Sauer introduced the term to the English-speaking world, defining the cultural landscape as an area successively altered by humans and fashioned by different cultural groups from the preceding natural landscape (Sauer 1925). Sauer and the Berkeley school were influential in promoting research into the ecological consequences of the human transformation of the environment (Thomas 1956; Turner et al. 1990).

There were other schools of landscape science, exemplified by the work of the Finnish geographer J. G. Granö, who identified and analysed landscape regions in Estonia and Finland in the 1920s and 1930s (Granö 1979; Jones 2003a). In his book 'Pure Geography", published in German in 1927 but not translated into English until 1997, he presented an anthropocentric and phenomenological approach to the study of landscape, defined as the environment as perceived by humans through all their senses (Granö 1929, 1997). His pioneering mappings of landscape included aspects not mapped previously, such as colour, lighting, sound and smells. His study of rhythms and mobility were a precursor to the Swedish geographer T. Hägerstrand's notion of the landscape of flows and movement (Hägerstrand 1993). Whereas in 'Pure Geography', Granö aimed to describe the phenomena of the perceived environment as objectively as possible, his study of the Altai Mountains was in contrast an artistic and poetic presentation of his own subjective experiences along with those of the people he met (Granö 1919-21).

Humanistic approaches to landscape from the 1960s onwards complemented studies of landscape morphology (Rowntree 1996; Jones 2003a, b; Wylie 2009). The British geographer D. Cosgrove distinguished between ecological and semiotic approaches to landscape research. Whilst ecology is concerned with the complex interactions of natural processes with each other and with humans, the semiotic approach considers the meanings people attach to landscapes through human cognition and representation, symbolically mediated (Cosgrove 2003). Influential studies of cultural perceptions and visual preferences for landscape covered topics such as landscape tastes (Lowenthal and Prince 1965), landscape as scenery (Lowenthal 1968), landscape values (Tuan 1974; Penning-Rowsell 1981; Penning-Rowsell and Lowenthal 1986), landscape as heritage (Lowenthal 1985, 1996), iconography of landscape (Daniels and Cosgrove 1988), landscape of the mind (Porteous 1990), and landscape and memory (Schama 1997). Much of this research into *landscape as a way of seeing* examined the ideological function of landscape images as reflecting and representing the values and norms of socio-economic elites (Cosgrove 1984; Daniels and Cosgrove 1988). In contrast, the North American landscape architect J. B. Jackson engaged with *vernacular landscape*, a landscape less looked at than lived in and shaped by ordinary people (Jackson 1986). The North American geographer D. Mitchell's materialist approach examines landscape as a site of production as well as of imagery, reflecting ongoing conflict and struggle between different socio-economic groups (Mitchell 1996, 2003).

The Danish–North American geographer K. R. Olwig has shown that the conceptualisation of *landscape as scenery* emerged in the seventeenth century. Taking an etymological and legal–historical approach, he has traced the origins of the landscape concept to the German and Scandinavian lands in the Middle Ages, where it referred a political–administrative entity in which the conditions and character of the land were shaped by customs and laws (Olwig 1996, 2002). This continues today as one of a diversity of landscape meanings, as illustrated by recent research on Nordic landscapes (Jones and Olwig 2008).

The terms *landscape science* and *landscape research* continue in use. M. Antrop has defined fundamentals and addressed the main topics of contemporary landscape research in Europe (Antrop 2000; Antrop and van Eetvelde 2017). The Landscape Research Group, existing since 1968, issues the international scientific journal 'Landscape Research', (now published by the Taylor and Francis Group) (Journal Landscape Research 2018). It has a broad view on landscapes with a focus on case studies and landscape reflections. The most cited paper in the journal is M. Tveit, Å. Ode and G. Fry's 'Key concepts in a framework for analysing visual landscape character' (Tveit et al. 2006).

5.3 Global Players in Landscape Research

Scientific institutions of the economically strong countries have adopted and promoted the term *landscape research*. A simple frequency distribution of literature reveals that research institutions from the EU, USA and China dominate recent publication activities in landscape research (Fig. 5.2). The term *ecosystem* needed to be included into this search in order to exclude the great number of publications

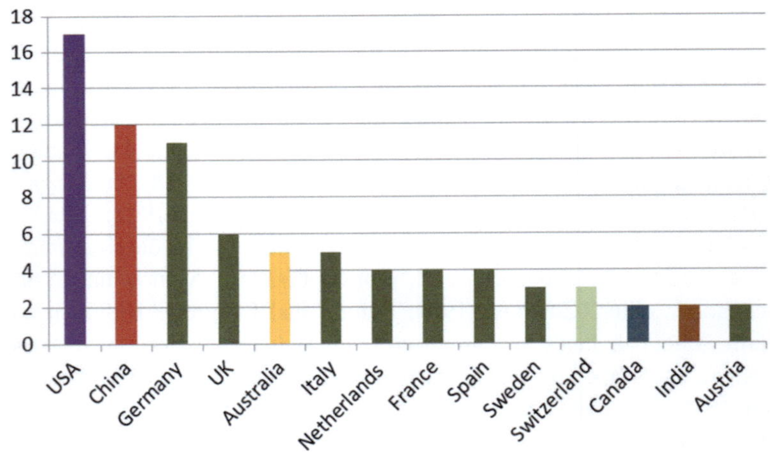

Fig. 5.2 Percentage of recent publications on landscape research grouped by country. The country of the first author's institution was counted. Search terms were 'Landscape + Research + Ecosystem', year = 2018, search engine Google Scholar. Sample of $n = 623$ publications having best matches, remainder of 20% distributes over 47 countries having a share of less than 2% each

where the term *landscape* is used figuratively, as in medical research. As all publications are in English, this analysis can provide a rough orientation only, not more.

It is noticeable that in these recent publications of 2018, the *landscape* term in the USA is mainly used in association with *landscape conservation.* In Germany, it is most closely related to *ecosystem services* and in regions of the former Roman Empire most frequently to *cultural landscapes.*

5.4 Landscapes in Permanent Transformation: The Case of Central Europe

A characteristic of landscapes is that they undergo continuous transformation due to natural processes as well as to the historical development of human society (Antrop 1998; Lipský 1995; Plit and Myga-Piatek 2014).

5.4.1 Neolithic to Medieval Times

The Neolithic Revolution, which started at the geological epoch of the Holocene around 12 thousand years ago, created settlements, agricultural fields, domesticated plants and animals and rural landscapes (Zeder 2008). Soil tillage and irrigation converted land around settlements to agricultural land. This process expanded slowly northwards from the Fertile Crescent and the Mediterranian area. From about 5 thousand years BC, equestrian nomads migrated from Central Asia and South-east Europe to Central Europe, for example, the Yamnaya culture (Mathieson et al. 2015; Bower 2017). Besides the Indo-European languages, they brought domesticated horses to Central Europe and promoted Bronze Age cultures (Bower 2017). The combination of sedentary arable farming with horse-based mobility became a successful basis for food production, trade and military activities (McNeill 1987).

A landscape structure consisting of forests, pastures, arable lands, open waters and interconnecting transport routes at community or regional levels was advantageous for the development of local rural cultures of Central Europe, for example, the Únětice culture (Meller and Michel 2018). They were largely confined to the few existing fertile forest steppe landscapes of Central Europe in the vicinity of the Hartz Mountains, cultivating fertile Loess and Alluvial soils.

Landscapes east of the Rhine River remained generally less cultivated than landscapes west of the Rhine. During the period of the Roman Empire at its largest extent about 117 AD (Kelly 2007), the *Limes Germanicus* (German frontier) divided two very differently cultivated regions in Central Europe along a line from Castra Regina (Regensburg) to Colonia and further north along the Rhine River. Landscapes located to the south and west were characterised by the Roman high culture: cities and villages, urban and rural infrastructure, sophisticated arable plant cropping, horticulture and viticulture. East and north of the Limes line fortfications, and especially east of the Elbe River, many landscapes remained more natural: sparsely populated and largely covered by virgin forests and natural peatlands.

In North-Central and North Europe in the period between 500 and 1000 AD, land cultivation and settlement became completely sedentary and concentrated, and changes in settlement location and fields became less common (Widgren and Pedersen 2011).

In the European zone of natural forests, cultivation of land for agriculture was commonly by slash and burn (Rösch et al. 2017). Later, it was recognised that agricultural soils suffer leaching and depletion of nutrients in the forest zone. Land management required stepwise measures of soil improvement by organic fertilisation (Baum et al. 2016) and by technologies such as plaggen cultivation (Giani et al. 2014; Acksel et al. 2017; Kern et al. 2019; Fig. 5.3). Lifestock husbandry provided dung, liquid manure and nutrient-enriched grass sods (plaggen) for the improvement of soils. Silvo-pastoral practices were also important in livestock farming (Vera 2000; Emanuelsson 2009).

Besides the cultivation of forested areas for food production, charcoal demand for smelting of iron became increasingly a major cause of deforestation in Central Europe (Iles 2016). Expanding trade routes across Europe and Eurasia permitted the

Fig. 5.3 Plaggen soil (German: Plaggenesch) in North Germany. These fertile Anthrosols are the result of humus accumulating soil cultivation from the Middle Ages onwards

exchange of goods and knowledge from the Neolithic and Bronze Ages (Zinkina et al. 2017), promoted technological progress and contributed to landscape transformations.

Ridge and furrow fields (German: Wölbäcker) resulted from the availability of mechanical soil ploughs with ploughshares and mouldboards (Küster 1997). A similar surface structure could also result from hand digging. This provided land drainage and increased stability of crop yields in humid landscapes. These often hidden structures of medieval fields still influences pathways of runoff and soil erosion in agricultural landscapes (Deumlich 2012).

The use of water and wind power for water management and improved rural land drainage through field drains and ditches have changed features and functions of agricultural landscapes from the end of the Middle Ages onwards.

5.4.2 *Modern Times*

Since the beginning of modern times, associated with the renaissance, the reformation and the age of discoveries, the speed of landscapes transformations has accelerated. In the Dutch Golden Age (roughly spanning the seventeenth century), the economy of this country boomed due to political circumstances, immigration of skilled workers, development of corporate finance, fertile soils and favourable climate. Culture, arts and education blossomed; multifunctional cultural landscapes evolved (Swart 1969).

Setbacks due to wars, epidemics and famines, associated with climate cooling of the Little Ice Age, were typical for the transition from medieval to modern Europe (Pfister 2007; Alfani 2011; Tagungsbericht 2015; Baten et al. 2018). Long wars such as the Thirty Years' War destroyed not only human life but also cultural achievements, especially on the territory of today's Germany. Villages fell desolate and infrastructure of cultural landscapes broke down. Forest vegetation overgrew debris and fields and formerly ploughed soils became forest soils again but kept the former anthropogenic processes in mind in terms of diagnostic properties. As soil erosion from cropping land is greater than from soil under natural vegetation (Nearing et al. 2017) landscape erosion processes and the sediment load of brooks, rivers and lakes decreased during landscape depopulation and forest succession.

As with other catastrophic events affecting nature and humans, the way was opened for new developments, induced by another population and their socio-economic conditions and constraints. Labour and scientific-technical innovations improved living conditions. This again was a precondition for a renaissance of culture, arts, science, and development of new cultural landscapes.

From the end of medieval times about 500 years ago, landscapes became increasingly multifunctional but maintained a rural character. The need for buffering against famines and seeking national food security were reasons for those with power to support agriculture. In the early nineteenth century peasants in the German kingdom of Prussia received personal freedom based on agrarian reforms

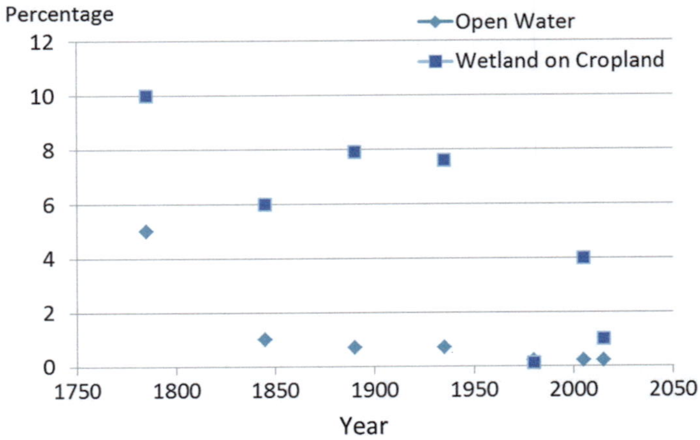

Fig. 5.4 Decline of wet areas on croplands in the Oderbruch region, located about 70 km east of Berlin, Germany. The fertile river lowland underlaid enhanced land reclamation since the mid-eighteenth century. Today, due to lack of funding for water management, 1–4% of croplands are too wet for permanent regular farming. This seems to be an acceptable compromise between farmers and conservationists. Data before 2000 according to Frielinghaus and Mueller (2003)

(Hubatsch 1989). Land consolidation was initiated in some regions such as in the Oderbruch landscape, located about 70 km east of Berlin (Mengel 1930, 1934). In order to improve both the land productivity and management, fields were re-arranged and grouped around single peasant's homesteads and villages. This significantly altered the landscape structure in this poldered lowland and developed this region (Frielinghaus and Mueller 2003, Fig. 5.4) to be the main food supplier of Berlin and other cities. River floods were diminished by construction of dikes and wetlands were drained by ditches. About 90% of the land became cropland. The proportion of planted line elements in the landscape, mainly hedgerows decreased from 51 m/ha in 1844 to 14 m/ha in 1980 (Frielinghaus and Mueller 2003).

Mechanisation allowed agricultural intensification. Consequently, on sloping land soil erosion from croplands has increased particularly fast over the past 200 years (Kappler et al. 2018).

At the beginning of the nineteenth century, agricultural science became an acknowledged scientific discipline in Prussia based on the research and education work of A. D. Thaer and promoted by W. von Humboldt (Frielinghaus 2004). Pioneers of agricultural science such as A. D. Thaer and Thünen (1826) developed the basics of land evaluation, land taxation, crop rotations, agricultural systems and rural economy in Prussia and Europe.

Food insecurity remained a serious issue in Europe. Poor social conditions among the lower class and famines promoted emigration to the New World, displacing and eliminating the native North American population and exploring and reclaiming extended landscapes for agriculture and other branches of the economy.

5.4.3 Industrial Age

The industrial epoch led to increased reclamation and exhaustion of natural resources, and to accelerated urbanisation. Due to population increase, the demand for food increased. Industrial development paved the way for the introduction of agricultural machinery, application of artificial fertilisers and soil and water management in agriculture. The landscape structure became adapted to these management processes (Blackbourn 2006; Poschold 2017).

The industrial age drastically altered the hydrologic regime of landscapes (Driescher 1974; Quast 2011; Bjornlund and Bjornlund 2019). Enhanced stream regulation for transportation and land reclamation, peat mining and wetland reclamation took place. This enhanced agriculture and industry but damaged the yields of freshwater fish and diminished landscape biodiversity.

In the second half of the nineteenth century, the transportation sector developed on the basis of steam engines. The electrification of rural landscapes began with the twentieth century. New landscape elements and structures such as railways, roads and power lines fragmented the landscape. They became habitats for successor species of flora and fauna. At the beginning of the twentieth century, agricultural landscapes in northeast Germany showed their greatest structural diversity (Frielinghaus and Mueller 1994) and thus high biodiversity.

Agricultural productivity shifted with the increased application of mineral fertilisers in the twentieth century (van der Ploeg et al. 2001; Antonkiewicz and Łabętowicz 2016). Their use was still moderate until World War II. The use of mineral fertiliser increased in the 1960s and later became excessive in Western Europe. This occurred alongside the development and application of pesticides and other agrochemicals. Growing awareness of their possible negative impacts on water, soil and biodiversity arose (Carson 1964; van der Ploeg et al. 2001). In Europe, generally, performance of agricultural landscapes and resulting food problems remained a serious issue in the industrial age due to political circumstances, wars and climate (Dronin and Bellinger 2005; Trentmann and Just 2006; Uskov and Uskov 2014).

Two devastating world wars and other technical wars of the twentieth century affected landscapes directly in new dimensions (Steinweg and Kerth 2018, Fig. 5.5). They created debris landscapes. Besides soil pollution of classical battlefields by explosives and other war material, bombing and widespread destruction of cities and industrial infrastructure occurred. A new post-war industry had to be created to deal with war legacies. More than 70 years after the end of World War II, many bombs are still explosive and need to be detected and defused.

Fig. 5.5 Landscapes influenced by wars and military activities over the past centuries are typical for Central Europe. The photograph shows a former military training area in an artificial heather and forest landscape in North Germany. It is now a habitat of the European wolf (*Canis lupus*) and other protected species

5.4.4 Post-World War II Period

The post-war period in Europe gave birth to an economic upswing in some countries of the later European Union (Abelshauser 1983; Temin 2002). All spheres of society and most regions in Europe have profited from sustained economic growth, rural landscapes included. Stimulated by national and European agricultural policies, land consolidation, agricultural soil and water management, implementation of modern crop management technologies and broad application of agrochemicals have become common. Lack of food was no longer an issue. On the contrary: permanent overproduction required market regulation. However, new environmental threats to soils, waters and landscapes emerged. Examples are: compaction of soils by agricultural machinery; pollution of groundwater by nitrate; acceleration of soil erosion through elimination of terraces, hedges and buffer strips in the course of land consolidation for industrial-style agriculture; loss of wetland habitats through land drainage; diseased forests and bogs due to air pollution; and high land consumption by road and highway construction.

Economic mergers accompanied the post-war agricultural development in European countries. The Rome treaties of 1957 (CVCE 2016) aimed to feed the population through a common agricultural market. At the same time, imports of agricultural products from the USA had to be ensured in return for the sale of European industrial products, via the GATT negotiations since 1947 (Goldstein et al. 2007). This contributed to areas of intensive animal production in north-western Europe. Imports of animal feed led to extreme nutrient surpluses in the Netherlands, northwest Germany and Denmark, with nitrate problems in groundwater and phosphorus transport into surface waters with sediment transport as part of soil erosion.

In countries under the patronage of the Soviet Union, these economic developments were weaker, substantially delayed and without public debates about negative side effects. Due to the collectivisation of agriculture in accordance with the Soviet Russian model, agricultural landscapes experienced their largest ever transformation. Complex measures of land consolidation (Russia: мелиорация (Melioratsya), East Germany: Melioration, West Germany: Flurbereinigung) were conducted from the mid-1960s to the 1980s in many countries. They were largely financed by the state and consisted of land drainage, irrigation, agricultural road construction, enlargement and re-arrangement of agricultural fields (German: Flurneuordnung) elimination of hedges and fruit trees along fields and dirt roads, and plantation of windbreaks and hedges along the borders of large rectangular fields (Schmidt 1961; Mueller et al. 1988). Large fields of 20–150 ha size and accordingly large agricultural units from 3 to 8 thousand hectares were created for the industrial production of agricultural commodities in East Germany. They were the basis for a high and stable agricultural production. This landscape structure consisting of large agricultural fields is still intact and has become subject to permanent debates about the nexus between productivity, sustainability and biodiversity.

At the end of this section, it should be noted that historical interactions of humans with landscapes could be demonstrated for a very small region of Central Europe only. More comprehensive and broader explanations covering further regions would have blown up the scope of this chapter. There is a huge number of special publications on historical landscape development in Europe and on other continents. This section should demonstrate that information about landscape transformations of the past helps to understand feature and functions of current and future landscapes.

5.5 Current Landscapes of the Anthropocene

Characteristics of the early Anthropocene, the current era of information technologies and globalisation, are human overpopulation in some regions, social conflicts, global instability of monetary and political systems, and unsolved environmental problems, putting pressure on all landscapes of our planet. The global

Fig. 5.6 Urbanisation is a dominating process and key driver of landscape alterations. The photograph shows a peri-urban landscape in Northern China. In the foreground is seen an irrigated field with maize. Soils are highly productive, and cropping systems are extremely intensive

economy has created a rural exodus and rapid urbanisation (Fig. 5.6). Globalisation and urbanisation are key driving forces of landscape change (Barau and Ludin 2012). More than half of the world's population now lives in cities (Haase and Schwarz 2009; Taylor 2016). Urban and peri-urban areas (Fig. 5.6) often include industrial structures that create legacies such as polluted soils. The high consumption in urban agglomerations affects neighbouring and remote rural landscapes through teleconnections (Seto et al. 2012).

War activities, migration of people due to wars, political instability and social inequality such as in Africa and the Near East puts pressure not only on directly affected regions and landscapes, but on those of migration routes and destinations. The millions of displaced persons are transforming the landscape by introducing new patterns, new process, new rhythms that require responses to the new conditions (Trovato 2015). Besides fundamental political decisions, landscape-based analyses and solutions for improving the social conditions for the affected people and their environment are required (Trovato 2019, Fig. 5.7).

The increasing global demand for energy and minerals leaves wounded landscapes for which rehabilitation solutions must be found (Sumina 1998; Douley and Audet 2016; Haubold-Rosar 2018; Knoche 2018, Fig. 5.8). In fact, notably large-scale opencast mining turns the landscape upside down, from a natural or pre-industrial rural area in a large-scale technogenic living space after mining. Therein rehabilitation of productive farmland and afforestation by guided succession and planting plays a key role to minimise the environmental footprint of mining (Schwarzer 2014). Even more upgrowing ecosystems on new ground are of overall importance for the revitalisation of the disturbed landscape, due to their multiple ecological functions, goods and public services (Schlenstedt et al. 2014; Knoche and Schlenstedt 2018).

Fig. 5.7 Informal tent settlements of refugees from Syria in the Lebanon. Refugees require space to meet their housing needs. The tent settlements do not meet the required standards for human dwellings and are causing ecological damage. The increasing number of people living in precarious conditions and scattered across the land is determining a loss of landscape character and quality. *Photographs* Maria G. Trovato

Fig. 5.8 Lignite opencast mining Welzow-Süd (in the background) and new farmland on dumped sites (with a vineyard in front) in the Lusatian Mining District, Germany. Photograph: Research Institute for Post-Mining Landscapes (FIB), Finsterwalde

Major objective of reclamation is to re-establish quite fertile soils as essential for a diverse food, fodder or feedstock production. Site adapted ecosystems should meet the very special soil properties but also be in good agreement with the traditional land use of the region. In a common way, mine site reclamation ends, when re-vegetation is achieved as intended and ecosystems come into a regular post-mining after-use (von Bismarck et al. 2014). However, from the ecological point of view and besides re-vegetation, restoring fully sustainable, integer and healthy ecosystems is a long-term process, taking several decades (Jordan et al. 1987). Because soil dynamics are quite high in the initial stage, it very often turns out rather difficult to predict the long-term ecosystem behaviour.

Rural landscapes have undergone a rapid transformation within only a few years (Tarolli et al. 2014). In Europe, viable infrastructure has enabled a good quality of life for the rural population. Land consolidation still occurs to varying degrees with the aim of adding value to landscapes and stabilising rural infrastruture. Land consolidation is governed by strict acts and rules in Europe (Vitikainen 2004; Thomas 2006) and is based on comprehensive participation of the rural community.

Despite these efforts, a breakdown of rural infrastructure and rapid landscape transformation towards agro-industrial and agro-energy landscapes are typical for many regions in Europe and Asia. Loss of biodiversity including insects and plants has accelerated due to elimination of field margins, reduced crop rotation systems, monocultures and large-scale application of effective pesticides such as neo-nicotinoids, herbicides such as glyphosates and other agrochemicals (Stenoien et al. 2018; Schäffer et al. 2018).

The globalisation of all spheres of the economy has created a global food market and increase of the flow of agricultural capital into profitable regions and centres outside the traditional rural areas (Robinson 2018a, b). This makes agricultural landscapes vulnerable to global financial and economic crises (Tarolli et al. 2014; Pinto-Correia et al. 2018). Large-scale land acquisitions (land grabbing), a highly controversial and debated issue (Dell'Angelo et al. 2017; Zoomers et al. 2017; Robinson 2018a; De Maria 2019), is not solely a typical trend in developing countries and transitional economies. It is also common in West and Central Europe (Bunkus and Theesfeld 2018).

Production-oriented, specialised, industrial-style farming can have both positive and negative consequences (Robinson 2018b). In the case of large agro-corporations and agro-holdings as practised in East Germany and other countries over the past twenty years, the consequences for agricultural landscapes and the rural population are clearly negative. Large fields and farms, implemented during the period of the Soviet Russian patronage in East Europe, are still very productive, but have paved the way for an accelerated loss of biodiversity, infrastructure and culture in these agricultural landscapes.

Profits from farming flow into other regions and promote the breakdown of rural infrastructure. The Common Agricultural Policy of the EU and other governance regulations were not able to stop those processes. On the contrary, regulations such as carbon crediting, the biofuel economy and the sale of state and federal lands to the highest bidder have promoted unsustainable development. Increasingly land is owned by people who feel no responsibility for living rural landscapes (Agrarbündnis 2015-2018). These land privatisation processes are associated with the alienation of people from the land, nature and landscapes and from the principles of sustainability. They need to be stopped and cultural landscapes maintained and sustainably developed (Succow 2012).

Forest landscapes are a biome-based landscape category. Forest ecosystems cover roughly one-third of the global land area and are among the most biologically rich and genetically diverse ecosystems on earth. They are important sources of livelihood to millions of people and contribute to national economic development in many countries (Köhl et al. 2018). Globally, boreal forests and tropical forests are widespread and play a key role for global cycles and biodiversity.

Very productive agricultural lands located in the broad-leaf forest zone and in the forest steppe zone are used for cropping. Artificial forests imbedded in those landscapes are part of agricultural landscapes in many regions. Conversion of forests to cropland is profitable in many regions, and hence, forest stocks are declining in Africa and South America (Köhl et al. 2018). Similarly, large-scale acquisitions of forest lands occur in Africa (Conigliani et al. 2018), south-east Asia and South America, as well as in Europe.

5.6 Challenges and Opportunities for Landscape Research in the Anthropocene

5.6.1 General Circumstances and Goals

The sustainable development goals of the United Nations are a vision for living conditions and behaviour of the global population to be achieved by 2030. Principal goals are ending poverty and hunger and ensuring healthy lives. Environmental issues such as water pollution, soil degradation and climate change are also explicitly addressed (United Nations 2015). Climate change and loss of biodiversity at the landscape scale combine to give alarming signals of planet-altering activity. The great challenge in the twenty-first century is to stop the decay and degradation of cultural and semi-natural landscapes, in other words, to achieve landscape sustainability. Essential attributes of landscapes for the survival and well-being of humans are clean water and air, fertile and healthy soils for food and other ecosystem services, and a green and biodiverse environment. Many people living in an urban environment, shaped in the virtual world of their smartphones and buying cheap food in the supermarket, have other perceptions about landscapes. The perception of landscapes and key environmental issues varies between people of different countries, regions and landscapes (Kienast et al. 2015; Dronin and Bychkova 2018) and needs to be considered when coming to definitive solutions for evolving cultural landscapes.

To improve the living conditions of people, and to stop soil, water and air pollution and degradation, better cycling of energy and matter fluxes is needed. Novel methods and technologies for understanding, monitoring and managing landscape processes must be developed and set into operation. Because of the multitude of issues to be addressed when planning and managing cultural landscapes, interdisciplinary and transdisciplinary approaches and decision tools are needed alongside disciplinary innovations. People must handle these tools ethically and responsibly.

5.6.2 *Towards Landscape Research in the Age of Information Technologies*

Research is a scientific activity that generates knowledge about any subject. Landscape research is research about landscapes and for landscapes, their inhibitants included. Landscape research can lead in understanding and monitoring landscape processes and managing landscapes in a sustainable way (Naveh 2007). Landscapes have attracted the interest of researchers of different disciplines (Antrop and van Eetvelde 2017). Cooperative work should become the rule. Landscape research has become a unifying area of research (Fry 2001; Barau and Ludin 2012; Kienast et al. 2015) because people still positively perceive landscapes and cherish their values for personal well-being. It can build bridges between researchers of different disciplines and between researchers and citizens. When coming to concrete decisions about the present and future of landscapes, researchers can objectify the debates between decision-makers and stakeholders, between productivists and environmentalists.

First, mechanisms in highly complex landscape systems need to be understood, and scientific knowledge generated. For example, more attention should be given to anthropo-landscapes, i.e. completely humanly transformed or degraded landscapes. This start with Anthroposols, highly humanly modified soils, ranging from highly productive soils due to past cultural activity (Anthrosols according to IUSS Working Group WRB 2015), to polluted and technogenic soils (Technosols according to IUSS Working Group WRB 2015). The huge volume of knowledge gained in the twentieth century concerns native soils, whilst the anthropogenic soils have new properties that should be explored thoroughly and managed carefully within the framework of their landscapes (Blum 2017; Gerasimova 2018; Gerasimova and Bezuglova 2019).

Social sciences and humanities are fundamental, to bring territorial approaches which take into account landscape quality. Moreover, in regards to such transformative knowledge and science, citizens should play an increasing role in research, monitoring and governing of landscape processes (Couvet 2018). The monitoring of environmental processes occurring in landscapes is a potential broad area of Citizen Science (Sayer et al. 2015; Couvet 2017; Wende and Walz 2019). Well-informed and enthusiastic people can create spatial data for research. Mobile sensor technologies such as smartphones and unmanned aerial vehicles can be used for efficient data gathering in the frame of landscape monitoring.

Amongst the multitude of scientific disciplines contributing to landscape research, two groups of scientists from natural sciences could play a greater role in future landscape research. These are landscape modellers and microbiologists.

5.6.3 Contributions of Landscape Modelling

The generation of knowledge in a globalised world of the Fourth Paradigm (Hey et al. 2009) is based on gathering and processing big data and scenario modelling (Barau and Ludin 2012; Sang and Ode-Sang 2015; Mirschel et al. 2016a, b; Karampiperis et al. 2019), particularly using GIS for adaption to locations, landscapes and regions (Trovato et al. 2017; Topaj et al. 2018, Fig. 5.9). Experiments at different scales, field observatories, and remote sensing methods need to deliver reliable data. Landscape models and scenarios allow a constricted extrapolation of knowledge in space and time. A main objective of these system-based models is to evaluate quantitatively relationships between landscape components and units and to predict future changes and trends resulting from external impacts or self-development. Some landscape models are used for the identification of emergent effects generated by intercations among spatial units (e.g. microclimate, runoff and populations viability). They vary widely in complexity, flexibility and spatial, temporal and taxonomic resolutions (Scheller 2013).

All spheres of landscape research are covered by models. They range from gene flow, species distribution and landscape functioning (Dyer et al. 2010; Braunisch

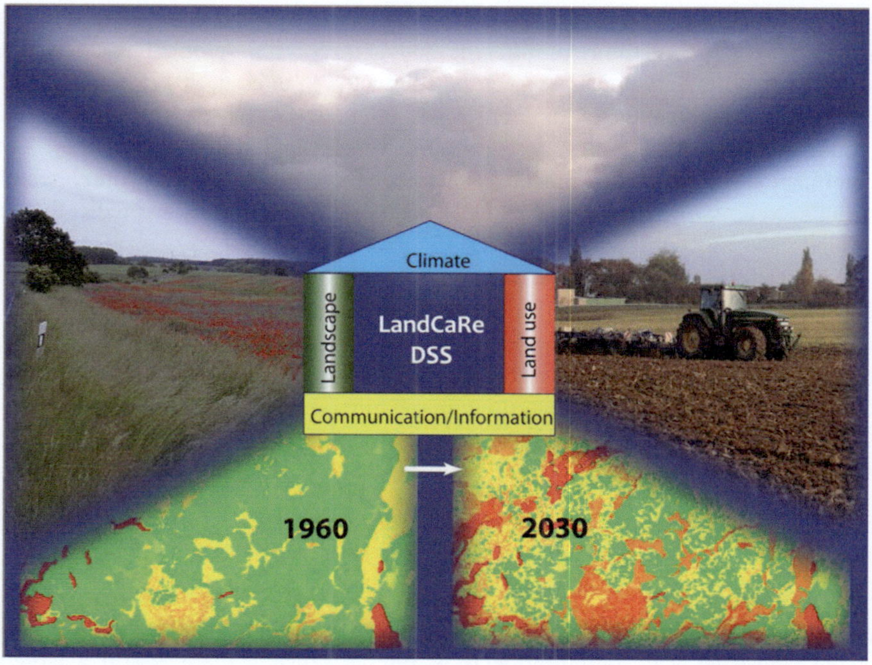

Fig. 5.9 Ecosystem models and decision support systems (DSS) are powerful tools for forecasting landscape developments. Graph: Leibniz Centre for Agricultural Landscape Research Muencheberg

et al. 2010; Wieland et al. 2017; Eaton et al. 2018) to virtual flows of landscape commodities and ecosystem services, and economic and social models of landscape development (Maldonado et al. 2018; Martellozzo et al. 2018; Schulp et al. 2019; Koch et al. 2019). Hydrological models are very commonly used to predict or reconstruct crucial impacts of weather alterations and climate change on hydrological processes and extremes like permafrost melting in the Arctics and floods (Thober et al. 2018; Victorov et al. 2019). Vegetation models (Foley et al. 1996; Cramer et al. 2001; Tchebakova et al. 2016), agro-ecosystem models (Kersebaum et al. 2007; Caubel et al. 2018; Nendel et al. 2019) and other models based on GIS and statistical approaches (Ruhovich 2018) enable prediction of crop productivity (Mirschel et al. 2014) and main nutrient fluxes in agricultural fields and landscapes (Kersebaum 2007).

Germany has developed the LandCare-DSS and other models. Scientists of this research centre explore the future of landscapes in several international cooperation projects. LandCaRe-DSS is a model-based interactive spatial information and decision support system for developing climate change adaptation strategies for the sustainable use of agro-landscapes. Graph: ZALF Muencheberg (Mueller et al. 2016b).

Many landscape-related decisions, for example, the risk assessment of pesticides (Schäffer et al. 2018), erosion risk (Deumlich et al. 2006) or environmental hazards (Svalova 2019), are based on modelling results. This raises questions about adequacy and reliability of models (Kersebaum et al. 2016; Getz et al. 2018; Pontius et al. 2018). The availability of different models for a certain problem is a desirable status so that model comparisons can be conducted (Yin et al. 2017; Pontius et al. 2018). The combination of different types of models, such as physically based models with statistical models, could improve their resolution and reliability. The combination of statistical models, including Bayesian belief networks, which include expert knowledge (e.g. Troldborg et al. 2013), can be used to predict threats to landscape function.

More and more complex models up to decision support systems have been developed (Mirschel et al. 2016a). In future, a further explosion in developing landscape models can be expected. They need to be permanently updated and re-calibrated with monitoring data (Kersebaum et al. 2015; Baatz et al. 2018). Big data and data mining will play a progressive role.

5.6.4 Contributions of Microbiology

Scientific enquiry dealing with landscapes implies the view over macroscopic elements, properties and processes. Besides what is recognisable by the naked eye at the landscape scale or by-products of remote sensing, there is still a largely unexplored microcosm of landscapes drivers. Microbiology has ascertained its structure and function and developed powerful tools of detection and manipulation of molecular structures. The microcosm, both the naturally existent and the anthropogenically overprinted or the potentially manipulated one, has crucial influence on functioning of landscapes and on the existence of all their inhabitants of macro-flora and macro-fauna, humans included (Martin 2008; Rotshild 2014; Behrendt et al. 2016; Müller et al. 2018; Reverey et al. 2018; Steinmuller et al. 2019). This microcosm underlies not only a rapid exploration process but is the subject of definition as landscapes of the microscale (Hanson et al. 2012). Recent outbreaks of food-borne infectious diseases such as bird flu and bovine spongiform encephalopathy (mad cow disease) have demonstrated significant knowledge gaps (Bhunia 2018; Figuié 2018). If science-based solutions for their prevention, healing and mitigation are not yet available, uncertainty of citizens and unpredictable actionism of authorities can be the consequence. Successfully addressing, monitoring and tackling crucial problems of landscape sustainability and the survival of inhabitants, including the spread and resistances of pathogens, prevention of epidemics, sustainable intensification of agriculture, future lifestock husbandry, waste processing and genetic modification of organisms, are impossible without further innovations in microbiology. Landscape genetics has developed as a special field of research. It integrates data and analysis methods from landscape ecology, spatial statistics, geography and population genetics to understand the spatial distribution of genetic variation (Manel et al. 2003; Storfer et al. 2010).

5.6.5 The Emergence of New Scientific Disciplines

More and more landscape-related scientific disciplines are emerging at the interface between existing ones. Examples are hydropedology (Lin 2012) and soil hydrology (Vereecken et al. 2015), hydroecology and ecohydrology (Wood et al. 2008), landscape sustainability science (Wu 2013; Opdam et al. 2018) and landscape genetics (Manel et al. 2003). Accelerated specialisation of scientists is an inherent characteristic of science. This is the basis for progress and innovation, but raises questions about their ability and responsibility for the implementation of novelties into complex landscape systems. Besides specialists, well-educated and responsible generalists are needed. Classical disciplines of landscape science and related disciplines such as geography or soil science could educate these generalists. When coming to complex solutions for future landscapes, landscape research may serve as an umbrella for disciplinary, interdisciplinary and transdisciplinary work of scientists (Zonneveld 1995).

5.6.6 Landscape Planning

Landscape planning based on comprehensive knowledge and advanced technologies plays a central and increasing role in evolving cultural landscapes (Antipov et al. 2006). The concepts of ecosystem services (Grunewald and Bastian 2013;

Bartlett et al. 2017; Bukvareva et al. 2015, 2018), impact assessment procedures (Kravchenko et al. 2008; Helming 2014; Rozas-Vásquez et al. 2018; Gutzler et al. 2015), life-cycle assessment (LCA) (ISO 14040 2006) and ecological footprint accounting (Mancini et al. 2018) are important modules for the evaluation of the sustainability of solutions and decisions. Decisions about land for food versus land for bioenergy or for nature and biodiversity are typical application areas of these research and decision tools. If agricultural lands are subject to reclamation, degradation, conversion, compensation measures or owner change, the resulting effects on soil quality and crop yield potential may be reliably estimated (Mueller et al. 2010, 2014a, b, 2016a).

The term *landscape approach* is frequently used in landscape planning, particularly in landscape conservation planning. 'Landscape approaches seek to provide tools and concepts for allocating and managing land to achieve social, economic, and environmental objectives in areas where agriculture, mining, and other productive land uses compete with environmental and biodiversity goals' (Sayer et al. 2013). Landscape approaches are especially valuable for taking into account the value of biodiversity as the basis of ecosystem functioning and for the optimising tasks of ecosystem services use and biodiversity conservation (Bukvareva 2018).

Digital data of earth observation, mapping and other spatial databases (Khoroshev 2016; Trovato et al. 2017; Hardt et al. 2018) are important for landscape researchers, decision-makers and other stakeholders. The landscape approach and the ecosystem services concept are applicable to real-world planning processes and integrated landscape management (Bastian et al. 2015; Miklos 2019). It is aimed at ecologically sustainable, economically beneficial and socially eligible adaptation of spatial land use decisions to landscape pattern. Ecologically oriented landscape planning takes into consideration both internal properties of a unit and geographical context as well as contribution of each unit to matter flows and emergent effects. Transferable and standardised evaluation methods for these data need to be developed for different scales of application (Mueller et al. 2016b; Eulenstein et al. 2016; Hoffmann et al. 2018a, b, c).

The evaluation of landscapes regarding their values for well-being and spirituality of people (Ode et al. 2008; Ball 2015), awareness of landscape values (Jones 1993, 2009; Walz and Stein 2018; Radford et al. 2019), critical examination of research terminology (Jones and Daugstad 1997; Jones 2003b) and the intersection of landscape science with landscape art remain important topics in landscape planning and decision processes. Landscape architecture will face new challenges in designing cultural landscapes and mediating between productivists and naturalists by adding components of fine arts and aesthetics into landscape projects and plans. Landscape architecture is 'the design of outdoor areas, landmarks, and structures to achieve environmental, social-behavioural, or aesthetic outcomes' (Jellicoe and Jellicoe 1987; Dramstad et al. 1996).

Sustainable development goal no. 11 of the United Nations states: 'Make cities and human settlements inclusive, safe, resilient and sustainable' (United Nations 2015). Urban areas need to be developed as cultural landscapes (Taylor 2016) through a bundle of measures (Regia 2014), including urban greening and promotion of biodiversity (Kowarik and Körner 2005; Kowarik et al. 2016).

Simensen et al. (2018) demand that landscape characterisation and mapping should be more purpose-orientated. To characterise the state of cultural landscapes and heritage sites in the United Kingdom (UK) and some other countries (Bartlett et al. 2017), a *Landscape Character Assessment* has become common (Fairclough et al. 2018). It provides a basis for spatial planning, management and decision-making for districts and heritage sites in the UK. If provided by standardised methodologies (Tudor 2014; Fairclough et al. 2018), it can be used for monitoring landscape changes over time (García-Martín et al. 2018) and improving the operationality of ecosystem approaches in landscape planning (Morrison et al. 2018). Key observation points can improve the validity and reliability of visual impact assessments significantly (Palmer 2019).

Landscape planning should experience new inputs from socio-economic research and citizens in terms of new lifestyle approaches and model projects for specific landscapes. As well as landscape planners, many people care about conservation and sustainable practices for landscapes. *Landscape stewardship* conceptualises the steering of landscapes towards sustainability by including all relevant stakeholders at the landscape level (García-Martín et al. 2018). It comprises all 'efforts to create, nurture and enable responsibility in landowners and resource users to manage and protect land and its natural and cultural heritage' (Brown and Mitchell 2000; Plieninger and Bieling 2017).

The role of public participation in accordance with the European Landscape Convention remains challenging. In research, there is need to refine methods for participation and communication between different actors in landscape matters. There is need for more knowledge about landscape perceptions as the landscape becomes increasingly a subject of public discussion. This necessitates further investigation of the meanings of landscape for different groups in increasingly multicultural societies. Another task for research is to examine critically participatory approaches and ways in which they might be biased or manipulated (Stenseke and Jones 2011). Examples and experiences of implementation of the ELC need critical examination (Jones and Stenseke 2011; Jørgensen et al. 2016).

Public participation is one aspect of a wider dimension of landscape democracy (Egoz et al. 2018). The Centre for Landscape Democracy, established in 2014 at the Department of Landscape Architecture, Norwegian University of Life Sciences, has the following research objectives: to promote national and international critical discourse on the relationship between landscape and democratic society; to explore and examine the linkages between human rights, landscape, democracy and public policy interventions (legislation, policy and planning and design practice); and to produce and collate theoretical, methodological and applied knowledge on landscape democracy from a variety of disciplines and policy perspectives. There is need for more discourse and knowledge on the landscape as a spatial materialisation of democracy, and on the role of democratic values in protecting, managing and planning landscape (CLaD 2016).

The relationship between landscape and social justice is an emerging field of research (Peil and Jones 2005; Jones 2006; Olwig and Mitchell 2009). This entails the development of a landscape concept that is sensitive to issues of social justice. There is need for research on ways in which mobilisation of representational and material landscapes creates and sustains social injustice and how legislation, policies, and management and planning practices relating to landscape issues can promote social justice (Setten et al. 2019).

5.6.7 The Role of Scientific Societies for Landscape Research

Research must lead to the qualification of public debates by delivering objective results. Powerful scientific societies such as the International Association for Landscape Ecology (IALE 2018) and the International Geographical Union (IGU 2018) are providing significant contributions to landscape research and transforming the results into public awareness. Examples of further international organisations active in landscape research are the International Union for Conservation of Nature (IUCN 2018), the Society for Urban Ecology (SURE 2018), the International Federation of Landscape Architects (IFLA 2018) and the International Council on Monuments and Sites (ICOMOS 2018). At European level, the Permanent European Conference for the Study of the Rural Landscape (PECSRL 2018), the European Network of Universities for the implementation of the European Landscape Convention (UNISCAPE 2018) and the European Council of Landscape Architecture Schools (ECLAS 2018) are examples of engagement for landscapes. Based on thorough research, other international scientific bodies and organisations such as the International Union of Soil Sciences (IUSS 2018) could play a greater role in studies and decision support concerning landscape sustainability. Further societies where scientists and practitioneers work together, such as the World Association of Soil and Water Conservation (WASWAC 2018), the International Soil Conservation Organisation (ISCO 2018), the International Soil and Tillage Research Organisation (ISTRO 2018) and many others, can provide valuable contributions towards landscape research for innovations and sustainability, too.

5.7 Some Landscapes of Particular Need for Better Research and Monitoring

Population growth, globalisation and climate change are altering the structure and functions of some regions and landscapes particularly fast. Considerable knowledge gaps exist. These are aquatic landscapes (seascapes) and the Arctic. Other landscapes face increasing human pressure such as those designated for food production (agricultural landscapes) and protected landscapes. Landscape development in Russia, the world's largest country in area, is influencing global cycles and facing particular challenges of sustainable development (Mueller et al. 2016b; Fan et al. 2018).

5.7.1 Seascapes

Seascape is a term sometimes used in marine geography and ecology (Falconer et al. 2013; Musard et al. 2014). Seascapes and their ecosystems cover more than two-third of the global surface. As compared with landscapes, they are not yet well explored and understood. This will change because of the pressure on natural resources and awareness of sea level rise due to climate change. The interface between sea and land—shallow waters, tidal flats and coasts, which are most productive and highly vulnerable ecosystems—will become subject to increased efforts of monitoring, risk assessment and management (Allen et al. 2018; Cottrell et al. 2018; Micallef and Rangel-Buitrago 2019). Research on seascapes investigates both resource use and ecological aspects (Musard et al. 2014; Kavanaugh et al. 2016; Pittman 2018; Mendel et al. 2019). Sustainable development goal no. 14 of the United Nations states 'Conserve and sustainably use the oceans, seas and marine resources' (United Nations 2015).

Seascapes are in part underwater landscapes. Landscape approaches can be applied to seascapes as well. The use of the term seascape in this sense is not yet very common. The scope of the European Landscape Convention includes inland water and marine areas (COUNCIL OF EUROPE 2000).

5.7.2 Landscape Research and Conservation in the Arctic

The rush to appropriate and exploit all parts of the globe has reached the Arctic, and this process will be accelerated by climate change (Kröger 2019) because of easier access. Landscapes in these regions have become multifunctional. Natural scientific research and monitoring will provide better understanding of Arctic landscape geosystems and human impacts (Sumina 2000; Masyagina et al. 2019). Monitoring of the permafrost zone (cryolithic zone) includes some fundamental processes such as groundwater hydrology, the thermal regime of the active layer and numerous cryogenic geological processes such as ground deformation and translocation, solifluction, thermoabrasion of shorelines, thermokarst, thermoerosion and ice dynamics (Pavlov 2008). Thermokarst processes have implications for ecosystems (Sumina 1998; McPartland et al. 2019; Claudino-Sales 2019) and for construction of infrastructure (Stanilovskaya 2019).

Conservation of Arctic landscapes cannot stop these processes but prevent them from becoming worse due to human-induced additional pressure. Strategies for the cultural development of cities and settlements in Arctic regions inhibited by recently sedentary indigenious peoples are challenging as ethno-cultural specifics need to be considered (Sergunin 2019; Landauer and Juhola 2019, Fig. 5.10). Arctic and other landscapes inhibited by indigenious pastoralists face a number of threats and conflicts with local industries (Lipski and Storozhenko 2019; Landauer and Juhola 2019). Those regions will require particular landscape stewardship for their maintenance and preservation.

Fig. 5.10 Landscapes of the Arctic are threatened by climate warming and mining industries. They are inhabited by indigenious peoples. For many of them, reindeer herding is their way of life. In these landscapes, there are extra difficulties for monitoring landscape processes and stewardship for sustainable landscape management and conservation. The photographs in the upper line show the Arctic grass tundra (left) and shrub tundra (right). The photograph below shows a little boy of the Yamal Nenets'ethnic group with his favourite husky. Photograph was taken on Yamal Peninsula in August 2011. Yamal is a particular hotspot of conflict between the gas industry and use and conservation of landscapes by nomadic pastoralists. Photograph with kind permission of Victor Slodkevich, Yamal, Russia

5.7.3 *Progress in Landscape Conservation Needed: The Case of Russian Zapovedniks*

State nature reserves (zapovedniks) in the Russian Federation are the strictest protected areas in the world (IUCN category I, IUCN 2018). They were created for nature and ecosystem research almost about 100 years ago. Russia has 103 zapovedniks with a total area of 34.9 million hectares, representing about 2% of Russian territory. The purpose of zapovendniks is to protect native biodiversity and to research ecosystem processes (Spetich et al. 2009). Whilst their protection status and management are ruled by law, goals and reality have differed. Their performance for meeting conservation objectives and research programmes has been insufficient (Colwell et al. 1997; Newell and Henry 2016). The monitoring and protection of zapovedniks cannot be guaranteed because of financial hardship and a local population which does not support the concept (Oostergreen and Shvarts 2000; Lieske 2011). One of the basic purposes of zapovedniks—research for understanding natural landscape processes as a baseline of landscape research—is still underdeveloped. The existing geographical network of zapovedniks of Russia needs further development in order to achieve optimal representativeness (Chibilev 2017).

5.7.4 *Paradigm Shifts for Agricultural Landscapes*

Agriculture must satisfy a basic human right: having enough food. Sustainable development goal no. 2 of the United Nation formulates this as follows: 'End hunger, achieve food security and improved nutrition and promote sustainable agriculture' (United Nations 2015). The focus of agriculture must be food security (Lal and Stewart 2011; Mueller et al. 2010; Dronin and Kirilenko 2011). Agricultural production is based on human impacts on biological cycles. Agricultural and associated forested and aquatic landscapes are ecosystems and thus providers of ecosystem services (DeClerck et al. 2016; Köppke and Schnug 2017). They are not only a source of food, fibre, timber and other materials but important secondary natural landscapes for leisure activities of the urban population. Agriculture is also a provider of valuable non-tradable goods or ecosystem services (Van Huylenbroeck et al. 2007) such as clean groundwater, fresh air and the beauty of wildflowers and butterflies. The current trends in many agricultural landscapes, e.g. their transformation to agro-landscapes, agro-energy and agro-industrial landscapes, designed for maximum productivity at high input rates of agro-chemicals, is a critical issue requiring better planning, monitoring and management decisions (Fig. 5.11). Pollinator decline in agricultural landscapes (Holzschuh et al. 2016; Kovács-Hostyánszki et al. 2019) is a clear indicator of unsustainable agriculture. There is a need for the development of improved and comprehensive ecotoxicological monitoring systems for agrochemicals (Schäffer et al. 2018). Agricultural

Fig. 5.11 Two diverging examples of farming in the twenty-first century. The left-hand photograph shows a farmer in the Chilenean Andes. He prepares the land for potato growing. The right-hand photograph shows an industry-style farm in Germany. The farmer is fertilising maize with liquid manure from a pig factory. Selling the energy produced by a biogas plant, wind turbines and solar batteries is a basis for a good income. Both agricultural systems require research on their sustainability, the future of landscapes and all their inhibitants

intensification has led not only to high productivity, but also to disturbances, damage and decline of ecosystems and rural cultural landscapes in some regions (Succow 2012; Stöven et al. 2015; Eulenstein et al. 2016; Schnug and de Kok 2016). This might be avoided through *Sustainable Intensification*, which involves simultaneously improving productivity and environmental management of agricultural land (Buckwell et al. 2014; Rockström et al. 2017; FAO 2017). *Agroecology* (Altieri 1995, 2018) is another approach towards sustainable agriculture. Wezel et al. (2009) characterise it as a science, a movement and a practice. There is a need for more conceptual work, knowledge and data in these important fields of research.

Along with a stable and economic agricultural production from their farms, farmers become landscape stewards (Raymond et al. 2016), ensuring ecosystem services through managing agricultural landscapes. Innovations in future agricultural landscapes should include the re-exploration of lost knowledge and experience. For example, simple measures of restructuring agricultural landscapes in the framework of land consolidation such as planting of hedgerows, grassland buffer strips, planting fruit trees or coppicing along dirt roads and ditches are proper measures of innovative restoration. These measures benefit ecosystem functioning (Arnaiz-Schmitz et al. 2018; Gontijo 2018; Holden et al. 2019; Jiang et al. 2019a, b).

The viability and feasibility of *Sustainable Intensification* must be proven. This will be important for the future of agricultural production systems and landscapes. Organic farming (Federal Ministry of Food and Agriculture 2017), which is more sustainable than conventional or controlled integrated farming, and other systems such as urban agriculture (Lohrberg et al. 2015) will probably not be able to feed the global population on current diets (Yadav et al. 2013; Reganold and Wachter 2016; Meemken and Qaim 2018).

A problem of all approaches of agricultural management seems to be their claim to work everywhere. In this connection, traditional approaches, which have worked well in some landscapes, are included in a global blame for environmental problems. An example is the debate about conservation tillage versus ploughing technologies (Mueller et al. 2018). Landscape approaches for land management, for example, local nitrogen management and planning (Andersen et al. 2019), could help to improve this. It is important especially in the case of bioavailable compounds from artificial fertilisers, a major factor responsible for accelerating the eutrophication of inland waters (Sobczyński and Joniak 2009; Joniak 2018).

Controlled local and regional food systems (production, consumption and waste recycling) based on landscape-adapted agriculture (Kiryushin 1995) are a promising approach towards strengthening rural communities and landscape sustainability (Polis et al. 2004; Tregear et al. 2007; Low et al. 2015; Schaller et al. 2018). International long-term agricultural experiments have shown that the highest and most stable crop yields and highest soil fertility, in the long run, were achieved through application of organo-mineral fertilisation (Körschens et al. 2014). Soil organic carbon management, using dung from animal husbandry, straw and stubble of cereals and other plant residues within the crop rotation, needs to be part of the local and regional nutrient cycle of soil (Lal 2008). Local agricultural food chains and nutrient cycles are worth being re-installed. The Common Agricultural Policy (CAP) of the European Union, as a tool of state support aimed at maintaining a viable farm population and increasing output (Robinson 2018a), needs to be developed with more emphasis on landscape stewardship.

5.7.5 Agri-Environmental Monitoring Systems

Networks of agri-environmental monitoring systems deliver important data about the status of agricultural ecosystems (Brandt et al. 2002; Schindler et al. 2010; Körschens et al. 2014; Saparov 2014; Sychev et al. 2016; McKenzie et al. 2017, Figs. 5.12 and 5.13). Many countries and their federal units have built up agri-environmental monitoring systems or general environmental monitoring systems including monitoring of agricultural lands and soils (Sychev et al. 2016; Glante et al. 2018; Gubler et al. 2018). Research institutions contribute to improving the methodology by finding the best indicators (Dannowski et al. 2014; Hoffmann and Kratz 2018; Römbke 2018), detecting causal relationships of ecosystem components (Couvet et al. 2011; Eisendle-Flöckner and Hilberg 2014; Hilberg and Eisendle-Flöckner 2016; Hoffmann et al. 2018a, b, c). Soils as the basis of agriculture and associated water bodies and wetlands as possible sinks for pollutants play a particular role in environmental monitoring of agricultural landscapes (Saljnikov et al. 2014a, b; Römbke et al. 2016; Joniak et al. 2017; Lischeid et al. 2018; Antić-Mladenović et al. 2018, Fig. 5.13).

They need to be better interlinked with models and decision support systems (Romanenkov et al. 2007; Mirschel et al. 2016a; Baatz et al. 2018) for more reliable and faster results and decisions about landscape sustainability.

Fig. 5.12 Advanced technical systems for monitoring the status of soils and vegetation in agricultural landscapes. The left-hand photograph shows a gas-flux measurement station on cropland operated by the Thuringian State Institute of Agriculture, Germany. The station is combined with lysimeters. Researchers aim to understand processes of plant growth and cycles of water and fertilisers in soil, plant and ground-level air layer. The right-hand photograph shows part of forest measurement station operated by the Thünen Institute of Forest Ecosystems in Eberswalde, Germany

Fig. 5.13 Evaluation of grassland ecosystems by indicator plants, soil survey and field measurement of mechanical, hydro-physical and chemical parameters. The photographs show a multispecies grazing experiment on large plots of the ZALF research station in Paulinenaue, Germany

5.8 Holistic Landscape Observations and Experiments as Drivers of Technological Progress: A Few Examples from a German Perspective

As landscapes are multifunctional and holistic (Antrop and van Eetvelde 2017), research projects and their infrastructure need to be adapted to the complexity resulting from both the diversity of geosystems and ecosystems diversity and the different views and claims of stakeholders. Isolated solutions of landscape monitoring focusing on single aspects of landscape sustainability and short-term case studies are still possible, but not longer the rule. Research projects and networks are increasingly multidisciplinary and their complexity is also increasing. Five examples of landscape experiments of differing and increasing complexity from the German perspective are presented below:

(1) The CarboZALF project (Augustin et al. 2011; Deumlich et al. 2017) is a natural science-based interdisciplinary landscape project aimed at understanding and quantificating the soil carbon balance in Pleistocene landscapes. It is financed by the budget of the Leibniz Centre for Agricultural Landscape Research (ZALF) in Muencheberg, Germany, an institution of the Leibniz Association. Such projects provide new insights into the interactions between abiotic and biotic landscape components. The potential of transferring the project results into other landscapes is still somewhat unclear. However, innovative methods developed in the framework of projects like this, such as field measurement systems for gas fluxes (Hoffmann et al. 2018a, b, c), are clearly transferable to other regions.

(2) The international project 'Water Monitoring in Central Asia' (CAWa) (CAWa Flyer 2017; Schöne et al. 2018) was a project of high complexity. Covering different countries and landscapes, it aimed at reliable transboundary water resource management in Central Asia. It was financed by the German Federal Foreign Office and coordinated by the German Research Centre for Geosciences (GFZ) in Potsdam Germany. The project run from 2008 to 2017 and included more than 20 scientific institutions. Within the CAWa project, a network of automatic monitoring stations was installed in Central Asia, especially in remote areas and at higher altitudes, providing continuously meteorological and hydrological parameters (Schöne et al. 2018). Third-party funded international multidisciplinary projects like CAWa are very important for the transfer of knowledge and technologies. They provide impulses for innovations in landscape research through installed technologies and monitoring systems. The WUEMoCA tool for monitoring irrigated cropland use and water use efficiency at the landscape level (Schönbrodt-Stitt et al. 2018) is an example for an applicable research product. Due to the limited duration of these projects, long-term data cannot be guaranteed.

(3) The KULUNDA project (Meinel et al. 2014; Kulunda Portal 2018) was an international multi- and transdisciplinary project, running 2011–2016, funded

by the German Federal Ministry of Education and directed by the Martin-Luther-University Halle-Wittenberg. Goals of the KULUNDA project were to mitigate soil degradation and desertification processes, to stabilise soil carbon, to increase crop yield by development and implementation of sustainable land management practices and to promote rural and regional development in Siberia (Kulunda Portal 2018). Experiments were conducted in different biogeographic regions: the dry steppe, the typical steppe and the forest steppe. Eight German and three Russian scientific organisations and a company that produces landscape-adapted agricultural machinery were included. The work packages considered social and institutional factors of implementing new technologies. Ecological and economic strategies for sustainable and management in the Russian steppes were developed. New lysimetric measurement methods for the soil water balance (Meissner et al. 2014; Balykin et al. 2016) and technologies for soil conservation tillage (Meinel et al. 2014; Grunwald et al. 2016) were novel tools implemented. As the study region belongs to the Eurasian steppe belt, the largest potential 'bread basket' of the world, the interest in follow-up projects is obvious.

(4) Terrestrial environmental observatories (TERENO) form an interdisciplinary and long-term research programme involving six research centres of the Helmholtz Association and several collaborating research institutes and universities. This large-scale project aims to study the long-term ecological, social and economic impact of global change at regional level (TERENO 2018). Fluxes of water, matter and energy within the continuum of the groundwater–soil–vegetation–atmosphere system, long-term changes of the composition and functioning of micro-organisms and plants and fauna are measured at key observation sites across Germany. The German Lysimeter network TERENO SOILoilCan is a key component of TERENO. Within the framework of TERENO SOILCan, 132 fully automated lysimeter systems (Hertel and von Unold 2014) were installed at 14 highly equipped experimental field sites across the four TERENO observatories. Relevant state variables of grassland and arable ecosystems are monitored since 2010 (Pütz et al. 2016).

(5) The European Long-Term Ecosystem, critical zone and socio-ecological Research Infrastructure (eLTER RI) research is an example of a comprehensive research programme (eLTER 2018). It has roots in other long-term ecological research programmes and platforms which have gained particular momentum in Europe during the past 20 years. It is a network of scientists, disciplines, institutions, data and metadata, and research projects (Mirtl 2010; Mirtl et al. 2018). The Helmholtz Association is particularly engaged in eLTER programmes and the German Federal Ministry of Education and Research is a main supporter. The eLTER research infrastructure comprises long-term ecosystem, critical zone and socio-ecological research. It covers 250 research sites representing major European terrestrial, freshwater and transitional water ecosystems. eLTER RI explores interactions between abiotic and biotic ecosystem components at multiple scales, including human–environment interactions (eLTER 2018). Recently, it became part of the European Strategy

Forum on Research Infrastructures (ESFRI) roadmap, containing research projects of pan-European priority (eLTER 2018). eLTER has a European and global focus. The conjunction of eLTER with other international ecosystem networks and observatories offers new opportunity for advancing Earth system dynamics modelling (Baatz et al. 2018). The landscapes of the globe will benefit from this work, but indirectly.

These few projects demonstrate the dynamic in the community of landscape research under the pressure of globalisation and imbedded in the need for permanent growing economies. Projects 2-5 mentioned above, and 4-5 in particular, seem not to directly link with landscape research in a narrow sense as they do not focus on the landscape scale. However, they are inextricably linked with landscapes, both in methodologies (discrete observation systems in typical landscapes) and in results. Whilst delivering innovations, they can initiate and push solutions at landscape level.

5.9 Innovations and Knowledge Transfer for Improving Landscape Sustainability

The great challenges of the Anthropocene are achieving landscape sustainability with high productivity, developing cultural landscapes and preserving largely natural landscapes. Landscape research can play a leading role for understanding and monitoring landscape processes, thus creating a basis for managing many landscapes in a sustainable way.

How to tackle the challenges for landscape research in detail? We start with information about the *status quo* and about available and potentially transferrable approaches for understanding, monitoring and managing landscapes.

A new book series 'Innovations in Landscape Research' will provide this information. It emphasises the role of research for improving landscape sustainability. There is great potential to develop and apply innovative research methods. These range from single high-tech measurement devices to complex forecasting and decision tools. The book series will provide a knowledge exchange between the English- and Russian-speaking scientific communities as a first step towards comparable methods and standards for monitoring and managing landscape processes. The following questions are among those that should be answered in detail:

How do landscapes function?

What will future landscapes look like?

How can we sustainably develop intensively used and stressed landscapes?

Scientific innovations and decision-making tools are the key for answering and solving these challenging questions. The series will provide information on advanced methods and results of disciplinary, interdisciplinary and transdisciplinary work in landscape research. A broad array of methods to measure, assess, forecast, utilise, and control landscapes and their components will be presented in this book series. They include field and laboratory measurement methods, methods of resource evaluation, functional mapping and risk assessment, and sensing methods for landscape monitoring. They contain methods for data analysis and ecosystem modelling, methods and technologies for optimising the use of multifunctional landscapes, for the bioremediation of soil and water and basics and procedures of landscape planning. The series 'Innovations in Landscape Research' will provide a new view on landscapes with some focus on scientific and technological innovations, on soils and problems of optimising agricultural landscapes under conditions of progressive urbanisation. Edited volumes of the book series 'Innovations in Landscape Research' will address the following topics:

- Status and trends of landscape research
- Exploring and optimising agricultural landscapes
- Landscape modelling and decision support
- Basics and tools for landscape planning
- Assessing soil resources and quality
- Water resource and quality monitoring
- Tools for water and wetland management
- Assessing soil degradation
- Rehabilitation of degraded landscapes
- Forest management and agroforestry
- Understanding and monitoring landscape biodiversity
- Understanding key landscapes processes
- Agricultural field experiments
- Landscape sensor and monitoring technologies
- Advances in soil hydrology and lysimetry
- Advances in soil monitoring
- Land improvement and irrigation
- Agricultural soil and plant management
- Landscape monitoring concepts and studies

Landscape research is benefiting from innovations and scientific tools within several disciplines. The books in this series present status analyses, methodical chapters and case studies showing the practical relevance and feasibility of novel decision tools and technologies. The books are intended as a source of information and communication for researchers, teachers, students, stakeholders and other citizens interested in the topics of landscape science and related disciplines (Fig. 5.14). Stakeholders such as landscape planners, advisors, farmers, environmentalists, and local authorities, who are interested in information about novel methods and their application in the concrete landscape they are responsible for, should benefit from the knowledge presented in these books. The books of this series will contain highly valuable information for forming the basis for managers and decision-makers at all levels, from local bodies to the highest international panels.

Fig. 5.14 International cooperation and knowledge transfer are keys for making innovations in landscape research applicable and operable. The photographs show a Russian–German team on the groundwater lysimeter station of the ZALF research station Paulinenaue, Germany, and a Kazakh–German team at work of soil sampling for special analyses in a semi-desert landscape (Saksaul vegetation on Takyr soils)

Particular attention is paid to trends and scientific–technological innovations, agricultural landscapes and agri-environmental problems, the inclusion of still unknown studies from Russia and Central Asia, the mitigation of human-induced processes that destroy the functioning of landscapes, as well as the remote-sensed monitoring of landscape changes, ecosystem modelling and forecasting of landscape processes and the creation of knowledge, rules and tools for handling the multifunctionality of landscapes.

5.10 Conclusions

1. Landscapes of the globe provide food security and welfare of the population. Many are productive but lack sustainability.
2. Landscapes are experiencing huge transformations with increasing speed. In the past, they have been associated with improvements of the living conditions of many people, but also with increasing environmental problems.
3. Our short review revealed great challenges and opportunities for landscapes research in the Anthropocene. Landscape research in terms of knowledge generation and transfer of ideas and technologies for landscape development and conservation can help to initiate landscape sustainability in many cases.
4. Directions of landscape development and technical innovations must be embedded in ethical guidelines of cultural development.
5. The novel book series 'Innovations in Landscape Research' presents an array of advanced methods for understanding, monitoring and managing landscapes. It may serve as a first knowledge, data and communication basis for wise decisions regarding the development of landscapes.

References

Abelshauser W (1983) Wirtschaftsgeschichte der Bundesrepublik Deutschland (1945–1980) Edition Suhrkamp 1241 = NF 241 Neue historische Bibliothek). Suhrkamp, Frankfurt am Main 1983, ISBN 3-518-11241-4

Acksel A, Kappenberg A, Kühn P, Leinweber P (2017) Human activity formed deep, dark topsoils around the Baltic Sea. Geoderma Regional 10:93–101. https://doi.org/10.1016/j.geodrs.2017.05.005

Agrarbündnis (2015-2018) Der kritische Agrarbericht 2015 https://www.kritischer-agrarbericht.de/2015.346.0.html, Der kritische Agrarbericht 2016 https://www.kritischer-agrarbericht.de/2016.357.0.html, Der kritische Agrarbericht 2017 https://www.kritischer-agrarbericht.de/2017.368.0.html, Der kritische Agrarbericht 2018 https://www.kritischer-agrarbericht.de/2018.382.0.html. Accessed on 19 Dec 2018

Ahern K, Cole L (2012) Landscape scale—towards an integrated approach. ECOS 33(3/4): 7–11. https://www.banc.org.uk/wp-content/uploads/2015/05/ECOS-33-3-4-6-Landscape-scale-integrated-approach.pdf. Accessed on 19 Dec 2018

Alfani G (2011) The Famine of the 1590s in Northern Italy. An Analysis of the Greatest "System Shock" of Sixteenth Century. Histoire & Mesure. Edition EHESS, pp 17–50. https://journals.openedition.org/histoiremesure/4119

Allen TR, Crawford T, Montz B, Whitehead J, Lovelace S, Hanks AD, Christensen AR, Kearney GD (2018) Linking water infrastructure, public health, and sea level rise: integrated assessment of flood resilience in coastal cities. Public Work Manag & Policy, 1–30. https://doi.org/10.1177/1087724X18798380

Altieri M (1995, 2018) Agroecology. The science of sustainable agriculture. CRC Press, Boca Raton. 448 p. ebook version Taylor & Francis 2018. https://www.taylorfrancis.com/books/9780429964015

Andersen PS, Andersen E, Graversgaard M, Christensen AA, Vejre H, Dalgaard T (2019) Using landscape scenarios to improve local nitrogen management and planning. J Environ Manage 232(15):523–530. https://doi.org/10.1016/j.jenvman.2018.11.023

Antić-Mladenović S, Kresović M, Čakmak D, Perović V, Saljnikov E, Ličina V, Rinklebe J (2018) Impact of a severe flood on large-scale contamination of arable soils by potentially toxic elements (Serbia). Environ Geochem Health, pp 1–18. https://doi.org/10.1007/s10653-018-0138-4

Antipov AN, Kravchenko VV, Semenov YuM et al (2006) Landscape planning: Tools and experience in implementation. Irkutsk: V.B. Sochava Institute of Geography SB RAS Publishers, 2006. 149 p. http://www.irigs.irk.ru/docs/lndscpln/Landscapeplannnig_Irkutsk_2006.pdf. Accessed on 19 Dec 2018

Antonkiewicz J, Łabętowicz J (2016) Chemical innovation in plant nutrition in a historical continuum from ancient Greece and Rome until modern times. Chem Didact Ecol Metro 21(1–2):29–43. https://doi.org/10.1515/cdem-2016-0002

Antrop M (1998) Landscape change: plan or chaos? Landsc Urban Plann 41:155–161

Antrop M (2000) Geography and landscape science. Belgeo (Online), pp 9–36 https://doi.org/10.4000/belgeo.13975. Accessed on 19 Dec 2018

Antrop M, Van Eetvelde V (2017) Landscape perspectives. The holistic nature of landscape. Springer Science + Business Media B.V. Landscape Series, vol 23, 436 p

Arend J (2017) Russlands Bodenkunde in der Welt. Eine ost-westliche Transfergeschichte 1880-1945, Vandenhoeck and Ruprecht, Göttingen, 308 p

Arnaiz-Schmitz C, Herrero-Jáuregui C, Schmitz MF (2018) Losing a heritage hedgerow landscape. Biocultural diversity conservation in a changing social-ecological Mediterranean system. Sci Total Enviro 637–638:374–384. https://doi.org/10.1016/j.scitotenv.2018.04.413

Augustin J, Hierold W, Sommer M, Deumlich D, Gerke HH, Koszinski S, Schröder B, Hufnagel J, Verch G, Wirth S, Jochheim H, Nendel C, Lischeid G, Schindler U, Siemens J, Lang F, Kaupenjohann M, Kirchner G, Schwank M (2011) CarboZALF–ein interdisziplinäres Landschaftsexperiment zum Kohlenstoffhaushalt von Agrarlandschaften. In: DBG Exkursionsführer 114/2011, pp 60–68 http://publ.ext.zalf.de/publications/a3625df3-001e-4ae7-b321-2e03d1fbfb4d.pdf. Accessed on 19 Dec 2018

Baatz R, Sullivan PL, Li L, Weintraub SR, Loescher HW, Mirtl M, Groffman PM, Wall DH, Young M, White T, Wen H, Zacharias S, Kühn I, Tang J, Gaillardet J, Braud I, Flores AN, Kumar P, Lin H, Ghezzehei T, Jones J, Gholz HL, Vereecken H, Van Looy K (2018) Steering operational synergies in terrestrial observation networks: opportunity for advancing earth system dynamics modeling. Earth Syst Dyn 9(2):593–609. https://doi.org/10.5194/esd-9-593-2018

Ball BC (2015) The landscape below: soil, spirituality and agriculture. Wild Goose Publications, Glasgow, 168 p

Balykin D, Puzanov A, Stephan E, Meissner R (2016) Using the innovative lysimeter technology in the German–Russian Research Project "KULUNDA". In: Mueller L, Sheudshen A, Eulenstein F (eds) Novel methods for monitoring and managing land and water resources in Siberia. Springer Water. Springer, Cham, pp 387–399. https://doi.org/10.1007/978-3-319-24409-9_16

Barau AS, Ludin ANM (2012) Intersection of landscape, Anthropocene and fourth paradigm. Living Rev Landscape Res 6:1. http://lrlr.landscapeonline.de/Articles/lrlr-2012-1/download/lrlr-2012-1Color.pdf. Accessed on 19 Dec 2018

Bartlett D, Gomez-Martin E, Milliken S, Parmer D (2017) Introducing landscape character assessment and the ecosystem service approach to India: a case study. In: Landscape and urban planning. Elsevier. pp 257–266. ISSN 0169-2046, ISSN 0169-2046

Bastian O, Grunewald K, Khoroshev AV (2015) The significance of geosystem and landscape concepts for the assessment of ecosystem services: exemplified on a case study in Russia. Landscape Ecol 30(7):1145–1164

Baten J, Steckel R, Larsen C, Roberts C (2018) Multidimensional patterns of European health, work, and violence over the past two Millennia. In Steckel R, Larsen C, Roberts C, Baten J (eds) The backbone of Europe: health, diet, work and violence over two Millennia (Cambridge Studies in Biological and Evolutionary Anthropology, pp 381–396). Cambridge University Press, Cambridge. https://doi.org/10.1017/9781108379830.015

Baum T, Nendel C, Jacomet S, Colobran M, Ebersbach R (2016) "Slash and burn" or "weed and manure"? A modelling approach to explore hypotheses of late Neolithic crop cultivation in pre-alpine wetland sites. Veget Hist Archaeobot 25:611. https://doi.org/10.1007/s00334-016-0583-x

Behrendt U, Kämpfer P, Glaeser SP, Augustin J, Ulrich A (2016) Characterisation of the N_2O producing soil bacterium *Rhizobium azooxidifex* sp. nov. Int J Syst Evol Microbiol 66 (6):2354–2361. https://doi.org/10.1099/ijsem.0.001036

Berg LS (1915) The objectives and tasks of geography (Predmet i zadachi geografii). Proc Imp Russ Geogr Soc 51(9):463–475 (in Russian)

Berg LS (1930) Landscape-geographical zones of the USSR, 28 p. (in Russian, Берг Л.С. Ландшафтно-географические зоны СССР. Л. 1930. с. 28)

Bhunia AK (2018) Introduction to foodborne pathogens. pp 1–23. In: Foodborne microbial pathogens. Food science text series. Springer, New York, NY. https://doi.org/10.1007/978-1-4939-7349-1_1

Bjornlund V, Bjornlund H (2019) Understanding agricultural water management in a historical context using a socioeconomic and biophysical framework. Agric Water Manag 213:454–467. https://doi.org/10.1016/j.agwat.2018.10.037

Blackbourn D (2006) The conquest of nature. Water, landscape and the making of modern Germany. Jonathan Cape. 320 p. ISBN-10: 0224060716; ISBN-13: 978-0224060714

Blum WEH (2017) Soils within cities: global approaches to their sustainable management—composition, properties, and functions of soils of the urban environment. Die Bodenkultur 68 (1):71–72; ISSN 0006-5471

Bower B (2017) Science news. Magazine 192(9):16. https://www.sciencenews.org/article/how-asian-nomadic-herders-built-new-bronze-age-cultures. Accessed on 2 Feb 2019

Brandt JJE, Bunce RGH, Howard DC, Petit S (2002) General principles of monitoring land cover change based on two case studies in Britain and Denmark. Landsc Urban Plann 62:37–51

Braunisch V, Segelbacher G, Hirzel AH (2010) Modelling functional landscape connectivity from genetic population structure: a new spatially explicit approach. Mol Ecol 19(17):3664–3678. https://doi.org/10.1111/j.1365-294X.2010.04703.x

Brown J, Mitchell B (2000) The stewardship approach and its relevance for protected landscapes. The George Wright Forum 17(1):70–79. http://citeseerx.ist.psu.edu/viewdoc/download?doi=10.1.1.450.6279&rep=rep1&type=pdf. Accessed on 19 Dec 2018

Buckwell A, Heissenhuber A, Blum WEH (2014) The sustainable intensification of European Agriculture. A review. Sponsored by the Rise Foundation, 96 p. http://www.risefoundation.eu/images/files/2014/2014_%20SI_RISE_FULL_EN.pdf. Accessed on 19 Dec 2018

Bukvareva E (2018) The optimal biodiversity–a new dimension of landscape assessment. Ecol Indic 94(2):6–11. https://doi.org/10.1016/j.ecolind.2017.04.041

Bukvareva E, Grunewald K, Bobylev S, Zamolodchikov D, Zimenko A, Bastian O (2015) The current state of knowledge of ecosystems and ecosystem services in Russia: a status report. AMBIO. 44(6):491–507. https://doi.org/10.1007/s13280-015-0674-4

Bukvareva E, Zamolodchikov D, Grunewald K (2018) National assessment of ecosystem services in Russia: methodology and main problems. Sci Total Environ 655:1181–1196. https://doi.org/10.1016/j.scitotenv.2018.11.286

Bunkus R, Theesfeld I (2018) Land grabbing in Europe? Socio-cultural externalities of large-scale land acquisitions in East Germany. Land 7:98. https://doi.org/10.3390/land7030098

Carson R (1964) Silent Spring, 1st edn. Fawcett Crest. 304 p

Caubel J, de Cortazar-Atauri IG, Vivant AC, Launay M, de Noblet-Ducoudré N (2018) Assessing future meteorological stresses for grain maize in France. Agric Syst 159:237–247. https://doi.org/10.1016/j.agsy.2017.02.010

CAWa Flyer (2017) Regional research network "Central Asian Water". https://www.cawa-project.net/fileadmin/cawa/00_home/Flyer_November17-Webversion.pdf. Accessed on 19 Dec 2018

Chibilev AA (2017) The origins and development paths of zapovednik management in Russia. Geogr Nat Resour 38(3):211–216. https://doi.org/10.1134/S1875372817030015
CLaD (2016) About CLaD. Centre for Landscape Democracy (CLaD), Norwegian University of Life Sciences. https://www.nmbu.no/en/faculty/landsam/research/centers/clad/about. Accessed on 21 Feb 2019
Claudino-Sales V (2019) Natural system of Wrangel Island Reserve, Russia. In: Coastal world heritage sites. Coastal Research Library, vol 28. Springer, Dordrecht, pp 209–214. https://doi.org/10.1007/978-94-024-1528-5_31
Colwell MA, Dubynin AV, Koroliuk AY, Sobolev NA (1997) Russian nature reserves and conservation of biological diversity. Nat Areas J 17(1):56–68. https://www.jstor.org/stable/43911637?seq=1#page_scan_tab_contents. Accessed on 19 Dec 2018
Conigliani C, Cuffaro N, D'Agostino G (2018) Large-scale land investments and forests in Africa. Land Use Policy 75:651–660. https://doi.org/10.1016/j.landusepol.2018.02.005
Cosgrove D (1984) Social formation and symbolic landscape. University of Wisconsin Press, Madison
Cosgrove D (2003) Landscape: ecology and semiosis. In: Palang H, Fry G (eds) Landscape interfaces. Cultural heritage in changing landscapes. Landscape series, vol 1. Kluwer Academic Publishers, Dordrecht/Boston/London, pp 15–20
Cottrell RS, Fleming A, Fulton EA, Nash KL, Watson RA, Blanchard JL (2018) Considering land–sea interactions and trade-offs for food and biodiversity. Glob Chang Biol 24(2):580–596. https://doi.org/10.1111/gcb.13873
COUNCIL OF EUROPE (2000) European Landscape Convention. Committee of Ministers of the Council of Europe. Florence. www.coe.int/EuropeanLandscapeConvention; https://rm.coe.int/CoERMPublicCommonSearchServices/DisplayDCTMContent?documentId=09000016802f80c6. Accessed on 19 Dec 2018
Couvet D (2017) From biodiversity to policies to politics (Chap. 28). In: Peterson SA, Somit A (eds) Handbook of biology and politics. Edward Elgar Publications, Cheltenham, pp 487–512
Couvet D (2018) Citizen and biodiversity governance (Chap. I/9). In: Sychev VG, Mueller L (eds) Novel methods and results of landscape research in Europe, Central Asia and Siberia. Monograph in 5 Volumes, Vol I Landscapes in the 21th Century: Status Analyses, Basic Processes and Research Concepts. © «FSBI "VNII Agrochemistry» 2018, pp 71–76. https://doi.org/10.25680/1757.2018.97.98.009, http://vniia-pr.ru/monografii/pdf/tom1-9.pdf. Accessed on 19 Dec 2018
Couvet D, Devictor V, Jiguet F, Julliard R (2011) Scientific contributions of extensive biodiversity monitoring. C R Biol 334:370–377. https://doi.org/10.1016/j.crvi.2011.02.007
Cramer W, Bondeau A, Woodward FI, Prentice IC, Betts R, Brovkin V, Cox PM, Fischer V, Foley JA, Friend AD, Kucharik C, Lomas MR, Ramankutty N, Sitch S, Smith B, White A, Young-Molling C (2001) Global response of terrestrial ecosystem structure and function to CO_2 and climate change: results from six dynamic global vegetation models. Glob Change Biol 7:357–373. https://doi.org/10.1046/j.1365-2486.2001.00383.x
Crutzen PJ (2002) Geology of mankind. Nature 415:23. https://doi.org/10.1038/415023a
CVCE (2016) European NAvigator. Etienne Deschamps. 1958–1968 Successes and crises. http://www.cvce.eu/obj/1958_1968_successes_and_crises-en-5136b72a-0de2-4636-bda0-27e58b6c83e8.html. Accessed on 2 Feb 2019
Daniels S, Cosgrove D (eds) (1988) The iconography of landscape. Essays on the symbolic representation, design and use of past environments. Cambridge University Press, Cambridge
Dannowski R, Schindler R, Cremer N, Eulenstein F (2014) Methods of in situ groundwater quality monitoring: basis for the efficiency survey of agricultural groundwater conservation measures. In: Mueller L, Saparov A, Lischeid G (eds) Novel measurement and assessment tools for monitoring and management of land and water resources in agricultural landscapes of Central Asia. Environmental Science And Engineering. Springer, Cham, pp 275–288. https://doi.org/10.1007/978-3-319-01017-5_16

De Maria M (2019) Understanding Land in the Context of large-scale land acquisitions: a brief history of land in economics. Land 8:15. https://doi.org/10.3390/land8010015, https://www.mdpi.com/2073-445X/8/1/15

DeClerck FAJ, Jones SK, Attwood S, Bossio D, Girvetz E, Chaplin-Kramer B, Enfors E, Fremier AK, Gordon LJ, Kizito F, Lopez Noriega I, Matthews N, McCartney M, Meacham M, Noble A, Quintero M, Remans R, Soppe R, Willemen L, Wood SLR, Zhang W (2016) Agricultural ecosystems and their services: the vanguard of sustainability?. Curr Opin Environ Sustain 23:92–99. https://doi.org/10.1016/j.cosust.2016.11.016

Dell'Angelo J, D'Odorico P, Rulli MC (2017) Threats to sustainable development posed by land and water grabbing. Curr Opin Environ Sustain 26–27:120–128. https://doi.org/10.1016/j.cosust.2017.07.007

Deumlich D (2012) Struktur und Prozess - Einfluss historischer Landbewirtschaftung auf lineare Fließwege bei Starkregen in Brandenburg (D) Landscape Online 31:1–19. https://doi.org/10.3097/lo.201231, https://www.landscapeonline.de/archive/2012/31/Deumlich_LO31_2012.pdf

Deumlich D, Kiesel J, Thiere J, Reuter HI, Völker L, Funk R (2006) Application of the SIte COMparison Method (SICOM) to assess the potential erosion risk: a basis for the evaluation of spatial equivalence of agri-environmental measures. CATENA 68(2–3):141–152

Deumlich D, Rogasik H, Hierold W, Onasch I, Völker L, Sommer M (2017) The CarboZALF-D manipulation experiment–experimental design and SOC patterns. Int J Environ Agric Res (IJOEAR) 3(1):40–50. https://ijoear.com/Paper-January-2017/IJOEAR-JAN-2017-3.pdf, ISSN: 2454-1850

Dokuchaev VV (1883) Russian black earth (Russkii chernozem) St. Peterburg Imperatorskoe Vol'noe Ekonomicheskoe Obshchestvo 1883, 384 p (in Russian)

Doyle-Capitman CE, Decker DJ, Jacobson CA (2018) Toward a model for local stakeholder participation in landscape-level wildlife conservation. Hum Dimens Wildl 23(4):375–390. https://doi.org/10.1080/10871209.2018.1444215

Douley D, Audet P (2016) What part of mining are ecosystems? Defining success for the 'restoration' of highly disturbed Landscapes. In: Squires VR (ed) Ecological restoration. Global challenges, social aspects and environmental benefits. Nova Science Publishers, New York, pp 57–88

Dramstad W, Olson JD, Forman RT (1996) Landscape ecology principles in landscape architecture and land-use planning. Island press, Washington DC. 80 p. ISBN-13: 978-1559635141, ISBN-10: 1559635142

Driescher E (1974) Veränderungen an Gewässern Brandenburgs in historischer Zeit. In: Landesumweltamt Brandenburg, Studien und Tagungsberichte Nr. 47, Oct. 2003 (in German) https://mlul.brandenburg.de/media_fast/4055/q_bd47a.pdf. Accessed on 19 Dec 2018

Dronin NM, Bellinger E (2005) Climate dependence and food problems in Russia (1900–1990). CEU Press, New York, Budapest, p 360

Dronin N, Bychkova A (2018) Perceptions of American and Russian environmental scientists of today's key environmental issues: a comparative analysis. Environ Dev Sustain 20(5):2095–2105. https://doi.org/10.1007/s10668-017-9979-8

Dronin N, Kirilenko A (2011) Climate change, food stress, and security in Russia. Reg Environ Change 11(Suppl 1):167–178. https://doi.org/10.1007/s10113-010-0165-x

Dyer RJ, Nason JD, Garrick RC (2010) Landscape modelling of gene flow: improved power using conditional genetic distance derived from the topology of population networks. Mol Ecol 19:3746–3759. https://doi.org/10.1111/j.1365-294X.2010.04748.x

Eaton S, Ellis C, Genney D, Thompson R, Yahr R, Haydone DT (2018) Adding small species to the big picture: species distribution modelling in an age of landscape scale conservation. Biol Conserv 217:251–258. https://doi.org/10.1016/j.biocon.2017.11.012

ECLAS (2018) European council of landscape Architecture Schools. http://www.eclas.org/index.php/about. Accessed on 19 Dec 2018

Egoz S, Jørgensen K, Ruggeri D (eds) (2018) Defining landscape democracy. A path to spatial justice. Edward Elgar Publishing, Cheltenham, Northampton

Eisendle-Flöckner U, Hilberg S (2014) Hard rock aquifers and free-living nematodes—an interdisciplinary approach based on two widely neglected components in groundwater research. Ecohydrology 8:368–377. https://doi.org/10.1002/eco.1516

eLTER (2018) Long-Term Ecosystem Research in Europe. http://www.lter-europe.net/elter-esfri. Accessed on 19 Dec 2018

Emanuelsson U (2009) The rural landscapes of Europe. How man has shaped European nature. The Swedish Research Council. Formas, 383 p

Eulenstein F, Saparov A, Lukin S, Sheudshen AK, Mayer WH, Dannowski R, Tauschke M, Rukhovich OV, Lana M, Schindler R, Pachikin K, Drechsler H, Cremer N (2016) Assessing and controlling land use impacts on groundwater quality. In: Mueller L, Sheudshen A, Eulenstein F (eds) Novel methods for monitoring and managing land and water resources in Siberia. Springer Water. Springer, Cham, pp 635–666. https://link.springer.com/chapter/10.1007/978-3-319-24409-9_29

Fairclough G, Herlin IS, Swanwick C (2018) Routledge handbook of landscape character assessment: current approaches to characterisation and assessment, 1st edn. Routledge, 294 p

Falconer L, Hunter DC, Telfer TC, Ross LG (2013) Visual, seascape and landscape analysis to support coastal aquaculture site selection. Land Use Policy 34:1–10

Fan P, Chen J, Ouyang Z, Groisman P, Loboda T, Gutman G, Prishchepov AV, Kvashnina A, Messina J, Moore N, Qi J (2018) Urbanization and sustainability under transitional economies: a synthesis for Asian Russia. Environ Res Lett 13(9). IOP Publishing Ltd. http://iopscience.iop.org/article/10.1088/1748-9326/aadbf8/pdf

FAO (2017) Voluntary guidelines for sustainable soil management. Food and agriculture organization of the United Nations. Rome, Italy. http://www.fao.org/3/a-bl813e.pdf. Accessed on 19 Dec 2018

Federal Ministry of Food and Agriculture (BMEL) (2017) Organic Farming in Germany, as of: January 2017. http://www.bmel.de/SharedDocs/Downloads/EN/Agriculture/OrganicFarming/Organic-Farming-in-Germany.pdf;jsessionid=BB4C7A098D6FCFF68C51D48D2B00FC2E.1_cid385?__blob=publicationFile. Accessed on 19 Dec 2018

Figuié M (2018) Collective action in response to emerging zoonotic diseases . In: Serge Morand MF (eds) Emergence of infectious diseases. Risks and issues for societies, pp 75–97. http://agritrop.cirad.fr/587548/1/ID587548.pdf. Accessed on 19 Dec 2018

Foley JA, Prentice IC, Ramankutty N, Levis S, Pollard D, Sitch S, Haxeltine A (1996) An integrated biosphere model of land surface processes, terrestrial carbon balance, and vegetation dynamics. Global Biogeochem Cycles 10(4):603–628. https://doi.org/10.1029/96GB02692

Frielinghaus M (2004) Albrecht Daniel Thaer in Brandenburg und Berlin. Fördergesellschaft Albrecht Daniel Thaer, Agrarhistorischer und kulturhistorischer Reiseführer. Findling, Neuenhagen 2004, ISBN 3-933603-28-5

Frielinghaus M, Mueller L (1994) Ableitung von Zielen fuer die Landschaftsgestaltung eines Flussniederungsstandortes aus alten Kartenunterlagen. Archives of Agronomy and Soil Science Arch. Acker- Pflanzenbau Bodenkd, Berlin 38:115–126 (in German, Engl. summary)

Frielinghaus M, Mueller L (2003) Landnutzung im Flusspolder Oderbruch in den letzten 250 Jahren (in German). Wasser & Boden 55(6):4–8

Fry GLA (2001) Multifunctional landscapes—towards transdisciplinary research. Landsc Urban Plann 57(3):159–168. https://doi.org/10.1016/s0169-2046(01)00201-8

García-Martín M, Plieninger T, Bieling C (2018) Dimensions of landscape stewardship across Europe: landscape values, place attachment, awareness, and personal responsibility. Sustainability 10(1):263. https://doi.org/10.3390/su10010263

Gerasimova MI (2018) Updates of the Russian soil classification system (Chap. II/1). In: Sychev VG, Mueller L (eds) Novel methods and results of landscape research in Europe, Central Asia and Siberia. Monograph in 5 Volumes. Vol II Understanding and monitoring processes in soils and water bodies, pp 12–16. https://doi.org/10.25680/9874.2018.63.37.098, http://vniia-pr.ru/monografii/pdf/tom2-1.pdf. Accessed on 19 Dec 2018

Gerasimova MI, Bezuglova O (2019) Functional-Environmental and properties-oriented approaches in classifying urban soils (In Memoriam Marina Stroganova) In: Vasenev II et al (eds) Urbanization: challenge and opportunity for soil functions and ecosystem services, Proceedings of the 9th SUITMA Congress, January 2019, Springer series Geography, pp 4–10. https://doi.org/10.1007/978-3-319-89602-1_2

Getz WM, Marshall CR, Carlson CJ, Giuggioli L, Ryan SJ, Romañach SS, Boettiger C, Chamberlain SD, Larsen LL et al (2018) Making ecological models adequate. Ecol Lett 21 (2):153–166. https://doi.org/10.1111/ele.12893

Giani L, Makowsky L, Mueller K (2014) Plaggic Anthrosol: Soil of the Year 2013 in Germany: An overview on its formation, distribution, classification, soil function and threats. J Plant Nutr Soil Sci 177(3):320–329. https://doi.org/10.1002/jpln.201300197

Glante F, Marx M, Römbke J (2018) Soil monitoring in Germany (Chap. II/18). In: Sychev VG, Mueller L (eds) Novel methods and results of landscape research in Europe, Central Asia and Siberia. Monograph in 5 Volumes.Vol II Understanding and Monitoring Processes in Soils and Water Bodies, pp 89–94. https://doi.org/10.25680/7493.2018.26.90.115, http://vniia-pr.ru/monografii/pdf/tom2-18.pdf. Accessed on 19 Dec 2018

Glinka KD (1927) The great soil groups of the world and their development (Translated from the German by CF Marbut.) Edwards Brothers, Ann Arbor, MI, 235 p

Goldstein JL, Rivers D, Tomz M (2007) Institutions in international relations: understanding the effects of the GATT and the WTO on world trade. Int Org 61(1):37–67

Gontijo LM (2018) Engineering natural enemy shelters to enhance conservation biological control in field crops. Biol Control. Available online 28 Oct 2018. https://doi.org/10.1016/j.biocontrol.2018.10.014

Granö JG (1919-21) Altai. Vaellusvuosina nähtyä ja elettyä, 2 Vols. WSOY, Porvoo

Granö JG (1929) Reine Geographie. Eine methodologische Studie beleuchtet mit Beispielen aus Finnland und Estland. Acta Geographica 2:2

Granö O (1979) Johannes Gabriel Granö 1882–1956. Geographers. Biobibliogr Stud 3:73–84

Granö JG (1997) Pure geography. In: Granö O, Paasi A (eds) The John Hopkins University Press, Baltimore and London

Grunewald K, Bastian O (eds) (2013) Ökosystemdienstleistungen – Konzept, Methoden und Fallbeispiele. Springer-Spektrum, Heidelberg, 332 p. https://www.springer.com/de/book/9783827429865

Grunwald L-C, Belyaev VI, Hamann M, Illiger P, Stephan E, Bischoff N, Rudev NV, Kozhanov NA, Schmidt G, Frühauf M, Meinel T (2016) Modern cropping systems and technologies for soil conservation in Siberian agriculture. In: Mueller L, Sheudshen A, Eulenstein F (eds) Novel methods for monitoring and managing land and water resources in Siberia. Springer Water. Springer, Cham, pp 681–715. https://doi.org/10.1007/978-3-319-24409-9_31

Gubler A, Wächter D, Schwab P, Hug A, Meuli R, Keller A (2018) Long-term observation of soils within the Swiss soil monitoring network NABO (Chap. II/19). In: Sychev VG, Mueller L (eds) Novel methods and results of landscape research in Europe, Central Asia and Siberia. Monograph in 5 VolumesVol II Understanding and Monitoring Processes in Soils and Water Bodies, pp 94–99. https://doi.org/10.25680/4039.2018.68.24.116, http://vniia-pr.ru/monografii/pdf/tom2-19.pdf. Accessed on 19 Dec 2018

Gutzler C, Helming K, Balla D, Dannowski R, Deumlich D, Glemnitz M, Knierim A, Mirschel W, Nendel C, Paul C, Sieber S, Stachow U, Starick A, Wieland R, Wurbs A, Zander P (2015) Agricultural land use changes—a scenario-based sustainability impact assessment for Brandenburg, Germany. Ecol Indic 48(2015):505–517

Haase D, Schwarz N (2009) Simulation models on human–nature interactions in urban landscapes: a review including spatial economics, system dynamics, cellular automata and agent-based approaches. Living Rev Landsc Res 3:2. http://www.livingreviews.org/lrlr-2009-3. Accessed on 19 Dec 2018

Hägerstrand T (1993) Samhälle och natur. In: Region och miljö – ekologiska perspektiv på den rumsliga närings- och bosättningsstrykturen. NordREFO 1:14–59

Hanson CA, Fuhrman JA, Horner-Devine MC, Martiny JBH (2012) Beyond biogeographic patterns: processes shaping the microbial landscape. Nat Rev Microbiol 10:497–506. https://doi.org/10.1038/nrmicro2795

Hardt E, de Pablo CL, de Agar PM, Dos Santos RF, Pereira-Silva EF (2018) GIS-based detection and quantification of patch-boundary patterns for identifying landscape mosaics. Appl Ecol Environ Res 16(2):1381–1398. ISSN 1785 0037. http://dx.doi.org/10.15666/aeer/1602_13811398. Accessed on 19 Dec 2018

Haubold-Rosar M (2018) Methods of agricultural recultivation of post-mining landscapes in Northeastern Germany (Chap. V/38). In: Sychev, VG, Mueller, L (eds) Novel methods and results of landscape research in Europe, Central Asia and Siberia, Vol V Landscape planning, management and rehabilitation, pp 178–183. http://vniia-pr.ru/monografii/pdf/tom5-38.pdf. Accessed on 19 Dec 2018

Helming K (2014) Impact assessment for multifunctional land use. In: Novel measurement and assessment tools for monitoring and management of land and water resources in agricultural landscapes of Central Asia. Springer Environment, pp 223–234. https://link.springer.com/chapter/10.1007/978-3-319-01017-5_12

Hertel C, von Unold G (2014) Third-generation lysimeters: scientific engineered monitoring systems. In: Mueller L, Saparov A, Lischeid G (eds) Novel measurement and assessment tools for monitoring and management of land and water resources in agricultural landscapes of Central Asia. Environmental science and engineering. Springer, Cham, pp 175–184. https://doi.org/10.1007/978-3-319-01017-5_9

Hettner A (1905) Das Wegen und Methoden der Geographie. Geographische Zeitschrift H 11:615–629

Hettner A (1927) Die Geographie, ihre Geschichte, ihr Wesen und ihre Methoden. Hirt Verlag Breslau, 463 p

Hey T, Tansley S, Tolle K (2009) The fourth paradigm: data intensive scientific discovery, Published by Microsoft Research, October 2009 ISBN: 978-0-9825442-0-4

Hilberg S, Eisendle-Flöckner U (2016) About faunal live in Austrian Aquifers—historical background and current developments. Austrian J Earth Sci AJES special issues Hydrogeology 109/1:119–134. http://dx.doi.org/10.17738/ajes.2016.0009. Accessed on 19 Dec 2018

Hoffmann J, Kratz W (2018) Nationale Indikatoren mit Bezug zu Pflanzenschutz und Biodiversität in der Agrarlandschaft. Natur Landschaft 93(6):273–279

Hoffmann C, Golozubov O, Alyabina I, Heinrich U (2018) Standards for the collection, management and provision of soil research data: two national case studies (Chap. I/12). In: Sychev VG, Mueller L (eds) Novel methods and results of landscape research in Europe, Central Asia and Siberia. Monograph in 5 Volumes. Vol I Landscapes in the 21th Century: Status Analyses, Basic Processes and Research Concepts. © «FSBI "VNII Agrochemistry» 2018, pp 85–89. https://doi.org/10.25680/6422.2018.85.59.012, http://vniia-pr.ru/monografii/pdf/tom1-12.pdf. Accessed on 19 Dec 2018

Hoffmann J, Wittchen U, Berger G, Stachow U (2018) Moving window growth-A method to characterize the dynamic growth of crops in the context of bird abundance dynamics with the example of Skylark (*Alauda arvensis*). Ecol Evol 8(17): 8880–8893. https://doi.org/10.1002/ece3.4398, https://onlinelibrary.wiley.com/doi/full/10.1002/ece3.4398

Hoffmann M, Wirth S, Beßler H, Engels C, Jochheim H, Sommer M, Augustin J (2018c) Combining a root exclusion technique with continuous chamber and porous tube measurements for a pin-point separation of ecosystem respiration in croplands. J Plant Nutr Soil Sci 181(1):41–50. https://doi.org/10.1002/jpln.201600489

Holden J, Grayson RP, Berdeni D, Bird S, Chapman PJ, Edmondson JL, Firbank LG, Helgason T, Hodson ME, Hunt SFP, Jones DT, Lappage MG, Marshall-Harries E, Nelson M, Prendergast-Miller M, Shaw H, Wade RN, Leake JR (2019) The role of hedgerows in soil functioning within agricultural landscapes. Agric Ecosyst Environ 273:1–12. https://doi.org/10.1016/j.agee.2018.11.027

Holzschuh A, Dainese M, González-Varo JP, Mudri-Stojnić S, Riedinger V, Rundlöf M, Scheper J, Wickens JB, Wickens VJ, Bommarco R, Kleijn D, Potts SG, Roberts SPM, Smith HG, Vilà M, Vujić A, Steffan-Dewenter I (2016) Mass-flowering crops dilute pollinator abundance in agricultural landscapes across Europe. Ecol Lett 19:1228–1236. https://doi.org/10.1111/ele.12657

Hubatsch W (1989) Die Stein-Hardenbergschen Reformen. Darmstadt: Wissenschaftliche Buchgesellschaft, 1989 ISBN 3-534-05357-5

IALE (2018) International Association for Landscape Ecology. http://www.landscape-ecology.org/home.html. Accessed on 19 Dec 2018

ICOMOS (2018) International Council on Monuments and Sites. https://www.icomos.org/fr/. Accessed on 19 Dec 2018

IFLA (2018) International Federation of Landscape Architects. http://iflaonline.org/. Accessed on 19 Dec 2018

IGU (2018) International Geographical Union. https://igu-online.org/. Accessed on 19 Dec 2018

Iles L (2016) The role of metallurgy in transforming global forests. J Archaeol Method Theory 23 (4):1219–1241. ISSN 1072-5369. https://doi.org/10.1007/s10816-015-9266-7

ISCO (2018) International Soil Conservation Organization. https://www.tucson.ars.ag.gov/isco/. Accessed on 19 Dec 2018

ISO 14040:2006 (2006) Environmental management—life cycle assessment—principles and framework. International Organisation for Standardisation (ISO), Geneve. https://www.iso.org/standard/37456.html/. Accessed on 19 Dec 2018

ISTRO (2018) International Soil and Tillage Research Organization. https://www.istro.org/. Accessed on 19 Dec 2018

IUCN (2018) International Union for the Conservation of Nature. https://www.iucn.org/. Accessed on 19 Dec 2018

IUSS (2018) International Union of Soil Sciences. https://www.iuss.org/. Accessed on 19 Dec 2018

IUSS Working Group WRB (2015) World Reference Base for Soil Resources 2014, update 2015. International soil classification system for naming soils and creating legends for soil maps. World Soil Resources Reports No. 106. FAO, Rome, 192 p. http://www.fao.org/3/i3794en/I3794en.pdf. Accessed on 19 Dec 2018

Jackson JB (1986) The vernacular landscpe. In: Penning-Rowsell EC, Lowenthal D (eds) Landscape meanings and values. Allen and Unwin, London, pp 65–81

Jellicoe G, Jellicoe S (1987) The landscape of man: shaping the environment from prehistory to the present day. Thames & Hudson Ltd, 2nd revised edition, 400 p. ISBN 9780500274316

Jiang C, Liu J, Zhang H, Zhang Z, Wang D (2019a) China's progress towards sustainable land degradation control: insights from the northwest arid regions. Ecol Eng 127:75–87. https://doi.org/10.1016/j.ecoleng.2018.11.014

Jiang L, Bao A, Jiapaer G, Guo H, Zheng G, Gafforov K, Kurban A, De Maeyer P (2019b) Monitoring land sensitivity to desertification in Central Asia: convergence or divergence? Sci Total Environ 658:669–683. https://doi.org/10.1016/j.scitotenv.2018.12.152

Jones M (1993) Landscape as a resource and the problem of landscape values. In: Rusten C, Wøien H (eds) The politics of environmental conservation. Proceedings from a Workshop in Trondheim March 26, 1993. Report no. 6/93. University of Trondheim, Centre for Environment and Development, Trondheim, pp 19–33

Jones M (2003a) The concept of cultural landscape: discourse and narratives. In: Palang H, Fry G (eds) Landscape interfaces. Cultural heritage in changing landscapes. Landscape Series Vol 1. Kluwer Academic Publishers, Dordrecht/Boston/London, pp 21–51. https://doi.org/10.1007/978-94-017-0189-1

Jones M (2003b) Human geographical landscapes. J. G. Granö's approach to landscape as scientist and artist. In: Granö O (ed) Origin of landscape science. J.G. Granö and a New Pure Geography for a New State. A Collection of Papers. Turun yliopiston maantietieteen laitoksen julkaisuja – Publicationes Instituti Geographici Universitatis Turkuensis 167: 71–98

Jones M (ed) (2006) Essays on landscape law and justice. Norsk Geografisk Tidsskrift–Norwegian. J Geogr 60(1):1–127
Jones M (2009) Analysing landscape values expressed in planning conflicts over change in the landscape. In: Van Eetvelde V, Sevenant M, Van De Velte L (eds) Re-Marc-able Landscapes. Marc-ante Landschappen. Liber Amicorum Marc Antrop. Academia Press, Gent, pp 193–205
Jones M, Daugstad K (1997) Usages of the "cultural landscape" concept in Norwegian and Nordic landscape administration. Landsc Res 22(3):267–281. https://doi.org/10.1080/01426399708706515
Jones M, Olwig KR (eds) (2008) Nordic Landscapes. Region and Belonging on the Northern Edge of Europe. University of Minnesota Press, Minneapolis 628 p. ISBN: 978-0-8166-3915-1
Jones M, Stenseke M (eds) (2011) The European landscape convention. Challenges of participation. Landscape Series Vol 13. Springer, Dordrecht. https://doi.org/10.1007/978-90-481-9932-7
Joniak T (2018) Concentration of nutrients and N:P ratio in surface sediments of small water bodies of agricultural and forest areas. In: SGEM 2018 GeoConference Proceedings 18, vol 51, pp 213–220
Joniak T, Kuczyńska-Kippen N, Gąbka M (2017) Effect of agricultural landscape characteristics on the hydrobiota structure in small water bodies. Hydrobiologia 793(1):121–133. https://doi.org/10.1007/s10750-016-2913-5
Jordan WR, Gilpin ME, Aber JD (eds) (1987) Restoration ecology: a synthetic approach to ecological research. Cambridge University Press, Cambridge. 347 p. ISBN 0521331102
Jørgensen K, Clemetsen M, Thorén KH, Richardson T (eds) (2016) Mainstreaming landscape through the European landscape convention. Routledge, Abingdon and New York
Journal Landscape Research (2018) http://www.tandfonline.com/loi/clar20. Accessed on 19 Dec 2018
Kappler C, Kaiser K, Tanski P, Klos F, Fülling A, Mrotzek A, Sommer M, Bens O (2018) Stratigraphy and age of colluvial deposits indicating Late Holocene soil erosion in northeastern Germany. CATENA 170:224–245. https://doi.org/10.1016/j.catena.2018.06.010
Karampiperis P, Lokers R, Neveu P, Hologne O, Kakaletris G, Candela L, Filter M, Manouselis N, Stavrakaki M, Zervas P (2019) Big data in agricultural and food research: challenges and opportunities of an integrated big data E-infrastructure. In: Emrouznejad A, Charles V (eds) Big data for the greater good. Studies in big data. Springer, Cham, Vol 42, pp 129–150. https://doi.org/10.1007/978-3-319-93061-9_6
Kavanaugh MT, Oliver MJ, Chavez FP, Letelier RM, Muller-Karger FE, Doney SC (2016) Seascapes as a new vernacular for pelagic ocean monitoring, management and conservation. ICES J Mar Sci 73(7):1839–1850. https://doi.org/10.1093/icesjms/fsw086
Kazakov LK, Chizhova VP (2001) Engineering geography. Moscow, Landros, 268 p. (in Russian, Казаков Л.К., Чижова В.П. Инженерная география.- М.: Лэндрос, 2001.– 268с)
Kelly C (2007) The roman empire: a very short introduction. Oxford University Press, 168 p. ISBN 0192803913
Kern J, Giani L, Teixeira W, Lanza G, Glaser B (2019) What can we learn from ancient fertile anthropic soil (Amazonian Dark Earths, shell mounds, Plaggen soil) for soil carbon sequestration? CATENA 172:104–112. https://doi.org/10.1016/j.catena.2018.08.008
Kersebaum KC (2007) Modelling nitrogen dynamics in soil-crop systems with HERMES. In: Kersebaum KC, Hecker J-M, Mirschel W, Wegehenkel M (eds) Modelling water and nutrient dynamics in soil-crop systems: proceedings of the workshop on "Modelling water and nutrient dynamics in soil-crop systems" held on 14–16 June 2004 in Müncheberg, Germany, 2007, pp 59–73. Springer, Dordrecht
Kersebaum KC, Hecker J-M, Mirschel W, Wegehenkel M (eds) (2007) Modelling water and nutrient dynamics in soil-crop systems: proceedings of the workshop on "Modelling water and nutrient dynamics in soil-crop systems" held on 14–16 June 2004 in Müncheberg, Germany, 2007, 271 p. Springer, Dordrecht

Kersebaum KC, Boote KJ, Jorgenson JS, Nendel C, Bindi M, Frühauf C, Gaiser T, Hoogenboom G, Kollas C, Olesen JE, Rötter RP, Ruget F, Thorburn PJ, Trnka M, Wegehenkel M (2015) Analysis and classification of data sets for calibration and validation of agro-ecosystem models. Env Model Software 72:402–417. https://doi.org/10.1016/j.envsoft.2015.05.009

Kersebaum KC, Kroes J, Gobin A, Takáč J, Hlavinka P, Trnka M, Ventrella D, Giglio L, Ferrise R, Moriondo M, Dalla Marta A, Luo Q, Eitzinger J, Mirschel W, Weigel H-J, Manderscheid R, Hoffmann M, Nejedlik P, Hösch J (2016) Assessing the uncertainty of model based water footprint estimation using an ensemble of crop growth models on winter wheat. Water 8:571. https://doi.org/10.3390/w8120571

Khoroshev AV (2016) Modern directions of structural landscape science. Proceedings of the Russian Academy of Sciences. Geographical Series, Science Publishing Moscow № 3, pp 7–15 (in Russian, Хорошев А.В. 2016 Современные направления структурного ландшафтоведения. Известия Российской академии наук. Серия географическая, издательство Наука (М.), № 3, с. 7–15)

Kienast F, Frick J, van Strien MJ, Hunziker M (2015) The Swiss landscape monitoring program — a comprehensive indicator set to measure landscape change. Ecol Modell 295:136–150. https://doi.org/10.1016/j.ecolmodel.2014.08.008

Kiryushin VI (1995) Methodology for the development of adaptive-landscape farming systems and technologies for the cultivation of agricultural crops, Moscow, Timiryasev Academy 1995, 81 p (in Russian) (Кирюшин В.И. Методика разработки адаптивно-ландшафтных систем земледелия и технологий возделывания сельскохозяйственных культур.-М. МСХА им.К.А.Тимирязева, 1995. - 81с.)

Knoche D (2018) Forest reclamation in post-mining landscapes—The Lusatian Lignite District, Eastern Germany (Chap. V/39). In: Sychev VG, Mueller L (eds) Novel methods and results of landscape research in Europe, Central Asia and Siberia, Vol. V Landscape planning, management and rehabilitation, pp 183–188. http://vniia-pr.ru/monografii/pdf/tom5-39.pdf. Accessed on 19 Dec 2018

Knoche D, Schlenstedt J (2018) Forest reclamation in the Lusatian Lignite District - A wounded landscape heading for new horizons. In: Proceedings of the 12th international conference on mine closure, 03.-07.09.2018, Leipzig, pp 677–688

Koch J, Dorning MA, Van Berkel DB, Beck SM, Sanchez GM, Shashidharan A, Smart LS, Zhang Q, Smith JW, Meentemeyer RK (2019) Modeling landowner interactions and development patterns at the urban fringe. Landsc and Urban Plann 182:101–113. https://doi.org/10.1016/j.landurbplan.2018.09.023

Köhl M, Lasco R, Cifuentes M, Jonsson Ö, Korhonen KT, Mundhenk P, Navar Jde J, Stinson G (2018) Changes in forest production, biomass and carbon: results from the 2015 UN FAO Global forest resource assessment. For Ecol Manage 352(2015):21–34. https://doi.org/10.1016/j.foreco.2015.05.036

Köppke U, Schnug E (2017) Biolandwirtschaft: Angewandter Hochwasserschutz. Ökologie & Landbau 34–35

Körschens M, Albert E, Baumecker M, Ellmer F, Grunert M, Hoffmann S, Kismanyoky T, Kubat J, Kunzova E, Marx M, Rogasik J, Rinklebe J, Rühlmann J, Schilli C, Schröter H, Schroetter S, Schweizer K, Toth Z, Zimmer J, Zorn W (2014) Humus und Klimaänderung - Ergebnisse aus 15 langjährigen Dauerfeldversuchen. Arch Agron Soil Sci 60(11):1485–1517

Kovács-Hostyánszki A, Földesi R, Báldi A, Endrédi A, Jordán F (2019) The vulnerability of plant-pollinator communities to honeybee decline: a comparative network analysis in different habitat types. Ecol Indic 97:35–50. https://doi.org/10.1016/j.ecolind.2018.09.047

Kowarik I, Körner S (eds) (2005) Wild urban woodlands. New perspectives for urban forestry. Springer, Berlin/Heidelberg. https://www.springer.com/de/book/9783540239123

Kowarik I, Fischer LK, Honold J (2016) Beeinflusst Artenvielfalt die Wertschätzung der Stadtnatur? In: Kowarik I, Bartz R, Brenck M (eds) Naturkapital Deutschland – TEEB DE (2016): Ökosystemleistungen in der Stadt – Gesundheit schützen und Lebensqualität erhöhen. Technische Universität Berlin, Helmholtz-Zentrum für Umweltforschung – UFZ. Berlin,

Leipzig, 300 p. https://www.agrar.hu-berlin.de/de/institut/departments/daoe/bk/forschung/klimagaerten/weiterfuehrende-materialien-1/2016_teeb-oekosystemleistungen-in-der-stadt.pdf. Accessed on 19 Dec 2018

Kravchenko VV, Wende W, Ignatov AV et al (2008) Assessment of environmental impacts and ecological expertise: professional experience of EIA issues in Russia and Germany (in Russian and German) (Кравченко, В. В., Венде, В., Игнатов, А. В. и др. Оценка воздействия на окружающую среду, и экологическая экспертиза / Изд-во Института географии СО РАН: Иркутск - Берлин – Бонн, 2008. – 199 с.) http://www.irigs.irk.ru/files/EE%20und%20OVOS%20russian%20version.pdf. Accessed on 19 Dec 2018

Krebs N (1922) Natur und Kulturlandschaft. Zeitschrift der Gesellschaft für Erdkunde zu Berlin 3–4:81–94

Kröger M (2019) The global land rush and the arctic. In: Finger M, Heininen L (eds) The global arctic handbook. Springer, Cham, pp 27–43. https://doi.org/10.1007/978-3-319-91995-9_3

Kulunda Portal (2018) How to prevent the next "Global Dust Bowl"? Ecological and Economic Strategies for Sustainable Land Management in the Russian Steppes: A Potential Solution to Climate Change. http://www.kulunda.eu/?q=en/home. Accessed on 19 Dec 2018

Küster HJ (1997) Geschichte der Landschaft in Mitteleuropa. C. H. Beck, München 1997, ISBN 3-406-45357-0

Lal R (2008) Soils and sustainable agriculture. A review. Agron Sustain Dev 28:57–64. https://doi.org/10.1051/agro:2007025

Lal R, Stewart BA (2011) World soil resources and food security. CRC Press, Boca Raton, 574 p

Landauer M, Juhola S (2019) Loss and damage in the rapidly changing Arctic. In: Mechler R, Bouwer L, Schinko T, Surminski S, Linnerooth-Bayer J (eds) Loss and damage from climate change. Climate risk management, policy and governance. Springer, Cham. https://doi.org/10.1007/978-3-319-72026-5_18

Lewis SL, Maslin MA (2015) Defining the anthropocene. Nature 519:171–180. https://doi.org/10.1038/nature14258

Lieske B (2011) Selbstverständnis des russischen Naturschutzes und die Entwicklung der Sapowedniks – eine Interviewstudie. Waldökologie, Landschaftsforschung und Naturschutz Heft 11 (2011) pp 63–79. http://www.afsv.de/download/literatur/waldoekologie-online/waldoekologie-online_heft-11-2.pdf. Accessed on 19 Dec 2018

Lin H (2012) Hydropedology. Synergistic integration of soil science and hydrology, 1st edn. Academic Press, Cambridge, 858 p

Lipski S, Storozhenko O (2019) Russia's circumpolar territories inhabited by indigenous peoples: economic and legal aspects of development, pp 470–494. https://doi.org/10.4018/978-1-5225-6954-1.ch022. In: Erokhin V, Gao T, Zhang X (eds) Handbook of research on international collaboration, economic development, and sustainability in the Arctic. 703 p. https://doi.org/10.4018/978-1-5225-6954-1

Lipský Z (1995) The changing face of the Czech rural landscape. Landsc Urban Plann 31(1):39–45

Lischeid G, Kalettka T, Holländer M, Steidl J, Merz C, Dannowski R, Hohenbrink T, Lehr C, Onandia G, Reverey F, Pätzig M (2018) Natural ponds in an agricultural landscape: external drivers, internal processes, and the role of the terrestrial-aquatic interface. Limnologica 68:5–16. https://doi.org/10.1016/j.limno.2017.01.003

Lohrberg F, Lička L, Scazzosi L, Timpe A (eds) (2015) Urban agriculture Europe. Jovis. 232 p. ISBN 978-3-86859-371-6

Low SA, Adalja A, Beaulieu E, Key N, Martinez S, Melton A, Perez A, Ralston K, Stewart H, Suttles S, Vogel S, Jablonski BBR (2015) Trends in U.S. Local and Regional Food Systems. AP-068, U.S. Department of Agriculture, Economic Research Service, January 2015. 86 p. https://scholarship.sha.cornell.edu/cgi/viewcontent.cgi?referer=https://scholar.google.de/&httpsredir=1&article=2062&context=articles. Accessed on 19 Dec 2018

Lowenthal D (1968) The American scene. Geogr Rev 58:61–88

Lowenthal D (1985) The past is a foreign country. Cambridge University Press, Cambridge

Lowenthal D (1996) Possessed by the past. The heritage crusade and the spoils of history. The Free Press, New York

Lowenthal D, Prince H (1965) English landscape tastes. Geogr Rev 54:309–346

Maldonado AD, Aguilera PA, Salmerón A, Nicholson AE (2018) Probabilistic modeling of the relationship between socioeconomy and ecosystem services in cultural landscapes. Ecosyst Serv 33:146–164. https://doi.org/10.1016/j.ecoser.2018.04.007

Mancini MS, Galli A, Coscieme L, Niccolucci V, Lin D, Pulselli FM, Bastianoni S, Marchettini N (2018) Exploring ecosystem services assessment through ecological footprint accounting. Ecosyst Serv 30:228–235. https://doi.org/10.1016/j.ecoser.2018.01.010

Manel S, Schwartz MK, Luikart G, Taberlet P (2003) Landscape genetics: combining landscape ecology and population genetics. Trends Ecol Evol 18:189–197. https://doi.org/10.1016/S0169-5347(03)00008-9

Marsh GP (1864) Man and nature. Or, physical geography as modified by human action. Charles Scribner & Co., New York

Martellozzo F, Amato F, Murgante B, Clarke KC (2018) Modelling the impact of urban growth on agriculture and natural land in Italy to 2030. Appl Geogr 91:156–167. https://doi.org/10.1016/j.apgeog.2017.12.004

Martin P (2008) Epidemics: lessons from the past and current patterns of response. Comptes Rendus Geosci 340(9–10):670–678. https://doi.org/10.1016/j.crte.2007.12.005

Masyagina OV, Evgrafova SYu, Bugaenko TN, Kholodilova VV, Krivobokov LV, Korets MA, Wagner D (2019) Permafrost landslides promote soil CO_2 emission and hinder C accumulation. Sci Total Environ 657:351–364. https://doi.org/10.1016/j.scitotenv.2018.11.468

Mathieson I et al (2015) Eight thousand years of natural selection in Europe. https://doi.org/10.1101/016477, https://www.biorxiv.org/content/10.1101/016477v2. Accessed on 2 Feb 2019

McKenzie BM, Stobart R, Brown JL, George TS, Morris N, Newton AC, Valentine TA, Hallett PD (2017) Platforms to test and demonstrate sustainable soil management: integration of major UK field experiments, AHDB Final Report RD-2012-3786, 178 p

McNeill WH (1987) The eccentricity of wheels, or eurasian transportation in historical perspective. Am Hist Rev 92:1111–1126

McPartland MY, Kane ES, Falkowski MJ, Kolka R, Turetsky MR, Palik B, Montgomery RA (2019) The response of boreal peatland community composition and NDVI to hydrologic change, warming, and elevated carbon dioxide. Global Change Biol 25(1):93–107. https://doi.org/10.1111/gcb.14465

Meemken E-M, Qaim M (2018) Organic agriculture, food security, and the environment. Annual review of resource economics. Annu Rev Resour Econ 10:4.1–4.25 https://doi.org/10.1146/annurev-resource-100517-023252. Accessed on 19 Dec 2018

Meinel T, Grunwald LC, Akshalov K (2014) Modern technologies for soil management and conservation in Northern Kazakhstan. In: Mueller L, Saparov A, Lischeid G (eds) Novel measurement and assessment tools for monitoring and management of land and water resources in agricultural landscapes of Central Asia. Environmental science and engineering. Springer, Cham, pp 455–464. https://doi.org/10.1007/978-3-319-01017-5_27

Meissner R, Rupp H, Seyfarth M (2014) Advanced technologies in lysimetry. In: Mueller L, Saparov A, Lischeid G (eds) Novel measurement and assessment tools for monitoring and management of land and water resources in agricultural landscapes of Central Asia. Environmental science and engineering. Springer, Cham, pp 159–173. https://doi.org/10.1007/978-3-319-01017-5_8

Meller H, Michel K (2018) Die Himmelsscheibe von Nebra. Der Schlüssel zu einer untergegangenen Kultur im Herzen Europas, Propyläen Verlag, 384 p, ISBN-13 9783549076460

Mendel B, Schwemmer P, Peschko V, Müller S, Schwemmer H, Mercker M, Garthe S (2019) Operational offshore wind farms and associated ship traffic cause profound changes in distribution patterns of Loons (Gavia spp.). J Environ Manage 231:429–438. https://doi.org/10.1016/j.jenvman.2018.10.053

Mengel P (1930) Das Oderbruch, vol 1. Verlagsgesellschaft R. Müller m.b.H, Eberswalde, p 1930

Mengel P (1934) Das Oderbruch, vol 2. Verlagsgesellschaft R. Müller m.b.H, Eberswalde, p 1934

Meynen E, Schmithüsen J (eds) (1953-1962) Handbuch der naturräumlichen Gliederung Deutschlands. Bundesanstalt für Landeskunde, 8 Vol

Micallef A, Rangel-Buitrago N (2019) The management of coastal landscapes. In: Rangel-Buitrago N (eds) Coastal scenery. Coastal research library. Springer, Cham, vol 26, pp 211–247. https://doi.org/10.1007/978-3-319-78878-4_7

Miklos L, Kočická E, Izakovicova Z, Esprit D, Špinerová A, Diviaková A, Miklósová V (2019) Landscape as a geosystem. Springer International Publishing, 161 p. https://www.springer.com/us/book/9783319940236

Millennium Ecosystem Assessment (2005) Ecosystems and human well-being: synthesis. Island Press, Washington, DC. 137 p. https://www.millenniumassessment.org/documents/document.356.aspx.pdf. Accessed on 19 Dec 2018

Mirschel W, Wieland R, Wenkel K-O, Nendel C, Guddat C (2014) YIELDSTAT - a spatial yield model for agricultural crops. Eur J Agron 52(2014):33–46. https://doi.org/10.1016/j.eja.2013.09.015

Mirschel W, Wenkel K-O, Berg M, Wieland R, Nendel C, Köstner B, Topazh AG, Terleev VV, Badenko VL (2016a) A spatial model-based decision support system for evaluating agricultural landscapes under the aspect of climate change. In: Müller L, Sheudshen AK, Eulenstein F (eds) Novel methods for monitoring and managing land and water resources in Siberia. Springer International Publishing, Cham, pp 519–540. https://link.springer.com/chapter/10.1007/978-3-319-24409-9_23

Mirschel W, Barkusky D, Hufnagel J, Kersebaum K-C, Nendel C, Laacke L, Luzi K, Rosner G (2016b) Coherent multi-variable field data set of an intensive cropping system for agro-ecosystem modelling from Müncheberg, Germany. Open Data J Agric Res 2(1):1–10. https://odjar.org/article/view/15412

Mirtl, M (2010) Introducing the next generation of ecosystem research in Europe: LTER-Europe's Multi-Functional and Multi-Scale Approach. In: Müller F, Baessler C, Schubert H, Klotz S (eds) Long-term ecological research: between theory and application, pp 75–93. https://doi.org/10.1007/978-90-481-8782-9_6

Mirtl M, Borer E, Burns E, Djukic I, Forsius M, Haubold H, Hugo W, Jourdan J, Lindenmayer D, McDowell WH, Muraoka H, Orenstein D, Pauw J, Peterseil J, Shibata H, Wohner C, Yu X, Haase P (2018) Genesis, goals and achievements of long-term ecological research at the global scale: a critical review of ILTER and future implications. Sci Total Environ 626:1439–1462. https://doi.org/10.1016/j.scitotenv.2017.12.001

Mitchell D (1996) The lie of the land. Migrant workers and the California landscape. University of Minnesota Press, Minneapolis

Mitchell D (2003) The right to the city. Social justice and the fight for public space. Guilford Press, New York

Mitchell N, Rössler M, Tricaud P-M (eds) (2009) World heritage cultural landscapes. A handbook for conservation and management. Published by the UNESCO World Heritage Centre. http://whc.unesco.org/documents/publi_wh_papers_26_en.pdf. Accessed on 19 Dec 2018

Mitscherlich EA (1931) Der Boden als Vegetationsfaktor (pflanzenphysiologische Bodenkunde). In: Handbuch der Bodenlehre. Hrs. E. Blanck, Springer Verlag Berlin 1931, Bd. 9, S. 497–541

Moon DG, Landa E (2017) The centenary of the journal soil science: reflections on the discipline in the United States and Russia around a hundred years ago. Soil Sci 182(6):203–215

Morrison R, Barker A, Handley J (2018) Systems, habitats or places: evaluating the potential role of landscape character assessment in operationalising the ecosystem approach. Landsc Res 43(7):1000–1012. https://doi.org/10.1080/01426397.2017.1415314

Mueller L, Tille P, Hartleb R, Homer A (1988) Preparation and performance of drainage constructions in the Oderbruch region (Vorbereitung und Durchführung von Entwässerungsmaßnahmen im Oderbruch). (in German) Melioration und Landwirtschaftsbau, Berlin, 22(7):318–321

Mueller L, Schindler U, Mirschel W, Shepherd TG, Ball B, Helming K, Rogasik J, Eulenstein F, Wiggering H (2010) Assessing the productivity function of soils: a review. Agron Sustain Dev 30(3):601–614. https://doi.org/10.1051/agro/2009057

Mueller L, Schindler U, Ball BC, Smolentseva E, Sychev VG, Shepherd TG, Qadir M, Helming K, Behrendt A, Eulenstein F (2014a) Productivity potentials of the global land resource for cropping and grazing. In: Mueller L, Saparov A, Lischeid G (eds) Novel measurement and assessment tools for monitoring and management of land and water resources in agricultural landscapes of Central Asia. Springer International Publishing, Cham, pp 115–142. Environmental science and engineering. https://doi.org/10.1007/978-3-319-01017-5_6

Mueller L, Behrendt A, Shepherd TG, Schindler U, Ball BC, Khudyaev S, Kaiser T, Dannowski R, Eulenstein F (2014b) Simple field methods for measurement and evaluation of grassland quality. In: Mueller L, Saparov A, Lischeid G (eds) Novel measurement and assessment tools for monitoring and management of land and water resources in agricultural landscapes of Central Asia. Springer International Publishing, Cham, pp 199–222, Environmental Science and Engineering. https://doi.org/10.1007/978-3-319-01017-5_11

Mueller L, Schindler U, Hennings V, Smolentseva EN, Rukhovich OV, Romanenkov VA, Sychev VG, Lukin S, Sheudshen AK, Onishenko L, Saparov A, Pachikin K, Behrendt A, Mirschel W, Eulenstein F (2016a) An emerging method of rating global soil quality and productivity potentials. In: Mueller L, Sheudshen AK, Eulenstein F (eds) Novel methods for monitoring and managing land and water resources in Siberia, Springer Water. Springer International Publishing, Cham, pp 573–595. http://link.springer.com/chapter/10.1007%2F978-3-319-24409-9_26

Mueller L, Sheudshen AK, Sychev VG, Romanenkov VA, Dannowski R, Eulenstein F (2016b) Potential of applying novel monitoring and management methods to Siberian landscapes. In: Mueller L, Sheudshen AK, Eulenstein F (eds) Novel methods for monitoring and managing land and water resources in Siberia, Springer Water. Springer International Publishing, Cham, pp 719–760. http://link.springer.com/chapter/10.1007/978-3-319-24409-9_32

Mueller L, Eulenstein F, McKenzie BM, Schindler U, Mirschel W (2018) Tillage depth and crop yields: re-evaluation of Late Holocene soil tillage trials in Eastern and Central Europe (Chap. IV/64). In: Sychev VG, Mueller L (eds) Novel methods and results of landscape research in Europe, Central Asia and Siberia. Monograph in 5 Volumes. Vol IV Optimising Agricultural Landscapes «FSBI "VNII Agrochemistry» 2018, pp 312–319. https://doi.org/10.25680/1929.2018.23.16.329, http://vniia-pr.ru/monografii/pdf/tom4-64.pdf

Müller T, Ruppel S, Behrendt U, Lentzsch P, Müller M (2018) Antagonistic potential of fluorescent pseudomonads colonizing wheat heads against mycotoxin producing Alternaria and Fusaria. Front Microbiol 9, Article 2124. https://doi.org/10.3389/fmicb.2018.02124

Musard O, Le Dû-Blayo L, Francour P, Beurier J-P, Feunteun E, Talassinos L (2014) Underwater seascapes: from geographical to ecological perspectives. Springer International Publishing, 291 p https://doi.org/10.1007/978-3-319-03440-9, https://www.springer.com/la/book/9783319034393

Naveh Z (2007) Transdisciplinary challenges in landscape ecology and restoration ecology—an Anthology. Landsc Ser 6(XVII):423 p

Nearing MA, Xie Y, Liu B, Ye Y (2017) Natural and anthropogenic rates of soil erosion. Int Soil Water Conserv Res 5(2):77–84. https://doi.org/10.1016/j.iswcr.2017.04.001

Neef E (1967) Die theoretischen Grundlagen der Landschaftslehre. Haack, Gotha, 1. Edt. 152 p

Nendel C, Rötter RP, Thorburn PJ, Boote KJ, Ewert F (2019) Editorial Introduction to the Special Issue "Modelling cropping systems under climate variability and change: impacts, risk and adaptation". Editorial Agric Syst 159:139–143. https://doi.org/10.1016/j.agsy.2017.11.005

Newell JP, Henry LA (2016) The state of environmental protection in the Russian Federation: a review of the post-Soviet era. Eurasian Geogr Econ 57(6):779–801 (Russian Geographies and Economics) https://doi.org/10.1080/15387216.2017.1289851

Ode A, Tveit MS, Fry G (2008) Capturing landscape visual character using indicators: touching base with landscape aesthetic theory. Landsc Res 33:89–117. http://www.tandfonline.com/doi/full/10.1080/01426390701773854. Accessed on 19 Dec 2018

Olwig KR (1996) Recovering the substantive nature of landscape. Ann Am Assoc Geogr 86:630–653

Olwig KR (2002) Landscape, nature and the body politic. From Britain's Renaissance to America's New World. University of Wisconsin Press, Madison

Olwig KR, Mitchell D (eds) (2009) Justice. Power and the political landscape, Routledge, Abingdon and New York

Oostergreen D, Shvarts E (2000) Russian Zapovedniki in 1998: recent progress and new challenges for Russia's Strict Nature Preserves. USDA Forest Service Proceedings RMRS-P-14, pp 209–213. http://www.fs.fed.us/rm/pubs/rmrs_p014/rmrs_p014_209_213.pdf. Accessed on 19 Dec 2018

Opdam P, Luque S, Nassauer J, Verburg PH, Wu J (2018) How can landscape ecology contribute to sustainability science? Landsc Ecol 33(1):1–7. https://doi.org/10.1007/s10980-018-0610-7

Palmer JF (2019) The contribution of key observation point evaluation to a scientifically rigorous approach to visual impact assessment. Landsc Urban Plann 183:100–110. https://doi.org/10.1016/j.landurbplan.2018.11.001

Passarge S (1919) Die Grundlagen der Landschaftskunde. 3 Bände. Vol I, Beschreibende Landschaftskunde, L. Friederichsen & Co. Hamburg, 222 p

Pavlov AV (2008) Monitoring of the cryolithic zone (in Russian, Engl subtitles) Novosibirsk Acad. publisher Geo, 229 p. (Павлов А.В. "Мониторинг криолитозоны" Новосибирск Академическое издательство «Гео») 2008 г. 229 с), ISBN 978-5-9747-0128-3

PECSRL (2018) Permanent European conference for the study of the rural landscape. http://www.pecsrl.org. Accessed on 19 Dec 2018

Peil T, Jones M (eds) (2005) Landscape, law and justice. Proceedings of a Conference Organised by the Centre for Advanced Study at the Norwegian Academy of Science and Letters, Oslo 15–19 June 2003. Instituttet for sammenlignende kulturforskning, Serie B, Skrifter CXVIII. Novus forlag, Oslo

Penning-Rowsell EC (1981) Fluctuating fortunes in gauging landscape value. Prog Hum Geogr 5 (1):25–41

Penning-Rowsell EC, Lowenthal D (eds) (1986) Landscape meanings and values. Allen and Unwin, London. 172 p. ISBN 0047100036

Pfister C (2007) Climatic extremes, recurrent crises and witch hunts: strategies of European societies in coping with exogenous shocks in the late sixteenth and early seventeenth centuries. Medieval History J 10(1&2)37, 33–73. http://www.hist.unibe.ch/unibe/portal/fak_historisch/dga/hist/content/e11168/e52524/e69145/e186327/e188624/22_Pfister-Witches-07_ger.pdf Accessed on 19 Dec 2018

Pinto-Correia T, Primdahl J, Pedroli B (2018) European Landscapes in Transition - Implications for Policy and Practice. Cambridge University Press, 330 pp

Pittman SJ (ed) (2018) Seascape ecology, 1st edn. Wiley, 501 pp

Plieninger T, Bieling C (2017) The emergence of landscape stewardship in practice, policy and research, pp 10–24. In: Bieling C, Plieninger T (eds) The science and practice of landscape stewardship. Cambridge University Press, Cambridge, 402 p. ISBN: 9781107142268, https://doi.org/10.1017/9781316499016.002

Plit J, Myga-Piątek U (2014) The degree of landscape openness as a manifestation of cultural metamorphose. Quaestiones Geogra 33(3):145–154. http://geoinfo.amu.edu.pl/qg/archives/2014/QG_33_3_145-154.pdf. Accessed on 19 Dec 2018

Polis GA, Power ME, Huxel GR (Eds) (2004) Food webs at the landscape level. The University of Chicago press. Chicago and London 2004, 528 p. https://www.press.uchicago.edu/ucp/books/book/chicago/F/bo3631997.html, accessed on Dec 19, 2018

Porteous JD (1990) Landscapes of the mind. Worlds of Sense and metaphor. University of Toronto Press, Toronto

Poschold P (2017) Geschichte der Kulturlandschaft. Entstehungsursachen und Steuerungsfaktoren der Entwicklung der Kulturlandschaft, Lebensraum- und Artenvielfalt in Mitteleuropa. 2. aktualisierte Auflage, Ulmer Verlag, 320 p. ISBN 978-3-8001-0926-5

Potthoff K (2013) The use of 'cultural landscape' in 19th century German geographical literature. Norsk Geografisk Tidsskrift–Norwegian J Geogr 67(1): 49–54

Pütz T, Kiese R, Wollschläger U, Groh J, Rupp H, Zacharias S, Priesack E, Gerke HH, Gasche R, Bens O, Borg E, Baessler C, Kaiser K, Herbrich M, Munch J-C, Sommer M, Vogel HJ, Vanderborght J, Vereecken H (2016) TERENO-SOILCan: a lysimeter-network in Germany observing soil processes and plant diversity influenced by climate change. Environ Earth Sci 75:1242. https://doi.org/10.1007/s12665-016-6031-5

Quast J (2011) Towards an integrated land and water resources management in Brandenburg's Oderbruch-Havelland Wetland Belt: History and Strategies for the Future. Die Erde 142(1–2):141–162. https://www.die-erde.org/index.php/die-erde/article/view/46/0. Accessed on 19 Dec 2018

Radford SL, Senn J, Kienast F (2019) Indicator-based assessment of wilderness quality in mountain landscapes. Ecol Indic 97:438–446. https://doi.org/10.1016/j.ecolind.2018.09.054

Ratzel F (1893) Die Vereinigten Staaten von Nord-Amerika, vol. 2, Politische und Wirtschafts-Geographie, 2nd edn. R. Oldenburg, München

Ratzel F (1895-96) Die deutsche Landschaft. Halbmonatshefte der Deutsche Rundschau 4:407–428

Raymond CM, Bieling C, Fagerholm N, Martin-Lopez B, Plieninger T (2016) The farmer as a landscape steward: comparing local understandings of landscape stewardship, landscape values, and land management actions. Ambio 45(2):173–184. https://doi.org/10.1007/s13280-015-0694-0, https://www.ncbi.nlm.nih.gov/pmc/articles/PMC4752565/

Reganold JP, Wachter JM (2016) Organic agriculture in the twenty-first century. Nature Plants 2, Article number: 15221. https://doi.org/10.1038/nplants.2015.221

Regia C (ed) (2014) Landscape planning and rural development. Key issues and options towards integration. Springer Briefs in Geography 147 p. https://doi.org/10.1007/978-3-319-05759-0

Reverey F, Ganzert L, Lischeid G, Ulrich A, Premke K, Grossart HP (2018) Dry-wet cycles of kettle hole sediments leave a microbial and biogeochemical legacy. Sci Total Environ 627:985–996. https://doi.org/10.1016/j.scitotenv.2018.01.220

Pontius Jr. RG, Castella J-C, de Nijs T, Duan Z, Fotsing E, Goldstein N, Kok K, Koomen E, Lippitt CD, McConnell W, Sood AM, Pijanowski B, Verburg P, Veldkamp AT (2018) Lessons and challenges in land change modeling derived from synthesis of cross-case comparisons. Trends in spatial analysis and modelling. In: Behnisch M, Meinel G (eds) Trends in spatial analysis and modelling. Geotechnologies and the environment, vol 19. Springer, Cham, pp 143–164. https://doi.org/10.1007/978-3-319-52522-8_8

Ritter C (1832) Die Erdkunde von Asien, vol 1, Der Norden und Nord-Osten von Hoch-Asien. Reimer, Berlin

Robinson G (2018a) New frontiers in agricultural geography: transformations, food security, land grabs and climate change. Boletin de la Asociacion de Geografos Espanoles (78):1–48. http://dx.doi.org/10.21138/bage.2710

Robinson G (2018b) Globalization of agriculture. Ann Rev Resour Econ 10(1):133–160. https://www.annualreviews.org/doi/10.1146/annurev-resource-100517-023303, https://doi.org/10.1146/annurev-resource-100517-023303

Rockström J, Williams J, Daily G, Noble A, Matthews N, Gordon L, Wetterstrand H, DeClerck F, Shah M, Steduto P, de Fraiture C, Hatibu N, Unver O, Bird J, Sibanda L, Smith J (2017) Sustainable intensification of agriculture for human prosperity and global sustainability. Ambio 46(1):4–17. https://doi.org/10.1007/s13280-016-0793-6

Romanenkov VA, Smith JU, Smith P, Sirotenko OD, Rukhovitch DI, Romanenko IA (2007) Soil organic carbon dynamics of croplands in European Russia: estimates from the "model of humus balance'. Reg Environ Change 7:93–104. https://doi.org/10.1007/s10113-007-0031-7

Römbke J (2018) Monitoring the biological quality of soil based on the structure and functions of soil organism communities (Chap. I/71). In: Sychev VG, Mueller L (eds) Novel methods and results of landscape research in Europe, Central Asia and Siberia. Monograph in 5 Volumes. Vol I Landscapes in the 21th Century: Status Analyses, Basic Processes and Research Concepts. © «FSBI "VNII Agrochemistry» 2018, pp 367–372. https://doi.org/10.25680/1244.2018.27.50.071, http://vniia-pr.ru/monografii/pdf/tom1-71.pdf. Accessed on 19 Dec 2018

Römbke J, Gardi C, Creamer R, Miko L (2016) Soil biodiversity data: actual and potential use in European and national legislation. Appl Soil Ecol 97:125–133

Rösch M, Biester H, Bogenrieder a, Eckmeier E, Ehrmann O, Gerlach R, Hall M, Hartkopf-Fröder C, Herrmann L, Kury B, Lechterbeck J, Schier W, Schulz E (2017) Late neolithic agriculture in temperate Europe—a long-term experimental approach. Land 6:11. https://doi.org/10.3390/land6010011

Rotshild EV (2014) Infections in nature. Ecological conception. Environ Epidemiol 8(2):7–279 (in Russian) (Ротшильд Е.В. Инфекции в природе. Экологическая концепция. 3-е издание книги// Энвайронментальная эпидемиология, 2014, том 8, № 2, С. 7 – 279.)

Rowntree LB (1996) The cultural landscape concept in American human geography. In: Earle C, Mathewson K, Kenzer MS (eds) Concepts in human geography. Rowman & Littlefield, Lanham, pp 127–159

Rozas-Vásquez D, Fürst C, Geneletti D, Almendra O (2018) Integration of ecosystem services in strategic environmental assessment across spatial planning scales. Land Use Policy 71:303–310. https://doi.org/10.1016/j.landusepol.2017.12.015

Ruhovich, O (2018) Methods for forecasting winter wheat crop productivity in the Oka River Basin with the Use of Relief, Climate and Soil Data (Chap. II/10) (in Russian, Глава II/10: методы прогнозирования урожайности озимой пшеницы в бассейне р. оки с использованием рельефа, климата и почв). In: Sychev VG, Mueller L (eds) Novel methods and results of landscape research in Europe, Central Asia and Siberia. Monograph in 5 Volumes.Vol. II Understanding and Monitoring Processes in Soils and Water Bodies, pp 51–56. https://doi.org/10.25680/7431.2018.70.89.107, http://vniia-pr.ru/monografii/pdf/tom2-10.pdf. Accessed on 19 Dec 2018

Saljnikov E, Cakmak D, Muhanbet A, Kresovic M (2014a) Biological indices of soil organic matter in long term fertilization experiment. Zemljiste i Biljka –Soil and Plant 63(2):11–20 http://www.sdpz.rs/images/casopis/2014/ZIB_vol63_no2_2014_pp11-20.pdf. Accessed on 19 Dec 2018

Saljnikov E, Saljnikov A, Rahimgalieva S, Cakmak D, Kresovic M, Mrvic V, Dzhalankuzov T (2014b) Impact of energy saving cultivations on soil parameters in northern Kazakhstan. Energy 77:35–41. https://doi.org/10.1016/j.energy.2014.03.042

Sang N, Ode-Sang AA (2015) Review on the state of the art in scenario modelling for environmental management. Swedish Environmental Protection Agency report 6695. Nov. 2015, 275 p. https://www.naturvardsverket.se/Documents/publikationer6400/978-91-620-6695-6.pdf?pid=16935. Accessed on 19 Dec 2018

Saparov A (2014) Strategy of sustainable soil and plant resource management in the Republic of Kazakhstan. In: Mueller L, Saparov A, Lischeid G (eds) Novel measurement and assessment tools for monitoring and management of land and water resources in agricultural landscapes of Central Asia. Environmental Science and Engineering. Springer, Cham, pp 611–619. https://doi.org/10.1007/978-3-319-01017-5_38

Sauer CO (1925) The morphology of landscape. Univ California Publ Geogr 2(2):19–53

Sayer J, Sunderland T, Ghazoul J, Pfund J-L, Sheil D, Meijaard E, Venter M, Boedhihartono AK, Day M, Garcia C, van Oosten C, Buck LE (2013) Ten principles for a landscape approach to reconciling agriculture, conservation, and other competing land uses. PNAS 110(21):8349–8356. https://doi.org/10.1073/pnas.1210595110

Sayer J, Margules C, Bohnet I, Boedhihartono A, Pierce R, Dale A, Andrews K (2015) The role of citizen science in landscape and seascape approaches to integrating conservation and development. Land 4:1200–1212. https://doi.org/10.3390/land4041200, https://espace.cdu.edu.au/eserv/cdu:58924/Dale_58924.pdf

Schäffer A, Filser J, Frische T, Gessner M, Köck W, Kratz W, Liess M, Nuppenau E-A, Roß-Nickoll M, Schäfer R, Scheringer M (2018) Der stumme Frühling - Zur Notwendigkeit eines umweltverträglichen Pflanzenschutzes. Diskussion Nr. 16. Nationale Akademie der Wissenschaften - Leopoldina, Halle (Saale). https://www.boeblingen-sindelfingen.greenpeace.de/sites/www.boeblingen-sindelfingen.greenpeace.de/files/2018_diskussionspapier_pflanzenschutzmittel.pdf. Accessed on 19 Dec 2018

Schaller L, Targetti S, Villanueva AJ, Zasada I. Kantelhardt J, Arriaza F, Bal T, Fedrigotti BV, Giray FH, Häfner K, Majewski E, Malak-Rawlikowska A, Nikolov D, Paoli J-C, Piorr A, Rodríguez-Entrena M, Ungaro F, Verburg PH, Viaggi D (2018) Agricultural landscapes, ecosystem services and regional competitiveness - Assessing drivers and mechanisms in nine European case study areas. Land Use Policy Vol 76, July 2018, p 735-745 https://doi.org/10.1016/j.landusepol.2018.03.001

Schama S (1997) Landscape and memory. Alfred A. Knopf, New York

Scheller RM (2013) Landscape modeling. Encyclopedia of Biodiversity (Second Edition) 2013:531–538. https://doi.org/10.1016/B978-0-12-384719-5.00387-7

Schenk W, Weizenegger S (2006) Cultural landscape management in Europe and Germany. In: Agnoletti M (ed) The conservation of cultural landscapes. Wallingford (CABI), pp 183–196

Schindler U, Müller L, Dannowski R, Barkusky D, Francis G (2010) Long-term measurements to quantify the impact of arable management practices on deep seepage and nitrate leaching. In: Müller F, Baessler C, Schubert H, Klotz S (eds) Long-term ecological research. Springer, Dordrecht, pp 243–252. https://doi.org/10.1007/978-90-481-8782-9_17

Schlenstedt J, Brinckmann A, Häfker U, Haubold-Rosar M, Kirmer A, Knoche D, Landeck I, Lorenz A, Rümmler F, Stärke M, Tischew S, Wiedemann D (2014) Rekultivierung. In: Drebenstedt C, Kuyumcu M (eds) Braunkohlensanierung – Grundlagen, Geotechnik, Wasserwirtschaft, Brachflächen, Rekultivierung, Vermarktung. Springer Vieweg, Berlin, Heidelberg, pp 487–578

Schlüter O (1903) Die Siedlungen im nordöstlichen Thüringen. Ein Beispiel für die Behandlung Siedelungsgeographischer Fragen, Costenoble, Berlin, p 453

Schmidt G (1961) Zur Organisation des Meliorationswesens in der DDR. Agrartechnik Heft 11. November 1961, 501–505 (in German) http://440ejournals.uni-hohenheim.de/index.php/de_agrartechnik/article/view/6474. Accessed on 19 Dec 2018

Schnug E, De Kok LJ (eds) (2016) Phosphorus in agriculture: 100% Zero. Dordrecht, 353 p. https://www.springer.com/de/book/9789401776110. Accessed on 19 Dec 2018

Schönbrodt-Stitt S, Conrad C, Dimov D, Ergashev I, Löw F, Morper-Busch L, Muminov S, Ruziev I, Schorcht G, Solodky G, Sorokin A, Sorokin D, Stulina G, Toshpulatov R, Zaitov S, Kitapbayev A, Unger-Shayesteh K (2018) The WUEMoCA tool for monitoring irrigated cropland use and water use efficiency at the landscape level of the Aral Sea basin (Chap. IV/72). In: Sychev VG, Mueller L (eds) Novel methods and results of landscape research in Europe, Central Asia and Siberia. Monograph in 5 Volumes. Vol. IV Optimising Agricultural Landscapes « FSBI "VNII Agrochemistry » 2018, pp 351–356. https://doi.org/10.25680/9880.2018.75.14.337. http://vniia-pr.ru/monografii/pdf/tom4-72.pdf. accessed on 19 Dec 2018

Schöne T, Zubovich A, Zech C, Illigner J, Sharshebaev A, Mandychev D, Shakirov A, Moldobekov B, Lauterjung J (2018) Water Monitoring in Central Asia – The Central Asian Water Project (CAWa) (Chapter I/13:). In: Sychev VG, Mueller L (eds) Novel methods and results of landscape research in Europe, Central Asia and Siberia. Monograph in 5 Volumes. Vol I Landscapes in the 21th Century: Status Analyses, Basic Processes and Research Concepts. © « FSBI "VNII Agrochemistry » 2018, pp 90–94. https://doi.org/10.25680/1757.2018.97.98.009, http://vniia-pr.ru/monografii/pdf/tom1-13.pdf. Accessed on 19 Dec 2018

Schulp CJE, Levers C, Kuemmerle K, Tieskens KF, Verburg PH(2019) Mapping and modelling past and future land use change in Europe's cultural landscapes. Land Use Policy Vol 80, January 2019, p 332-344. https://doi.org/10.1016/j.landusepol.2018.04.030

Schwarzer M (2014) Von Mondlandschaften zur Vision eines neuen Seenlandes. Der Diskurs über die Gestaltung von Tagebaubrachen in Ostdeutschland, Springer VS, Wiesbaden. https://www.springer.com/la/book/9783658056391. Accessed on 29 Jan 2019

Selman P (2006) Planning at the landscape scale. Routledge, 213 p

Sergunin A (2019) Russian Arctic Cities' Sustainable Development Strategies, pp 495–511. https://doi.org/10.4018/978-1-5225-6954-1.ch023. In: Erokhin V, Gao T, Zhang X (eds) Handbook of research on international collaboration, economic development, and sustainability in the Arctic. 703 p. https://doi.org/10.4018/978-1-5225-6954-1

Seto KC, Reenberg A, Boone CG, Fragkias M, Haase D, Langanke T, Marcotullio P, Munroe DK, Olah B, Simon D (2012) Urban land teleconnections and sustainability. Proc Natl Acad Sci USA 109(20):7687–7692

Setten G, Brown KM, Rørtveit HN (2019) Landscape and social justice. In: Howard P, Thompson I, Waterton E, Atha M (eds) The Routledge companion to landscape studies, 2nd edn. Routledge, Abingdon, pp 418–428

Shaw DJB, Oldfield J (2007) Landscape science: a Russian geographical tradition. Ann Assoc Am Geogr 97(1):111–126. ISSN 0004-5608. https://core.ac.uk/download/pdf/1396599.pdf

Simensen T, Halvorsen R, Erikstad L (2018) Methods for landscape characterisation and mapping: a systematic review. Land Use Pol 75:557–569. https://doi.org/10.1016/j.landusepol.2018.04.022

Sobczyński T, Joniak T (2009) Vertical changeability of physical-chemical features of bottom sediments in three lakes in aspect type of water mixis and intensity of human impact. Pol J Environ Stud 18(6):1091–1097

Spetich MA, Kvashnina AE, Nukhimovskya YD, Rhodes OE Jr (2009) History, administration, goals, values, and long-term data of Russia''s strictly protected scientific nature reserves. Nat Areas J 29(1):71–78. https://www.srs.fs.usda.gov/pubs/ja/ja_spetich010.pdf

Stanilovskaya JS (2019) Landslides in permafrost zone of Russia. In: Pradhan S, Vishal V, Singh T (eds) Landslides: theory, practice and modelling. Advances in natural and technological hazards research, vol 50. Springer, Cham, https://doi.org/10.1007/978-3-319-77377-3_14

Steffen W, Broadgate W, Deutsch L, Gaffney O, Ludwig C (2015) The trajectory of the Anthropocene: the great acceleration. Anthropocene Rev 2(1):81–98. https://doi.org/10.1177/2053019614564785

Steinmuller HE, Dittmer KM, White JR, Chambers LG (2019) Understanding the fate of soil organic matter in submerging coastal wetland soils: a microcosm approach. Geoderma 337:1267–1277. https://doi.org/10.1016/j.geoderma.2018.08.020

Steinweg B, Kerth M (2018) Soils as witnesses of wars: an overview and further research needs (Chap. I/10). In: Sychev VG, Mueller L (eds) Novel methods and results of landscape research in Europe, Central Asia and Siberia. Monograph in 5 Volumes, vol I Landscapes in the 21th Century: Status Analyses, Basic Processes and Research Concepts. © «FSBI "VNII Agrochemistry» 2018, pp 76–80. https://doi.org/10.25680/9680.2018.48.98.010

Stenoien C, Nail KR, Zalucki JM, Parry H, Oberhauser KS, Zalucki MP (2018) Monarchs in decline: a collateral landscape-level effect of modern agriculture. Insect Sci 25(4):528–541. https://doi.org/10.1111/1744-7917.12404

Stenseke M, Jones M (2011) Conclusion: benefits, difficulties, and challenges of participation under the European Landscape Convention. In: Jones M, Stenseke M (eds) The European landscape convention. Challenges of participation. Landscape Series, vol 13. Springer, Dordrecht, pp 295–309

Storfer A, Murphy MA, Spear SF, Holderegger R, Waits LP (2010) Landscape genetics: where are we now? Mol Ecol 19(17):3496–3514. https://doi.org/10.1111/j.1365-294X.2010.04691.x

Stöven K, Jacobs F, Schnug E (2015) Mikroplastik: Ein selbstverschuldetes Umweltproblem im Plastikzeitalter. Journal für Kulturpflanzen 67(7):241–250

Succow M (2012) Positionspapier Agrarwirtschaft. http://www.succow-stiftung.de/tl_files/pdfs_downloads/Hintergrundmaterialien_Weltbeweger/Michael%20Succow_Postionspapier_Agrarwirtschaft.pdf. Accessed on 19 Dec 2018

Sumina OI (1998) The taxonomic diversity of quarry vegetation in North-West Siberia and Chukotka. Polar Geogr 22(1):17–55. https://doi.org/10.1080/10889379809377635

Sumina OI (edt) (2000) Research on anthropogenic impacts in the Russian Arctic: review and bibliography guide to Russian Arctic science. Arctic Centre of University of Lapland. Rovaniemi.2000. p 8-43

SURE (2018) Society for urban ecology http://www.society-urban-ecology.org/. Accessed on 19 Dec 2018

Svalova V (ed) (2019) Natural hazards and risk research in Russia. Springer Earth and Environmental Science, 399 p. https://doi.org/10.1007/978-3-319-91833-4

Swart KW (1969) The miracle of the Dutch Republic as seen in the seventeenth century. London: H.K.Lewis & Co Ltd., cited by Appelbaum 2012 http://www.dianamuirappelbaum.com/?p=583#.XB3fmGkxmpo. Accessed on 19 Dec 2018

Sychev VG, Yefremov EN, Romanenkov VA (2016). Monitoring of Soil Fertility (Agroecological Monitoring) In: Novel Methods for Monitoring and Managing Land and Water Resources in Siberia. Springer Water p 541-561, https://link.springer.com/chapter/10.1007/978-3-319-24409-9_24

Tagungsbericht (2015) Famines During the 'Little Ice Age' (1300-1800). Socio-natural Entanglements in Premodern Societies, 19.02.2015–20.02.2015 Bielefeld. In: H-Soz-Kult, 25.04.2015, www.hsozkult.de/conferencereport/id/tagungsberichte-5939. Accessed on 19 Dec 2018

Tarolli P, Vanacker V, Middelkoop H, Brown AG (2014) Landscapes in the anthropocene: state of the art and future directions. Anthropocene 6:1–2. https://doi.org/10.1016/j.ancene.2014.11.003

Taylor K (2016) The historic urban landscape paradigm and cities as cultural landscapes. Challenging orthodoxy in urban conservation. Landsc Res 41(4):471–480

Tchebakova NM, Parfenova EI, Soja AJ (2016) Significant Siberian vegetation change is inevitably brought on by the changing climate. In: Novel methods for monitoring and managing land and water resources in Siberia Springer Water, pp 269–285. https://link.springer.com/chapter/10.1007/978-3-319-24409-9_10

Temin P (2002) The golden age of European growth reconsidered. Eur Rev Econ History 6(1):3–22

TERENO (2018) Terrestrial environmental observatories. http://www.tereno.net/overview-en?set_language=en. Accessed on 19 Dec 2018

Thober S, Kumar R, Wanders N, Marx A, Pan M, Rakovec O, Samaniego L, Sheffield J, Wood EF, Zink M (2018) Multi-model ensemble projections of European river floods and high flows at 1.5, 2, and 3 degrees global warming. Environ Res Lett 13(1):014003. https://doi.org/10.1088/1748-9326/aa9e35

Thomas WL (ed) (1956) Man's role in changing the face of the earth. University of Chicago Press, Chicago

Thomas J (2006) What's on regarding land consolidation in Europe? In: Shaping the Change. Proceedings of the XXIII FIG Congress, Munich, Germany, October 8–13, 2006, pp 1–16. https://www.fig.net/resources/proceedings/fig_proceedings/fig2006/papers/ts80/ts80_03_thomas_0311.pdf. Accessed on 19 Dec 2018

Thünen JH (1826) Der isolirte Staat in Beziehung auf Landwirthschaft und Nationalökonomie. Hamburg, 1826. (in German) Deutsches Textarchiv http://www.deutschestextarchiv.de/thuenen_staat_1826. Accessed on 19 Dec 2018

Topaj A, Badenko V, Medvedev S, Terleev V (2018) Dynamically adjusted forecasting of agro-landscape productivity using massive computations of crop model in GIS environment (Chap. III/53). In: Sychev VG, Mueller L (eds) Novel methods and results of landscape research in Europe, Central Asia and Siberia. Monograph in 5 Volumes. vol III Landscape Monitoring and Modelling © « FSBI "VNII Agrochemistry » 2018, pp 253–257, https://doi.org/10.25680/3309.2018.28.99.246, http://vniia-pr.ru/monografii/pdf/tom3-53.pdf. Accessed on 19 Dec 2018

Tregear A, Arfini F, Belletti G, Marescotti A (2007) Regional foods and rural development: the role of product qualification. J Rural Stud 23(1):12–22. https://doi.org/10.1016/j.jrurstud.2006.09.010

Trentmann F, Just F (eds) (2006) Food and conflict in Europe in the age of the two world wars. Palgrave Macmillan, 296 p. https://doi.org/10.1057/9780230597495

Troldborg M, Aalders I, Towers W, Hallett PD, McKenzie BM, Bengough AG, Lilly A, Ball BC, Hough RL (2013) Application of Bayesian Belief Networks to quantify and map areas at risk to soil threats: using soil compaction as an example. Soil Tillage Res 132:56–68

Troll C (1939) Luftbildforschung und Landeskundige Forschung. Erdkundliches Wissen. Schriftenreihe für Forschung und Praxis, Heft 12, F. Steiner Verlag, Wiesbaden
Troll C (1971) Landscape ecology (geoecology) and biogeocenology–a terminology study. Geoforum 2(8):43–46
Trovato MG (2015) E-scape: landscape and refugee influx. Landsc Architect Australia 147:74–79
Trovato MG (2019) A landscape perspective on the impact of Syrian refugees in Lebanon. In: Asgary A ed) Resettlement challenges for displaced populations and refugees. Springer International Publishing AG, part of Springer Nature January 2019, pp 41–64. https://doi.org/10.1007/978-3-319-92498-4_4
Trovato MG, Ali D, Nicolas J, El Halabi A, El Meouche S (2017) Landscape risk assessment model and decision support system for the protection of the natural and cultural heritage in the eastern mediterranean area. Land 6(4):76. https://doi.org/10.3390/land6040076
Tuan Y-F (1974) Topophilia. A study of environmental perception, attitudes, and values. Prentice-Hall, Englewood Cliffs
Tudor C (2014) An approach to landscape character assessment. October 2014. https://assets.publishing.service.gov.uk/government/uploads/system/uploads/attachment_data/file/691184/landscape-character-assessment.pdf. Accessed on 19 Dec 2018
Turner II BL, Clark WC, Kates RW, Richards JF, Mathews JT, Meyer WB (eds) (1990) The earth as transformed by human action. Global and regional changes in the biosphere over the last 300 years. Cambridge University Press, Cambridge
Tveit M, Ode Å, Fry G (2006) Key concepts in a framework for analysing visual landscape character. Landsc Res 31(3):229–255
UNISCAPE (2018) European network of universities for the implementation of the European landscape convention. http://www.uniscape.eu/. Accessed on 19 Dec 2018
United Nations (2015) Agenda 2030: sustainable development goals. 17 goals to transform our world https://www.un.org/sustainabledevelopment/. Accessed on 19 Dec 2018
Uskov IB, Uskov AO (2014) Bases for adaptation of agriculture to climate change (reference). Russian Academy of Science, Saint-Petersburg 2014, 384 p. (in Russian: Основы адаптации земледелия к изменениям климата (спавчное издание).Санкт Петербург. 2014г. 384 с.). ISBN 978-5-4469-0469-3
Van der Ploeg R, Schweigert P, Bachmann J (2001) Use and misuse of nitrogen in agriculture: the German story, scientific world journal. 1:737–744. Published online 2001 Oct 30. https://doi.org/10.1100/tsw.2001.263, PMCID: PMC6084271 PMID: 12805882 https://www.ncbi.nlm.nih.gov/pmc/articles/PMC6084271/. Accessed on 9 Dec 2018
Van Huylenbroeck G, Vandermeulen V, Mettepenningen E, Verspech A (2007) Multifunctionality of agriculture. A review of definitions, evidence and instruments. Living Rev. Landscape Res 1:3. http://www.livingreviews.org/lrlr-2007-3. Accessed on 19 Dec 2018
Vera FWM (2000) Grazing ecology and forest history. CABI Publishing, Wallingford, UK, p 528
Vereecken H, Huisman JA, Hendricks Franssen HJ, Brüggemann N, Bogena HR, Kollet S, Javaux M, van der Kruk J, Vanderborght J (2015) Soil hydrology: recent methodological advances, challenges, and perspectives. Water Resour Res 51(4):2616–2633. https://doi.org/10.1002/2014WR016852
Victorov AS, Orlov TV, Kapralova VN, Trapeznikova ON, Sadkov SA, Zverev AV (2019) Stochastic modeling of natural lacustrine thermokarst under stable and unstable climate. In: Svalova V (eds) Natural hazards and risk research in Russia. Innovation and discovery in Russian science and engineering. Springer, Cham, pp 241–267. https://doi.org/10.1007/978-3-319-91833-4_18
Vitikainen A (2004) An overview of land consolidation in Europe. Nordic J Survey Real Estate Res 1:25–44. ftp://fgg.uni-lj.si/Sendable/MFERLAN/ZEMLJISKI%20KATASTER%20II/8_NALOGA/arvo.pdf. Accessed on 19 Dec 2018
Vogel C (1851) Geographische Landschaftsbilder. Handbuch zur Belebung geographischer Wissenschaft für Lehrer und Gebildete überhaupt, Hinrichs, Leipzig

Von Bismarck F, Andrich A, Berkner A, Boldorf K, Dallhammer W-D, Drebenstedt C, Freytag K, Kadler A, Meyer H-D, Schlenstedt J, Schmidt R, Strzodka M, Weymanns K-O (2014) Rechtliche, finanzielle und organisatorische Grundlagen. In: C Drebenstedt, M Kuyumcu (eds) Braunkohlensanierung - Grundlagen, Geotechnik, Wasserwirtschaft, Brachflächen, Rekultivierung, Vermarktung. Springer Vieweg, Berlin, Heidelberg, pp 73–129

Walz U, Stein C (2018) Indicator for a monitoring of Germany's landscape attractiveness. Ecol Ind 94(2):64–73. https://doi.org/10.1016/j.ecolind.2017.06.052

WASWAC (2018) World Association of Soil and Water Conservation. http://www.waswac.org/waswac/index.htm. Accessed on 19 Dec 2018

Wende W, Walz U (2019) Landschaft im Wandel. https://www.buergerschaffenwissen.de/projekt/landschaft-im-wandel. Accessed on 29 Jan 2019

Wezel A, Bellon S, Doré T, Francis C, Vallod D, David C (2009) Agroecology as a science, a movement and a practice. A review. Sustain Dev 29(4):503–515. https://doi.org/10.1051/agro/2009004

Widgren M, Pedersen EA (2011) Agriculture in Sweden: 800 BC–AD 1000. In: Myrdal J, Morell M (ed) The agrarian history of Sweden: from 4000 BC to AD 2000, pp. 46–71. Nordic academic press, Lund. http://urn.kb.se/resolve?urn=urn:nbn:se:su:diva-57991

Wieland R, Kerkow A, Früh L, Kampen H, Walther D (2017) Automated feature selection for a machine learning approach toward modeling a mosquito distribution. Ecol Modell 352:108–112. https://doi.org/10.1016/j.ecolmodel.2017.02.029

Wimmer J (1885) Historische Landschaftskunde. Verlag der Wagner'schen Universitäts-Buchhandlung, Innsbruck

Wood PJ, Hannah DM, Sadler JP (eds) (2008) Hydroecology and ecohydrology past, present and future. Wiley, 464 p, ISBN 978-0-470-01017-4

Wu J (2013) Landscape sustainability science: ecosystem services and human well-being in changing landscapes. Landsc Ecol 28(6):999–1023. https://doi.org/10.1007/s10980-013-9894-9

Wylie J (2009) Landscape. In: Gregory D, Johnston R, Pratt G, Watts MJ, Whatmore S (eds) The dictionary of human geography, 5th edn. Wiley-Blackwell, Chichester, pp 409–411

Yadav SK, Babu S, Yadav MK, Singh K, Yadav GS, Pal S (2013) A review of organic farming for sustainable agriculture in Northern India. Int J f Agron 2013, Article ID 718145, 8 p. http://dx.doi.org/10.1155/2013/718145. Accessed on 19 Dec 2018

Yin X, Kersebaum KC, Kollas C, Manevski K, Baby S, Beaudoin N, Öztürk I, Gaiser T, Wu L, Hoffmann MP, Armas-Herrera CM, Charfeddine M, Conradt T, Constantin J, Ewert F, de Cortazar-Atauri IG, Giglio L, Hlavinka P, Hoffmann H, Launay M, Louarn G, Manderscheid R, Mary B, Mirschel W, Nendel C, Pacholski A, Palosuo T, Ripoche-Wachter D, Rötter R, Ruget F, Sharif B, Trnka M, Ventrella D, Weigel H-J, Olesen JE (2017) Performance of process-based models for simulation of grain N in crop rotations across Europe. Agric Syst 154:63–77. https://doi.org/10.1016/j.agsy.2017.03.005

Zalasiewicz J, Williams M, Haywood A, Ellis M (2011) The anthropocene: a new epoch of geological time? Phil Trans R Soc Lond A 369:835–841

Zeder MA (2008) Domestication and early agriculture in the Mediterranean Basin: Origins, diffusion, and impact. PNAS August 19, 2008 105 (33) 11597-11604; https://doi.org/10.1073/pnas.0801317105

Zinkina J, Ilyin IV, Korotayev A (2017) The early stages of globalization evolution: networks of diffusion and exchange of domesticates, technologies, and luxury goods. Socionauki. 16(1). https://www.socionauki.ru/journal/articles/875359/

Zonneveld IS (1995) Land ecology. SPB Academic Publishing bv, Amsterdam, p 199

Zoomers A, van Noorloos F, Otsuki K, Steel G, van Westen G (2017) The rush for land in an urbanizing world: from land grabbing toward developing safe, resilient, and sustainable cities and landscapes. World Dev 92:242–252. https://doi.org/10.1016/j.worlddev.2016.11.016

Part II
Concepts for Landscape Assessment and Evolvement

Chapter 6
Concepts and Measures for Maintaining Wilderness and Landscape Biodiversity in the Anthropocene

Vladimir Bocharnikov

Abstract "Wilderness" is a term used in everyday language with different culturally influenced meanings. Wilderness is largely perceived fundamentally as areas of untouched, or so-called "virgin," countryside, which in ecological terms contain relatively undisturbed natural animal and plant populations. In areas where urban lifestyles dominate in Russia, there is a lower likelihood of daily interactions with nature and a greater need for intentionality before such interactions occur. We understand wilderness (*dikaya priroda*) as a largely undisturbed place (region, landscape) and as a subject of human recognition and legislative protection. Starting out from this pragmatic point of view, we developed an index of wildernesses in Russia for mapping and planning purposes. Roadless areas, settlements, industrial infrastructure, and powerlines are important input parameters for this index. Small-scale GIS maps were created of wildernesses and their level of protection. Three large Federal districts in the Asian part of Russia contain most remaining large areas of wilderness, consisting of undisturbed flora and fauna; a total of more than nine million square kilometers (54.6% of Russia). We emphasize the particular role of Arctic and sub-Arctic regions as wilderness and biodiverse ecosystems.

Keywords Wilderness · Russia · National inventory · Anthropogenic disturbance · Landscape

6.1 Introduction: Concepts of Wilderness

The current Anthropocene epoch is characterized by humans exerting accelerating effects on the environment, including all regions, landscapes, and ecosystems on earth (Crutzen 2002; Lewis and Maslin 2015; Mueller et al. 2016, 2018). Throughout the world, urban culture has emerged and developed over the past centuries. It is an ideal

V. Bocharnikov (✉)
Far-Eastern Branch, Pacific Geographical Institute,
Russian Academy of Sciences, 7 Radio Street, Vladivostok 690041, Russia
e-mail: vbocharnikov@mail.ru

L. Müller and F. Eulenstein (eds.), *Current Trends in Landscape Research*,
Innovations in Landscape Research, https://doi.org/10.1007/978-3-030-30069-2_6

Fig. 6.1 Brown bear (*Ursus arctos arctos* L.) in the secondary small-leaved forest of Southern Siberia. Photograph with kind permission of Svetlana Babina

way of life that is systematic, orderly, defined, and called safe by man, without contact with nature. More than half of the world's population lives in cities (Taylor 2016), far from any genuine wilderness (Harvey 2015). Landscapes with little human impact and habitats for wildlife (Fig. 6.1) are decreasing.

Wilderness is often idealized as an undisturbed space, a place where we can escape the spaces otherwise dominated by humans and their activities. In Western philosophical literature, concern about wilderness appeared almost two centuries ago in North America, when the scale of the extermination of wild animals and the transformation of their habitats by European settlers and colonists became obvious and caused concern (Kormos et al. 2015). While wilderness is thought of as a place or places, today it is also a legal or zoning designation (i.e., a sociopolitical overlay) of a biophysical space.

The reason for this image is a modern Western culture that has been built as a dichotomy between humans and nature (culture and nature). Roderick Nash (Nash 2001) suggests that wilderness is best understood using the conception of a spectrum of conditions or environments ranging from the purely wild at one end to civilized or industrial/urban/rural surroundings at the other.

Wilderness means areas shaped largely by natural forces, in contrast to an area shaped by humans, though most parts of the world are now believed to be affected by climate-induced disturbance. The classic American environmental strategist

Aldo Leopold noted that "wilderness" provides an understanding of a certain standard thought of as corresponding to natural "healthy" land, and which should be the standard for the organization of "restoration" projects to restore human-disturbed areas.

Wilderness can be a place only in our imagination. It is usually a place truly experienced by only a small section of society who actually can either afford to travel to undeveloped places or who lives there in isolated conditions. It means areas shaped largely by natural forces in contrast to an area shaped by humans, though most parts of the world are now believed to be affected by climate-induced disturbance. The traditional concept of wilderness views it as an objective place separate from humans, primarily shaped by natural forces and free from human inhabitants and structures (McCloskey 1966).

However, it took more than a century and a half for this concept to become ingrained in the world's environmental community and to be accepted as one of the most strictly protected natural areas, i.e., those areas which are relatively free from significant anthropogenic impact but also protected for human experiences of the high-integrity countryside. Wildland preservation, however, is motivated by a variety of ethical, biological, cultural, and recreational concerns.

Those working on wildland preservation issues have been forced to consider the issue of local economic impacts, as those supporting the commercial development of these wild natural landscapes emphatically asserts that wildland preservation damages the local and national economies by restricting access to valuable natural resources and constraining commercial economic activity that would otherwise take place. It can be recognized and legislatively protected by governments. While the official line (see the IUCN PA Database) is that the level of global nature protection is increasing, there is a decreasing trend in such natural places on the planet, where the unknown is protected as more valuable than what is already familiar. Wilderness instantiates a unique form of human–land relations, and wilderness boundaries reveal their permeable and fluid nature.

In the whole world, and to a very large extent in Russia, the Arctic, the Taiga, and the high mountains and deserts are often perceived as the most pristine wilderness on earth due to their very low level of anthropogenic influence (Fig. 6.2).

There have been several global, continental, and national wilderness assessments (McCloskey and Spalding 1989; Lesslie et al. 1988; Sanderson et al. 2002; Cao et al. 2017). A spatial assessment of the character of wilderness is based on ecosystem integrity and, around the world, wilderness is commonly defined by the absence of permanent habitation and roads, and, most importantly, the lack of forms of economic activity which cause large-scale transformations and violate natural ecosystems (Casson et al. 2016).

This chapter illustrates the especially low level of human influence in the Russian Arctic, though we are given a wider picture of Siberia, on a broad scale from the Ural Mountains in the west to the Pacific in the east (Fig. 6.3). The efforts described in this paper focused on obtaining a highly generalized but realistic representation of the current degree of anthropogenic disturbance in the Russian Federation.

Fig. 6.2 Sikhote-Alin intact southern oak forest landscapes. Photograph with kind permission of Ivan Seryodkin

6.2 Wilderness Concepts and Practice in Russia

"Wilderness" is an idea, concept, system, theory, and practice of activities for the protection of natural functions and processes; the human environment (Bocharnikov 2016). Because the modern concept of wilderness is primarily a cultural or political one, the scientific foundations for wilderness are still being established. Rarely are efforts to protect wildlands motivated by an interest in promoting economic growth. It is acknowledged that wilderness and definitions of wilderness are complex and somewhat problematic across cultures, and we should look to the past for the reason for that.

The first step proposed here is to understand wilderness as one part of larger landscapes of inhabitation. Worldwide, the category "wilderness" refers to the global network of protected areas and, though according to the well-known ecologist Dezhkin (2005), this category could be a valuable reserve of wilderness conservation in Russia, so far this has not happened. We introduce into scientific circulation the concept of *dikaya priroda* ("wilderness") as a term signifying all the large areas of natural landscapes that are undisturbed by significant human impact (Bocharnikov 2016; Bocharnikov and Egidarev 2015).

We understand *dikaya priroda* or "wilderness" as a new environmental index, designed to generally assess the degree of preservation of natural geosystems, depending on the degree of anthropogenic impact, but which, in contrast to the

Fig. 6.3 The author in Kamchatka with the Northern Pacific in the background. Photograph by Vladimir Bocharnikov

traditional focus, displays a variety of parameters of anthropogenic disturbances to the natural environment of the developed areas. Here, we briefly consider and compare the history of conservation in Russia and the USA. In the USA, in the second half of the nineteenth century, a popular romantic notion of wild countryside was "transplanted to the soil" of the North American continent by European settlers. Later, there was a narrower focus on functional, pragmatic activities for the organization and development of national parks on public and Federal lands for recreation and to a lesser extent for environmental purposes.

In Russia, the idea of law based on the primacy of creating and maintaining state nature reserves as standards of nature, allowed a correct understanding to develop the natural properties of natural systems whose key functions of monitoring and research require minimal human impact (Shtilmark 2001). If we turn to Russian environmental history, we can see a big difference between the USA and Russia.

At the beginning of late Tsarist Russian, then Soviet history, the founders of the national nature protection strategy, A. G. Kozhevnikov and A. P. Semenov-Tyan-Shansky, were well aware of the development of the idea of wild countryside, and they made some real steps in this direction, similar to those in North America. But because of major cultural, conceptual, and political differences in the early days of the process of organizing the first reserves, the rationalism of the scientific component for nature preservation was unconditionally given.

Major conservation gains achieved by the Soviet state were made by effectively protecting nature reserve systems within inhabited areas but not in these remote zones, with only a few exceptions. The period of the prerevolutionary "spiritual" attitude to nature came to an end in the "flames" of the civil war, and the basic ideas of "freedom representation" and "nature rights" were ruthlessly uprooted in the era of Soviet industrialization and collectivization. It is enough to recall and reproduce the characteristic appeals: "… we will declare a battle on nature," "we do not expect favors from nature," etc. (Lavrenko et al. 1958; Zapovedniki 1977).

The preservation of nature and the eternal inviolability of nature reserves were laid out and defined by Soviet Russia government decrees of 1920–1924. By 1950, there were 128 strict nature reserves in Russia; the total area was then approximately 0.56% of the country (Sokolov 2006). The eternal inviolability of nature reserves gradually came under pressure. Pragmatic environmental ideas lost sight of their original principles and basic beliefs in the face, for example, of campaigns by most Soviet leaders such as wind walls to stop erosion, plowing the steppe, irrigation, etc.

Obviously, the further development of the network of reserves and national parks in Russia—clearly one of the largest and last wilderness areas in the world—needs serious correction. However, the biggest flaw in the existing system of wilderness and nature protection areas is currently its obviously insufficient geographical representation in terms of both the zone and specific types of landscapes (Figs. 6.4 and 6.5).

An analysis of the existing network of protected areas (Prisyazhnaya et al. 2016) shows that, despite all its undoubted merits and value, this network is very unevenly distributed across the country in a spatial sense and is also very uneven in terms of the size of the protected areas, as well as having poor infrastructure.

However, the network of nature reserves continues to grow, and they effectively perform environmental functions, significantly helping to develop science and solve problems related to the rational use of natural resources. Even if, earlier, publications by Soviet scholars condemned the fact that in international practice, particularly in the USA, "natural areas allocated as national parks are created for the state or purchased from landowners primarily for the recreation of the urban population and serve not only nature conservation but also, to a large extent, the tourism industry" (Zapovedniki Dalnego Vostoka SSSR 1985, p. 5).

In the context of the continuing huge global losses of natural wildlife, habitat areas, such as in Russia, could be a more realistic environmental contribution by the state, but they have not yet received a proper socioeconomic and geopolitical assessment in the world community. Unlike the rest of the world, Russia is

Fig. 6.4 The autumn at the Sikhote-Alin Nature Reserve (*Sikhote-Alinskiy zapovednik*) Photograph with kind permission of Ivan Seryodkin

Fig. 6.5 Middle stretch of the Aldan River (Yakutya, August 2008). Photograph by Vladimir Bocharnikov

experiencing "demographic desertification" not only in the vast expanses of Siberia and the Far East, but also in the European part of the country. These are, in essence, from an economic point of view, "empty territories" which can rightly be considered useful resources, conditions, and factors.

Moscow and St. Petersburg are the main centers of attraction for people, and about 80% of Russian residents tend to live or work in "profitable" cities. But throughout the Asian North, the fact should be taken into account that in the official reserves and the main sites of oil and gas production, pipeline routes now call into serious question whether any of the large areas of primarily unprotected, official natural countryside or wilderness will remain in the long term (Fig. 6.6).

In other words, we need a clear demonstration of the populated and uninhabited territories of the Russian Federation which allows us to understand in fact what part is played by their economic and social use by humankind. We examine preserved natural landscapes that are still very large, even by world standards, and far from the main zone of settlement of the territory of the Russian Federation.

Many practical considerations, obtained over decades in which wildlife areas have existed and been organized, have been considered in the development of programs, justifying the environmental frameworks and creating environmental networks. They have also been included in the strategy for implementing recreational activities, conducting environmental education and developing ecotourism. But we urgently need to develop a strategy for conserving the remaining huge areas of wilderness which have not been given official conservation status.

At the same time, what is particularly important here is the clarity and simplicity of the identification and the implementation of the calculations of their units (Isakov 1983). The reason for that is that Russia, a huge northern country located on a third

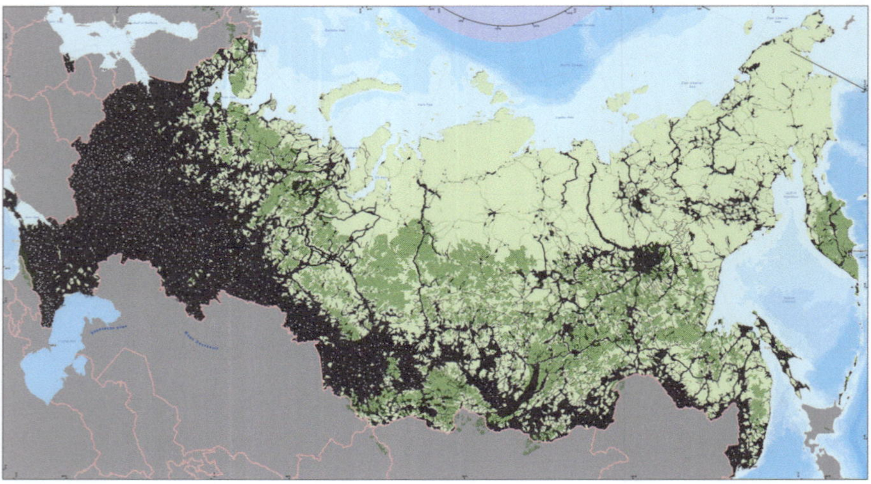

Fig. 6.6 The map shows the largest areas of wilderness (light green), landscapes without human transformation (green) and what are called man-made geosystems, surrounded by big urban and rural sites, for the entire territory of Russia

of the largest continent of the planet, contains the world's natural heritage, which still allows the global ecological balance to be maintained (Reimers and Shtilmark 1978). It is well-known that the modern strategy of wilderness protection should be based on preserving the ecological balance while maintaining existing types of environmental management, which in the theoretical sense should be based on geosystems' sustainability as a result of transformation serving as the basis for development.

It is obvious that there is an urgent need to improve and modernize, and this is considered in the provisions of both the existing "concept for the development of the system of specially protected natural areas of Federal importance for the period up to 2020" and the recent plans and programs of the Ministry of Natural Resources and Environment (Minprirody). Most likely, the further development of the network of reserves and national parks in Russia—clearly one of the largest and last wilderness areas in the world—needs a serious correction, even though strict nature reserves (*zapovednik*) have become a reality and their use is restricted to acclimatization in reserves of economically useful species, etc.

The best way to identify anthropogenic impacts has been to map all types of roads serving mining companies, although we also "translate" the results of our calculations within the boundaries of official administrative-territorial divisions.

The specificity of geographical space is determined by the hierarchy, which allows different classifications to be made, with a simultaneous display of patterns of discontinuity and continuality, where each object affects its environment, and is present and interacts with other objects in a fixed manner. Our geospatial analysis of Russian wilderness suggests that it is "empty" from an economic point of view, but that such areas should be preserved because they support the ecological integrity of the planet.

6.3 Wilderness Index

Against this background, it is important to understand the role of the existing network of protected areas, in comparison with the vast areas that have been identified by us as large areas of wilderness. Our highlighted, generalized display of different types of transport routes on the map gives us a clear view of the huge scale of transformation of geosystems, primarily in the context of how they are linked to the centers of transport and economic development in the territory. We present original results gained by assessing the degree of anthropogenic disturbance in landscapes of the Russian Federation on the basis of GIS technology.

We have developed a wilderness index for mapping and planning purposes. The methodology for this assessment conducted for Russia starts with the recognition of a unified spatiotemporal continuum, one pole of which is represented by a virgin natural environment (a wilderness unaffected by human management and impact).

while the other pole is a cultural landscape, developed territories, and urbanized areas. We identified "off-road" and "uninhabited" human territory, facilitating binary encoding ("humanized nature"/"wilderness"), with a high degree of accuracy to formalize and calculate the probability of the existence of areas of wilderness. This may present particular difficulties for mapping and GIS, both of which depend on carefully defined attributes and discrete criteria.

Methodologically, we used widely applied standard quantitative geoinformation analyses showing (in absolute terms and as a percentage) large tracts of off-road and uninhabited (or nearly uninhabited) areas, as well as the opposite; anthropogenically disturbed areas.

To identify the share of those areas of wilderness that have an official conservation status, the work included a spatial analysis of how these areas intersected with protected areas (state nature reserves, including biosphere reserves; national parks; nature reserves of Federal importance, UNESCO world natural heritage sites; Ramsar lands and key ornithological areas (ABA). For this comprehensive assessment, we also used digital data on the borders of specially protected natural areas of international and Federal importance in the Russian Federation (http://gis-lab.info/qa/oopt.html; http://oopt.kosmosnimki.ru/).

A "wilderness index" was calculated by the simple formula:

$$\text{Rast} = \sum\left(\text{SPrdor} + \sum \text{SPlpos}\right)^{*} + \sum \text{Znbf} + \sum \text{SPlpin/OPtre},$$

where:

Rast = the current state (an area or percentage thereof) of the remaining large wilderness areas in relation to a part or the whole of the Russian Federation;
$\sum$SPrdor is the total area of roadless areas;
$\sum$SPlpos is the total area of settlements of all types;
$\sum$Plpin is the total area of all types of industrial infrastructure and linear industrial facilities (power lines, pipelines, etc.);
OPtre is the total area of the administrative-territorial unit or physiographic country or landscape and ecological macroregion for example, according to the Global 200 (Olson and Dinerstein 2002).

The theoretical justification of this concept has already been published by us (Bocharnikov 2012, 2014), as have some of the algorithms for calculations and data on nature throughout Russia (Bocharnikov and Egidarev 2017a, b), although the main information is contained in the works not yet published to date and presented here.

6.4 Mapping Wilderness for Regions of Russia

The idea of protecting wilderness, with reference to the international experience and with adjustments for Russian realities, has been expressed before. However, the idea of wilderness has not found much popularity, remaining little-known in most Russian scientific and conservation circles (Bocharnikov 2017). Yet the overwhelming evidence from many case studies and reviews around the country demonstrates that protecting wilderness is an overall positive net economic benefit to communities and regional economies, as well as contributing to an enhanced quality of life.

We calculated the areas of wilderness within the boundaries of the "net" administrative-territorial divisions under official acts. In the small-scale cartographic analysis of the territory of Russia presented here, carried out on the basis of GIS technology (software: ESRI ArcGIS10.2), the levels of intact landscapes and their protection are described. The working scale of our research is based on vector topographic maps of Digital Chart of the World (DCW) data on a 1:1,000,000 scale, which was upgraded by the Russian company DATA+ in 2002.

However, when we map out very large areas of wilderness, along with the impacts of economic development, patterns are clear, not only for Southern Siberia and the south of the Far East, but also in Central Siberia and the northeast. A correct visualization of the socioeconomic situation of the country on the basis of our proposed integral indicator offers a new view of the specifics of the spatial distribution of resources, labor, services, and culture that will help to understand the real role of Russian regions in the development of economic space (Fig. 6.7).

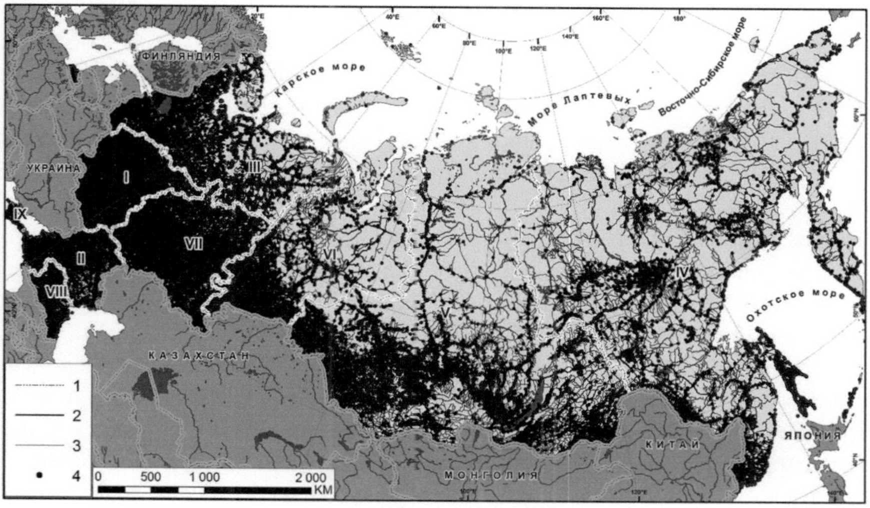

Fig. 6.7 Current distribution of large natural Russian roadless areas. Legend: 1—border of the Russian Federation; 2—Russian Federal District (*okrugs*) borders; 3—regional administrative borders; 4—cities; 5—low-integrity human-influenced areas (blank areas); 6—hard anthropogenically altered and human-influenced areas (black)

The methodological component of the work on wilderness assessment is based on the identification of five main features that are considered to be most characteristic of wilderness. These indicators characterize a minimal, noneconomic impact on the territory. The information comes from plots of no less than 50 thousand ha (500 km^2), making it possible to paint a very detailed picture of economic development throughout Russia.

The area was calculated and contours drawn using standard GIS software functions (Fig. 6.7). The first step consists of the calculation of sites which are permanently uninhabited and unused by humans. Settlements, industrial areas, and infrastructure, including all forms of human development (land and inland water transport, communication means of all types, power lines, pipelines, etc.) are excluded.

The second component for identifying wilderness areas is the mapping of all those territories that have an official conservation status (reserve, national park or reserve of federal significance). Such areas often have a higher level of preservation of natural functions than the surrounding areas that do not have such formal recognition. The algorithm used in this work to approximately calculate or apply the "coarse filter" used to display the "unpopulated" territory shows land plots which contain no permanent settlements, industrial enterprises (including extracting and processing), transport, means of communication of any kind, highways, railways, power lines, oil and gas pipelines, and other large objects testifying to economic activity.

We need to make a more accurate assessment in the near future, especially to divide remote historically populated areas from lands under agriculture use in the last two centuries. These influences are not commonly understood in ordinary general education maps, in educational atlases, in many thematic maps, or in reference books.

Most importantly, in our calculations of wilderness integrity, attention must be drawn to the fact that most important areas of wilderness are found in the territory of Asian or "Eastern" Russia, with only very fragmented natural landscapes within the "northern territories" of the European part (Bocharnikov and Egidarev 2017a, b).

Figures 6.7 and 6.8 show that human influence within Russia is far from uniform: the widest area of dense settlement is in the European part of Russia (from 60° north to the southern borders of the country), to the east; the band of intense settlement somewhat narrows toward the Urals and beyond, and widens again with more intense settlement from Novosibirsk to Tomsk, Kuzbass, and Altai, with further narrowing along the Trans-Siberian zone and the industrial zones. Even with this preliminary comparison of the territories, we are considering in the large grid of administrative and territorial divisions in the Federal districts, it is obvious that in the European part of Russia, the only area of wild countryside that is significant and comparable to the Asian part of the country is the North-West district (Fig. 6.7).

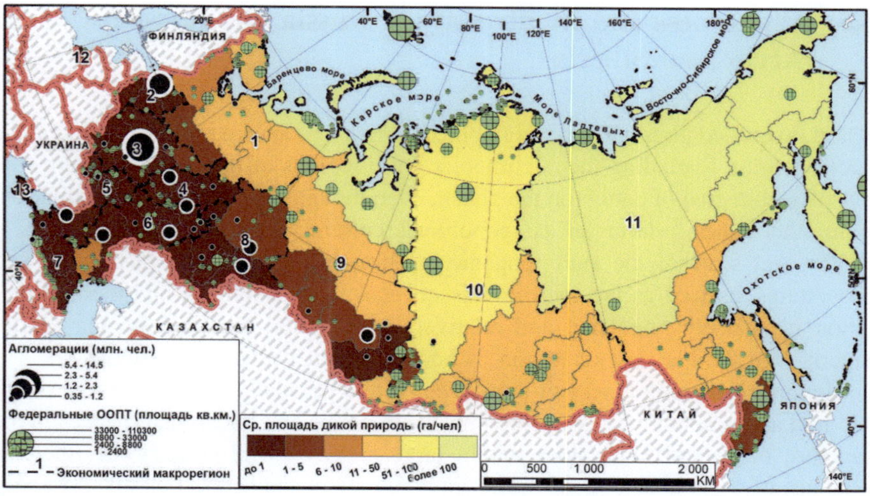

Fig. 6.8 Map of Russia illustrating differences between European and south Siberian areas with high population density and prosperous industrial areas. The black circles are large Russian agglomerations; green circles are protected areas at Federal level (*zapovedniks* and national parks), the average percentages of remaining wilderness areas are shown from yellow (high wilderness rate) to brown (low wilderness rate) with an index calculated as the area of wilderness (in hectares) per 1 resident. *Source* Bocharnikov and Egidarev (2016)

In the European part of Russia, the North-West Federal Okrug (No. 1) has preserved almost half of its territory in a state with a low anthropogenic impact (48.1% of the total area), but in the Central, Southern and Privolzhsky districts, only 3% remain. The North Caucasian Federal District and what was until recently the Crimean Federal District both contain areas of intact wildlife populations (generally preserved in remote mountain areas) of only about 1.9 and 6.7% of the total area, respectively. Most of the existing agglomerations are found here, with a small number of protected areas.

Among Russian Arctic residents, the differences in the relationships between people and high-integrity countryside can differ by many orders of magnitude. For example, in Karelia, there are 8.5 ha of wild countryside per person (Arkhangelsk—15, Murmansk—12, Komi—33), but in Chukotka, this indicator of a personal relationship with nature reaches a value of 1351 ha per person. The Nenets Autonomous District is also among the Arctic leaders in intact countryside: it has 540 ha per capita. Yakutia has half that level, at 269 ha, and the Yamal-Nenets district half again, with 114 ha per person. The Krasnoyarsk Territory has about 87 ha of wild countryside per person (Bocharnikov and Egidarev 2016).

6.5 Wilderness and Biodiversity of Russia, the Russian Arctic and Siberia

Based on unified physical and geographical terms, we can support the claim that the Arctic Region (including the circumpolar north regions of the planet) contains the most intact degree of nature on the planet, even though it is fundamentally different from other zones in terms of its geopolitical, cultural, and economic conditions.

The real scale of the anthropogenic transformation of natural landscapes and ecosystems, as well as the critical consequences of geochemical and thermal pollution in sparsely populated and relatively undisturbed areas of the Far North, Siberia, and the Far East of Russia, is poorly documented and monitored.

However, in remote areas of the Arctic, where populations are relatively small, and foci and areas of anthropogenic impact are points and linear disturbances in the structure of natural landscapes within large areas of wild countryside, there is evidence of a significant negative impact both locally and across borders, primarily in areas subject to industrial exploitation and the transport of minerals. Abundant natural resources and large, uninhabited territories make the Arctic unique, as does the animal world of the Arctic, though it has the limited species composition typical of northern biogeocenoses (Fig. 6.9).

Fig. 6.9 The fox (*Vulpes vulpes* L.) often lives close to human settlements and working places. Northern Sakhalin Island. Photograph with kind permission of Ivan Tiunov

Continuous changes in the natural environment are occurring. Oil and gas production and supplies of these raw materials coming from abroad have a favorable effect on the environmental safety of many countries in Europe and Asia, but at the same time, current activities are negatively affecting the environment in sparsely populated and relatively underdeveloped regions of Russia. It can be stated that the value of these territories is growing rapidly at the same time as a massive anthropogenic impact on nature, especially due to the cardinal changes as a result of global climate change. The Arctic owes its exceptionally high integrity and importance to conservation activities at the end of the twentieth century.

The dictionary written by N. F. Reimers and A. V. Yablokov in 1982 defined *dikaya priroda* as “the nature of the “wild”: areas of nature, not disturbed by human economic activity, which a person affects only as a biological being or only indirectly through global changes in the world; the concept of subjective: from the point of view of the citizen.” And, although the general spatial characteristics of Eastern Russia do not change, there is a big difference between the developed areas of Siberia and such areas at European part of Russia.

Let us consider in detail the Siberia. In mostly within Siberia, there is the area whose development remains most closely connected to the exploitation of the richness of nature. The landscapes of the forest steppes, as well as small areas of the subtropical zone of the south of European Russia, are those which have been transformed to the greatest extent through industrialization, and only in the polar territories are there still a high proportion of conserved natural landscapes (Table 6.1).

At the beginning of the third millennium, mankind is still not fully aware of the exceptional importance of nature and has not found a harmonious way to maintain human health, while at the same time preserving the natural appearance of landscapes, the functionality and productivity of ecosystems, and the structure of the human environment. What is the reason?

On the basis of our calculations of wild countryside, the relevant map layers were “cut out” to show the distribution of remaining wilderness areas on the territory of the Siberian Federal district. Siberia is like the Amazon and the largest comparable green and white landmasses on the planet; it is part of the planet’s lungs

Table 6.1 The degree of preservation of natural environmental conditions in large remote landscapes of Russia

Physical geographical regions and countries (type of landscape)	Area of the landscape (km^2)	Wilderness area within landscape (km^2)	Wilderness (%)
Arctic deserts	75,641.6	72,825.5	96.3
Arctic Tundra	721,073.0	653,907.0	90.7
Mountain sparse forests and dwarf forests	1,842,033.5	1,638,462.5	88.9
Forest tundra and woodlands	448,524.7	408,540.2	91.1
Taiga	4,575,734.2	3,385,917.7	74.0

and kidneys. It is the guardian of freshwater reserves of global importance (think for example of the Baikal Lake).

The advantageous economic and geographical position of Western Siberia, due to its location at the crossroads of the railway lines and the great Siberian rivers Ob and Irtysh, right by the industrial Urals, developing Kazakhstan and Eastern Siberia, is enhanced by the monopolist role of the "owner" of the main fuel resources of northern Eurasia. Western Siberia accounts for 4/5 of Russian and more than 1/4 of global reserves of natural gas, and 70% of all oil reserves in the country (more than a third in the world), which determines the region's fate of intensive development.

Nonetheless, on the map (Fig. 6.6), it is clearly seen that the entire southern part of the district has been completely changed by anthropogenic effects. However, the plans by the Ministry of Natural Resources and Environment of the Russian Federation (MPR) and the development strategies of domestic protected areas, developed with the help of scientific and public environmental organizations, do not include the official "legalization" of protected areas in this category in Russia in the near future.

As a first step, our pioneering project was carried out in 2015–18 to assess the current degree of anthropogenic disturbance of the entire territory of Russia based on a calculated index of wilderness area integrity (see technical details in Bocharnikov and Egidarev 2015). More attention and effort are needed to achieve more exact and scientific data. We can already report some rethinking on the real role which protected areas can play.

The demonstration of the value of large protected areas shows that in most regions there are only fragments/islands of wild countryside left, which are not sufficient either to preserve biospheric functions or to preserve the populations of many terrestrial vertebrates (large mammals, birds, reptiles, and amphibians). These well-preserved natural landscapes are close to meeting the definition of Category 1b protected areas as described by the International Union for Conservation of Nature (IUCN).

This territorial assessment of the Arctic, based on a wilderness integrity index, allows us to understand the importance and urgency of conservation measures for both flora and fauna and the bioproductivity of ecosystems and biological diversity throughout Russia. The main significance of these "naturalness" indices is that the higher scored areas may be able to maintain a high level of integrity in terms of biospheric function due to their size and naturalness, but their wildlife is still subjected to a variety of influences. Particular attention should be paid to the protection of rare and endangered species in the face of increasingly intensified anthropogenic impacts and the very real effects of global climate change.

About 4.8 million km^2 of area remains relatively intact in the Far Eastern Federal District, where the share of natural areas is 78.3%. And although locally, and sometimes within the larger region, there are many consequences of all types of nature use, including hunting, fishing, and the harvesting of wild plants, as well as recreational activities and tourism, the naturalness of wilderness has been maintained. Thus, the vast territories of wild countryside that are preserved in Russia for the whole planet are not so much resource riches as a vital condition for maintaining the global ecological balance; a fact not yet recognized by the world community.

With this in mind, the role of wilderness within protected areas is even more important, as we have tried to understand by special calculations.

6.6 Transborder Issues in the Amur River Basin

Another example of the difference between the developed areas of Siberia and wilderness areas is the transborder issue, especially in the south of the Russian Far East, for instance, the Amur River Basin (Fig. 6.10). Natural and semi-natural ecosystems cover approximately half (1 million km^2) of the Amur-Heilong basin. Thus, the basin as a whole still has potential to sustain biodiversity and natural processes on large geographic scales. In Russia, approximately 75% of arable land is located in the southern riverine plains. This is driven in part by differences in topography, climate, and soil conditions, but also in part by social and political processes.

The Amur-Heilong is the largest river in northeast Asia. It is one of the longest undammed rivers in the world. The largest portion of the Amur-Heilong river basin, about 1 million km^2, lies within the Russian Federation. Mountain ranges, ridges, foothills, and plateaus cover two-thirds of the region. Most mountains are low, ranging from 300 to 1000 m in elevation The Amur River Basin Wilderness and Ecological Network should be developed as soon as possible as a conceptual framework for organizations to cooperate and set priorities at a pan-Asian level.

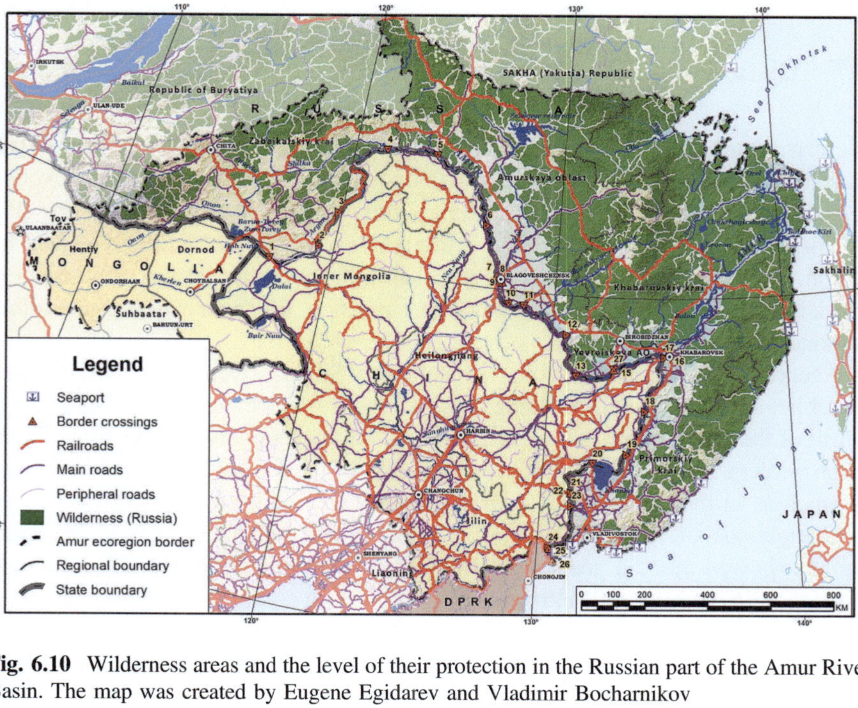

Fig. 6.10 Wilderness areas and the level of their protection in the Russian part of the Amur River Basin. The map was created by Eugene Egidarev and Vladimir Bocharnikov

Important tasks for biodiversity conservation in this zone will include setting aside intact ecosystems in protected areas and supporting traditional uses of natural resources. The main elements of the Amur River Basin Wilderness Biological and Landscape Diversity Strategy should be:

- a description of how the ecosystem in Eastern Asia functions or fails to function, in terms of causes and effects;
- a vision of ecological targets for the intermediate and long term, respecting ecological connections and geological, scenic and cultural values, acknowledging the biogeographic regions of Asia, its surrounding seas and intercontinental ecological relations;
- guidance on the full integration of nature conservation into other policies, in particular, agriculture, forestry, fisheries, land use and water planning and management and regional development, based on the principle of environmental sustainability;
- an East Asian ecological network, maintaining and enhancing the conservation and coherence of natural and semi-natural habitats and natural processes of Asian importance, paying particular attention to characteristic threatened and endangered species;
- priorities for Asian actions, including the more effective implementation of international nature conservation instruments and funds, and identifying where urgent action is needed by national authorities;
- public awareness, education, and information programs.

6.7 Distances from Settlements to Wilderness

We carried out special calculations based on the assumption that for people who want to go to the countryside, up to 10 km is possible for walking and cycling access; 11–25 km for day trips to the countryside, returning home by car; 25–50 km when the possibility of visits is limited to some extent by the time spent on arrival and fuel consumption; while at distances from 50 to 100 km or more, only specially organized and/or targeted trips are practical for those wishing to explore "untouched countryside."

According to our calculations, in Southern Siberia (Kemerovo oblast, Altai Republic, and Khakassia), the share of wilderness of protected areas comes close to a fifth of the overall area (23, 7, 19.1, and 22.4%, respectively). In the Republic of Altai, wilderness is preserved in more than 40% of the territory, or more than twice as much as in the territory of existing protected areas (Fig. 6.11).

Comparison of the availability of wilderness and protected areas in the same regions showed that the areas with well-preserved natural landscapes (conditional size 20 × 25 km), significantly more accessible (average) for the population in such areas as the Irkutsk region, the Krasnoyarsk region, the Republics of Tuva and Buryatia, or all autonomous districts, which is a good reflection of the relatively lower degree of their anthropogenic disturbance.

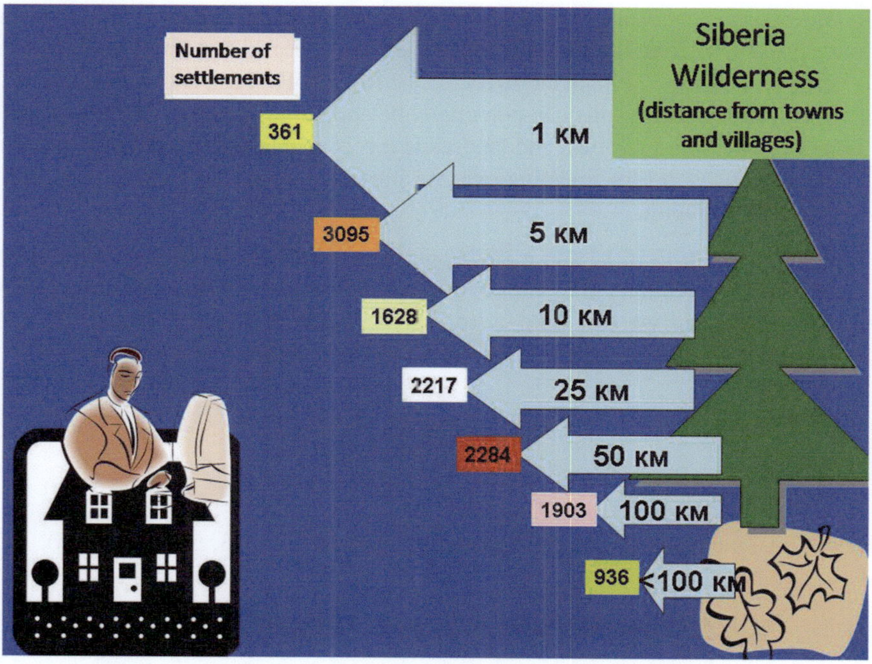

Fig. 6.11 Actual accessibility of large (more than 50 km^2) wilderness areas for residents of Siberia

Our calculation of the population's access to protected areas showed the following results: the remoteness of settlements varies significantly in various areas of the Siberian Federal District (SFD), averaging from 31.7 (Republic of Khakassia) to 184.0 km in the Irkutsk region. Almost half (49%) of villages, towns, and cities are at a distance of 25 km; 23% of settlements are from 50 to 100 km away from available areas of wilderness, and only 3% of residents have to travel a distance of 100 km or more. When updating existing nature protection areas in different categories and planning new ones, these results could be taken into consideration.

6.8 Conclusions

1. Wild countryside (wilderness, Russian: *dikaya priroda*) is a fundamental category for the development of human civilization.
2. It is largely associated with remoteness from disturbances caused by human economic activity, with broad and impressive biodiversity, and with high ecosystem services and landscape values.
3. The current geological epoch of the Anthropocene is characterized by accelerated human influence on wilderness areas. This requires their monitoring and assessment, and finally appropriate planning and conservation measures.

4. As a basis for this, we developed a wilderness index and conducted a GIS-based status analysis of wilderness areas in Russia. The latter revealed that the most extensive areas of wilderness were in the Arctic, in the Russian Far East, and in Siberia.
5. Our approach and results can potentially be applied to monitor wilderness areas at different scales, regional and landscape levels included. They could also serve as a starting point for updating existing protected areas and future nature conservation planning.

Figures 6.12, 6.13, 6.14 and 6.15 (Appendix) show some impressive species of fauna in the wild countryside. We need to pay more attention to public awareness and education, and we can learn a lot if we consider the actual role of existing protected areas and opportunities for the public to use them.

Appendix

Faunal inhabitants of wilderness landscapes (Figs. 6.12, 6.13, 6.14 and 6.15).

Fig. 6.12 Oriental honey-buzzard (*Pernis ptilorhyncus*). Terney town, Primorsky Krai. October 8, 2012. Photograph with kind permission of Ivan Seryodkin

Fig. 6.13 Domestic reindeer (*Rangifer tarandus*) at Tofalarya (Irkutskaya oblast). Photograph with kind permission of Konstantin Klokov

Fig. 6.14 Elk (*Alces alces*) on mesotrophic peatland in Western Siberia. Photograph with kind permission of Eugene Strelnikov

Fig. 6.15 Female Wapiti Red Deer (*Cervus elaphus xanthopygus*) in winter (Sikhote-Alin Nature Reserve). Photograph with kind permission of Ivan Seryodkin

References

Bocharnikov VN (2012) Tsennosti v dikoy prirode sovremennogo cheloveka – novy aktualny fokus meghdistsiplinarnykh issledovani (Wilderness value of modern humans as new current focus of interdisciplinary research). Astrakhanski vestnik ecologicheskogo obrazovania 1:72–87

Bocharnikov VN (2014) O meghdunarodnoi terminologii, zarueghnykh predstavlenniyakh I geographicheskom smysle konseptstii dikoi prirody (About international terminology, foreign understanding and geographical understanding of wilderness). In: Geographicheskie issledovanyua vostochnykh raionov Rossii: etapy osvoenyua I perspektivy razvitiya. Pacific Geographical Institute, Vladivostok, pp 32–36

Bocharnikov VN (2016) Geoprostranstvo "dikoy prirody" Rossii (Wild Russia Geospace). ARGO Bull 5:165–175

Bocharnikov VN (2017) Dikaya priroda I osobookhranyaemyye prirodnye territorii (OOPT) Sibiri (Wilderness and nature protected areas (PAs) of Siberia). In: Kotlyakov VM, Chibilev AA, Tishkov AA (eds) Voprosy geografii. Sb. 143. Geograficheskiye osnovy zapovednogo dela (k 100-letiyu zapovednoy sistemy Rossii) (Questions of geography. Collection of papers. 143. Geographical bases of nature reserve management (on the 100th anniversary of the Russian nature reserve system). Kodeks Publication House, Moscow, pp 106–133

Bocharnikov VN, Egidarev EG (2015) "Dikaya priroda" – novyy prirodookhrannyyi ndeks dlya territorii Rossii (Wilderness: a new indicator for the assessment of existing anthropogenic transformation on the territory of Russia). Problemy regional'noy ekologii (Probl Reg Ecol) 5:75–80

Bocharnikov VN, Egidarev EG (2016) Ecologo-geographicheskoye kartographirovaniye ekonomicheskogo prostranstva Rossii na osnove technology GIS (Ecological and geographical mapping of Russian economic regions based on GIS technologies). Vestnik Volgogrskogo gos. un-ta. Ser. 3, Ekon. Ekol. (Bull Volgograd State Univ Ser 3 Econ Ecol) 3:163–176

Bocharnikov VN, Egidarev EG (2017a) Dikaya priroda v landshaftakh I ekoregionakh Rossii (Wilderness in landscapes and ecoregions of Russia). Geografiya I prirodnyye resursy (Geogr Nat Res) 4:38–49

Bocharnikov VN, Egidarev EG (2017b) "Dikaya priroda" kak strategicheski element prostranstva Rossii (Wilderness as a strategic element of Russian territory). Astrakhanski vestnik ecologicheskogo obrazovania 2:11–21

Cao Y, Long Y, Rang R (2017) Research on the identification and spatial distribution of wilderness areas at the national scale in mainland China. Chin Landscape Architect 33 (258):26–33

Casson SA, Martin VG, Watson A, Stringer A, Kormos CF (Volume Editors) (2016). No 25: Wilderness protected areas: management guidelines for IUCN Category 1b protected areas. IUCN World Commission on Protected Areas (WCPA)

Crutzen PJ (2002) Geology of mankind. Nature 415:23. https://doi.org/10.1038/415023a

Dezhkin VV (2005) Osobo okhranyaemye prirodnye territorii mira I nekotorye Rossiiskogo zapovednogo dela (Specially protected natural territories of the world and some problems of the Russian nature reserve). In: Dezhkin VV (ed) Use and protection of natural resources in Russia recreational resources and protected areas, vol 3, pp 142–151

Harvey MWT (2015) Wilderness forever. University of Washington Press, Seattle. 328 p

Isakov YuA (1983) Printsipy planirovaniya osobo okhranyaemykh prirodnykh territory v SSSR (Principles of planning for the protected areas in USSR) In: Landscape conservation and planning. M. P. 128–140

Kormos CF, Bertzky B, Jaeger T, Shi Yu, Badman T, Hilty J, Mackey BG, Mittermeier RA, Locke H, Osipova E, Watson JEM (2015) A Wilderness Approach under the World Heritage Convention. Conserv Lett 9:228–235

Lavrenko EM, Heptner VG., Kirikov SV, Formozov AV (1958) Perspektivnyy plan geograficheskoy seti zapovednikov (proyekt) (Perspective planning geographic network of reserves (project). Okhranyayemoye prirody I zapovednoye delo v SSSR (Protected by nature and protected in the USSR). Bul. 3. Moscow: an SSSR, pp 3–92

Lesslie RG, Mackey BG, Preece KM (1988) A computer based methodology for the survey of wilderness in Australia. Environ Conserv 15:225–232

Lewis SL, Maslin MA (2015) Defining the anthropocene. Nature 519:171–180. https://doi.org/10.1038/nature14258

McCloskey JM (1966) The Wilderness Act of 1964: its background and meaning. Oregon Law Rev 45:288–321

McCloskey JM, Spalding H (1989) A reconnaissance-level inventory of the amount of wilderness remaining in the world. Ambio. https://doi.org/10.1073/sf3280-015-0693-1

Mueller L, Sheudshen AK, Syso A, Barsukov P, Smolentseva EN, Khodzher T, Sychev VG, Romanenkov VA, Rukhovich OV, Puzanov A, Dannowski R, Schindler U, Eulenstein F (2016) Land and water resources of Siberia, their functioning and ecological state. In: Mueller L, Sheudshen AK, Eulenstein, F (eds) Novel methods for monitoring and managing land and water resources in Siberia. Springer Water. Springer International Publishing, Cham, pp 3–74. http://link.springer.com/chapter/10.1007%2F978-3-319-24409-9_1

Mueller L, Blum WEH, Schenk W, Romanenkov V, Sychev VG, Jones M, Kowarik I, Ball BC, McKenzie BM, Gerasimova M, Dronin NN, Kazakov LK, Kravchenko VV, Bukvareva E, Couvet D, Sheudzhen AK, Saparov A, Pachikin K, Abuduwaili J, Kosolapov VM, Schindler U, Mirschel W, Schnug E, Lischeid G, Eulenstein F (2018). About landscapes and their utilisation: status and trends of landscape research. In: Sychev VG, Mueller L (eds) Novel methods and results of landscape research in Europe, Central Asia and Siberia. Monograph in 5 Volumes. Vol I Landscapes in the 21th century: status analyses, basic processes and research concepts, 502 p. Chapter I/1: pp 16–25. http://vniia-pr.ru/monografii/pdf/tom1-1.pdf. Accessed 30 Jan 2019

Nash RF (2001) Wilderness and the American mind. Yale University Press, New Haven, CO, 413 pp

Olson D, Dinerstein E (2002) The global 200: priority ecoregions for global conservation 1. Ann Missouri Bot Gard 89:199–224

Prisyazhnaya AA, Chernova OV, Snakin VV (2016) Razvitiye systemy osobo okhranyaemykh prirodnykh territorye (OOPTP – osnova sokhranenuya biologicheskogo raznoobraziya prirodnykh kompleksov (Development of a network of nature reserves: the basis for conserving the biological diversity of natural systems). Space and time. vol 11, issue 1: The system of the planet Earth. Fixed network address: 2227-9490e-aprovr_east11-1.2016.41

Reimers NF, Shtilmark FR (1978) Osobo okhranyaemye prirodnue territorie (Protected areas). Moscow Mysl, 296 p

Reimers NF, Yablokov AV (1982) Slovar terminov I ponyatyi, svyazannykh s okhranoi ghivoi prirody (Glossary of terms and concepts related to wildlife conservation). Nauka, Moscow, 144 p

Sanderson EW, Jaiteh M, Levy MA, Redford KH, Wannebo AV, Woolmer G (2002) The human footprint and the last of the wild: the human footprint is a global map of human influence on the land surface, which suggests that human beings are stewards of nature, whether we like it or not. Bioscience 52(10):891–904

Shtilmark FR (2001) Absolyutnaya zapovednost' – posledniy oplot real'noy okhrany dikoy prirody (Absolute reserves—the last stronghold of real wildlife protection). Gumanitarny y ekologicheskiy zhurnal (Humanitarian Ecol J) 3:111–113

Sokolov VYe (2006) Biosfernyye zapovedniki: tseli I problemy (Biosphere reserves: goals and problems). Lektsii laureatov Demidovskoy premii (1993–2004) (Lectures of the winners of the Demidov Prize (1993–2004). Ekaterinburg. http://www.aif.ru/dontknows/file/zapovednaya_sistema_rossii_spravka

Taylor K (2016) The Historic Urban Landscape paradigm and cities as cultural landscapes. Challenging orthodoxy in urban conservation. Landscape Res 41(4):471–480

Zapovedniki (1977) Zapovedniki Sovetskogo Soyuza (Natural Reserves of the Soviet Union) Moscow. Lesnaya promyshlennost' Publishing House, 88 p

Zapovedniki Dalnego Vostoka SSSR (1985) (Natural Reserves of the Russian Far East). In: Sokolov VYe, Syroechkovsky YeYe (eds). M: Mysl, 1980, 264 p

Chapter 7
Towards an Interdisciplinary Understanding of Landscape Qualities: Wilderness, Tranquillity and Dark Skies

Flurina M. Wartmann, William A. Mackaness, Nicole Bauer, Janine Bolliger and Felix Kienast

Abstract The need to assess landscape qualities has become increasingly important over the past decades, with landscapes being continuously shaped and re-shaped through dynamic natural and anthropogenic processes. It is now widely recognised that landscapes need to be studied both in terms of their physical and ecological elements as well as how people living in or visiting landscapes perceive and interact with them. Different approaches have been developed over time for assessing these variegated aspects of landscape qualities, which range from methods in the natural sciences to the social sciences and humanities. Using the three examples of wilderness, tranquillity and dark skies as landscape qualities, we review existing quantitative and qualitative approaches to illustrate the potential of interdisciplinary landscape research. Furthermore, we highlight the potential of novel methodologies and data sources to study these landscape qualities, including the use of machine learning for automated image recognition, analysis of social media data (tags, location and image content), as well as citizen science approaches.

Keywords Landscape qualities · Landscape research · Landscape perception · Wilderness mapping · Tranquillity mapping · Nightscapes · Dark skies · Soundscapes

7.1 Introduction

Landscapes are rapidly changing across the world. For Western Europe, two simultaneous but contrasting trends are observed: firstly, abandonment of agricultural lands followed by spontaneous reforestation in marginal areas, and secondly,

F. M. Wartmann (✉) · N. Bauer · J. Bolliger · F. Kienast
Swiss Federal Research Institute for Forest, Snow and Landscape WSL, Zürcherstrasse 111, 8903 Birmensdorf, Switzerland
e-mail: flurina.wartmann@wsl.ch

W. A. Mackaness
School of Geosciences, Drummond Street, Edinburgh EH8 9XP, UK

L. Müller and F. Eulenstein (eds.), *Current Trends in Landscape Research*, Innovations in Landscape Research, https://doi.org/10.1007/978-3-030-30069-2_7

high socio-economic pressure on urban and peri-urban areas leading to conversion of cultivated and natural land into built-up areas. The first trend leads to wilderness areas with high recreational quality, while the second trend results in increasing numbers of recreationists from growing population centres seeking landscapes exhibiting qualities such as wilderness, tranquillity or dark skies. However, assessing and maintaining these landscape qualities is challenging. Consequently, these qualities run the risk of being neglected in policies and decision-making impacting future landscapes. In this chapter, we focus on these 'neglected' landscape qualities of *wilderness*, *tranquillity* and *dark skies*. We first introduce general approaches for 'sensing landscapes' using different sources of information and analysis methods. Using the three examples of wilderness, tranquillity and dark skies, we then describe state-of-the-art methods for assessing and modelling landscape qualities through qualitative and quantitative approaches, as well as highlight innovative new methodologies and data sources that are emerging. Finally, we discuss the challenges and opportunities of modelling perceived landscape qualities in the context of landscape research.

7.1.1 Capacities in the Sensing of Landscapes

Landscapes can be sensed by the individual (experiential), through digital sensors (often as earth observations), and they can be modelled using surrogate variables. In combination with a wide variety of qualitative and quantitative techniques, it is possible to classify, detect change and generally model different characteristics of the landscape. The free sharing of global datasets (of increasing recency and often of high resolution), coupled with bespoke, intuitive to use, mobile apps able to record many of the attributes of landscape, has facilitated a significant growth in citizen science. Thus human and digital sensor networks coupled with analytical techniques can produce a wide range of outputs and perceptions that are critical to public engagement and policy-making (Fig. 7.1).

Well established is the combination of remote sensing (RS), geographic information systems (GIS) and spatial analysis techniques to characterise landscape, monitor change and model landscape ecologies (Yang et al. 2013). Remote sensing gives emphasis to the physical properties of landscape, but the human dimension to experiencing landscape is equally important (Cosgrove 1984). In this respect, citizen science has been transformative to research, a myriad of projects are testament to the importance of user-generated content in experiencing and monitoring landscape. The potential and benefit of public participation in modelling and mapping qualities of the landscape is well understood (Dickinson and Bonney 2012). Bespoke mobile apps, open-source software and data portals have hugely facilitated the gathering and sharing of geographical data in support of the earth observation agenda (Fritz et al. 2017). This direct and active involvement of a large number of contributors has been termed *active crowdsourcing*. In addition to these directed citizen projects, imagery and related user-generated text-data from social media has

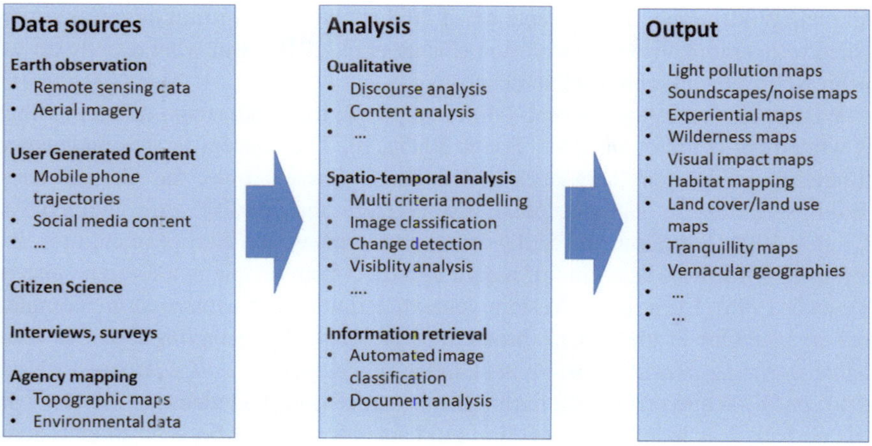

Fig. 7.1 Overview of examples of data sources, analysis methods and outputs to assess landscapes

been used as a data source (for example from platforms including Flickr, Instagram, Geograph and Pinterest). The analysis and use of such data has been termed *passive crowdsourcing*. Notwithstanding issues of data quality (in particular, recorder bias, completeness and currency), user-generated data have enabled the creation of vernacular geographies with an emphasis on understanding the perception and experience of 'place'—important in understanding how we perceive and value landscape (Purves et al. 2011). In summary, what Fig. 7.1 seeks to convey is a multi-methodological approach to understanding qualities of landscapes that draws on a fusion of human and digital sensor networks that results in a deeper comprehension of what landscapes mean—culturally and ecologically. In the following sections, we explore these ideas with reference to three important qualities of landscape (wilderness, tranquillity and dark skies), and discuss approaches of how we might model and characterise these qualities.

7.2 Wilderness

Wilderness has become an important concept in nature conservation, as well as in tourism and outdoor recreation. On the one hand, the continued expansion of human infrastructure and intensive land use means we continue to loose wilderness as the human footprint on the land is growing larger (Sanderson et al. 2002). On the other hand, many rural areas, for instance, in Europe, face agricultural land abandonment, and lacking intervention, will revert to some form of 'wilderness' (Höchtl et al. 2005). The need to conserve wilderness areas has been recognised in politics and nature conservation (Jones-Walters and Čivić 2010), and the European parliament put wilderness on its political agenda (European Parliament 2009). In parallel, WildEurope, a conglomerate of NGOs and other institutions, was established to operationalise the resolution (Fisher et al. 2010). But what exactly do we mean when we talk about wilderness?

Wilderness is a vague concept—different people may understand different things as wilderness (Lutz et al. 1999; Bauer 2005a, b). The concept is also understood differently in different languages and cultures. For example, the German term 'Wildnis' differs in its conceptual history from the English term 'wilderness' (Cronon 1996; Stremlow and Sidler 2002) and in many non-European cultures, the notion of wilderness as a kind of nature alienated from people is a foreign concept (Descola 1996). Even in the Western context, definitions of wilderness are complex and diverse. One of the perhaps broadest is by Nash: '*It is tempting to let the term define itself: to accept as wilderness those places people call wilderness* (Nash 2014, p. 5)'. While this phrase highlights the social constructedness of the concept, it is perhaps less helpful in providing guidelines for those who need to take decisions on the management and protection of wilderness.

If we want to provide tangible evidence for policy-making and planning that will aid in protecting wilderness as an important quality of landscapes, it becomes necessary to define what wilderness is, how people perceive it and where we can still find wild places. A growing body of literature in a range of different research areas deals with these questions. One strand of research is concerned with producing robust and repeatable models in geographic information systems (GIS) that tell us where we can still find wilderness (Kliskey and Kearsley 1993; Fritz et al. 2000; Carver et al. 2012, 2013; Orsi et al. 2013; Carver and Fritz 2016b). Another strand of research investigates public perception of wilderness (Habron 1998; Bauer 2005a, b; Mc Morran et al. 2008; Bauer et al. 2018). In the following, we provide an overview of these two different strands, acknowledging that they are not distinct areas of research but can be seen as a spectrum from more perception-focused research to GIS modelling. We also include more critical views of the concept of wilderness as a cultural invention (Cronon 1996). Finally, we introduce novel potential approaches for wilderness mapping using automated image classification approaches that allow the detection of potential 'wilderness' from photographs.

7.2.1 Mapping Wilderness

Nash's definition of wilderness being what people think it serves to illustrate wilderness as a sociocultural construct, but mapping wilderness characteristics in GIS requires operationalisable definitions and selecting criteria and attributes (Carver and Fritz 2016a). In the USA, for example, the Wilderness Act of 1964 defines wilderness through terms such as '*untrammeled*', '*undeveloped*', '*natural*', '*opportunities for solitude and unconfined recreation*' (US Congress 1964). The federal agencies involved in managing wilderness areas in the National Wilderness Preservation System published a conceptual framework that links management actions directly to these qualities described in the Wilderness Act (Landres et al. 2008).

At the global level, the IUCN defines Wilderness areas (IUCN Category Ib) as: '*Protected areas that are usually large unmodified or slightly modified areas, retaining their natural character and influence, without permanent or significant human habitation, which are protected and managed so as to preserve their natural condition*' (IUCN 2018). In most wilderness definitions, particularly those that became the basis for mapping, we can identify an ecological component of the degree of naturalness or modification, as well as a perceptual component (Lesslie and Taylor 1985). In the literature, an important distinction is made between wilderness quality and wilderness areas. Wilderness qualities are defined as '*the extent to which any specified unit area is remote from and undisturbed by the impacts and influence of modern technological society*' (Lesslie 2016, p. 21). Wilderness areas are defined as '*relatively large intact natural areas—places where wilderness quality is defined using agreed thresholds recognised by society*' (Lesslie 2016, p. 21). In their guidelines for the management of wilderness in Natura 2000 areas, the European Commission defines wilderness as: '*A wilderness is an area governed by natural processes. It is composed of native habitats and species, and large enough for the effective ecological functioning of natural processes. It is unmodified or only slightly modified and without intrusive or extractive human activity, settlements, infrastructure or visual disturbance* (European Commission 2013, p. 10).

Different definitions and notions of wilderness exist, and are still being debated. The issue of definition is not merely an academic concern, but is of direct consequence for mapping, as we can only map what we have clearly defined. It is also paramount for policy-making and planning (Mc Morran et al. 2008), as a lack of clear definitions may hamper the effectiveness of protecting wilderness. In this respect, mapping wilderness quality is considered an important endeavour to assess the status of this landscape quality. Carver and Fritz (2016b) highlight the paradox of mapping wilderness and elaborate why it is better to map than not to map: mapping wilderness helps to localise where more or less wild places exist, providing hard proof that is needed to convince decision-makers and politicians of the need to protect wilderness. Even before GIS became mainstream technology, McCloskey and Spalding (1989) mapped the world's remaining wilderness areas using two criteria: areas more than 4000 km^2 in extent and more than 6 km from recorded human features were identified based on Jet Navigation Charts at a scale of 1:2 million. The resulting map showed that around a third of the world's area could be seen as wilderness, with the largest areas located at high latitudes and in desert areas (McCloskey and Spalding 1989).

Most wilderness mapping conceptualises the quality of wilderness as a spectrum of the 'degree to which a place is remote from and undisturbed by the influences of modern technological society', which Lesslie and Taylor (1985) and Nash (2014) described as the 'wilderness continuum'. This continuum avoids drawing sharp distinctions between what is, and what is not wilderness, acknowledging that the drawing of such an (imaginary) boundary is highly dependent on social and cultural norms. It also incorporates the notion of 'modern technological society's impact', and not of indigenous and traditional societies. In inventorying wilderness in South

Australia, Lesslie and Taylor (1985) implemented the wilderness continuum concept in a spatial manner, defining wilderness as 'four wilderness quality indicators (Lesslie and Taylor 1985). Using similar indicators for the national scale map of Australia (remoteness from settlement, remoteness from access, distance from human artefacts and naturalness of the land at national scale), the *Australian National Wilderness Inventory* was arguably the first comprehensive use of GIS to map wilderness quality (Lesslie and Maslen 1995). Many wilderness mapping initiatives have followed, but the two basic factors of the Australian Inventory, *naturalness* and *remoteness* are, in some form, used in almost all models of wilderness quality today (Carver and Fritz 2016a). Some models extended the criteria to four components: naturalness, remoteness, ruggedness (as a measure of the terrain) and human impact (Fritz et al. 2000; Carver et al. 2002; Carver and Fritz 2016b). Different methods have been applied to wilderness mapping, with Comber et al. (2010) empirically demonstrating how different models lead to different wilds, highlighting the importance of selecting criteria and weightings. For example, for the UK, Fritz et al. (2000) implemented weighted distance decay models (e.g. of population) to avoid defining rigid thresholds and included differential weighting of wilderness quality data layers through multi-criteria evaluation (Fritz et al. 2000). Weighting of GIS layers has been done through expert-consultation (Radford et al. 2019), and public participation GIS has also been tested as a means to gage different perspectives and provide weightings for wilderness models in a participatory manner (Carver et al. 2002).

The challenges involved in mapping wilderness quality notwithstanding, mapping wilderness have become an important tool for conservation, and a range of wilderness mappings have been conducted across different scales, from global to local. At a global level, Lesslie (1998), in work for the World Conservation Monitoring Centre, extended the Australian concept to cover the entire world. Other global mapping initiatives include the Global Methodology for Mapping Human Impacts on the Biosphere (GLOBIO, UNEP 2002) and the Human Footprint and the 'Last of the Wild' map (Sanderson et al. 2002). Recent global wilderness mappings highlight the decline in wilderness areas over the past decades (Watson et al. 2016) and how the world's largest remaining wilderness areas are concentrated in a handful of countries, including Russia, Canada, Australia, the USA and Brazil (Watson et al. 2018). For Europe, a Wilderness Quality Index has been published that maps wilderness quality as a continuous surface (EEA 2011), highlighting Northern Scandinavia, Iceland and Northern Scotland as areas where pockets of wilderness remain. Fritz et al. (2000) mapped wilderness quality across the UK. Examples of wilderness maps at national scale include the aforementioned example of Australia (Lesslie and Taylor 1985), Austria (Plutzar et al. 2016), Iceland (Ólafsdóttir et al. 2016), Scotland (SNH 2013) and Switzerland (Radford et al. 2019). The Swiss case study confirmed the important role of mountains, showing how the largest contiguous areas with high wilderness quality are to be found in the Swiss alps (Fig. 7.2).

In another case study for two mountainous regions in Southern Switzerland, the focus was on mapping the quality of remoteness rather than wilderness through the

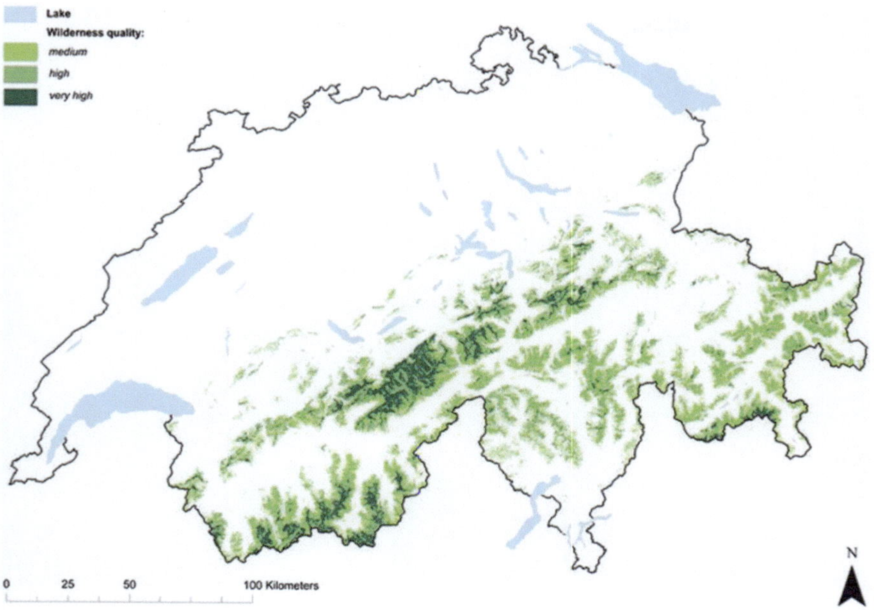

Fig. 7.2 Last remaining continuous areas with high wilderness quality are to be found in the alpine areas of Switzerland (Own illustration. Data sources: basemap data from BFS GEOSTAT/ Federal Office of Topography swisstopo, data on wilderness quality from Radford et al. 2019)

use of a remoteness measure developed specifically for that study (Boller et al. 2010). Boller et al., argued the concept of remoteness was better suited to mountainous regions of Europe as areas that have long been settled and managed by alpine communities, and where the cultural heritage and traditional land use form an important part of the experience of alpine landscapes.

Case studies of wilderness mappings at local scale that can guide protected area management include, for instance, the Death Valley National Park in the USA (Carver et al. 2013), the Cairngorms and the Loch Lomond and The Trossachs National Park in Scotland (Comber et al. 2010; Carver et al. 2012) or the Carpathian mountains in Romania (Măntoiu et al. 2016).

One of the areas where wilderness quality has been most comprehensively mapped is Scotland, which has been mapped at global, regional and local scales. Comparing these different maps highlights how wilderness is, to some extent, a scale-dependant concept. On global wilderness maps, Scotland does not show up as an area with high wilderness quality (UNEP 2002). On a wilderness map of Europe (EEA 2011), we start to see small pockets in Scotland emerging as some of the wildest areas, the largest areas within Europe being located in Iceland and Scandinavia (Kuiters et al. 2013). For Great Britain, wilderness mapping with public participation GIS showed that Scotland contained most of the top 10% areas considered 'wildest' (Carver et al. 2002). At the national scale, the Scottish Natural Heritage mapping highlighted areas of Northern and Northwestern Scotland as

Fig. 7.3 Cairngorms is one of Scotland's largest Wild Land Areas and protected through the Cairngorms National Park. Image: 'Munros Trip 9/08 (ct28)' by Ted and Jen on Flickr (CC BY 2.0)

areas with high wilderness quality, whereas the Southern part of Scotland showed low levels (SNH 2013). At the local scale, wilderness maps for the two national parks in Scotland (Cairngorms and the Loch Lomond and The Trossachs) highlight differences in wilderness quality between two protected areas (Carver et al. 2012), with generally higher levels of wilderness quality in the Cairngorms (Fig. 7.3). Such wilderness maps at the local level can form an empirical basis for managing development within a protected area, and guide recreational planning.

The Scottish case also illustrates how wilderness mapping can have an influence on policy-making. The data from the national mapping (SNH 2013) was used to directly inform Scottish policy (SNH 2014a). In a later phase, the largest and most wild areas were identified, which are now listed as 42 'Wild Land Areas' (SNH 2014b). The list includes small areas, e.g., around single mountains such as Ben More on Mull, or parts of the island of Rum. Larger Wild Land Areas are the Cairngorms (Fig. 7.3) and parts of the Scottish Highlands.

7.2.2 *Wilderness as a Cultural Notion*

It is important to note that some of the areas now identified as Wild Land Areas in Scotland were once populated and the basis for rural farming communities. This changed during the Clearances in the late eighteenth and early nineteenth century.

Fig. 7.4 Suisnish ruins on the Isle of Skye in Scotland. During the Clearances, thousands of families were forcefully removed from their homes to make way for commercial sheep farming. Image: 'Suisnish Ruins' by Ross Collins on Flickr (CC BY 2.0)

Landowners, sometimes forcefully and involving human atrocities, evicted rural communities from the land they were living on to make way for more lucrative commercial sheep farming (Richards 2012). It is this human suffering of highland communities that eventually resulted in what many people today perceive as wilderness (Fig. 7.4). Perhaps unsurprisingly, notions of wilderness and wild land are still controversially discussed in Scotland (Mc Morran et al. 2008). The continued erasure of people from the land through environmental narratives of nature is criticised, with writers such as James Hunter advocating for acknowledging people's history and connection to place (Hunter 2014).

The case of erasing people from the landscape is not unique to the Highlands of Scotland. Other examples include conservation initiatives across the globe that, with the aim of protecting idealised (Western) notions of wilderness, displaced indigenous people from the land they inhabited (Neumann 1998; Agrawal and Redford 2009). Acknowledging that wilderness is a social and cultural construct makes it possible to also critically engage with this concept. In his essay 'The Trouble with Wilderness, or, Getting Back to the Wrong Nature', Cronon (1996) investigates wilderness from a historical perspective, critical of the separation of people and nature, and calling for a rethinking of the human–nature relationship. Stremlow and Sidler (2002) explore the concept of wilderness ('Wildnis' in German) as a cultural phenomenon through its use in modern literary and journalistic texts to study societally rooted perceptions of wilderness, showing how wilderness is constructed conflictingly, as a place of fear, and as a place of the idyllic. In more applied research, a case study on the Isle of Harris in Scotland shows how cultural notions of wilderness, which are related to the exclusion of crofting communities, can be re-thought as a more inclusive concept that integrates conservation and crofter's stewardship of the land that creates possibilities for more socially just future landscapes (Mackenzie 2006). To shape landscapes where the importance of people and their ties to place as well as nature conservation are acknowledged, it is important to understand and take into account public opinion towards nature and wilderness.

7.2.3 The Perception of Wilderness

In the highly urbanised and developed continent of Europe, only few areas of wilderness remain, not all of which are formally protected and thus risk being encroached upon by infrastructure development. At the same time, we observe continuing agricultural abandonment creating new possibilities for (secondary) wilderness areas. In order to safeguard the continued existence of wilderness areas in the future and potentially enable some formerly used areas to revert to secondary wilderness, it is important to understand how people who live in and around such areas perceive wilderness and nature to design appropriate landscape management policies that will be supported by the public.

People's attitudes towards wilderness have been studied with respect to their expectations and experiences in specific areas (Durrant and Shumway 2004; Wallner et al. 2007; Bauer et al. 2018), as well as independently of whether they are visitors or residents in a particular area (Lutz et al. 1999; Cordell et al. 2003; Johnson et al. 2004; Bauer 2005a, b; Bauer et al. 2009).

Regarding general opinions about wilderness, a study in Canada found that respondents to a questionnaire survey had relatively positive opinions and favoured wilderness protection, irrespective of whether they lived in rural or urban areas (Lutz et al. 1999). However, a photograph-rating task in the same study revealed that rural dwellers had higher thresholds for what they considered as wild than urban dwellers. For instance, only 11% of rural dwellers considered a photograph of a valley with agriculture in the foreground as wilderness compared to 43% of the urban dwellers (Lutz et al. 1999). This shows that while opinions about wilderness were overall positive, people had different understandings about what wilderness is. Based on the US National Survey on Recreation and the Environment (NSRE) in the year 2000, a study on the influence of ethnic background showed that while there were differences in visitation rates and use-values between different groups, less differences were observed for intrinsic and non-use values attributed to wilderness (Johnson et al. 2004). In the same survey younger people, residents in the East of the USA and ethnic minorities were more likely to be in favour of expanding the system of wilderness areas (Cordell et al. 2003). Results from a comparison of the NSRE survey results between 1994 and 2000 indicated that awareness of wilderness and the importance of non-use values had increased in the overall population, but this did not translate to an increased acceptance of expanding wilderness areas (Cordell et al. 2003).

In a comprehensive study on the public understanding of wilderness in Switzerland, Bauer (2005b) investigated public attitudes towards wilderness with over 1500 written questionnaires sent out to a representative sample of the population. The public understanding of wilderness in Switzerland was relatively congruent with scientific criteria, including the lack of visible human impact and the degree of naturalness. A difference in public and scientific understanding was that for the public, areas with dense vegetation are seen as wild, irrespective of their development history (Bauer 2005b). Several socio-demographic factors influenced opinions about wilderness, including age, place of residence (rural, urban) and general attitudes towards nature (e.g. nature conservationists). Critical opinions towards wilderness were more likely from people living in rural areas, older people and people who grew up in rural areas. More positive opinions were likely from nature conservationists, younger people and city dwellers (Bauer 2005b). Proponents of wilderness were less tolerant of visible human influences in wilderness areas. Opponents were more tolerant of the presence of infrastructure and perceived areas with higher human influence as wilderness than the proponents. People in the French- and Italian-speaking parts of Switzerland also held more critical opinions about wilderness, whereas people in the German-speaking part were more positive, which may reflect sociocultural differences in wilderness perception between regions but could also be linked to more general differences with the terms used in German, French and Italian. Based on these findings, Bauer (2005b) presents region-specific management suggestions that include the creations of different 'wilds' to satisfy the needs and wishes of different groups.

In a survey about attitudes towards wilderness in the South-Western Carpathians in Romania, attitudes of people living within protected areas were divided into two groups (Bauer et al. 2018). One group that had a more utilitarian attitude towards nature and wished for nature to stay unchanged and be aesthetically pleasing while another group favoured unrestricted access to wilderness and considered it important for wilderness to be remote. While attitudes towards nature and wilderness were positive for both groups, they were more critical of protected areas, highlighting how the local population differentiates between wilderness in general and protected areas, a finding similar to the study on wilderness areas in Utah (Durrant and Shumway 2004).

Research into perception of wilderness thus brought forth the various and diverse factors that influence people's perception of and attitudes towards wilderness, ranging from factors at societal to more individual level.

7.2.4 *New Methods for Wilderness Research: Automated Image Annotation*

Wilderness is perception-based—it is what people perceive it to be. For instance, the absence of infrastructure, wide open lands, or rough terrain. This makes wilderness an interesting concept to be studied through the use of methods such as automated image annotation, a process through which keywords are automatically assigned to an image. This is possible through machine learning, where a computer algorithm 'learns' to recognise certain elements in a picture, say the presence of people, or a dog, and then scans new pictures for being more, or less likely, to also depict these elements it has been trained to detect. It does so through a training dataset that is manually annotated, where human annotators have decided whether a picture depicts people, dogs or any other elements that are of interest to users (e.g. cars, houses, etc.). This annotated dataset is then provided as an input for the algorithm for learning to recognise similar pictures and label them. Some companies are offering pre-trained algorithms that can be applied to any images and will automatically annotate them with tags. One such example is the Google Vision API, where interestingly, the algorithm has been trained to use the label 'wilderness'. For instance, it labelled Figs. 7.3 and 7.4 as 'wilderness'. Automated image recognition thus opens new research avenues, where landscape characteristics based on perception can be inferred from images. In combination with user-generated images from social media platforms such as Instagram or Flickr, where users often add coordinates to an image uploaded (i.e. geotags), it is now possible to study spatial patterns of wilderness through user-generated image content.

7.3 Tranquillity

Today, the term tranquillity is widespread, including for marketing and place promotion. Tranquillity is therefore something that people seek and value, but what does tranquillity mean? Dictionaries provide definitions such as 'the quality or state of being tranquil' and synonyms include 'peacefulness', 'quietness', 'serenity'. The Roman philosopher Seneca (ca. 4 BC–65 AD) wrote in his work 'de tranquillitate animi' (c.f. Reynolds 1998) about tranquillity as a state of mind that is the opposite to anxiety and worry. Achieving such a state of mind is a challenge that can be made easier through the presence of certain environmental or landscape characteristics. For example, in Japanese aesthetics, high importance is ascribed to harmonious environments promoting a tranquil state of mind, as illustrated, for example, in design principles of Japanese gardens (van Tonder and Lyons 2005).

Although tranquillity is something that is individually perceived, many would agree that some landscape characteristics are more likely to instil tranquillity. Seeking tranquillity, peace and quietness away from the hustle and bustle of everyday life was shown to be a driver of recreation behaviour (Beard and Ragheb

1983; Frick et al. 2007), with recreationists visiting areas that exhibit these qualities. In the UK, for example, tranquillity was found to be among the main reasons why people visited the countryside (NBS 2004). Apart from attracting visitors, there are positive health effects associated with tranquil areas (Ulrich et al. 1991; De Vries et al. 2003; Velarde et al. 2007; Shepherd et al. 2013; Seresinhe et al. 2015; Vienneau et al. 2017).

Given its importance for health and recreation, tranquillity has become recognised as an important quality of landscapes that is reflected at the policy level. The UK, for instance, included the protection of tranquillity in its National Planning Policy Framework, stating that: '*Planning policies and decisions should aim to: […] identify and protect areas of tranquillity which have remained relatively undisturbed by noise and are prized for their recreational and amenity value for this reason*' (National Planning Policy Framework 2012, p. §123). At the European level, the focus has been on environmental noise, which can be seen as an important aspect of the more holistic notion of tranquillity. The Environmental Noise Directive (END2002/49/EC) (END 2002) relates to the assessment and management of environmental noise, and requires member states to prepare noise maps and action plans, including the identification of noise exposure levels and preserving areas with low environmental noise levels. We see tranquillity as a more all-encompassing concept because we can imagine an area that is completely silent (e.g. an abandoned industrial area), but this area does not necessarily instil tranquillity.

From a research perspective, tranquillity is a multifarious concept that has been investigated in different fields, using various approaches. In the following, we provide a brief overview of strands of research and different methodologies relating to the exploration and study of tranquillity. We highlight how user-generated content in the form of geotagged social media photographs provide a novel source of information for investigating tranquillity as a landscape quality.

7.3.1 Image Rating Experiments About Tranquillity and Landscape Preference

In environmental psychology, the attention restoration theory by Kaplan and Kaplan (1989) states that natural environments facilitate restoration through 'soft fascination', a combination of capturing our attention effortlessly ('moderate fascination') and pleasure (preference or aesthetic quality). Based on this theoretical framework, Herzog and Bosley (1992) conducted image rating experiments to investigate the relationship between landscape preference and tranquillity. Images with different landscape settings, including fields and forests, mountains and water bodies were shown to participants. Results showed a high correlation of preference and tranquillity ratings, but for some landscape settings such as standing bodies of water, tranquillity was rated higher than preference. These experiments established tranquillity and landscape preference as two related, but distinct concepts (Herzog and Bosley 1992). In a follow-up study, Herzog and Barnes (1999) investigated the influence of the factors 'unstructured openness', 'mistiness' and 'surface calmness' in predicting tranquillity and preference ratings in three landscape settings (desert, large water body and field/forest), showing that surface calmness was a predictor in some settings (including large water bodies), but not all. Another study investigated differences of tranquillity ratings for images with urban and natural scenes, where tranquillity was rated significantly higher for natural than urban scenes (Herzog and Chernick 2000). Factors that were positively correlated with tranquillity in natural surroundings were how well a setting appeared to be cared for (neatness), the amount of foliage visible (nature) and the amount of visible open space (Herzog and Chernick 2000). In these experiments, the focus had been on rating visual stimuli. In the next section, we summarise research on soundscapes that takes into account noise levels in addition to visual stimuli.

7.3.2 Measuring Soundscapes

Various efforts have been made to develop tools that predict tranquillity based on visual landscapes and their associated soundscapes. Pheasant et al. (2008) investigated the relationship between perceived naturalness of a scene, sound levels and tranquillity ratings by presenting visual and acoustic data of 11 English rural and urban landscapes to 44 subjects. Based on their findings, they developed a tranquillity rating prediction tool (TRAPT) that consists of an equation combining sound measures and visual stimuli (Pheasant et al. 2009). To validate and further develop this prediction tool, 18 subjects were presented with various imagery and pre-recorded sounds from three locations (a city garden, an urban park and a rural churchyard). The subsequent analysis revealed a linear relationship between tranquillity ratings with naturalness and recorded sound levels. Tranquillity ratings increasing with naturalness (percentage of natural features present in a scene) and decreased with increasing sound levels (Pheasant et al. 2009). This work demonstrated the possibility of predicting tranquillity based on features perceived in the surroundings (including natural features such as water, foliage, rock, but also including cultural heritage and traditional farm buildings) and sound levels from human sources. Arguing that due to their accessibility and population numbers in urban settings, tranquil urban spaces may be relatively more important to the overall population than tranquil rural spaces, Watts et al. (2011) investigated three urban recreation areas, where levels of tranquillity were estimated based on the prediction tool and compared to ratings from visitors in field-based surveys. In a subsequent study, eight urban green areas were examined and visitor questionnaires with 252 respondents were used to explore potential refinements to the prediction tool (Watts et al. 2013). Though able to produce good predictions, potential refinements to the model include giving penalty points for parks where graffiti and litter, is noticeably present, as this was shown to influence visitors' sense of tranquillity and improved predictions (Pheasant et al. 2009). Other applications of the tranquillity rating prediction tool include exploring its relevance to wind farm siting and discusses the impact of differing expectations about tranquillity in urban areas and the countryside, highlighting the need for high levels of precision in defining anthropogenic disturbance both visual and acoustic (Watts and Pheasant 2015a). Further exploring the utility of the prediction tool in natural surroundings, a set of controlled experiments were used in which video footage was captured for 46 different natural scenes in Scotland and various noise sources were added to study the effect of different sources of noise and tranquillity and wildness ratings (Watts and Pheasant 2015b). Ratings from respondents show that tranquillity is a different construct from wilderness, and that tranquillity was much more impacted by acoustic variables, including human noise. Interestingly, adding natural sounds did not significantly improve tranquillity ratings. Overall, their analysis led Watts and Pheasant to argue that soundscape is an integral factor in people's sense of tranquillity and should therefore be an integral part of the assessment process (Watts and Pheasant 2015b).

Collectively, research on soundscapes shows that sound needs to be included as part of any assessment process while also revealing the complex and subtle nature of anthropogenic and natural sound in terms of its frequency, level and people's expectations.

7.3.3 Mapping Tranquillity

As one of the most active institutions for raising awareness about tranquillity and the protection of tranquil areas since the 1990s, the non-governmental organisation 'Campaign to Protect Rural England' (CPRE) commissioned a series of research and mapping projects (CPRE and the Countryside Commission 1995; Levett 2000; MacFarlane et al. 2004; CPRE 2005, 2007, Jackson et al. 2008). In his pioneering work for ASH consulting, Simon Rendel conducted the first mapping of tranquil areas in 1991 in a study for the Department of Transport. Rendel's map showed how significant areas so far unaffected by development, but undesignated and unprotected, were vulnerable to a proposed transportation corridor. Based on this work, the CPRE and the Countryside Agency (1995) produced a set of 'Tranquil Area' maps for England. This approach to tranquillity mapping was based on modelling the impact of audio-visual disturbances in the form of roads, railways, power stations and settlements, defining thresholds to these sources of noise. One of the main critiques of this early approach was that fixed thresholds from noise sources were used, and that the definitions of the thresholds and sources of disturbance were based on expert-opinions, rather than public consultations (Levett 2000). In aiming to address these critiques and to include the views of the public, a subsequent project in the Northumberland National Park and West Durham Coalfield in England conducted a comprehensive participatory appraisal to determine what people considered as tranquil (MacFarlane et al. 2004). In a second step, the researchers operationalised the factors contributing to and diminishing tranquillity in a GIS model that included as layers such as remoteness, naturalness of land cover, openness and noise sources. The different layers were weighted again through a public consultation exercise. In a subsequent study, a similar approach was also applied to the Chilterns in the South East of England (Jackson et al. 2008), and then scaled-up to produce a tranquillity map for England (CPRE 2007). To produce the tranquillity map for England, over 1000 people were consulted and the thresholds applied to the GIS layers (for instance, the distance at which the impact of a road on tranquillity decreases) were determined through photograph-rating tasks with the public (Jackson et al. 2008). The importance of taking into account public opinion was also recognised in a study in the Dorset Area of Outstanding Beauty, which incorporated views from different stakeholders, including parts of the population considered hard to reach. In total, views of over 800 participants were collected through participatory appraisals including focus groups, household and on-site surveys (Hewlett et al. 2017). The criteria that people deemed important were open landscape, the absence of human-made noise, traffic and the presence of other people. Based on this information a GIS model was created, using over 70 input layers (Terradillos and Wilkinson 2015).

7.3.4 New Methods for Tranquillity Research: Analysing Tags from Geotagged Social Media Photographs

One of the challenges in research on landscape qualities such as tranquillity or wilderness is that these qualities are based on people's perception, and such perceptions have commonly been cost-intensive to assess across larger areas. If national surveys are used, sample sizes that can be achieved are limited given common financial constraints. As a consequence, the data are often not available at spatial granularities that allow detailed spatial modelling. In this respect, social media data generated by users is a novel form of data that enables researchers to study place-based experiences of people in landscapes (Guerrero et al. 2016; van Zanten et al. 2016; Chesnokova and Purves 2018; Wartmann et al. 2018). This body of work is based on people contributing content to social media platforms such as Instagram, Flickr or Twitter in the form of photographs, associated text (i.e. 'tags' or 'hashtags') and often also provide coordinates for their photographs ('geotags').

Through the tags associated with images, we can, for instance, select images that have been tagged with words such as 'tranquillity', 'tranquil', 'peaceful' or 'quiet'—which provide us with a dataset of images that are likely to have been taken at locations where people experienced tranquillity (Fig. 7.5).

Mapping the spatial distribution of these locations while controlling for the underlying distribution of all images provides us with an assessment of areas where people experience tranquillity based on social media data such as Flickr images (Wartmann et al. 2019; Wartmann and Mackaness, in review).

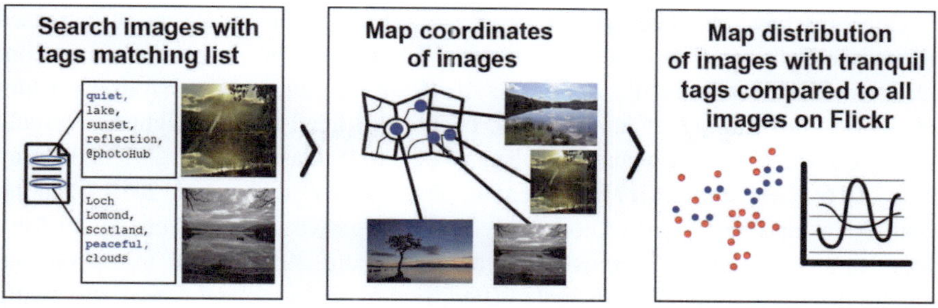

Fig. 7.5 Methodology for assessing tranquillity based on selecting geotagged photographs from social media through keywords associated with tranquillity

7.4 Dark Skies

Artificial light at night is an important driver of global change in the twenty-first century (Davies and Smyth 2018). In European cities, artificial light at night has become omnipresent: from streetlamps, illuminated buildings and soccer courts, restaurants, shop windows, and blinking billboards to our homes, where we illuminate front doors with security lights or decorate lawns and windows with artificial lighting. Artificial light at night provides a range of benefits, including safety on roads at night for traffic and pedestrians, as well as aspects of convenience. The unprecedented increase in lighting also has negative impacts, including the disappearance of dark skies (e.g. Fig. 7.6), which makes it impossible for people to observe the stars at night (CfDS 2009).

Fig. 7.6 Night sky above Domat/Ems and Tamins in the Canton of Grisons, Switzerland (image courtesy of B. and C. Wartmann)

Concerns have also been raised about ecological and health effects of increased night-time lighting (Smith 2008). The term 'light pollution' is often used in connection with dark skies and is defined as '*artificial light that is excessive or intrudes where it is not wanted*' (Claudio 2009, p. 29). Artificial light at night has largely negative effects on biodiversity (Hölker et al. 2010; Gaston et al. 2013; Brüning et al. 2016; Knop et al. 2017), organisms' physiology (Hölker et al. 2010; Brüning et al. 2016) and behaviour (Knop et al. 2017). Negative impacts were found for a large variety of animal species (Rich and Longcore 2013). Both terrestrial and aquatic insects are negatively affected (Perkin et al. 2014). Street lamps effectively act as light traps and can kill billions of insects throughout the year, in turn reducing food availability for predators, which affects entire ecosystems (Rich and Longcore 2013). In sea turtles, artificial light was shown to impact nesting behaviour and to be negatively related to nest densities (Silva et al. 2017; Hu et al. 2018), as well as reduce survival rates of turtle hatching by disorienting them away from the sea towards illuminated beach infrastructure (Lorne and Salmon 2007). Negative effects were also demonstrated for nocturnal mammals such as beach mice (Bird et al. 2004) and bats (Polak et al. 2011; Russo et al. 2017; Rydell et al. 2017; Azam et al. 2018).

Furthermore, there is now increasing empirical evidence that light pollution also has detrimental and potentially far-reaching effects on people's health (Navara and Nelson 2007; Cho et al. 2015). Artificial light at night may be associated with increased risk of different types of cancer (Kloog et al. 2008; Stevens 2009), obesity (Fonken et al. 2010) and diabetes (Spiegel et al. 2005). The internal (circadian) clock regulates various bodily functions and is closely linked to the daily change from light to dark. The hormone melatonin, for example, is only produced in dark phases. If melatonin production is disrupted, sleep disturbances or changes in the immune system can occur (Chepesiuk 2009). An economic study estimates the damage caused by light pollution to the USA alone at US$ 7 billion per year (Gallaway et al. 2010).

From a landscape perspective, the loss of dark skies changes not only the way organisms and ecosystems function, but also the way people (can't) experience dark skies, which is considered a cultural loss. Across the world, large natural areas with dark skies are disappearing and in many cities and nearby areas, the night skies are lightened-up so much that a fifth of all people in the world can no longer observe the milky way with their naked eyes (Cinzano et al. 2001). Exceptions are some areas in highly developed European nations, where night skies in some (small) areas have become darker again. This development is linked to the decline in primary industries and the reduced need for lighting linked to industrial processes, which lead to less artificial light emissions at night in some parts of northern Germany or South-West England (Bennie et al. 2014). The prevailing trend, however, continues to be an increase in night lighting, particularly in urbanised areas (Fig. 7.7).

With increased awareness about light pollution, actions are being taken to reduce artificial light emissions (Gaston et al. 2012). Initiatives to mitigate light emissions related to public infrastructure include advanced street-light technologies. These

Fig. 7.7 'Earth's City Lights'. Marc Imhoff/NASA GSFC, Christopher Elvidge/NOAA NGDC, Image: Craig Mayhew and Robert Simmon/NASA GSFC (Creative Commons CC0)

may rely on energy-efficient, long-lived and robust LED illuminants, which have been shown to considerably reduce light pollution (Bennie et al. 2014; Poiani et al. 2015; Shahzad et al. 2016). Despite some promising examples, light pollution levels are still increasing across most European countries, which brings the monitoring of nightscapes and their perception by the public onto the research agenda.

7.4.1 The Perception of Artificial Light Pollution and Dark Skies

For example, the reduction of light pollution is an ambition of the Campaign to Protect Rural England—their interactive map revealing England's darkest and most light-polluted skies (CPRE 2018). In Switzerland, the public perceives light emissions as a new source of environmental pollution that has to be limited (Zumthor et al. 2005). In areas such as the Val Müstair in the Canton of Grisons that are still relatively little affected by night pollution, dark skies are perceived positively by inhabitants as well as visitors (Mazenauer 2015). Despite positive associations with dark skies in Val Müstair, no consensus was found to further enhance this landscape quality through specific measures, such as turning off street lighting between one and five in the morning every day or reducing the availability of artificially led ski slopes and sledging trails. Interview participants were in favour of such artificial lighting for snowsport activities, but also supported guided star gazing walks as touristic activity (Mazenauer 2015). However, with increasing light pollutions across Europe, the future of localised initiatives related to dark skies remains unsure. According to a survey by the Swiss Federal Office for the Environment (FOEN) among the Cantons of Switzerland, there is a demand for standards about light emissions, but so far, only few cantons have provided guidelines on reducing artificial light emissions, for example, Basel-Landschaft (AuE 2004).

In order to inform the public and to provide baseline measures of the landscape quality of dark skies, it is important for environmental monitoring programmes to include observations of light pollution. The Swiss landscape monitoring program (LABES) provides such time series from 1992 onwards (FOEN and WSL 2013; Kienast et al. 2015). The indicator light emissions aggregate many processes such as urbanisation, land abandonment or remoteness in one measure, without being able to distinguish between the processes. From a landscape monitoring perspective, on the one hand, the comprehensiveness of this indicator is welcome and cost-efficient new measure. On the other hand, it does not allow evaluations of the effectiveness of specific environmental policies.

7.4.2 Modelling Light Emissions

Light emissions can be modelled using satellite imagery (Cinzano et al. 2001). For Switzerland, light emission modelling showed that these emissions have increased by 70% between 1994 and 2009, with fewer areas being dark at night. For example, on the Swiss plateau and in the Jura mountains, there is no cell that is entirely dark at night (Fig. 7.8), the last dark 1×1 km cell in the Jura mountain disappeared. The total area with night-darkness decreased significantly between 1992 and 2000, but afterwards, the curve flattens out considerably. Such a flattening is not observed in the values of night brightness/total light emission (not shown) for the last ten years, which increases continuously. This indicates a certain concentration of light-lit areas since the turn of the millennium.

The importance of protecting dark skies from becoming 'extinct' with increasing use of, often unnecessary, artificial lighting is increasingly being recognised. Apart from environmental benefits in protecting insects, birds, bats and many other species, reducing artificial lighting helps to save energy. From a landscape perspective, we do not yet know enough how people experience the loss of dark skies, which is a landscape quality that forms part of an extended understanding of wilderness or wild areas. Such dark sky areas in connection with wilderness may become an important economic resource that generates income through tourism. For instance, the International Dark Sky Association has the ambition of creating 'Dark Sky Parks' such as the one designated in the Cairngorms, Scotland. Other examples include the US National Park service launching a successful series of dark sky protection initiatives, such as the designation of the Petrified Forest National

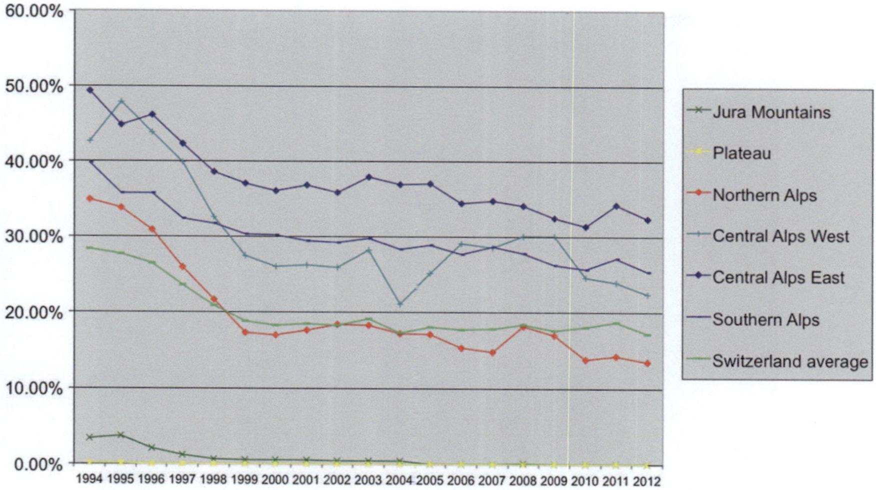

Fig. 7.8 Relative area of complete night-darkness over the years 1992–2012 in the biogeographic regions of Switzerland (definition night-darkness: Radiance < 1 $(W * cm^{-2} * sr^{-1} * \mu m^{-1}) * 10^{-10}$). Three years are averaged in each case. The spatial resolution is 1 km^2 (own illustration from data source: Kienast et al. 2015)

Park in Arizona as an International Dark Sky Park, and the Yellow Stone National Park launched activities to reduce artificial light (US National Park Service 2018).

7.4.3 *New Methods for Dark Sky Research: Citizen Science*

Citizen science—often defined as science conducted by citizens who aren't professional scientists—has recently gained popularity not only as a way to engage people in diverse topics, but also as a form of collecting data that would otherwise be unfeasible, be it for bird conservation, galaxy classification or protein structure modelling. Examples of citizen science projects on dark skies show how research on landscape qualities can benefit from engaged people collecting data on the ground.

So far, dark sky research has primarily focused on measuring light emissions through satellite data. However, satellites measure light visible from space, not the brightness on the ground that affects people and other organisms. This information needs to be modelled using the satellite measures as input data and calibrating the model with ground truth data—data which would be impossible to collect without the help of citizen scientists. Several projects concerning dark skies have been launched, which can produce important datasets that complement existing research on dark skies (Kyba et al. 2013). Examples include the 'Globe at Night' web platform (https://www.globeatnight.org/webapp/) where you select how your sky

Fig. 7.9 Globe at Night App allows citizen scientists to report which star constellations they can see at night at a particular location and time, allowing scientists to estimate sky brightness. Image: 'Light pollution: It's not pretty' by jpstanley on Flickr (CC BY 2.0)

looks compared to a series of star charts (Fig. 7.9). The data collected through this platform then allows researchers to track changes in sky brightness over time at the global or national level.

Because 'Globe at Night' provides a distinct number of options showing different levels of sky constellations visible, it does not provide a very exact measurement of sky brightness. Therefore, the 'Loss of the Night' app for mobile phones (http://lossofthenight.blogspot.com/) was designed to complement 'Globe at Night'. Based on the 'Sky Map' app by Google, citizen scientists are asked how many individual starts they can recognise in the night sky. Determining the least bright scar identified enables researchers to determine more accurately the levels of sky brightness as ground truth data. Another approach is the 'The Dark Sky Meter' app (https://www.darkskymeter.com/), which makes use of the smartphone camera to measure sky brightness. All these citizen-collected datasets are made available to the public where anyone interested can view the data (http://www.myskyatnight.com/#map).

Citizen science projects are not without caveats, including the disproportionately large contribution of few participants, difficulties of maintaining participant's motivation and questions revolving around data quality (Cohn 2008; Eveleigh et al. 2014; Lukyanenko et al. 2016). These challenges notwithstanding, citizen science offers a new tool to collect data on landscape qualities, and while we have focused on citizen science for dark skies, such an approach is feasible to apply to other landscape qualities.

7.5 Conclusion

Landscape qualities including tranquillity, wilderness and dark skies are ideal topics through which to illustrate the very essence of landscape research as a multidisciplinary and therefore multi-perspectival research. On the one hand, these concepts can be investigated through a positivistic or essentialist perspective, measuring the physical, in other words, the more tangible aspects of landscapes from a natural science perspective. For instance, we can measure noise levels, light emissions seen from space, and distances from human infrastructure as indications of landscape qualities people perceive. On the other hand, we can assess these landscape qualities from a social science perspective, asking about attitudes towards certain landscape qualities such as wilderness, or how people perceive tranquillity or dark skies in certain landscape settings, and, last but not least, what values and meanings groups of people or societies associate with these qualities. In this chapter, we aimed to illustrate that multiple forms of research can be combined to create a more holistic view on landscape qualities, including physical aspects, as well as how they are perceived and evaluated by individuals and societies. We introduced novel approaches and methodologies to investigate landscape qualities, using, for instance, automated image recognition, analysis of geotagged social media imagery and texts, as well as citizen science approaches, highlighting emerging trends and potential new research directions.

This chapter is not meant as a thorough review of the concepts of wilderness, dark skies and tranquillity and associated methodologies for their assessment. Rather, we see it as an illustration of the potential for landscape research to engage multiple perspectives from diverse research fields and serve as the basis for discussions about multiple and diverse approaches to assess landscapes and their qualities.

References

Agrawal A, Redford K (2009) Conservation and displacement: an overview. Conserv Soc 7(1):1

AuE BL (2004) Stopp der Lichtverschmutzung. http://www.darksky.ch/dss/wp-content/uploads/2004/01/BL_flyer.pdf. Last accessed 15 March 2019

Azam C, Le Viol I, Bas Y, Zissis G, Vernet A, Julien J-F, Kerbiriou C (2018) Evidence for distance and illuminance thresholds in the effects of artificial lighting on bat activity. Landscape Urban Plann 175:123–135

Bauer N (2005a) Attitudes towards wilderness and public demands on wilderness areas. In: Kowarik I, Körner S (eds) Wild urban woodlands. New perspectives for urban forestry. Springer, Berlin, pp 47–66

Bauer N (2005b) Für und wider Wildnis: Soziale Dimensionen einer aktuellen gesellschaftlichen Debatte. Bristol Sc. Bern: Haupt, 185 p

Bauer N, Wallner A, Hunziker M (2009) The change of European landscapes: human-nature relationships, public attitudes towards rewilding, and the implications for landscape management in Switzerland. J Environ Manage 90(9):2910–2920

Bauer N, Vasile M, Mondini M (2018) Attitudes towards nature, wilderness and protected areas: a way to sustainable stewardship in the South-Western Carpathians. J Environ Plann Manage 61 (5–6):857–877

Beard JG, Ragheb MG (1983) Measuring leisure motivation. J Leisure Res 15(3):219–228

Bennie J, Davies TW, Duffy JP, Inger R, Gaston KJ (2014) Contrasting trends in light pollution across Europe based on satellite observed night time lights. Sci Rep 4:3789

Bird BL, Branch LC, Miller DL (2004) Effects of coastal lighting on foraging behavior of beach mice. Conserv Biol 18(5):1435–1439

Boller F, Hunziker M, Conedera M, Elsasser H, Krebs P (2010) Fascinating remoteness: the dilemma of hiking tourism development in peripheral mountain areas. Mt Res Dev 30(4):320–331

Brüning A, Hölker F, Franke S, Kleiner W, Kloas W (2016) Impact of different colours of artificial light at night on melatonin rhythm and gene expression of gonadotropins in European perch. Sci Total Environ 543:214–222

Carver S, Fritz S (2016a) Introduction. In: Carver S, Fritz S (eds) Mapping wilderness: concepts, techniques and applications. Springer, Dordrecht, pp 1–16

Carver S, Fritz S (2016b) Mapping wilderness: concepts, techniques and applications. Springer, Dordrecht, 204 p. ISBN 978-94-017-7397-3. https://doi.org/10.1007/978-94-017-7399-7

Carver S, Evans AJ, Fritz S (2002) Wilderness attribute mapping in the United Kingdom. Int J Wilderness 8(1):24–29

Carver S, Comber A, McMorran R, Nutter S (2012) A GIS model for mapping spatial patterns and distribution of wild land in Scotland. Landscape Urban Plann 104(3):395–409

Carver S, Tricker J, Landres P (2013) Keeping it wild: mapping wilderness character in the United States. J Environ Manage 131:239–255

CfDS (2009) Blinded by the Light? A Handbook on Light Pollution. BAA, Hampshire, 40 p. http://www.britastro.org/dark-skies/pdfs/HANDBOOKTEXT.pdf. Last accessed 18 Mar 2019

Chepesiuk R (2009) Missing the dark: health effects of light pollution. Environ Health Perspect 117(1):A20

Chesnokova O, Purves RS (2018) From image descriptions to perceived sounds and sources in landscape: analyzing aural experience through text. Appl Geogr 93:103–111

Cho Y, Ryu S-H, Lee BR, Kim KH, Lee E, Choi J (2015) Effects of artificial light at night on human health: a literature review of observational and experimental studies applied to exposure assessment. Chronobiol Int 32(9):1294–1310

Cinzano P, Falchi F, Elvidge CD (2001) The first world atlas of the artificial night sky brightness. Mon Not R Astron Soc 328(3):689–707

Claudio L (2009) Switch on the night: policies for smarter lighting. Environ Health Perspect 117 (1):A28–A31

Cohn JP (2008) Citizen science: can volunteers do real research? Bioscience 58(3):192

Comber AE, Carver S, Fritz S, McMorran R, Washtell J, Fisher P (2010) Different methods, different wilds: evaluating alternative mappings of wildness using fuzzy MCE and Dempster-Shafer MCE. Comput Environ Urban Syst 34(2):142–152. https://doi.org/10.1016/j.compenvurbsys.2009.10.006

Cordell HK, Tarrant MA, Green GT (2003) Is the public viewpoint of wilderness shifting. Int J Wilderness 9(2):27–32

Cosgrove D (1984) Social formation and symbolic landscapes. The University of Wisconsin Press, Madison, WI. ISBN 0-299-15514-5, 332 p

CPRE (2005) Mapping tranquillity. Defining and assessing a valuable resource. Campaign to Protect Rural England, Northumberland. http://www.cpre.org.uk/resources/countryside/tranquil-places/item/download/369. Last accessed 15 Mar 2019

CPRE (2007) Tranquillity map: England, campaign to protect rural England. http://www.cpre.org.uk/resources/countryside/tranquil-places/item/download/352. Last accessed 15 Mar 2019

CPRE (2018) England's light pollution and dark skies. https://www.nightblight.cpre.org.uk/maps. Last accessed 15 Mar 2019

CPRE and the Countryside Commission (1995) Tranquil areas: England map. https://www.thenbs.com/PublicationIndex/documents/details?Pub=CPRE&DocID=273255. Last accessed 15 Mar 2019

Cronon W (1996) The trouble with wilderness, or, getting back to the wrong nature. In: Cronon W (ed) Uncommon ground: rethinking the human place in nature. W.W. Norton and Company, New York, pp 69–90

Davies TW, Smyth T (2018) Why artificial light at night should be a focus for global change research in the 21st century. Glob Change Biol 24(3):872–882

De Vries S, Verheij RA, Groenewegen PP, Spreeuwenberg P (2003) Natural environments—healthy environments? An exploratory analysis of the relationship between greenspace and health. Environ Plann A 35(10):1717–1731

Descola P (1996) In the society of nature, a native ecology in Amazonia. Cambridge University Press, Cambridge, 372 p. https://doi.org/10.2458/v6i1.21503. ISBN-13: 978-0521574679, ISBN-10: 0521574676

Dickinson JL, Bonney R (2012) Citizen science: Public participation in environmental research. Cornell University Press, Ithaca, NY, 240 p

Durrant JO, Shumway JM (2004) Attitudes toward wilderness study areas: a survey of six southeastern Utah counties. Environ Manage 33(2):271–283

EEA (2011) Wilderness quality index Europe. https://www.eea.europa.eu/data-and-maps/figures/wilderness-quality-index/wilderness-quality-index-including-terrain. Last accessed 15 Mar 2019

European Commission (2013) Guidelines on Wilderness in Natura 2000. Management of terrestrial wilderness and wild areas within the Natura 2000 Network. Technical report-2013-069. http://ec.europa.eu/environment/nature/natura2000/wilderness/pdf/WildernessGuidelines.pdf. Last accessed 15 Mar 2019

European Parliament (2009) Wilderness in Europe European Parliament resolution of 3 February 2009 on Wilderness in Europe (2008/2210(INI)). https://publications.europa.eu/en/publication-detail/-/publication/5bc06340-9f99-4071-9e26-a4abfc34ce6e/language-en. Last accessed 15 Mar 2019

Eveleigh A, Jennett C, Blandford A, Brohan P, Cox AL (2014) Designing for dabblers and deterring drop-outs in citizen science. In: Proceedings of the SIGCHI conference on human factors in computing systems, pp 2985–2994

Fisher M Carver S, Kun Z, McMorran R, Arrell K, Mitchell G (2010) Review of status and conservation of wild land in Europe. Project commissioned by the Scottish government. Leeds, UK

FOEN and WSL (2013) Neue Ansätze zur Erfassung der Landschaftsqualität. Zwischenbericht Landschaftsbeobachtung Schweiz (LABES). Berne. https://www.bafu.admin.ch/dam/bafu/de/dokumente/landschaft/uw-umwelt-wissen/neue_ansaetze_zurerfassungderlandschaftsqualitaet.pdf.download.pdf/neue_ansaetze_zurerfassungderlandschaftsqualitaet.pdf. Last accessed 15 Mar 2019

Fonken LK, Workman JL, Walton JC, Weil ZM, Morris JS, Haim A, Nelson RJ (2010) Light at night increases body mass by shifting the time of food intake. Proc Natl Acad Sci 107 (43):18664–18669

Frick J, Degenhardt B, Buchecker M (2007) Predicting local residents' use of nearby outdoor recreation areas through quality perceptions and recreational expectations. Forest Snow Landscape Res 81(1–2):31–41

Fritz S, Carver S, See L (2000) New GIS approaches to wild land mapping in Europe. In: McCool SF, Cole DN, Borrie WT, O'Loughlin J (eds) Wilderness science in a time of change conference—Volume 2: Wilderness within the context of larger systems, 1999 May 23–27, Missoula, MT. Proceedings RMRS-P-15-VOL-2. Ogden, UT: US Department of Agriculture, Forest Service, Rocky Mountain Research Station, pp 120–127

Fritz S, Fonte CC, See L (2017) The role of citizen science in earth observation. Remote Sens 9 (4):357

Gallaway T, Olsen RN, Mitchell DM (2010) The economics of global light pollution. Ecol Econ 69(3):658–665

Gaston KJ, Davies TW, Bennie J, Hopkins J (2012) Reducing the ecological consequences of night-time light pollution: options and developments. J Appl Ecol 49(6):1256–1266

Gaston KJ, Bennie J, Davies TW, Hopkins J (2013) The ecological impacts of nighttime light pollution: a mechanistic appraisal. Biol Rev 88(4):912–927

Guerrero P, Møller MS, Olafsson AS, Snizek B (2016) Revealing cultural ecosystem services through Instagram images: the potential of social media volunteered geographic information for urban green infrastructure planning and governance. Urban Plann 1(2):1–17

Habron D (1998) Visual perception of wild land in Scotland. Landscape Urban Plann 42(1):45–56

Herzog TR, Barnes GJ (1999) Tranquility and preference revisited. J Environ Psychol 19 (2):171–181

Herzog TR, Bosley PJ (1992) Tranquility and preference as affective qualities of natural environments. J Environ Psychol 12(2):115–127

Herzog TR, Chernick KK (2000) Tranquility and danger in urban and natural settings. J Environ Psychol 20(1):29–39

Hewlett D, Harding L, Munro T, Terradillos A, Wilkinson K (2017) Broadly engaging with tranquillity in protected landscapes: a matter of perspective identified in GIS. Landscape Urban Plann 158:185–201

Höchtl F, Lehringer S, Konold W (2005) "Wilderness": what it means when it becomes a reality—a case study from the southwestern Alps. Landscape Urban Plann 70(1–2):85–95

Hölker F, Wolter C, Perkin EK, Tockner K (2010) Light pollution as a biodiversity threat. Trends Ecol Evol 25(12):681–682

Hu Z, Hu H, Huang Y (2018) Association between nighttime artificial light pollution and sea turtle nest density along Florida coast: a geospatial study using VIIRS remote sensing data. Environ Pollut 239:30–42

Hunter J (2014) On the other side of sorrow: nature and people in the Scottish Highlands. Birlinn Ltd., Edinburgh, UK, p 272

IUCN (2018) Category Ib: wilderness area. https://www.iucn.org/theme/protected-areas/about/protected-areas-categories/category-ib-wilderness-area. Last accessed 17 Oct 2018

Jackson S, Fuller D, Dunsford H, Mowbray R, Hext S, MacFarlane R, Haggett C (2008) Tranquillity mapping: developing a robust methodology for planning support. Report to the Campaign to Protect Rural England, Centre for Environmental & Spatial Analysis, Northumbria University, Bluespace environments and the University of Newcastle upon on Tyne. Retrieved from http://www.cpre.org.uk/resources/countryside/tranquil-places/item/download/542. Last accessed 15 Mar 2019

Johnson CY, Bowker JM, Bergstrom JC, Ken Cordell H (2004) Wilderness values in America: does immigrant status or ethnicity matter? Soc Nat Res 17(7):611–628

Jones-Walters L, Čivić K (2010) Wilderness and biodiversity. J Nat Conserv 18(4):338–339

Kaplan R, Kaplan S (1989) The experience of nature: a psychological perspective. Cambridge University Press, Cambridge, p 352

Kienast F, Frick J, van Strien MJ, Hunziker M (2015) The Swiss Landscape Monitoring Program—a comprehensive indicator set to measure landscape change. Ecol Model 295:136–150

Kliskey AD, Kearsley GW (1993) Mapping multiple perceptions of wilderness in southern New Zealand. Appl Geogr 13(3):203–223

Kloog I, Haim A, Stevens RG, Barchana M, Portnov BA (2008) Light at night co-distributes with incident breast but not lung cancer in the female population of Israel. Chronobiol Int 25 (1):65–81

Knop E, Zoller L, Ryser R, Gerpe C, Hörler M, Fontaine C (2017) Artificial light at night as a new threat to pollination. Nature 548(7666):206

Kuiters AT, van Eupen M, Carver S, Fisher M, Kun Z, Vancura V (2013) Wilderness register and indicator for Europe. ec.europa.eu/environment/nature/natura2000/wilderness/pdf/Wilderness_register_indicator.pdf. Last accessed 15 Mar 2019

Kyba CCM, Wagner JM, Kuechly HU, Walker CE, Elvidge CD, Falchi F, Ruhtz T, Fischer J, Hölker F (2013) Citizen science provides valuable data for monitoring global night sky luminance. Sci Rep 3:1835

Landres P, Barns C, Dennis JG, Devine T, Geissler P, McCasland CS, Merigliano L, Seastrand J, Swain R (2008) Keeping it wild: an interagency strategy to monitor trends in wilderness character across the National Wilderness Preservation System. General technical report RMRS-GTR-212. US Department of Agriculture, Forest Service, Rocky Mountain Research Station, Fort Collins, CO, 77 P., 212

Lesslie R (1998) Global wilderness, Cambridge: UNEP-WCMC. Dataset derived using the Digital Chart of the World 1993 version and methods based on the Australian National Wilderness Inventory. https://www.unep-wcmc.org/resources-and-data/global-wilderness. Last accessed 15 Mar 2019

Lesslie R (2016) The wilderness continuum concept and its application in Australia: lessons for modern conservation. In: Carver S, Fritz S (eds) Mapping wilderness: concepts, techniques and applications. Springer, Dordrecht, pp 17–33

Lesslie R, Maslen M (1995) National wilderness inventory handbook of procedures, content and usage, 2nd edn. Commonwealth Government Printer, Canberra, 98 pp

Lesslie R, Taylor SG (1985) The wilderness continuum concept and its implications for Australian wilderness preservation policy. Biol Conserv 32(4):309–333

Levett R (2000) A headline indicator of tranquillity: definition and measurement issues. Interim report to CPRE. CAG Consultants for CPRE, London

Lorne JK, Salmon M (2007) Effects of exposure to artificial lighting on orientation of hatchling sea turtles on the beach and in the ocean. Endangered Species Res 3(1):23–30

Lukyanenko R, Parsons J, Wiersma YF (2016) Emerging problems of data quality in citizen science. Conserv Biol 30(3):447–449

Lutz AR, Simpson-Housley P, Deman AF (1999) Wilderness: rural and urban attitudes and perceptions. Environ Behav 31(2):259–266

MacFarlane R, Haggett C, Fuller D, Dunsford H, Carlisle B (2004) Tranquillity mapping: developing a robust methodology for planning support. Newcastle upon Tyne: Report to the Campaign to Protect Rural England, Countryside Agency, North East Assembly, Northumberland Strategic Partnership, Northumberland National Park Authority and Durham County Council, Centre for Environmental & Spatial Analysis, Northumbria University

Mackenzie AFD (2006) A working land: crofting communities, place and the politics of the possible in post-Land Reform Scotland. Trans Inst Br Geogr 31(3):383–398

Măntoiu DŞ, Nistorescu MC, Şandric IC, Mirea IC, Hăgătiş A, Stanciu E (2016) Wilderness areas in Romania: a case study on the South Western Carpathians. In: Carver S, Fritz S (eds) Mapping wilderness. Springer, Dordrecht, pp 145–156

Mazenauer C (2015) Nächtliche Dunkelheit im Val Müstair. Eine Untersuchung über die Wahrnehmung und das touristische Potenzial der nächtlichen Dunkelheit. Master thesis at the Geography Department, University of Zurich, Switzerland. https://tinyurl.com/y79qyrcy. Last accessed 15 Mar 2019

Mc Morran R, Price MF, Warren CR (2008) The call of different wilds: the importance of definition and perception in protecting and managing Scottish wild landscapes. J Environ Plann Manage 51(2):177–199

McCloskey JM, Spalding H (1989) A reconnaissance-level inventory of the amount of wilderness remaining in the world. Ambio 221–227

Nash RF (2014) Wilderness and the American mind, 5th edn. Yale University Press, New Haven, CT, USA, p 413

National Planning Policy Framework (2012) Conserving and enhancing the natural environment. Government of the United Kingdom. https://www.gov.uk/guidance/national-planning-policy-framework/11-conserving-and-enhancing-the-natural-environment. Last accessed 15 Mar 2019

Navara KJ, Nelson RJ (2007) The dark side of light at night: physiological, epidemiological, and ecological consequences. J Pineal Res 43(3):215–224

NBS (2004) Quality of life counts. Indicators for a strategy for sustainable development for the United Kingdom. 2004 Update. Updating the baseline assessments made in 1999. http://www.nies.go.jp/db/sdidoc/qolc2004.pdf. Last accessed 15 Mar 2019

Neumann RP (1998) Imposing wilderness, Struggles over livelihood and nature preservation in Africa. University of California Press, Berkeley (California studies in critical human geography) 268 pp. ISBN: 9780520234680

Ólafsdóttir R, Sæþórsdóttir AD, Runnström M (2016) Purism scale approach for wilderness mapping in Iceland. In: Carver S, Steffen F (eds) Mapping wilderness. Springer, Dordrecht, pp 157–176

Orsi F, Geneletti D, Borsdorf A (2013) Mapping wildness for protected area management: a methodological approach and application to the Dolomites UNESCO World Heritage Site (Italy). Landscape Urban Plann 120:1–15

Perkin EK, Hölker F, Tockner K (2014) The effects of artificial lighting on adult aquatic and terrestrial insects. Freshw Biol 59(2):368–377

Pheasant R, Horoshenkov K, Watts G, Barrett B (2008) The acoustic and visual factors influencing the construction of tranquil space in urban and rural environments tranquil spaces-quiet places? J Acoust Soc Am 123(3):1446–1457

Pheasant RJ, Watts GR, Horoshenkov KV (2009) Validation of a tranquillity rating prediction tool. Acta Acustica United Acustica 95(6):1024–1031

Plutzar C, Enzenhofer K, Hoser F, Zika M, Kohler B (2016) Is there something wild in Austria? In: Carver S, Fritz S (eds) Mapping wilderness. Springer, Dordrecht, pp 177–189

Poiani S, Dietrich C, Barroso A, Costa-Leonardo AM (2015) Effects of residential energy-saving lamps on the attraction of nocturnal insects. Lighting Res Technol 47(3):338–348

Polak T, Korine C, Yair S, Holderied MW (2011) Differential effects of artificial lighting on flight and foraging behaviour of two sympatric bat species in a desert. J Zool 285(1):21–27

Purves RS, Edwardes A, Wood J (2011) Describing place through user generated content. First Monday 16(9). https://ojphi.org/ojs/index.php/fm/article/view/3710/3035. Last accessed 15 Mar 2019

Radford SL, Senn J, Kienast F (2019) Indicator-based assessment of wilderness quality in mountain landscapes. Ecol Ind 97:438–446

Reynolds LD (1998) de tranquillitate animi. dial. 9 in: L. Annaei Senecae dialogorum libri duodecim; Reynolds LD (ed). Clarendon, Oxford. ISBN: 9780198146599. https://doi.org/10.1093/actrade/9780198146599.book1

Rich C, Longcore T (2013) Ecological consequences of artificial night lighting. Island Press, Washington, p 480

Richards E (2012) The highland clearances. Birlinn Ltd., Edinburgh, UK. ISBN:978-1-84158-542-0, eBook ISBN: 978-0-85790-524-6

Russo D, Cistrone L, Libralato N, Korine C, Jones G, Ancillotto L (2017) Adverse effects of artificial illumination on bat drinking activity. Anim Conserv 20(6):492–501

Rydell J, Eklöf J, Sánchez-Navarro S (2017) Age of enlightenment: long-term effects of outdoor aesthetic lights on bats in churches. R Soc Open Sci 4(8):161077

Sanderson EW, Jaiteh M, Levy MA, Redford KH, Wannebo AV, Woolmer G (2002) The human footprint and the last of the wild. Bioscience 52(10):891–904

Seresinhe CI, Preis T, Moat HS (2015) Quantifying the impact of scenic environments on health. Sci Rep 5(Article number 16899)

Shahzad G, Yang H, Ahmad AW, Lee C (2016) Energy-efficient intelligent street lighting system using traffic-adaptive control. IEEE Sens J 16(13):5397–5405

Shepherd D, Welch D, Dirks KN, McBride D (2013) Do quiet areas afford greater health-related quality of life than noisy areas? Int J Environ Res Publ Health 10(4):1284–1303

Silva E, Marco A, da Graca J, Perez H, Abella E, Patino-Martinez J, Martins S, Almeida C (2017) Light pollution affects nesting behavior of loggerhead turtles and predation risk of nests and hatchlings. J Photochem Photobiol B Biol 173:240–249

Smith M (2008) Time to turn off the lights. Nature 457(7225):27

SNH (2013) Mapping Scotland's wildness, non-technical methodology. https://www.nature.scot/mapping-scotlands-wildness-and-wild-land-non-technical-description-methodology. Last accessed 15 Mar 2019

SNH (2014a) Core Areas Of Wild Land 2013 map Scottish Natural Heritage's advice to government—16th June 2014. https://www.nature.scot/sites/default/files/2018-02/CAWL2013map-AdvicetoGovernment-RevisedFinal-17June2014.pdf. Last accessed 15 Mar 2019

SNH (2014b) Wild land areas. https://www.nature.scot/wild-land-area-descriptions. Last accessed 15 Mar 2019

Spiegel K, Knutson K, Leproult R, Tasali E, Van Cauter E (2005) Sleep loss: a novel risk factor for insulin resistance and Type 2 diabetes. J Appl Physiol 99(5):2008–2019

Stevens RG (2009) Light-at-night, circadian disruption and breast cancer: assessment of existing evidence. Int J Epidemiol 38(4):963–970

Stremlow M, Sidler C (2002) Schreibzüge durch die Wildnis: Wildnisvorstellungen in Literatur und Printmedien der Schweiz, 1st edn. Haupt, Bern

Terradillos A, Wilkinson K (2015) Broadly Engaging with Tranquility Project Report IV: GIS. reshare.ukdataservice.ac.uk/851934/197/PRJ4-Final-10-September-2015.pdf. Last accessed 15 Mar 2019

Ulrich RS, Simons RF, Losito BD, Fiorito E, Miles MA, Zelson M (1991) Stress recovery during exposure to natural and urban environments. J Environ Psychol 11(3):201–230

UNEP (2002) Globio: global methodology for mapping human impacts on the biosphere. United Nations Environment Programme. www.globio.info. Last accessed 15 Mar 2019

US Congress (1964) Wilderness act, Public law 88-577. http://www.gpo.gov/fdsys/pkg/STATUTE-78/pdf/STATUTE-78-Pg890.pdf. Last accessed 15 Mar 2019

US National Park Service (2018) Night skies. https://www.nps.gov/subjects/nightskies/stargaze.htm. Accessed 4 Oct 2018

van Tonder GJ, Lyons MJ (2005) Visual perception in Japanese rock garden design. Axiomathes 15(3):353–371

van Zanten BT, van Berkel DB, Meentemeyer RK, Smith JW, Tieskens KF, Verburg PH (2016) Continental-scale quantification of landscape values using social media data. Proc Natl Acad Sci 113(46):12974–12979

Velarde MD, Fry G, Tveit M (2007) Health effects of viewing landscapes—landscape types in environmental psychology. Urban Forest Urban Greening 6(4):199–212

Vienneau D, de Hoogh K, Faeh D, Kaufmann M, Wunderli JM, Röösli M, Group SNCS (2017) More than clean air and tranquillity: residential green is independently associated with decreasing mortality. Environ Int 108:176–184

Wallner A, Bauer N, Hunziker M (2007) Perceptions and evaluations of biosphere reserves by local residents in Switzerland and Ukraine. Landscape Urban Plann 83(2–3):104–114

Wartmann FM, Acheson E, Purves RS (2018) Describing and comparing landscapes using tags, texts, and free lists: an interdisciplinary approach. Int J Geogr Inf Sci 32(8):1572–1592

Wartmann FM, Tieskens KF, van Zanten BT, Verburg PH (2019). Exploring tranquillity experienced in landscapes based on social media. Appl Geogr

Watson JEM, Shanahan DF, Di Marco M, Allan J, Laurance WF, Sanderson EW, Mackey B, Venter O (2016) Catastrophic declines in wilderness areas undermine global environment targets. Curr Biol 26(21):2929–2934

Watson JEM, Venter O, Lee J, Jones KR, Robinson JG, Possingham HP, Allan JR (2018) Protect the last of the wild. Nature 563:27–30

Watts GR, Pheasant RJ (2015a) Identifying tranquil environments and quantifying impacts. Appl Acoust 89:122–127

Watts GR, Pheasant RJ (2015b) Tranquillity in the Scottish Highlands and Dartmoor National Park—the importance of soundscapes and emotional factors. Appl Acoust 89:297–305

Watts GR, Pheasant RJ, Horoshenkov KV (2011) Predicting perceived tranquillity in urban parks and open spaces. Environ Plann B Plann Des 38(4):585–594

Watts GR, Miah A, Pheasant R (2013) Tranquillity and soundscapes in urban green spaces—predicted and actual assessments from a questionnaire survey. Environ Plann B Plann Des 40 (1):170–181

Wartmann FM, Mackaness WA (in review) Mapping tranquillity based on interviews and social media data. Submitted to Landscape Research

Yang X, Fu B, Chen L (2013) Remote sensing and geospatial analysis for landscape pattern characterization. In: Fu B, Jones K (eds) Landscape ecology for sustainable environment and culture. Springer, Dordrecht, pp 205–221

Zumthor P, Beer I, Mathieu J (2005) Wieviel Licht braucht der Mensch, um leben zu können, und wieviel Dunkelheit?/Di quanta luce ha bisogno l'uomo per vivere e di quanta oscurità? vdf-Hochschulverlag und Editrice Compositori, Zurich and Bologna

Chapter 8
Tools for Landscape Science: Theory, Models and Data

Marcel van Oijen

Abstract We review the different roles of theory, models and data in landscape science. The need for science at the landscape scale is argued. Landscape theory is considered as a repository of probabilistic patterns rather than as a collection of laws of nature. We present a typology of such patterns for five distinct landscape features: land cover, land use, patch properties, patch interactions and exogenous influences. We show how theory for these features can support landscape modelling, and we provide a checklist of questions for model developers. The limited availability of data on landscapes is discussed, and how that leads to uncertainties in theoretical patterns as well as models. We analyse how probability theory can be used to account for these uncertainties, strengthening the links between theory, models and data, and facilitating decision support.

Keywords Agricultural landscapes · Bayesian methods · Data · Ecosystem services · Landscape theory · Models · Probability theory · Uncertainties

Abbreviations

ALS	Agricultural landscapes
ES	Ecosystem services
FLS	Forest landscapes

8.1 Introduction

8.1.1 Terminology

The study of landscapes has a long history. Biophysical and cultural aspects of landscapes have been studied in a range of disciplines including geography,

M. van Oijen (✉)
Centre for Ecology & Hydrology, Bush Estate, Penicuik EH26 0QB, UK
e-mail: mvano@ceh.ac.uk

L. Müller and F. Eulenstein (eds.), *Current Trends in Landscape Research*, Innovations in Landscape Research, https://doi.org/10.1007/978-3-030-30069-2_8

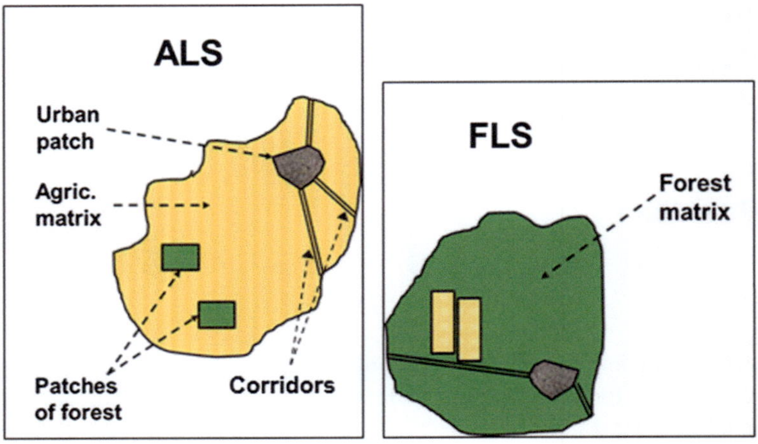

Fig. 8.1 Landscape elements: matrix, patches and corridors. Left: Agricultural landscape (ALS). Right: Forest landscape (FLS)

biogeochemistry and landscape ecology. This chapter does not review that history, but focuses on landscapes from the perspective of a scientist who wants to predict the impact of environmental change on landscape processes and the services that landscapes provide. From this pragmatic perspective, a *landscape* is any area that consists of multiple different ecosystem types that interact with each other. To describe landscapes, we use the "matrix-patch-corridor" terminology of Forman (1995) (see Fig. 8.1). The *matrix* is the dominant ecosystem type that covers most of the area within the landscape. *Patches* are smaller areas of different ecosystem types within the matrix, and *corridors* are linear landscape elements that separate the ecosystems. The corridors can function as conduits as well as barriers. When over time patches become larger or more numerous, the matrix will diminish in size until it no longer covers the majority of the land, and the term matrix loses it meaning. This can continue until a different ecosystem type is seen as the matrix. We shall often refer to the matrix as just another patch, albeit the largest one. *Agricultural landscapes* (ALS) have a matrix of managed farmland (crops or grassland), and in *forest landscapes* (FLS), the matrix is forest.

8.1.2 The Need for Landscape Science

Much of the literature in the ecological, environmental and agricultural sciences focuses on single ecosystem types, many of which even have their own dedicated journals. However, concepts and methods are converging. A common concept is that of ecosystem services (ES), which encapsulates the many different ways in which ecosystems are important to mankind. Any single ecosystem such as a forest or field of wheat can be productive, have biodiversity, contribute to the greenhouse gas balance, intercept air pollution, be appreciated for its beauty and provide other services. But we can ask the question: Are these ES properly studied at the level of a single ecosystem, or should we study them as part of the larger landscape?

Many ES depend on connections between ecosystems. The biodiversity of a patch is replenished by the influx of species from the surrounding landscape. Air pollution may or may not reach a patch depending on the presence of tree barriers. Biogeochemists study the flow of water, carbon and nutrients between different landscape elements. Agronomists track the dispersion of genetically modified organisms. ES such as aesthetic value and tourism are properties of a landscape, not of any single ecosystem within it. People aim to visit the 'Lake District' in England rather than any particular lake, village, forest or mountain. On the other hand, for ES such as productivity, the field scale may suffice. Farmers decouple the productivity of their fields from the landscape by controlling the availability of water (irrigation, drainage), nutrients (fertilization), weeds, pests and diseases (biocides). But this decoupling is never complete, and the required intensity of management partly depends on the landscape, not just the field. For example, field hydrology management will require more effort when upstream forests are cut. We conclude that landscapes are more than the sum of their constituent ecosystems, implying a need to consider problems at the landscape scale. This is the justification for landscape science, and the term *landscape services* is increasingly used.

8.1.3 Tools for Landscape Science

Many process-based field-scale models exist, for crops, grasslands and forests, that can be used to simulate processes in individual patches. However, these models are not equipped to simulate landscape processes (Van Oijen et al. 2018). For specific ecosystem combinations, landscape models exist as well (Baker 1989; Baker and Cai 1992; Baker et al. 1991; Duretz et al. 2011), but their development is still in its infancy. This shows that, despite the long history of landscape study, landscape science is still a young and developing discipline. This chapter discusses the tools that this developing discipline requires. In particular, we aim to clarify the roles of theory, models and data in landscape science. Our examples shall focus on ALS, but much of the discussion will be more generally valid.

8.2 Landscape Theory

8.2.1 Theory in the Applied Sciences

Many of the applied sciences underwent large changes in the 1960s. After the publication of Von Bertalanffy's 'General Systems Theory' (1968), any object of study came to be seen as a system whose behaviour depended on internal feedbacks as well as exogenous influences. In the environmental and ecological sciences, the systems approach led to the development of complex process-based dynamic models. This lowered the status of 'theory' (Van Oijen 2009). Models were no longer seen as implementations of a discipline-specific theory; they were representations of known processes and feedbacks. The diminished role of theory implied that models no longer needed to be 'correct', they needed to be useful. In landscape science, debates about the definition of 'landscape'—as a primarily human or natural construct—died down.

These developments reflected the realization that the applied sciences cannot produce any simple 'laws', comparable to those of theoretical physics. Organisms, ecosystems, landscapes and economies are complex systems, and their behaviour shows regularities but is not fully predictable from any deterministic laws. The conservation laws for matter and energy hold for all systems, and they underpin element budgets of catchments (Bormann and Likens 1967) and calculations of energy flows in ecosystems (Odum 1968), but they leave many degrees of freedom for most ES. The role of theory therefore has to be modest: summarizing the likelihood of different kinds of system behaviour under different conditions. Theory can only provide probability distributions for the properties and performance of systems. So what kinds of probabilistic patterns can landscape theory capture? One example is the observation that ALS of higher spatial complexity tend to be richer in biodiversity, including natural enemies of crop pests and weeds, than simpler ALS (Begg et al. 2017; Chaplin-Kramer et al. 2011; Roschewitz et al. 2005). Another example is that grassland-dominated ALS in Europe tend to lose between 70 and 100% of their nitrogen imports to the environment, with fairly regular proportions of losses to soil, groundwater and atmosphere (Dalgaard et al. 2012). Knowledge of such theoretical patterns helps modellers develop and test their models. But a daunting variety of patterns can be detected in landscapes. Regularities can be found in space, in time and in relationships between landscape elements. To bring order into this study, we need a typology of landscape patterns; one is proposed in the following section (Table 8.1).

Table 8.1 A typology of landscape patterns

Feature	Examples of patterns
Land cover	Deforestation, urbanization, desertification, Von Thünen's 'Isolated State'
Land use	Intensified crop management in some areas, more organic agriculture in others, increase in agroforestry
Properties of individual patches	Variable emission factors for N_2O, loss of soil organic matter
Interactions between patches	Introduction of shelter belts, ecoducts
Exogenous influences	Climate change, increased tourism, increasing impact of global markets

8.2.2 The Five Types of Pattern in Landscape Theory

Landscapes show regular patterns in: (1) land cover, (2) land use, (3) properties of individual patches, (4) interactions between patches and (5) exogenous influences on the landscape. We mention examples for each of these five types of pattern.

What kinds of *land cover* can we expect in our landscapes? Are there any obvious trends in time of land cover change? Are there standard spatial configurations of the different patches in a landscape? Where can we expect corridors? There are not many answers to these questions: theory regarding land cover dynamics is not well developed. Worldwide, several trends are noticeable: increasing use of land for agriculture, increasing urbanization, denser transport networks and desertification. At finer scales, we see an increase in parcel size in agriculture. As for the spatial configuration of landscape elements, some theoretical patterns were already proposed two centuries ago. Von Thünen, in his “Isolated State” concept, suggested that agricultural landscapes would self-organize in rings around a central market town (von Thünen 1826). Vegetation requiring the least effort in maintenance and transport would be found furthest from the centre. Some recent observations on real ALS in Brittany (France) show similar circular patterns around farms (Thenail and Baudry 2004). There is also theory for the spatial arrangement of river networks, which often follows a fractal pattern that can be explained using thermodynamics (Kleidon et al. 2013). It is to be expected that the identification of temporal and spatial patterns of land cover and land cover change will benefit from the increasing availability of remote sensing data.

Land use and land-use change have received increasing attention in recent decades because of their role in greenhouse gas emissions. A common trend in many ALS is increasing intensity of management. In general, information about land cover needs to be combined with other sources of information to reliably assess how land is used. Levy et al. (2018) combined eight different data sets—including self-reporting by farmers, afforestation records and remotely sensed land cover (Corine)—to estimate annual land-use patterns in Scotland from 1969 to 2015 at a spatial resolution of 0.1×0.1 km. They accounted for uncertainty in each of the data sets, using a Bayesian method to combine all information into probability distributions for land-use change. The work showed that the carbon balance for land-use change in Scotland was dominated by losses in conversion to intensively managed cropland and gains due to expansion of timber forestry.

Regularities in the *properties of individual patches* are studied by disciplines that underpin landscape science, such as ecophysiology and soil science. These patterns are important to landscape scientists—they need to know that grassland soils tend to be richer in organic matter than crop lands—but information about such patterns is inherited from the other disciplines, and not exclusive to landscape theory.

Interactions between patches, on the other hand, belong to the core domain of landscape theory. However, knowledge is still limited. There are many ideas about how movement of pollutants, pollen and seeds through the landscape can be controlled through the judicious placement of tree shelter belts and green strips, based on models from epidemiology, hydrology and atmospheric dynamics (e.g. Bealey et al. 2014). However, the efficacy of such measures is still under debate. For landscape theory, this means that probability distributions for interactions between patches are currently wide, acknowledging this large uncertainty.

As with all systems, landscapes are subject to many *exogenous influences*. Market constraints affect the scope for export of products and for tourism—and these are also affected by the accessibility of the landscape: the proximity of airports, railway stations and motorways. Climatic trends alter the delivery of ES, as do the regional levels of air pollution, invasive species, pests and diseases.

8.2.3 The Development of LS Theory

The preceding brief overview has shown that landscape theory can be seen as the collection of common patterns among landscapes for land cover, land use, connectivity, internal constraints and exogenous drivers. Such patterns are being detected and reported in the literature, but the research effort is still limited. Most of environmental and ecological science focuses on smaller spatial scales and/or single ecosystems or habitats rather than multi-ecosystem landscapes. How can landscape theory progress? When identifying patterns, it will be important to draw on landscape typologies. Patterns can more easily be detected for specific landscape types than across all landscapes. Fortunately, efforts to develop comprehensive landscape typologies are underway. An example is the recent map of Norwegian landscape types produced by Erikstad et al. (2017). Their map reveals common landscape gradients with altitude and distance from the sea. Research can be targeted at specific landscape types. The already mentioned regularities in nitrogen losses observed in European grassland ALS (Dalgaard et al. 2012) may not hold in the same way for other types of landscape.

However, landscape theory should remain open to similarities between different landscape types and to similarities with theory in other disciplines. Progress towards landscape theory for ALS can be made by exchanging ideas with FLS scientists and ecologists, as they bring complementary expertise. The study of ALS tends to focus on productivity and the mitigation of short-term impacts of human actions through environmental offsets, management of fauna and flora, hedgerows and other measures. FLS science adds to this the elements of long-term dynamics, climate change impacts and individual-based modelling. And ecology brings focus on biodiversity, genetics, traits and diseases.

But how should landscape theory be developed concretely? How exactly can we make any general statements about landscape patterns? We already mentioned that the patterns are of necessity probabilistic; they are not laws of nature. So the patterns—such as the results from meta-analyses and inventories—must be described as probability distributions rather than as deterministic functions. And when new information becomes available, the distributions can be updated using the methods from probability theory. This implies the use of Bayesian methods to reduce uncertainties (Van Oijen 2017). These methods also offer a natural place to grand theories and concepts for complex adaptive systems, such as scaling theory (West 2018), panarchy theory (Holling et al. 2002), the principle of maximum entropy (Bertram and Dewar 2013), criteria for resilience of agro-ecosystems (Cabell and Oelofse 2012) and the already mentioned thermodynamical explanation of fractal river flow networks (Kleidon et al. 2013). None of these theories and principles can predict anything with great accuracy for any specific organism or system, but they give us—within wide uncertainty bounds—a first idea of what to expect, a first quantification of our probability distributions.

Having concluded that landscape theory provides probabilistic patterns rather than deterministic laws, we now discuss how such theory can be of use to landscape modelling and data collection.

8.3 Landscape Models

Landscape models are more than a loose collection of models for specific ecosystems or farm systems. Their defining characteristic is that they represent interactions between different patches either explicitly (by simulating transport processes) or implicitly (e.g. by including dependencies on 'landscape complexity'). The practice of 'tiling', which is common in vegetation modelling as part of global climate simulation and involves different ecosystem-specific models being run independently for different patches of land, is not landscape modelling, as it ignores the interactions between ecosystems.

8.3.1 Designing Landscape Models: A Checklist

In contrast to landscape theory of which the primary purposes are to summarize patterns and support modelling, landscape models are desired to be predictive. This requires the models to target well-defined landscapes. Key questions are the following: (1) Which landscapes do we want to model? (2) Why do we want to model them? (3) At which spatiotemporal scale should our model operate? (4) Which land cover types, land uses, patch properties, interactions between patches, and exogenous drivers must our model represent? We briefly discuss this checklist of four questions (Table 8.2).

The aim of the model can be to simulate a specific individual landscape (or set of landscapes) that we can locate on a map. Alternatively, the model may be targeted at a class of landscapes, wherever they happen to be, such as 'ALS in temperate climates'. The latter requires a much wider range of model applicability, i.e. broader knowledge of relevant mechanisms.

Our reasons for being interested in certain landscapes affect the amount of detail that must be represented in the model. If we are interested in the impact of

Table 8.2 A checklist for landscape model development

Question	Examples of answers
Which landscapes do we want to model?	The English lake District, Grassland-based ALS in NW-Europe
Why do we want to model them?	To predict the impact of greening measures on biodiversity, to predict the impact of climate change on productivity and other ES
At which spatiotemporal scale should the model operate?	Catchments of 1–10 km^2 (at a resolution of 100 m) for the years 2000–2100 (at an annual time step)
Which land cover types, land uses, patch properties, interactions between patches, and exogenous drivers must the model represent?	Grassland and forest, intensive management, etc.

management decisions or environmental change, then these must be represented as input variables or parameters. And the range of ES that we are interested in defines a minimal set of output variables that must be calculated.

The spatiotemporal scale of the model refers to both *extent* (time period of interest, area covered by the landscape) and *resolution* (time step and spatial grain). Both should match the spatial heterogeneity and the length scales of connecting processes. However, there is a great variety of connections in landscapes. Transport can take place via atmosphere, land, soil and groundwater. A key problem in landscape modelling is that there is no 'natural' length scale to these connections: process length scales vary by orders of magnitude. Atmospheric transport of ammonia is much more localized than that of nitrogen oxides, dispersal of pollen and seeds varies between meters and kilometres depending on particle size etc. Atmospheric transport models exist, but they are still in their infancy for topologically complex terrain, and similar problems exist for hydrological models.

The final question of our checklist provides a link between model development and theory. The land cover types, land uses, patch properties, interactions between patches, and exogenous drivers that our model must represent are of course also the five types of landscape pattern that are studied by landscape theory. Theory thus provides the first ideas about which processes and feedbacks to include in the model, and initial estimates for variables and parameters. In this way, landscape theory informs landscape model design without prescribing model structure in any detail.

8.4 Data on Landscapes

As in all sciences, data underpin both theory development and modelling (Van Oijen 2002). Theory is formed by detecting patterns in data. Landscape model design is inspired by this data-based theory, but data are also used more directly in modelling as external driving variables (e.g. weather conditions, market prices), as information for parameter calibration, and as test-cases for the predictive capacity of the models. However, data availability is still limited in landscape science. A common problem is that driver data tend to be available only at coarse spatial resolution. When such spatially averaged data are used as input to field or landscape models, which tend to be nonlinear, outputs are biased because of aggregation error. There are ways to correct for this error (Van Oijen et al. 2017), but these are not easily applicable in the case of complex models. Improvements in downscaling of driver data are thus required. In contrast to driver data, data on the ecosystems that make up landscapes are often only available for small areas within individual patches. This hampers any testing of model predictions at the landscape scale. A notable exception to this is the classical work on element budgets in catchments, where a single water outflow point can act as a natural integrator of nutrient losses in the whole landscape (Bormann and Likens 1967). Similar methods are now used to quantify losses of dissolved organic and inorganic carbon from landscapes, a hitherto poorly known component of the carbon balance (Dinsmore et al. 2010).

An additional obstacle in landscape science is the difficulty of carrying out experiments at the landscape scale. It is therefore hard to distinguish causal relationships from mere correlations. However, the consequences of natural large-scale perturbations and disturbances (e.g. fires, epidemics, droughts) can be compared for different landscapes. Information on such extreme events will become increasingly available through the further improvement of remote sensing techniques and the increasing placement of small sensors in the landscape [see, e.g., the Cyber-Physical Systems landscapes of Lee et al. (2015)]. Further understanding of the processes underlying the spread of organisms through the landscape will be derived from the increasingly refined techniques for landscape genomics (Whipple and Holeski 2016).

Irrespective of any expected improvements in measurement methods, data on landscapes will—just as landscape theory and models—have a degree of uncertainty attached to them. The question thus becomes: How should landscape science combine theory, models and data when all have uncertainties?

8.5 Towards a Probabilistic Synthesis

At various points in this text, we have identified links between the three tools of landscape science: theory, models and data. A summary of these links is provided in Fig. 8.2. We alluded to the fact that uncertainties are ubiquitous, and rigorous book-keeping of these uncertainties is essential. This implies a need for using probability theory, in particular Bayesian probabilistic methods (Van Oijen 2017), in all areas of landscape science. We see this first in the derivation of theory from data, not as laws, but as probability distributions. The resulting theory, but also theory from underpinning 'small-scale' disciplines (crop science, grassland science, soil science, etc.), provides prior probability distributions for the parameter values of landscape models. Data to compare with model predictions further allow Bayesian calibration of the parameter distributions, thereby reducing uncertainty, and testing of model quality. Monte Carlo sampling from the parameter distributions, and from uncertain input variables, is used to generate stochastic model predictions.

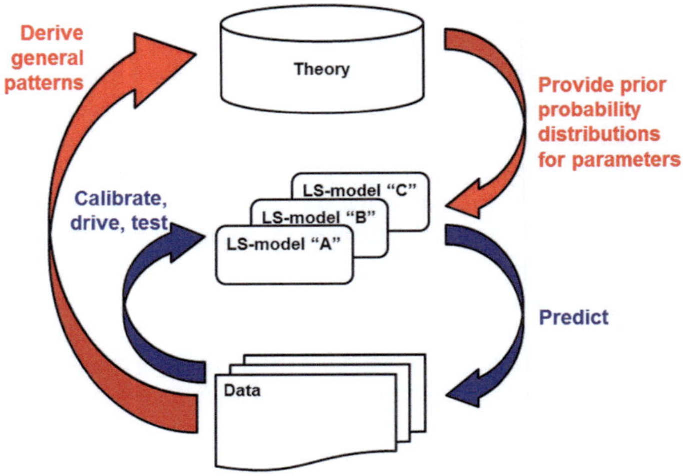

Fig. 8.2 Landscape science: The connections between theory, models and data

If this probabilistic approach to landscape science is adopted throughout, other benefits follow. The design of landscape monitoring programmes can be targeted at reducing key uncertainties. Landscape theory, landscape models and probability theory together can identify which measurements will reduce uncertainty in model predictions the most (as was shown for forest models, Van Oijen et al. 2005). Secondly, probabilistic modelling facilitates decision support, where managers need to assess uncertainties associated with different courses of action, and risk analysis (Van Oijen et al. 2013). To this end, the landscape modelling can be embedded in simple graphical models, such as spatial Bayesian Belief Networks (BBN). BBN are already being used in FLS to decide on the spatial allocation of nature conservation areas, and the process of building the graphical model has been shown to facilitate stakeholder involvement (Gonzalez-Redin et al. 2016). Finally, if a landscape model becomes too large and unwieldy for Bayesian calibration, it can be approximated using a stochastic emulator (Kennedy and O'Hagan 2001).

8.6 Conclusions

We have outlined a view of landscape science in which theory is a repository of existing knowledge, expressed as a collection of conditional probability distributions. Such theory will always remain 'in progress' as new data are being gathered, more appropriate landscape typologies defined, and more robust patterns of landscape behaviour identified. The search for such patterns can be organized around the five types of landscape pattern that we distinguished. However, hard laws will not be found because landscapes are complex systems. The theory will support the development and application of landscape models that are evaluated not for 'correctness' but for predictive capacity in tests against data.

We emphasized the use of probability theory in this endeavour. It can be employed in developing theory, providing prior distributions for model parameters, quantifying uncertainty in data, calibrating and evaluating models, correcting upscaling errors, providing decision support, facilitating risk analysis and replacing complex models with stochastic emulators.

Acknowledgements I thank the organizers of the Landscape 2018 meeting in Berlin for their invitation to participate, and for the stimulating discussions about the future of landscape science

References

Baker WL (1989) A review of models of landscape change. Landsc Ecol 2:111–133

Baker WL, Cai Y (1992) Ther.Le programs for multiscale analysis of landscape structure using the GRASS geographical information system. Landsc Ecol 7:291–302. https://doi.org/10.1007/BF00131258

Baker W, Egbert SL, Frazier GF (1991) A spatial model for studying the effects of climatic change on the structure of landscapes subject to large disturbances. Ecol Model 56:109–125. https://doi.org/10.1016/0304-3800(91)90195-7

Bealey WJ, Loubet B, Braban CF, Famulari D, Theobald MR, Reis S, Reay DS, Sutton MA (2014) Modelling agro-forestry scenarios for ammonia abatement in the landscape. Environ Res Lett 9:125001. https://doi.org/10.1088/1748-9326/9/12/125001

Begg GS, Cook SM, Dye R, Ferrante M, Franck P, Lavigne C, Lövei GL, Mansion-Vaquie A, Pell JK, Petit S, Quesada N, Ricci B, Wratten SD, Birch AE (2017) A functional overview of conservation biological control. Crop Prot 97:145–158. https://doi.org/10.1016/j.cropro.2016.11.008

Bertalanffy LV (1968) General system theory: foundations, development, applications. George BrazillerInc, New York

Bertram J, Dewar RC (2013) Statistical patterns in tropical tree cover explained by the different water demand of individual trees and grasses. Ecology 94:2138–2144

Bormann FH, Likens GE (1967) Nutrient cycling. Science 155:424–429. https://doi.org/10.1126/science.155.3761.424

Cabell J, Oelofse M (2012) An indicator framework for assessing agroecosystem resilience. Ecol Soc 17. https://doi.org/10.5751/ES-04666-170118

Chaplin-Kramer R, O'Rourke ME, Blitzer EJ, Kremen C (2011) A meta-analysis of crop pest and natural enemy response to landscape complexity. Ecol Lett 14:922–932. https://doi.org/10.1111/j.1461-0248.2011.01642.x

Dalgaard T, Bienkowski JF, Bleeker A, Dragosits U, Drouet JL, Durand P, Frumau A, Hutchings NJ, Kedziora A, Magliulo V, Olesen JE, Theobald MR, Maury O, Akkal N, Cellier P (2012) Farm nitrogen balances in six European landscapes as an indicator for nitrogen losses and basis for improved management. Biogeosciences 9:5303–5321. https://doi.org/10.5194/bg-9-5303-2012

Dinsmore KJ, Billett MF, Skiba UM, Rees RM, Drewer J, Helfter C (2010) Role of the aquatic pathway in the carbon and greenhouse gas budgets of a peatland catchment. Glob Chang Biol 16:2750–2762. https://doi.org/10.1111/j.1365-2486.2009.02119.x

Duretz S, Drouet J, Durand P, Hutchings N, Theobald M, Salmon-Monviola J, Dragosits U, Maury O, Sutton M, Cellier P (2011) NitroScape: a model to integrate nitrogen transfers and transformations in rural landscapes. Environ Pollut 159:3162–3170. https://doi.org/10.1016/j.envpol.2011.05.005

Erikstad L et al (2017) LandskapstyperiNorge: Nymetodikk for kartleggingavlandskap. https://www.regjeringen.no/contentassets/8eebe3a246ee4baa8b6eb94ef9827eab/erikstad.pdf. Accessed on 17 Mar 2019

Forman RTT (1995) Land mosaics: the ecology of landscapes and regions. Cambridge University Press, Cambridge, New York, p 632

Gonzalez-Redin J, Luque S, Poggio L, Smith R, Gimona A (2016) Spatial Bayesian belief networks as a planning decision tool for mapping ecosystem services trade-offs on forested landscapes. Environ Res 144(Part B):15–26. https://doi.org/10.1016/j.envres.2015.11.009 (The provision of ecosystem services in response to global change)

Holling C, Gunderson L, Ludwig D (2002) In quest of a theory of adaptive change. In: Gunderson L, Holling C (eds) Panarchy: understanding transformations in human and natural systems. Island Press, Washington, pp 3–22

Kennedy MC, O'Hagan A (2001) Bayesian calibration of computer models. J R Stat Soc: Ser B (Stat Methodol) 63:425–464

Kleidon A, Zehe E, Ehret U, Scherer U (2013) Thermodynamics, maximum power, and the dynamics of preferential river flow structures at the continental scale. Hydrol Earth Syst Sci 17:225–251. https://doi.org/10.5194/hess-17-225-2013

Lee J, Bagheri B, Kao H-A (2015) A cyber-physical systems architecture for industry 4.0-based manufacturing systems. Manuf Lett 3:18–23. https://doi.org/10.1016/j.mfglet.2014.12.001

Levy P, Van Oijen M, Buys G, Tomlinson S (2018) Estimation of gross land-use change and its uncertainty using a Bayesian data assimilation approach. Biogeosciences 15:1497–1513. https://doi.org/10.5194/bg-15-1497-2018

Odum EP (1968) Energy flow in ecosystems: a historical review. Am Zool 8:11–18

Roschewitz I, Gabriel D, Tscharntke T, Thies C (2005) The effects of landscape complexity on arable weed species diversity in organic and conventional farming. J Appl Ecol 42:873–882. https://doi.org/10.1111/j.1365-2664.2005.01072.x

Thenail C, Baudry J (2004) Variation of farm spatial land use pattern according to the structure of the hedgerow network (bocage) landscape: a case study in northeast Brittany. Agr Ecosyst Environ 101:53–72. https://doi.org/10.1016/S0167-8809(03)00199-3

Van Oijen M (2002) On the use of specific publication criteria for papers on process-based modelling in plant science. Field Crop Res 74:197–205

Van Oijen M (2009) Theory and models for managed ecosystems: from confusion to certainty and back again. 40 Years Theory and Model at Wageningen UR 25

Van Oijen M (2017) Bayesian methods for quantifying and reducing uncertainty and error in forest models. Curr For Rep 3:269–280. https://doi.org/10.1007/s40725-017-0069-9

Van Oijen M, Rougier J, Smith R (2005) Bayesian calibration of process-based forest models: bridging the gap between models and data. Tree Physiol 25:915–927. https://doi.org/10.1093/treephys/25.7.915

Van Oijen M, Beer C, Cramer W, Rammig A, Reichstein M, Rolinski S, Soussana J-F (2013) A novel probabilistic risk analysis to determine the vulnerability of ecosystems to extreme climatic events. Environ Res Lett 8:015032. https://doi.org/10.1088/1748-9326/8/1/015032

Van Oijen M, Cameron D, Levy PE, Preston R (2017) Correcting errors from spatial upscaling of nonlinear greenhouse gas flux models. Environ Model Softw 94:157–165. https://doi.org/10.1016/j.envsoft.2017.03.023

Van Oijen M, Bellocchi G, Höglind M (2018) Effects of climate change on grassland biodiversity and productivity: the need for a diversity of models. Agronomy 8:14. https://doi.org/10.3390/agronomy8020014

von Thünen J (1826) Die isolierte Staat in Beziehung auf Landwirtshaft und Nationalökonomie. Perthes, Hamburg

West G (2018) Scale: the universal laws of life and death in organisms, cities and companies. W&N.320 pp, ISBN 9781780225593

Whipple AV, Holeski LM (2016) Epigenetic inheritance across the landscape. Front Genet 7. https://doi.org/10.3389/fgene.2016.00189

Chapter 9
Analyzing Vegetation–Soil–Topography Relationships of Landscapes: A Multiscale Geosystem Approach and Its Application

Alexander V. Khoroshev

Abstract Most landscape-ecological models deal with either between-component relationships at a single hierarchical level or with hierarchy of a single geocomponent, often ignoring variety of types of between-component relationships and density of linkages in space. We developed the multiscale geosystem approach to clarify the relations between landscape attributes, governing ecological processes, and relevant scale levels of processes description. We treat landscape as a multi-pattern geosystem that is composed by superposition of spatial relationships formed at several scale levels and reflecting various contemporary ecological processes as well as legacy of former processes and geographical phenomena. We propose the procedure that allows comparison of a series of hypotheses about the appropriate spatial extent of the higher-order system, i.e., the extent to which the focal system responds to constraint. Constraints from the higher-order system are interpreted as an emergent effect evolving as a result of interactions between neighboring spatial units. The properties of a landscape unit are treated as a product of both intra-level interactions between geocomponents and inter-level relations with a set of surrounding units. The proposed statistical procedure is aimed at splitting spatial variance of geocomponents' (soils and vegetation) properties to effects of inter-level and intra-level interactions. Non-metric multidimensional scaling was applied to reduce dimensionality of raw field data collected in Republic of Udmurtia (mixed-forests zone) and to range ecological factors affecting vegetation and soil properties. Response surface regression models were applied to relate properties of soils and vegetation to each other and to a set of relief morphometric properties measured from digital elevation model in a range of square neighborhoods. Comparison of quality of equations built for various neighborhoods allowed establishing the size of holistic higher-order geosystems that impose frame conditions on properties of the focus-level units. We revealed contemporary and former-time processes that shaped patterns of vegetation and soil cover and their

---

A. V. Khoroshev (✉)
Faculty of Geography, Lomonosov Moscow State University,
Leninskiye gory 1, 119991 Moscow, Russia
e-mail: avkh1970@yandex.ru

L. Müller and F. Eulenstein (eds.), *Current Trends in Landscape Research*,
Innovations in Landscape Research, https://doi.org/10.1007/978-3-030-30069-2_9

characteristic space scale. The proposed approach is scale-invariant and can be applied to study of relations between the focus-level units and the higher-order units at any map resolution.

Keywords Geosystem · Landscape · Hierarchy · Response surface regression · Neighborhood · Morphometrical features · Forest · Udmurtia

9.1 Introduction: Modern Challenges for Studying Spatial Relations and Hierarchy in Landscapes

Modern landscape ecology acknowledges that types of spatial relations vary from place to place (Cushman and Huettmann 2010). Combinations, proportions, and configuration of patches, landforms, communities, etc., may replace each other in space with either discrete boundaries or non-discrete ecotones. Hierarchical organization of a landscape requires finding approaches to adequate identification and description of holistic units with close interactions among the geocomponents such as vegetation, soil, water, bedrocks, and topography. However, it is well known that each unit as a holistic entity is embedded into a higher-order system and may respond to dynamic events in the lower-order units (Wu and David 2002; Turner and Gardner 2015). Since landscape is treated as a complex system and has ability for self-organization which is indicated by spatial patterns and temporal rhythms, identification of relevant space scale for each process or phenomenon is critically needed (O'Neill 1988; Phillips 1999; Hay et al. 2002; Ben Wu and Archer 2005; McMahon et al. 2004; Khoroshev et al. 2007).

Landscape-ecological studies face a set of challenges that restrict correct reflection of hierarchical organization in forecasting future states of a landscape and land-use decisions. First, most models deal with either between-component relationships on a single hierarchical level or hierarchy of a single geocomponent (in most cases, vegetation cover or relief). However, the privilege of landscape science is the study of relationships between groups of correlating properties. Second, landscape mapping is poorly supplied with quantitative approaches to comparative evaluating contributions of abiotic template, self-organization of geocomponents, and intra-phytocoenotic relations to spatial heterogeneity. This results in incorrect delineation of units based on inaccurate determination of principal driving factors. Third, we face the lack of quantitative methods that would enable us to evaluate contributions of the higher-level geosystems to spatial variability of the lower-level ones that is to consider top-down constraints for local ecological processes. Fourth, types of between-components relationships and density of linkages can vary in space. Ignorance for these phenomena restricts opportunities for extrapolation of plot-related information for the needs of territorial planning.

Modern landscape research needs deep insight into relations between biotic and abiotic geocomponents as well as between hierarchical levels. The useful tools for the three-dimensional view of landscape with due consideration for the spatial organization of relief and geological patterns were developed within the framework of the geosystem concept. This concept was introduced to geography by Victor Sochava in 1963 (Sochava 1978). He initiated treating a set of interrelated geo-components on the Earth's surface irrespective of scale as a "geosystem". Recently, Bastian et al. (2015) analyzed contemporary state of the art in the geosystem concept. Since the early years of the concept, the principal challenge for the identification of complex geographical systems has been almost unlimited number of attributes and linkages between them. In 1970s, this fact inspired certain pessimism in feasibility of studying territories as physical systems (Reteyum 1972). The cited author insisted on limiting the "geosystem" notion by the single kind of geographical objects, namely, that with unidirectional matter flow or, later, by centered systems with a field of influence around the core. Nowadays, explicit statistical tools allow reducing dimensionality, and we have opportunities to replace infinite number of attributes by a small number of virtual variables which integrate interdependent attributes and are easily interpretable as ecological factors (Khoroshev and Merekalova 2006). Furthermore, methods like principal components analysis or multidimensional scaling allow ranging virtual variables (factors) by their contributions to spatial heterogeneity. This is not a purely technical advance but, more important, a promising opportunity to derive rationales for revealing hierarchies in nature which is one of the most critical challenges for landscape studies.

So far, the geosystem approach has been developed mainly in Eastern and Central Europe (Angelstam et al. 2013; Bastian et al. 2015). Landscape ecology, nowadays, follows the general trend to three-dimensional study of landscape rather than traditional commitment to "flat" representation of relationships among patches, corridors, and matrix. Relief patterns may be detected and formalized relatively easily due to availability of SRTM models or lidar data. One of the main points of departure for the geosystem concept is more or less dense linkages between abiotic template and soil and vegetation patterns. These relationships are believed to be governed both intra-level and inter-level dependencies.

9.2 What Is the Essence of the Multiscale Geosystem Approach?

In this paper, we apply the geosystem concept to study relationships between vegetation, soil, and topography. In our opinion, the geosystem approach should satisfy the following conditions.

- Interrelated elements are defined and quantitatively described on a set of spatial objects.
- Type of dependency between elements is statistically described, and significance is evaluated.

- System-forming linkages are established; their physical meaning is determined.
- Common properties of elements imposed by external factors and/or higher-order systems (frame conditions) are established.
- Functional role of a focus-level geosystem in a higher-order geosystem is identified.
- Geosystems under investigation are classified.
- Geographical conditions that permit the existence of a system are revealed.
- Boundaries of a geosystem are mapped.

We treat landscape as a multipattern geosystem that is composed by superposition of spatial relationships formed at several scale levels and reflecting various contemporary ecological processes as well as legacy of former processes and geographical phenomena. We distinguish three types of systems and apply these concepts consequently aimed at multiscale description of landscape structure.

The first group of systems (A-systems below) is formed by attributes of this or that geocomponent. The system emerges if some attributes can be treated as competing elements. For example, humus accumulation in taiga soils can replace podzolization under favorable conditions of nutrients supply (e.g., in Umbrisols[1]) or vice versa (e.g., in Retisols). However, in a certain range of landscape conditions, both processes can be manifested in Umbric Retisols with various ratios of humus and podzolic horizons. By reducing dimensionality of raw field data (e.g., thickness of genetic horizons), a researcher has the opportunity to switch to new variables which values reflect coordinates of a sample plot on an axis of some ecological factor. The sensitivity to this factor is treated below as a *property* of a geocomponent, namely ratio of competing attributes which can replace each other depending on intensity of ecological factor manifestation. When a researcher deals with abundances of plant species, the axes reflect competitive relations literally.

The second group (B-systems below) consists of interacting properties of several geocomponents (soil, vegetation cover, groundwater, sediments) that cannot exist separately from each other or, more precisely, their properties emerge as a result of interactions. The abovementioned properties are treated as elements of B-systems, the latter being treated as geographical individuals. The properties of the geocomponents are related by radial matter and/or energy flows. Each combination of the correlating properties is driven by the common ecological factor and produces emergent effect in a system, e.g., biological productivity, chemical composition of groundwater, and soil profile. Of course, we need to determine the spatial limits where each combination of properties exists. Since this kind of systems involves relations between abiotic template and biota at a certain geographical space, they may be referred to as geosystems. Theoretically, the fact of strong correlations in a group of properties may indicate the positive feedback loop that in certain sites generates separate individual geosystem with well-manifested boundaries. In contrast, if one fails to reveal such groups on a territory, this may indicate continuum with highly independent species.

[1]Soils here and below are named according to IUSS (2015).

The third group (C-systems below) is formed by territorial conjugations of elementary operational territorial units (OTUs). Theoretical foundation involves dependence of a system operation on its structure, i.e., on manner of linkage between subsystems (Lange 1965). Results of information which transfer from OTU to the neighboring ones are believed to be a frame condition for the OTU located in the geometrical center of this group that is contextual effects *sensu* Fotheringham et al. (2002). Common "efforts" of spatial elements in a C-system generate a field of a certain attribute, change of any element results in change of a field, and its potential in each OTU. For example, field of groundwater level is determined by combination of landforms: Progressive down-cutting of one of ravines induces lowering of groundwater level in all the surrounding units within a C-system. Elements in this kind of system are not independent on each other. Though extinction of a single element does not destroy the whole system, it affects properties of the other elements. However, the function of a system as a whole, e.g., ability to provide a set of habitats, is preserved since disappearance of a single element can be compensated by similar elements. When modification of structural properties does not entail changes in functional properties, this is referred to as structural adaptation of a system to external impacts (van Gigch 1974). If the holistic properties of a mosaic territory are determined by uniform effect of parent rocks, the area of such rocks, at least within the layer of seasonal fluctuations of temperature and circulation of groundwater, corresponds to the area of a three-dimensional higher-order geosystem in relation to spatial elements. System-forming flows of air, water, heat, etc., create frame conditions for any spatial element.

Principles of the general systems theory require that inherent properties of a landscape unit be related to constants imposed by a higher-order system. Elementary homogeneous landscape unit ("focus-level" unit below) is hypothesized to be a holistic entity united by elemental cycle, water cycle, and other forms of matter and energy exchange between the geocomponents. Before we evaluate constraints from the higher-order system, we should determine its spatial extent. Below we propose the procedure that allows comparison of a series of hypotheses about the appropriate spatial extent of the higher-order system, i.e., the extent to which the focal system responds to constraint. Constraints from the higher-order system are interpreted as an emergent effect evolving as a result of interactions between neighboring spatial units. Thus, the properties of a landscape unit are treated as a product of both interior interactions between geocomponents and spatial relations with surrounding units. The task of crucial importance is to clarify whether elementary landscape unit interacts with environment by a whole set of properties or by groups of properties separately. If the latter holds true, this would indicate that focus-level geosystem is a superposition of effects generated by independent processes. In other words, it is assumed that different processes operate at different spatial scales (Fotheringham et al. 2017). A researcher tests the hypothesis that the geosystem has multistructural and multiscale organization (Wu et al. 2000).This paper is aimed at establishing the relations between landscape attributes, governing ecological processes, and relevant scale levels of processes description.

9.3 Application of Multiscale Geosystem Approach

9.3.1 Splitting the Sources of Spatial Variability in Geosystems

We propose the model of splitting the sources of spatial variability in geosystems. Comparing the contributions of individual and contextual effects allows deep insight into the reasons for spatial heterogeneity (Fotheringham et al. 2002). Suppose, X is a property of a geocomponent, e.g., a sensitivity of a herb layer to water supply manifested in ratio of mesophytes and hygrophytes. A, B, and C are some properties of the other geocomponents (e.g., tree layer, soil color, texture of soil-forming sediments). The scheme in Fig. 9.1 illustrates the following hypotheses aimed at explaining sources of spatial variability of the property X.

1. The values of property X (coordinates of the factor axis) are determined both by intra-level relationships with properties of the other geocomponents (A, B, C—digit 1 on Fig. 9.1) and by inter-level relationships, i.e., frame conditions imposed by the higher-level geosystems (digit 2 on Fig. 9.2). Values of X may vary in space independently of some other properties D.

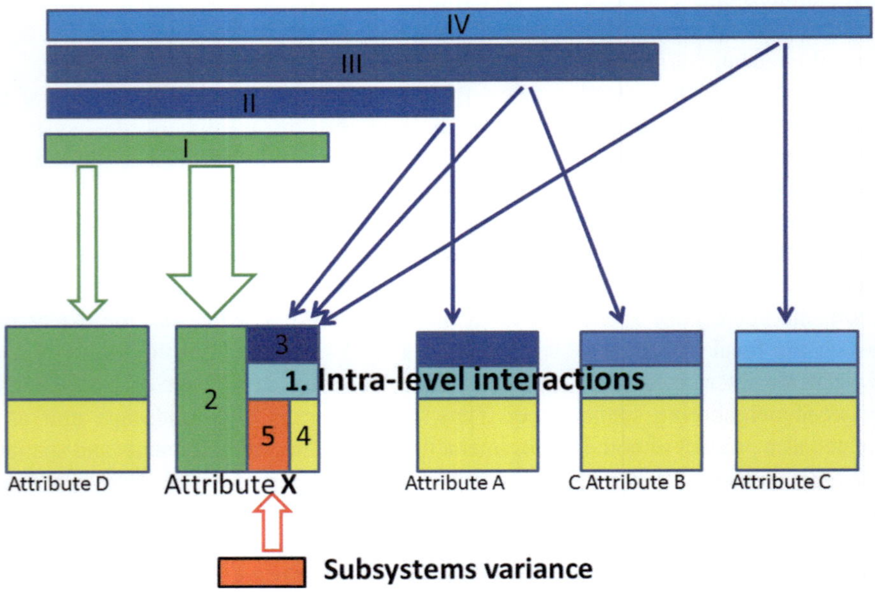

Fig. 9.1 Model of splitting the sources of spatial variability of the geosystem property X at the focus hierarchical level. 1–5—sources of spatial variability. A, B, C, D—other geosystem properties at the focus hierarchical level I–IV—hypothetic higher-order geosystems. Any property of geocomponent may indicate processes at various scale levels as well as results of interactions with other geocomponents at the focus level. For detailed explanations, see in text

Fig. 9.2 Study area in Republic of Udmurtia in Russia, mixed-forests landscape zone (rectangle with black outline)

2. Correlation of properties X, A, B, C, D can be influenced by similar frame conditions imposed by the embracing higher-order geosystem (I in Fig. 9.1) or even by geosystems of several higher orders (I, II, III, IV) simultaneously (digit 3 in Fig. 9.1)
3. Part of property X variability can be determined by random processes independently on both other geocomponents and higher-order geosystems (digit 4 on Fig. 9.1)
4. Part of property X variability unexplained by neither intra-level nor inter-level relationships nor random processes can be determined by spatial variability at the lower hierarchical level, i.e., by variance of subsystems (digit 5 on Fig. 9.1)

Spatial properties of a higher-order geosystem can be described in a number of ways. Below we use a combination of landforms as a surrogate for characteristic of spatial heterogeneity and describe it by morphometric values: standard deviation of elevations, summary length of thalwegs, vertical curvature, and horizontal curvature. However, in general, a higher-order geosystem can be characterized by different sets of properties depending on the research focus. The greatest experience in assessing emergent effects of spatial pattern has been accumulated in landscape-ecological studies of how spatial pattern of landscape cover affects viability of populations (Forman 2006). Another example involves studying runoff regime as an emergent property of a basin geosystem. In this case, a researcher assumes that plant cover influences time and rate of snowmelt and ratio of surface and subsurface flows. Hence, he/she can evaluate proportions of forested and deforested areas, small-leaved and coniferous stands, mature and young stands as the surrogates for plant cover pattern. If one studies microclimatic controls of agricultural productivity, it is relevant to consider share of ravines as accumulators of cold air masses as well as share and connectivity of wind shelterbelts.

9.3.2 Study Area and Measurements

The approach was tested in the mixed-forests zone of eastern East-European plain, Russia, Republic of Udmurtia. The study area is located in the Kama River basin on the Nysha-Vala interfluve within the elevated denudation plain shaped by erosion (Fig. 9.2). The interfluves are asymmetric with steeper south-facing slopes which are deeply dissected by ravines and small river valleys. Solid Permian sedimentary rocks, mainly marls and clays, emerge on valley slopes and contribute to the development of calcium rich soils. Though glaciation did not entered the area, the legacy of the Pleistocene is manifested as a mantle of loess-like loams on flat interfluves and slopes and a mantle of fluvioglacial sands in the valleys bottoms. Nutrient-rich Umbrisols, Umbric Retisols develop on loess-like loams and Permian rocks under mixed forests (*Picea abies*, *Abies sibirica*, *Querqus robur*, *Tilia cordata*) while podzols prevail on sands under the coniferous forests (*Pinus sylvestris*, *Picea abies*). Secondary forests are dominated by *Betula pendula*, *Populus tremula*, and *Tilia cordata*. In wet habitats at the footslope forests with domination of *Alnus glutinosa* and nitrophilous herbs occur. Small bogs with *Pinus sylvestris* and *Betula pubescens* occur locally on the flat interfluves.

We collected 171 integrated field descriptions representative of the characteristic landforms, geological conditions, and forest types. The raw field data involved measurements of plant species abundance (separately for trees, shrubs, low shrubs, herbs, and mosses), thickness of soil horizons, texture, and Munsell colors (hue, value, and chroma) measured at 5 cm interval.

9.3.3 Multiscale Statistical Algorithm for Delineation of Geosystems and Building Hierarchy

We developed the stepwise procedure that allows revealing geosystems and comparing hypotheses about the appropriate spatial extent of the higher-order geosystem, i.e., the extent to which the focal geosystem responds to constraint.

To describe systems of competing elements (the A-systems), we used non-metric multidimensional scaling (NMDS) to calculate position of each sample plot on axes. For each group of attributes, we calculated nonparametric Gamma correlations (r) and converted correlation matrix to distance (d) matrix by equation $d = 1 - r$. Then we derived sensitivity ($a_{1i}, \ldots, a_{4i}$) of each raw variable to NMDS axes and calculated coordinates of sample plots on the NMDS axes by means of equations system:

$$y_i{}^j = a_{1i}x_1{}^j + a_{2i}x_2{}^j + a_{3i}x_3{}^j + a_{4i}x_4{}^j \tag{9.1}$$

where $y_i{}^j$—known value of variable i measured in sample plot j, $a_{1i}, \ldots, a_{4i}$—sensitivity coefficients for variable i in relation to axes 1, 2, 3 or 4, $x_1{}^j, \ldots, x_4{}^j$—coordinate value for the sample plot j.

The axes were rationalized as sensitivity to ecological factors in four ways. First, opposite values of sensitivity coefficients for species or soil horizons suggest interpretation if we know contrasting ecological requirements of these species or horizons. Second, coordinate values can be related to the other well-known ecological attributes by means of ANOVA, regression equations, correlations, etc. Third, spatial distribution of values on a map can suggest solution if we know principal physiographic contrasts in the study area. Fourth, it is easy to interpret values of the axis if we have already obtained physical interpretation of any other strongly correlating axis. Relating the axes values to raw field data we interpreted physical meaning of axes. For example, if the thickness of peat horizon is scoring high on the axis and that of podzolic horizon scoring low, the axis was interpreted as mobility of soil water. At the next steps, axes values were used as dependent variables in equations.

B-systems were composed of the sets of correlating properties. These sets were derived from the results of calculating nonparametric Spearman's correlations and verified by principal components analysis (PCA) from NMDS axes. Positively and negatively correlating properties have the opposite values of PCA factor loadings. To describe relations between the geocomponents, we composed Response Surface Regression (RSR) models for each property:

$$Y = b_0 + \Sigma b_n X_n + \Sigma b_k X_n^2 + \Sigma b_z X_n X_m \pm \varepsilon,$$

where Y—property of a geocomponent (e.g., tree layer), i.e., the coordinate of a sample plot (OTU) on the axis of ecological factor, $X_{n(m)}$—property of another geocomponent (e.g., soils) on the axes that are significant for it, $b_{n(k,l)}$—regression coefficients, and ε—standard error of estimate.

To describe C-systems, we used operational territorial units (OTUs) of the same size formally defined as a pixel of digital elevation model (DEM). Linear dimension of OTU was chosen as 400 m in order to focus on rather large landforms such as flat interfluves area, valley slope, terrace, and floodplain. and to omit deliberately unnecessary topographic details. Sample plots were located in dominant position in the center of OTU and provided with integrated field description. Each OTU was considered to be representative of the natural geosystem defined by common genesis of abiotic template and topography. For the square neighborhood (with linear dimensions varying from 1200 to 6000 m) of each out, we calculated four morphometric attributes: vertical dissections (as a standard deviation of elevations), horizontal dissection (sum of thalwegs lengths), vertical curvature, and horizontal curvature. The obtained values were referred to the OTU in the center of the area used for calculation. We composed RSR model similar to (9.1) but with different variables:

$$Y = b_0 + \Sigma b_n X_n + \Sigma b_k X_n^2 + \Sigma b_l X_n X_m \pm \varepsilon \tag{9.2}$$

where Y—property of a geocomponent (e.g., tree layer), i.e., the coordinate of a sample plot (OTU) on the axis of ecological factor, $X_{n(m)}$—morphometrical attributes of relief in the square neighborhoods of various linear dimensions varying from 1200 to 6000 m, $b_{n(k,l)}$—regression coefficients, ε—standard error of estimate.

The size of square neighborhood reflected our hypotheses about the spatial parameters of the higher-order geosystem (below "1200 m geosystems", "2000 m geosystems" etc.) and contextual effects. Comparing coefficients of determination r^2, we determined the peak value among a series of equations for the various neighborhoods. The peak value was interpreted as the linear dimension of a geosystem that imposed the strongest constraints on the state of the focus-level unit. Considering the linkage between the characteristic space scale of a process and the size of emerging heterogeneous geosystem, we could explain which kind of contemporary of former-time landscape process shaped the geosystem and created frame conditions for the certain property of the geocomponents in OTUs.

To determine spatial limits for various types of intra-level and inter-level relationships, we verified the quality of the statistical models (9.1) and (9.2) for the various combinations of landscape attributes. Analysis of variance (ANOVA) and discriminant analysis were applied to relate residuals from the models to types of landforms, forest types, rank of anthropogenic changes, and types of sediments. The type of landscape conditions with the narrowest range of residuals close to zero was considered to be the most favorable conditions for a certain type of relationships.

9.3.4 Soil-Vegetation Intra-level Relationships as a Reflection of Ecological Drivers

The A-systems were revealed by means of NMDS. This allowed us to range ecological factors by their contributions to spatial variability of each geocomponent. The composition of competing groups of plant species in relation to each factor indicated that nutrients supply was the most significant factor for trees, low shrubs, and herbs (base saturation of soils, in particular), while water supply was ranked as the second factor. Nitrogen supply turned out to the most significant factor for herbs. In contrast, water supply contributed most for shrubs, composition of soil profile and soil colors. Stage of succession was ranked as the third factor for trees. These findings suggest that soil properties could be particularly sensitive to drainage conditions depending on relief, and most vegetation layers (except for shrubs) are controlled mainly by types of sediments and/or transport of nutrients by groundwater. Hence, we needed to establish which hierarchical level of relief organization was responsible for the spatial distribution of soil and vegetation properties. Of course, soils and plants are linked by close interactions and some of their properties may response to relief similarly. At the next step, we tested the hypothesis that geocomponent properties are united in several combinations (B-systems) which vary in space in agreement with each other.

RSR models for B-systems (intra-level relations) showed that the species composition of an herb layer was almost independent on the other vegetation layers. Coefficients of determination were insignificant ranging from 0.03 to 0.19 for any NMDS axis. Evidently, this fact indicates that characteristic time and/or space scale of herbs differ much from that of the other geocomponents. However, structure of soil profile was significant for the 2D, 3D, and 4th axes which are sensitive to nutrients supply and texture of soil-forming sediments, r^2 accounting for, respectively, 0.20, 0.17, and 0.15.

Tree and shrub layers are more closely related to soil profile and soil colors with r^2 reaching 0.24–0.32. Nutrient-poor well-drained soils favor high abundance of *Pinus sylvestris, Juniperus communis*, and *Rosa majalis*. Nutrient-rich and humus-rich soils are covered with communities dominated by *Alnus glutinosa* or combination of *Querqus robur*, *Acer platanoides*, *Corylus avellana*, *Sambucus racemosa*, *Padus aviculare*.

It is obvious that inter-geocomponent intra-level relations explain rather low proportion of variance. Hence, similar communities may develop on different soils and vice versa. This could be influenced by the contextual effects that is contribution of frame conditions from the higher levels of landscape organization, i.e., by processes of another origin and characteristic space scale.

9.3.5 Topographic Effects on Soils and Vegetation as the Indicators of Inter-level Relationships

Now we turn to the results obtained in investigation of inter-level relationships, that is of emergent effects resulting from interactions of neighboring spatial elements. These interactions were believed to be responsible for the part of variance of the OTUs properties. We classified OTUs (pixels of DEM) by morphometric attributes in the square neighborhoods with linear dimensions 1200, 2000, 2800, 3600, 6000 m (Fig. 9.3). Visualization of classification showed evidence that the "1200 m geosystems" reflect contemporary erosion-induced patterns. The "6000 m geosystems" correspond to contrasts induced by the various rates of neotectonic uplift which resulted in the variety of valleys widths and slopes dissection. Intermediate sizes of geosystems are indicative of relief pattern formed by the Pleistocene sedimentation, mainly the development of fluvioglacial sandy terraces, and the Holocene denudation at slopes resulting in either exposure of solid rocks on steep slopes or preservation of periglacial loess-like loams on the gentle ones.

The properties of the geocomponents formed two distinct groups in relation to the hierarchical level of relief organization they "prefer", that is the level that imposes the most significant constraints on spatial variability. Properties of the tree layer (the 1st, 2D, 3D, and 4th axes) were the most sensitive to the relief conditions in the "6000 m geosystems," i.e., to rather large patterns with obvious neotectonic genesis (Fig. 9.4). The same is true for most properties of shrubs, some properties of herbs (the 2D and 3D axes), soil color (the 2D axis) and soil profile (the 3D axis). Almost all of these axes were sensitive to nutrients supply, to contrast between loess-like loams and sands in particular. Soil profile (thicknesses of horizons) and soil colors are partly controlled by the closest environment in "1200 m geosystems." Hypothetic "2800 m geosystems" were the least significant and could hardly be treated as real holistic ones. Soil texture does not exhibit clear relation to relief of any size of hypothetic geosystems.

For example, the 2D and the 3D axes for the herb layer reflect sensitivity to nutrients supply and soil texture, respectively. As the linear size of hypothetic square neighborhood increases from 1200 to 6000 m, coefficient of determination increases monotonously within the interval 0.20–0.34 for the 2D axis and in the interval 0.10–0.27 for the 3D axis. Maximum values on the 3D axis for soil horizons and the 4th axis for soil colors correspond to slowed down transformation of litter to humus under the slightly abundant humidity. Munsell values are high in the 5–15-cm-thick topsoil; Munsell chroma is high at the depth of 30–40 cm and low at the depth of 50–60 cm. This indicates poor humus accumulation and iron illuviation. Such combination of soil attributes occurs under either minimum or maximum vertical dissection in the neighborhood 2000 m (i.e., parabolic pattern), that is in moist habitats of either flat interfluves or steep deeply dissected slopes, respectively. The latter option occurs due to emergence of groundwater at the footslopes or in ravines. Minimum values of the abovementioned axes correspond

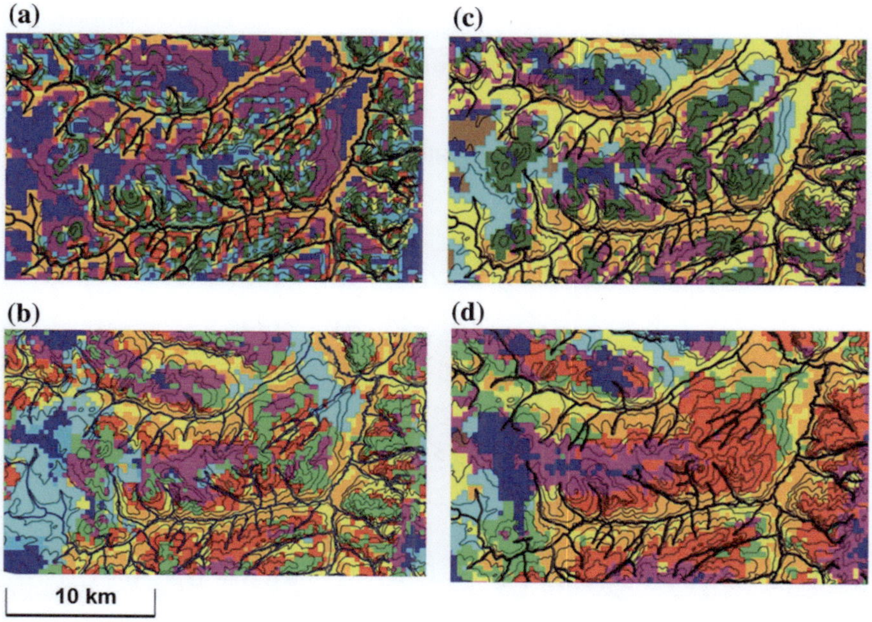

Fig. 9.3 Classification of relief patterns by morphometric attributes in square neighborhood with linear dimension 1200 m (**a**), 2000 m (**b**), 3600 m (**c**), and 6000 m (**d**). Red color depicts the most deeply dissected steep slopes, purple—flat interior areas of the wide interfluves, orange—concave areas of river valley bottoms and lower slopes, light blue—flat or undulating areas, green—narrow convex ridges

to elevated humus accumulation and low thickness of litter on narrow convex interfluves and gentle well-drained slopes.

The group of properties (the 4th axis for trees, the 1st axis for shrubs, and the 1st axis for soil color) is sensitive to redox conditions and vary in agreement with relief patterns with linear size 3600 m, with vertical curvature in particular. Classification of OTUs by morphometric features of relief in the neighborhood of this size ignores individual small valleys and gullies but distinguishes slopes of large valleys differing in degree of dissection, flat interfluve areas, and river terraces. Localities with zero or low vertical curvature are dominated by species with high tolerance to elevated water supply (e.g., *Betula pubescens*, *Alnus incana*, *Ribes nigrum*) and gleization in soils often with Histic horizon.

To identify the boundaries of the holistic geosystems, it was necessary to reveal a range of conditions which allowed similar response of various geocomponents to frame conditions imposed by the higher-order geosystems. We compared quality of the equations and residual values for the hypothetic high-order geosystems of various sizes. We verified the quality of equations for various types of landforms and determined which landforms ensure the least possible (close to zero) residuals. In other words, we established abiotic conditions in which inter-level relations are

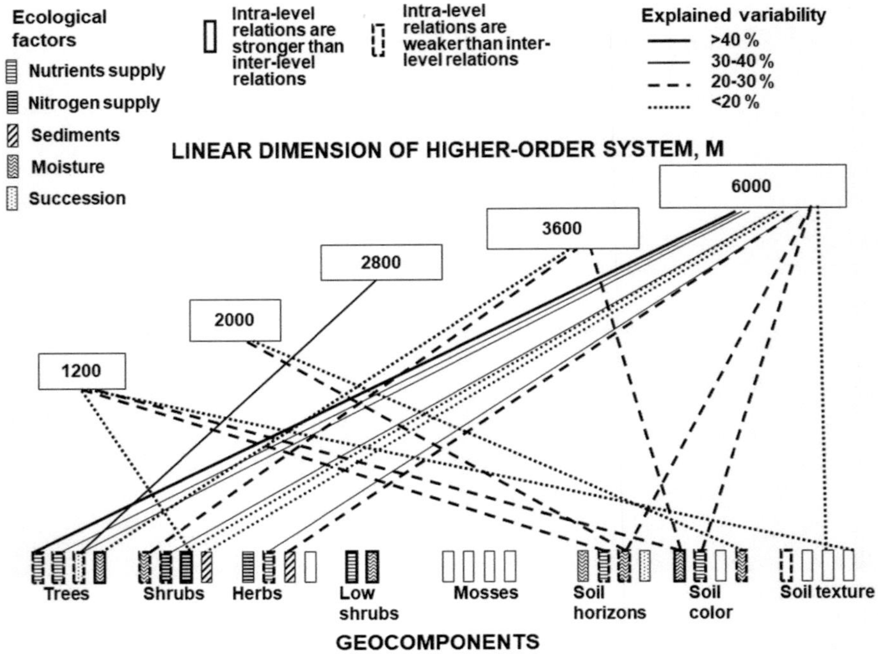

Fig. 9.4 Sensitivity of the geocomponents to relief-driven processes in the higher-order geosystems. The graph shows that geocomponents exhibit various space scales of relationships with topography-driven processes. Most properties of trees, shrubs, and herbs obey the heterogeneity of the highest hierarchical level (6000 m), while soils may reflect patterns at the lower levels (1200 and 2000 m)

the most well-pronounced. This allowed us to classify the properties of the geocomponents by their sensitivity to scale levels.

The first group of properties is not related to relief and topography-mediated processes in any surrounding of an OTU. Species composition of tree layer in peatlands is an example of such properties.

The second group of properties obeys the processes of a single level of the higher-order geosystems. Relief pattern either indicates paleo-process or controls contemporary process. The other scale levels are insignificant, and residuals in the corresponding equations deviate from zero much. For example, species composition of tree layer in flat interfluves highly depends on morphometric features of relief in the geosystems with linear dimension 6000 m.

The third group of properties obeys the processes operating within the geosystems of several rank orders. The ratio of coniferous and broad-leaved trees (the 1st axis) is driven by both “1200” and “6000” geosystems. The highest quality of equation is characteristic of the steep southwest-facing slopes of the largest valleys. In the closest environment (“1200 m geosystems”), the most significant predictors are vertical and horizontal relief dissection which indicate geological contrasts, more precisely the varying influence of Permian carbonate rocks being locally covered by the Quaternary sediments. Broad-leaved trees prefer habitats with the closest to surface carbonate rocks in the upper slopes, while coniferous ones—the habitats with sandy fluvioglacial sands where the influence of carbonate rocks is not manifested. In contrast, in the widest environment ("6000 m geosystems") vertical and horizontal curvature are the most significant predictors which are responsible for dispersion and accumulation of flows. For the tree layer, this means contrasting habitats in convex ridges and concave valleys. The difference between the best predictors evidences difference in the nature of driving factors in small and large higher-order geosystems.

Among the RSR models that relate properties of trees, shrubs, and herbs to the relief of "6000 m geosystems" the minimal (i.e., close to zero), residuals were detected for 64 units out of 171 examined ones. Another 21 units had low residuals for trees and herbs. Then we tested the hypothesis that close to zero residuals occur within certain type of landforms and at certain stage of recovery succession. The model for tree layer was successful in units with high canopy cover, thick litter, high abundance of fir, spruce, and aspen, and low abundance of pine. Evidently, such a set of attributes distinguishes units with zonal undisturbed vegetation from the secondary forests. Shrubs and herbs exhibit stronger dependence on the relief of the higher-order geosystems if the thickness of sands over loams does not exceed 20–30 cm. On sandy soils, intra-level linkages contribute to spatial variability more than inter-level ones.

Humus accumulation exhibits the best dependence on relief of "1200 m geosystems" on the interfluves rather than on slopes. The thicker humus horizons are typical for convex or concave surfaces. Reduced thickness occurs in flat areas where humus accumulation is commonly replaced by litter accumulation and podzolization. Humus accumulation decreases as the width of flat interfluve increases. The influence of "6000 m geosystems" is more well-manifested in inclined areas. Increase in vertical dissection results in decrease of podzolization due to reduced duration of the period with temporal excessive humidity in spring which is critical for this soil-forming process. The other reason is strong influence of solid carbonate rocks which are closer to surface at almost any position within slopes of ridges despite local peculiarities. Obviously, common legacy of the neotectonic uplift create powerful frame conditions within each type of the "6000 m geosystems."

9.3.6 *Nature of Geosystems Hierarchy Based on Vegetation–Soil–Topography Relationships*

In this research, we applied the approach allowing distinction among contributions of inter-level and intra-level relations to spatial heterogeneity of the focus-level landscape units. We did not postulate the sizes of the higher-level geosystems a priori but revealed them based on statistical models. Comparing response surface regression equation for various neighborhoods, we managed to reveal the neighborhood that imposes the most significant constraint over several combinations of the correlated vegetation and soil properties at the lowest scale level.

Ranging ecological factors using NMDS showed that their significance differs for various layers of vegetation and soils. Territory of interest contains features of three neighboring landscape zones which are manifested in different landscape units depending on mineral nutrition and water supply. Relief patterns detected by a set of morphometric features indicate geological structures and neotectonic trends that affect occurrence or co-occurrence of features of taiga, mixed-forests, and broad-leaved forests zones.

Most vegetation layers were highly sensitive to patterns dictated by the levels of nutrient-supply in soil-forming sediments. Soils are more sensitive to water supply determined by relief. Sensitivity of herb layer to a combination of soil horizons, though not very high, reflects high geochemical contrasts of soils on sands, loess-like-loams and solid carbonate bedrocks. Trees and shrubs layers are better related to soil conditions than herbs. All of them co-vary in space in concordance with frame conditions imposed by relief of the higher-order geosystems. However, the sensitivity of correlating vegetation and soil properties to the combined effects of neighboring landforms evidences interdependencies among intra-level and inter-level relations in a landscape. The sensitivity of soil and vegetation properties to relief differs much.

Geosystems of different rank orders were generated by processes acting at different space and time scales and themselves generate contemporary processes affecting soils and vegetation. Increase in relief dissection makes inter-level relations more complex due to the growing effect of fine-scale processes.

Relief patterns with approximate linear sizes 2000 and 6000 m provided integrated information on vegetation, soil and geochemical flows and, hence, can be treated as landscape patterns. Geosystems with linear size 6000 m play the most significant role for the spatial variability of most geocomponents properties. Macro-landforms correspond to either wide and narrow sectors of the convex asymmetric interfluves composed of Permian solid sedimentary rocks or river valleys along neotectonic joints. This major topographic contrast forms the most important feature of landscape heterogeneity, namely the difference between nutrient-rich units on loess-like loams and marls and, on the other hand, nutrient-poor units on fluvioglacial sands which cover the lowest concave parts of slopes and high terraces. The first variant corresponds to forests with high abundance of *Tilia cordata*, *Ulmus glabra*, *Acer platanoides*, *Abies sibirica*, *Corylus avellana*, *Viburnum opulus*, *Padus aviculare*, *Asarum europaeum*, *Melica nitens*, *Asperula odorata*, *and Pulmonaria officinalis*. These communities are formed on Umbric Retisols or Umbrisols which provide favorable conditions for the species typical of the broad-leaved forests. In contrast, nutrient-poor podzols on sands ensure dominance of boreal species (*Picea abies*, *Juniperus communis*, *Vaccinium myrtillus*, *Vaccinium vitis-idaea*) and similarity of plant communities with taiga zone. Thus, contrasting sedimentation in Pleistocene was the reason for close neighborhood of "islands" of taiga and broad-leaved communities in space. Land-use decisions should use 6.0-km scale level for rational distribution of agricultural, forestry, and nature protection activity, e.g., relatively nutrient-poor sandy soils on long gentle valley slopes are much better for forest recovery than for plowing (Khoroshev 2010).

Our research showed evidence that the phytocoenoses (tree layer, in particular) may vary in concordance with spatial pattern of neighboring units within the higher-order geosystem despite similar inherent abiotic template of OTU. It means that despite similar types of landforms and sediments, the OTUs have dissimilar tree layer driven by contrasts in the neighborhood (3000 m in each direction) induced by neotectonic processes. It is worth noting that size of landform may act as a separate ecological factor which regulates a lot of ecological processes. We found evidence that some types of inter-level relations are manifested differently on wide and narrow interfluve areas with similar interior topography and geology. In other words, the size of landform does matter for vegetation.

The properties that exhibited no response to relief pattern in any higher-order geosystem (e.g., tree layer in peatlands), most likely, have another characteristic space scale than those considered. Obviously, the low-resolution DEM used in our research omitted topographic details that could be more closely connected to significant contemporary ecologically processes. Hence, such properties of OTU may be driven by variability in lower-order geosystems to a great extent (variance source 5 in Fig. 9.1). Another explanation involves real ignorance of relief and high contribution of self-development of the geocomponent.

The proposed procedure allows revealing the range of scale levels that impose frame conditions on the focus-level units by comparing sets of significant predictors in the equations and signs of regression coefficients. If the parameters of the equations are similar for various scale levels (i.e., various neighborhoods in the case under consideration), it means that the close environment (e.g., 1200 m in our example for tree layer) characterize only a small part of the wider area (6000 m in our example) where the driving factor operates. If at the intermediate scale levels any significant relations are not manifested, we face the phenomenon of self-similarity. In this case, composition of tree layer differs in ravines and local interfluve between them following the same rule as in large valleys and a ridge between them. The rule may be as follows: The more deeply dissected relief is, the lower is the proportion of hygrophytes. If the parameters of the statistically significant equations differ at various scale levels, we conclude that the driving process changes at the higher scale level.

In the region of interest, mutual adaptation of the geocomponents and inter-level relations depend on the stage of development of phytocoenosis. In units with mature and overmature forests, vegetation layers are strongly adapted to frame conditions of the higher-order geosystems. In contrast, in secondary disturbed forests inter-level linkages are not well manifested. Hence, anthropogenic disturbance forces landscape units to "switch of" temporarily their relations with the higher scale level and behave more independently following self-development regularities. Geological conditions that support zonal vegetation, such as loams or thin sandy layer over loams, may ensure stronger dependence on inter-level relations as compared to the nutrient-poor sandy deposits. The latter support vegetation communities that resemble taiga, that is landscape more inherent to regions located further to the north (so-called extra-zonal landscapes). Hence, the geosystems with approximate linear dimension 6000 m impose well-pronounced frame conditions on a set of geocomponents only in landscape conditions that ensure intra-level relations inherent for the mixed-forests zone in its natural state.

9.4 Conclusions

Application of the geosystem approach and statistical procedures provide opportunities to establish hierarchical organization of a landscape based on interaction of geocomponents. When landscape ecologists proceed from the concept that soils and vegetation follow topographic patterns at multiple scales, he or she acquires greater chances to relate ecological biota-regulated processes both to contemporary matter flows and to remedy of past geological epochs. The approach was tested in the mixed-forest landscape zone, but it is applicable to any zone with humid climate where water flows can link remote units easily. In arid regions, the approach should be tested separately. The proposed approach is scale-invariant and can be applied to study of relations between the focus-level units and the higher-order units at any map resolution. Furthermore, combination of landforms used in our example is only a particular case of a much wider range of possible attributes of hypothetic higher-order systems. The approach allows relating properties of focus-level units (soils, vegetation, animals, groundwater, etc.), for example, to the combinations of land-use types, microclimate conditions, and connectivity of patches. Comparing the models built for the various hypothetic surroundings, it is easy to reveal the relevant set of higher hierarchical levels. We believe that this opportunity may be helpful in landscape planning which needs to find rationales for the optimal scale of land-use decisions.

Acknowledgements The study was supported by Russian Foundation for Basic Research (RFBR projects 17-05-00447). The contribution of Ksenia Merekalova and Roman Bekkiev to field research is greatly acknowledged.

References

Angelstam P, Grodzynskyi M, Andersson K, Axelsson R, ElbakidzeM Khoroshev A, Kruhlov I, Naumov V (2013) Measurement, collaborative learning and research for sustainable use of ecosystem services: landscape concepts and Europe as laboratory. AMBIO 42:129–145

Bastian O, Grunewald K, Khoroshev AV (2015) The significance of geosystem and landscape concepts for the assessment of ecosystem services: exemplified on a case study in Russia. Landscape Ecol 30(7):1145–1164

Ben Wu X, Archer SR (2005) Scale-dependent influence of topography-based hydrologic features on patterns of woody plant encroachment in savanna landscapes. Landscape Ecol 20:733–742
Cushman SA, Huettmann F (eds) (2010) Spatial complexity, informatics, and wildlife conservation. Springer, Tokyo, p 458
Forman RTT (2006) Land mosaics. Cambridge University Press, Cambridge, p 632
Fotheringham AS, Brunsdon S, Charlton M (2002) Geographically weighted regression: the analysis of spatially varying relationships. Wiley, 269 p
Fotheringham AS, Yang W, Kang W (2017) Multiscale geographically weighted regression (MGWR). Ann Am Assoc Geogr 107(6):1247–1265
Hay GJ, Dube P, Bouchard A, Marceau DJ (2002) A scale-space primer for exploring and quantifying complex landscapes. Ecol Model 153:27–49
IUSS Working Group WRB (2015) World Reference Base for Soil Resources 2014, update 2015. International soil classification system for naming soils and creating legends for soil maps. World Soil Resources Reports No. 106. FAO, Rome, 192 p
Khoroshev A (2010) Multilevel analysis of landscape structure for land use decisions. In: Barančoková M, Krajčí J, Kollár J, Belčáková I (eds) Landscape ecology—methods, applications and interdisciplinary approach. Bratislava, Institute of Landscape Ecology, Slovak Academy of Sciences, pp 99–112
Khoroshev AV, Merekalova KA (2006) Uncertainty of relations between landscape components—a tool for modelling evolution of spatial pattern. Ekológia (Bratislava) 25(Suppl. 1):122–130
Khoroshev AV, Merekalova KA, Aleshchenko GM (2007) Multiscale organization of intercomponent relations in landscape. In: Dyakonov KN, Kasimov NS, Khoroshev AV, Kushlin AV (eds) Landscape analysis for sustainable development. Theory and applications of landscape science in Russia. Alex Publishers, Moscow, pp 93–103
Lange O (1965) Wholes and parts: a general theory of system behavior. Pergamon Press, 74 p
McMahon G, Wiken EB, Gauthier DA (2004) Toward a scientifically rigorous basis for developing mapped ecological regions. Environ Manage 34(1):111–124
O'Neill RV (1988) Hierarchy theory and global change. In: Rosswall T, Woodmansee RG, Risser PG (eds) SCOPE 35. scales and global change: spatial and temporal variability in biospheric and geospheric processes. Wiley, UK, pp 29–45
Phillips JD (1999) Divergence, convergence and self-organization in landscapes. Ann Am Assoc Geogr 89(3):466–488
Reteyum AYu (1972) Physical-geographical studies and system approach. In: System studies. Nauka, Moscow, pp 90–110 (in Russian)
Sochava VB (1978) Introduction to the theory of geosystems. Nauka, Novosibirsk, 320 p (in Russian)
Turner M, Gardner RH (2015) Landscape ecology in theory and practice. Pattern and process. Springer, Berlin, p 482
van Gigch JP (1974) Applied general systems theory. Harper & Row Publishers, New York, p 439
Wu J, David JL (2002) A spatially explicit hierarchical approach to modelling complex ecological systems: theory and applications. Ecol Model 153:7–26
Wu J, Jelinski DE, Luck M, Tueller PT (2000) Multiscale analysis of landscape heterogeneity: scale variance and pattern metrics. Geogr Inf Sci 6:6–19

Chapter 10
Ecosystem Services of Russian Landscapes

Elena Bukvareva, Dmitry Zamolodchikov and Karsten Grunewald

Abstract Russian landscapes provide important ecosystem services (ES) of local, regional and global scale and are crucially important for the economy and people of the country. The Project TEEB-Russia is the first attempt at a nation-wide ES assessment in Russia. The result of the first phase of the project (2013–2015) was the "Prototype National Report on Ecosystem Services of Russia, Volume 1, Terrestrial Ecosystems Services." A methodology for ES assessment was developed with allowance for the current status of the national public statistics. ES volumes supplied by ecosystems and consumed by humans were assessed. The degree of ES use was assessed by the ratio of supplied ES to consumed ES. These methodological approaches allowed to compare the regions of Russia and define regions, which are ES donors and ES consumers. However, further progress in defining the principles of ES management requires moving from the interregional to landscape scale. In particular, the optimization of the tasks of biodiversity conservation and ES use can be effectively solved at the landscape level.

Keywords Ecosystem services · Supplied · Demanded · Consumed services · National assessment · Regional comparison · Management of ES · Biodiversity conservation

E. N. Bukvareva (✉)
Biodiversity Conservation Center, Vavilova Street 41, Office 2, 117312 Moscow, Russia
e-mail: bukvareva@gmail.com

D. Zamolodchikov
Biological Faculty, Lomonosov Moscow State University, Leninskie gory, d. 1, str. 12, 119234 Moscow, Russia
e-mail: dzamolod@mail.ru

Centre for Ecology and Productivity of Forests of Russian Academy of Sciences, Profsouznaya Street, 84/32, 117997 Moscow, Russia

K. Grunewald
Leibniz Institute of Ecological Urban and Regional Development, Weberplatz 1, 01217 Dresden, Germany
e-mail: k.grunewald@ioer.de

L. Müller and F. Eulenstein (eds.), *Current Trends in Landscape Research*, Innovations in Landscape Research, https://doi.org/10.1007/978-3-030-30069-2_10

10.1 Introduction

The concept of ecosystem services (ES) is one of the most rapidly developing fields of up-to-date ecological research, aiming at maintaining life-supporting functions of biodiversity and sustainable development of the biosphere. The international TEEB process (The Economics of Ecosystems and Biodiversity; TEEB 2019; Hedden-Dunkhorst et al. 2015) encourages studies on the national and regional levels. National ES assessments are an important tool for maintaining ES and have already been carried out by many countries, including national projects within the TEEB framework (TEEB 2019).

Russian landscapes, diverse and occupying large areas, provide important ecosystem services of local, regional and global scale and are crucially important for the economy and people of the country. The importance and necessity of the ES concept development in Russia were proclaimed by the scientific community and a number of pioneering works on ES assessments were made in Russia and other post-soviet new independent states (Bastian et al. 2015; Bukvareva et al. 2015; Grunewald et al. 2014a, b). However, these work samples had a local or regional scale.

To start a national ES valuation and accounting process the Russian–German project "TEEB-Russia. Ecosystem Services Evaluation in Russia: First Steps" was initiated in 2013 by the Biodiversity Conservation Center (Moscow) in cooperation with the Leibniz Institute of Ecological Urban and Regional Development (Dresden) (Grunewald et al. 2014c). The TEEB-Russia project is commissioned by the German Federal Agency for Nature Conservation (BfN) with funds from the German Federal Ministry for the Environment, Nature Conservation and Nuclear Safety (BMU) and was supported by the Ministry of Natural Resources and Environment of the Russian Federation. The result of the first phase of the project "TEEB-Russia 1" (2013–2015) was the first volume of the Prototype National Report on Ecosystem Services of Russia (Bukvareva and Zamolodchikov 2018), hereinafter Prototype Report. This report pursues methodological goals and shows possible approaches to non-monetary estimation of terrestrial ES at the national level on the data available today.

In the present publication, we discuss the methodology of national non-monetary assessment of ES in Russia explaining indicators of supplied, demanded and consumed ES and providing examples of ES valuation by different methods. After that we discuss the main problems encountered during the assessment and future assessment tasks including analysis of interrelations between biodiversity and ES at interregional and landscape scales.

10.2 Materials and Methods

10.2.1 Data Sources and Assessment Units

The primary data sources for assessing ES were open-state databases, published maps and statistical digests including, first and foremost, public databases of the Federal State Statistics Service (FSSS) and other federal agencies, digital cartographic materials from "Land Resources of Russia" (Stolbovoi and McCallum 2002), a map of the terrestrial ecosystems of Northern Eurasia (Bartalev et al. 2003) and National Atlas of Russia (2004–2008).

The assessment area was the territory of Russian Federation. The main national-level sources of socio-economic data, as well as some environmental indicators in Russia, are the public databases of FSSS and other federal agencies, which produce data for the constituents of the Russian Federation. Thus, constituents of the Russian Federation—oblasts, krais, republics, etc., (hereinafter the regions) were used as assessment units. The use of administrative regions as assessment units corresponds well to the state statistics, but this approach does not fit well with ecosystem processes. However, we believe this approach was the most appropriate for the national system of ES assessment in Russia because it corresponds to the national statistical system and allows a comparison of ES between the regions. Similar approach was implemented for some European sub-continental assessments using NUTS (EU statistical areas) as a spatial mapping unit for the ES evaluation (Maes et al. 2011; Schulp et al. 2014; Zulian et al. 2014).

10.2.2 Classification of Ecosystem Services

Classification of terrestrial ES adopted in the Prototype Report combines the approaches of the Millennium Ecosystem Assessment (2005), CICES (Haines-Young and Potschin 2013) and the National Strategy of Biodiversity Conservation in Russia (2001). A comparison of ES classifications in MEA, CICES and in the Prototype Report is presented in Bukvareva et al. (2019). In total, 31 ES were considered (Table 10.1). ES were grouped into four categories: productive (provisioning), i.e., production of biomass that is removed from ecosystems by people; environment-forming (regulating), i.e., the establishment and maintenance of environmental conditions conducive to human life and economic development; informational (cultural), i.e., all kinds of information that is contained in natural ecosystems and can be used by people; recreational, i.e., formation and maintenance of natural conditions for different types of recreation. The names "productive", "environment-forming" and "informational" retain the designations of the main groups of life-supporting functions of biodiversity in the National Strategy of Biodiversity Conservation in Russia (2001).

Table 10.1 Methods for assessing ES in the prototype report

ES	Methods*			
	1	2	3	4
Productive (provisioning)				
Wood production	X			
Non-wood production of forest and other terrestrial ecosystems	X			
Production of fodder on natural pastures (hayfields were not taken into consideration)		X		
Production of freshwater ecosystems, primarily fish				X
Game production	X			
Production of honey in natural areas				X
Environment-forming (regulating)				
Climate and atmosphere regulation				
Biogeochemical climate regulation				
Carbon storage	X			
Regulation of greenhouse gas flows (only CO_2 was considered)	X			
Biogeophysical climate regulation				X
Air purification by vegetation (absorption of pollutants by suburban forests)		X		
Hydrosphere regulation				
Water protection and water regulation				
Regulation of runoff volume		X		
Regulation of runoff variability (runoff stabilization)		X		
Assurance of water quality by terrestrial ecosystems		X		
Assurance of water quality by freshwater ecosystems (water self-purification and dilution)		X		
Soil formation and protection				
Soil protection from erosion				
Soil protection from water erosion			X	
Soil protection from wind erosion			X	
Prevention of damage from soil washing into water bodies				X
Prevention of damage from landslides and mudflows				X
Establishment of soil bioproductivity				X
Self-purification of soils			X	
Regulation of cryogenic processes		X		
Regulation of biological processes important for the economy and for security				
Ecosystem regulation of species with economic importance (agricultural and forest pests, invasive and synanthropic species)				X
Pollination of farm crops			X	
Ecosystem regulation of species with medical, biomedical and veterinary importance				X
Informational (cultural)				
Genetic resources of wild species and populations			X	

(continued)

Table 10.1 (continued)

ES	Methods*			
	1	2	3	4
Information on structure and functioning of natural systems that can be used by humans			X	
Aesthetic and educational importance of natural systems			X	
Ethical, spiritual and religious importance of natural systems				X
Recreational				
Formation of natural conditions for daily recreation near home, weekend recreation, recreation at summer cottages			X	
Formation of natural conditions for educational and active tourism in the nature			X	
Formation of natural conditions for resort recreation (except seacoasts)				X

*Numbers correspond to the assessment methods noted in the text

With regard to the assessment of landscape ES, the following differences between our ES classification and other ES classifications should be explained. A significant part of the territory of Russia, especially in the European part of the country, is occupied by agricultural and cultural landscapes that combine natural and anthropogenic elements. For these landscapes, approaches to evaluating recreational and provisioning ES are especially important. In contrast to the CICES, we do not incorporate recreational and information ES in a single category of cultural ES. Instead, we consider provisioning, regulating and informational ES as the main categories and separately identify recreational category as integrative ES that are coupled to all of the first three ES categories to various extents depending on the type of recreation. For example, regulating ES which provide good quality of environment is the most important for recreation at summer cottages and resorts; productive ES such as game and fish production are the most important for sporting, hunting and fishing; informational ES such as aesthetic, educational, ethical and spiritual importance of natural systems are the most important for educational tourism in nature, etc. (Bukvareva et al. 2019). Another important feature of our ES classification applies to agricultural production. CICES includes ES related to food production by humans (crops, livestock, aquaculture), but we do not consider them as ES since we believe that ecosystems create natural conditions for these industries (soils, water, climate), which is taken into account in regulating ES. Moreover, ES related to food production by humans include a certain amount of purely human resources (machines, technologies, fuel, labor of humans and domestic animals). Generally, agricultural production negatively depends on the share of the area of natural ecosystems in the region and, in fact, is in direct opposition to other ES (Sect. 5.4). Accounting for agricultural production as ES can lead to conflicting results when assessing ES, and especially when assessing the relationship between ES and the state of ecosystems and biodiversity.

10.2.3 Methods of ES Assessment

The following methods of ES assessment were used depending on the data availability and methodological clarity.

1. **Direct quantitative ES valuation** was possible if values of supplied, consumed and demanded ES are presented in the public statistical databases and reports. Today, possibilities for this valuation method are extremely limited. In total, supplied and consumed volumes only for five out of 31 ES were directly evaluated to a relatively complete extent (Table 10.1): the ES wood production (Fig. 10.3)—on the base of data of FSSS (Rosstat 2013) and the Federal Agency for Forestry of the Russian Federation, the ES of game production—on the base of statistical digest of the Department of State Policy and Regulation of Hunting and Conservation of Hunting Resources of the Ministry of Natural Resources and Environment of the Russian Federation (Lomanova et al. 2011), the ES of non-wood production of terrestrial ecosystems (on the base of data on harvest and stocks of mushrooms and berries in the second half of the 1980s from statistical compilation by Egoshina (2005), the ES of carbon storage and regulation of CO_2 flows (Fig. 10.4)—on the base of data from national reports due to Russia's obligations to implement the UN Framework Convention on Climate Change (National Report 2013).
2. **Indirect quantitative evaluation** of ES was applied in case of a lack of direct statistical data but in the presence of cartographic and statistical data that allowed us to evaluate the desired indicators. This method of ES evaluation corresponds to the indirect measurement of ES as defined in Vihervaara et al. (2017) and to the extrapolation of primary data as defined in Martínez-Harms and Balvanera (2012) and consists in the transformation of cartographic and statistical data using known coefficients and simple equations that can be classified as conceptual and deterministic physical and chemical models (Dunford et al. 2017). Six ES were assessed by this method: one provisioning ES and five regulating ES (Table 10.1; Figs. 10.5, 10.6, 10.7).
3. **Estimation of ES score** was applied if there was no data to evaluate the ES themselves and it was only possible to estimate factors affecting them. We believe that supplied ES is determined by natural factors (e.g., the area of natural ecosystems) while consumed and demanded ES are determined by socio-economic factors (including polluting emissions as a result of human activity). The range of values of the selected factor was divided into 10 classes with a score from 1 to 10 points assigned to each class (smaller scores correspond to lower values of the factor). To combine several factors their scores in a region were summed up and the resulting total value was translated into a 10-point scale. Nine ES were estimated by scores: four regulating ES, three informational ES and two recreational ES (Table 10.1; Fig. 10.8).
4. **Statement of the task** of ES assessment, if methodological approaches for ES assessment are not ready or data were not available.

In total, about one-third of considered ES was quantitatively evaluated (five ES were directly evaluated and six ES were indirectly evaluated), one-third (nine ES) was assessed with scores and one-third (ten ES) was not assessed (Table 10.1, for details, see Bukvareva et al. 2019).

10.3 Assessment of Supplied, Demanded and Consumed ES

The extreme diversity of natural and socio-economic conditions in Russia requires estimation of ES, which are supplied by ecosystems and ES, which are demanded and consumed by people. The supplied ES are generally correlated with the area of ecosystems. The demanded and consumed ES are linked to population density, economic development and transport accessibility of the regions. The most common pattern is an inverse relationship between the area of natural ecosystems and the density of ES consumers. This pattern is evident almost everywhere because economic activity in general leads to disturbance of natural ecosystems, and it is most clearly manifested in the large and diverse territory of Russia (Fig. 10.1, Bukvareva et al. 2015; Bukvareva and Zamolodchikov 2018). In order to have the possibility of an adequate comparison of very heterogeneous regions, we used "supplied-demanded-consumed" approach for the national ES assessment in Russia.

Supplied ES were defined as ES provided by ecosystems regardless of the presence or absence of people, for example, annual allowable cut (Fig. 10.3a), abundance of game animals, biomass and productivity of mushrooms and berries, productivity of natural pastures, carbon content in ecosystems (Fig. 10.4a), amount of pollutants that could potentially be neutralized by ecosystems, volume of water that could potentially be purifies by ecosystems (Figs. 10.6a and 10.7a), runoff provided due to regulation by ecosystems.

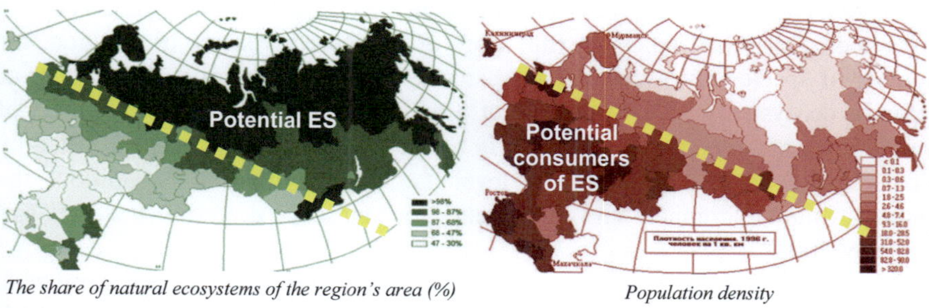

Fig. 10.1 Comparison of the distribution of potential ES provided by ecosystems and potential consumers of ES in different regions of Russia

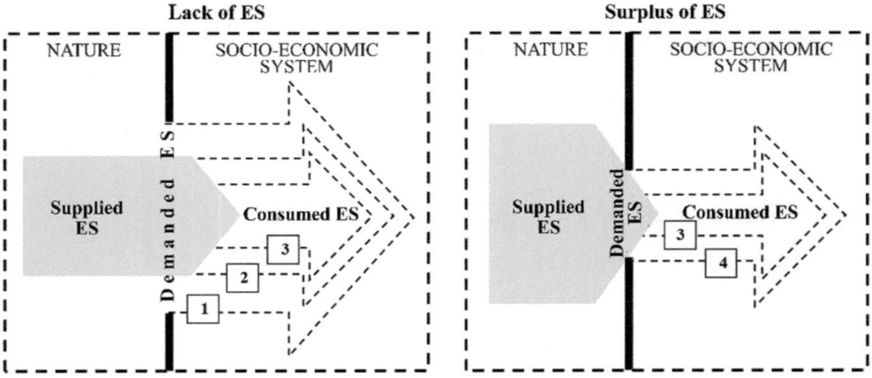

Fig. 10.2 Possible ratios of volume of supplied, demanded and consumed ES: 1—consumed ES is equal to demanded ES and exceeds supplied ES (possible for provisioning and recreational ES); 2—consumed ES is less than demanded ES because of lack of provided volume (all ES); 3—consumed ES is less than demanded ES because of lack of technological, legal or economic means of ES use (all ES); 4—consumed ES is less than supplied ES because of low demand for a service when demanded ES is less than supplied ES (all ES)

Indicators of supplied ES should be assessed taking into account the sustainable use of ecosystems and their components, i.e., it is equal to the volume of provisioning and recreational ES that can be used by people without disturbance of ecosystem structure and functioning (regulating and informational ES cannot be overused as discussed below). However, data from open databases allowed us to correctly estimate the supplied volume for only one ES—wood production (annual allowable cut indicator, Fig. 10.3a), while other supplied provisioning ES were estimated by proxy indicators such as the total population number of game species and productivity of natural pastures.

When evaluating of ES score, the indicator of supplied ES reflects the capacity of natural factors that form ES. For example, supplied volume of the ES of soil self-purification was assessed on the basis of a map of soil capacity for self-purification from the National Atlas of Russia (2004–2008); supplied volume of the ES of aesthetic and educational importance of ecosystems was estimated by a combination of three indicators (the share of natural ecosystems in a region, the number of species of vascular plants per unit of area of a region, the number of types of ecosystems per unit of area of a region); supplied volume of the ES of forming natural conditions for tourism in nature (Fig. 10.8a) was also estimated by a combination of three indicators (the level of comfort of the natural conditions, the quality of the environment and an indicator for landscape diversity (Basanets and Drozdov 2006).

Some authors include anthropogenic inputs (e.g., energy, machinery, fertilizers, pesticides, labor, etc.) in the volume of ES supply (Burkhard et al. 2014; Martínez-Harms and Balvanera 2012). When physical ES evaluation in the Prototype Report, we did not consider anthropogenic inputs as part of ES. As noted

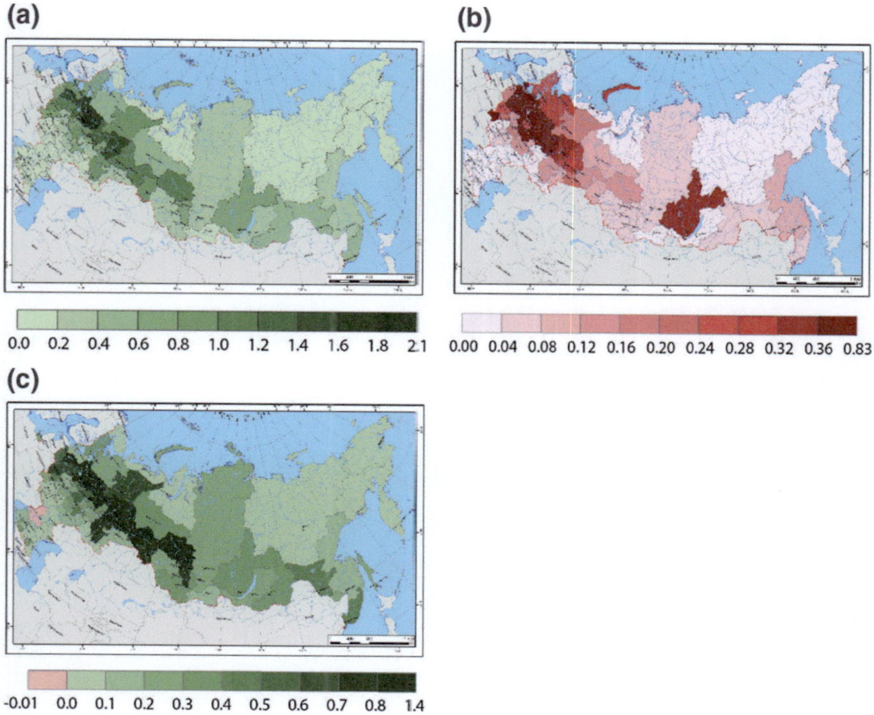

Fig. 10.3 ES of wood production: **a** supplied ES-annual allowable cut (m^3/ha/yr), **b** consumed ES-timber felling (m^3/ha/yr), **c** the degree of ES use-unused residual of allowable cut (m^3/ha/yr) (negative values indicate that timber felling exceeded the allowable cut)

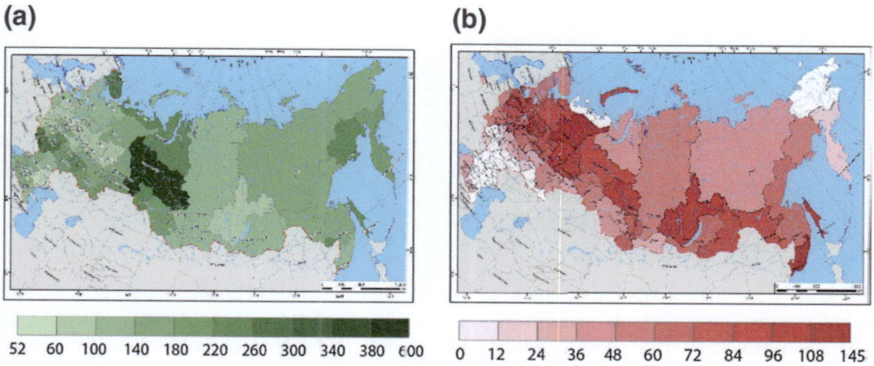

Fig. 10.4 ES of the carbon storage: **a** supplied volume-total carbon content in phytomass and soil (tC/ha), **b** consumed volume-carbon content in managed forests (tC/ha)

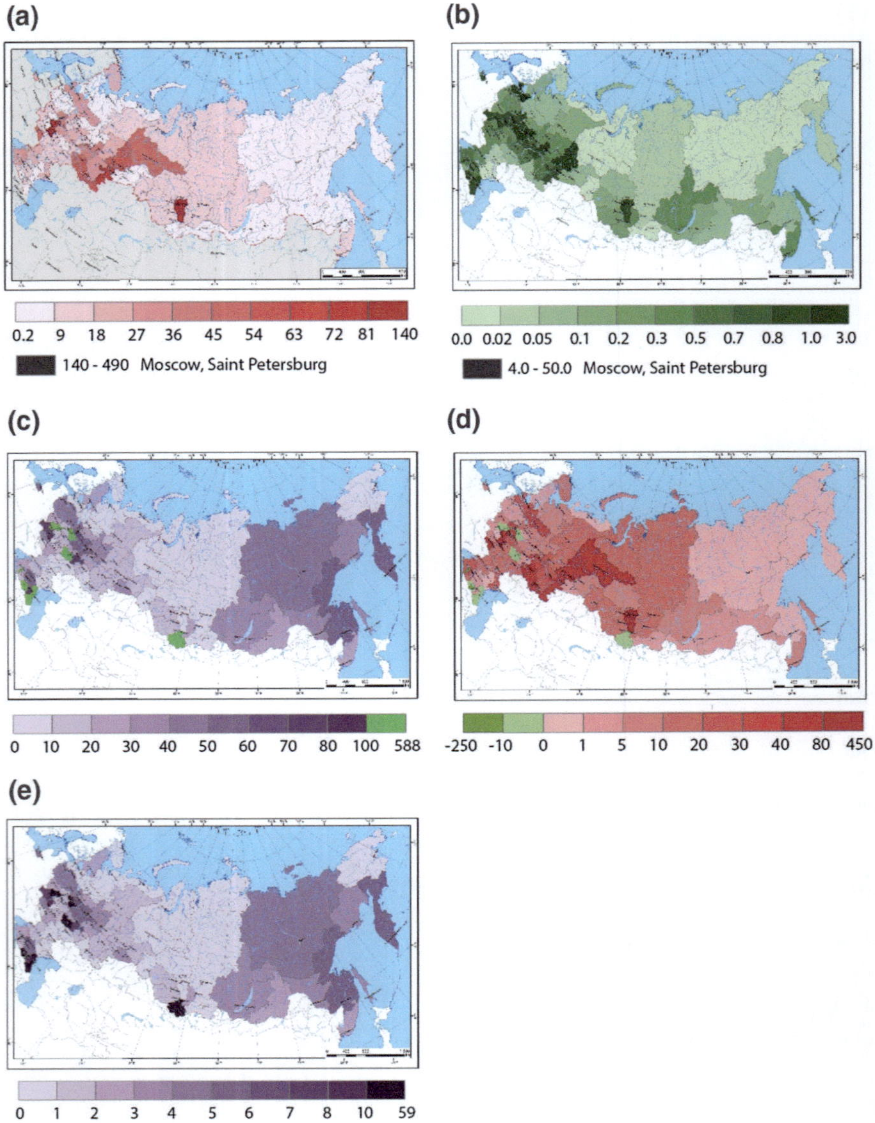

Fig. 10.5 ES of air purification by suburban forests: **a** the demanded ES: the amount of emissions of toxic gases (t/ha/yr), **b** the consumed ES: the amount of toxic gases actually absorbed by suburban forests (kg/ha/yr), **c** the degree of potential meeting the demand for ES: the share of toxic gases that can potentially be absorbed (violet) or the excess of toxic gases, which can be absorbed by suburban forests over real emissions (green) (%), **d** the excess or deficit of the ES: the residual of toxic gases which cannot be absorbed by suburban forests (red), or the excess of toxic gases over real emissions, which can be absorbed by forests (green) (kg/ha/yr), **e** the degree of actual meeting the demand for ES: the share of toxic gases absorbed by suburban forests (%)

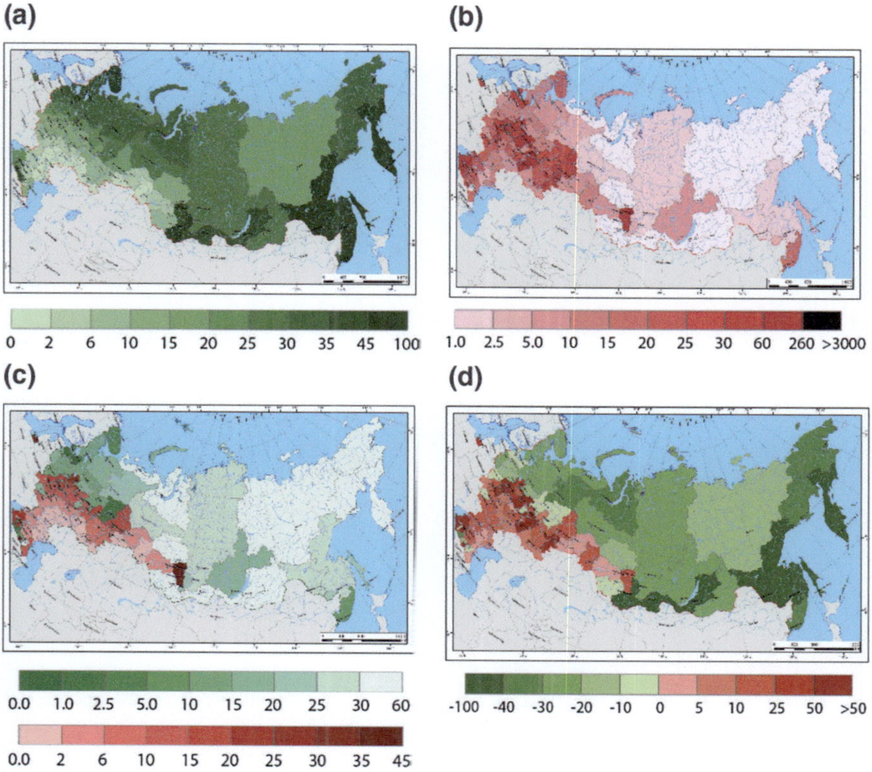

Fig. 10.6 ES of assurance of water quality by freshwater ecosystems: **a** supplied ES-the amount of wastewater that can potentially be purified to a safe concentration of pollutants due to dilution and transformation of pollutants (m^3/ha/yr), **b** demanded ES volume-discharge of polluted wastewater (m^3/ha/yr), **c** consumed ES volume-volume of actually purified wastewater (m^3/ha/yr), green spectrum—regions where the volume of wastewater discharge is less than the capacities of ecosystems to purify it, the consumed ES volume equals the demanded ES volume; red spectrum—regions where the volume of wastewater discharge exceeds the capacities of ecosystems to purify it, the consumed ES volume equals the supplied ES volume, **d** deficit or excess of the ES—untreated wastewater remainder or unused ES volume (m^3/ha/yr)

above, the inclusion of anthropogenic inputs in ES evaluation is a questionable approach. This issue can be resolved in the future comprehensive economic evaluation of ES of Russia.

Demanded ES were defined as ES that are necessary to fulfill needs of the population and economy of a region, for example, logging volume, fish take, hunting production, amount of natural fodder, etc., that are necessary for regional business and household welfare, volume of runoff needed for the population and the economy, amount of pollutant emissions that be neutralized by ecosystems, the number of tourists in nature that is necessary for regional business.

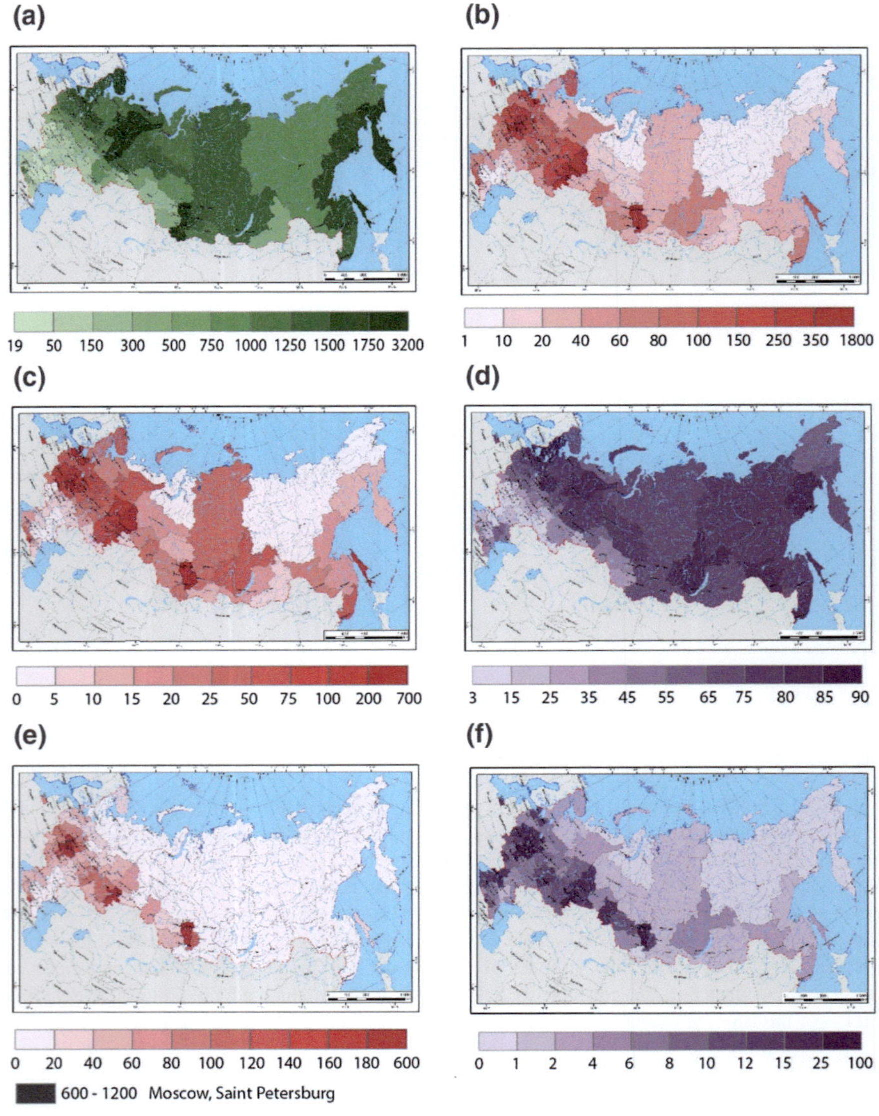

Fig. 10.7 ES of assurance of water quality by terrestrial ecosystems: **a** the supplied ES: potentially purified runoff (m^3/ha/yr), **b** the demanded ES: polluted runoff (m^3/ha/yr), **c** the consumed ES: purified runoff (m^3/ha/yr), **d** the degree of actual meeting the demand for ES: the share of polluted runoff purified by ecosystems (%), **e** the volume of unmet need for the ES: the residual of polluted runoff unpurified by ecosystems (m^3/ha/yr), **f** the degree of ES use: the share of actually purified runoff in the potentially purified runoff (%)

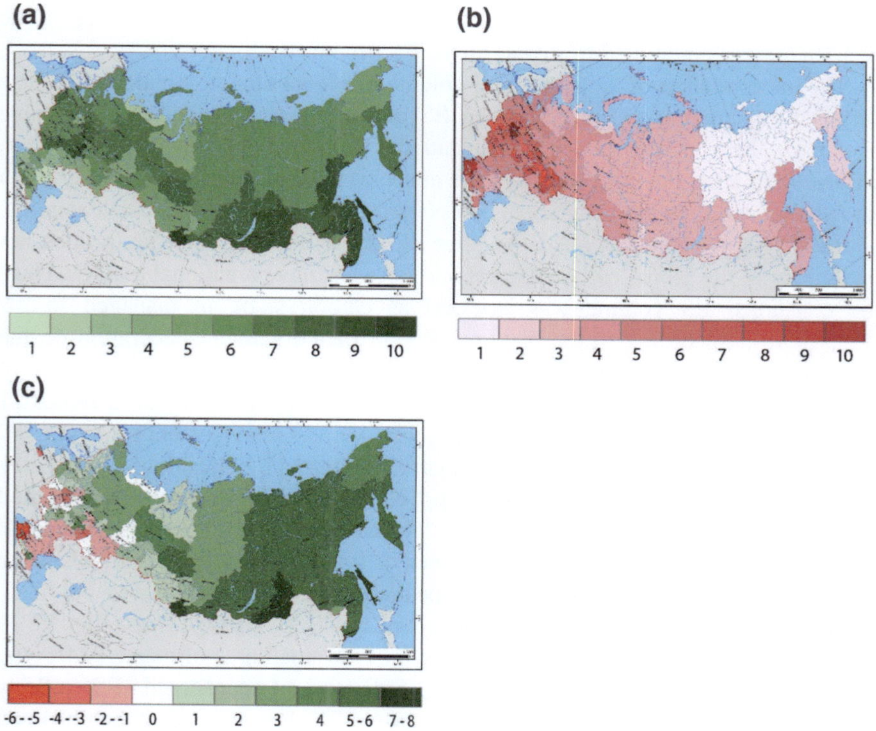

Fig. 10.8 Estimation of the score of the ES of forming natural conditions for tourism in nature: **a** supplied ES, **b** consumed ES, **c** comparison of the natural and socio-economic factors determining the supplied and consumed ES

In the Prototype Report demanded ES were assessed only for ES related to the purification of the natural environment from pollutants. In these cases, the amount of pollutants that must be neutralized by ecosystems may be used as the indicator for demanded ES. In the Prototype Report, annual pollutant emissions (Fig. 10.5a) and annual discharge of polluted wastewater (Fig. 10.6b) and volume of polluted runoff (Fig. 10.7b) were used as a proxy of this indicator. However, maximum permissible concentrations were taken into account in the indicator of supplied ES of assurance of water quality by freshwater ecosystems (Fig. 10.6a). In the future, demanded ES will have to be calculated more accurately as the difference between the amount of pollutant emissions and their maximum permissible concentrations.

For most other ES, demanded volumes are determined primarily by socio-economic features of a region and were not assessed in the first phase of the project since the economic ES assessment was not the task of it.

Consumed ES were defined as ES that are actually used by people, for example, actual logging volume (Fig. 10.3b), hunting production (e.g., the number of elk shot by hunters), mushroom and berry harvest, amount of fodder eaten by cattle in natural pastures, amount of pollutants actually neutralized by ecosystems

(Fig. 10.5b), volumes of runoff and wastewater purified by ecosystems (Figs. 10.6c and 10.7c) and amount of freshwater used by people.

The interpretation of the consumed volume of global climate-regulating ES is a certain problem. For example, for the ES of carbon storage, consumed volume should be the benefit that the people and the economy of a region derive from climate regulation due to carbon storage in ecosystems of that region. Today, both knowledge and data are insufficient to assess these indicators. A simple economic approach could be to calculate the price of stored carbon, but Russia does not participate in the global carbon market and does not have a national market. Thus, the consumed ES volume was assessed as carbon stock in managed forests (Fig. 10.4b). According to the National Report of the Russian Federation on the Cadastre of Anthropogenic Emissions by Sources and Removals by Sinks of Greenhouse Gases not Controlled by the Montreal Protocol for 1990–2011 (National Report 2013),[1] since Russia officially declares management for UNFCCC purposes in this forest category. This evaluation revealed a tremendous discrepancy between the spatial distribution of the supplied ES and the consumed ES attributed to managed forests. The supplied ES, i.e., total carbon stocks in phytomass and soil (according to the database "Land Resources of Russia", Stolbovoi and McCallum 2002) is the highest in regions with vast peat ecosystems (West Siberia) and black earth regions (southern part of European Russia) (Fig. 10.4a). Thus, this ES is poorly recorded just in these regions because of the small area of managed forests.

When evaluating of ES score, the indicator of consumed ES reflects socio-economic factors that determine the use of ES. For example, the consumed volume of the ES of soil self-purification was assessed by a combination of three indicators (population density, the share of croplands in the regions and the share of the polluted area in the regions); consumed volume of the ES of aesthetic and educational importance of ecosystems was estimated by a combination of indicators of population density and transport accessibility; consumed volume of the ES of forming natural conditions for tourism in nature (Fig. 10.8b) was estimated by a combination of three indicators (the investment appeal of the regions, population health and potential tourist demand (Basanets and Drozdov 2006).

The definitions of supplied, demanded and consumed ES adopted in the Prototype Report lie in the mainstream of a common ES understanding, but in some details differ from approaches of other authors (Bukvareva et al. 2017). Short reviews of indicators of supplied, demanded and consumed ES adopted in the Prototype Report is presented in Bukvareva et al. (2017, 2019).

The ratios of supplied, demanded and consumed ES differ between regions, as they depend on properties of ecosystems and socio-economic conditions, which are highly heterogeneous across Russia. A physical analogy to supplied ES would be the total amount of potential ES in a room of nature (the gray arrow in

[1]The cadastre is the report of Russia on greenhouse gas emission/absorption and the GHG balance in managed forests. Carbon stocks in forests are intermediate estimates of the cadaster (National report 2013).

Fig. 10.2). The demanded ES would be compared with a window between the natural and socio-economic rooms, which allows a certain amount of ES to pass from nature to the socio-economic system. The consumed ES corresponds to forces that draw ES through the window from the natural to the socio-economic room (Bukvareva et al. 2017). Supplied and consumed ES form the real flow of ES, while the demanded ES volume can only be considered a condition or restriction affecting this flow.

In densely populated regions, demanded and consumed ES could exceed supplied ES (Fig. 10.2, left). Such regions are located in the central and southern parts of European Russia and in patches in the south of Siberia. By contrast, in remote regions where the population density is lower, demanded and consumed ES could be smaller than supplied ES (Fig. 10.2, right). Such regions are found in the major part of Siberia and in the North of Russia.

Demanded volume may hypothetically exceed *supplied* volume for all ES categories. The Prototype Report contains examples of such situations for ES related to the purification of the natural environment from pollutants. Supplied ES of air purification by suburban forests is less than demanded ES in most regions (Fig. 10.5c, d), although for the preliminary assessment, we considered that supplied ES is determined by the maximum physiological absorptive capacity of trees, the excess of which leads to their death (this amount is ten times the amount of toxic gases that are actually absorbed by trees in cities, as measured). Supplied ES of assurance of water quality by freshwater ecosystems is less than demanded ES in densely populated regions with developed industry and agriculture (red spectrum in Fig. 10.6d). Such cases should be recognized as evidence of a high environmental hazard, when even the maximum capabilities of ecosystems are no longer able to provide acceptable environmental quality.

Interrelations between *consumed* and *supplied* ES volumes are specific to ES categories. Consumed volume can exceed supplied volume of provisioning and recreational ES (number 1 in Fig. 10.2). That leads to overexploitation (overfishing, overhunting, etc.) or disturbance by excessive recreational load. ES estimations based on state statistics basically do not reveal overexploitation (e.g., is in Fig. 10.3c); however, data on illegal, unreported and unregulated (IUU) harvesting (Sect. 5.3) may change this result.

Consumed ES volume cannot exceed the supplied volume of regulating and cultural ES because the overuse is impossible for these. People can exist only in the given environment and draw benefit or harm from it. If the demanded ES volume exceeds the supplied ES volume, ecosystems cannot maintain acceptable parameters of the environment and people have to live in an unfavorable environment. For example, if amount of wastewater exceeds purifying ability of freshwater ecosystems, then consumed ES is equal to supplied ES, since ecosystems purify only the amount of water that they can (Fig. 10.6c, red spectrum). In that case, the quality of the environment deteriorates. Similarly, it is impossible to overexploit informational ES, since it is impossible to use more information than there is in nature. This information can be lost due to the degradation of ecosystems or the extinction of species, but it is obviously impossible to overuse it.

Consumed ES volume could be less than supplied volume of all ES categories. The primary reason for this would be low demand for a service, when the demanded volume is less than the supplied ES volume (number 4 in Fig. 10.2). A lack of technological, legal or economic means for ES use was another widespread cause of insufficient consumption of supplied ES volume (numbers 3 in Fig. 10.2). For example, a lack of logging equipment or roads prevents cutting down the amount of timber required for normal operation of wood-processing enterprises in a region. Potentially useful natural genetic resources might not be used due to lack of theoretical knowledge and technologies.

Relationships between *consumed* and *demanded* ES volumes are also specific to ES categories. Consumed volume of provisioning and recreational ES can exceed required volume in case of an extremely inefficient planning and management system. For example, if biological resources are harvested in excessive amounts that cannot be processed or transported to another region or if the number of people who relax in nature is excessively high, this reduces the quality of their rest and the economic profit from recreation (Bukvareva and Zamolodchikov 2018). Consumed ES volume could be less than demanded volume due to lack of supplied ES (number 2 in Fig. 10.2), for example, when the emission of pollutants exceeds the capacity of forests to absorb them. It is a case of the ES of air purification by suburban forests when pollution is not completely absorbed in any region (Fig. 10.5e). For the ES of assurance of water quality by terrestrial ecosystems, consumed ES is always less than demanded ES (Fig. 10.7b, c). The demand for this ES cannot be entirely satisfied because terrestrial ecosystems are not capable of completely purifying polluted runoff, especially during the snow melt period. All regions therefore have residual unpurified runoff (Fig. 10.7e), i.e., the demand for the service is not satisfied anywhere.

Supplied, demanded and consumed ES volumes can be compared if they are measured in the same units. Maes et al. (2011) suggest expressing ES capacity in total ecosystem area or biomass, whereas ES flow (the partial analog of consumed ES) in units per time period. This approach does not allow them to be compared. The cited authors propose to solve this problem using bundles of ES that include information of both ES capacity and flow. However, direct quantitative comparison may be more efficient.

Ratios and differences of supplied, demanded and consumed ES allowed us to estimate the degree of ES use and degree of meeting the demand for ES, which may be useful for decision-makers. Examples of management interpretation of indicators are shown in Table 10.2.

10.4 Comparison of the Regions

One of the main advantages of assessment of supplied, demanded and consumed ES is the ability to compare regions that are donors and consumers of ES, which is crucially important at national-level ES assessment. Donor–consumer relationships

Table 10.2 Indicators of the degree of ES use and degree of meeting the demand for ES and corresponding messages for decision-makers (V_s—supplied ES, V_c—consumed ES, V_d—demanded ES)

Indicators	Ratios and differences of ES volumes	Examples	Messages for decision-makers
The degree of ES use	V_c/V_s $V_c/V_s \times 100\%$	The ratio of actually purified runoff to the potential purification capacity of terrestrial ecosystems (Fig. 10.7f)	In all regions (except cities of Moscow and St. Petersburg), the potential purifying capacity of ecosystems, on average, significantly exceeds actual purification of polluted runoff
Unused (if positive) or overdrawn (if negative) ES volume	$V_s - V_c$	The unused residual of the annual allowable cut (Fig. 10.3c)	According to official data, wood resources are underused in most regions, especially in forest regions of European part of Russia and West Siberia
The degree of potential meeting the demand for ES	V_s/V_d $V_s/V_d \times 100\%$	The share of toxic gases that can potentially be absorbed by suburban forests (Fig. 10.5c)	The maximum gas-absorbing capacity of suburban forests is insufficient to neutralize toxic emissions in the majority of regions. Many regions have a maximum capacity below 50% of annual emissions
Excess (if positive) or deficit (if negative) of ES	$V_d - V_s$	The residual volume of polluted runoff which cannot be neutralized by water ecosystems or untapped opportunities of ecosystems for wastewater treatment (Fig. 10.6d)	Industrially and agriculturally developed regions in the central and southern European part of Russia and southern West Siberia experience a deficit of this ES that indicates overexploitation of freshwater ecosystems
		The residual of toxic gases that cannot be absorbed by suburban forests, or the excess of forest absorption capacity over real emissions (Fig. 10.5d)	The maximum gas-absorbing capacity of suburban forests is insufficient to neutralize toxic emissions in the majority of regions. Many regions have a significant amount of unabsorbed pollutants under any conditions

(continued)

Table 10.2 (continued)

Indicators	Ratios and differences of ES volumes	Examples	Messages for decision-makers
The degree of actual meeting the demand for ES	V_c/V_d $V_c/$ $V_d \times 100\%$	The share of polluted runoff that is actually purified by ecosystems (Fig. 10.7 d)	Terrestrial ecosystems in industrially developed regions in the central European part of Russia and southern Ural and Siberia cannot cope with severe pollution and purify no more than a third of polluted of runoff
		The share of toxic gases absorbed by suburban forests (Fig. 10.5e)	Toxic gases are not completely absorbed in any region. In the majority of regions, less than 10% of emissions are absorbed
Volume of unmet need for ES	$V_d - V_c$	The residual of polluted runoff unpurified by terrestrial ecosystems (Fig. 10.7e)	Terrestrial ecosystems in industrially developed regions in the central European part of Russia and southern Ural and Siberia cannot cope with severe pollution of runoff. A significant part of the pollution falls into the waterbodies

between regions are important for ES of regional and interregional scale and are not essential for local and in situ ES, for example, pollination or soil formation. In Russia, the overwhelming majority of ES is important at the regional and interregional scales (Bukvareva and Zamolodchikov 2018). Examples of interregional donor–recipient relationships and a preliminary set of indicators for detecting ES donor regions in Russia are presented in BfN scripts (Bukvareva 2014a, b).

The indicators of the degree of ES use (Fig. 10.8f), the degree of actual and potential meeting the demand for ES (Figs. 10.6d, f and 10.8d) and the volume of unmet need for ES (Fig. 10.8e) highlight environmentally hazardous and relatively safe regions and are important primarily for regional government. For example, the indicator of the use of ES of assurance of water quality by terrestrial ecosystems, i.e., the share of actually purified runoff in the potentially purified runoff (Fig. 10.7f), highlights regions where the ES is used almost entirely. The indicators for air purification showed that in many regions even the potential absorption capacity of suburban forests is less than half of the toxic emissions (Fig. 10.5c), and actual absorption everywhere is less than the emissions (Fig. 10.5e). This means that in most regions a large amount of toxic gases remains unabsorbed under any weather conditions (red spectrum in Fig. 10.5d).

The indicators of unused/overdrawn ES volume (Figs. 10.3c and 10.9c) and excess/deficit of ES (Fig. 10.7d) show regions, which tend to be consumers or donors of ES, which are of interregional scale and their supplied volumes can move between regions. For example, uneven use of annual allowable cut in the regions (Fig. 10.3c) can be compensated by transporting excess wood to timber processing enterprises in other regions. The excess of the ES of assurance of water quality by freshwater ecosystems (Fig. 10.6d) and the ES of runoff regulation can be used in other regions if they are located in the proper direction of water flow. The excess of the ES of air purification by suburban forests (green spectrum in Fig. 10.5d) cannot be moved to other regions, so this ES works at a local scale, thus, there are not donor–consumer relationships between regions for this ES. However, ES assessment methodology applied in the Prototype Report does not take into account the relative spatial location of the regions (see Sect. 5.2) that should be corrected in the future evaluations.

As mentioned earlier, only one-third of the ES considered was quantitatively estimated. The inclusion of ES scores in the comparison of regions allowed as to extend the comparison of the regions to two-thirds of considered ES. For this, quantitative indicators were also converted into scores. Comparative matrices show the distribution of supplied and consumed ES scores across regions and differences of these scores (Vsupplied–Vconsumed). Such a matrix is presented in Fig. 10.9. It reflects the qualitative balance of natural factors determining ES supply by ecosystems and socio-economic factors determining ES consumption by the people and the economy in the regions. These matrices are analogous to potential/flow and flow/demand matrices of land cover types (Burkhard et al. 2012, 2014) but estimate regions, not land cover types. Each column of the matrix in Fig. 10.9 represents the balance of factors determining ES supply and consumption by region, that is, corresponds to the maps, examples of which are shown in Fig. 10.8c.

The values close to zero and light colors indicate that in corresponding regions, natural and socio-economic factors have close scores of intensity in comparison with other regions. Green color and positive scores indicate relative predominance of ES supply factors and red color and negative scores indicate relative predominance of ES consumption factors. For example, this matrix shows that supply and consumption factors for provisioning ES are more balanced than for other ES categories (except for a few regions where their consumption prevails). For most of the regulating ES, we see a strong prevalence of supply factors in the North-European and Asian parts of Russia (federal districts 1, 7 and 8), while consumption factors prevail in the remaining territory (except for "carbon" ES due to ambiguities in determining their consumed volume discussed in Sect. 10.3). More detailed analysis can reveal the relative intensity of ES within their categories and for regions within federal districts (Bukvareva and Zamolodchikov 2018).

Region	Productive (Provisioning): Wood production	Non-wood production	Production of natural fodder	Game production	Environment-forming (Regulating): Carbon storage	Regulation of CO_2 flows	Air purification by suburban forests	Regulation of water runoff volume	Regulation of runoff variability	Water purification by terr. ecosyst.	Water purification in aq. ecosyst.	Agric. soil protect. from water eros.	Agric. soil protect. from wind eros.	Self-purification of soils	Pollination	Informational (Cultural): Natural genetic resources	Information on structure and funct.	Aesthetic and educational value	Recreational: Daily and weekend recreation	Tourism in nature
1 North West Federal District																				
Arkhangelsk Oblast	-4	1	0	0	-5	2	1	5	2	5	4	1	1	4	0	2	0	3	2	3
Vologda Oblast	-1	1	0	0	-6	1	-3	5	0	5	3	0	1	4	0	7	1	5	5	2
Leningrad Oblast	-4	0	0	-1	-5	-6	6	-3	-3	-1	0	1	0	2	0	0	-3	2	3	1
Kaliningrad Oblast	-1	-2	0	0	-1	-3	5	3	-4	-3	-2	-2	-3	-2	-1	0	2	1	1	-3
Murmansk Oblast	-1	2	0	0	2	1	-1	4	1	6	3	1	1	7	0	-3	-1	2	-1	3
Nenets Autonomous Okrug	0	0	1	0	5	2	0	7	1	7	6	0	0	2	0	0	2	3	1	0
Novgorod Oblast	-2	-4	0	-2	-7	-3	2	6	0	4	3	0	0	5	1	2	0	2	4	2
Pskov Oblast	-1	-3	0	0	-2	-4	2	4	2	3	4	1	1	3	3	6	4	2	4	3
Republic of Karelia	-3	0	0	1	-3	-1	1	6	1	7	4	1	1	7	0	3	0	2	0	1
Komi Republic	0	6	0	0	-6	3	0	8	1	7	5	1	1	3	0	1	-1	3	2	3
2 Central Federal District																				
Belgorod Oblast	-1	1	0	-3	7	2	-4	-1	-8	-2	-5	1	2	-5	1	-6	-3	-8	-1	0
Bryansk Oblast	-5	-3	0	-1	-3	-1	5	1	6	-1	-2	2	2	0	6	3	3	-2	2	-1
Vladimir Oblast	-6	-3	0	-3	-5	-3	7	2	0	-6	-5	0	1	-4	4	-2	-1	-2	-1	0
Voronezh Oblast	-2	-1	0	-3	6	1	1	-1	2	-2	-5	2	2	-7	0	-7	-6	-7	2	2
Ivanovo Oblast	-2	-4	0	-1	-5	-3	5	2	-1	-4	-4	2	2	2	5	1	3	1	1	0
Kaluga Oblast	-2	1	0	1	-4	-1	7	2	1	-3	-4	1	1	2	4	-3	-3	0	2	-1
Kostroma Oblast	0	2	0	1	-8	0	1	-2	1	4	2	0	0	4	1	7	6	5	6	4
Kursk Oblast	-1	1	0	-1	6	5	4	-1	-4	-2	-1	0	2	-4	3	-6	-1	-7	1	-1
Lipetsk Oblast	-2	0	0	1	4	6	-5	0	-1	-5	-5	2	2	-3	1	1	4	-8	1	0
Moscow Oblast	-2	-3	0	-3	-3	-9	0	-6	-5	-6	-5	1	1	-3	4	-3	-4	-3	-2	-4
Oryol Oblast	-1	0	0	-2	3	6	3	1	4	-2	-3	0	1	-4	5	1	7	-5	2	0
Ryazan Oblast	-3	-3	0	-2	3	3	-1	1	3	-7	-3	2	2	-5	6	-2	-3	-4	1	0
Smolensk Oblast	0	-1	0	-1	-3	-4	1	3	2	-4	-1	1	1	2	4	2	0	0	3	0
Tambov Oblast	-2	-1	0	1	5	3	0	2	3	-1	-3	3	2	-4	0	-4	-3	-6	3	3
Tver Oblast	-3	-1	0	-4	-6	-4	2	1	0	-2	2	0	0	2	2	1	-4	1	4	1
Tula Oblast	-1	2	0	-3	2	2	-2	-1	1	-9	-6	0	2	-7	5	-3	-3	-5	-1	-1
Yaroslavl Oblast	-2	2	0	-1	-5	-2	3	3	-3	-3	-2	0	1	2	3	-1	-1	1	3	-2
3 Volga Federal District																				
Kirov Oblast	-3	2	0	0	-6	1	2	5	2	2	1	1	1	3	3	3	-1	2	4	3
Nizhny Novgorod Oblast	-4	0	0	-1	-4	0	6	-1	-2	-1	-6	2	2	-2	4	-4	-5	-2	1	2
Orenburg Oblast	-1	0	1	1	4	0	-7	-4	4	-1	-3	0	1	-8	4	-1	-1	-4	3	0
Penza Oblast	-2	1	0	-2	2	1	5	0	-2	-1	-5	2	3	-2	5	-7	-5	-6	3	1
Perm Krai	0	4	0	0	-7	0	0	2	-2	2	-1	0	1	3	1	0	-5	3	4	3
Republic of Bashkortostan	1	2	-1	1	-1	1	0	3	-3	-3	-4	-2	1	-4	4	-2	-4	-2	5	-2
Mari El Republic	-6	0	0	1	-4	1	4	2	2	2	-3	1	2	1	4	4	6	3	5	-1
Republic of Mordovia	-2	-1	0	1	0	2	1	2	4	-1	-2	2	3	0	7	-2	1	-3	3	3
Republic of Tatarstan	0	0	0	-1	1	3	0	-1	-3	-3	-6	3	3	-7	5	-6	-6	-5	2	-1
Samara Oblast	-1	1	1	-2	4	-2	-2	-2	-1	-4	-9	1	3	-8	0	-7	-4	-6	1	-3
Saratov Oblast	-1	1	0	0	4	0	0	0	-2	-1	-3	2	0	-6	0	-3	-3	-5	2	-1
Udmurt Republic	-5	1	0	-2	-5	-3	4	2	-2	1	-3	2	2	2	5	2	3	0	3	1
Ulyanovsk Oblast	-1	-3	0	1	2	-2	4	0	6	-2	-6	3	3	-3	3	-7	-2	-2	4	4
Chuvash Republic	-1	-1	0	2	-1	-2	5	1	-1	-2	0	0	2	-4	6	-3	1	-4	1	3
4 South Federal District																				
Astrakhan Oblast	0	0	-5	0	0	2	-2	-5	-1	0	-4	-2	-6	1	2	3	4	3	1	-2
Volgograd Oblast	-1	1	0	-3	2	1	-1	0	-2	0	-4	1	-2	-6	1	-3	-2	-4	-1	-2
Krasnodar Krai	-1	0	0	-3	1	4	1	-6	2	-2	-4	2	2	-8	0	-4	1	-5	-1	-6
Republic of Adygea	0	0	-1	-2	0	0	8	-2	4	-1	0	2	2	0	-3	1	5	-2	-1	2
Republic of Kalmykia	-1	0	-5	0	0	4	0	-1	0	0	-2	-5	-8	0	0	4	6	4	1	-2
Rostov Oblast	-1	1	-2	-3	3	-1	1	-6	-4	-1	-1	2	1	-9	2	-7	-4	-6	1	0
5 Northern Caucasus FD																				
Republic of Ingushetia	0	0	-1	-2	1	-4	2	-5	5	-1	0	-1	2	-3	4	0	5	-2	0	4
Kabardino-Balkar Republic	0	0	-6	-1	2	-2	7	-2	-1	-2	2	-2	1	-2	2	0	3	0	0	1
Karachay-Cherkess Republic	0	0	-7	-5	0	-5	0	6	0	3	1	-4	-1	2	2	3	4	4	0	1
Republic of Dagestan	-1	1	-6	0	2	6	6	-8	-4	0	-2	-3	-4	-1	4	-4	4	-2	-1	-1
Republic of North Ossetia-Alania	0	0	-3	2	1	-1	9	-1	-3	-5	-1	-3	1	0	4	0	3	-1	-1	1
Stavropol Krai	-1	1	-3	0	3	4	4	-9	-4	0	-5	0	-2	-5	1	-2	3	-2	2	-1
Chechen Republic	0	0	-2	1	0	2	6	-6	-3	-1	-2	-4	-1	-2	4	2	6	-2	-1	4
6 Ural Federal District																				
Kurgan Oblast	-4	-1	5	0	2	3	2	0	0	-1	-2	3	4	-6	8	3	3	2	6	2
Sverdlovsk Oblast	-2	1	0	1	-5	-2	0	3	-3	-3	-4	1	1	0	1	-1	-6	1	0	0
Tyumen Oblast	2	0	1	-1	4	0	1	2	-6	1	-2	3	3	-3	2	1	-3	5	6	2
Khanty-Mansi Autonom. Okrug	0	4	0	0	2	-1	-5	5	-1	7	5	1	1	0	0	2	-2	4	3	4
Chelyabinsk Oblast	-1	0	3	6	1	-4	-3	1	-5	-7	-7	1	1	-7	5	-4	-2	0	4	-2
Yamalo-Nenets Autonom. Okrug	-1	3	0	0	2	0	-1	5	1	6	7	0	0	3	0	1	0	3	3	1
7 Siberia Federal District																				
Altai Krai	-1	0	1	1	1	-1	1	3	1	-1	2	1	-2	-4	6	0	2	-1	4	1
Zabaykalsky Krai	0	3	1	2	-5	-3	1	3	1	2	6	0	1	5	1	6	1	2	5	7
Irkutsk Oblast	-4	6	0	1	-7	-1	2	5	0	4	3	1	1	3	0	2	-3	4	4	5
Kemerovo Oblast	1	0	2	0	-4	-7	-3	3	-4	0	-1	-2	1	0	3	4	2	2	2	2
Krasnoyarsk Krai	-1	4	0	0	-2	1	0	7	1	6	5	1	1	6	0	-1	-2	2	2	3
Novosibirsk Oblast	-1	2	4	2	1	3	1	0	3	0	-2	3	0	-6	6	-5	-4	-1	6	1
Omsk Oblast	1	7	3	0	1	1	-1	1	4	-2	-4	3	-1	-7	6	-1	1	1	4	1
Altai Republic	0	1	0	3	-4	0	2	9	0	8	9	-4	0	7	0	9	8	8	4	8
Republic of Buryatia	0	2	0	1	-3	-2	1	6	1	6	8	1	2	6	1	6	0	4	4	6
Tyva Republic	-1	5	0	2	0	4	1	6	1	6	8	-2	-5	7	1	9	4	7	5	6
Republic of Khakassia	0	0	0	2	-4	-1	1	9	0	-1	7	-1	-5	1	2	8	8	5	4	2
Tomsk Oblast	2	-4	0	0	1	-2	0	3	3	4	3	1	1	0	0	-1	-5	3	2	5
8 Far East Federal District																				
Amur Oblast	0	-1	0	1	-4	1	1	6	0	4	8	1	1	3	0	4	0	4	6	5
Jewish Autonomous Oblast	-1	0	0	3	-1	-4	3	6	1	4	7	0	0	2	0	5	6	6	2	3
Kamchatka Krai	-1	0	0	0	3	-2	1	7	1	9	9	0	0	6	0	-1	0	3	5	5
Magadan Oblast	-1	0	0	0	1	1	1	8	0	8	7	0	0	5	0	0	-1	2	5	5
Primorsky Krai	0	-1	0	1	-7	-4	2	6	1	5	4	1	1	3	0	2	-3	5	2	6
Republic of Sakha (Yakutia)	-1	7	0	0	0	0	1	5	1	4	4	1	1	4	0	1	0	3	5	5
Sakhalin Oblast	0	-4	0	0	-3	-6	2	9	-4	5	7	1	1	6	0	2	4	6	3	8
Khabarovsk Krai	-1	2	0	0	-2	-1	1	8	1	8	8	1	1	5	0	3	-1	4	5	5
Chukotka Autonomous Okrug	0	0	0	0	3	1	0	8	0	5	7	0	0	7	0	0	0	2	3	4

Fig. 10.9 Difference between supplied and consumed ES scores by region

10.5 The Main Problems of Assessment and Future Tasks

The first national ES assessment in Russia demonstrated that ES of terrestrial ecosystems is critical for the well-being of the population and economy of Russia. The volume of the most important ES provided by ecosystems is comparable to the amount of basic needs of the population and economy of the Russian regions for regulation of the environment and natural bioproduction. A number of the most important life-supporting ES are fully used, or they are already not sufficient to meet the needs of people and the economy. This is true for ecosystem regulation of runoff, ensuring water quality by terrestrial ecosystems, water purification in aquatic ecosystems, and absorption of air pollutants by suburban forests.

Thus, the results of the "TEEB-Russia 1" project prove that it is necessary to immediately start forming a national system of ES monitoring and assessment, as well as mechanisms of integrating ES values in decision-making. However, for this it is necessary to solve a number of methodological problems, the most important of them are briefly discussed below.

10.5.1 Selection of Assessment Units

A significant problem inherent in the use of administrative units in Russia is the extremely unequal area of them. Single values for vast areas such as Krasnoyarsk Krai, Yakutia and other large regions in Siberia, the North and the Far East of Russia could not adequately describe the diversity of natural and socio-economic conditions inside these regions. The territorial subdivisions of the largest regions are required for future ES evaluations.

Another important problem is the incorporation of data organized according to river basins into the general ES assessment performed according to administrative regions. This is necessary because water-related governmental agencies and scientific institutes operate according to river basins. Moreover, water-related ES requires modeling at the basin scale for their adequate evaluation. Freshwater fishery management and partly water management are carried out at the basin scale. However, a comparison of ES and regions requires a uniform grid of valuation units. Therefore, it is necessary to find the most effective algorithm for translating data on basins into administrative regions and management recommendations, and back from the grid of administrative units to the basins.

10.5.2 Spatial Interrelations Between Regions

Evaluation of supplied, demanded and consumed ES allows to compare regions that are donors and consumers of ES. For a more accurate detection of donor–recipient relationships ES spatial characteristics, directional ES flow and user movement should be considered (Costanza 2008) as well as distance between regions and their position relative to the water flows, currents, prevailing winds and animal migrations. Methods of ES flows spatial analysis are developed primarily at the regional level (e.g., Bagstad et al. 2013; Nedkov and Burkhard 2012; Stürck et al. 2014; Syrbe and Walz 2012), but the principles can be further used for interregional comparisons at the national level. The concept of "providing/benefiting areas" (Syrbe and Walz 2012) should be extremely useful in the future ES assessments in Russia, given the highly uneven distribution of ecosystems and population in the territory.

10.5.3 The Lack of Data

National ES assessment and monitoring can be based on a nation-wide regularly updated system of data collection. Today, such a system is mostly absent. Annually updated governmental databases allowed us to completely quantitatively evaluate only one ES (wood production) as well as separate indicators for three regulating ES (air purification by suburban forests, regulation of runoff volume and assurance of water quality by freshwater ecosystems). In total, more than 30% of the indicators (9 out of 25 quantitative indicators and 6 out of 21 indicators, which were used for evaluating ES scores) were obtained from governmental databases and statistical compilations. 70% of the indicators were either obtained directly or calculated using data from statistical compilations, analytical reviews, and cartographic and remote sensing materials that are produced by various institutions and are not updated at a fixed frequency.

A significant problem for assessment of consumed volumes of provisioning ES is the large amount of illegal, unreported and unregulated (IUU) bioresource harvesting in Russia. For example, with regard to ES of wood production, illegal forest harvesting in some regions makes up tens of percent above official data (Gryaznov et al. 2011; Kotlobay et al. 2006; Ptichnikov and Kuritsyn 2011). IUU harvesting remains an important obstacle for evaluation ES of game and freshwater fish production in Russia (Bukvareva and Zamolodchikov 2018).

A potential source of data that can be regularly updated in the future is the map of terrestrial ecosystems obtained from satellite imagery. In the Prototype Report, four quantitative indicators were calculated using this map. Two more indicators based on this map were used for estimating ES scores. Regular update of this map and development of a set of algorithms for indirect ES quantification on the basis of vegetation cover can make this map the basis for assessing the significant part of ES that cannot be directly evaluated on the basis of statistical data.

With regard to biodiversity assessment and monitoring, the current national system of data collection is able to make it only fragmentarily. Most of the data needed for biodiversity monitoring are not available at the national level, as they are present only in individual studies for individual regions, or none at all. Only changes in exploited species are formally tracked by federal bioresource-related agencies (Bukvareva et al. 2019).

10.5.4 The Need for a Comprehensive Assessment of All ES Categories

The priority management of only one part of ES can lead to wrong and harmful decisions due to ES trade-offs (e.g., Allan et al. 2015; King et al. 2015; Maes et al. 2012; Raudsepp-Hearne et al. 2010; Turkelboom et al. 2018). In the most striking form, ES trade-off manifests itself showing negative correlation between indexes of ES which are provided by natural ecosystems (e.g., regulating ES) and agriculture production which CICES also considers as ES (e.g., Casalegno et al. 2013; Felipe-Lucia and Comín 2015; Qiu and Turner 2013).

As discussed above (Sect. 2.2), we do not consider agriculture production as ES. Indicators of agricultural production in the regions are determined by the share of agriculture land and negatively depend on the share of natural ecosystems (Fig. 10.10a). For the vast majority of ES, which are provided by natural ecosystems, the opposite is true (e.g., see in Fig. 10.10b). Generally, agricultural production and most regulating ES are in direct opposition to each other.

With regard to possible trade-offs between ES considered in the Prototype Report, the conflict between the use of productive and recreational ES, on the one hand, and the maintenance of environment-forming and informational ES, on the other hand, is extremely relevant in Russia. Exploitation of bioresources, primarily forest, is a significant sector of the economy. In the overwhelming majority of cases, the highest priority is to maximize the product that can be extracted

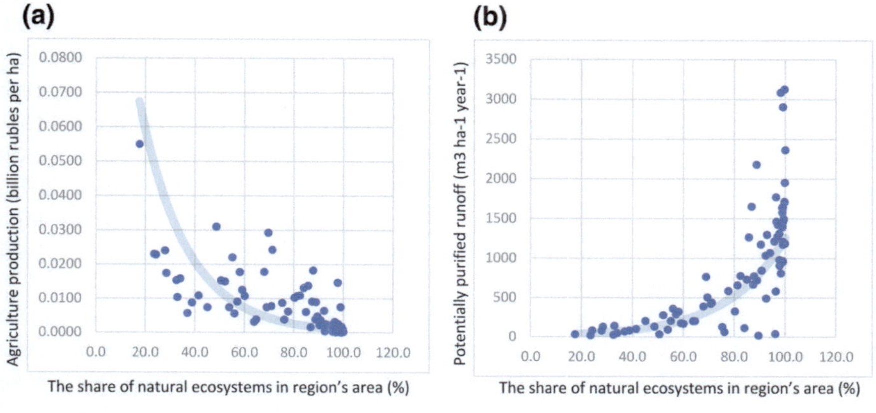

Fig. 10.10 Dependence of **a** agriculture production and **b** supplied ES of assurance of water quality by terrestrial ecosystems on the share of natural ecosystems in region's area

sustainably from ecosystems (timber, seafood, game production). Harvesting of bioresources inevitably disturbs natural ecosystems, their functions and biodiversity, that is, environment-forming and informational ES. This is an old and well-known conflict. A relatively new conflict in ES management goals arose between recreational and informational ES in strictly protected areas called in Russian “zapovedniks” (Concept of the development of the system of nature-protected areas of federal importance until 2020). Traditionally, the priority task of zapovedniks was the preservation and study of natural complexes undisturbed by man, that is, the maintenance and use of information ES for the preservation of information of natural ecosystems. However, the presence of tourists, even for educational purposes, inevitably disrupts the functioning of natural populations and ecosystems, that is, it conflicts with this information ES.

Based on current state statistics, productive ES can be assessed better than all other categories of ES. However, prioritizing only productive ES will lead to an inadequate understanding of the whole value of ES and biodiversity and to the wrong management decisions.

10.5.5 Landscape Approach to Optimization of Tasks of Biodiversity Conservation and ES Maintenance

Biodiversity is the structural basis of ecosystem functioning (EF) and is a key factor in determining the mean level and stability of ecosystem functioning (Cardinale et al. 2012; Hooper et al. 2012; Tilman et al. 2012, 2014). Positive effects of species diversity and intraspecific diversity on EF (productivity, biomass, rate of nutrient cycling, invasion resistance, stability, etc.) were confirmed by hundreds of experiments (Bardgett and van der Putten 2014; Gross et al. 2014; Handa et al. 2014; Forsman 2014; Forsman and Wennersten 2016; Hughes et al. 2008) as well as surveys of real-world systems (Lewandowska et al. 2016). The evidence obtained for grasslands (Grace et al. 2016; Maestre et al. 2012) and forests (Baruffo et al. 2013; Cavanaugh et al. 2014; Nadrowski et al. 2010; Paquette and Messier 2011; Vilà et al. 2013,; Wang et al. 2011) may be the most interesting for landscape research in Russia. Relationships between biodiversity and ES are still not so clear. However, despite negative or mixed relationships between some ES and biodiversity the majority of ES positively depend on biodiversity (Cardinale et al. 2012; Harrison et al. 2014) and sustaining the long-term flow of many ES require high levels of biodiversity (Science for Environment Policy 2015). Changes in biodiversity inevitably lead to ES change.

Thus, the tasks of biodiversity conservation and ES use and maintenance cannot be separated from each other. This issue is one of the main topics in the next phase of the project (TEEB-Russia 2). However, the work done to date allows us to make some preliminary judgments on this issue today and outline further studies.

The principle of optimum biodiversity (Bukvareva 2014c, 2018; Bukvareva and Aleshchenko 2013) may be one of possible theoretical approaches to this issue. According to this principle, species diversity and intrapopulation diversity are inseparable adaptive characteristics of interacting hierarchical biodiversity levels—communities and populations—to environmental conditions. The optimum values of diversity provide maximum efficiency and survivability of communities and populations and thus, the maximum EF. The optimal diversity values depend on the degree of environmental stability and the amount of available resource. The optimal values of intrapopulation diversity decrease in more stable environments. The optimal values of species richness increase in more stable and rich environments. Thus, natural undisturbed communities that are adapted to rich and stable conditions tend to consist of a large number of species with low intrapopulation diversity, that is, specialists with narrow ecological niches. Communities that are adapted to scarce unstable conditions tend to consist of a small number of species with high intrapopulation diversity, that is, generalists with wide ecological niches. In rich unstable and scarce stable environments, we may expect some intermediate optimal diversity values.

The crucial question is on what scale is the relationship between diversity and functioning a significant factor for decision-making.

Comparisons of indicators of biodiversity and EF/ES on a large scale (global, continental, national) only show that communities adapted to different climatic and geographical conditions differ in EF and capacity to produce ES. For example, negative correlation between species richness and carbon content in ecosystems revealed for Britain (Anderson et al. 2009) does not mean that we need to reduce species diversity for better carbon storage. This means that ecosystems that store carbon are located in the north of the country and their low species diversity is optimal in those climatic conditions.

Preliminary results obtained in the project "TEEB-Russia 1" showed that relationship between ES indicators and species richness should be considered as a correlation, and not as a causal relationship. The average values of ES indicators in the regions depend primarily on climatic conditions, relief and the share of natural ecosystems in regions' area. Species richness that is typical for regional ecosystems is also determined by climate and relief. Thus, we have correlation but not causal relationship between species richness and EF/ES (Fig. 10.11). Examples of such

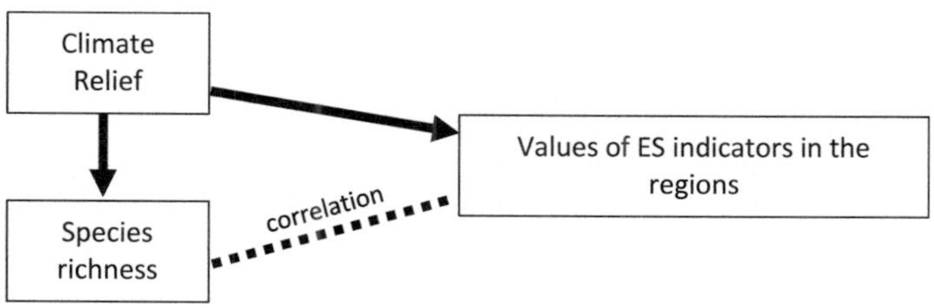

Fig. 10.11 Correlation and causal relationship between species richness and ES at national and landscape assessment scales (solid arrows show causal relationships and dotted lines show correlation

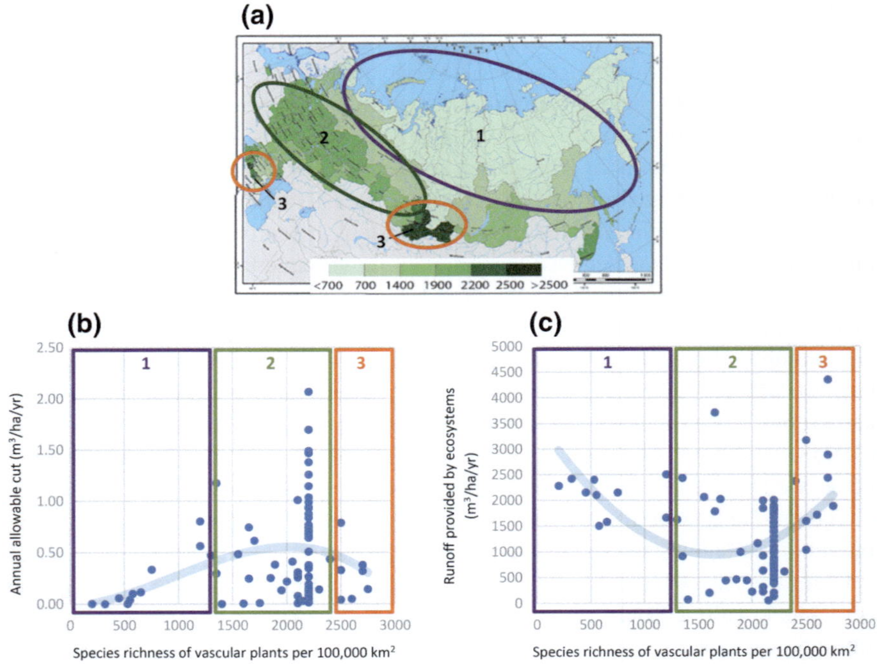

Fig. 10.12 Correlation between species richness of vascular plants and supplied ES: **a** number of species of vascular plants per 100,000 km^2 (according to the National Atlas of Russia); **b** correlation between species richness and supplied ES of wood production, **c** correlation between species richness and supplied ES of runoff regulation. 1—northern regions with a cold harsh climate, 2—temperate lowland regions, 3—temperate mountain regions

correlations are shown in Fig. 10.12. The relationship between supplied ES of wood production and species richness of vascular plants per 100,000 km^2 has unimodal humpback form with maximum ES values when diversity values are of around 2000 (Fig. 10.12b). The relationship between supplied ES of runoff regulation and species richness has unimodal U-shaped form with minimum ES values when diversity values are of around 1800 (Fig. 10.12c). These are only preliminary data and a comprehensive analysis with the identification of significant variables and dependencies will be a matter for future project phases, however, some of the most likely explanations can be made already now. In this example, climatic conditions and relief determine both species richness and supplied ES volumes. Species richness has minimum values in the northern regions (1 in Fig. 10.12a) and medium values in the temperate regions (2 in Fig. 10.12a), which reflects the adaptation of communities to the regional climatic conditions. The highest species richness characterizes mountain regions (3 in Fig. 10.12a) due to the large diversity of communities in altitude gradients. Supplied ES volumes also depend on climate and relief. Maximum wood production is a feature of lowland temperate regions and decreases both in northern and mountain regions. Runoff regulation by ecosystems, in opposite, is minimal in lowland temperate regions because ecosystems provide the highest volume of runoff in sufficiently moistened northern regions and in mountain regions (Bukvareva and Zamolodchikov 2018).

A causal relationship between species richness and EF/ES occurs at the landscape level when comparing plots of communities (habitats) of the same type that are disturbed to varying degrees. On a landscape, anthropogenic and natural disturbances transform the mosaic of communities adapted to natural conditions (relief, soils, water supply, etc.) pushing biodiversity away from the optimal state. The optimal biodiversity values can be broken due to anthropogenic changes of environmental conditions and because of direct disturbance of populations and communities (Bukvareva 2018). The general direction of anthropogenic changes of the environment is destabilization. Direct anthropogenic impact on populations and communities is expressed primarily in reduction of species richness and intrapopulation diversity. As a result, populations and communities leave their optimal state and move to suboptimal state. The further they move away from the optimal state, the weaker and more unstable is their EF and ES. Thus, in the landscape scale, interconnections between indicators of biodiversity and EF/ES can be useful background for decision-making. With regard to species diversity, two main management consequences can be formulated.

First, the optimal species diversity can be relatively low in unstable or scarce conditions. Despite this, it provides the maximal effectiveness of a community under these conditions. Thus, the criterion for the choice of conservation priorities should be the distance of anthropogenic shift away from the optimal diversity, but not high formal diversity indexes (e.g., species richness). The ultimate goal should be preservation of diversity of typical communities for a given landscape or region, including natural communities with low species diversity.

Second, the optimal biodiversity concept may be an additional approach to resolve ES trade-offs. For example, intensive use of provisioning services, especially food, fiber and biofuel production, greatly simplified ecosystem structure. This simplification enhanced certain provisioning services, but reduced others, particularly regulating services (Cardinale et al. 2012). One of the reasons for this trade-off is the different response of biodiversity to management for regulating and provisioning ES. While the first requires conservation of the optimal diversity values, the latter push populations and communities away from the optimal state. Thus, landscape planning should consider the conflict between goals of biodiversity management for different ES.

10.6 Conclusion

Russian landscapes provide important ES and are crucially important for the economy and people of the country. It is necessary to immediately start forming a national system of ES monitoring and assessment. The main methodological approaches to national ES assessment were proposed in the Prototype National Report "Ecosystem Services of Russia. Volume 1. Terrestrial Ecosystem Services" (Bukvareva and Zamolodchikov 2018).

Quantitative evaluation of supplied, demanded and consumed ES allows to assess the degree of ES use and the degree of satisfaction of the demand for ES in the regions and to reveal environmentally hazardous and relatively safe regions as well as regions, which tend to be consumers or donors of ES. ES assessment methodology applied in the Prototype Report does not take into account the relative spatial location of the regions that should be corrected in the future ES assessments.

However, only one-third of the services reviewed were quantified by the currently available data. In order to expand the range of ES when comparing regions, we used ES scores. Scores of supplied ES reflect natural factors that determine the capacity of ecosystems to perform ES. Scores of consumed and demanded ES reflect socio-economic factors that determine the need for ES and their use by humans. The comparison of the regions by the balance of supplied and consumed ES scores shows where natural factors dominate, and where socio-economic factors prevail.

Despite a significant data gap for a full ES evaluation the formation of a national system of ES assessment can be started on the basis of the current federal system of data collection. Priority attention should be paid to solving in the short term the following methodological problems.

- elaborate methods allowing (a) to take into account spatial diversity of natural and socio-economic conditions within the largest constituent entities of the Russian Federation and (b) to translate data for ES evaluation and estimates of ES between river basins and administrative regions,
- determine the methods of accounting for ES spatial characteristics, directional ES flow and user movement as well as spatial interrelations between the regions,
- ensure the availability and regular updating of data related to ES assessment, which are already collected by federal agencies and of the map of terrestrial ecosystems of Russia based on satellite imagery, develop a set of algorithms and models for indirect quantitative ES assessment,
- ensure a comprehensive assessment of all ES categories, and avoidance of priority accounting of the only easily calculated economic value of provisioning and recreational ES.

An important task to be addressed in the future is to assess the links between ES and biodiversity in order to optimize the objectives of biodiversity conservation and ES maintenance and use. This task should be addressed primarily at the landscape level, taking into account the adaptation of populations and communities to environmental conditions.

Acknowledgements The TEEB-Russia project was commissioned by the German Federal Agency for Nature Conservation (BfN) with funds from the German Federal Ministry for the Environment, Nature Conservation, and Nuclear Safety (BMU) and was supported by the Ministry of Natural Resources and Environment of the Russian Federation.

References

Allan E, Manning P, Alt F, Binkenstein J, Blaser S, Blüthgen N, Böhm S, Grassein F, Hölzel N, Klaus VH, Kleinebecker T, Morris EK, Oelmann Y, Prati D, Renner SC, Rillig MC, Schaefer M, Schloter M, Schmitt B, Schöning I, Schrumpf M, Solly E, Sorkau E, Steckel J, Steffen-Dewenter I, Stempfhuber B, Tschapka M, Weiner CN, Weisser WW, Werner M, Westphal C, Wilcke W, Fischer M (2015) Land use intensification alters ecosystem multifunctionality via loss of biodiversity and changes to functional composition. Ecol Lett 18(8):834–843. https://doi.org/10.1111/ele.12469. Accessed 15 Jan 2019

Anderson BJ, Armsworth PR, Eigenbrod F, Thomas CD, Gillings S, Heinemeyer A, Roy DB, Gaston KJ (2009) Spatial covariance between biodiversity and other ecosystem service priorities. J Appl Ecol 46:888–896. https://doi.org/10.1111/j.1365-2664.2009.01666.x. Accessed 15 Jan 2019

Bagstad KJ, Johnson GW, Voigt B, Villa F (2013) Spatial dynamics of ecosystem service flows: A comprehensive approach to quantifying actual services. Ecosyst Serv 4:117–125. https://doi.org/10.1016/j.ecoser.2012.07.012. Accessed 15 Jan 2019

Bardgett RD, van der Putten WH (2014) Belowground biodiversity and ecosystem functioning. Nature 515: 505–511. http://dx.doi.org/10.1038/nature13855. Accessed 15 Jan 2019

Bartalev SA, Belvard AS, Ershov DV, Isaev AS (2003) A new SPOT4- VEGETATION derived land cover map of Northern Eurasia. Int J Remote Sens 24(9):1977–1982. https://doi.org/10.1080/0143116031000066297. Accessed 15 Jan 2019

Baruffo M, Schmid B, Bruelheide H, Chi X, Hector A, Ma K, Michalski S, Tang Z, Niklaus PA (2013) Biodiversity promotes tree growth during succession in subtropical forest. PLoS One 8 (11):e81246. http://dx.doi.org/10.1371/journal.pone.0081246. Accessed 15 Jan 2019

Basanets LP, Drozdov AV (2006) Nature management for tourism, ecological imperative and prospects of Russia. In: Nature management and sustainable development. World Ecosystems and Problems of Russia. KMK—Association of The Scientific Publications, Moscow (in Russian)

Baruffo M, Schmid B, Bruelheide H, Chi X, Hector A, Ma K, Michalski S, Tang Z, Niklaus PA (2013) Biodiversity promotes tree growth during succession in subtropical forest. PLoS One 8 (11):e81246. http://dx.doi.org/10.1371/journal.pone.0081246. Accessed 15 Jan 2019

Bukvareva E (2014a) Die globale Bedeutung der russischen Ökosystemfunktionen und das Problem der unterschiedlichen Skalen von Ökosystemdienstleistungen, pp 93–126. In: Grunewald K, Bastian O, Drozdov A (eds) TEEB-Prozesse und Ökosystem-Assessment in Deutschland, Russland und weiteren Staaten des nördlichen Eurasiens. BfN-Skripten 372, Bonn (in German and Russian)

Bukvareva E (2014b) Ein erster Indikatorensatz zur Bestimmung des Status von Regionen als Bereitsteller bzw. Empfänger von ÖSD in verschiedenen räumlichen Maßstäben, pp 50–65. In: Grunewald K, Bastian O, Drozdov A, Grabowsky V (eds) Erfassung und Bewertung von Ökosystemdienstleistungen (ÖSD)—Erfahrungen, insbesondere aus Deutschland und Russland. BfN-Skripten 373, Bonn (in German and Russian)

Bukvareva E (2014c) The summary of the principle of optimal diversity of biosystems. LAP Lambert Academic Publishing, 52 p

Bukvareva E (2018) The optimal biodiversity—a new dimension of landscape assessment. Ecol Indic 94(2):6–11. http://dx.doi.org/10.1016/j.ecolind.2017.04.041. Accessed 15 Jan 2019

Bukvareva E, Aleshchenko G (2013) The principle of optimum diversity of biosystems. KMK—Association of The Scientific Publications, Moscow, 522 p (In Russian)

Bukvareva E, Zamolodchikov D, Grunewald K (2019) National assessment of ecosystem services in Russia: Methodology and main problems. Sci Total Environ 655(2019):1181–1196. https://doi.org/10.1016/j.scitotenv.2018.11.286. Accessed 15 Jan 2019

Bukvareva E, Zamolodchikov D, Kraev G, Grunewald K, Narykov A (2017) Supplied, demanded and consumed ecosystem services: Prospects for national assessment in Russia. Ecol Indic 78:351–360. https://doi.org/10.1016/j.ecolind.2017.03.034. Accessed 15 Jan 2019

Bukvareva EN, Grunewald K, Bobylev SN, Zamolodchikov DG, Zimenko AV Bastian O (2015) The current state of knowledge of ecosystems and ecosystem services in Russia: a status report. AMBIO 44(6):491–507. https://doi.org/10.1007/s13280-015-0674-4. Accessed 15 Jan 2019

Bukvareva EN, Zamolodchikov DG (eds) (2018) Ecosystem services of Russia: Prototype National Report, vol. 1. Terrestrial ecosystems services. Adapted English version of the report, originally published in Russian in 2016. BCC Press, Moscow, 148 p. Available at: http://www.biodiversity.ru/programs/ecoservices/first-steps/Ecosystem-Services-Russia_V1_eng_web.pdf. Accessed 15 Jan 2019

Burkhard B, Kroll F, Nedkov S, Muller F (2012) Mapping ecosystem service supply, demand and budgets. Ecol Indicators 21:17–29

Burkhard B, Kandziora M, Hou Y, Müller F (2014) Ecosystem service potentials, flows and demands—Concepts for spatial localisation, indication and quantification. Landscape Online 34:1–32. https://doi.org/10.3097/lo.201434. Accessed 15 Jan 2019

Cardinale BJ, Duffy JE, Gonzalez A, Hooper DU, Perrings C, Venail P, Narwani A, Mace GM, Tilman D, Wardle DA, Kinzig AP, Daily GC, Loreau M, Grace JB, Larigauderie A, Srivastava DS, Naeem S (2012) Biodiversity loss and its impact on humanity. Nature 486 (7401):59–67

Casalegno S, Inger R, Desilvey C, Gaston KJ (2013) Spatial covariance between aesthetic value & other ecosystem services. PLoS One 8(6):e68437. https://doi.org/10.1371/journal.pone.0068437. Accessed 15 Jan 2019

Cavanaugh KC, Gosnell JS, Davis SL, Ahumada J, Boundja P, Clark DB, Mugerwa B, Jansen PA, O'Brien TG, Rovero F, Sheil D, Vasquez R, Andelman S (2014) Carbon storage in tropical forests correlates with taxonomic diversity and functional dominance on a global scale. Glob Ecol Biogeogr 23:563–573. http://dx.doi.org/10.1111/geb.12143. Accessed 15 Jan 2019

Concept of the development of the system of nature protected areas of federal importance until 2020. Approved by the Government of the Russian Federation of December 22, 2011 No. 2322 p. Moscow (In Russian). Available at: http://government.ru/docs/4281/ Accessed 15 Jan 2019

Costanza R (2008) Ecosystem services: multiple classification systems are needed. Biol Conserv 141:350–352

Dunford RW, Harrison PA, Bagstad KJ (2017) Computer modelling for ecosystem service assessment. In: Burkhard B, Maes J (eds) Mapping ecosystem services. Pensoft Publishers, Sofia, pp 124–135

Egoshina TL (2005) Non-wood plant resources of Russia. NIA-Nature, Moscow, 80 p (in Russian)

Felipe-Lucia MR, Comín FA (2015) Ecosystem services–biodiversity relationships depend on land use type in floodplain agroecosystems. Land Use Policy 46:201–210. https://doi.org/10.1016/j.landusepol.2015.02.003. Accessed 15 Jan 2019

Forsman A, Wennersten L (2016) Inter-individual variation promotes ecological success of populations and species: evidence from experimental and comparative studies. Ecography 39:630–648. https://doi.org/10.1111/ecog.01357. Accessed 15 Jan 2019

Forsman A (2014) Effects of genotypic and phenotypic variation on establishment are important for conservation, invasion and infection biology. Proc Natl Acad Sci US 111:302–307. https://doi.org/10.1073/pnas.1317745111. Accessed 15 Jan 2019

Grace JB, Anderson TM, Seabloom EW, Borer ET, Adler PB, Harpole WS, Hautier Y, Hillebrand H, Lind EM, Pärtel M, Bakker JD, Buckley YM, Crawley MJ, Damschen EI, Davies KF, Fay PA, Firn J, Gruner DS, Hector A, Knops JM, MacDougall AS, Melbourne BA, Morgan JW, Orrock JL, Prober SM, Smith MD (2016) Integrative modelling reveals mechanisms linking productivity and plant species richness. Nature 529:390–393. http://dx.doi.org/10.1038/nature16524. Accessed 15 Jan 2019

Gross K, Cardinale BJ, Fox JW, Gonzalez A, Loreau M, Polley HW, Reich PB, van Ruijven J (2014) Species richness and the temporal stability of biomass production: a new analysis of recent biodiversity experiments. Am Nat 183:1–12. http://dx.doi.org/10.1086/673915. Accessed 15 Jan 2019

Grunewald K, Bastian O, Drozdov A (eds) (2014a) TEEB-Prozesse und Ökosystem-Assessment in Deutschland, Russland und weiteren Staaten des nördlichen Eurasiens. BfN-Skripten 372, Bonn (in German and Russian)

Grunewald K, Bastian O, Drozdov A, Grabowsky V (eds) (2014b) Erfassung und Bewertung von Ökosystemdienstleistungen (ÖSD) – Erfahrungen, insbesondere aus Deutschland und Russland. BfN-Skripten 373, Bonn (in German and Russian)

Grunewald K, Zimenko A, Bastian O, Bukvareva E, Grigorian A, Wende W (2014c) Konzeption des Projektes, pp 146–161. In: Grunewald K, Bastian O, Drozdov A (eds) TEEB-Prozesse und Ökosystem-Assessment in Deutschland, Russland und weiteren Staaten des nördlichen Eurasiens. BfN-Skripten 372, Bonn (in German and Russian)

Gryaznov SE, Kuzminyh YV, Bogachev YK (2011) Assessment and measurement of illegal logging in the forest sector of the Russian Federation. News of Higher education institutions. Forest Journal 5:124–130 (In Russian)

Haines-Young R, Potschin M (2013) Common international classification of ecosystem services (CICES): consultation on Version 4, August–December 2012. EEA Framework Contract No. EEA/IEA/09/003

Handa IT, Aerts R, Berendse F, Berg MP, Bruder A, Butenschoen O, Chauvet E, Gessner MO, Jabiol J, Makkonen M, McKie BG, Malmqvist B, Peeters ET, Scheu S, Schmid B, van Ruijven J, Vos VC, Hättenschwiler S (2014) Consequences of biodiversity loss for litter decomposition across biomes. Nature 509:218–221. http://dx.doi.org/10.1038/nature13247. Accessed 15 Jan 2019

Harrison PA, Berry PM, Simpson G, Haslett JR, Blicharska M, Bucur M, Dunford R, Egoh B, Garcia-Llorente M, Geamănă N, Geertsema W, Lommelen E, Meiresonne L, Turkelboom F (2014) Linkages between biodiversity attributes and ecosystem services: A systematic review. Ecosyst Serv 9:191–203. https://doi.org/10.1016/j.ecoser.2014.05.006. Accessed 15 Jan 2019

Hedden-Dunkhorst B, Braat L, Wittmer H (2015) TEEB emerging at the country level: Challenges and opportunities. Ecosyst Serv 14:37–44. https://doi.org/10.1016/j.ecoser.2015.03.007. Accessed 15 Jan 2019

Hooper DU, Adair EC, Cardinale BJ, Byrnes JEK, Hungate BA, Matulich KL, Gonzalez A, Duffy JE, Gamfeldt L, O'Connor MI (2012) A global synthesis reveals biodiversity loss as a major driver of ecosystem change. Nature 486:105–108. https://www.nature.com/articles/nature11118Accessed 15 Jan 2019

Hughes AR, Inouye BD, Johnson MTJ, Underwood N, Vellend M (2008) Ecological consequences of genetic diversity. Ecol Lett 11:609–623. https://doi.org/10.1111/j.1461-0248.2008.01179.x. Accessed 15 Jan 2019

King E, Cavender-Bares J, Balvanera P, Mwampamba TH, Polasky S (2015) Trade-offs in ecosystem services and varying stakeholder preferences: Evaluating conflicts, obstacles, and opportunities. Ecol Soc 20(3). http://dx.doi.org/10.5751/ES-07822-200325 Accessed 15 Jan 2019

Kotlobay A, Lopina O, Kharchenkov Yu, Bryukhanov A, Schegolev A, Smirnov D (2006) Evaluation of timber of doubtful origin and analysis of the practice of introducing systems for tracing the origin of wood in a number of multi-forest regions of the North-West, Siberia and the Far East of Russia. WWF, Moscow, 56 p (In Russian)

Lewandowska AM, Biermann A, Borer ET, Cebrián-Piqueras MA, Declerck SA, De Meester L, Van Donk E, Gamfeldt L, Gruner DS, Hagenah N, Harpole WS, Kirkman KP, Klausmeier CA, Kleyer M, Knops JM, Lemmens P, Lind EM, Litchman E, Mantilla-Contreras J, Martens K, Meier S, Minden V, Moore JL, Venterink HO, Seabloom EW, Sommer U, Striebel M, Trenkamp A, Trinogga J, Urabe J, Vyverman W, Van de Waal DB, Widdicombe CE, Hillebrand H (2016) The influence of balanced and imbalanced resource supply on biodiversity functioning relationship across ecosystems. Philos Trans R Soc B 371:20150283. https://doi.org/10.1098/rstb.2015.0283. Accessed 15 Jan 2019

Lomanova NV et al (eds) (2011) The state of game resources in the Russian Federation in 2008–2010: Information-analytical materials. Game animals of Russia (biology, protection, resource management, rational use). Issue 9. Phys. Culture, Moscow, 219 p (in Russian)

Maes J, Paracchini ML, Zulian GA (2011) European assessment of the provision of ecosystem services. Towards an atlas of ecosystem services, Publ Office EU, Luxembourg

Maes J, Paracchini ML, Zulian G, Dunbar MB, Alkemade R (2012) Synergies and trade-offs between ecosystem service supply, biodiversity, and habitat conservation status in Europe. Biol Conserv 155:1–12. https://doi.org/10.1016/j.biocon.2012.06.016. Accessed 15 Jan 2019

Maestre FT, Quero JL, Gotelli NJ, Escudero A, Ochoa V, Delgado-Baquerizo M, García-Gómez M, Bowker MA, Soliveres S, Escolar C, García-Palacios P, Berdugo M, Valencia E, Gozalo B, Gallardo A, Aguilera L, Arredondo T, Eldridge DJ, Espinosa CI, Florentino A, Gaitán J, Gatica MG, Ghiloufi W, Gómez-González S, Gutiérrez JR, Hernández RM, Huang X, Huber-Sannwald E, Jankju M, Miriti M, Monerris J, Mau RL, Morici E, Naseri K, Ospina A, Polo V, Prina A, Pucheta E, Ramírez-Collantes DA, Romao R, Tighe M, Torres-Díaz C, Val J, Veiga JP, Wang D, Zaady E (2012) Plant species richness and ecosystem multifunctionality in global drylands. Science 335:214–218. https://doi.org/10.1126/science.1215442. Accessed 15 Jan 2019

Martínez-Harms M-J Balvanera P (2012) Methods for mapping ecosystem service supply: a review. Int J Biodivers Sci Ecosys Serv Manage 8:1–2, 17–25. https://doi.org/10.1080/21513732.2012.663792. Accessed 15 Jan 2019

Millennium Ecosystem Assessment: Synthesis (2005) Island Press, Washington

Nadrowski K, Wirth C, Scherer-Lorenzen M (2010) Is forest diversity driving ecosystem function and service? Curr Opin Environ Sustain 2:75–79. https://doi.org/10.1016/j.cosust.2010.02.003. Accessed 15 Jan 2019

National Atlas of Russia. Volumes 1–4. (2004–2008). Federal Agency for Geodesy and Cartography of Russia, Moscow (in Russian)

National Report (2013) National Report of the Russian Federation on the Cadastre of Anthropogenic Emissions by Sources and Removals by Sinks of Greenhouse Gases Not Controlled by the Montreal Protocol for 1990–2011 (2013) Moscow, 421 p (in Russian)

National Strategy of Biodiversity Conservation in Russia (2001) RAS, Ministry of Natural Resources of Russia, GEF Project "Conservation of Biodiversity", Moscow, 75 p (in Russian)

Nedkov S, Burkhard B (2012) Flood regulating ecosystem services – Mapping supply and demand, in the Etropole municipality, Bulgaria. Ecol Indic 21:67–79. https://doi.org/10.1016/j.ecolind.2011.06.022. Accessed 15 Jan 2019

Paquette A, Messier C (2011) The effect of biodiversity on tree productivity: from temperate to boreal forests. Glob Ecol Biogeogr 20:170–180. https://doi.org/10.1111/j.1466-8238.2010.00592.x. Accessed 15 Jan 2019

Ptichnikov A, Kuritsyn A (2011) Systems for tracing wood in Russia: the experience of timber companies and forest management bodies. Analytical report. WWF of Russia, Moscow, 116 p (In Russian)

Qiu J, Turner MG (2013) Spatial interactions among ecosystem services in an urbanizing agricultural watershed. PNAS 110(29):12149–12154. https://doi.org/10.1073/pnas.1310539110. Accessed 15 Jan 2019

Raudsepp-Hearne C, Peterson GD, Bennett EM (2010) Ecosystem service bundles for analyzing tradeoffs in diverse landscapes. Proc Natl Acad Sci USA 107(11):5242–5247. https://doi.org/10.1073/pnas.0907284107. Accessed 15 Jan 2019

Rosstat (2013) Regions of Russia. Socioeconomic indicators. Statistical Bull. Moscow (in Russian). Available at: www.gks.ru/wps/wcm/connect/rosstat_main/rosstat/ru/statistics/publications/catalog/doc_1138623506156. Accessed 15 Jan 2019

Schulp CJE, Burkhard B, Maes J, Van Vliet J, Verburg PH (2014) Uncertainties in ecosystem service maps: a comparison on the European scale. PLoS One 9(10):e109643. https://doi.org/10.1371/journal.pone.0109643. Accessed 15 Jan 2019

Science for Environment Policy (2015) Ecosystem Services and the Environment. In-depth Report 11 produced for the European Commission. DG Environment by the Science Communication Unit, UWE, Bristol. https://doi.org/10.2779/57695. Accessed 15 Jan 2019

Stolbovoi V, McCallum I (2002). Land Resources of Russia (CD-ROM). International Institute for Applied Systems Analysis and the Russian Academy of Science, Laxenburg, Austria: Available at: http://webarchive.iiasa.ac.at/Research/FOR/russia_cd/guide.htm. Accessed 15 Jan 2019

Stürck J, Poortinga A, Verburg PH (2014) Mapping ecosystem services: The supply and demand of flood regulation services in Europe. Ecol Indic 38:198–211. https://doi.org/10.1016/j.ecolind.2013.11.010. Accessed 15 Jan 2019

Syrbe R-U, Walz U (2012) Spatial indicators for the assessment of ecosystem services: Providing, benefiting and connecting areas and landscape metrics. Ecol Indicators 21:80–88. https://doi.org/10.1016/j.ecolind.2012.02.013. Accessed 15 Jan 2019

TEEB (2019) http://www.teebweb.org/areas-of-work/country-studies-home/. Accessed 15 Jan 2019

Tilman D, Reich PB, Isbell F (2012) Biodiversity impacts ecosystem productivity as much as resources, disturbance, or herbivory. Proc Natl Acad Sci USA 109:10394–10397. https://doi.org/10.1073/pnas.1208240109. Accessed 15 Jan 2019

Tilman D, Isbell F, Cowles JM (2014) Biodiversity and Ecosystem Functioning. Ann Rev of Ecology, Evolution, and Systematics 45(1):471–493. https://doi.org/10.1146/annurev-ecolsys-120213-091917. Accessed 15 Jan 2019

Turkelboom F, Leone M, Jacobs S, Kelemen E, García-Llorente M, Baró F, Termansen M, Barton DN, Berry P, Stange E, Thoonen M, Kalóczkai Á, Vadineanu A, Castro AJ, Czúcz B, Röckmann C, Wurbs D, Odee D, Preda E, Gómez-Baggethun E, Rusch GM, MartínezPastur G, Palomo I, Dick J, Casaer J, Dijk J, Priess JA, Langemeyer J, Mustajoki J, Kopperoinen L, Baptist MJ, Peri PL, Mukhopadhyay R, Aszalós R, Roy SB, Luque S, Rusch V (2018) When we cannot have it all: Ecosystem services trade-offs in the context of spatial planning. Ecosyst Serv 29:566–578. https://doi.org/10.1016/j.ecoser.2017.10.011. Accessed 15 Jan 2019

Vihervaara P, Mononen L, Santos F, Adamescu M, Cazacu C, Luque S, Geneletti D, Maes J (2017) Biophysical quantification. In: Burkhard B, Maes J (eds) Mapping Ecosystem Services. Pensoft Publishers, Sofia, pp 95–101

Vilà M, Carrillo-Gavilán A, Vayreda J, Bugmann H, Fridman J, Grodzki W, Haase J, Kunstler G, Schelhaas MJ, Trasobares A (2013) Disentangling biodiversity and climatic determinants of wood production. PLoS ONE 8(2):e3530. http://dx.doi.org/10.1371/journal.pone.0053530. Accessed 15 Jan 2019

Wang W, Lei X, Ma Z, Kneeshaw DD, Peng C (2011) Positive relationship between aboveground carbon stocks and structural diversity in spruce-dominated forest stands in New Brunswick, Canada. Forest Sci 57(6):506–515

Zulian G, Polce C, Maes J (2014) ESTIMAP: A GIS-based model to map ecosystem services in the European Union. Ann Bot (Roma) 4:1–7. https://doi.org/10.4462/annbotrm-11807. Accessed 15 Jan 2019

Chapter 11
Citizen and Landscape Governance

Denis Couvet

Abstract Biodiversity governance requires relevant biodiversity data and interpretation, policy proposals. Essential Biodiversity Variables are a major concept to identify priorities in regards to these data, how to organize and relate the different kind of biodiversity data. A major type of biodiversity variables to develop to examine and improve human–biodiversity relationships concern ecosystems, where their diversity ought to be related to human livelihoods diversity. In this regard, beyond 'historical' and 'modern' landscapes, 'ecological' and 'post-modern' landscapes ought to be characterized and promoted, a way to improve biodiversity–society relationships; in particular the quality of democratic procedures in regards to (i) the diversity of biodiversity-society models, landscapes, and options proposed, (ii) the number of biological, social, and cultural dimensions explored.

Keywords Biodiversity · Governance · Essential biodiversity variables · Landscapes · Teleconnection · Biosphere people

11.1 Introduction

Associated to the wide diversity of biodiversity values, from intrinsic to different types of instrumental values, landscape dynamics, hence governance, concerns every citizen, social group. Being a fuzzy scientific concept, there are different ways to conceptualize landscapes. Emerge at least two major different ones, a biophysical and a cultural viewpoint. We will examine successively these two viewpoints, complementary. We will first examine how the landscape stands, compared to biodiversity concepts and variables, especially species ones, and how citizen contribute to their characterization. We will then examine how people stand and

D. Couvet (✉)
UMR MNHN-CNRS-SU 'CESCO', 36, rue Geoffroy Saint-Hilaire (cp 41), 75231 Paris Cedex 05, France
e-mail: couvet@mnhn.fr

L. Müller and F. Eulenstein (eds.), *Current Trends in Landscape Research*, Innovations in Landscape Research, https://doi.org/10.1007/978-3-030-30069-2_11

connect to different landscapes, a question which matters especially when landscapes become unfavorable to human activities. We will finally examine the relation between these two viewpoints.

11.2 Landscapes from a Biophysical Viewpoint: Biodiversity

From a biophysical viewpoint, there are two ways to examine landscapes. We will not consider the physicochemical viewpoint, with the concept of geodiversity, where there can be geodiversity hotspots (e.g., Tasmania), and with a corresponding geo-ethics. A point we will not explore further. We will here examine the biological viewpoint, with the concept of biodiversity.

11.2.1 Essential Biodiversity Variables (EBV)

To determine the biodiversity variables that should be examined, in particular in regards to landscapes, the concept of EBV considers the different levels of biological organization, scientific disciplines concerned, from the gene to the ecosystem (Pereira et al. 2013, and Table 11.1). EBV classes identify six types of entities which should be monitored (Schmeller et al. 2017); the hierarchy between EBV classes, in case of limiting possibilities to monitor biodiversity, being a lagging question. For example, the species concept is a major tool for biodiversity public policies, represented by the first three EBV classes. However, the higher biological organization levels, communities, ecosystems, and landscapes, represented by the following three EBV classes, are also most relevant; ecosystem diversity being an objective in itself in biodiversity preservation, necessary to preserve species diversity, and also ecosystem services (and see below).

Table 11.1 Six different classes of essential biodiversity variables (EBV), from Pereira et al. (2013)

EBV class	Major observation techniques	Type of biodiversity assessment
Genetic variability	Genomics	Adaptation potential of species
Populations, species	Citizen extensive monitoring	Species conservation status
Species characters	Museum collections	Ecological functions
Community composition	Citizen extensive monitoring	
Ecosystems functions	Intensive monitoring, remote sensing	Ecosystem services
Ecosystems structure	Remote sensing	Landscape quality

To hierarchize biodiversity variables to monitor, another difficulty is that many variables can exist for each class. For example, community composition can be described through the number of species, species diversity, weighted by their phylogenetic proximity, and also community abundance, e.g., number of birds... (Couvet et al. 2011; Hill et al. 2016).

Biodiversity data are especially necessary to relate biological diversity and ecological functions, hence ecosystem services, to document the effect of different human pressures, public policies, on biodiversity. Assessing biodiversity adaptive potential to global change is a growing challenge. It concerns threatened species as well as pests and pathogens, whose adaptation is costly to humans, and also 'ordinary' biodiversity whose adaptation is necessary to biodiversity and ecological functions to adapt to global change, avoiding ecosystem disruption. Adaptive management should integrate such adaptation, requiring closer time-scales in regards to observations, to know how biodiversity reacts on short time-scales, to be able to modify management in consequence.

Such needs have to be confronted with observational possibilities, focusing on techniques and human abilities available, including social and environmental costs. To develop monitoring, biodiversity observation networks (BON) build capacities, investigating what should be monitored, identifying observation methods, managing and interpreting data. Observation methods include extensive and intensive monitoring, ecological field studies, and remote sensing (Schmeller et al. 2017). Some methods require rather professionals, whereas others are better performed by citizens, especially extensive monitoring (Schmeller et al. 2017; Couvet et al. 2011).

Citizen, as para-taxonomists or para-ecologists, plays an important role in BON, documenting especially species-traits and community composition. Monitoring citizens demonstrated major variations of farmland and butterflies communities, in terms of abundance and composition, relative abundance of specialist species (Julliard et al. 2004). At the opposite, more classical observations, on species number and diversity, made by professionals, failed to detect changes (Dornelas et al. 2014), both results being compatible (Hill et al. 2016). Indeed, relating different biodiversity classes dynamics is a major scientific task, to assess contradiction, to get a better image of biodiversity dynamics (Hill et al. 2016).

11.2.2 The Case of Microorganisms

Recent progresses in the characterization of microorganisms communities are deeply affecting our biodiversity representations. Defined as bacteria, fungi, a large part of the protists, microorganisms are the majority of living individuals, in terms of number of individuals and probably as number of species (Locey and Lennon 2016), thus in terms of diversity, moreover in terms of phylogenetic differences, a trivial result when one considers that microorganisms diversity is a lot older than plants and animals diversity. Microorganisms dynamics has been neglected because they were not readily accessible through human senses, visions…, although their human importance in terms of diseases, allergies, human uses (e.g., making of wine and cheese), is fundamental.

The conservation status of most microorganisms is not available; for the ones which are obligatory host of threatened macro species, the future could be very bleak. At the opposite, some phyla, like toxic cyanobacteria, benefit from human disturbance, proliferating in overexploited and polluted waters. Climate change could also lead to noticeable changes in their relative proportion, for example, between diatoms and foraminifers.

Much of their importance for humans could go through the holobiont, the way they interact with macro-organisms, humans, and mammals, for example, being inhabited by several very large communities of microorganisms, necessary for many physiological functions, our health, accounting for the effect of plant diversity on minimization of allergies, or the beneficial effects of organic agriculture on plant composition (Carey 2016). Microorganisms are probably most important for several ecosystems services, an important issue being how to monitor such communities, to make sense of their variation, and moreover how ecosystem management affects microorganisms fate.

Emphasizing the hidden importance of microorganisms helps to mind the gap between the scientific representation of biodiversity and its biological reality. Indeed, most scientific biodiversity representations barely integrate microorganisms, although they are the majority, in terms of number of individuals and probably number of species. Moreover, their functional importance is major. Microorganism's state and dynamics might reflect the quality of human practices in a landscape, especially in regards to agriculture, and addressing soil communities.

11.2.3 Landscape and EBV Variables

Related to EBV, a landscape can be defined as a set of neighboring ecosystems, which can be viewed from a human viewpoint. Thus, landscape states and dynamics can be significantly documented with EBV. Ecosystems functions and structures (Table 11.1) classes are especially relevant. Nevertheless, intimately related to humans, social, and cultural variables ought also to be documented, to get a complete landscape assessment.

In this regard, different landscapes can be recognized with the notion of anthropobiome. Such notion distinguishes six kinds, in three sets. (1) High human density, with towns and villages, where are 80% of humans but represent only 6% of the non-icy terrestrial area. (2) Low human density but with high human pressure represented by agriculture, crops, and prairies. (3) Low human density and human pressure, often called 'natural', and represent 40% of the planet.

Urban and rural landscapes structures vary, along many different dimensions, with human practices, independently of any biodiversity conservation programs. Landscape makers, stakeholders are numerous and diverse: farmers, architects, and territorial managers. Indeed, even when strong direct ties between humans and species are considered—whether these are charismatic, domesticated, exploited, noxious, etc.—landscape variables also a determine relationship with these species. In other terms, direct relations between humans and genes or species are difficult to conceptualize, manage, independently of landscapes practices.

11.2.4 Ecosystems Red-Lists

Ecosystems as systems organizing interactions between living organisms and their physic-chemical substrate, are a level of biological organization important to preserve, and noticed as such according to CBD (Convention on Biological Diversity). Besides species red-list, UICN (International Union for Conservation of Nature) proposed an ecosystems red-list, including criteria for identification and threat level assessment (Keith et al. 2013). A major difficulty remains to establish a consensual typology for ecosystems, based on biophysical and/or social criteria. Such typology is a necessary step to identify a '*threatened ecosystem*,' and ought to be devised according to a relevant landscape typology.

11.3 Landscapes Diversity from a Human Viewpoint

We will now consider how considering landscape human uses should lead to consider four meaningful types of landscapes, combining biophysical and social questions.

11.3.1 Hidden Teleconnections: 'Historical' and 'Modern' Landscapes

In regards to landscape diversity, assuming mostly a biophysical viewpoint, a typology was recently proposed, distinguishing 'modern' and 'historical' ecosystems (Barnosky et al. 2017), the scale considered corresponding also to two types of landscapes. Such biophysical viewpoint corresponds to the 'productionist' perspective (Fouilleux et al. 2017), where human livelihoods are considered as an outcome of the biophysical possibilities.

In such typology, 'historical' landscapes prioritize the intrinsic value of biodiversity. Considered as reservoirs of biodiversity, they are also associated to the option value of biodiversity. They are rather populated with anthropophobic species, like large carnivores, having difficulties to coexist with humans. Such species are often threatened, and as a matter of fact need protection. Conservation objectives are usually considered as independent of any utilitarian perspective, of any

Table 11.2 Different landscapes, EBV preferentially associated

Landscapes	EBV privileged
Historical	Threatened biological diversity, especially at the species level
Modern	Ecosystem services (community composition, ecosystem functions, and structure), possibly associated with synthetic organisms (GMOS, gene editing …), to maximization of labor productivity, related to landscape uniformity
Ecological	– Biodiversity variables associated with landscape resilience, hence to biological and social diversity. Local and/or specialist species, every organism involved in ecological networks, from microorganisms (holobionts) to pollinators – Ecological functions (community composition, ecosystem functions, and structure), their adaptability (functional and phylogenetic diversity) – Landscape structural heterogeneity (associated in particular to agro-ecology)
Post-modern	'*Nature contribution to people*' Biological diversity relevant for social minorities (from cultivated to symbolic species): small farming, peasants, indigenous people, and environmental refugees Involves traditional ecological knowledge

necessity for humans, except for cultural and ethical needs, although preservation of ecosystem services has been recently proposed as a motive of advocacy for such areas. Associated to historically low human impacts, social acceptability to restrain human activity should be higher in such landscapes.

'Modern' landscapes host essential human activities, especially agriculture. Production is most important, hence the preservation of ecological functions involved, like soil fertility. The objective is to optimize goods and services production, to decrease their environmental impact. Ecological engineering is a proposed method to improve ecological functioning, exotic and synthetic biodiversity being a possible component, depending on their performances in regards to ecosystem functioning (Barnosky et al. 2017). Being heavily used by humans, preserving biodiversity in such landscapes requires deep changes of human activities, like going from industrial agriculture to agro-ecology.

As a result, relevant EBV classes differ for these two landscapes (Table 11.2).

11.3.2 Teleconnection Between Landscapes: Global Commodity Chains

A caveat with this dual, biophysical, typology is that relations between remote landscapes are hidden. Indeed, most often modern landscapes require, from a human viewpoint, other, often remote landscapes, through teleconnections, to be functional. Such dependency is in terms of natural, human, and social resources.

Fluxes mechanisms associated with such dependency can be described with global commodity chains (or 'GCC'), where the chain going from production to

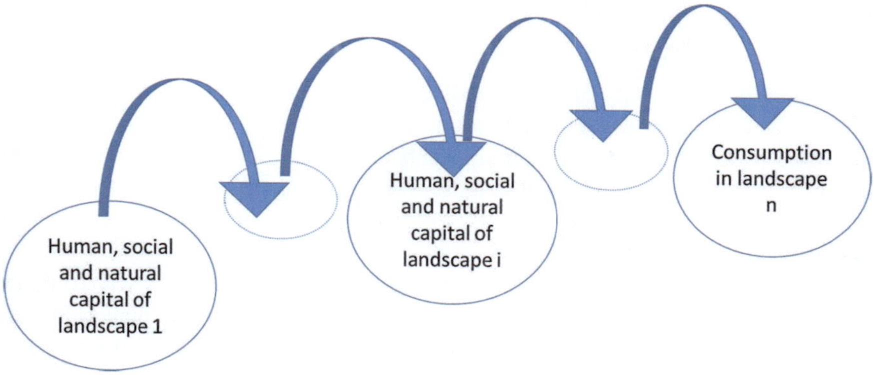

Fig. 11.1 GCC as relations between landscapes, through infrastructures, necessary for teleconnections between remote landscapes

consumption can function at the world level (Fig. 11.1). As a result, production and consumption sustainability cannot be locally assessed, but require to take into account GCC logics (Gereffi 1996).

LCA are a major tool to quantify teleconnections through GCC, assessing GCC logics. LCA can be assessed for a product, or at the level of territory–aggregating LCAs of all products consumed at the scale of this territory, corresponding to the ecological footprint of people at such spatial scale, which leads to the possibility to compare such 'territorial' LCA to the biophysical capacity of the same territory.

However, considering such LCA, also called attributional LCA, has major conceptual limits, disregarding, when LCA are minimized, probable rebound effects associated with lifestyles. That is, replacement of products with lower environmental impacts can stimulate consumptions, new infrastructures, leading to the so-called rebound effect (Alcott 2005). As a result, consequential LCA (CLCA), or prospective life-cycle analyses, allows the possibility to anticipate and reduce possible rebound effects (Chaplin-Kramer 2017). CLCA require defining (i) the market, (ii) demand and offer, their relationships, (iii) the functional unit, the practices involved, relating production and consumption modes. Indeed, there are tight relationships between production and consumption in the case of food and agriculture, affecting biodiversity (Desquilbet et al. 2016). For example, organic consumers are lower meat consumers, production and consumption modes of corn are related (Sukhedv et al. 2016). In other words, livelihoods, consumption modes, of citizen, are a critical matter for landscape dynamics. In regards to agriculture, the most important human pressure on biodiversity, different types of agriculture, from 'intensive' to 'organics,' correspond to different teleconnections with landscapes, associated with different pressures on landscapes, livelihoods. In other words, to be compared, the types of agriculture ought to be assessed through GCC. For example, to assess the environmental effects of GM soybean production, the complete food chain it belongs to, the corresponding GCC, need to be considered.

11.3.3 'Biosphere' and 'Ecosystem' People

LCA analyses exemplify unequal ecological exchange between low and high 'Human Development Index' (HDI) countries—HDI integrating wealth, health and education (Cumming and von Cramon-Taubadel 2018), richer countries importing environmental goods from poorer countries. That means that the so-called modern landscapes (see above) are significantly different variable on ecological grounds, at least depending on wealth.

To describe more precisely such variability, the notion of ecological footprint (EF) is useful. EF indicates the possible ecological overshoot of the planet. Present overshoot increases with the 'Human Development Index', HDI (Cumming and von Cramon-Taubadel 2018). Such EF variations correspond to different ways for people to relate to world economy. The higher the HDI, the more local people depend on teleconnected, often remote, landscapes. In reverse, the lower the HDI, the more local landscapes are overexploited for export; in such landscapes, exportation toward wealthier landscapes adds to local consumption. Indeed, most degraded landscapes would belong to lower HDI countries (Sutton et al. 2016). As a matter-of-fact, Gadgil (2013) characterize two types of people, biosphere and ecosystem people, depending if their well-being depends on local or global landscapes (Fig. 11.2), corresponding to societies interactions.

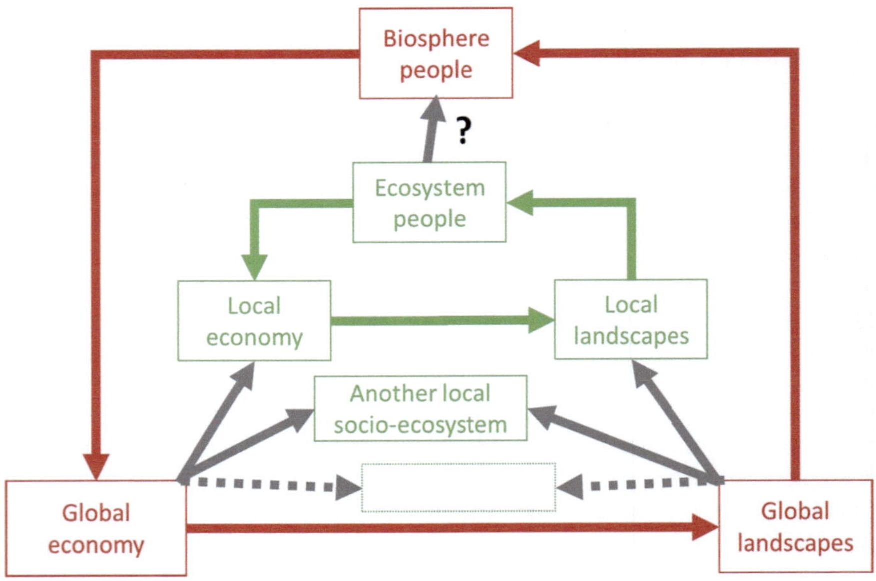

Fig. 11.2 Relations and interdependence of biosphere and ecosystem people with local and global landscapes and economies? indicates possible migrations (adapted from Cumming and von Cramon-Taubadel 2018)

A consequence could be that, beyond unsustainable production in many landscapes, hidden teleconnections constrain choices in every landscape, in different ways depending on the predominance of exports or imports. The notion of red and green loop can characterizes differences that modern landscapes establish with their local (global) landscapes (Fig. 11.2), and see (Cumming and von Cramon-Taubadel 2018).

11.3.4 Degraded Landscapes and Refugees

Associated to ecological overshoot and hidden relationships with remote landscapes, landscape degradation can occur, in different ways. Natural resources can be exhausted, barely renewed, and/or life conditions become not anymore suitable for humans. For example, some climate scenarios anticipate that 3.5 billion people leave in landscapes that might lose 50% of their productivity (Mora et al. 2015). That correspond also to the notion of 'Ghosts of the Anthropocene,' a notion to characterize landscapes where major degradation from a human origin arose (Tsing et al. 2017). For example, Chernobyl and Fukushima areas, abandoned nuclear site, Minimata bay, Aral Sea ….

Taking into account teleconnections, between biosphere and ecosystem people, their landscapes, management ought to integrate, beyond human preferences, biophysical limits, an inescapable consequence of the overshoot of planetary limits in modern landscapes, environmental refugees. Indeed, degraded landscapes and refugees are a testimony of the urgency of such considerations. That is, locally, beyond a certain level of degradation, people cannot anymore leave in degraded landscapes. People are then dispossessed from their landscapes, becoming environmental refugees, leaving unwillingly their landscape. The United Nations counted 30 millions of refugees in 2016, for which environmental degradation as a cause should interact with other social causes. The historical case of the Shtetl suggests that there might not be happy conclusion for environmental refugees.

11.4 'Citizen' Governance: 'Ecological' and 'Post-modern' Landscapes

Such degradations, future plausible ones associated to ecological overshoot, lead to major difficulties for biosphere people and environmental refugees to land. Beyond difficult, perhaps impossible top-down global responses, the power of people initiatives, the citizen perspective becomes a major matter of concern. Citizen initiatives, changing lifestyles, relationships between production and consumption modes, for example, with the 'slow food' movement (Schneider 2008) are numerous. A typology of world-views diversity (De Witt et al. 2017) can help to understand, acknowledge, such diversity of responses.

World-view diversity should include at least four types of world-views. With the 'modern' world-view, associated to dependence on global ecosystems (see above), natural science and associated techniques are considered as the major tools to improve humans–biodiversity relationships, lifestyles being not questioned. With the 'traditional' world-view, 'historical' landscapes host traditional values, in particular spiritual ones, necessary for human–biodiversity relationships, avoiding the great divide between nature and culture.

Finally, in a pluralistic world, 'post-modern' and 'ecological' world-views (De Witt et al. 2017) are important alternatives, associated to corresponding 'ecological' and 'post-modern' landscapes. A major advance with these two last categories is to integrate the hidden relationships with remote landscapes. We will now present these two categories.

11.4.1 'Post-modern' Landscapes: (Dominated) Ecosystem People

Such landscapes are seen through the lenses of 'ecosystem people' (Fig. 11.2), in particular dominated ones, for which living requires exporting their natural capital (Gadgil 2013). As a result, in such landscape, social issues become of major importance. That concerns social minorities, depauperate and/or deprived ones, gender issues…. More generally, the post-modern world-view considers that social uses of biodiversity, ecosystems, hence environmental questions, priorities, and urgencies, vary significantly with social groups. That concerns in particular farmers, leading to proposals to favor small farming, 'peasant' agriculture, corresponding to a certain kind of agriculture, based on agro-ecology, organics. Such perspective concerns a large majority of farmers, probably more than 90% of the farmers in the planet, based primarily in Asia and Africa; at the opposite, modern farmers being a tiny minority (Dorin 2017). Improving perspectives for post-modern landscapes probably correspond to ecological landscapes, which we will consider now.

11.4.2 'Ecological' Landscapes: A Necessary Perspective for Biosphere People?

In 'ecological' landscapes, the primary attempt is to combine the biophysical and human viewpoints, their retroactions; landscapes and societies resilience being a major principle. Accordingly, priority objectives should be the maintenance of local biological and social diversity, landscape structural heterogeneity, the diversity of ecological functions, their adaptability, involving functional and phylogenetic species diversity.

To reach such objective, major changes are necessary from the biophysical and human viewpoints. Minimizing human ecological footprint should be considered as a prerequisite to preserve biodiversity. A consequence being that human productivity is not any more maximized, such landscape concern at first biosphere people in 'modern' landscapes, especially the ones who 'need to land', that is, to have an EF compatible with planetary limits. Landscapes associated to vegans, low-energy lifestyles, including 'transition towns', might be good examples of such projected landscapes.

Ecological landscapes are also an interesting perspective for post-modern landscapes, for ecosystem people, in general, for territories having difficulties with the modern development model. For example, in high HDI countries, shrinking cities, especially present in high HDI countries might be favorable landscapes to try such new ecological landscapes, offering the possibility for these to become at the forefront of the ecological transition rather than to be the bandwagon of modernity.

Production modes should correspond to agro-ecology in regards to agriculture, where the primary issue is the use of ecological functions involving local and/or specialist's species, to minimize agriculture environmental impacts (Table 11.2). That concerns small farming, in general farmers belonging to post-modern landscapes. The way farmers manage biodiversity, what matters to them, for example, in regards to biological control, becomes a central issue to prioritize EBV to monitor in such 'ecological' landscapes.

11.5 Further Considerations on These Four Kinds of Landscapes

Having described what would be the major features of the typology we propose, we can now make further comparisons between these four kinds of landscapes, which combine biophysical and human considerations.

The notion of 'human footprint' (HF), which measures the human pressure of a given landscape—disregarding to whom such footprint benefits, local or remote people—can complete the information brought by the EF, exemplifying differences between these four kinds of landscapes. In landscapes in 'high HDI' countries, EF should be large and HF low. At the opposite, in landscapes of 'low HDI' countries, EF should be comparatively lower and HF higher (see Fig. 11.3). Traditional landscapes should have both low footprints, while ecological landscapes have the objective to lower both footprints compared to modern and post-modern landscapes, combining lower consumption, and improved relations between human modes of production and consumption, with the support of the functioning of ecosystem, in particular through agro-ecology. The type of agriculture developed might vary with the landscape (Table 11.3). Notice that post-modern and ecological agricultures might converge, as suggested above, although the pathway and origins might differ.

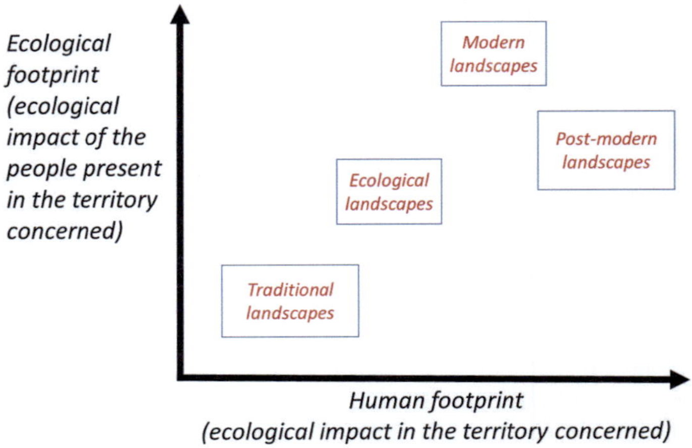

Fig. 11.3 Suggested classification of different landscapes, according to ecological footprint and human pressure

11.5.1 Landscapes and 'Shared Socio-Economic Pathways' (SSP)

To consider the possible future of the planet, 'shared socio-economic pathways' (SSP) have been constructed. Five major SPP are identified (O'Neill et al. 2017), varying in terms of expected human demography, social inequality, economic growth, human environmental impacts… (see Fig. 11.4). SSP2 is at the middle between the four other SSPs.

As landscapes differ in terms of ecological and human footprint, SSP also exhibits the same type of variations. Relating landscapes to SSP—which are constructed at a world level—offers a possibility to relate processes occurring at the level of local territories with regulation occurring at the planetary level, that is to relate top-down and bottom-up processes. Based on simulation results on SSP

Table 11.3 Different kind of environmentalities for the different types of landscapes

Type of landscape	Type of agriculture	Type of environmentality
Historical	No (or traditional types)	Bio power, sovereignty, and truth
Modern	Industrial, intensive (associated to land-sparing, Schneider 2008)	Discipline and neoliberalism
Ecological	Agro-ecology, including organic (associated to land-sharing, Schneider 2008)	Geared around deliberative citizen democracy
Post-modern	Small farming, peasant agriculture	Discipline and neoliberalism

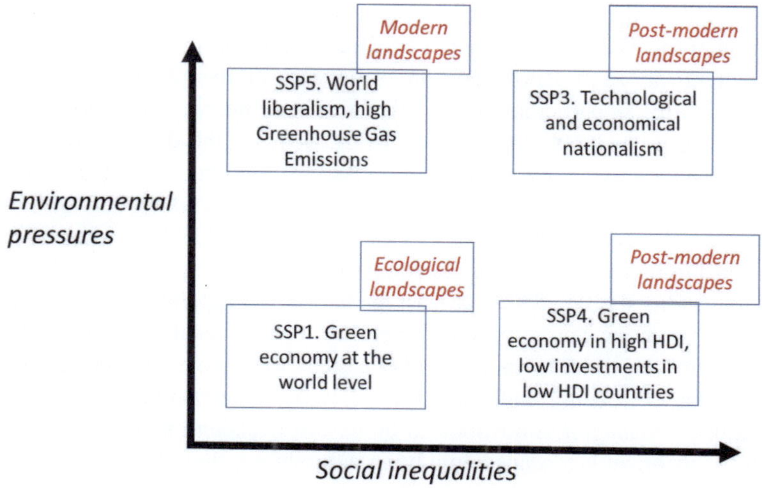

Fig. 11.4 Correspondence between SSP, inequalities and planetary limits and type of landscapes

(O'Neill et al. 2017), one can propose indeed preferential associations, along two axes that differentiate SSP, environmental pressure and social inequalities (Fig. 11.4). Historical and ecological landscapes can be related to SSP1, the sustainable pathway. Modern and post-modern landscapes are rather associated with other pathways, the conflictual intensity varying between among pathways, being more intense in SSP3 and SSP4, corresponding especially to post-modern landscapes. At the opposite, SSP5 assume, implicitly, a certain landscape convergence at the world level, corresponding to the 'modern' landscape perspective.

11.5.2 *Relationships Between Landscapes and Environmentalities*

Beyond footprints variations, diversity of relations, accordance, with SSP, the different landscapes might correspond to different norms. Modern landscapes would be rather associated with industrial norms, efficiency. Ecological landscapes are rather associated with environmental norms, post-modern landscapes being dominated by social inequalities. Overall, norms, speeches, discourses, philosophies, and scientific analyses should differ, corresponding to different governmentalities operating in the domain of the environment, or environmentalities (Fletcher 2010).

As Fletcher (2010) identified five kinds of environmentalities, one can indeed preferentially associate these to the different landscapes. For example, internalization of norms and rules, and top-down ruling are two kinds of environmentalities probably preferentially associated with 'modern' and 'post-modern' landscapes (Table 11.3).

Ecological landscapes could correspond to the perspective of liberal environmentalism proposed by Fletcher (2010), for the following reasons. Major improvements in such landscapes are necessary in regards to governance, or environmentality. In particular, deliberative citizen democracy ought to improve the quality of the procedures. As proposed by A. Sen, for the deliberation to be satisfactory, requires two conditions; (1) a diversity of options, so that each actor can have a multiplicity of choices, are not confronted with false dilemmas with a TINA ('There is no alternative') discourse. (2) Each option should take into account many social and biological dimensions, beyond wealth, health, quality of life. Overall, these dimensions should address at least the ten central capabilities proposed by M. Nussbaum, based on Sen's work (Couvet 2017). As a result, citizen science will have an increasing role, to multiply alternatives, to explore different dimensions. Citizen science is being understood in a wide understanding from gathering information to managing innovation to be part of decision-making.

Finally, historical landscapes, involving naturalness, should rather be associated with environmentalities involving biopower, sovereignty, and truth (Table 11.3).

Further questions should be considered in regards to the relation between landscapes and environmentalities. (1) A same landscape might be seen from different perspectives, for example, from a post-modern and an ecological viewpoint. The interactions between these different perspectives might bring new perspectives. (2) The interactions between these different kinds of landscapes might affect their dynamics, as they should coevolve, in particular, due to their interactions through GCC. (3) The relation that people establish with these different landscapes might change with time, a question associated with the previous one.

11.5.3 Traditional Ecological Knowledge and Environmentalities

In regards to environmentalities, traditional ecological knowledge (TEK) can have a major importance in post-modern and ecological landscapes. TEK are supposed to take into account all kind of knowledge, especially vernacular knowledge, in particular of indigenous people. In complement, the notion of 'nature contribution to people' (NCP) should help to analyze the complex, multiple connections, of local people with nature (Díaz et al. 2018)—although, from a scientific analysis viewpoint, the notion of nature can be considered as rather vague, and should be related to the notion of biodiversity and geodiversity. The notion of NCP is close to the notion of ecosystem services, developed by conservation biology to pursue a comparable objective, to integrate more closely biodiversity and geodiversity in human logics and decisions. A shared advantage of these notions, NCP or ecosystem services, is to go beyond the notion of natural resources. These concepts emphasize that biodiversity, nature, are not only resources for humans, but are necessary for every aspect of human life, have cultural, spiritual value. These notions also underline there can be different human viewpoints of a same landscape, in particular of what would be preferable for humans.

11.6 Mandatory Conclusions

1. Building citizen capacities in regards to landscape management, in terms of information and objectives
2. Taking into account the way different citizen, stakeholders relate to landscapes, citizens having a critical role in terms of expertise capacity, scrutinizing public policies.
3. Recognizing different types of landscapes, depending on the state of biodiversity, ecological and human footprints, social inequalities, so that people can understand local issues, how these local issues are related to global ones.
4. Improving the quality of democratic procedures in regards to landscapes policies, improving environmentalities.
5. Proposing a diversity of options, each taking into account many social and biological dimensions, beyond wealth, health, quality of life, integrating all central capabilities (Couve 2017).

References

Alcott B (2005) Jevons' paradox. Ecol Econ 54(1):9–21

Barnosky AD, Hadly EA, Gonzalez P, Head J, Polly PD, Lawing AM, Blois J (2017) Merging paleobiology with conservation biology to guide the future of terrestrial ecosystems. Science 355(6325):eaah4787

Carey J (2016) News feature: crucial role of belowground biodiversity. Proc Natl Acad Sci 113 (28):7682–7685

Chaplin-Kramer, R, Sim S, Hamel P, Bryant B, Noe R, Mueller C, Rigarlsford G, Kulak M, Kowal V, Sharp R, Clavreul J, Price E, Polasky S, Ruckelshaus M, Dailyet G (2017) Life cycle assessment needs predictive spatial modelling for biodiversity and ecosystem services. Nat Commun 8. https://doi.org/10.1038/ncomms1506

Couvet D, Devictor V, Jiguet F, Julliard R (2011) Scientific contributions of extensive biodiversity monitoring. C R Biol. 334:370–377

Couvet D (2017) From biodiversity to policies to politics. In: Peterson SA, Somit A (eds) Handbook of biology and politics. Edward Elgar pub

Cumming GS, von Cramon-Taubadel S (2018) Linking economic growth pathways and environmental sustainability by understanding development as alternate social–ecological regimes. Proc Natl Acad Sci 115(38):9533–9538

De Witt A, Osseweijer P, Pierce R (2017) Understanding public perceptions of biotechnology through the "Integrative Worldview Framework". Public Underst Sci 26(1):70–88

Desquilbet M, DorinB Couvet D (2016) Land sharing vs land sparing to conserve biodiversity: how agricultural markets make the difference. Environ Model Assess. https://doi.org/10.1007/s10666-016-9531-5

Díaz S, Pascual U, Stenseke M, Martín-López B, Watson RT, Molnár Z, Polasky S (2018) Assessing nature's contributions to people. Science 359(6373):270–272

Dorin B (2017) India and Africa in the global agricultural system (1960–2050): towards a new sociotechnical regime? Econ Polit Wkly LII(25–26):5–13

Dornelas M, Gotelli NJ, McGill B, Shimadzu H, Moyes F, Sievers C, Magurran AE (2014) Assemblage time series reveal biodiversity change but not systematic loss. Science 344 (6181):296–299

Fletcher R (2010) Neoliberal environmentality: towards a poststructuralist political ecology of the conservation debate. Conserv Soc 8(3):171–181

Fouilleux E, Nicolas B, Alpha A (2017) 'Feeding 9 billion people': global food security debates and the productionist trap. J Eur Public Policy 1–20

Gadgil M (2013) Poverty and biodiversity. In: Levin SA (ed) Encyclopedia of biodiversity. Academic Press, London

Gereffi G (1996) Global commodity chains: new forms of coordination and control among nations and firms in international industries. Competition Chang 1(4):427–439

Hill SLL, Harfoot M, Purvis A, Purves DW, Collen B, Newbold T, Burgess ND, Mace GM (2016) Reconciling biodiversity indicators to guide understanding and action. Conserv Lett 9(6):405–412

Julliard R, Jiguet F, Couvet D (2004) Common birds facing global changes: what makes a species at risk? Glob Change Biol 10:148–154

Keith DA, Rodríguez JP, Rodríguez-Clark KM, Nicholson E, Aapala K, Alonso A, Asmussen M, Bachman S, Basset A, Barrow EG, Benson JS, Bishop MJ, Bonifacio R, Brooks TM, Burgman MA, Comer P, Comín FA, Essl F, Faber-Langendoen D, Fairweather PG, Holdaway RJ, Jennings M, Kingsford RT, Lester RE, Mac Nally, R, McCarthy MA, Moat J, Oliveira-Miranda MA, Pisanu P, Poulin B, Regan TC, Riecken U, Spalding MD, Zambrano-Martínez S (2013) Scientific foundations for an IUCN red list of ecosystems. PLOS one 8(5):e62111 (2013)

Locey KJ, Lennon JT (2016) Scaling laws predict global microbial diversity. Proc Natl Acad Sci 2016:2015–21291

Mora C, Caldwell IR, Caldwell JM, Fisher MR, Genco BM, Running SW (2015) Suitable days for plant growth disappear under projected climate change: potential human and biotic vulnerability. PLoS Biol 13(6):e1002167

O'Neill BC, Kriegler E, Ebi KL, Kemp-Benedict E, Riahi K, Rothman DS, Levy M (2017) The roads ahead: narratives for shared socioeconomic pathways describing world futures in the 21st century. Glob Environ Change 42:169–180

Pereira HM, Ferrier S, Walters M, Geller GN, Jongman RHG, Scholes RJ, Coops NC (2013) Essential biodiversity variables. Science 339(6117):277–278

Schmeller DS, Böhm M, Arvanitidis C, Barber-Meyer S, Brummitt N, Chandler M, Chatzinikolaou E, Costello MJ, Ding H, García-Moreno J, Gill M, Haase P, Jones M, Juillard R, Magnusson WE, Martin CS, McGeoch M, Mihoub J-B, Pettorelli N, Proença V, Peng C, Regan E, Schmiedel U, Simaika JP, Weatherdon L, Waterman C, Xu H, Belnap J (2017) Building capacity in biodiversity monitoring at the global scale. Biodivers Conserv 2017:1–26. https://doi.org/10.1007/s10531-017-1388-7

Schneider S (2008) Good, clean, fair: the rhetoric of the slow food movement. Coll Engl 70 (4):384–402

Sukhdev P, May P, Müller A (2016) Fix food metrics. Nat News 540(7631):33

Sutton PC, Anderson SJ, Costanza R, Kubiszewski I (2016) The ecological economics of land degradation: impacts on ecosystem service values. Ecol Econ 129:182–192

Tsing AL, Swanson HA, Gan E, Bubandt N (eds) (2017) Arts of living on a damaged planet: ghosts and monsters of the Anthropocene. University Minnesota Press, 376 p

Chapter 12
Learning Landscape Approach Through Evaluation: Opportunities for Pan-European Long-Term Socio-Ecological Research

Per Angelstam, Michael Manton, Fatima Cruz, Mariia Fedoriak and Yurij Pautov

Abstract Sustainable development as a societal process aimed at securing sustainability is challenging. To encourage the necessary knowledge production and learning in different social-ecological contexts requires a place-based networking research infrastructure that involves multiple academic disciplines and non-academic actors. Long-term socio-ecological research (LTSER) platform is one approach with $\sim$80 initiatives globally. To encourage transdisciplinary learning through evaluation we defined a normative model for ideal performance at both local platform and network levels. Four surveys were then sent out to 67 self-reported LTSER platforms. Focusing on the network level, we analyzed the spatial distribution of both long-term ecological monitoring sites within LTSER platforms, and LTSER platforms across

P. Angelstam (✉)
School for Forest Management, Faculty of Forest Sciences, Swedish University of Agricultural Sciences (SLU), PO Box 43, 73921 Skinnskatteberg, Sweden
e-mail: per.angelstam@slu.se

M. Manton
Faculty of Forest Science and Ecology, Vytautas Magnus University, Studentu g. 13, 53362 Akademija, Kaunas, Lithuania
e-mail: michael.manton@vdu.lt

F. Cruz
Sustainable Forest Management Research Institute, University of Valladolid, Campus de la Yutera. Av. de Madrid 44, 34071 Palencia, Spain
e-mail: fatimaregina.cruz@uva.es

M. Fedoriak
Department of Ecology and Biomonitoring, Chernivtsi National University, 2 Kotsyubynski street, Chernivtsi 58029, Ukraine
e-mail: m.m.fedoriak@gmail.com

Y. Pautov
Silver Taiga Foundation for Sustainable Development, PO Box 810, 167000 Syktyvkar, Russia
e-mail: ypautov@komimodelforest.ru

L. Müller and F. Eulenstein (eds.), *Current Trends in Landscape Research*, Innovations in Landscape Research,
https://doi.org/10.1007/978-3-030-30069-2_12

the European continent. Finally, narrative biographies about 18 LTSER platforms in different stages of development were analyzed. While the siting of LTSER platforms represented biogeographical regions well, variations in land-use history and governance arrangements were poorly represented. Ecosystem research (72%) dominated social system research (28%). Maintenance of a platform required 3–5 staff members, was based mainly on national funding and had 1–2 years of future funding secured. Networking with other landscape approach concepts was common. Individually, and as a network, LTSER platforms have good potential for transdisciplinary knowledge production and learning about sustainability challenges. To benefit from the large range of variation among Pan-European social-ecological systems, we encourage collaboration among different landscape approach concepts such as LTSER platform and Model Forest, ecological reference landscapes like zapovedniks as well as traditional systems for landscape stewardship.

Keywords European continent · Landscape approach · Learning through evaluation · LTSER platform · Model forest · Social-ecological system · Stakeholder engagement · Transdisciplinary research

12.1 Introduction

As a complement to the ecosystem services approaches in land-use policy, governance and planning, implementation on the ground requires skills to navigate the complexity of interactions within landscapes as social-ecological systems. It is essential to focus both on sustainable development as a societal process (Baker 2006), and on ensuring sustainability in social-ecological systems. Landscape is a well-established concept that can aid knowledge production and learning by fostering transdisciplinary knowledge production and learning (Angelstam et al. 2013). This requires integration of researchers and other knowledge producers representing different disciplines, as well as stakeholders representing different sectors at multiple levels (Termorshuizen and Opdam 2009).

To maintain natural capital through functional green (ecological) infrastructure (e.g., European Commission 2013), thereby enhancing human well-being, modified landscapes often require capacity-building in social systems, and action through conservation, management and restoration in ecological systems. To scale up research and development in support of sustained delivery of ecosystem services is a challenging task (Angelstam et al. 2017a). It requires identification of the acceptable level of modification of the biophysical environment (e.g., Manton and Angelstam 2018), place-based co-ordination of human management of land and water resources, as well as engaging and incentivizing stakeholders and actors to act sustainably (e.g., Dawson et al. 2017). The general term landscape approach captures this complex web of interactions (Axelsson et al. 2011; Sayer et al. 2015; Angelstam et al. 2019a). A wide range of landscape approach concepts aimed toward place-based knowledge production and engaged stakeholder collaboration have emerged (Angelstam and Elbakidze 2017). One such concept is the long-term socio-ecological research (LTSER) platform (e.g., Haberl et al. 2006, 2009; Singh et al. 2013; Gingrich et al. 2016; Bretagnolle et al. 2018). Currently, there are ~80 LTSER platform initiatives globally (Mirtl et al. 2018). The LTSER network emerged as a bottom-up process, where existing local and national initiatives became recognized at the European level (Singh et al. 2013). Enhancing collaboration among LTSER platforms at the international level is the next desirable level of ambition toward using multiple landscapes as a laboratory (Angelstam et al. 2019a, b; Holzer et al. 2018).

While landscape approach is commonly advocated, and implementation of initiatives is often highlighted as success stories, formal audits against a norm that states what should be delivered are rarely made. Hence, it is difficult to assess the contributions of different landscape approach concepts in terms of what they actually deliver on the ground. The term learning through evaluation concept captures this challenge (Lähteenmäki-Smith 2007).

The aim of this chapter is to report on a recent audit by Angelstam et al. (2019a) about the extent to which European LTSER platform initiatives live up to the LTSER platform concept's own norms developed for place-based knowledge production and learning toward sustainable landscapes. First, we defined a normative model for the ideal performance of LTSERs at platform and network levels. Second, we analyzed the location of platforms across Pan-European gradients of biophysical, anthropogenic and intangible interpretations of landscape. Third, we sent out four surveys with increasing complexity to 67 self-reported LTSER platforms. Fourth, we compiled narrative biographies for 18 LTSER platforms in different development stages. The discussion focuses on two dimensions. First, how can landscape approach concepts such as LTSER can be sustained as local hubs of problem-solving landscape laboratories, and how they can form a research infrastructure. Second, we advocate learning among different landscape approach concepts such as LTSER platform and Model Forest (Angelstam et al. 2019b), ecological reference landscapes like zapovedniks (Shtilmark 2003) and traditional approaches to landscape stewardship in rural landscapes (Angelstam and Elbakidze 2017; Fedoriak et al. 2019).

12.2 Methods and Materials

12.2.1 Normative Model

To assess performance of individual LTSER platforms, we developed a normative model by integrating Grove's et al. (2013) architectural metaphor of "siting, construction and maintenance" of individual LTSER platforms, and Mirtl's et al. (2013) triangle of region and actors (i.e., landscape as a coupled social-ecological system), research, infrastructure and co-ordination (Fig. 12.1). This approach

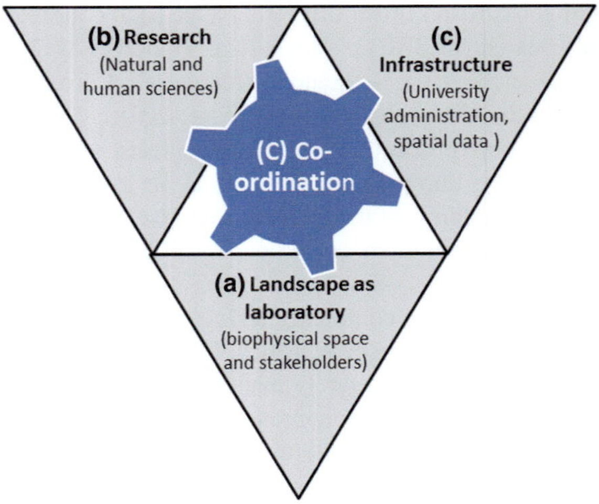

Fig. 12.1 Landscape approach according to the architectural metaphor of the LTSER platform concept (Grove et al. 2013; Mirtl et al. 2013) involves **a** siting a landscape as a socio-ecological system laboratory and engaging stakeholders in knowledge production and learning, **b** constructing by integrating researchers from different disciplines and securing an infrastructure for collecting and analyzing quantitative and qualitative data and **c** maintaining continuous facilitation and co-ordination to sustain transdisciplinarity (see Table 12.1)

Table 12.1 Criteria and indicators developed by Angelstam et al. (2019a) for the assessment of LTSER platforms as (1) a place-based research infrastructure based on Grove's et al. (2013) architectural metaphor "siting-construction-maintenance" (A, B, C) for individual platforms and (2) for LTSER platforms as a distributed network (D)

Criterion	Indicator
Siting (A)	1. Ecoregional representation
	2. Representation of anthropogenic change
	3. Representation of intangible interpretations
Construction (B)	4. Human versus natural science research
	5. LTER sites in LTSER platforms
	6. Stakeholder structure (at least five partners)
	7. Land ownership structure
Maintenance (C)	8. Number of full-time workers
	9. How funding is spent on main functions
	10. Funding sources
	11. Duration of secured future funding
	12. Past survival
Network (D)	13. Reaction frequency
	14. Response time
	15. Opportunity for socio-ecological analyses
	16. National support

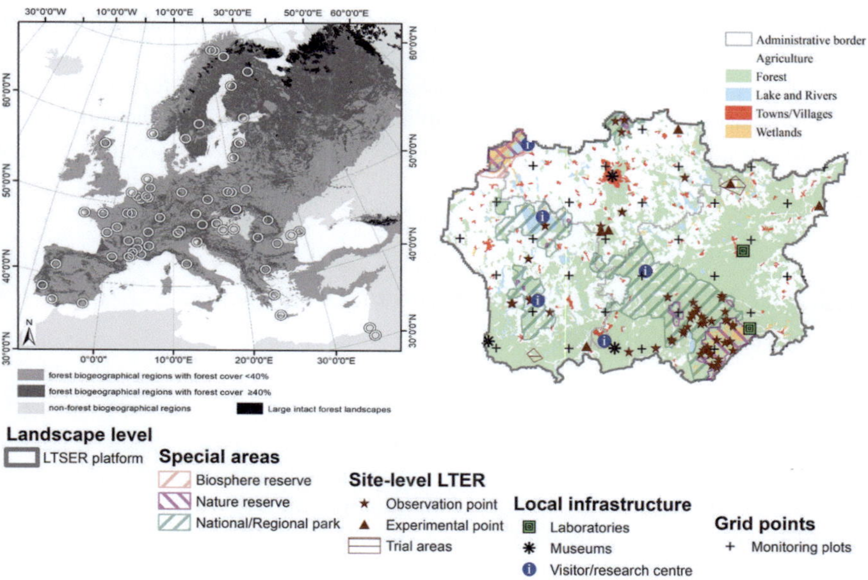

Fig. 12.2 LTSER platforms provide opportunity to address large spatial extents and both ecological and social system research topics (Metzger et al. 2010), as well as how they interact (Singh et al. 2013). The figure illustrates how 67 LTSER platforms form a multi-level place-based research infrastructure in Europe. Nested within LTSER platforms (from left to right) in a fictive LTSER platform there are special areas such as biosphere reserve and national park, LTER (Long-Term Ecological Research) sites that focus on ecosystem monitoring comprising highly instrumented Master Sites, Regular Sites and Satellite Sites, as well as local infrastructure and monitoring grid points (e.g., Mirtl et al. 2013 p. 417). Globally, this research infrastructure comprises ~80 long-term socio-ecological research (LTSER) platforms and ~700 LTER sites (Mirtl et al. 2018)

resulted in four criteria and generation of 16 indicators for which verifier variable data were collected (Table 12.1).

To visualize the locations of LTSER platforms, we created a standardized platform area of 10,000 km^2 (Fig. 12.2). The rationale is three-fold. First, this size is indicated in LTSER guidelines (100–10,000 km^2, Mirtl et al. 2008). Second, from an ecological point of view, we relied on the area requirements of focal species (Lambeck 1997), meaning that a planning unit should be in the order of 1000–10,000 km^2 (Angelstam et al. 2004 p. 435). Third, from a social system perspective people generally do not commute more than 1.5 h per day, i.e., corresponding to ca 50–60 km one-way travel distance by car or train (e.g., Lindelöw 2018), hence also ca. 10,000 km^2.

To collect verifier variable data that matched the 16 indicators in Table 12.1 we sent out four surveys. Survey-1 aimed at identifying the individuals responsible for LTSER platform co-ordination, ecological system research and social system research in each platform. Survey-2 focused on characterizing the construction and maintenance of an LTSER platform. Survey-3 was designed as an online Web tool which LTSER platforms could use to check that the GIS polygon of their platform was correct, and if needed draw or adjust its shape directly. Finally, Survey-4 focused on evaluating the extent to which and how LTSER platforms work with green infrastructure as a key transdisciplinary topic to address the supply and provision of ecosystem services in the LTSER platform areas as social-ecological systems.

12.2.2 LTSER Platforms in Different Development Stages as Case Studies

We also collected case study narratives for 18 LTSER platforms (Table 12.2). Having emerged as bottom-up initiatives in different settings, today's LTSER platforms represent a wide gradient from those just interested in embarking on the LTSER concept, and to those that have been active for >15 years. The case study narratives were structured by the four assessment criteria (1) siting, (2) construction and (3) maintenance of individual platforms on one hand and cross-platform (4) networking on the other (Table 12.1). From these structured narratives, we extracted and summarized the key themes for each criterion. In the results section quantitative indicators are reported first, and then the results from the case study narratives.

12.3 Results

12.3.1 Siting

The 67 LTSER platforms represented 23 countries. In terms of biophysical interpretation of landscape, there was good coverage of LTSER platforms in the Alpine, Boreal, Atlantic, Continental and Mediterranean biogeographical regions (Indicator A1). Gradients of anthropogenic land cover change are an important feature allowing design of comparative studies of the effects of anthropogenic factors on ecosystem services, such as across LTSER platform areas. Forest is the most widespread potential natural land cover in Europe and ranges from lost to present but modified and intact forest landscapes. The location of LTSER platforms only in the western half of the European continent means that intact forest landscapes that can be used as reference landscapes (e.g., Naumov et al. 2018) are missing (Indicator A2). Similarly, an example of an intangible interpretation of the landscape concept, countries with the full range of governance arrangements values were not represented, thus missing important constituent social system variables (Indicator A3).

Table 12.2 Case study narratives were collected for 18 LTSER platforms in different stages of development and analyzed with respect to the four LTSER platform criteria siting, construction, maintenance and networking (see Table 12.1)

LTSER platform (code according https://data.lter-europe.net/deims/site/)	Country	Years of operation as a LTSER platform (up to 2019)	Area (km^2)	Number of local administrative units
Waddensee (lter_eu_nl_001)	Netherlands	3	6155	17
Engure (lter_eu_lv_01)	Latvia	9	644	5
Oracle/BVRE Orgeval (Seine river basin), (rbv_fr_05; part of lter_eu_fr_002)	France	30	1200/ 78,000	16/8400
Negev Highlands (lter_eu_il_017)	Israel	5	1700	2
Roztochya (lter_eu_ua_004)	Ukraine	Potential	280	10
Tovel lake (lter_eu_it_090)	Italy	Potential	90	1
Doñana LTSER Platform (lter_eu_es_001)	Spain	11	2736	12
Plaine and Val de Sevre (lter_eu_fr_009)	France	19	435	24
Poloniny National Park (lter_eu_sk_010)	Slovakia	2	342	10
Braila Island (lter_eu_ro_006)	Romania	18	2597	25
Bergslagen (lter_eu_se_001)	Sweden	8	44,000	40
Eizenwurzen (lter_eu_at_001)	Austria	15	5904	91
Lithuanian coastal site (lter_eu_lt_004)	Lithuania	Abandoned	NA	NA
Helsinki (lter_eu_fi_002)	Finland	13	745	3
Arava (lter_eu_il_016)	Israel	10	1650	1
Baixo Sabor (lter_eu_pt_002)	Portugal	10	1590	5
Montado-Alentejo (lter_eu_pt_001)	Portugal	8	32,700	53
Chernivtsi region	Ukraine	Potential	8097	11

The qualitative results from the 18 narratives of LTSER platform initiatives illustrate the European diversity of local and regional social-ecological contexts. Three groups of landscape types had LTSER platforms. The first was rural agricultural landscapes in different development stages. The second involved river catchments and coastal areas, both of which forming gradients from urban settings including urbanization and industrial decline via agricultural areas in different socio-ecological transition, and to protected areas of different kinds. The third group was formed by historic informal regions in steep socio-ecological gradients with complex governance legacies linked to different land ownerships and landscape histories, both within countries and in cross-border regions.

12.3.2 Construction

Survey-1 showed that the three LTSER platform functions co-ordination, responsibility for ecological and social science research, respectively, was served by one person [39% of the platforms (n = 28)], two persons (50% of the platforms) and three persons (11% of the platforms). Research on ecological systems (73%) outnumbered research on social systems (27%). All LTSER platforms had at least one LTER site. Concerning the profile of stakeholder participation in spatial planning for biodiversity conservation and human well-being (indicator B6) and land ownership (indicator B7) there was a clear focus on the local and regional levels.

The narratives showed that the construction of LTSER platforms included both top-down to bottom-up approaches. The first came out of national-level competitive initiatives to develop LTSER platforms with the aim to enable integrative research about ecosystem services together with stakeholders. The second group was formed by national parks, municipalities and regional planning units that realized the need for integrated land planning to cope with socio-economic pressures on landscapes as social-ecological systems, and biodiversity conservation through promotion of sustainable landscape development and integrated planning. The third group was LTSER platforms the establishment of which was triggered by concrete drivers for knowledge production and learning bottom-up, such as declining human population and demographic challenges in rural areas, need for landscape restoration, a severe flooding event, securing water quality, river damming for hydroelectric use, decline of charismatic focal farmland birds and threats to beekeeping. These initiatives led to monitoring projects, later evolving into research projects at local, regional and international levels, which over time did or may transform into transdisciplinary research gathering ecologists, economists and social scientists as well as stakeholders.

12.3.3 Maintenance

The LTSER platforms' most frequently mean number of full-time workers (indicator C8) was 3–5 persons, but almost the same proportion of the respondents said 1–2 persons (n = 29). Research (ecological 29%, social 11%) accounted for the largest funding expenditure for the platforms (indicator C9). Other expenses included data collection (26%), co-ordination (16%), travels in the field (12%) and stakeholder engagement (7%) (n = 29). Regarding funding sources (indicator C10) almost 90% of the LTSER platforms relied on national grants, and about 50% of the platforms were supported by EU grants as well as from regional sources (n = 29). The "duration of secured funding" (indicator C11) was short-term. Most commonly funding was secured for only 1–2 years (37%) in advance. About 30% of the LTSER platforms reported funding for the next 3–5 years (n = 29). "Long-term survival" (indicator C12) was assessed by comparing data for 2010 with those for 2016. Of the 30 platforms listed in 2010, only three had disappeared by 2016.

Analyzing the narratives, we identified three mechanisms to sustain a more or less loose researcher-stakeholder network as a key foundation for a LTSER platform. First, some platforms had permanent staff based at a national research institute or university, with a desire to do LTSER platform work. National and regional funding in successive shorter periods was frequently complemented with mainly disciplinary short-term projects. Long-term monitoring of biodiversity and socio-economic data were key assets in emerging, young and long-lived LTSER platforms. Second, skills to identify key topics, and to write proposals to secure and sustain multiple minor short-term sources were combined with a patchwork of disciplinary research, post-graduate and consulting projects. Wise integration of funding for research and stakeholder engagement can transition into transdisciplinarity. Third, to survive some LTSER platforms exercise opportunistic use of short-term research funding through participation in national and EU-projects, however, this may be insufficient to sustain desired monitoring initiatives and to allow time for preparing peer-review publications.

12.3.4 Networking

To assess the reaction frequency to the four surveys (indicator D13) at the network level we analyzed their response rates. Surveys 1–4 were answered by 28, 29, 21 and 14 respondents, respectively. In total, 43 LTSER platforms responded to at least one of the four surveys. LTSER platforms that had delivered polygons in Survey-3 had a significantly higher response rate. To assess LTSER platforms as a communicating network of place-based research, we used as a proxy the frequency of occurrence of LTSER platforms that responded to four different surveys and how fast they responded to them in relation to the requested 14 day limit (Indicator D14). In the first survey, all but one of the 28 platforms that responded met this requirement. With increasing survey complexity response times became longer. The opportunity for analyzing socio-economic data collected at the level of public administrative units (Indicator D15) was estimated by comparing our estimates of how large (i.e., 10,000 km^2) a sufficiently sized LTSER platform ought to mirror the size of a sufficiently large areas that reflects both ecological and social system analyses comprehensively on one hand, and the size of administrative units at different levels of governance on the other. Of the 43 LTSER platforms for which a polygon could be attributed, a total of 18 were 1000–10,000 km^2 in size and 8 met the requirement of 10,000 km^2. For those 43 platforms with boundaries in DEIMS-SDR, a total of 50% were supported by their respective host countries.

Finally, networking with other LTSER platforms, but also with other landscape approach concepts, was widespread. This included Model Forest, EU LEADER, Ramsar, UNESCO Biosphere Reserve, UNESCO Global Geopark, Zone Atelier, World Heritage Site, zapovednik and, as well as a wide range of professional and researcher networks.

12.4 Discussion

12.4.1 Comparisons with the Normative Model for LTSER Platforms

Long-term socio-ecological research (LTSER) emerged in response to the recognition of increased effects of human activities on sustainability at local to global levels (Singh et al. 2013; Mirtl et al. 2018). These challenges can often be considered as wicked (Duckett et al. 2016), which calls for transdisciplinary knowledge production and learning (Angelstam et al. 2013; Holzer et al. 2018). The LTSER platform concept aims at being a place-based infrastructure that supports collaborative knowledge production and learning by academic and non-academic participants (Haberl et al. 2006; Singh et al. 2013). As a network, the ambition is to develop context-specific solutions by drawing upon multiple LTSER platforms representing biophysical, anthropogenic and intangible properties of landscapes as social-ecological systems (e.g., Matthews and Selman 2006; Metzger et al. 2010). Based on mixed quantitative and qualitative methods Table 12.3 provides an overview of the compliance with the normative model.

The siting of LTSER platforms affects the opportunity to design macroecological research, natural experiments and comparative studies of government, governance and politics. From this point-of-view, the European LTSER platforms represent the socio-ecological diversity within the EU well (Metzger et al. 2010). However, parts of important socio-economic, landscape history and governance gradients that exist on the entire Pan-European continent are missing from this network. For example, many post-Soviet countries are missing (e.g., Russia, Belarus, Ukraine), and some platforms are no longer functioning (e.g., Lithuania). The absence of large intact

Table 12.3 Opportunities for improvement of LTSER platforms' performance as place-based research infrastructure

Criterion	Indicator	Opportunities for improvement
Siting (A)	1. Ecoregional representation	–
	2. Representation of anthropogenic change	Include reference areas representing ecological integrity and resilience
	3. Representation of intangible interpretations	Include wider gradients of governance and political cultures
Construction (B)	4. Human versus natural science research	Strengthen the role of humanities and social sciences
	5. LTER sites in LTSER platforms	Encourage macroecological approaches
	6. Stakeholder structure	–
	7. Land ownership structure	–
Maintenance (C)	8. Number of full-time workers	–
	9. How funding is spent on main functions	Increase proportion funding spent for humanities and social sciences
	10. Funding sources	Funding at EU-level need to support also local LTSER platforms
	11. Duration of secured future funding	Need to encourage longer-term funding
	12. Past survival	–
Network (D)	13. Reaction frequency	Develop incentives for cross-platform collaboration
	14. Response time	Develop incentives for cross-platform collaboration
	15. Opportunity for regional meta-analyses	Compare macroecological and socio-economic data
	16. National support	Increased support from north and east European countries

forest landscapes as reference areas for ecological integrity, and of regions with clearly top-down governance with low levels of democracy are two examples. The first example can be illustrated by studies on the effect of habitat amount and configuration on biodiversity (e.g., Roberge et al. 2008), and the effects of large herbivores on trees (Angelstam et al. 2017b). Similarly, comparative studies of different governance arrangements illustrate that different societal trajectories require solutions that are adapted to both stakeholder engagement patterns and spatial planning legacies (e.g., Elbakidze et al. 2010, 2013).

At the network level, a critically important dimension of a landscape approach research infrastructure is that it covers wide gradients in all dimensions of landscape. First, this involves capturing the full gradient of ecosystem state (Naumov et al. 2018; Manton and Angelstam 2018). Second, the network needs to cover a wide range of social systems, such as from bottom-up democratic governance to top-down authoritarian (e.g., Elbakidze et al. 2010). To achieve this, the LTSER network needs to establish platforms outside the EU as well as collaborate with other concepts.

The Soviet zapovednik system for strictly protected areas, including monitoring data and phenological letters of nature (Shtilmark 2003), is a grand infrastructure to build on. However, problems with funding of both place-based research and other research are widespread in post-Soviet countries. For example, in Ukraine, the situation deteriorated significantly in the years after independence in 1991, Slovakia was in a similar situation before entrance to EU in 2004, and Lithuania's LTSER platform no longer functions due to the withdrawal of governmental support, although there is interest from the communities of international landscape approach concepts and local research sector to reinvigorate. Although EU funds helped to improve the situation, impacts of the previous regime can be felt. Thus, the most efficient way to develop a network in such countries with limited financial opportunities would be to implement EU-projects that are dedicated to support the establishment of a LTSER platforms network in both EU and former post-Soviet countries.

Concerning the construction of LTSER platforms, the addition of the "socio"-component to already established long-term ecological monitoring/research sites is not straightforward. This is reflected both by a dominance of ecological research according to our survey work, and a dominance of ecological research in peer-review publications (Dick et al. 2018). Nevertheless, the same study demonstrated that the trend over time is positive for social science contributions from LTSER platforms. Moving from research restricted to natural science or human science research toward transdisciplinary knowledge production through collaborating researchers, practitioners and citizens means a radical change in the way knowledge production is carried out and how infrastructure for this is built (Holzer et al. 2018). There is often epistemological and methodological friction when engaging in transdisciplinary research (e.g., Furman and Peltola 2013; Mirtl et al. 2013).

So far, the LTSER platform concept has been viewed through an ecosystem and natural science lens. To balance this, the LTSER platform concept also needs to incorporate the perspectives of social sciences and the humanities. To achieve this, mutual respect from both cultures of human and natural science research [sensu "two cultures" of Snow (1959)], respectively, is required. Developing a transdisciplinary research agenda overarching multiple LTSER platforms, and facilitating researchers' and stakeholders' ability to participate in processes of knowledge production and learning would be an important contribution.

The qualitative approach based on 18 narratives corroborates the quantitative analysis. The narratives demonstrate the long period from the appearance of a transdisciplinary idea to its realization, and that collaboration among different landscape approach concepts was common (e.g., Angelstam and Törnblom 2004; Axelsson et al. 2013; Bretagnolle et al. 2018). While on the one hand this can be an obstacle for establishing a place-based distributed network of landscape approach initiatives as research infrastructure (see http://www.lter-europe.net/elter-esfri), a positive pragmatic approach is to encourage collaboration in different constellations of LTSER platforms based on their characteristics. Finally, the size of LTSER platform areas matters. Addressing interactions between macroecological patterns and processes on the one hand and governance, political cultures and socio-economic factors on the other, requires platform areas that are sufficiently large to contain multiple territorial units that can provide such data.

12.4.2 Landscape Approach and Traditional Landscape Stewardship

Landscape approach entails a collaborative effort of researchers, stakeholders, practitioners and policy-makers toward bottom-up projects and actions to promote a sustainable development process and sustainability in their own place and region (Axelsson et al. 2011; Sayer et al. 2015). This can be called social innovation (Moulaert et al. 2005). Creative actions, social participation, collaboration among different levels of decision-making and different sectors of society are all common features of social innovation. The necessary conditions for developing place-based transdisciplinary research representing different social-ecological contexts include: (1) existence of long-term data about ecological and social systems, "compass" sensu Lee (1993), (2) sufficient time for developing collaborative capacity, "gyroscope" sensu Lee (1993) and (3) sufficient co-ordination (Angelstam and Elbakidze 2017).

The main goal of social innovation from the perspective of landscape approach is to facilitate that a place-based, permanent and renewable change takes place toward a more equitable and sustainable society. Neumeier (2012:55) defined social innovation as "changes in attitudes, behaviors or perceptions of a group of people joined in a network of aligned interests that in relation to the group's horizon of experience lead to new and improved ways of collaborative action within the group and beyond". It can thus help address important challenges for local communities and stakeholder groups. This is urgently required to address the interconnected wicked challenges of economic development, ecological integrity, and social justice that are essential components of human well-being through a stronger territorial basis (e.g., Duckett et al. 2016). This calls for revival of collective action (Primdahl et al. 2018), which can be sought both through analyses of past local collective systems for landscape stewardship, and where they remain in terms of for example different types of landscape stewardship in traditional village systems (Angelstam and Elbakidze 2017; Fedoriak et al. 2019, see Fig. 12.3).

Fig. 12.3 Our vision is that place-based initiatives working with different landscape approach concepts, and traditional forms for landscape stewardship, would exchange experiences about knowledge production and learning toward sustainable landscapes. The authors of this chapter present (1) Komi Model Forest (IMFN 2008) in Russia (initiative emerged 1997), (2) Bergslagen Model Forest and LTSER platform in Sweden (initiative emerged 1999), (3) Palencia Model Forest candidate in Spain (initiative emerged 2015), (4) Beekeeping as a form of collaborative learning in Chernivtsi, Ukraine (initiative emerged 2014/15) and (5) the Lithuanian State Forestry Enterprise, which was established in 2018 (https://www.vivmu.lt/en/), is looking for approaches to stakeholder collaboration

There is, thus, potential for integration among different landscape approach concepts and initiatives as a research infrastructure (Fig. 12.3). This would enhance the use of Pan-European gradients in biophysical, anthropogenic and intangible interpretations of landscapes for knowledge production and learning toward sustainable social-ecological systems. This needs to be matched by effective bridging of barriers in terms of competition between organizations and concepts that focus only on their own version of what a landscape approach means. We therefore encourage wide use of our systematic approach to learning through evaluation (see also Angelstam et al. 2019a, b, c).

Acknowledgements We acknowledge funding from EU Horizon 2020 for the research infrastructure project eLTER, FORMAS (project number 2017:1342) and to the Lithuanian Science Council [grant number P-MIP-17-210] for the FunGILT project.

References

Angelstam P, Barnes G, Elbakidze M, Marsh A, Marais C, Mills A, Polonsky S, Richardson DM, Rivers N, Shackleton R, Stafford W (2017a) Collaborative learning to unlock investments for functional ecological infrastructure: bridging barriers in social-ecological systems in South Africa. Ecosyst Serv 27:291–304

Angelstam P, Elbakidze M (2017) Forest landscape stewardship for functional green infrastructures in Europe's West and East: diagnosing and treating social-ecological systems. In: Bieling C, Plieninger T (eds) The science and practice of landscape stewardship. Cambridge University Press, Cambridge, pp 124–144

Angelstam P, Elbakidze M, Axelsson R, Khoroshev A, Tysiachniouk M, Pedroli B, Tysiachniouk M, Zabubenin E (2019a) Model forests in Russia as landscape approach: demonstration projects or initiatives for learning towards sustainable forest management? Forest Policy Econ 101:96–110

Angelstam P, Grodzynskyi M, Andersson K, Axelsson R, Elbakidze M, Khoroshev A, Kruhlov I, Naumov V (2013) Measurement, collaborative learning and research for sustainable use of ecosystem services: landscape concepts and Europe as laboratory. Ambio 42(2):129–145

Angelstam P, Manton M, Elbakidze M, Sijtsma F, Adamescu M, Avni N, Beja P, Bezak P, Zyablikova I, Cruz F, Bretagnolle V, Díaz-Delgado R, Ens B, Fedoriak M, Flaim G, Gingrich S, Lavi-Neeman M, Medinets S, Melecis V, Muñoz-Rojas J, Schäckermann J, Stocker-Kiss A, Setälä H, Stryamets N, Taka M, Tallec G, Tappeiner U, Törnblom J, Yamelynets T (2019b) LTSER platforms as a place-based transdisciplinary research infrastructure: learning landscape approach through evaluation. Landscape Ecol On-line 34:1461–1484. https://doi.org/10.1007/s10980-018-0737-6

Angelstam P, Munoz-Rojas J, Pinto-Correia T (2019c) Landscape interpretations and landscape approach initiatives can foster knowledge production and learning that sustain multiple ecosystem services. Landsc Ecol 34(7): 1445–1460. https://doi.org/10.1007/s10980-019-00866-z

Angelstam P, Törnblom J (2004) Maintaining forest biodiversity in actual landscapes—European gradients in history and governance systems as a "landscape lab". In: Marchetti M (ed) Monitoring and indicators of forest biodiversity in Europe—from ideas to operationality. European Forest Institute, symposium No 51. pp 299–313

Angelstam P, Manton M, Pedersen S, Elbakidze M (2017b) Disrupted trophic interactions affect recruitment of boreal deciduous and coniferous trees in northern Europe. Ecol Appl 27 (4):1108–1123

Angelstam P, Roberge J-M, Lõhmus A, Bergmanis M, Brazaitis G, Dönz-Breuss M, Edenius L, Kosinski Z, Kurlavicius P, Lārmanis V, Lūkins M, Mikusinski G, Račinskis E, Strazds M, Tryjanowski P (2004) Habitat modelling as a tool for landscape-scale conservation—a review of parameters for focal forest birds. Ecol Bull 51:427–453

Axelsson R, Angelstam P, Elbakidze M, Stryamets N, Johansson K-E (2011) Sustainable development and sustainability: landscape approach as a practical interpretation of principles and implementation concepts. J Landscape Ecol 4(3):5–30

Axelsson R, Angelstam P, Myhrman L, Sädbom S, Ivarsson M, Elbakidze M, Andersson K, Cupa P, Diry C, Doyon F, Drotz MK, Hjorth A, Hermansson JO, Kullberg T, Lickers FH, McTaggart J, Olsson A, Pautov Yu, Svensson L, Törnblom J (2013) Evaluation of multi-level social learning for sustainable landscapes: perspective of a development initiative in Bergslagen, Sweden. AMBIO 42(2):241–253

Baker S (2006) Sustainable development. Routledge, London/New York

Bretagnolle V, Berthet E, Gross N, Gauffre B, Plumejeaud C, Houte S, Badenhausser I, Monceaua K, Allier F, Monestiez P, Gaba S (2018) Towards sustainable and multifunctional agriculture in farmland landscapes: lessons from the integrative approach of a French LTSER platform. Sci Total Environ 627:822–834

Dawson L, Elbakidze M, Angelstam P, Gordon J (2017) Governance and management dynamics of landscape restoration at multiple scales: learning from successful environmental managers in Sweden. J Environ Manage 197:24–40

Dick J, Orenstein DE, Holzer JM, Wohner C, Achard AL, Andrews C, Avriel-Avni N, Beja P, Blond N, CabelloJ Chen C, Díaz-DelgadoR Giannakis GV, Gingrich S, Izakovicova Z, Krauze K, Lamouroux N, LecaS MelecisV, Miklós K, Mimikou M, Niedrist G, Piscart C, Postolache C, Psomas A, Santos-Reis M, Tappeiner U, Vanderbilt K, Van Ryckegem G (2018) What is socio-ecological research delivering? A literature survey across 25 international LTSER platforms. Sci Total Environ 622:1225–1240

Duckett D, Feliciano D, Martin-Ortega J, Munoz-Rojas J (2016) Tackling wicked environmental problems: the discourse and its influence on practice in Scotland. Landscape Urban Plann 154:44–56

Elbakidze M, Angelstam P, Sandström C, Axelsson R (2010) Multi-stakeholder collaboration in Russian and Swedish Model Forest initiatives: adaptive governance towards sustainable forest management? Ecol Soc 15(2):14

Elbakidze M, Hahn T, Mauerhofer V, Angelstam P, Axelsson R (2013) Legal framework for biosphere reserves as learning sites for sustainable development: a comparative analysis of Ukraine and Sweden. Ambio 42(2):174–187

European Commission (2013) Green Infrastructure (GI)—Enhancing Europe's Natural Capital. COM 249. European Commission, Brussels

Fedoriak MM, Angelstam PK, Kulmanov OL, Tymochko LI, Rudenko SS, Volkov RS (2019) Ukraine is moving forward from 'Undiscovered honey land' to active participation in international monitoring of honey bee colony losses. BeeWorld. https://doi.org/10.1080/0005772X.2018.1554279

Furman E, Peltola T (2013) Developing socio-ecological research in Finland: challenges and progress towards a thriving LTSER network. In: Singh SJ, Haberl H, Chertow M, Mirtl M, Schmid M (eds) Long term socio-ecological research (LTSER). Springer, Dordrecht, pp 443–459

Gingrich S, Schmid M, Dirnböck T, Dullinger I, Garstenauer R, Gaube V, Haberl H, Kainz M, Kreiner D, Mayer R, Mirtl M, Sass O, Schauppenlehner T, Stocker-Kiss A, Martin W (2016) Long-term socio-ecological research in practice: lessons from inter-and transdisciplinary research in the Austrian Eisenwurzen. Sustainability 8(8):743

Grove JM, Pickett STA, Whitmer A, Cadenasso ML (2013) Building and urban LTSER: the case of the Baltimore ecosystem study and the D.C./B.C., ULTRA-Ex project. In: Singh JS, Haberl H, Schmid M, Mirtl M, Chertow M (eds) Long term socio-ecological research studies in society nature interactions across temporal and spatial scales. Springer, Dordrecht, pp 369–408

Haberl H, Gaube V, Díaz-Delgado R, Krauze K, Neuner A, Peterseil J, Plutzar C, Singh SJ, Vadineanu A (2009) Towards an integrated model of socioeconomic biodiversity drivers, pressures and impacts. A feasibility study based on three European long-term socio-ecological research platforms. Ecol Econ 68:1797–1812

Haberl H, Winiwarter V, Andersson K, Ayres RU, Boone C, Castillo A, Cunfer G, Fischer-Kowalski M, Freudenburg WR, Furman E, Kaufmann R, Krausmann F, Langthaler E, Lotze-Campen H, Mirtl M, Redman CL, Reenberg A, Wardell A, Warr B, Zechmeister H (2006) From LTER to LTSER: conceptualizing the socioeconomic dimension of long-term socioecological research. Ecol Soc 11(2):13

Holzer JM, Carmon N, Orenstein DE (2018) A methodology for evaluating transdisciplinary research on coupled socio-ecological systems. Ecol Ind 85:808–819

IMFN (2008) Model forest development guide. International Model Forest Network Secretariat, Ottawa, Canada

Lähteenmäki-Smith K (ed) (2007) Learning through evaluation: the Nordic experience. Nordregion Report 2007:3

Lambeck RJ (1997) Focal species: a multi-species umbrella for nature conservation. Conserv Biol 11(4):849–856

Lee KN (1993) Compass and gyroscope: integrating science and politics for the environment. Island Press, Covelo

Lindelöw D (2018) Running to stand still—the role of travel time in transport planning. SWECO, Gothenburg

Manton M, Angelstam P (2018) Defining benchmarks for restoration of green infrastructure: a case study combining the historical range of variability of habitat and species' requirements. Sustainability 10:326

Matthews R, Selman P (2006) Landscape as a focus for integrating human and environmental processes. J Agric Econ 57(2):199–212

Metzger MJ, Bunce RGH, Van Eupen M, Mirtl M (2010) An assessment of long term ecosystem research activities across European socio-ecological gradients. J Environ Manage 91(6):1357–1365

Mirtl M, Borer ET, Djukic I, Forsius M, Haubold H, Hugo W, Jourdan J, Lindenmayer D, McDowell WH, Muraoka H, Orenstein DE, Pauw JC, Peterseil J, Shibata H, Wohner C, Yu X, Haase P (2018) Genesis, goals and achievements of long-term ecological research at the global scale: a critical review of ILTER and future directions. Sci Total Environ 626:1439–1462

Mirtl M, Frenzel M, Furman E, Ohl C, Krauze K, Grünbühel C (2008) LTER-EUROPE: Criteria and recommendations. Version 5.2, 2008-05-27. Retrieved from http://www.lter-europe.net/document-archive/central/I3034v02-LTER-Europe-Criteria.pdf

Mirtl M, Orenstein DE, Wildenberg M, Peterseil J, Frenzel M (2013) Development of LTSER platforms in LTER-Europe: challenges and experiences in implementing place-based long-term socio-ecological research in selected regions. In: Singh SJ, Haberl H, Chertow M, Mirtl M, Schmid M (eds) Long term socio-ecological research. Springer, Dordrecht, pp 409–442

Moulaert F, Martinelli F, Swyngedouw E, Gonzales S (2005) Towards alternative model(s) of local innovation. Urban Stud 42(11):1969–1990

Naumov V, Manton M, Elbakidze M, Rendenieks Z, Priedniek J, Uglyanets S, Yamelynets T, Zhivotov A, Angelstam P (2018) How to reconcile wood production and biodiversity conservation? The Pan-European boreal forest history gradient as an "experiment". J Environ Manage 218:1–13

Neumeier S (2012) Why do social innovations in rural development matter and should they be considered more seriously in rural development research?—Proposal for a stronger focus on social innovations in rural development research. Sociologia Ruralis 52(1):48–69

Primdahl J, Kristensen L, Arler F, Angelstam P, Aagaard Christensen A, Elbakidze M (2018) Rural landscape governance and expertise—on landscape agents and democracy. In: Egoz S, Jorgensen K, Ruggeri D (eds) Defining landscape democracy: a path to spatial justice. Edward Elgar Publishing, Cheltenham, UK

Roberge J-M, Angelstam P, Villard M-A (2008) Specialised woodpeckers and naturalness in hemiboreal forests—deriving quantitative targets for conservation planning. Biol Cons 141:997–1012

Sayer J, Margules C, Boedhihartono AK, Dale A, Sunderland T, Supriatna J, Saryanthi R (2015) Landscape approaches, what are the pre-conditions for success? Sustain Sci 10(2):345–355

Shtilmark F (2003) History of the Russian zapovedniks. Russian Nature Press, Exeter

Singh SJ, Haberl H, Chertow M, Mirtl M, Schmid M (eds) (2013) Introduction. In: Singh SJ, Haberl H, Chertow M, Mirtl M, Schmid M (Eds) Long term socio-ecological research: studies in society-nature interactions across spatial and temporal scales. Springer Science and Business Media, Berlin, pp 1–26

Snow CP (1959) The two cultures. Cambridge University Press, Cambridge

Termorshuizen JW, Opdam P (2009) Landscape services as a bridge between landscape ecology and sustainable development. Landscape Ecol 24(8):1037–1052

Chapter 13
Regenerative Landscape Development: A Transformational Methodology for Thrivability of Landscapes

Leah V. Gibbons

Abstract Although the integration of science, practice, culture, and spirituality is recognized as necessary to move toward sustainability, most transdisciplinary frameworks are not inclusive of the necessary worldviews, paradigms, aims, processes, and components. Landscape sustainability science focuses on a pivotal level for scientific, practitioner, and stakeholder efforts toward sustainability, yet collaboration and progress have been slow. Regenerative development, a development and design methodology based on a holistic worldview, has potential as an integrating and transformational methodology to fill these gaps. A new paradigm of regenerative landscape development could shift the aim from sustainable social–ecological systems to thriving living systems in which health, well-being, and happiness increase continually across scales. This potential of regenerative landscape development in practice is with two case studies of projects in Viña del Mar, Chile and Juluchuca, Guerrero, Mexico. Finally, recommendations moving forward in constructing regenerative landscape development as a new paradigm are proposed. If fully understood, embraced, and realized, regenerative development holds incredible potential for a future that is not just sustainable but is thriving (This text is adapted with permission from Gibbons et al. in Sustainability 10:1910, 2018).

Keywords Regenerative development · Sustainable development · Landscape sustainability science · Ecological design · Ecological planning · Urban design · Urban planning · Landscape design · Landscape planning · Social–ecological systems · Complex adaptive systems

L. V. Gibbons (✉)
School of Sustainability, Arizona State University, Tempe, AZ 85287, USA
e-mail: leah.gibbons@asu.edu

L. Müller and F. Eulenstein (eds.), *Current Trends in Landscape Research*, Innovations in Landscape Research, https://doi.org/10.1007/978-3-030-30069-2_13

13.1 Introduction

Sustainability of landscapes is pivotal for linking local and global sustainability efforts (Forman 2008; Opdam et al. 2013; Wu 2013). Landscapes are the foundation of critical life-support systems. They are the scale at which scientists and practitioners blend natural and social scientific knowledge and practice and the scale at which inhabitants of a place most directly affect and connect with the land and each other (Nassauer 2012; Opdam et al. 2013; Wu 2013). Yet, landscape sustainability remains an elusive goal. Scientists, scholars, practitioners, and stakeholders still struggle to collaborate in impactful and transformative ways toward sustainable processes and outcomes.

Landscape sustainability science calls for "a place-based, use-inspired science of understanding and improving the dynamic relationship between ecosystem services and human well-being with spatially explicit methods" (Wu 2013, p. 1014), yet it has not provided holistic guidance for the transformational regeneration of multi-scalar landscapes. Sustainability research and action in landscapes should integrate ecological, geophysical, social, cultural, and spiritual components and their temporal and spatial dynamics (Fischer et al. 2012; Opdam et al. 2013; Wu 2013). However, current frameworks and methodologies for sustainable development in landscapes are incomplete, fragmented, rudimentary, and neglect important foundational elements necessary for landscape sustainability (Gibbons et al. 2018).

Regenerative development (RD) is a methodology with the potential to fill this gap. RD answers recent calls by scientists and practitioners alike for a reconceptualization of sustainability from reductive and mechanistic to holistic, from sustaining or improving health and well-being to creating abundant, thriving social–ecological systems (i.e., living systems) (duPlessis 2012; Olsson et al. 2010; Russell 2013; Ziervogel et al. 2016). RD integrates aspects of science, practice, culture, and spirituality that have thus far been segregated, rendering sustainability an unachievable ideal. By integrating these necessary components through a holistic paradigm and developmental change process methodology, RD makes transformational change toward not just sustainability but *thrivability* of living systems a real possibility. When used as a platform that also integrates landscape sustainability science, thrivability of landscapes may be achieved.

In this chapter, the theoretical and practical development of the relatively new articulation and practice of regenerative sustainability and regenerative development is introduced. Regenerative development's strengths and limitations are discussed, and it is suggested that integrating landscape sustainability science for a new paradigm of regenerative landscape development can lead to thriving landscapes. The potential of applying such a paradigm is illustrated with two case studies of regenerative development at the landscape scale. Needs for developing the regenerative landscape development paradigm are discussed, and a call to action for scientists, practitioners, and inhabitants of communities as foundational components of landscapes is put forth.

13.2 Regenerative Sustainability and Development

Amidst a predominantly descriptive–analytical science and reductive, mechanistic discourse and practice around sustainability, there are calls and efforts for *regenerative sustainability* to move beyond improving human well-being within ecological limits toward creating abundant social–ecological systems (i.e., living systems) in which all life thrives (duPlessis 2012; Olsson et al. 2010; Russell 2013; Ziervogel et al. 2016). Regenerative sustainability is based on recent scientific understandings in ecology, quantum physics, systems theory, and developmental change theory as well as more ancient ways of knowing and being in the world (e.g., indigenous knowledge and practices) (duPlessis 2012). It understands that the worldviews we hold underlie the (un)sustainability and health of living systems and intentionally adopts a holistic/ecological worldview (hereafter referred to as "holistic worldview") that considers all life (duPlessis and Brandon 2015; Meadows 1999). This worldview sees humans and nature as part of one autopoietic system. Concepts of self and extended-self, interrelationship and interdependency, reciprocity and responsibility are key.

Through understanding the world as a whole living system that is dynamic and relational, regenerative sustainability suggests that effective and ethical action will follow the rules of nature, build healthy, mutually beneficial interdependent relationships, accept change as necessary for life, and contribute to the well-being of the whole system (duPlessis and Brandon 2015). Reintegrating humans with nature as well as the physical, mental, emotional and spiritual aspects of existence is a vital step (Berejnoi et al. 2019; duPlessis and Brandon 2015; Wilber 2000). In this context, spirituality is defined as the awareness and experience of belonging to the larger, interconnected community of life, of the purpose and meaning of life within this context, and the development of personal and community values and actions out of these. From this regenerative sustainability paradigm, new relationships between nature and society, humans and the built environment, and humans and each other emerge. These can be manifested through regenerative development.

RD is a developmental change methodology that has been practicing and evolving regenerative sustainability for more than 20 years. It translates regenerative sustainability principles into design and development frameworks, context-specific designs and plans, locally adapted technologies, and capacities in social–ecological (i.e., living systems) systems that manifest continually higher levels of health, well-being, and vitality across scales (Benne and Mang 2015; duPlessis and Brandon 2015; Mang and Reed 2012). These include many accepted approaches, such as multifunctional landscapes, design experiments, safe-to-fail experiments, adaptive design, parametric design, biomimicry, biophilia, and permaculture. It also uses technologies specific to regenerative development, such as living systems thinking, story of place, and integral assessment (Ahern 2013; Lovell and Johnston 2009; Mang et al. 2016; Nassauer 2012).

RD bridges the science–practice–culture–spirituality gap, integrating and operationalizing necessary elements for sustainability that thus far have gone unintegrated and often unrecognized by sustainability scholars and practitioners (Gibbons et al. 2018). RD rests upon a foundation of systemic changes that include the power to transcend paradigms, the mindset out of which the system arises, the aims of the system, and the capacity of the system to self-organize (Abson et al. 2017; Mang and Reed 2012; Meadows 1999). Specifically, RD deeply engages all stakeholders with a place and works to create a "new mind" with which to see, understand, and interact with place and the world (Mang and Reed 2012). It seeks to create not just residents of a place, but inhabitants, who are deeply connected to and care for the place (i.e., social–ecological system) of which they are an integral part (Cole 2012). Thus, continuous learning and participation of stakeholders and inhabitants through action, reflection, and dialogue are key to manifesting the long-term aspirations of any RD project (Reed 2007). The cultivation and integration of our inner intangible reality—how individuals and, by extension, the social collective perceive and experience their world—with our outer tangible reality are a key component for the creation and regeneration of healthy communities and landscapes. Regenerative development creates positive feedbacks between outer and inner dimensions of existence, the constructed/created environment, and worldviews/paradigms. Regenerative capacity can only occur in perpetuity when human beings live our interconnectedness with all of existence, experiencing the meaning, purpose, and responsibility that arises from it.

RD includes the following key principles for sustainability of landscapes that other frameworks neglect: (1) manifesting potential-inherent future states of being for a system that is health-promoting and value-adding to the larger systems of which it is a part—rather than focusing on problems–solutions; (2) addressing the root causes of (un)sustainability by shifting worldviews to holistic ones through a process of deeply engaging stakeholders, inhabitants of a place, and practitioners in a collaborative, co-creative process; (3) working in whole systems rather than fragments; (4) creating mutually beneficial relationships among the sociocultural and ecological components of systems that evolve through time; (5) working across scales to add integral, life-conducive value to systems; (6) growing the capacity of the living systems to increase viability, vitality, and evolutionary capacity (i.e., regenerative capacity) (Gibbons et al. 2018). The goal is nothing less than catalyzing the transformation of social–ecological systems across scales into thriving, regeneratively sustainable states (Benne and Mang 2015; Mang and Reed 2012). By comparison, other sustainable design and development approaches and conceptualizations are anthropocentric, fragmented, prescriptive; such approaches and conceptualizations focus on the symptoms of unsustainability and aim for, among other things, incremental improvements, doing less harm, mitigating damage, or managing both humans and nature (duPlessis 2012; Kopnina 2015; Reed 2007) (Fig. 13.1).

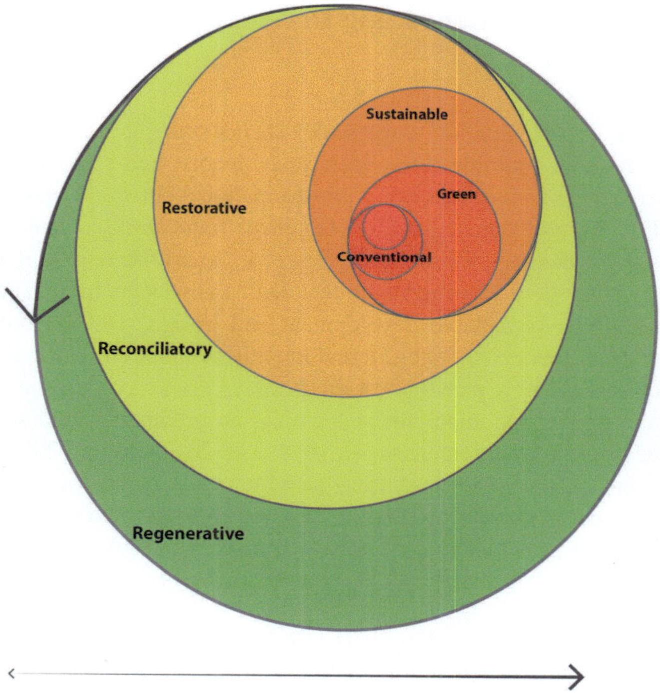

	Worldview	Paradigm	Values	Ideas	Actions	Aim
Regenerative	Holistic	Holistic /Regenerative	Collective individualism Interbeing	Wholeness Humans as nature Potential Co-creation	Developmental change processes	Regenerate
Reconciliatory			Interdependence	Humans as integrated with nature		
Restorative			Stewardship	Humans healing nature		Improve
Sustainable				Ecosystem services/ human well-being Engagement	Sustainable development goals	Maintain
Green				Problems-Solutions	LEED, BREEAM, indicators, checklists	Operate
Conventional	Mechanistic	Reductionistic	Independence	Fragmentaion Extraction	Codes, regulations, laws	Stabilize

Fig. 13.1 Spectrum from conventional to regenerative development and design methodologies. Each methodology builds upon and incorporates the last in an ascending spiral, representing an evolution from simple to more complex and inclusive ways of being and interacting in the world. The arrows in both the spiral and to the right of the chart show the direction of this evolution. The properties of lower methodologies are inherently present and available in upper methodologies, if needed. Further, each level is based upon a supporting worldview and paradigm, out of which specific values, ideas, levels of work, and actions develop. These properties influence one another to vary degrees, with worldviews exerting the strongest influence, as the arrow across the top indicates. The integration of all methodologies can result in a regenerative process to create whole, thriving, and healthy living systems. From Gibbons et al. (2018)

13.3 Regenerative Landscape Development—A New Paradigm

The inherent complexity of large-scale social–ecological systems, such as landscapes, requires multiple disciplines, practitioners, and stakeholders to effectively understand, envision, and enact transformational change toward regenerative sustainability. Current regenerative development frameworks are not specific for landscapes (e.g., Mang and Reed 2012; Plaut et al. 2016). There is an opportunity to further integrate regenerative development with knowledge from landscape sustainability science, sustainable development and design, ecology, and complex adaptive systems theory to enable transformational research and action in landscapes to construct a new paradigm of regenerative landscape development (RLD). For example, knowledge about social learning, co-production, and social change could be incorporated more intentionally (Bos et al. 2013; Holling 2004; Reed et al. 2010; Watson 2014). RLD as a new paradigm and methodology could unite science, practice, culture, and spirituality into one coherent field that overcomes the challenges typically encountered in interdisciplinary and transdisciplinary work; it could also affect significant transformational sustainability change from local to global scales.

In constructing a new paradigm, RD brings a holistic worldview, frameworks, methodologies, technologies, and methods to operationalize many scientific principles. RLD could utilize concepts and tools from ecology, such as pattern–process–scale interactions, diffusion, metastability, and incorporation, to aid in understanding and shaping landscape relationships, flows, and dynamics. It could explicitly include multifunctionality, redundancy, heterogeneity, diversity, modularity, connectedness, local interactions, and tight feedbacks as critical components of healthy, self-organizing systems (Holling 2004; Wallner et al. 1996; Wu 2008; Wu and Wu 2013). Knowledge and tools from complex adaptive systems science could aid in understanding when and where to implement specific projects and in what relationships to one another to leverage transformational change (Abson et al. 2017; Gunderson and Holling 2002; Wallner et al. 1996). Cultural and spiritual practices could help people create meaningful, motivational, and positively reciprocal personal and societal connections to each other and nature. Finally, scientists and practitioners could intentionally use RD projects as designed experiments, testing ecological, design, and sustainability hypotheses while promoting discourse and allowing time for change to occur. They could implement adaptive design and learn by doing, incorporating new knowledge and adjusting efforts based on experiments (Ahern 2011, 2013; Felson and Pickett 2005; Holling 2004).

Understanding landscapes as holarchical living systems, in which change cascades upscale from lower levels as well as across scales, landscapes can be transformed using RLD in networked, nested communities comprising landscapes (Daly and Cobb 1994; Gibbons et al. 2018; Gunderson and Holling 2002). Interestingly, in addition to transformations in non-human dominated ecosystems, the development of altruistic human societies also follows this pattern

(Gunderson and Holling 2002; Gutenschwager 2013). This suggests that significant social–ecological transformation is possible following an RLD paradigm and methodology.

13.4 Regenerative Development Case Studies

The following case studies illustrate regenerative development theory and practice at landscape scales. They integrate science, practice, culture, and spirituality within a regenerative development framework to create regenerative capacity and move toward regenerative sustainability at a landscape scale.

13.4.1 Las Salinas Project, Viña del Mar, Chile

Las Salinas is a 40-acre brownfield site located in Viña del Mar, Chile. It is owned by the Chilean energy company, COPEC SA and has been used as a petroleum fuel distribution site for decades. In 2015, COPEC SA created a detailed redevelopment plan that maximized the development allowed there and presented it to the city as a "gift." This plan faced fierce backlash from the community, who feared it would bring similar problems that had overwhelmed the community in recent decades, namely increased traffic congestion, decreased quality of life, and decreased agricultural yields (Hennick 2018). The company decided to shift from a transactional approach to a reciprocal one. It enlisted Regenesis Group, a regenerative development consulting firm, to facilitate a collaborative relationship and regenerative development project with the Viña del Mar community (Reed 2018). Las Salinas embodies the progression from conventional to green to restorative to regenerative methodologies and the collective integration of each stage. It also demonstrates how a specific site can play a catalyzing regenerative role in the landscape (Box 13.1).

The design team conducted a detailed integral assessment considering geological, ecological, and human components of the living system as well as their interactions through time and space. They also conducted deep listening sessions with the community, who was considered part of the design team. They connected to a feel of nostalgia for the Viña del Mar of the past, which exemplified its moniker, "Garden City" (Hennick 2018; Reed 2007). Viña del Mar, which translates to "Vineyard of the Sea," was a place associated with gentility, abundance, diversity, social and ecological connectivity, beauty and vitality; these community elements had degenerated over the last several decades. Las Salinas sits between what was a biodiverse hillside and the sea—an important connective element in the landscape. The regenerative development concept that emerged was to co-create Las Salinas as a connecting place and hub for the regeneration of the social and ecological components of Viña del Mar and the region beyond (Hennick 2018).

Eight bridging concepts emerged for the site design, which directly and indirectly connects the following biogeophysical and social elements of the city: habitat connection, estuary health, mobility, meaningful public space, cultural centers, community centers, marketplace, food networks, and youth education (Reed 2018). The site design proposes an ecosystem regeneration on the hillside which connects Las Salinas to a greater ecological context via its streets (which function as ecological corridors), green roofs, courtyards (which function as ecological patches or stepping stones), and linear parks that offer diffused connectivity throughout the space (Sasaki 2018). Seascape views are preserved; an elevator takes people from the hillside to sea level, re-engaging the community with its seafront and public spaces. A fully accessible and interconnected public realm network promotes greater social integration (Fig. 3). In all, 4.5 billion square feet have been allocated for mixed-use LEED-ND certified development (Hennick 2018; Reed 2007).

In addition to ecological and social connections, the design for Las Salinas regenerates the complexity and health of the landscape in multiple ways. Stormwater is slowed and allowed to infiltrate the soil and recharge the aquifer using appropriate landscaping and design, pervious pavements, and the reuse of water (Reed 2018; Sasaki 2018). Sediments are filtered through the landscape before reaching the sea. Spaces for social integration and interaction with nature (e.g., parks, boardwalks, beaches) are abundant (Sasaki 2018).

Collaborations between regional stakeholders have already resulted in urban forestry and watershed regeneration initiatives, and more mutualistic relationships continue to emerge. Shifting from a conventional and transactional development and design proposal to a regenerative, reciprocal, and co-creative one has garnered the support of eighteen activist groups, including municipal planners who had initially opposed the development. It has integrated and built upon Las Salinas' LEED-ND certification to play a key role in manifesting the potential of Viña del Mar by working across scales to add value and manifest potential (Reed 2018).

There is still a long way to go from the planning and remediation phases to construction and there is no guarantee that the regenerative direction of the project will continue. However, development and design team members are confident that the collaborative atmosphere they established during the planning process will continue through the construction phase and beyond. They feel that the energy field of will and caring that has been developed will continue to grow and evolve. They see this manifesting in the collaborative relationships that are forming in the community (Hennick 2018; Reed 2018). Regenesis Group's involvement in the Las Salinas project continues, focusing on the development and design team members' inner/personal development that is necessary to keep the project moving forward in a regenerative direction (Reed 2018). Further, team members understand that the project must keep evolving to stay relevant (Hennick 2018).

Box 13.1. Regenerative development outcomes and indicators in Las Salinas

Manifesting potential:

- Connecting social and ecological elements of the landscape, catalyzing a return of abundance, diversity, and vitality to Viña del Mar.

Shifting worldviews:

- Shift from transactional to reciprocal relationships
  - Shift from developer-led gifted plan to co-created, collaborative plan.
- Shift from mechanistic worldviews (conventional plan) to ecological worldviews (regenerative plan).

Working in whole systems:

- Plan returns ecological and social diversity, abundance, connectivity, and vitality to landscape.

Creating mutually beneficial, co-evolving relationships:

- Regional urban forestry and watershed regeneration initiatives developed; others continue to emerge
- Support from 18 activist groups that originally opposed development
- Collaborating with community groups to co-create a cultural foundation, recover natural beach formation dynamics, improve ocean access for neighboring communities.

Adding value across scales:

- Social and ecological connections with surrounding landscape via eight bridging concepts—habitat connection and ecosystem regeneration, estuary health, mobility, meaningful public space, cultural centers, community centers, marketplace, food networks, and youth education.

Growing regenerative capacity:

- Deep stakeholder and inhabitant collaboration has grown will of inhabitants in the larger system of Viña del Mar; community collaborations are forming outside of the Las Salinas project
- Continuing community dialogue and collaboration
- Continuing engagement in regenerative development at the individual and collective levels
- Recognition that the project must keep evolving
- Learning from mistakes and making adjustment.

13.4.2 Playa Viva, Juluchuca, Guerrero, Mexico

Playa Viva is a 200-acre regenerative ecoresort in Juluchuca, Guerrero, Mexico. Playa Viva's developers and owners wanted the resort to be more than green or sustainable. They wanted it to improve not only the land it was on but also the surrounding landscape and community. From 2006–2007, Regenesis Group facilitated a process of discovery and co-creation from which emerged a regenerative direction for the resort, which opened in 2009 (Benne and Mang 2015; Playa Viva 2018; Reed 2018). Playa Viva is a good example of a project that developed regenerative capacity and continues to evolve that capacity today (Box 13.2).

An integral assessment revealed that Playa Viva was once a small but important community that was part of a thriving regional population of 10,000 people (Reed 2018). As a result of the richness of the estuary, forest, and coast in this region, the community provided valuable goods to the surrounding areas. Deep listening and dialogue sessions with inhabitants revealed a story of abundance, dense biodiversity, trees bursting with fruit, lagoons overflowing with fish, and nature teaming with life (Playa Viva 2018). In the 1920s, a large portion of the coastal landscape of the state of Guerrero was slashed and burned and turned into coconut monocultures, degenerating the former vitality and abundance of this place. Rapid discharge of water attributable to the loss of key vegetation led to shortages in water supplies for Juluchuca. Young residents were leaving the community in search of better economic opportunities and quality of life. The estuary was slowly degenerating into a marsh (Playa Viva 2018; Reed 2018).

The regenerative concept that emerged for Playa Viva was that it could once again be a place of vitality and abundance, one that added value to the surrounding ecosystems, community of Juluchuca, and social–ecological systems further upstream. Playa Viva decided to focus on estuary regeneration, community coevolution, and transformational guest experiences. Before Playa Viva began development on the land it owned, it began working with the community of Juluchuca as well as communities throughout the watershed to co-create and sponsor education, health, and economic development initiatives. These initiatives now offer guests of Playa Viva the opportunity to become deeply involved in community and ecosystem experiences, ones that they can continue to be involved in following their return home via investment in and growth of local businesses and initiatives (Benne and Mang 2015).

Playa Viva began by establishing permaculture and biodynamic farming training programs for local farmers, helping to create a healthier watershed, agroecosystems, and livelihood opportunities for the community. Production expanded beyond the staple beans, corn, and squash to include fruits, vegetables, and tropical flowers. Playa Viva helped to create a community-supported agriculture cooperative for local farmers to have a market for their produce. The market has evolved and now includes a vibrant regional farmers' market (Playa Viva 2018).

Additionally, Playa Viva sponsored a detailed chemical analysis of this region's salt, leading to the discovery that this resource is distinctive, with desirable mineral content. The community sponsored a local fair trade cooperative that sells this unique salt to local resorts as well as internationally; this marks an evolution from previous methods, when visitors of the resort promoted the products by means of person-to-person marketing. Cooperative members continue to use traditional means of harvesting the salt, thus preserving ecosystem health and cultural heritage. Playa Viva also helped create a market for local coconut products and ecotourism. Further, it co-created and sponsors education and health initiatives that provide needed supplies for local schools and health clinics as well as English tutoring (Playa Viva 2018).

This holistic approach supports the local community with financial, intellectual, market, and social capital, creating local living economies. By engaging in a reciprocal relationship with these small industries through economic assistance, business training, access to resources, and access to markets, Playa Viva has helped increased their profitability and business viability. A supply and a demand for these activities have created a positive synergy between social and ecological components of this system and contributed to its vitality and abundance (Playa Viva 2018).

Playa Viva helped to develop a local turtle sanctuary that transformed poachers into turtle sanctuary employees and stewards. They collect and nurture turtle eggs as well as oversee the release of baby turtles into the wild. They have earned visibility and status within the community and now view themselves as defenders of indigenous turtles. They have become preservation experts and are important in the community's environmental regeneration. Additionally, Playa Viva is catalyzing collaborations that are regenerating estuaries along the coast of the region, reviving a critical landscape element; it is also regenerating marine life, creating a carbon sink, improving local fisheries and water quality, stabilizing the land, and providing storm surge protection (Benne and Mang 2015; Playa Viva 2018).

Visitors to Playa Viva can participate in many of the above-mentioned activities, leading to friendships with villagers and investments in community businesses and initiatives (Reed 2018). Guests also pay a 2% Regenerative Trust fee that is directed to local environmental and community efforts (Playa Viva 2018). These efforts have helped to increase economic opportunities, quality of life, and ecosystem health, drawing youth who now see a future here back to the community (Benne and Mang 2015; Reed 2018).

Playa Viva is regenerating ecosystems on its own property as well. There is a 160-acre nature preserve where coastal forest biodiversity is undergoing regeneration through the planting of over 10,000 native trees. A once-thriving lagoon that dried up because of invasive cattle grass is being restored. Its biodiversity is increasing and Playa Viva plans to regenerate the entire lagoon. Playa Viva is slowly expanding its mangrove ecosystem through restoring and extending waterways that once thrived. The resort's permaculture-designed landscaping, which balances native, drought-tolerant, and aesthetic/food-bearing species, attracts birds and beneficial insects while serving as a living classroom for guests, local farmers, and WWOOFers (participants in World Wide Opportunities on Organic Farms) (Playa Viva 2018).

In terms of the built environment, Playa Viva is also replenishing the local aquifer by using gray water on gardens and mini-living systems for blackwater. Water is reused whenever possible, and nutrients are extracted to enrich the soils. All energy production is solar and off-grid. Buildings are constructed of local, salvaged, and eco-friendly materials by local artisans, i.e., local craftspeople using traditional methods. They are designed according to natural cooling principles, with no need for air conditioning. Hotel materials (e.g., soaps, linens, dinnerware) and food are sourced both locally and on-site, where a wide variety of fruits, nuts, vegetables, seeds, chickens, and fish are available. As a reflection of Playa Viva's values of a strong community, there is a balance of private and public spaces. Further, Playa Viva serves as a teaching model for local contractors and other resort owners in the area (Playa Viva 2018).

Playa Viva has continued its commitment to regeneration. It is allowing itself time to adapt and expand gradually in a coevolving mutualism, receiving and responding to feedback from the community, ecosystem, and guests, with an increasingly beneficial effect on the surrounding community and ecosystems. Playa Viva continues to evolve because owners and employees are constantly reflecting on their value-adding role in this coevolving living system, striving to be beneficial components. They demonstrate a commitment to place and to themselves as regenerative agents (Reed 2018).

Box 13.2. Regenerative development outcomes and indicators in Playa Viva

Manifesting potential:

- Abundance and vitality returning to landscape—estuary regeneration, community co-evolution, and transformational guest experiences (see below).

Shifting worldviews:

- Collaborative, reciprocal relationships
  - Farmer trainings and products
  - Salt and coconut businesses
  - Ecotourism, turtle sanctuary.
- Continually asking how to continue co-evolving with place
- Understanding that whole living systems and their relationships must be healthy for their elements to be healthy and productive is reflected in actions (see below).

Working in Whole Systems:

- Increasing biodiversity through ecosystem restoration, resort Permaculture landscaping, upriver farming training that decreases toxic chemical inputs, and regional estuary regeneration

Creating mutually beneficial, co-evolving relationships:

- Reciprocal economic relationships–Permaculture and biodynamic farmers, salt cooperative, coconut and ecotourism businesses
- Reciprocal humanitarian relationships—educational and health initiatives
- Reciprocal ecosystems relationships—coastal forests, mangroves, lagoon, watershed health, increasing biodiversity
- Initiatives continue to emerge and grow
- On-site development continues at rate that allows co-creation, feedback, and adjustment.

Adding value across scales:

- Sponsoring education, health, and economic initiatives and support locally and regionally
- Estuary regeneration up and down coast
- Healthier watershed, agroecosystems, and livelihood opportunities for the community
- Increasing biodiversity
- Permaculture and biodynamic farmer trainings
- Replenishing aquifer.

Growing regenerative capacity:

- Beneficial initiatives continue to emerge and grow
- Ecosystem restoration:
  - Estuary
  - Mangrove swamps
  - Coastal forests
  - Lagoon.
- Evolving thriving local living economies:
  - Permaculture and biodynamic farmer trainings
  - Local and international markets for farming products, coconut products, salt,
  - Ecotourism
  - Guest investment in local businesses and initiatives
- Young people returning to area for good quality of life, livelihood opportunities.

13.5 Recommendations in Constructing Regenerative Landscape Development

To advance regenerative landscape development as a paradigm, we identify several needs. RLD needs conceptual and theoretical development, which should more fully integrate landscape sustainability science, ecology, complex adaptive systems science, practice, culture, and spirituality through the platform of regenerative development. Scientific research and evidence can be developed to enhance current practices and vice versa. Methodological frameworks should guide how design processes structure and integrate science, practice, knowledge, and action. These frameworks should be able to be locally adapted for a variety of contexts, scales, and cultures.

Assessment tools should move beyond typical prescriptive, fragmented, and deterministic checklists to monitor dynamic change, developmental trajectories, and the ongoing contribution of systems to the health and evolution of their larger wholes. Sociocultural and ecological processes and outcomes, as well as their relationships, can be assessed. To move toward sustainability, frameworks and assessment tools should allow for the flexibility, redundancy, and diversity found in living systems while also adequately addressing their complexity. They, and the humans using them, should begin to acknowledge that we cannot "solve" sustainability problems, but we can offer developmental pathways that can lead to greater health, vitality, and prosperity of the entire system (duPlessis and Brandon 2015). Frameworks and assessments should be applied and adaptable across scales, with larger scales providing guidelines for approaches based on lower working levels and smaller scales providing the mechanisms driving higher-level processes (Wu and Loucks 1995). Adaptive management and design experiments should be part of the strategies used to monitor and assess new and existing systems and projects (Felson and Pickett 2005). Assessment tools should also be applicable to new as well as existing systems and be able to guide their development toward regenerative sustainability. Such a process requires an ongoing participatory and reflective process that nurtures social learning and is part of a culture of regenerative sustainability (duPlessis and Brandon 2015; Reed et al. 2010; Wahl 2016).

Educational programs for practitioners, students, scientists, stakeholders, and inhabitants of a place should teach the theory and practice of regenerative landscape development. Lastly, design experiments at all scales, including regional areas, dense urban areas, rural areas, neighborhoods, communities, building sites, and even households need to be conducted to inform policy and land use regulations (Cole 2012; duPlessis 2012; Opdam et al. 2013). Case studies on regenerative development and other regenerative sustainability projects can be conducted and used to create a portfolio of transdisciplinary working methods, frameworks, technologies, and assessments as well as their outcomes that may be adapted and used in different specific and local situations. Additionally, we must find the most effective methods for mechanistic approaches to enhance and inform approaches based on an ecological worldview to enable successful shifts toward regenerative sustainability. Policy, governance, power, and funding shifts reflecting this expanded paradigm and new knowledge will be key to implementation. Policy must make it not only possible to implement regenerative development but also desirable and, potentially, even required.

Although regenerative development has tremendous potential as a transformational sustainability approach, it also faces challenges that could subvert it. For instance, we should use regenerative design as part of a larger regenerative development process; otherwise, systemic transformation toward thrivability will not occur. We should be cautious of RD being "greenwashed" and detached of its essential meaning and aims. It is also necessary to be aware of how the human ego may interact with regenerative development processes. It may be difficult for people to accept the uncertainty and maintain the patience necessary for the regenerative process to unfold in complex, dynamic, unpredictable living systems. Acting with timing not in alignment with the dynamics of holistic living systems can trigger degenerative instead of regenerative processes in living systems (Gunderson and Holling 2002; Mang et al. 2016).

We should be careful about when and how we use reductionistic methods in conjunction with holistic methods, striving for a respect and balance of each that is called for in each particular place and time. We should become comfortable with "failure," recognizing that a regenerative development approach requires an attitude of experimentation and learning as well as openness to systems behaving in ways we cannot predict (Ahern 2011; Felson et al. 2013; Holling 2004). We should be mindful that RD requires that we create processes and products that are amenable to change, adaptation, and evolution. It also continuous monitoring of processes capacity-building collaborative partnerships between inhabitants of a place, scientists, and stakeholders.

13.6 Conclusions

This chapter has proposed regenerative landscape development (RLD) as a means of transforming current landscape sustainability theory and practice to co-create thriving living systems that manifest increasingly higher levels of health, well-being, and happiness across scales. RLD would more fully integrate the methodology of regenerative development (RD) with landscape sustainability science. RD has been in practice for 20+ years and addresses gaps in current sustainability science and practice, integrating necessary aspects of science, practice, culture, and spirituality that have thus far been inadequately addressed. To reach its potential for co-creating transformational change toward thriving landscapes, RD should more fully integrate landscape sustainability science. The integration will be an iterative process of inner transformation at the personal and collective levels with the outer transformation of the physical world. Intentionally holding and operating from a holistic worldview are essential.

RD is not a natural practice for most people in the Western world. We are steeped in a mechanistic worldview, and our default beliefs, thinking mechanisms, and actions emerge from there. RD ultimately grows from individuals who consciously commit to changing their own worldviews and ways of being in the world. This is no small task; it takes constant commitment and effort. Learning how to be a RD practitioner or regenerative inhabitant of a living system is not easy. It is not as simple as implementing a formula or technologies or following a prescribed list of activities. It is as much an art as it is a science; it will take a complementary approach of disciplines, practices, and ways of knowing and being in the world to move forward. It will take continual effort to create the commitment and caring necessary to continue on a regenerative pathway. It will take a new kind of practitioner, scientist, and inhabitant of a place who possesses new skills, mindsets, and aspirations and constantly nurtures these. Initially, it may be difficult to find fully willing participants and to provide them with the training and support they need. Yet, it might just be the approach to regenerate humanity and all life beyond.

It is time to raise the aim of sustainability from improving the relationship between humans and ecosystems in landscapes to living in ways that nourish the perpetuation of well-being for all life in living systems. It is time to stop focusing on problems and instead focus on potential. It is time for humans to take responsibility for their co-creative role in the state of well-being of the living systems of which they are a part and live in ways that are full of purpose, meaning, and fulfillment. By fully integrating science, practice, culture, and spirituality in a new paradigm of regenerative landscape development, achieving these new aims for thrivability from local to global scales might just be possible. You are invited by all who do this work to be part of that process.

References

Abson DJ, Fischer J, Leventon J, Newig J, Schomerus T, Vilsmaier U, von Wehrden H, Abernethy P, Ives CD, Jager NW, Lang DL (2017) Leverage points for sustainability transformation. Ambio 46:30–39

Ahern J (2011) From fail-safe to safe-to-fail: Sustainability and resilience in the new urban world. Landscape Urban Plann 100:341–343

Ahern J (2013) Urban landscape sustainability and resilience: the promise and challenges of integrating ecology with urban planning and design. Landscape Ecol 28:1203–1212

Benne B, Mang P (2015) Working regeneratively across scales—insights from nature applied to the built environment. J Clean Prod 109:42–52

Berejnoi E, Ulluwishewa R, Cloutier S, Gibbons L, Puga S (2019) Spirituality: the missing link of sustainability and happiness: a framework for holistic development. In: Kee Y, Lee SJ, Phillips R (eds) Perspectives on community well-being, pp TBA. Springer, Berlin

Bos JJ, Brown RR, Farrelly MA (2013) A design framework for creating social learning situations. Glob Environ Change 23:398–412

Cole RJ (2012) Transitioning from green to regenerative design. Build Res Inf 40:49–53

Daly HE, Cobb JB Jr (1994) For the common good: redirecting the economy toward community, the environment, and a sustainable future. Beacon Press, 534 p. ISBN 0-8070-4705-8 duPlessis C (2012) Towards a regenerative paradigm for the built environment. Build Res Inf 40:7–22

duPlessis C, Brandon P (2015) An ecological paradigm as basis for a regenerative sustainability paradigm for the built environment. J Clean Prod 109:53–61

Felson AJ, Pickett STA (2005) Designed experiments: new approaches to studying urban ecosystems. Front Ecol Environ 3:549–556

Felson AJ, Bradford MA, Terway TM (2013) Promoting Earth Stewardship through urban design experiments. Front Ecol Environ 11:362–367

Fischer J, Dyball R, Fazey I, Gross C, Dovers S, Ehrlich PR, Brulle RJ, Christensen C, Borden RJ (2012) Human behavior and sustainability. Front Ecol Environ 10:153–160

Forman RTT (2008) Urban regions: ecology and planning beyond the city. Cambridge University Press, Cambridge, UK

Gibbons LV, Cloutier SA, Coseo PJ, Barakat A (2018) Regenerative development as an integrative paradigm and methodology for landscape sustainability. Sustainability 10:1910. https://doi.org/10.3390/su10061910

Gunderson LH, Holling CS (eds) (2002) Panarchy: understanding transformations in human and natural systems (synopsis). Island Press, 50 p. ISBN 978-1-55963-330-7

Gutenschwager G (2013) From epicurus to maslow: happiness then and now and the place of the human being in social theory. Cadmus 1:66–90

Hennick C (2018) Winter. Sustainable Chile. USGBC+, Transforming Our Built Environment, pp 38–47

Holling CS (2004) From complex regions to complex worlds. Ecol Soc 9:11 [online] URL: http://www.ecologyandsociety.org/vol9/iss1/art11

Kopnina H (2015) The victims of unsustainability: a challenge to sustainable development goals. Int J Sustain Dev World Ecol. https://doi.org/10.1080/13504509.2015.1111269

Lovell ST, Johnston DM (2009) Creating multifunctional landscapes: how can the field of ecology inform the design of the landscape? Front Ecol Environ 7:212–220

Mang P, Reed B (2012) Designing from place: a regenerative framework and methodology. Build Res Inf 40:23–38

Mang P, Haggard B, Regenesis (2016) Regenerative development and design: a framework for evolving sustainability. Wiley Inc, 235 p. ISBN 978-1-118-97286-1

Meadows D (1999) Leverage points: places to intervene in a system. The Sustainability Institute

Nassauer JI (2012) Landscape as medium and method for synthesis in urban ecological design. Landscape Urban Plann 106:221–229

Olsson P, Bodin Ö, Folke C (2010) Building transformative capacity for ecosystem stewardship in social-ecological systems. In: Plummer R, Armitage D (eds) Adaptive capacity and environmental governance (Chap. 13). Springer, pp 263–286. ISBN 978-3642121937

Opdam P, Nassauer JI, Wang Z, Albert C, Bentrup G, Castella JC, McAlpine C, Liu J, Sheppard S, Swaffield S (2013) Science for action at the local landscape scale. Landscape Ecol 28:1439–1445

Plaut J, Dunbar B, Gotthelf H, Hes D (2016) Regenerative development through LENSES with a case study of Seacombe West. Environ Des Guide 1–19

Playa Viva. Available online: http://www.playaviva.com Accessed 7 May 2018

Reed B (2007) Shifting from 'sustainability' to regeneration. Build Res Inf 35:674–680

Reed MS, Evely AC, Cundill G, Fazey I, Glass J, Laing A, Newig J, Parrish B, Prell C, Raymond C, Stringer LC (2010) What is social learning? Ecol Soc 15(4): r1. [online] URL: http://www.ecologyandsociety.org.ezproxy1.lib.asu.edu/vol15/iss4/resp1/

Reed B (2018) Regenesis group, Boston, NY, USA. Personal communication

Russell JM (2013) Thrivability: breaking through to a world that works. Triarchy Press, Devon United Kingdom. ISBN 978-1909470286

Sasaki, Las Salinas. Available online: http://www.sasaki.com/project/421/las-salinas/. Accessed 11 May 2018

Wahl DC (2016) Designing regenerative cultures. Triarchy Press, 287 p. ISBN 978-1909470774

Wallner HP, Naradoslawsky M, Moser F (1996) Islands of sustainability: a bottom-up approach towards sustainable development. Environ Plann 28:1763–1778

Watson V (2014) Co-production and collaboration in planning—The difference. Plann Theor Pract 15:62–76

Wilber K (2000) Waves, streams, states and self—further considerations for an integral theory of consciousness. J Conscious Stud 7:145–176

Wu J (2008) Toward a landscape ecology of cities: beyond buildings, trees, and urban forests. In: Carreiro MM, Song YC, Wu JG (eds) Ecology, planning, and management of urban forests: international perspectives (Chap. 2). Springer, pp 10-28. ISBN 978-0387714240

Wu J (2013) Landscape sustainability science: Ecosystem services and human well-being in changing landscapes. Landscape Ecol 28:999–1023

Wu J, Loucks OL (1995) From balance of nature to hierarchical patch dynamics: a paradigm shift in ecology. Q Rev Biol 70:439–466

Wu J, Wu T (2013) Ecological resilience as a foundation for urban design and sustainability. In: Pickett STA, Cadenasso M, McGrath BP (eds) Resilience in urban ecology and design: linking theory and practice for sustainable cities (Chap 10). Springer, pp 211–229. ISBN 978-9400753433

Ziervogel G, Cowen A, Ziniades J (2016) Moving from adaptive to transformative capacity: building foundations for inclusive, thriving, and regenerative urban settlements. Sustainability 8:955. https://doi.org/10.3390/su8090955

Chapter 14
The iCASS Platform: Nine Principles for Landscape Conservation Design

Robert M. Campellone, Kristina M. Chouinard, Nicholas A. Fisichelli, John A. Gallo, Joseph R. Lujan, Ronald J. McCormick, Thomas A. Miewald, Brent A. Murry, D. John Pierce and Daniel R. Shively

Abstract The Anthropocene presents society with a super-wicked problem comprised of multiple contingent and conflicting issues driven by a complex array of change agents. Super-wicked problems cannot be adequately addressed using siloed decision-making approaches developed by hierarchical institutions using science that is compartmentalized by discipline. Adaptive solutions will rest on human ingenuity that fosters transformation toward sustainability. To successfully achieve these objectives, conservation and natural resource practitioners need a paradigm that transcends

R. M. Campellone (✉)
National Wildlife Refuge System, U.S. Fish and Wildlife Service,
Mail-Stop: NWRS, 5275 Leesburg Pike, Falls Church, VA 22041-3803, USA
e-mail: rob_campellone@fws.gov

K. M. Chouinard
Migratory Bird Program/National Wildlife Refuge System, U.S. Fish and Wildlife Service,
Jackson, TN 38305, USA
e-mail: tina_chouinard@fws.gov

N. A. Fisichelli
Climate Change Response Program, National Park Service, Fort Collins, CO 80524, USA
e-mail: NFisichelli@schoodicinstitute.org

N. A. Fisichelli
Schoodic Institute at Acadia National Park, Winter Harbor, ME 04693, USA

J. A. Gallo
Conservation Biology Institute, Corvallis, OR 97333, USA
e-mail: john.gallo@consbio.org

J. R. Lujan
U.S. Fish and Wildlife Service, National Wildlife Refuge System, Albuquerque,
NM 87103, USA
e-mail: joseph_lujan@fws.gov

This chapter is a reprint of a chapter in Landscape and Urban Planning Volume 176, August 2018, Pages 64–74(https://doi.org/10.1016/j.landurbplan.2018.04.008). Reprint with kind permission of Elsevier. License order number4570171307086 of April 16, 2019.

L. Müller and F. Eulenstein (eds.), *Current Trends in Landscape Research*,
Innovations in Landscape Research, https://doi.org/10.1007/978-3-030-30069-2_14

single-institution interests and decision-making processes. We propose a platform for an emerging and evolutionary step change in sustainability planning: landscape conservation design (LCD). We use existing governance and adaptation planning principles to develop an iterative, flexible innovation systems framework—the "iCASS Platform." It consists of nine principles and five attributes—innovation, convening stakeholders, assessing current and plausible future landscape conditions, spatial design, and strategy design. The principles are organized around four cornerstones of innovation: people, purpose, process, and product. The iCASS Platform can facilitate LCD via processes that aim to create and empower social networks, foster stakeholder involvement, engender co-production and cross-pollination of knowledge, and provide multiple opportunities for deliberation, transparency, and collaborative decision-making. Our intention is to pivot from single-institution, siloed assessment and planning to stakeholder-driven, participatory design, leading to collaborative decision-making and extensive landscape conservation.

Keywords Adaptation · Design · Governance · Planning · Stakeholders · Sustainability

14.1 Introduction

The dawn of the Anthropocene—an era characterized by human-induced global ecological change and uncertainty—presents a preview of a possible future quite different from the environment that fostered the emergence and prosperity of present-day human societies. Adapting to the Anthropocene's complex array of

R. J. McCormick
Environmental Quality and Protection, Bureau of Land Management, Washington, DC 20036, USA
e-mail: rmccormi@blm.gov

T. A. Miewald
U.S. Fish and Wildlife Service, North Pacific Landscape Conservation Cooperative/National Wildlife Refuge System, Portland, OR 97232, USA
e-mail: thomas_miewald@fws.gov

B. A. Murry
U.S. Fish and Wildlife Service, Caribbean Landscape Conservation Cooperative, San Juan, PR 00913, USA
e-mail: brent_murry@fws.gov

D. J. Pierce
Washington Department of Fish and Wildlife, Wildlife Program, Olympia, WA 98501, USA
e-mail: John.Pierce@dfw.wa.gov

D. R. Shively
Watershed, Fish, Wildlife, Air & Rare Plants Program, U.S. Forest Service, Washington, DC 20250, USA
e-mail: dshively@fs.fed.us

change is a "super-wicked" problem (Levin et al. 2012, p. 2; Waddock 2013), comprised of multiple, contingent, and conflicting issues. Super-wicked problems cannot be fully assessed using siloed decision-making approaches developed by hierarchical institutions using disciplinary science (Norris et al. 2016). Finding adaptive solutions for how to thrive in the Anthropocene rests on human ingenuity fostering transformability toward social, economic, and ecological sustainability. To that end, we propose a platform for an emerging and evolutionary step change in sustainability planning: landscape conservation design (LCD) (see Table 14.1).

Our theory of change (Fig. 14.1) is grounded in the belief that just as Earth's biomes and human civilizations evolved during the Holocene and will continue to do so in the Anthropocene, so too must the governance structures and processes societies use to guide decision-making (Armitage et al. 2010; Voss et al. 2006). To be successful in tackling wicked problems, natural and cultural resource practitioners need a flexible, multi-stakeholder governance structure that transcends

Table 14.1 Glossary of terms

Term	Definition
Adaptive comanagement	A process by which institutional arrangements and ecological knowledge are tested and revised in a dynamic, ongoing, self-organized process of learning-by-doing (Folke et al. 2002)
Adaptation pathway	An analytical approach to planning that allows for uncertainty and change by encouraging consideration of multiple possible futures and the robustness and flexibility of options across these futures (Bosomworth et al. 2015)
Double- and triple-loop learning	Double-loop learning: revisiting assumptions about cause and effects; triple-loop learning: reassessing underlying values and beliefs, potentially resulting in changes to institutional norms (Argyris 1976; Butler et al. 2016; Pahl-Wostl 2009)
Governance	Sustaining coordination and coherence among a wide variety of stakeholders with different purposes and objectives (Pierre 2000) heuristic Involving or serving as an aid to learning, discovery, or problem-solving by experimental and especially trial-and-error methods (Merriam-Webster Dictionary, http://www.merriam-webster.com/dictionary/heuristic)
iCASS Platform	A heuristic for landscape conservation design. The *iCASS Platform* is an innovation systems framework consisting of five attributes and nine principles. The *iCASS* acronym stands for: i = innovation, iC = inclusiveness: convene stakeholders, iA = interdisciplinary assessment of current and plausible future conditions, iS^1 = interactive spatial design, and iS^2 = informative strategy design
Innovation systems framework	A holistic and dynamic approach to problem-solving that utilizes creative thinking to solve societal challenges (Iizuka 2013)
Landscape	A bounded area of indeterminate size that humans have an affinity for or connection to, and within which they assess appearance, quality, and function of the landscape based on social norms and interest (Termorshuizen and Opdam 2009)

(continued)

Table 14.1 (continued)

Term	Definition
Landscape conservation	Landscape conservation is the rapidly growing practice of people working together across large geographies, regardless of political boundaries, to conserve our natural and cultural heritage and ensure a sustainable future for both people and nature [Network for Landscape Conservation. (n.d.). Retrieved August 23 (2017)]. It connects wild lands, working lands, and urban areas into whole, healthy landscapes [or social-ecological systems], and enhances the conservation value of all lands [and waters] through the development of strategies that promote adaptation and resilience [Center for Large Landscape Conservation (n.d.)]
Landscape conservation design (LCD)	A stakeholder-driven, participatory process that: (1) integrates societal values and cross-jurisdiction, multisector interests with the best available interdisciplinary science and traditional knowledge; (2) assesses spatial and temporal patterns, vulnerabilities, risks, and opportunities for landscape elements valued by stakeholders; (3) results in a set of spatially explicit products and multi-objective adaptation strategies; and (4) which protects biodiversity, conserves ecosystem services, and promotes social-ecological systems (e.g., landscapes) that are resilient and sustainable for current and future generations
Silo	A system, process, department, etc. that operates in isolation from others (Oxford Dictionary, https://en.oxforddictionaries.com/definition/silo)
Social-ecological system	An integrated system of ecosystems and human society with reciprocal feedback and interdependence. The concept emphasizes the humans-in-nature perspective (Folke et al. 2010)
Social learning	Where multiple agents combine different values, experiences, and knowledge to identify issues and potential solutions, analyze alternatives, debate choices, and identify priorities through inclusive and deliberative processes (Ojha et al. 2013)
Stakeholder	All human agents and agencies, regardless of expertise, title, or role in the design process
Transformability	The capacity to create a fundamentally new system when ecological, economic, or social (including political) conditions make the existing system untenable (Walker et al. 2004)

single-institution interests and siloed decision-making (Knight et al. 2013; Toomey et al. 2016). To attain that vision, we briefly discuss the well-established concepts underlying landscape conservation design, and then using established adaptation principles, we develop an innovation systems framework—the iCASS Platform—and introduce its five attributes and nine principles. The purpose of this paper is to provide practitioners with a framework that can be used to guide, test, and evaluate landscape conservation design initiatives. Our intention is to ignite transformation

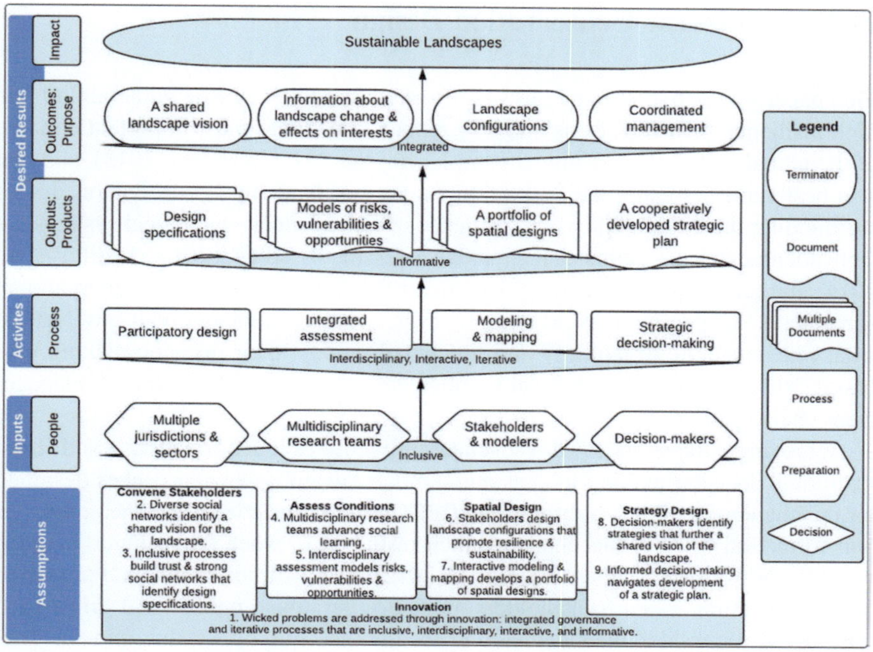

Fig. 14.1 Theory of change. The iCASS Platform, a set of attributes and principles for landscape conservation design, can result in effective inputs, activities, outputs, and outcomes that facilitate sustainable social–ecological systems

from single-institution, siloed assessment and planning to stakeholder-driven, participatory design, leading to collaborative decision-making and extensive landscape conservation.

It bears noting that this paper is unique among the many that tackle similar issues and scope: It is written by practitioners from federal/state agencies and nonprofit organizations that are enmeshed in the socio-political realities of conservation in the USA and, as such, provides a lens of pragmatism not necessarily present among scholarly papers. We acknowledge every country has a particular conservation context it must operate within, such as the challenges between private property rights versus the commons, economic constraints, etc. This paper, and the iCASS Platform it introduces, incorporates universal truths in adaptation planning and governance and, as such, can be applicable, not only within the US context, but the global community at large.

14.2 Landscape Conservation Design

This paper holds that sustainability—an overall condition of low vulnerability and high resilience (Wu 2013)—is and will remain, the single most important concern of the global community (Kuhlman and Farrington 2010; Waddock 2013), and that it is best achieved using the adaptation strategy of landscape conservation. Issues surrounding the vulnerability and resilience of linked social–ecological systems are well documented in global and national assessments (IPCC 2014; Melillo et al. 2014), national strategic planning documents (NFWPCAP 2012), and international agreements (United Nations 2015). However, questions remain about how, and at what rate, societies structurally and functionally adapt their decision-making processes when faced with either fully anticipated or completely unknown management challenges.

Achieving a trajectory toward sustainability requires transformation: a shift from traditional decision-making models and a move toward novel approaches designed for the challenges of the Anthropocene (IPCC 2014). Societies begin the process of transformation when events challenge their resilience or when their support systems begin to show vulnerabilities (Kates et al. 2012). Ideally, transformational processes gain momentum when stakeholders undergo iterative double- and triple-loop learning to reconsider their values and institutions, question status quo methodologies, and explore broad-scale intensive and extensive strategies (Argyris 1976; Butler et al. 2016; Pahl-Wostl 2009).

Sustainability science (Heinrichs et al. 2015) has broadened the discussion from incremental management actions focused on individual components of localized systems to holistic, transformational design processes that enhance the coupling of social and ecological systems (Berkes et al. 2003; Folke et al. 2010; Walker et al. 2004).

Transformational approaches, such as ecosystem management (Lackey 1998) and adaptive co-management (Folke et al. 2002; Plummer et al. 2012), have been developed and tested over recent decades. More recently, "large-scale," "landscape," and "large landscape" approaches (with or without the term "conservation" appended) have begun to be expressed in policy (DOI 2017) and explored in regional decision-making processes (Minnesota Prairie Plan Working Group 2011).

Erlhoff and Marshall (2008) note the design discipline includes the people, processes, and products that facilitate people's intention (or purpose) to transform their environment into a more desirable one. We refer to people, purpose, process, and product as cornerstones of innovation. Cope and Kalantzis (2011) describe design as "an engine of change" (p. 49), and Roozenburg and Eekels (1995) define it as "conceiving [of] an idea for [development of] an artifact or system, and expressing the idea in an embodiable form" (p. 53). Within the context of landscape conservation, design can be an adaptation pathway—a transformative decision-making approach that challenges, if not contributes to, the body of existing strategic frameworks. Our intention is to further landscape conservation through stakeholder-driven design processes, undertaken at regional scales, which identify desired landscape configurations and adaptation strategies that promote sustainability.

Therefore, we define LCD as a stakeholder-driven, participatory process that: (1) integrates societal values and cross-jurisdiction, multisector interests with the best available interdisciplinary science and traditional knowledge (*the people*); (2) assesses spatial and temporal patterns, vulnerabilities, risks, and opportunities for landscape elements valued by stakeholders (*the process*); (3) results in a set of spatially explicit products and multi-objective adaptation strategies (*the products*), and (4) which protects biodiversity, conserves ecosystem services, and promotes social–ecological systems (e.g., landscapes) that are resilient and sustainable for current and future generations (*its purpose*).

A proactive approach to landscape conservation that yields a robust, complex, interconnected network of protected, conserved, working, and developed lands can facilitate an adaptive response to emerging or unforeseen challenges and promote resilient societies within sustainable landscapes (Melillo et al. 2014; NFWPCAP 2012). The over-arching aim of LCD is to collaboratively develop innovative, coordinated strategies (*a strategic plan*) that facilitate adaptation across jurisdictions and sectors. In addition, it identifies landscape configurations (*blueprints*) that allow ecological systems to resist, recover, and adapt to changing conditions (Aplet and McKinley 2017; Fisichelli et al. 2016), typically by encouraging ecologically representative, redundant networks that work to restore or maintain ecosystem integrity, strengthen connectivity, and appropriately scale responses to disturbances (NFWPCAP 2012).

To be successful, LCD must be a bottom-up, collaborative approach to making decisions, as opposed to a top-down, single-institution planning process (Ansell and Gash 2008; NASEM 2015; Wyborn and Bixler 2013). LCD combines the concepts of social learning and spatial and strategic planning to come to understand the nature of ecological systems as the supporting context of social–ecological systems. LCD can be a crucial aspect of landscape conservation because of its participatory approach, and because successful landscape conservation rests on stakeholder choices (Nassauer 2012). Stakeholder participation in LCD can assure that products developed offer relevant guidance to those engaging in the stewardship and use of desired future landscapes (Nassauer 2012). LCD characteristics and how they differ from planning characteristics are presented in Table 14.2.

14.3 The iCASS Platform: An Innovation Systems' Framework For LCD

A single design process for landscape conservation would stifle the innovation needed to address the challenges of wicked problems (Erlhoff and Marshall 2008; Nassauer 2012). As an alternative, we propose a heuristic for LCD: the iCASS

Table 14.2 Landscape conservation design and equivalency with general planning characteristics organized by four cornerstones of innovation

Cornerstones of innovation	Planning characteristics (The current approach)	Landscape conservation design characteristics (The transformed approach)
People	Single institutions	Multiple institutions
	"Internal" coordination/ facilitation	Bridging organization coordinates/ facilitates
Purpose	Achieve the agent's mission, mandates, and goals	Achieve resilient and sustainable multifunctionality
	Focus is within jurisdictional boundaries	Focus is region-wide
	Agent-specific learning and decision-making	Social learning and decision-making
	Institutional success	Collective impact
Process	Institution-driven	Stakeholder-driven
	Rigid governance structure	Flexible governance structure
	Stakeholders provide input as part of the process	Stakeholders direct all aspects of the process
	Issue-driven	Value/interest-driven
	Atomistic	Holistic
	Disciplinary-based	Interdisciplinary-based
	Science informs single-agent policies, programs, strategies	Science informs multiple stakeholders' policies, programs, strategies
	Planning for historic/current conditions	Designing for potential futures
	Siloed decision-making	Horizontal decision-making
	Iterative	Iterative
Product	Determined by agent	Determined by stakeholders
	"Draft" until "Final"	Prototypes
	"Final" until revised	"Living"
	"Sit on the shelf"	Live in "the cloud"
	Text, maps, data	Same plus models, decision-support tools, strategic plans, cooperative agreements, etc.
	Directs institutional action at site-specific scale	Guides collective action at the regional-scale

Platform. It is an innovation systems' framework: a holistic, yet flexible system-based approach that encourages innovation to solve societal challenges (Iizuka 2013). It consists of *five attributes*—innovation, convening stakeholders, *a*ssessing conditions, spatial design, and strategy design (Fig. 14.2)—and *nine principles* (Fig. 14.3). To design it, we synthesized four sources: (1) international agreements on adaptation and biodiversity conservation (IPCC 2014; UNEP 2000, 2011), and US government adaptation policy (CEQ 2011) (Supplemental Material, Table S1); (2) input obtained during five practitioner forums hosted between

Table 14.3 This non-extensive list conveys case studies from the United States that illustrate attributes of the iCASS Platform

iCASS Platform: Attributes	Example	Summary/Purpose	References
Innovation	Carnegie Airborne Observatory	Using advanced imaging technology and data analytics to galvanize action that protects the environment over large geographic extents	Carnegie Airborne Observatory (2018). Who we are [HTML]. https://cao.carnegiescience.edu
	Conservation Innovation Center	Develops partnerships, processes, and products that maximize the efficiency and effectiveness of broad-scale conservation	Chesapeake Conservancy (2018). Conservation Innovation Center [HTML]. http://chesapeakeconservancy.org/conservation-innovation-center/
	Conservation Planning Atlas	Platforms that showcase spatial information and supporting documentation that technical experts, managers, and decision-makers can interact with	Conservation Biology Institute (n.d.). LCC Conservation Planning Atlases [HTML]. https://consbio.org/products/projects/conservation-planning-atlases
Inclusiveness: Convene Stakeholders	Landscape Conservation Cooperatives	A national network of bridging organizations that bring diverse stakeholders together to work collaboratively through conservation planning and design	Landscape Conservation Cooperative Network. (n.d.). About LCC [HTML]. https://lccnetwork.org/about/about-lccs
	National Wind Coordinating Collaborative	Identifies, defines, and discusses through broad stakeholder involvement and collaboration wind power-wildlife interactions	National Wind Coordinating Collaborative (2018). About NWCC [HTML]. https://www.nationalwind.org/about-nwcc
	Pajaro Compass Network	A diverse membership consisting of cross-jurisdiction, multisector interests working collaboratively to articulate collective values	The Pajaro Compass (2016). A Network for Voluntary Conservation [PDF]. http://www.pajarocompass.org/resources/documents/pdf/pajaro-report.pdf

(continued)

Table 14.3 (continued)

iCASS Platform: Attributes	Example	Summary/Purpose	References
Interdisciplinary Assessment of Current and Plausible Future Conditions	Integrated Landscape Assessment	Collaboratively exploring the dynamics of broad-scale, multi-ownership landscapes over time by evaluating and integrating information	Oregon State University (2018). Integrated Landscape Assessment Project [HTML]. http://inr.oregonstate.edu/ilap
	Landscape Conservation and Climate Change Scenarios for the State of Florida	A strategic instrument allowing exploration of potential future conditions which helps organizations make more informed choices in uncertain times	Vargas, J.C., Flaxman, M., & Fradkin, B. (2014). Landscape Conservation and Climate Change Scenarios for the State of Florida: A Decision Support System for Strategic Conservation. Summary for Decision Makers. GeoAdaptive LLC, Boston, MA and Geodesign Technologies Inc., San Francisco CA. Retrieved from http://peninsularfloridalcc.org/page/climate-change-scenarios
	Ecoregional Assessments	Identifies regionally important habitats and gauges potential changes to them	The Nature Conservancy (2017). Ecoregional Assessments [HTML]. https://www.conservationgateway.org/Files/Pages/ecoregional-assessment-to.aspx
Interactive Spatial Design	Adaptation Portfolio Spatial Decision Support System	An optimization algorithm integrated with priority mapping overlays that spatially allocates three management types of interest	Gallo, J. A. & Aplet, G. (2016). Allocating Land to a Three Zone Climate Adaptation Strategy Using a Spatial Decision Support System. Open Science Framework. May 14. https://osf.io/h5pa8/
	Arid Lands Initiative	A facilitated planning process that developed a priority areas portfolio that meets conservation goals for connectivity	Arid Lands Initiative (2014). Our Shared Priorities [HTML]. http://aridlandsinitiative.org/our-shared-priorities/

(continued)

Table 14.3 (continued)

iCASS Platform: Attributes	Example	Summary/Purpose	References
	Crucial Habitat Assessment Tool	Facilitates early non-regulatory planning efforts to reduce conflicts while ensuring wildlife values are better incorporated into land use planning	Western Association of Fish and Wildlife Agencies (2018). Crucial Habitat Assessment Tool: Mapping Fish and Wildlife Across the West [HTML]. http://www.wafwachat.org/
	Minnesota Prairie Conservation Plan	Identifies a landscape configuration and recommendations for protecting, enhancing, and restoring acreage goals	Minnesota Prairie Plan Working Group (2011)
Informative Strategy Design	Implementation of the Minnesota Prairie Conservation Plan	Identifies stakeholder management actions for prairie conservation that ensures no duplicative efforts, missed opportunities, or confusion	Minnesota Prairie Plan Working Group (2011). Implementing the Minnesota Prairie Conservation Plan. Minnesota Prairie Plan Working Group, Minneapolis, MN. 22p. Retrieved from http://www.mda.state.mn.us/~/media/Files/news/govrelations/pollinators/dnrprairieconsplan.ashx
	National Climate Adaptation Strategy	A comprehensive, multi-partner framework to guide responsible and effective actions by natural resource managers and other decision makers	National Fish, Wildlife and Plants Climate Adaptation Partnership (2012)
	Puget Sound Action Agenda	A diverse partnership effort, informed by qualitative and quantitative science, that identifies regional strategies and specific actions	Puget Sound Partnership., (n.d.). Action Agenda for Puget Sound [HTML]. http://psp.wa.gov/action_agenda_center.php

October 2014 and July 2015 in the USA (Supplemental Material); (3) governance and adaptation planning literature; and (4) over 225 years of combined professional experience in conservation research, planning, and delivery within the US context.

The iCASS Platform is a synthesis of adaptation planning concepts and methodologies presented as a set of attributes and principles organized around four cornerstones of innovation: people, purpose, process, and product (Dignan 2013). It emphasizes a design process that is inclusive, interdisciplinary, interactive, and informative. The iCASS Platform can facilitate LCD via processes that create and empower social networks, foster stakeholder involvement, engender co-production and cross-pollination of knowledge, and provide multiple opportunities for deliberation, transparency, learning, and collaborative decision-making (Iizuka 2013; Malerba 2002). By providing organizational guidance to stakeholders on the overall design process, while allowing space for local customization and innovation to unfold, the iCASS heuristic offers practical and flexible guidance for practitioners to follow and may contribute to the larger body of strategic decision-making approaches that exist. We provide case study examples from the authors' experience in the USA to demonstrate iCASS Platform attributes in Table 14.3.

Our intention is to pivot from single-institution, siloed assessment and planning to stakeholder-driven, participatory design, leading to collaborative decision-making and extensive landscape conservation. We understand that advocating a collaborative, multi-objective design approach to landscape conservation represents relatively new territory form any practitioners. While it is our assertion that the iCASS Platform provides a theoretical construct that facilitates making such a leap, only through monitoring many LCD processes, and evaluating if resilient and sustainable social–ecological landscapes are achieved as a result, can its success or failure be determined.

14.3.1 Attribute: i = Innovation

14.3.1.1 Principle #1: Wicked problems are addressed through innovation

Landscape conservation requires a design process that facilitates innovation: the exploration, development, and application of ideas that address wicked problems and improve human well-being (Brown 2009). Innovation emerges when a systems' framework is used to facilitate social learning among diverse agents (Iizuka 2013) who use that knowledge to question existing norms and design new concepts, services, and products (IPCC 2014; Malerba 2002; Ojha et al. 2013). Innovation is the central attribute for iCASS and underpins all other attributes.

Social learning is a foundational principle in innovation systems frameworks and stakeholder efforts to address change and uncertainty. It occurs when multiple agents combine their different values, experiences, and knowledge in order to identify issues and potential solutions, analyze alternatives, debate choices, and establish priorities through inclusive, deliberative processes (Ojha et al. 2013). It is expedited by iterative processes that provide opportunities for governance bodies, consisting of design practitioners, scientists, decision-makers, local experts, and other stakeholders, to engage in hypothesis building, experimentation, adaptive management, and deliberative forms of decision-making (Fulton Suri 2008; IPCC 2014; Jacobson and Robertson 2012). Social learning results in positive outcomes such as stronger relationships, fundamental changes in participant and organizational behavior, and development of innovative strategies (Berkes 2009; Salter et al. 2010).

LCD—similar to other research, planning, and design efforts—requires iterative processes to facilitate social learning in addressing complexity and uncertainty. It is a continuous cycle of conceptualization—an ongoing divergence and convergence of ideas (Roozenburg and Eekels 1995). The application of prototyping is integral to the iterative design process, foundational to experimentation, and key to social learning. Prototypes can inspire stakeholder engagement, visually represent and explore new ideas, produce tangible products that improve over time, and highlight design weaknesses or flaws prior to implementation (Erlhoff and Marshall 2008; Fulton Suri 2008).

Facilitating innovation capable of addressing landscape change requires integrated governance: stakeholder-driven, participatory decision-making grounded in social networks, experimentation, and empowerment (Crona and Parker 2012; Tress et al. 2005). Integrated governance is grounded in stakeholder representativeness and engagement, transparency, collaboration, and informed action (Carson 2009; Hartz-Karp 2007). It originates when cross-jurisdiction, multisector stakeholders progress from siloed decision-making approaches—which have traditionally promoted a polarity between conservation and resource use—to collaborative approaches to managing the flow of ecological services to society (Carcasson 2013; Hanleybrown et al. 2012). Integrated governance can foster innovation by promoting and facilitating transdisciplinary communication, expanding stakeholder knowledge by sharing ideas and perspectives across organizational boundaries, and considering multiple perspectives in making decisions (Erlhoff and Marshall 2008; Tress et al. 2005).

14.3.2 *Attribute: iC = Inclusiveness: Convening Landscape Stakeholders*

14.3.2.1 Principle #2: Diverse Social Networks Identify a shared Vision for the Landscape (Innovation Cornerstones: People, Purpose)

Convening a cross-jurisdiction, multisector body of stakeholders that represents the diversity of social values and interests in the landscape is central to the successful design and development of a sustainable social–ecological landscape (CEQ 2011; IPCC 2014; UNEP 2000, 2011). Inherent in this attribute is the recognition that no

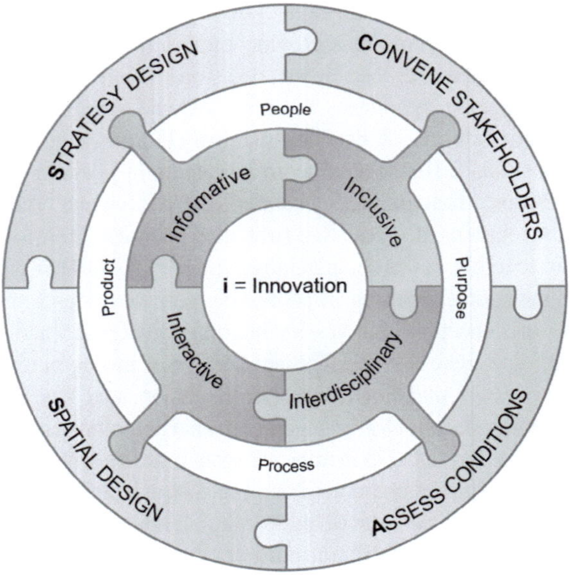

Fig. 14.2 iCASS Platform is an adaptive, innovative systems' framework for landscape conservation design organized around four cornerstones of innovation: people, purpose, process, and product. It includes five attributes: **a** innovation, **b** inclusiveness: convening stakeholders, **c** interdisciplinary assessment of current and plausible future conditions, **d** interactive spatial design, and **e** informative strategy design

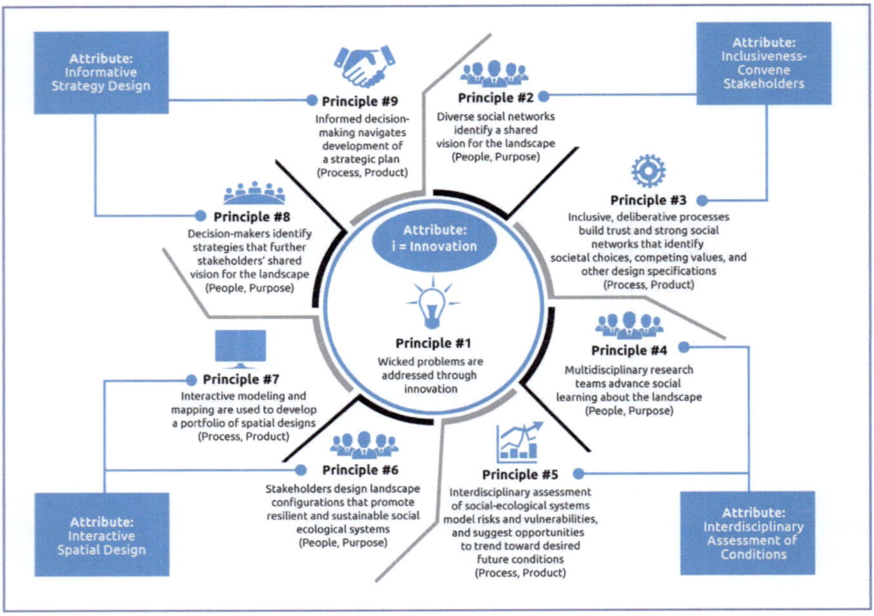

Fig. 14.3 The iCASS Platform consists of an adaptive set of nine principles organized around five attributes

single institution possesses the ability to respond to the complexity of the Anthropocene alone, and that an inclusive, stakeholder-driven approach to LCD relieves individual institutions from solving the sustainability conundrum alone. The purpose of being inclusive in LCD is to convene and empower abroad coalition of stakeholders to create a shared vision of a landscape responsive to societal perceptions of vulnerability and risk. Another objective is to design and deliver measures—a collective, stakeholder response—that promote a sustainable landscape (Roozenburg and Eekels 1995; Sayer et al. 2013).

A diversity of stakeholders in the design process increases opportunities to share knowledge, technical expertise, and other resources (Ansell and Gash 2008); assures that relevant science will be used to guide learning and decision-making (Funtowicz and Ravetz 1994; Pahl-Wostl 2009); and allows for individual and institutional champions to emerge in taking a collaborative decision-making approach to sustainability (Fulton Suri 2008). More importantly, it helps ensure design products are relevant to stakeholders (Nassauer and Opdam 2008; Salter et al. 2010). We know that identifying and integrating knowledge across disciplines and stakeholder groups can be challenging (Raymond et al. 2010), yet it is an essential step toward achieving inclusiveness. Bringing a broad range of stakeholders together is not, however, a panacea. Stakeholders often bring competing objectives and maladaptive agendas to the table (Butler et al. 2016), but working through disparate views and values is an important step in promoting inclusiveness, social learning, and coordinated management.

14.3.2.2 Principle #3: Inclusive, Deliberative Processes Build Trust and Strong Social Networks that Identify Societal Choices, Competing Values, and Other Design Specifications (Innovation Cornerstones: Process, Product)

Building effective partnerships capable of enduring the challenges of extended collaboration is difficult. Improved approaches to enhance the success of stakeholder participation and further transformation toward a desired future are needed (Hansen et al. 2012). Inclusive, decentralized decision-making processes, like LCD, are grounded in three core principles of open governance—participation, transparency, and collaboration (Pahl-Wostl 2009). Convening an inclusive set of cross-jurisdiction, multisector stakeholders requires unique and adaptive leadership models. Bridging organizations are key to facilitating inclusive, open governance processes that build trust among diverse stakeholders (Hanleybrown et al. 2012; Jacobson and Robertson 2012). Their role is to cultivate initial enthusiasm for the project, provide a neutral environment for stakeholders to deliberate, commence and coordinate work, and identify project champions to provide complementary leadership (Crona and Parker 2012).

Deliberation—the purposeful process of empowering stakeholders by providing facilitated forums that strive for transparency—is necessary in horizontal, participatory decision-making (Carcasson 2013; Ojha et al. 2013). Trust and collaboration are more likely to unfold and cultivate innovations that facilitate long-term outcomes when diverse stakeholders share views and information in open and transparent forums (Knight et al. 2013; Termorshuizen and Opdam 2009). Disagreement and conflict may arise at points throughout the deliberative process. Forward progress in LCD may be impeded until a bridging organization or stakeholder champion steps into resolve the conflict (Hartz-Karp 2007; Jacobson and Robertson 2012). Nevertheless, successful collaborations overcome such challenges, and combine expertise and resources to achieve synergistic effects in adaptation (Kania and Kramer 2011).

It is not enough to focus solely on participatory decision-making processes. Doing so can result in a stakeholder consensus for meeting their needs and interests without meeting ecosystem needs, especially over the long-term. What we strive for is a pragmatic, yet disciplined approach that produces substantive outcomes using the tenets of conservation science while empowering stakeholder social learning and negotiation (Groves and Game 2016). To that end, facilitated forums lead stakeholders to: identify design specifications; include values, interests and targets; present perspectives on vulnerabilities and risks; articulate assumptions; and develop scenarios, prototypes, and strategies for promoting sustainable social–ecological landscapes (Iversen et al. 2012; Jacobson and Robertson 2012). Design specifications guide compilation and co-production of scientific information—including interdisciplinary assessments of current and plausible future conditions—that supports decision-making.

14.3.3 Attribute: iA = Interdisciplinary Assessment of Current and Plausible Future Conditions

14.3.3.1 Principle #4: Multidisciplinary Research Teams Advance Social Learning About the Landscape (Innovation Cornerstones: People, Purpose)

Designing sustainable social–ecological systems requires fully realized narratives that express the social, economic, and ecological values held and desired by stakeholders (CEQ 2011; IPCC 2014; UNEP 2000, 2011). Assessing the current, plausible, and desired future conditions of a multifunctional landscape is a complex task suited to no single discipline (Salter et al. 2010). Multidisciplinary research teams consisting of technical and non-technical stakeholders, representing relevant social, economic, and ecological disciplines, and local, traditional, and indigenous groups are fundamental to landscape conservation and the design of sustainable social–ecological systems (IPCC 2014). Inherent in this attribute is the recognition that no single discipline can understand the uncertainties of how social–ecological systems respond to change; and that interdisciplinary assessment teams are needed to explore the transdisciplinary nature of, and issues associated with, those systems

(Erlhoff and Marshall 2008; NFWPCAP 2012; Norris et al. 2016). Interdisciplinary assessments compile and co-produce information that provides a holistic understanding of the complex spatial and temporal aspects of, and linkages among, social–ecological elements of interest to stakeholders. In addition, such assessments facilitate social learning about how environmental change and uncertainty affects those interests (NFWPCAP 2012; Salter et al. 2010).

14.3.3.2 Principle #5: Interdisciplinary Assessment of Social–Ecological Systems, Model Risks and Vulnerabilities, and Suggest Opportunities to Trend Toward Desired Future Conditions (Innovation Cornerstones: Process, Product)

Developing interdisciplinary assessments to provide a holistic, system-based perspective of stakeholder interests requires stakeholder engagement. Social, economic, and ecological conditions vary by region, and stakeholders are one of the most relevant sources of qualitative and quantitative information about those three elements (CEQ 2011; IPCC 2014; UNEP 2000, 2011). Computer models have proven valuable for systematizing and assembling quantitative information from various disciplines to better understand problems; however, they are less useful for assessing qualitative changes in societies and institutions, which can dramatically affect how individuals and groups interact with their landscape. These qualitative, normative aspects of decision-making have inspired movement toward participatory interdisciplinary assessments (Salter et al. 2010; UNEP 2011).

Bridging organizations can coordinate the development of participatory interdisciplinary assessments that seek to understand transdisciplinary challenges posed by complex societal interactions with a landscape. Design methodologies (Brown 2009; Erlhoff and Marshall 2008; Roozenburg and Eekels 1995) can be used to identify research questions from values and issues defined by stakeholders. Assessment teams should seek out and use existing quantitative data and qualitative information and knowledge relevant to stakeholder values and perceived risks; integrate local, traditional, and indigenous knowledge into all assessment aspects; use innovative technologies to model spatial and temporal patterns, vulnerabilities, and risks (IPCC 2014); and develop social, economic, and ecological endpoints as system-state indicators to understand the current condition of the landscape as well as for a range of plausible future scenarios (Allen and Hoekstra 2015; Groves and Game 2016; IPCC 2014; Tress et al. 2005).

Each scenario should be a coherent, internally consistent, plausible future condition (IPCC 2014; Melillo et al. 2014); and as a set of scenarios, they should collectively offer a divergent range of relevant possible futures (National Park Service 2013; Rowland et al. 2014). Useful scenarios allow decision-makers to craft experimental responses to multiple futures, comprehend the implications of specific uncertainties, and identify opportunities (National Park Service 2013; Rowland et al. 2014). Interdisciplinary approaches within this framework are needed to integrate information on historical, present-day, and possible future system states, including climate, land use, demographics, and other social–ecological drivers. Stakeholders can offer additional insights into the vulnerability, sensitivity, and adaptive capacity of valued social–ecological elements under different potential future system states. Participatory, interdisciplinary assessments can yield analysis of key information sources, ecological context, priority element rankings, data and knowledge uncertainty, vulnerable landscape elements, and potential climate refugia (Hansen et al. 2012). The scientific information co-generated from interdisciplinary assessments directly informs spatial design.

14.3.4 *Attribute: iS*[1] *= Interactive Spatial Design*

14.3.4.1 Principle #6: Stakeholders Design Landscape Configurations that Promote Resilient and Sustainable Social-Ecological Systems (Innovation Cornerstones: People, Purpose)

Spatial design uses the compiled and co-produced data, information, and knowledge from the assessment to identify locations best suited for the range of desired land uses (Groves and Game 2016). Technical and non-technical stakeholders and decision-makers who reflect the diversity of social values and interests on the landscape, and that have a thorough understanding of its vulnerabilities, risks, and potential opportunities, are essential when spatially designing configurations that promote sustainable social–ecological systems (CEQ 2011; IPCC 2014; UNEP 2000, 2011). Inherent in this attribute is a recognition that no single institution possesses the ability to model and map priority locations to retain the diversity of societal values and stakeholder interests, and that an interactive, participatory approach is more effective in visually representing desired future landscape conditions that are supported by stakeholders. The purpose of an interactive spatial design process is to produce landscape configurations that protect biodiversity and the delivery of ecosystem services, maintain and enhance ecosystem integrity, and promote sustainable social–ecological systems (Groves and Game 2016; IPCC 2014; NFWPCAP 2012).

A beneficial aspect of stakeholder engagement in spatial design is development of a deeper trust that the models used to identify priorities integrate their interests with other information and knowledge, which furthers social learning and collective agreement on resource allocation and landscape objectives (Melillo et al. 2014). Interactive spatial design helps stakeholders to organize and evaluate the costs and benefits of various options, communicate the uncertainty associated with various trade-offs, and identify priorities (Melillo et al. 2014; Moilanen et al. 2009). As they participate in spatial design, stakeholders learn that meeting one objective often makes it difficult or impossible to achieve others, requiring further negotiation. Overall, the co-development of a spatial design helps organize landscape elements while maintaining and improving stakeholder buy-in (De Groot et al. 2009; Melillo et al. 2014).

14.3.4.2 Principle #7: Interactive Modeling and Mapping are Used to Develop a Portfolio of Spatial Designs (Innovation Cornerstones: Process, Product)

Bridging organizations can coordinate development of a spatial design portfolio that promotes sustainable social–ecological systems. The use of decision support tools and applications, in conjunction with charrettes attended by stakeholders, facilitates development of complex spatial designs and visual communication of priority land-use decisions (De Groot et al. 2009; Moilanen et al. 2009). Two primary approaches to spatial modeling include: optimization models, which provide spatial analytical solution sets that strive for the biggest "bang for the buck" for one or a few objectives; and priority mapping overlay models, which assign a relative value of importance for a priority objective (e.g., biodiversity) or combination of objectives (e.g., social and ecological resilience) for every landscape analysis unit (Knight et al. 2013). For the latter approach, a large number of sub-objectives may be combined in a tree-based hierarchy. Both approaches have their benefits, and an integrated model using both may be desirable (Jankowski et al. 2014).

A spatial design portfolio contains model-based maps showing the potential locations of specific land-use or management activities, considering future scenarios (Moilanen et al. 2009). Design models attempt to maximize the size of core areas while minimizing perimeter and reducing edge effects. Other identified areas can include buffer zones and transition areas and important habitat connectivity areas that maintain landscape permeability and facilitate gene flow (Moilanen et al. 2009; NFWPCAP 2012; Schmitz et al. 2015). The portfolio can communicate spatial and temporal uncertainty by visually depicting variability and sensitivity of the modeling results (Jankowski et al. 2014) under various scenarios. Land-use prioritizations co-generated from spatial design directly inform strategy design.

14.3.5 *Attribute: iS^2 = Informative Strategy Design*

14.3.5.1 Principle #8: Decision-Makers Identify Strategies that Further Stakeholders' Shared Vision for the Landscape (Innovation Cornerstone: People, Purpose)

Strategy design builds upon the portfolio of products developed in spatial design to address the knowledge-action gap that occurs in many broad-scale, conservation planning efforts (Groves and Game 2016; Wyborn and Bixler 2013). Decision-makers—supported by scientists and planners—are critically important to successful development of coordinated adaptation strategies that promote sustainable social–ecological systems (IPCC 2014; Melillo et al. 2014; Watson et al. 2012; Wyborn 2015). Inherent in this attribute is a recognition that siloed decision-making models focused within institutional boundaries are unable to sufficiently respond to the complexity and rate of change in social–ecological systems. Horizontal decision-making models facilitate the development of information relevant across institutional boundaries and can inform collective action and have collective impact (Watson et al. 2012; Wyborn and Bixler 2013). The purpose of informed strategy design is to translate science products into mutually reinforcing strategies that identify stakeholder roles in fulfilling a shared vision for the landscape. A strategy can promote across jurisdiction, multisector approach to governance (Wyborn and Bixler 2013) that coordinates on-the-ground delivery, monitoring, evaluation, and an adaptive approach to design revision. Integrated governance encourages management across social and ecological boundaries and anticipates conditions that support viable and productive communities within social–ecological systems (NFWPCAP 2012). Strategy design maximizes synergies and minimizes trade-offs and confusion resulting from siloed decision-making and implementation.

14.3.5.2 Principle #9: Informed Decision-Making Navigates Development of a Strategic Plan (Innovation Cornerstones: Process, Product)

Bridging organizations bring stakeholders together around the assessment products and spatial prioritizations developed throughout the LCD process. In strategy design, stakeholders use that body of work to develop a common understanding of synergistic implementation approaches through the identification of high-level adaptation strategies. These strategies are then articulated as a collaboratively developed strategic plan that guides stakeholder activities within a multifunctional landscape (Melillo et al. 2014; Wyborn 2015).

The Collective Impact framework (Barberg 2015; Hanleybrown et al. 2012; Kania and Kramer 2011) sets an alignment goal where each stakeholder articulates their contribution to ensuring the sustainability of their supporting social–ecological system. As Naiman (2013) notes, when restoring large river systems, success depends "on a diverse group of stakeholders working together, not by requiring all participants do the same thing, but by encouraging each participant to undertake specific activities at which it excels in a way that supports and is coordinated with the actions of others" (p. 406). No single agent can deliver all the tools needed to implement a strategy. Strategy design identifies the full range of potential cross-jurisdiction, multisector tools that complement individual stakeholder efforts and make the collective whole more effective. This articulation of mutually reinforcing activities (Barberg 2015; Kania and Kramer 2011) lies at the core of strategy design.

Fostering agreements between stakeholders that have, at times, conflicting missions, mandates, and goals are difficult. Knowledge of and skill in conflict resolution, facilitation and negotiation techniques, and a diversity of tools such as structured decision-making, "robust decision-making," and trade-off analysis can help stakeholders get past road- blocks (Melillo et al. 2014). Best practices in co-governance suggest focusing first on low-hanging fruit—easy wins—to gain momentum (Ansell and Gash 2008). Early success builds trust, and stakeholders are more likely to attempt larger challenges (Berkes 2009). Overtime, continual action evolves into what Lauber et al. (2011) term the "partnership utilization" phase, where the stakeholders' implementation efforts mature.

14.4 Conclusion

Adapting to the Anthropocene's complex array of change agents is a "super-wicked" problem that cannot be fully assessed using siloed decision-making approaches developed by hierarchical institutions using disciplinary science. Developing solutions increase sustainability rests, in part, on the ability of natural and cultural resource practitioners to transform how they make decisions about land conservation and utilization.

LCD is a pathway to transforming sustainability planning. As a stakeholder-driven, participatory process, it identifies landscape configurations and adaptation strategies that promote sustainable social–ecological systems. We propose an innovation systems framework for LCD: the iCASS Platform—a flexible governance structure that transcends single-institution interests and siloed decision-making processes. The iCASS Platform is a set of attributes and principles organized around four cornerstones of innovation: people, purpose, process, and product. It emphasizes a design methodology that is inclusive, interdisciplinary, interactive, and informative. The iCASS Platform can facilitate LCD and expedite landscape conservation via processes that create and empower social networks, engender co-production and cross-pollination of data, information and knowledge, and provide multiple opportunities for deliberation, transparency, learning, and collaborative decision-making.

Although this paper focuses solely on the design component of the adaptive learning cycle (generally expressed as *plan* → *design* → *implement* → *monitor* → *evaluate* → *revise* → *repeat*) (Williams et al. 2009), we acknowledge the importance of testing the iCASS Platform (i.e., implementing, monitoring, and evaluating results) to determine its effectiveness furthering LCD. We argue that bridging organizations play a fundamental role in coordinating and facilitating LCD. We also contend that they are the appropriate entities to evaluate iCASS's effectiveness as a participatory design methodology—one that integrates societal values and cross-jurisdiction, multisector interests with the best available interdisciplinary science and traditional knowledge. Using iCASS to guide the assessment of spatial and temporal patterns, vulnerabilities, risks, and opportunities for landscape elements valued by stakeholders can produce a set of spatially explicit products and strategies that can achieve resilient and sustainable social–ecological systems for current and future generations.

Acknowledgements We wish to dedicate this article to the memory of our co-author, John Pierce, who passed away on February 23, 2018. John was the Chief Wildlife Scientist for the Washington Department of Fish and Wildlife, where he worked for over 30 years. John truly embodied the spirit of landscape conservation: passionate, innovative, and committed to collaboration. He had a unique ability to inspire others and will be greatly missed by all.

We thank, with great appreciation, the following for their contribution in development of this essay: Amanda Robertson, Anisa Romero, and the *Landscape Conservation Design Minimum Standards Working Group*—an ad hoc group of conservation professionals—for their assistance in developing an unpublished report (Campellone et al. 2014) that provided background information for this paper. The paper is benefited greatly from the thoughtful critical review of Catherine Doyle-Capitman, Sean Finn, Natalie Sexton, Scott Schwenk, and Steve Traxler. The authors acknowledge financial support from the U.S. Fish and Wildlife Service, Headquarters Office, National Wildlife Refuge System.

Disclaimer The findings and conclusions in this article are those of the authors and do not necessarily represent the views of the U.S. Fish and Wildlife Service or any other agency. Any use of trade, firm, or product names is for descriptive purposes only and does not imply endorsement by the US Government.

Appendix: Supplementary Data

Supplementary data associated with this article can be found, in the online version, at http://dx.doi.org/10.1016/j.landurbplan.2018.04.008.

References

Allen TFH, Hoekstra TW (2015) Toward a unified ecology, 2nd edn. Columbia University Press, New York, NY

Ansell C, Gash A (2008) Collaborative governance in theory and practice. J Public Adm Res Theor 18(4):543–571. https://doi.org/10.1093/jopart/mum032

Aplet GH, McKinley PS (2017) A portfolio approach to managing ecological risks of global change. Ecosyst Health Sustain 3(2):e01261. https://doi.org/10.1002/ehs2.1261

Argyris C (1976) Single-loop and double-loop models in research on decision making. Adm Sci Q 21(3):363–375. https://doi.org/10.2307/2391848

Arid Lands Initiative (2014) The arid lands initiative—Shared priorities for conservation at a landscape scale. Summary Prepared by Sonia A. Hall (SAH Ecologia LLC) and the Arid Lands Initiative Core Team. Wenatchee, Washington, 39 pp

Armitage D, Berkes F, Doubleday N (eds) (2010) Adaptive co-management: collaboration, learning, and multi-level governance. UBC Press, Vancouver, B.C

Barberg B (2015) Harnessing the power source for collective impact: mutually reinforcing activities. Retrieved 19 Dec 2015 from https://collectiveimpactforum.org/sites/default/files/Harnessing%20the%20Power%20Source%20for%20Collective%20Impact%20v4.pdf

Berkes F (2009) Evolution of co-management: role of knowledge generation, bridging organizations and social learning. J Environ Manage 90(5):1692–1702. https://doi.org/10.1016/j.jenvman.2008.12.001

Berkes F, Colding J, Folke C (2003) Navigating social-ecological systems: building resilience for complexity and change. Cambridge University Press, Cambridge, UK

Bosomworth K, Harwood A, Leith P, Wallis P (2015) Adaptation pathways: a playbook for developing options for climate change adaptation in natural resource management. Retrieved from https://terranova.org.au/repository/southern-slopes-nrm-collection/adaptation-pathways-a-playbook-for-developing-options-for-climate-change-adaptation-in-natural-resource-management/scarp-adaptation-pathways-a-playbook-final.pdf
Brown T (2009) Change by design: how design thinking transforms organizations and inspires innovation. HarperCollins Publishers, New York, NY
Butler JRA, Suadnya W, Yanuartati Y, Meharg S, Wise RM, Sutaryono Y et al (2016) Priming adaptation pathways through adaptive co-management: design and evaluation for developing countries. Clim Risk Manage 12:1–16. https://doi.org/10.1016/j.crm.2016.01.001
Campellone R, Bartuszevige A, Chouinard T, Lujan J, Miewald T, Murry B et al (2014) An analytical framework for conservation design at landscape-scales (landscape conservation design). Available at https://osf.io/vpmcj/
Carcasson M (2013) Tackling wicked problems through deliberative engagement. Colorado Municipalities, pp 9–13. Retrieved 14 Nov 2015 from http://www.academia.edu/4789230/Carcasson-_Tackling_wicked_problems_through_deliberative_engagement_Colorado_Municipalities
Carnegie Airborne Observatory (2018) Whoweare [HTML]. Retrieved from https://cao.carnegiescience.edu
Carson L (2009) Deliberative public participation and hexachlorobenzene stockpiles. J Environ Manage 90(4):1636–1643. https://doi.org/10.1016/j.jenvman.2008.05.019
Center for Large Landscape Conservation (n.d.) Donor Resource Document. Retrieved 5 Apr 2017 from http://largelandscapes.org
Chesapeake Conservancy (2018) Conservation Innovation Center [HTML]. Retrieved from http://chesapeakeconservancy.org/conservation-innovation-center/
Conservation Biology Institute (n.d.) LCC conservation planning atlases [HTML]. Retrieved from https://consbio.org/products/projects/conservation-planning-atlases
Cope B, Kalantzis M (2011) 'Design' in principle and practice: A reconsideration of the terms of design engagement. Des J 14(1):45–63. https://doi.org/10.2752/175630610X12877385838768
Council on Environmental Quality (2011) Implementing climate change adaptation planning in accordance with Executive Order 13514: Federal leadership in environmental, energy, and economic performance. Federal Agency Climate Change Adaptation Planning: Supporting Document. Washington, DC: The Executive Office of the President. Retrieved 14 Nov 2015 from https://www.whitehouse.gov/sites/default/files/microsites/ceq/adaptation_support_document_3_3.pdf
Crona B, Parker J (2012) Learning in support of governance: Theories, methods, and a framework to assess how bridging organizations contribute to adaptive resource governance. Ecol Soc 17 (1):32. https://doi.org/10.5751/ES-04534-170132
De Groot RS, Alkemade R, Braat L, Hein L, Willemen L (2009) Challenges in integrating the concept of ecosystem services and values in landscape planning, management and decision making. Ecol Complex 7(3):260–272. https://doi.org/10.1016/j.ecocom.2009.10.006
Department of the Interior, U.S (2017) Landscape-level management (604 DM 1). Washington, DC: Government Printing Office
Dignan A (2013) The operating model that is eating the world. The Ready. Retrieved 19 Dec 2015 from https://medium.com/the-ready/the-operating-model-that-is-eating-the-world-d9a3b82a5885#.ycnm0jodz
Erlhoff M, Marshall T (2008) Design dictionary: perspectives on design terminology. Berlin: Birkhäuser
Fisichelli N, Schuurman G, Hoffman C (2016) Is 'resilience' maladaptive? Towards an accurate lexicon for climate change adaptation. Environ Manage 57(4):753–758. https://doi.org/10.1007/s00267-015-0650-6
Folke C, Carpenter S, Elmqvist T, Gunderson L, Holling CS, Walker B (2002) Resilience and sustainable development: building adaptive capacity in a world of transformations. Ambio 31 (5):437–440. https://doi.org/10.1579/0044-7447-31.5.437

Folke C, Carpenter S, Walker B, Scheffer M, Chapin T, Rockström J (2010) Resilience thinking: Integrating resilience, adaptability and transformability. Ecol Soc 15(4), 20. Retrieved 19 Dec 2015 from http://www.ecologyandsociety.org/vol15/iss4/art20

Fulton Suri J (2008) Informing our intuition: design research for radical innovation. Rotman Mag. Retrieved 19 Dec 2015 from https://www.ideo.com/images/uploads/news/pdfs/Informing_Our_Intuition.pdf

Funtowicz SO, Ravetz JR (1994) Uncertainty, complexity and post-normal science. Environ Toxicol Chem 13(12):1881–1885. https://doi.org/10.1002/etc.5620131203

Groves CR, Game ET (2016) Conservation planning: Informed decisions for a healing the planet. Roberts and Company Publishers, Greenwood Village, Colorado

Hanleybrown F, Kania J, Kramer M (2012) Channeling change: making collective impact work. Stanford Soc Innov Rev 20, 1–8. Retrieved 19 Dec 2015 from http://ssir.org/articles/entry/channeling_change_making_collective_impact_work

Hansen L, Gregg RM, Arroyo V, Ellsworth S, Jackson L, Snover A (2012) The state of adaptation in the United States: an overview. Washington, DC: EcoAdapt. Retrieved 10 Oct 2015 from http://www.cakex.org/virtual-library/state-adaptation-united-states-overview

Hartz-Karp J (2007) How and why deliberative democracy enables co-intelligence and brings wisdom to governance. J Pub Deliberation 3(2), 1–9. Retrieved 19 Dec 2015 from http://espace.library.curtin.edu.au/R/?func=dbin-jump-full&object_id=201621&local_base=GEN01-ERA02

Heinrichs H, Martens P, Michelsen G, Wiek A (2015) Sustainability science: an introduction. Springer, Netherlands. https://doi.org/10.1007/978-94-017-7242-6

Iizuka M (2013) Innovation systems framework: still useful in the new global context? UNU-MERIT Working Paper Series, 2013–005, 20 pp. Retrieved 14 Nov 2015 from http://www.merit.unu.edu/publications/working-papers/abstract/?id=4868

IPCC (2014) Climate change 2014: impacts, adaptation, and vulnerability. Part A: global and sectoral aspects. In: Field CB, Barros VR, Dokken DJ, Mach KJ, Mastrandrea MD, Bilir TE (eds) Cambridge University Press, Cambridge, UK and New York, NY, USA, 1132 pp. Retrieved 20 June 2015 from http://www.ipcc.ch/pdf/assessment-report/ar5/wg2/WGIIAR5-PartA_FINAL.pdf

Iversen OS, Halskov K, Leong TW (2012) Values-led participatory design. CoDesign 8(2):87–103. https://doi.org/10.1080/15710882.2012.672575

Jacobson C, Robertson AL (2012) Landscape conservation cooperatives: bridging entities to facilitate adaptive co-governance of social-ecological systems. Hum Dimensions Wildlife 17 (5):333–343. https://doi.org/10.1080/10871209.2012709310

Jankowski P, Fraley G, Pebesma E (2014) An exploratory approach to spatial decision support. Comput Environ Urban Syst 45:101–113. https://doi.org/10.1016/j.compenvurbsys.2014.02.008

Kania J, Kramer M (2011) Collective impact. Stanford Soc Innov Rev, pp 36–41. Retrieved 19 Dec 2015 from http://ssir.org/articles/entry/collectiveimpact

Kates RW, Travis WR, Wilbanks TJ (2012) Transformational adaptation when incremental adaptations to climate change are insufficient. Proc Natl Acad Sci USA 109(19):7156–7161. https://doi.org/10.1073/pnas.1115521109

Knight AT, Rodrigues ASL, Strange N, Tew T, Wilson KA (2013) Designing effective solutions to conservation planning problems. In: Macdonald DW, Willis KJ (eds) Key topics in conservation biology, vol 2. Wiley, Oxford, UK, pp 362–383. https://doi.org/10.1002/9781118520178.ch20

Kuhlman T, Farrington J (2010) What is sustainability? Sustainability 2(11):3436–3448

Lackey RT (1998) Seven pillars of ecosystem management. Landscape Urban Plan 40(1):21–30. https://doi.org/10.1016/S0169-2046(97)00095-9

Lauber BT, Stedman RC, Decker DJ, Knuth BA, Simon CN (2011) Social network dynamics in collaborative conservation. Hum Dimensions Wildlife 16(4):259–272. https://doi.org/10.1080/10871209.2011.542556

Levin K, Cashore B, Bernstein S, Auld G (2012) Overcoming the tragedy of super wicked problems: constraining our future selves to ameliorate global climate change. Policy Sci 45:125–152. https://doi.org/10.1007/s11077-012-9151-0

Malerba F (2002) Sectoral systems of innovation and production. Res Policy 31(2), 247–264 Retrieved 14 Nov 2015 from https://doi.org/10.1016/S0048-7333(01)00139-1

Melillo JM, Richmond TC, Yohe GW (eds) (2014) Climate change impacts in the United States: The third national climate assessment. Global Change Research Program, Washington, DC, U. S. https://doi.org/10.7930/J0Z31WJ2

Minnesota Prairie Plan Working Group (2011). *Minnesota prairie conservation plan.* Minnesota Prairie Plan Working Group, Minneapolis, MN, 55p

Moilanen A, Wilson KA, Possingham HP (eds) (2009) Spatial conservation prioritization: quantitative methods and computational tools, vol 6. Oxford University Press, Oxford, UK

Naiman RJ (2013) Socio-ecological complexity and the restoration of river ecosystems. Inland Waters 3(4):391–410. https://doi.org/10.5268/IW-3.4.667

Nassauer JI (2012) Landscape as a medium and method for synthesis in urban eco- logical design. Landscape Urban Planning 106(3):221–229. https://doi.org/10.1016/j.landurbplan.2012.03.014

Nassauer JI, Opdam P (2008) Design in science: extending the landscape ecology paradigm. Landscape Ecol 23(6):633–644. https://doi.org/10.1007/s10980-008-9226-7

National Academies of Sciences Engineering and Medicine (2015) A review of the land- scape conservation cooperatives. The National Academies, Washington, DC. Advance online publication https://doi.org/10.17226/21829

National Fish, Wildlife and Plants Climate Adaptation Partnership (2012) National fish, wildlife and plants climate adaptation strategy. Washington, DC: Association of Fish and Wildlife Agencies, Council on Environmental Quality, Great Lakes Indian Fish and Wildlife Commission, National Oceanic and Atmospheric Administration, and U.S. Fish and Wildlife Service. http://dx.doi.org/%2010.3996/082012-FWSReport-1

National Park Service (2013) Using scenarios to explore climate change: A handbook for practitioners. National Park Service, Fort Collins, CO

National Wind Coordinating Collaborative (2018) About NWCC [HTML]. Retrieved from https://www.nationalwind.org/about-nwcc

Network for Landscape Conservation (n.d.) Retrieved 23 Aug 2017 from http://www.largelandscapenetwork.org

Norris PE, O'Rourke M, Mayer AS, Halvorsen KE (2016) Managing the wicked problem of transdisciplinary team formation in socio-ecological systems. Landscape Urban Planning 154:115–122. https://doi.org/10.1016/j.landurbplan.2016.01.008

Ojha HR, Hall A, Sulaiman R (2013) Adaptive collaborative approaches in natural resources governance. Routledge, New York, NY

Oregon State University (2018) Integrated landscape assessment project [HTML]. Retrieved from http://inr.oregonstate.edu/ilap

Pahl-Wostl C (2009) A conceptual framework for analyzing adaptive capacity and multi-level learning processes in resource governance regimes. Global Environ Change 19(3):354–365. https://doi.org/10.1016/j.gloenvcha.2009.06.001

Pierre J (2000) Debating governance: authority, steering, and democracy. Oxford University Press, New York, NY

Plummer R, Crona B, Armitage DR, Olsson P, Tengö M, Yudina O (2012) Adaptive comanagement: a systematic review and analysis. Ecol Soc 17(3):11. https://doi.org/10.5751/ES-04952-170311

Puget Sound Partnership (2014) The 2014/2015 action agenda for puget sound. Retrieved from http://www.psp.wa.gov/2014_action_agenda_download.php

Raymond CM, Fazey I, Reed MS, Stringer LC, Robinson GM, Evely AC (2010) Integrating local and scientific knowledge for environmental management. J Environ Manage 91(8):1766–1777. https://doi.org/10.1016/j.jenvman.2010.03.023

Roozenburg NFM, Eekels J (1995) Product design: fundamentals and methods. Wiley, New York, NY

Rowland ER, Cross MS, Hartmann H (2014) Considering multiple futures: scenario planning to address uncertainty in natural resource conservation. U.S. Fish and Wildlife Service, Washington, DC. Retrieved 16 Jan 2016 from http://www.fws.gov/home/climatechange/pdf/Scenario-Planning-Report.pdf

Salter J, Robinson J, Wiek A (2010) Participatory methods of integrated assessment—a review. Wiley Interdisc Rev Clim Change 1(5):697–717. https://doi.org/10.1002/wcc.73

Sayer J, Sunderland T, Ghazoul J, Pfund J, Sheil D, Meijaard E et al (2013) Ten principles for a landscape approach to reconciling agriculture, conservation, and other competing land uses. Proc Natl Acad Sci USA 110(21):8349–8356. https://doi.org/10.1073/pnas.1210595110

Schmitz OJ, Lawler JJ, Beier P, Groves C, Knight G, Boyce DA Jr et al (2015) Conserving biodiversity: Practical guidance about climate change adaptation approaches in support of land-use planning. Nat Areas J 35(1):190–203. https://doi.org/10.3375/043.035.0120

Termorshuizen JW, Opdam P (2009) Landscape services as a bridge between landscape ecology and sustainable development. Landscape Ecol 24(8):1037–1052. https://doi.org/10.1007/s10980-008-9314-8

The Nature Conservancy (2017) Ecoregional assessments [HTML]. Retrieved from https://www.conservationgateway.org/Search/Pages/results.aspx?k=Ecoregional%20Assessments&a=&r=

The Pajaro Compass (2016) Pajaro compass document and appendices. A network for voluntary conservation [PDF]. Retrieved from http://pajarocompass.org/resources/documents/

Toomey AH, Knight AT, Barlow J (2016) Navigating the space between research and implementation in conservation. Conserv Lett. https://doi.org/10.1111/conl.12315

Tress B, Tress G, Fry G, Opdam P (eds) (2005) From landscape research to landscape planning: Aspects of integration, education and application. Springer, Dordrecht, Netherlands

United Nations (2015) Adoption of the Paris Agreement: Proposal by the President (Draft Decision). FCCC/CP/2015/L.9, p 31. Retrieved 14 Nov 2015 from http://unfccc.int/resource/docs/2015/cop21/eng/l09.pdf

United Nations Environment Program (2000) Report of the fifth meeting of the conference of the parties to the convention on biological diversity (UNEP/CBD/COP/5). Decision V/6. Ecosystem Approach (CBD, 2000). Retrieved from https://www.cbd.int/decision/cop/?id=7148

United Nations Environment Program (2011) Report on how to improve sustainable use of biodiversity in a landscape perspective. (UNEP/CBD/SBSTTA/15/13), 14 p. Retrieved from https://www.cbd.int/doc/meetings/sbstta/sbstta-15/official/sbstta-15-13-en.pdf

Voss JP, Bauknecht D, Kemp R (eds) (2006) Voss JP, Bauknecht D, Kemp R (eds) (2006). Reflexive governance for sustainable development. Edward Elgar Publishing. Edward Elgar Publishing, Cheltenham

Waddock S (2013) The wicked problems of global sustainability need wicked (good) leaders and wicked (good) collaborative solutions. J Manage Glob Sustain 1(1):91–111. https://doi.org/10.13185/JM2013.01106

Walker B, Holling CS, Carpenter SR, Kinzig A (2004) Resilience, adaptability and transformability in social-ecological systems. Ecol Soc 9(2), 5. Retrieved 19 Dec 2015 from http://www.ecologyandsociety.org/vol9/iss2/art5/

Watson JEM, Rao M, Ai-Li K, Yan X (2012) Climate change adaptation planning for biodiversity conservation: a review. Adv Clim Change Res 3(1):1–11. https://doi.org/10.3724/SP.J.1248.2012.00001

Western Association of Fish and Wildlife Agencies (2018) Crucial habitat assessment tool: mapping fish and wildlife across the West [HTML]. Retrieved from http://www.wafwachat.org/

Williams BK, Szaro RC, Shapiro CD (2009) Adaptive management: The U.S. department of the interior technical guide. U.S. Department of the Interior, Washington, DC

Wu J (2013) Landscape sustainability science: ecosystem services and human well-being in changing landscapes. Landscape Ecol 28(6):999–1023. https://doi.org/10.1007/s10980-013-9894-9

Wyborn C (2015) Connectivity conservation: boundary objects, science narratives and the co-production of science and practice. Environ Sci Policy 51:292–303. https://doi.org/10.1016/j.envsci.2015.04.019

Wyborn C, Bixler RP (2013) Collaboration and nested environmental governance: scale dependency, scale framing, and cross-scale interactions in collaborative conservation. J Environ Manage 123(15):58–67. https://doi.org/10.1016/j.jenvman.2013.03.014

Chapter 15
Long-Term Experiences with the Action "Soil of the Year" in Germany

Monika Frielinghaus and Gerhard Milbert

Abstract Soils are valuable because they provide the basis for food and are the habitat for many animals, plants, and microorganisms. They are integral parts of landscapes, protect water resources, and store and transform nutrients. Soils act as a buffer against climatic effects and tell stories. This incomplete list already shows that soils are multifunctional and not replaceable. They are a limited resource whose protection has to be specially considered. Soils are threatened in their multi-functionality by pollutants and chemical substances, climate change, sealing, and erosion. All human activities use soils in a direct or indirect way. Therefore, everybody is a "soil stakeholder" and has to bear responsibility for soil protection. However, conservation cannot be successful without an awareness of and knowledge about soil valuation. This insight, which means understanding soils and their problems, has to be developed through education and the transfer of knowledge. This was the background against which the members of the German Soil Science Society (DBG) and the Federal Soil Association of Germany (BVB) initiated the action "Soil of the Year" in 2004. A committee prepared a concept for the launch and a guideline for the annual election procedure. The action is evaluated based on a presentation of 15 different soils.

Keywords Soil · Soil functions · Soil type · Landscape · Soil awareness · Soil of the year

M. Frielinghaus (✉) · G. Milbert
Kuratorium Boden des Jahres, Sprecher: Gerhard Milbert, Egmondstraße 37, 47608 Geldern, Germany
e-mail: frielinghaus@zalf.de
URL: http://www.boden-des-jahres.de

L. Müller and F. Eulenstein (eds.), *Current Trends in Landscape Research*, Innovations in Landscape Research, https://doi.org/10.1007/978-3-030-30069-2_15

15.1 Obliviousness Versus Awareness of Soil Value

In speeches by the chiefs of various Native American tribes, it says: "The earth is our mother, she nourishes us. What we put into the ground she returns to us" (Big Thunder, Wabanaki Algonkin), "We have not inherited the land from our parents, we have only borrowed it from our children" (Sammüller 2007).

There are good reasons for improving awareness of and responsibility for our soils. With the rapid development of engineering, technology, and trade, the basic attitude toward soils has changed in recent centuries. Individual scientists and politicians have repeatedly warned about natural catastrophes: "A nation that destroys its soils destroys itself" (Fallou 1862; Roosevelt 1936). However, this has hardly led to any changes. Thus, David Montgomery warns us, "Will modern soil conservation efforts prove too little and too late, like those of ancient societies? Or will we relearn how to preserve agricultural soils as we use them even more intensively? Extending the life span of our civilization will require reshaping agriculture to respect the soil not as an input to an industrial process, but as the living foundation for material wealth. As odd as it may sound, civilization's survival depends on treating soil as an investment, as a valuable inheritance rather than a commodity—as something other than dirt." (Montgomery 2007, p. 246).

What are the reasons why warnings are not taken seriously and our perception of the meaning and value of soils is currently warped?

Soils in industrialized countries are no longer perceived today as an indispensable basis of life. People's diet has largely become disconnected from food production. Much food is imported today from all over the world. This leads to "soil obliviousness" or to soil being treated too carelessly, since life-threatening food shortages are hard to be feared there (Rees 2015). This carelessness is all the more serious as it comes at the expense of developing countries. The increasing existential needs of large parts of the world's population in underdeveloped and developing countries have created a vicious circle: People there struggle daily with the limited resources of nature, and their fates suffer from the consumption of fertile soil (European Soil Charter 1972, 1996). This means that "forgotten land" in Europe has led to disinterest in the global "soil destruction." The purpose of the Soil of the Year action must be to bring new impetus into many strata of society that generates curiosity and interest in soils.

15.2 Aims of the Action Soil of the Year

Some special features of the medium of soil (as opposed to water or air) must be noted, to which attention must be paid when formulating the goals.

- Soils are largely out of our direct sensory experience; they are rare (e.g., on harvested arable land) and then only visible in two dimensions. They are not demarcated, clearly defined bodies. They form a continuum with very gradual transitions between the different soils. Soils are referred to both microscopically and globally.

- This uppermost part of the earth's crust, which we call "soil," formed over very long periods of time, often more than 10,000 years, so it is hard for a single generation to relate to it. In addition, very different soils were created, which can also change. Soil formation and alteration can rarely be traced (e.g., in case of erosion events). The negative changes are often very slow and therefore different from those of air or water (Frielinghaus and Sommer 2005).
- The functions of soils in landscapes and ecosystems are very complex and therefore difficult to understand.
- Soils are primarily associated with price and value from an economic point of view. They are perceived in the context of building land, settlement structures, motorway construction, commercial space, and related jobs. As yet, their value has not been assessed under the aspect of social and ecological functionality (Thoenes et al. 2004).
- As a result, soil-related environmental problems are scarcely communicated in society, e.g., only marginally and without continuity in schools.
- Knowledge about the very complicated circumstances found in the soil must be disseminated in a clearly understandable form. Thanks to the continuity of the action, recognition and "aha" effects are to be generated year by year in order to create a basic attitude shaped by emotions. This is regarded as a basis for positive behavior toward soil protection (Miehlich 2009).
- The action Soil of the Year is one of many activities designed to send out clear signals. These involve many actions in all German Federal States and even internationally. Awareness-raising communication is to be initiated in public. The different roles, awareness, and interests of the individual target groups must be observed by the multipliers. Knowledge transfer must be modified accordingly. This should support different networks over a lengthy period of time. Awareness does not arise at any particular place or time, but grows slowly, influenced by different, repetitive stimuli.

15.3 Experiences with the Presentation of the Soils of the Years 2005–2019

The International Soil Science Union launched "World Soil Day" on December 5, 2002, at its World Congress in Bangkok. In Germany, in 2004, the three Soil Science Associations German Soil Science Society, Federal Soil Association and Engineering Association for Contaminated Sites Management and Land Reclamation founded a "Soil of the Year" program together with the Federal Environment Agency. They commissioned a committee to steer and accompany the action "Soil of the Year," especially regarding public relations. Its members organize the presentation of the recommended soil of the next year. They publish posters and flyers and a "Soil of the Year" Web site (Boden des Jahres 2019). They organize events about soil protection. Together with members of the German soil agencies which propose the specific soil, they prepare a special event in the Federal Government Representation in Berlin. On December 5, 2004, the first Soil the Year 2005 was presented.

Experience shows that there have been very different expectations regarding the didactic suitability of the proposed soils. For the members of the committee, this required a learning process in order to achieve the formulated goals (Frielinghaus 2010). This will be explained using several examples. The following soils have been elected Soils of the Year in Germany (Table 15.1).

Some examples are described below to assess their suitability (Milbert 2018; Bodenwelten 2019; Boden des Jahres 2019)

Table 15.1 Soils of the Year 2005–2019

Year	Soil (German name)	WRB classification (closest name)	Didactic suitability[a]	Presenter of the soil (Country, Federal State)
2005	Schwarzerde	Chernozem	+++	Germany, Saxony-Anhalt
2006	Fahlerde	Albeluvisols/Retisols	+	Germany, Mecklenburg-Vorpommern
2007	Podsol	Podzols	+++	Germany, Lower Saxony
2008	Braunerde	Cambisols/Arenosols	+	Austria
2009	Kalkmarsch	Gleyic Calcaric Fluvisols	+++	Germany, Schleswig-Holstein
2010	Stadtboden	(Urban soils)/ Technosols	++	Germany, Berlin
2011	Brauner Auenboden	Fluvisols/Fluvic Cambisols	++	Germany, Baden-Württemberg
2012	Niedermoor	Histosols (Fens)	+++	Germany, Brandenburg
2013	Plaggenesch	Plaggic Anthrosols	+++	Germany, Lower Saxony
2014	Weinbergsboden	Regic Anthrosols	+++	Germany, Rhineland Palatinate
2015	Stauwasserboden	Stagnosols/Planosols	+++	Germany, North Rhine-Westphalia
2016	Grundwasserboden	Gleysols	++	Germany, Schleswig-Holstein
2017	Hortisol/ Gartenboden	Hortic Anthrosols	+++	Germany, Thuringia
2018	Felshumusboden	Folic Histosols/ Suprafolic Leptosols	++	Germany, Bavaria
2019	Kippenboden	(Dump Regosols)/ Technosols	++	Germany, Saxony

[a] The didactic suitability is based on a clear picture of the horizons, the recognition of the origin, the recognizable endangerment through use or climate change, or the possibility to tell stories: + moderate, ++ good, +++ very well suited

15.4 Chernozem—Soil of the Year 2005

Chernozems (black soils, Fig. 15.1) arise from calcareous parent material (mainly loess sediments) in regions with hot, dry summers, and cold winters (continental climate), preferably in the lee position of low mountain regions. The organic residues of grasses and herbs cannot be completely decomposed due to drought in summer and cold in winter. These humus-rich soils with a thick topsoil, usually 60–80 cm thick, were developed over a long time. Soil-dwelling animals such as hamsters, ground squirrels, and earthworms ensure that the soil is constantly mixed. The formation of black soils began during the end of the last Ice Age more than 10,000 years ago in a climate with predominantly steppe-like vegetation. Black soil areas were colonized very early on by people in the Stone Age for agriculture. In today's climate, no new black soil can arise in Germany. Black soils can store a lot of water and many nutrients and have therefore been particularly productive arable land for thousands of years. Large areas of black soils are to be found in the landscapes around Hildesheim (Lower Saxony), Magdeburg, Halle and Köthen

Fig. 15.1 Chernozem—Soil of the Year 2005. **a** Earthworm holes and filled holes dug by small mammals. *Photograph* Schulze-Kellinghaus, Soil of the Year committee. **b** High-yield winter wheat on Chernozem. *Photograph* I. Merbach, Soil of the Year committee. **c** Intensive animal life in black soils. *Photograph* I. Merbach, Soil of the Year committee

(Saxony-Anhalt) or in the Querfurt Basin (Thuringia). Black soils and black-like soils occupy about 3% of the soil surface and about 5% of agricultural land in Germany. The largest areas of black soil (56%) are in the Federal States of Saxony-Anhalt and Thuringia.

Although black soils are one of the best soils, they cannot compensate for the stresses of very intensive land use. The biggest danger is high surface sealing. They are also threatened by water and wind erosion, as well as increasing damage. Soil compaction is a significant risk factor. The best way of protecting black soils and the area in which the greatest experience has been gathered is environmentally sustainable agriculture.

15.5 Histosols (Fens)—Soil of the Year 2012

Histosols (Peat soils, Fig. 15.2) typically contain more than 30% organic material and are a dark brown to black color. Depending on their conservation status, fen-forming plant residues are visible to different extents. The subsoil of fen soils

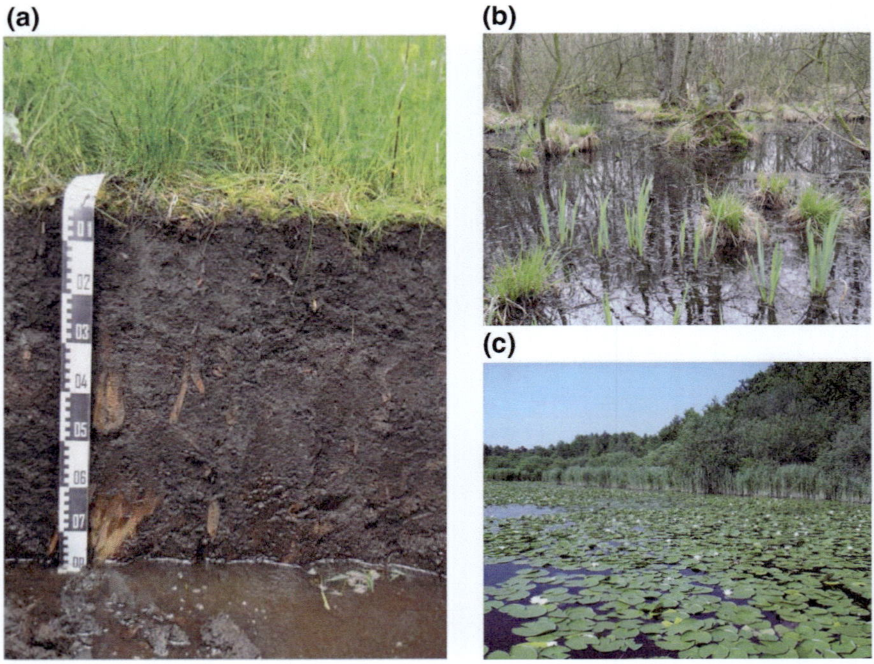

Fig. 15.2 Histosols (Fens)—Soil of the Year 2012. **a** Shallow drained fen soil. *Photograph* G. Milbert, Soil of the Year committee, **b** Terrestrialization process of a lake, leading to an eutrophic fen. *Photograph* Biologische Station Großes Heiliges Meer, Hopsten, Soil of the Year committee, **c** Eutrophic fen, alder swamp. *Photograph* M. Dworschak, Soil of the Year committee

often consists in glacial deposits such as sand, silt, loam and clay, or of lacustrine deposits. Depending on their parent material, lacustrine sediments are white (lime mud), olive green (algal gyttja), or dark brown (clay mud). Fen soils mainly develop in groundwater-influenced lowlands or along rivers and lakes. On a global scale, they are typical of the cool, humid climates of the northern hemisphere. In Germany, fen soils cover a surface of approx. 1 million hectares. The largest areas up to 30,000 ha are situated in Schleswig-Holstein, Lower Saxony, Mecklenburg-Western Pomerania, Brandenburg, Bavaria, and Baden-Württemberg. Usually, the development of fen soils is initiated by paludification at a high groundwater level or increasing sedimentation into lakes. Dead plant material accumulates under water saturation and air exclusion.

In siltation mires, the peat lies on the bottom of a water body above organic or mineral lacustrine sediments. The peats of fen soil areas are formed of dead roots, branches, leaves and sprouts of sedges, reeds, mosses, elders, willows, or other swamp plants. Decomposition of the organic material is slow and incomplete. A peat body only increases a few millimeters per year, directed to the water surface and/or the lake center. The peat increase occurs from the bottom to the top. If the peat layer thickness exceeds 30 cm, it is classified as fen soil (Histosols).

From an ecological point of view, natural fen soils are highly valuable. Only conformist, mostly rare, exceptional animal and plant specialists such as the large copper butterfly, cotton grass, and sedges are adapted to the high water content and special nutrient conditions.

Thick fen soils contain up to 2000 tonnes of carbon per hectare. Worldwide, Histosols store the most carbon per areal unit. Further, peat soils are important archives of nature and civilization. They conserve former vegetation and climate conditions as well as traces of settlements and former use.

For more than 1000 years, peat has been used as a solid fuel, medicine, and fertilizer. Until the 1950s, peat digging was carried out industrially. Limonite, a formation in fen soils with iron-rich groundwater infiltration, and lime mud were also excavated until the beginning of the twentieth century. Today organic material from peat is obtained for medical use on very few sites in Germany. Because of their rare occurrence, in Germany intact fen soils close to their natural state have been placed under nature conservation. Fen soils are often used for agriculture and forestry. In the past, they had to be drained by ditches or drainage systems which seriously and often irreversibly changed the peat properties. With different intensity, most fen soils in Germany are presently used as pasture land. Intensively used fen soils can release up to 40 tonnes of CO_2 per hectare and year. For the conservation and permanent protection of intact lowland peat soils, carefully elaborated development strategies are needed.

15.6 Stagnosols/Planosols—Soil of the Year 2015

Stagnosols/Planosols (Pseudogley soils, Fig. 15.3) form where precipitation water only drains away into the underground after long delay. Beneath a well permeable layer that is waterlogged after rainfall, there is a dense layer with low permeability. In the German soil classification, most soils affected by stagnating water belong to the Pseudogleys, and those with an extended wet phase to the Stagnogleys. The international classification allocates these soils primarily to the Planosols and Stagnosols.

Fig. 15.3 Stagnosols/Planosols (Pseudogleys)—Soil of the Year 2015. **a** Soil Profile Kottenforst Forest near Bonn. *Photograph* G. Milbert and M. Dworschak, Soil of the Year committee. **b** Wet bleached and orange mottling as well as black colored areas in the Pseudogley profile. *Photograph* G. Milbert, Soil of the Year committee. **c** Soil compaction after intensive soil traffic by agricultural machinery. *Photograph* G. Milbert, Soil of the Year committee

As a function of the soil's water permeability, the weather conditions and plant water use, surface water gley soils alternate between wet, moist, and dry phases. These phases may vary in length and may occur several times during the course of a year. This results in temporary excesses and temporary dearths of soil water. During the wet phases, a lack of oxygen adversely affects plant roots and soil organisms. On surface water above gley soils, only those plant species thrive that tolerate wet soil conditions. Little by little, a mottled to veined soil develops with spots bleached by water, side by side with rust-colored zones. Often, hard iron and manganese segregations form so-called concretions.

Pseudogley soils are unique natural bodies that often support forest plant communities that prefer changing moisture conditions, e.g., Common Oak-European Hornbeam forests. Markedly wet Pseudogley soils are—due to their extreme site conditions—well suited for rare animal and plant com-munities.

Pseudogley soils store precipitation water that evaporates with a time delay or is consumed by plants. In this way, they act as a buffer against precipitation peaks. The perched water drains slowly into the groundwater, possibly of nearby groundwater soils, and into watercourses or bodies.

Forests adapted to changing moisture conditions can be stable, productive ecosystems that are at the same time ecologically valuable. Tree species which tolerate perched soil water are the common oak, European hornbeam, ash, black alder, and downy birch. Those which are adversely affected include the Norway spruce, larch, and beech; they develop only shallow roots in perched water. In dry years, drought damage occurs. After several wet years in a row, roots suffer from lack of oxygen. During storm events, shallow root trees tend to be uprooted.

Timber harvesting does not only harm the soil during dry phases. Surface water gleys are sensitive to weather conditions and the climate. Increasing numbers of heavy rainfall events result in more frequent wet phases. If climate warming extends the vegetation period, plant water consumption will increase, and longer dry periods could occur.

Wet–dry Common Oak-European Hornbeam forests could develop into Beech forests in the long run. During the last 50 years, the vegetation period has already lengthened more than two weeks, and the number of heavy rainfall events has risen.

15.7 Hortisol—Soil of the Year 2017

The name Hortisol is of Latin origin and combines "hortus" (garden) and "solum" (soil). Hortisols (garden soils, Fig. 15.4) belong to the "terrestrial anthropogenic soils" of the German soil classification system. They have been so thoroughly modified by humankind that few characteristics resemble their original state. Similar soils are Plaggenesch (Plaggic Anthrosol) and vineyard soil.

The Hortisol has an active soil life with a particularly high number of earthworms that mix the soil material intensively.

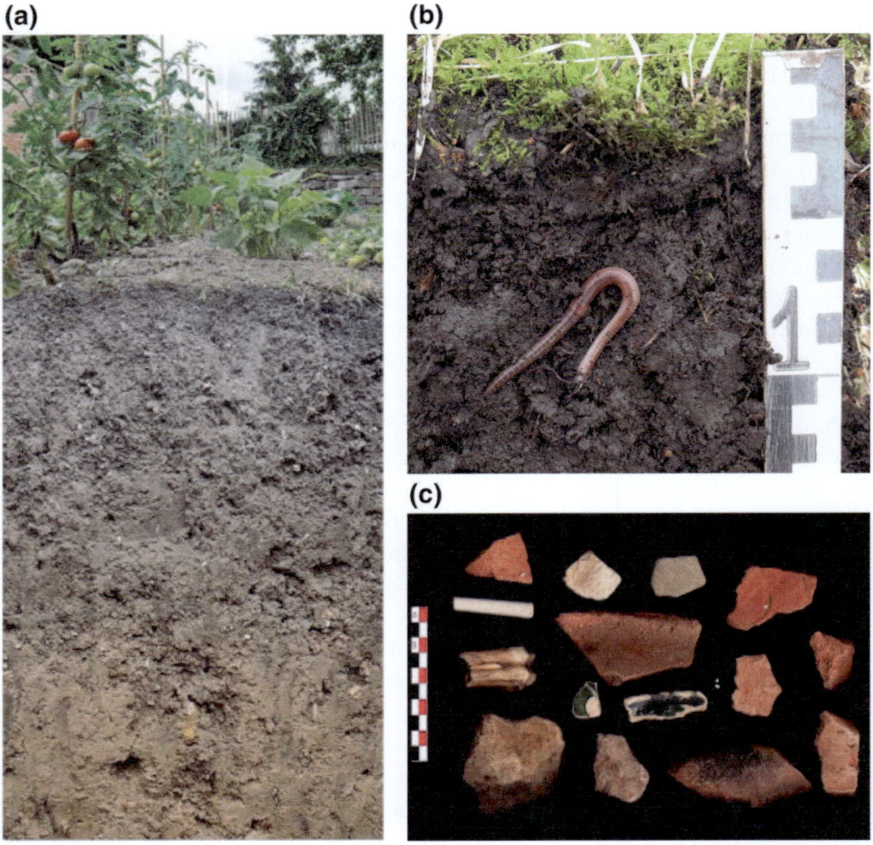

Fig. 15.4 Hortisol—Soil of the Year 2017. **a** Profile of a Hortisol (garden soil) *Photograph* Thüringer Landesanstalt für Umwelt und Geologie, Soil of the Year committee. **b** *Lumbricus terrestris* earthworm in a Hortisol profile. *Photograph* M. Dworschak/Geologischer Dienst NRW, Soil of the Year committee. **c** Ceramic shards and other historically interesting objects found in old garden soils. *Photograph* Geolog. Dienst NRW, Soil of the Year committee

The part of the soil that is subject to so-called bioturbation (Latin "turbare," to churn up) is a specific, humus-rich and often dark gray horizon. In combination with the also humus-enriched topsoil, this reaches down to more than 40 cm below the soil surface.

Why are garden soils so fertile and rich in humus? Human activity in combination with soil organisms, climate, and the parent material are the most important factors behind garden soil development. Garden soils literally grow on or in various soils or materials. Often, human beings have also brought in the material for the soil through laborious efforts, e.g., in old gardens on terraces on the slopes or in castle and cloister complexes. But it is only the gardener's cultivation over decades or even hundreds of years that makes the soil turn into a garden soil:

- The soil is regularly dug and turned, loosened and enriched with materials with spades, hoes and other tools.
- Compost, feces, ashes, sweepings, organic waste, lime, bones, sherds, etc., are put on the soil and mixed in.
- The soil is watered regularly during dry phases in summer and autumn.

Thanks to this work by the gardener, the soil grows by 40 to more than 100 cm, depending on how long the garden is in use. On the top, it develops a new soil layer that is particularly loose, fertile, and rich in humus and nutrients. This soil layer contains much more phosphorous and nitrogen than other soils. Both are important nutrients for crop plants.

Clay particles and humus retain a lot of rainwater that is then available to plants. The high humus content provides for high biological activity. In particular, the many earthworms help incorporate plant residues and compost into the soil, even down to a depth of more than 1 m. Roots, soil organisms and lime make for a soil structure composed of crumbs and consistent soil particles that keep the soil loose, but stable. As a result, garden soil can supply plant roots with oxygen and release carbon dioxide at its surface.

Wherever humans settled, they changed part of the land near their houses and cottages into gardens. This is where the oldest and deepest Hortisols can be found. Old villages contain sizeable areas of garden land even today, including land on their outskirts (Grabeland—land that is dug up). Various kinds of cabbages and root crops such as potatoes and beets were once cultivated there. Through long, intense use, typical Hortisols were developed.

Many innovations in gardening practice started out from cloisters, as they were once part of a pan-European network. In many cloister gardens, nuns and monks bred new kinds of fruit and vegetable. In Thuringia, there are more than 200 historical cloisters and parishes. Many of them still have old gardens.

In urban settlements, kitchen gardens became common from the twelfth and thirteenth centuries. The oldest gardens are situated in the core area developed in the High Middle Ages, and younger ones in the areas where the towns expanded within or outside their walls. These vegetable and kitchen gardens were very small. Cultivated for hundreds of years, they now contain typical Hortisols. They are rare in the city centers, where much of the soil has been removed or overbuilt. Furthermore, the soil is often polluted with fire debris, brick fragments, cement, and metal remnants. Urban areas (unlike rural areas) often used fertilizers such as liquid feces from the cesspools, organic household waste, and ash. Since the middle of the nineteenth century, sewage sludge, compost, and artificial fertilizers have been used. These gardens primarily contribute to the production of food, particularly in times of hardship, but also serve as a place to meet and communicate.

If soil is to be used carefully, it must be thought of as a living organism. Soil life needs to be preserved, and chemicals (mineral fertilizer, pesticides) used cautiously. Garden soils should not be sealed or heavily compacted, as they may lose their many functions for the ecosystem.

Unfortunately, many garden soils in the town centers and on the outskirts of villages are now being overbuilt and are no longer a "green lung." They are losing their social function.

In many cities, school garden projects play an important role in teaching children and young people about the value of and threats to soil.

15.8 Conclusions and Outlook

Experience with the Soil of the Year action confirms the difficulties associated with the perception of soils as a limited resource with essential functions. The action drew very different responses every year, with corresponding setbacks, but it can now be stated that it has taken a firm place in the soil-related activities of various institutions when it comes to improving society's awareness of soil. The continuity of naming a Soil of the Year and preparing interesting material, the continuous Internet presentation (Boden des Jahres 2019; Bodenwelten 2019), and the creation of a soil network have increasingly aroused national and international interest.

Under the patronage of the respective Federal State representation, the presentation of the next soil will take place on the World Soil Day. This annual event increases political interest in responsibility for soil protection.

Basic funding is essential to realize this action. For example, the German Soil Science Society, the Federal Association of Soil, and the Association for the Management of Contaminated Sites and Land Recycling assume responsibility on the committee. The Federal Environment Agency deals with the financing of printing and dispatching the materials. The committee members can only offer input to the "Soil of the Year campaign" for initiatives related to the Federal States' target groups. The regional soil specialists communicate with farmers, gardeners, foresters, museum staff, the staff of botanical and zoological gardens, landowners of all kinds, artists, parishes, association members and soil specialists employed by the authorities. These regional target groups have to find intelligent solutions and ask for political and technical decisions. Interested citizens in towns and rural areas and all types of media can only be reached effectively and with special regional or problem-oriented activities in the Federal States.

A memorandum from the Scientific Advisory Council on Soil Protection states: "We have discovered that citizens do not think as much about the use and protection of soils as scientists and soil protectors believed they did" (Wissenschaftlicher Beirat 2002).

The activities used to impart knowledge about soils as an essential basis for human existence and showing how endangered this limited resource is, unless action is taken to protect soils, have to be made even more goal-oriented and more effective. New knowledge of effective communication must lead to constant reviews and improvement of knowledge transfer methods. Various events (such as World Soil Day, Day of the Environment, Earth Day) and regional projects should take place regularly for all target groups. The exchange of experience between the actors should be encouraged. There is a very wide range of material available for each Soil of the Year from 2005 to 2019, enabling the various target groups to be reached.

References

Boden des Jahres (2019) www.boden-des-jahres.de. Accessed on 14 Jan 2019

Bodenwelten (2019) www.bodenwelten.de. Accessed on 14 Jan 2019

European Soil Charter (1972, 1996) European soil charter and revised European soil charter (ETS Nr.163), 1996. www.fao.org/soil-portal. Accessed on 14 Jan 2019

Fallou FA (1862) Pedologie oder allgemeine und besondere Bodenkunde. G. Schönfeld´s Buchhandlung Dresden 1862, 487, p 7

Frielinghaus M (2010) Aktion Boden des Jahres – ein Beitrag zur bewussten Wahrnehmung von Boden in der Gesellschaft. In Makki M, Frielinghaus M (eds.): Boden des Jahres 2010 – Stadtböden. Berliner Geographische Arbeiten 117, pp 1–5

Frielinghaus M, Sommer M (2005) Vorsorgender Bodenschutz – die Voraussetzung für eine nachhaltige Bodennutzung. In: Leibniz-Gemeinschaft (2005) (Eds): Zwischenruf-Umweltforschung für politische Praxis. Heft 2, pp 16–25

Miehlich G (2009) Bodenbewusstsein – ein Schlüssel zur Förderung des Bodenschutzes. In Alfred Toepfer Akademie für Naturschutz (Hrsg.) Bodenschutz im Spannungsfeld von Umwelt- und Naturschutz, Schneverdingen, pp 48–53

Milbert G (2018) Boden des Jahres - eine Initiative, damit wir nicht den Boden unter unseren Füßen verlieren. In: Vom Wert unserer Böden. Bund Heimat und Umwelt, S.54. ISBN:978-3-9253t4-50-0

Montgomery DR (2007) Dirt: the erosion of civilizations. University of California Press; 1st edition (13 April 2007) p 246

Rees WE (2015) Soils, Eco-footprints and Overshoot – can we still be sustainable? Oral presentation Berlin 3.12.2015, UBA Publikationen, www.uba.de. Accessed on Jan 14 2019

Roosevelt FD (1936) The great depression—the dust bowl. Performed by President, recorded 6 Sept 1936

Sammüller S (2007) Indianerweisheiten. Edition XXL, Fränkisch-Crumbach 2007, www.edition-xxl.de. Accessed on 14 Jan 2019

Thoenes HW, Lazar S, Huck S, Miehlich G (2004) Bodenbewusstsein – Wahrnehmung, Geschichte und Initiativen. In: D Rosenkranz, G Bachmann, W König, G Einsele (eds) Bodenschutz, ergänzbares Handbuch, 41, Lfg., VII 04, Erich Schmidt Verlag, Berlin

Wissenschaftlicher Beirat (2002) Wissenschaftlicher Beirat Bodenschutz beim BMU (2002) Ohne Boden - bodenlos. Eine Denkschrift zum Bodenbewusstsein, Berlin, p 56

Chapter 16
A Framework to Connect Biodiversity-Ecosystem Functioning Research to Habitat Fragmentation

Jiajia Liu, Lionel Hertzog, Guang Hu, Kris Verheyen and Mingjian Yu

Abstract The relationships between biodiversity and ecosystem functioning (BEF) are one of the key questions in ecological studies especially in the context of the current global decline in biodiversity. However, limited effort has been done to connect BEF relationships to habitat fragmentation while we are living in an age of global habitat fragmentation, especially forest fragmentation. In this chapter, we briefly discussed why such a connection is need. We follow by outlining the major mechanisms by which habitat fragmentation can affect BEF relationships: (i) fragmentation-driven non-random turnover of species, (ii) changes in species–species interactions affecting complementarity potential, and (iii) influences on insurance effects due to changes in environmental conditions and landscape patterns. We highlight the importance of considering spatiotemporal scales in studying BEF relationships. Finally, to promote further research in this area, we present the evidence currently available to science and outline major avenues for future studies.

Keywords Habitat fragmentation · Habitat loss · Biodiversity · Ecosystem functioning · Scales · Connectivity · Species interactions · Species turnover

---

Jiajia Liu and Lionel Hertzog have equal contributions in this chapter

---

J. Liu (✉) · L. Hertzog · K. Verheyen
Forest and Nature Lab, Ghent University, Gontrode, Geraardsbergsesteenweg 267, 9090 Melle-Gontrode, Belgium
e-mail: liujiajia.ecology@gmail.com

J. Liu · M. Yu (✉)
College of Life Sciences, Zhejiang University, Hangzhou, Zhejiang, China
e-mail: fishmj@zju.edu.cn

G. Hu
School of Civil Engineering and Architecture, Zhejiang Sci-Tech University, Hangzhou, China

L. Müller and F. Eulenstein (eds.), *Current Trends in Landscape Research*, Innovations in Landscape Research, https://doi.org/10.1007/978-3-030-30069-2_16

16.1 Introduction: Biodiversity and Ecosystem Functioning

About thirty years ago, ecosystem ecologists realized that ecosystem functions such as primary production or nutrient cycling were not only driven by abiotic conditions, but also by the biotic community composing ecosystems (Tilman et al. 2013). Joining forces with community ecologists, they derived a successful research program that explored the relationships between biodiversity and ecosystem functioning. The most famous research in this area was done based on grassland experiments such as the Cedar Creek Biodiversity Experiment in the USA (Isbell et al. 2013) and the Jena Experiment in Germany (Weisser et al. 2017). Results showed that biodiversity generally has a positive effect on ecosystem functioning (Liang et al. 2015; Ratcliffe et al. 2017; Huang et al. 2018).

Two main hypotheses were proposed to explain the positive biodiversity and ecosystem functioning (BEF) relationships: complementarity and selection effects. First, species within more diverse assemblages are expected to show higher complementarity (Cardinale et al. 2002) which may be driven by resource partitioning, such as complementarity light use strategies among neighboring plants (Jucker et al. 2014), abiotic facilitation by—for instance—the positive impact of nitrogen fixers (Kahmen et al. 2006) or by plant–soil feedbacks (Barry et al. 2018). Second, communities with more diverse assemblages are more likely to show selection effects which refer to the dominant role of a key species that contributes greatly to the ecosystem functioning. For example, large-diameter tree species play a disproportionally important role in carbon allocation and productivity (Slik et al. 2013; Lutz et al. 2018). Selection effects are considered to be more important at the early stage of succession (Tobner et al. 2016) while complementary effects will become more important at the late successional stages (Reich et al. 2012). The importance of these different mechanisms and their consequences on the relation between biodiversity and ecosystem functioning will be context-dependent. For instance, resource partitioning will be especially beneficial if resource supply is limited (Ratcliffe et al. 2017; Barry et al. 2018) and abiotic facilitation will become more important in stressed systems (Steudel et al. 2012). However, in a recent review Liu et al. (2018b) showed that while an extensive literature exists on BEF relationships on one side and on habitat fragmentation effects on biodiversity on the other side, limited effort has been done to connect the dots and look at fragmentation effect on BEF relationships (but see Hertzog et al. 2019).

16.2 What Is Habitat Fragmentation?

Habitat fragmentation is a process by which continuous and large habitats divided into smaller and isolated patches (Fahrig 2003). In landscapes under human pressure, habitat fragmentation often occurs together with habitat loss. Recent studies call for an explicit reorganization of habitat fragmentation studies to explore the effect of habitat fragmentation per se independently of the effects of habitat loss (Fahrig 2017). However, here we mainly discuss the effect of habitat fragmentation in general. An increasing numbers of studies have reported that global habitat fragmentation is a pervasive phenomenon (Haddad et al. 2015; Taubert et al. 2018), while intact large forests are rare (Watson et al. 2018). For example, 70% of the world' forest lies within 1 km of an edge (Haddad et al. 2015), and 19% of the tropical forests lie within 100 m of a forest edge (Brinck et al. 2017). Creation of edges after habitat fragmentation mainly affects ecosystems through environmental changes in humidity, fire frequency, nutrient, and light availability (Laurance et al. 2018; Smith et al. 2018). More isolated patches are less efficient in gene flow from a long-term perspective. As such, habitat fragmentation generally increases mortality of large old trees (Laurance et al. 2000), decreases species diversity, and leads to an erosion of functional traits (Magnago et al. 2014; Liu et al. 2018a). To sum up, both biodiversity and ecosystem processes can be strongly affected by habitat fragmentation.

16.3 A Need to Connect Habitat Fragmentation and BEF

Previous studies on biodiversity and ecosystem functioning relationships mainly focused on using a strong experimental design including a high level of replication and strictly controlled environmental conditions to explore biodiversity effect. However, these studies were criticized for their lack of realism and by extension their limited implications for real-world systems (Srivastava and Vellend 2005). First of all, these experiments assumed random species loss or gain a highly unlikely process in the real world (Wardle 2016). For example, differential responses to habitat fragmentation have been frequently reported within and among biological groups (Fontúrbel et al. 2015; Pfeifer et al. 2017). Second, environmental conditions have important effects on ecosystem functioning, both directly and indirectly via biodiversity effects, which has been supported by recent papers (De Laender et al. 2016). Third, previous studies ignored potential spatial effects such as metacommunity dynamics (Loreau et al. 2003). Therefore, in order to have a systematic understanding of diversity effects on ecosystem functioning in the real world under threat of habitat fragmentation and habitat loss, we here discuss how habitat fragmentation may affect BEF relationships and set up a tentative research agenda for future studies.

16.4 How Do Habitat Fragmentation and Habitat Loss Affect BEF Relationships?

16.4.1 Causes: Environmental Changes, Connectivity, and Habitat Loss

Habitat fragmentation leads to changes in environmental conditions, mainly through edge effects. Forest edges are, for instance, drier and warmer than forest interiors (Arroyo-Rodríguez et al. 2017; Laurance et al. 2018). These changes in environmental conditions will lead to shifts in the biotic community composition and in species–species interactions (Haddad et al. 2015; Wilson et al. 2016). Habitat fragmentation also leads to loss in connectivity between habitat patches reducing

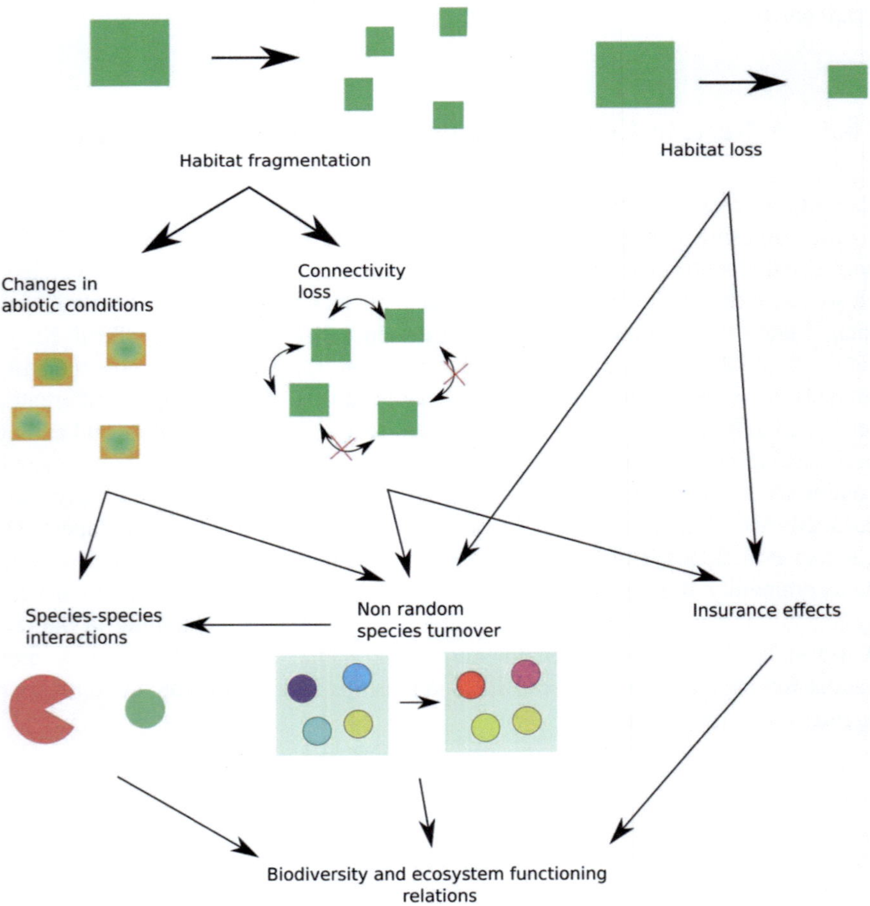

Fig. 16.1 A framework that explains how habitat fragmentation affects BEF relationships

movement and flow. These changes in connectivity will affect metacommunity dynamics, leading to shifts in the biotic community composition and in insurance effects, which proposes that functionally redundant species at a given time may show temporal complementarity (Thompson et al. 2017). Finally, habitat fragmentation will reduce average population sizes and species richness (Haddad et al. 2015), and this will affect biotic community composition and insurance effects (Fig. 16.1).

16.4.2 *Consequences for BEF Relationships: Non-random Species Turnover, Species–Species Interactions and Insurance Effects*

The biotic community with its set of traits affect ecosystem functioning (Hooper et al. 2005), and any changes in the functional composition of the communities driven by habitat fragmentation and habitat loss can potentially affect BEF relations. Most importantly, this occurs when species most vulnerable to habitat fragmentation and habitat loss have a disproportionally large contribution to ecosystem functions. For instance, large-diameter tree species contribute to 41% of aboveground live tree biomass (Lutz et al. 2018), and they have a higher carbon sequestration ability than smaller trees (Stephenson et al. 2014). However, they are vulnerable to habitat fragmentation via drier conditions induced by edge effects (Laurance et al. 2000). Similarly, large seed dispersers play key roles in the maintenance of forest carbon storage due to their important roles played in ecological interactions (Bello et al. 2015). However, large-bodied seed dispersers, such as hornbills and primates, are more sensitive to habitat fragmentation than small-bodied animals. These shifts could affect the selection effect which refers to the dominant role of a key species that contribute disproportionally to the ecosystem functioning, if these lost species exert an important impact on the BEF relationship. On the other hand, fragmentation-driven community shifts may also promote selection effect by changing the evenness and the dominance structure within the community (Hillebrand et al. 2008).

A second important consequence of fragmentation is the changes in species–species interactions such as resource use partitioning, competition, facilitation, and mutualism. These changes may be driven by shifts in the biotic communities. Fragmentation and habitat loss can, for instance, affect pollinator communities (Kramer et al. 2012), which will affect pollen dispersal and pollination efficiency (Rusterholz and Baur 2010; Hagen et al. 2012). Species–species interactions may also be driven by fragmentation-driven environmental changes such as abiotic stress or resource availability (Barry et al. 2018). Baert et al. (2016) showed that dominance increased under chemical stress in experimental microalgae communities leading to stronger BEF relations. Resource limitation may also promote resource use partitioning (Grace et al. 2016), a mechanism that might be behind the findings of Ratcliffe et al. (2016) that BEF relationships were stronger in water-limited conditions.

Finally, fragmentation and habitat loss will affect the potential for insurance effects, the fact that the functional consequence of species loss might be offset by other species present in the community (functional redundancy, Reich et al. 2012) or by immigration from other patches (metapopulation dynamics, Gonzalez et al. 2009). Habitat loss will reduce the species pool affecting the potential for functional redundancy. For example, in large forest fragments many species of top predators may be present such that the loss of any one of them will have little functional consequences. On the other hand, in small forest fragments where only one unique top predator species is present, the loss of one species may have strong functional consequences (Terborgh et al. 2001). Therefore, BEF relationships are expected to be stronger in smaller habitat patches (Liu et al. 2018b). In a recent study, Hertzog et al. (2019) revealed that the effect of tree richness on forest multifunctionality was stronger in small and isolated compared to large and well-connected fragments in mature temperate forests of Belgium. In addition, fragmentation will affect the connectivity between habitat forming a spatial insurance for ecosystem functioning in heterogeneous landscapes (Gonzalez et al. 2009). For example, diversity effects on ecosystem functions will be reduced if mobile grazer species were allowed to move among studied patches (France and Duffy 2006). Ecosystems tend to be more resilient to species loss when connected to other patches, and BEF relationships are stronger in more isolated fragments (Wu and Loucks 1995; Staddon et al. 2010). In contrast, connectivity loss erodes the stability of ecosystem functioning (Thompson et al. 2017).

16.5 The Importance of Spatiotemporal Scales from a Landscape Perspective

Ecosystem functioning can be enhanced by many different species across many times and from nearby places resulting in a spatiotemporal dependency of BEF relationships (Isbell et al. 2018).

16.5.1 Spatial Scales

BEF relationships are dependent on studied spatial scales (Thompson et al. 2018). For example, there was a significant BEF relationships in primary forests at very small spatial scales (0.04 ha), but it became insignificant at increasing spatial scales (0.25, 1 ha) (Chisholm et al. 2013). In addition, the strength of the BEF relationship can be different between ecosystems (Kardol et al. 2018).

However, there is a mismatch of between studied sample units in BEF studies and the size classes of habitat fragments in the real world. Sampling plots for BEF studies are typically 0.04 ha in forest systems (e.g., Chisholm et al. 2013; Ratcliffe et al. 2015; Liang et al. 2016) and varied considerably from 1 to more than 400 m^2 in grassland systems (Roscher et al. 2005). On the other hand, the mean size of forest fragments in tropical regions is 17 ha (Taubert et al. 2018), and most fragmentation studies used a gradient of habitat sizes ranged from 0.01 to more than 1000 ha (Brinck et al. 2017; Laurance et al. 2018; Liu et al. 2018a). Therefore, biased sampling toward certain sampling sizes (e.g., 0.04 ha for forest ecosystems) may not allow a full understanding of BEF relationship across spatial scales.

16.5.2 Temporal Scales

Many BEF experiments have focused on grasslands. For example, the longest term Cedar Creek Biodiversity Experiment was established in 1994 (Isbell et al. 2013), and the Jena Experiment was set up in 2002 (Weisser et al. 2017), and plots on islands in northern Sweden were established in 1996 (Fanin et al. 2018). However, other systems such as forests require a long term to reach a mature state while experimental forest plots have, until now, very short study periods. For example, the BEF-China experiment is only 10 years old (Huang et al. 2018), similar to other forest experiments in the world that are less than 20 years (http://www.treedivnet.ugent.be/). Interestingly, recent studies found that biodiversity effects are also influenced by successional stages, with a more profound positive diversity effect at the early successional stages (Lasky et al. 2014). Therefore, findings shown by short-term studies may be different from those shown by long-term studies (Reich et al. 2012). Similarly, Kardol et al (2018) showed that the effects of species loss on community biomass decreased over time, but the pattern was influenced by habitat size. Indeed, habitat fragmentation has long-term effects on biodiversity, which may persist for more than a century (Vellend et al. 2006; Jones et al. 2016). This may result in a time-delayed "functioning debt" by habitat fragmentation paired with an extinction debt (Liu et al. 2018b).

16.6 Conservation Implications from BEF Studies in Fragmented Landscapes

Habitat loss and fragmentation are a highly non-random process (Laurance 2008). Forest fragments are more likely to be located in relatively higher elevated, steep, and northern faced areas that are not suitable for agriculture (Sandel and Svenning 2013). This results in a pattern where areas with higher biodiversity and ecosystem functions are more readily converted to agricultural lands (Liu and Slik 2014).

Therefore, we may be unable to detect the full picture of BEF relationships due to lack of sufficient samples in extremely diverse systems. Indeed, BEF studies typically neglect tropical rainforests (Clarke et al. 2017), and recent studies found that more carbon did not equal more biodiversity in less disturbed tropical rain forests (Ferreira et al. 2018). Few studies have taken this into account, limiting the practical applications of BEF research to biological conservation (Srivastava and Vellend 2005).

16.7 Future Directions Under the Proposed Framework

The last two decades we have witnessed a shift of BEF studies from experimental studies to its applications in the real world (Steudel et al. 2012; Liang et al. 2015; Jucker et al. 2016), especially in heterogeneous and dynamic landscapes (Brose and Hillebrand 2016; Hertzog et al. 2019). Considering that habitat fragmentation is becoming a hugely important threat to global biodiversity and ecosystem functioning (Haddad et al. 2015; Wilson et al. 2016), future research is direly needed to further reveal how BEF relationships hold in fragmented landscapes. We therefore propose a research agenda based on a combination of different approaches.

16.7.1 Using Remote-Sensing Maps to Model Long-Term Fragmentation Effects

Large-scale fragmentation maps are increasingly available due to the rapid development of remote-sensing technology (Haddad et al. 2015; Taubert et al. 2018), which has promoted the global study on fragmentation effects on ecosystem functioning. For example, a study using global tropical forest fragmentation maps found that one-third of the carbon release was contributed by habitat fragmentation (Brinck et al. 2017). A recent study using large-scale fragmentation mapping found that habitat conditions were not improved in a nature reserve (Xu et al. 2017). Combining the available large-scale mapping information on habitat fragmentation and carbon release would allow the exploration of correlations between habitat fragmentation and ecosystem function at the global scale.

16.7.2 Empirical Studies in Plots Forming a Gradient of Fragmentation Intensity

Currently, several platforms have been developed to investigate long-term effects of habitat fragmentation on biodiversity and ecosystem functioning, such as BDFFP (Laurance et al. 2018), islands in the Thousand Island Lake (Wilson et al. 2016), islands in northern Sweden (Fanin et al. 2018), and the SAFE project (Ewers et al. 2011). These study sites have a gradient of fragmentation intensity, with systematically studied fragmentation effects, and would reveal how strongly the BEF relation responds to fragmentation intensity. However, more studies are needed to explore how widespread this response is, especially from a global network connecting and synthesizing the data from these fragmentation projects. Some studies have scaled up from survey plots to landscapes to assess the impacts of fragmentation on biodiversity (Lafortezza et al. 2010; Pfeifer et al. 2017), but further exploration on BEF relationships based on these fragmentation studies is required.

16.7.3 Long-Term Surveys on Biodiversity and Ecosystem Functioning

Although there are long-term projects that have studied the impact of fragmentation and biodiversity (Haddad et al. 2015), fragmentation effects on functioning (Kardol et al. 2018), and BEF relationships separately (e.g., http://www.treedivnet.ugent.be/), few have taken a long-term approach to study BEF relationships in fragmented landscapes. Fortunately, true islands in the Thousand Island Lake (Wilson et al. 2016), in the northern Sweden (Kardol et al. 2018), and habitat islands in BDFFP (Laurance et al. 2018), and the SAFE project (Ewers et al. 2011) can be considered as model system to study how habitat fragmentation affects ecosystem functioning. These systems can provide unusual research opportunities to address these questions, such as how ecosystem functioning responded to time-delayed effect of biodiversity loss caused by habitat fragmentation (Vellend et al. 2006) (Fig. 16.2).

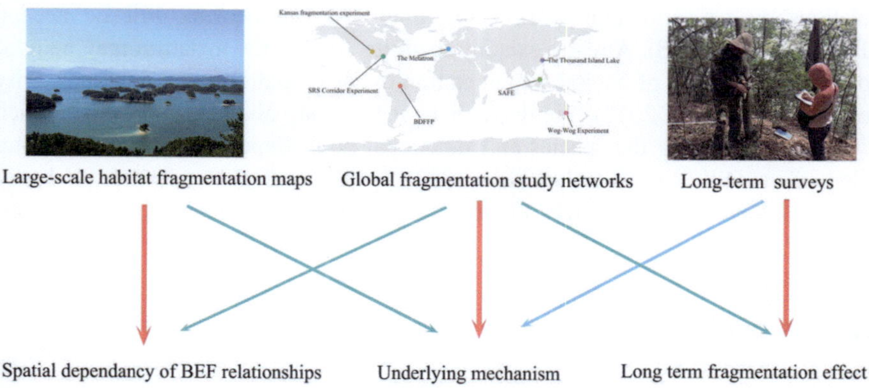

Fig. 16.2 A new agenda to explore BEF relationships in the fragmented landscapes combing data from large-scale fragmentation maps (e.g., Haddad et al. 2015), global fragmentation study networks (e.g., Wilson et al. 2016), and long-term surveys on biodiversity and ecosystem functioning (e.g., BDFFP project, see Laurance et al. 2018). Red lines indicate major approaches while blue lines indicate useful data to answer the linked questions

16.7.4 Modeling Work to Derive a Different Set of Predictions

In order to predict the relative importance of different mechanisms on the BEF relationship in fragmented habitats, mathematical models may be used. These models could be inspired by combining the approach outlined recently by De Laender (2016) on the effect of environmental conditions on BEF relationships together with the broader meta-ecosystem framework (Gounand et al. 2018). These models track the consequences of habitat fragmentation for functional diversity and the composition of the biotic communities and may provide powerful tools to predict dynamics to be further tested in experimental and real-world settings. Further inspiration to develop these models may come from the use of a limited number of functional "super-traits" such as body mass for consumers (Brose and Hillebrand 2016) in order to simplify model parameterization. These models, if proved to provide adequate fit to a priori expectation, may be applied to different scenarios of future land-use changes in specific landscapes in order to understand how will ecosystem functioning likely change and what mechanisms will be responsible for those (functional turnover, changes in species–species interaction or insurance effects) in order to design mitigation strategies.

16.7.5 Experimental Work to Disentangle the Different Mechanisms

To understand the relative importance of the different mechanisms through which fragmentation and habitat loss may affect biodiversity and ecosystem functioning relationships, experimental work using micro- or mesocosms will be needed (Hammill et al. 2018). Already existing facilities such as the metatron may be used to test for the effect of habitat fragmentation on BEF relation through connectivity loss (Legrand et al. 2012). Similarly, habitat loss may also be mimicked in such systems by reducing the species and functional pool. Experimentally reproducing edge effects will demand technical ingenuity but is a critical step in order to better understand and predict changes in BEF relationships under habitat fragmentation.

16.8 Conclusions

Natural habitats continue to be fragmented due to anthropogenic actions, and beyond affecting biodiversity, habitat fragmentation can alter the relationship between biodiversity and ecosystem functioning. In this chapter, we outlined the major mechanisms by which habitat fragmentation can affect BEF relations: (i) fragmentation-driven non-random turnover of species, (ii) changes in species interactions, and (iii) changes in insurance effects. We also highlighted the importance of considering spatiotemporal scales in studying BEF relationships. To promote BEF studies in fragmented landscapes, we presented the evidence currently available to science and outline major avenues for future research.

Acknowledgements The work was supported by the European Research Council [PASTFORWARD; grant no. 614839 attributed to KV] and the National Natural Science Foundation of China [grant no. 31361123001]. We thank Jianguo Wu, Maxwell Wilson, and Jinliang Liu for their helpful comments.

References

Arroyo-Rodríguez V, Saldaña-Vázquez RA, Fahrig L, Santos BA (2017) Does forest fragmentation cause an increase in forest temperature? Ecol Res 32:81–88. https://doi.org/10.1007/s11284-016-1411-6

Baert JM, Janssen CR, Sabbe K, De Laender F (2016) Per capita interactions and stress tolerance drive stress-induced changes in biodiversity effects on ecosystem functions. Nat Commun 7:12486. https://doi.org/10.1038/ncomms12486

Barry KE, Mommer L, van Ruijven J et al (2018) The future of complementarity: disentangling causes from consequences. Trends Ecol Evol in press. https://doi.org/10.1016/j.tree.2018.10.013

Bello C, Galetti M, Pizo MA et al (2015) Defaunation affects carbon storage in tropical forests. Sci Adv 1:e1501105. https://doi.org/10.1126/sciadv.1501105

Brinck K, Fischer R, Groeneveld J et al (2017) High resolution analysis of tropical forest fragmentation and its impact on the global carbon cycle. Nat Commun 8:14855. https://doi.org/10.1038/ncomms14855

Brose U, Hillebrand H (2016) Biodiversity and ecosystem functioning in dynamic landscapes. Philos Trans R Soc B Biol Sci 371:20150267. https://doi.org/10.1098/rstb.2015.0267

Cardinale BJ, Palmer MA, Collins SL (2002) Species diversity enhances ecosystem functioning through interspecific facilitation. Nature 415:426–429. https://doi.org/10.1038/415426a

Chisholm RA, Muller-Landau HC, Abdul Rahman K et al (2013) Scale-dependent relationships between tree species richness and ecosystem function in forests. J Ecol 101:1214–1224. https://doi.org/10.1111/1365-2745.12132

Clarke DA, York PH, Rasheed MA, Northfield TD (2017) Does biodiversity-ecosystem function literature neglect tropical ecosystems? Trends Ecol Evol 32:320–323

De Laender F, Rohr JR, Ashauer R et al (2016) Reintroducing environmental change drivers in biodiversity-ecosystem functioning Research. Trends Ecol Evol 31:905–915

Ewers RM, Didham RK, Fahrig L et al (2011) A large-scale forest fragmentation experiment: the stability of altered forest ecosystems project. Philos Trans R Soc B Biol Sci 366:3292–3302. https://doi.org/10.1098/rstb.2011.0049

Fahrig L (2003) Effects of habitat fragmentation on biodiversity. Annu Rev Ecol Evol Syst 34:487–515. https://doi.org/10.1146/annurev.ecolsys.34.011802.132419

Fahrig L (2017) Ecological responses to habitat fragmentation per se. Annu Rev Ecol Evol Syst 48:1–23. https://doi.org/10.1146/annurev-ecolsys-110316-022612

Fanin N, Gundale MJ, Farrell M et al (2018) Consistent effects of biodiversity loss on multifunctionality across contrasting ecosystems. Nat Ecol Evol 2:269–278. https://doi.org/10.1038/s41559-017-0415-0

Ferreira J, Lennox GD, Gardner TA et al (2018) Carbon-focused conservation may fail to protect the most biodiverse tropical forests. Nat Clim Chang 8:744. https://doi.org/10.1038/s41558-018-0225-7

Fontúrbel FE, Candia AB, Malebrán J et al (2015) Meta-analysis of anthropogenic habitat disturbance effects on animal-mediated seed dispersal. Glob Chang Biol 21:3951–3960. https://doi.org/10.1111/gcb.13025

France KE, Duffy JE (2006) Diversity and dispersal interactively affect predictability of ecosystem function. Nature 441:1139–1143. https://doi.org/10.1038/nature04729

Gonzalez A, Mouquet N, Loreau M (2009) Biodiversity as spatial insurance: the effects of habitat fragmentation and dispersal on ecosystem functioning. Biodiversity, Ecosyst Funct Ecosyst Serv 134–146

Gounand I, Harvey E, Little CJ, Altermatt F (2018) Meta-ecosystems 2.0: rooting the theory into the field. Trends Ecol Evol 33:36–46. https://doi.org/10.1016/j.tree.2017.10.006

Grace JB, Anderson TM, Seabloom EW et al (2016) Integrative modelling reveals mechanisms linking productivity and plant species richness. Nature 529:390–393. https://doi.org/10.1038/nature16524

Haddad NM, Brudvig LA, Clobert J et al (2015) Habitat fragmentation and its lasting impact on Earth's ecosystems. Sci Adv 1:e1500052. https://doi.org/10.1126/sciadv.1500052

Hagen M, Kissling WD, Rasmussen C et al (2012) Biodiversity, species interactions and ecological networks in a fragmented world. Adv Ecol Res 46:89–120. https://doi.org/10.1016/B978-0-12-396992-7.00002-2

Hammill E, Hawkins CP, Greig HS et al (2018) Landscape heterogeneity strengthens the relationship between β-diversity and ecosystem function. Ecology 99:2467–2475. https://doi.org/10.1002/ecy.2492

Hertzog LR et al (2019) Forest fragmentation modulates effects of tree species richness and composition on ecosystemmultifunctionality. Ecology. in press

Hillebrand H, Bennett DM, Cadotte MW (2008) Consequences of dominance: A review of evenness effects on local and regional ecosystem processes. Ecology 89:1510–1520. https://doi.org/10.1890/07-1053.1

Hooper DU, Chapin FS III, Ewel JJ (2005) Effects of biodiversity on ecosystem functioning: a consensus of current knowledge. Ecol Monogr 75:3–35. https://doi.org/10.1890/04-0922

Huang Y, Chen Y, Castro-Izaguirre N et al (2018) Impacts of species richness on productivity in a large-scale subtropical forest experiment. Science 362(80-):80–83. https://doi.org/10.1126/science.aat6405

Isbell F, Reich PB, Tilman D et al (2013) Nutrient enrichment, biodiversity loss, and consequent declines in ecosystem productivity. Proc Natl Acad Sci U S A 110:11911–11916. https://doi.org/10.1073/pnas.1310880110

Isbell F, Cowles J, Dee LE et al (2018) Quantifying effects of biodiversity on ecosystem functioning across times and places. Ecol Lett 21:763–778. https://doi.org/10.1111/ele.12928

Jones IL, Bunnefeld N, Jump AS et al (2016) Extinction debt on reservoir land-bridge islands. Biol Conserv 199:75–83. https://doi.org/10.1016/j.biocon.2016.04.036

Jucker T, Bouriaud O, Avacaritei D et al (2014) Competition for light and water play contrasting roles in driving diversity-productivity relationships in Iberian forests. J Ecol 102:1202–1213. https://doi.org/10.1111/1365-2745.12276

Jucker T, Avăcăriței D, Bărnoaiea I et al (2016) Climate modulates the effects of tree diversity on forest productivity. J Ecol 104:388–398. https://doi.org/10.1111/1365-2745.12522

Kahmen A, Renker C, Unsicker SB, Buchmann N (2006) Niche complementarity for nitrogen: an explanation for the biodiversity and ecosystem functioning relationship? Ecology 87:1244–1255. https://doi.org/10.1890/0012-9658(2006)87%5b1244:ncfnae%5d2.0.co;2

Kardol P, Fanin N, Wardle DA (2018) Long-term effects of species loss on community properties across contrasting ecosystems. Nature 557:710–713. https://doi.org/10.1038/s41586-018-0138-7

Kramer B, Poniatowski D, Fartmann T (2012) Effects of landscape and habitat quality on butterfly communities in pre-alpine calcareous grasslands. Biol Conserv 152:253–261. https://doi.org/10.1016/j.biocon.2012.03.038

Lafortezza R, Coomes DA, Kapos V, Ewers RM (2010) Assessing the impacts of fragmentation on plant communities in New Zealand: Scaling from survey plots to landscapes. Glob Ecol Biogeogr 19:741–754. https://doi.org/10.1111/j.1466-8238.2010.00542.x

Lasky JR, Uriarte M, Boukili VK et al (2014) The relationship between tree biodiversity and biomass dynamics changes with tropical forest succession. Ecol Lett 17:n/a-n/a. https://doi.org/10.1111/ele.12322

Laurance WF (2008) Theory meets reality: How habitat fragmentation research has transcended island biogeographic theory. Biol Conserv 141:1731–1744

Laurance WF, Delamônica P, Laurance SG et al (2000) Rainforest fragmentation kills big trees. Nature 404:836. https://doi.org/10.1038/35009032

Laurance WF, Camargo JLC, Fearnside PM et al (2018) An Amazonian rainforest and its fragments as a laboratory of global change. Biol Rev 93:223–247. https://doi.org/10.1111/brv.12343

Legrand D, Guillaume O, Baguette M et al (2012) The Metatron: An experimental system to study dispersal and metaecosystems for terrestrial organisms. Nat Methods 9:828–833. https://doi.org/10.1038/nmeth.2104

Liang J, Zhou M, Tobin PC et al (2015) Biodiversity influences plant productivity through niche–efficiency. Proc Natl Acad Sci. https://doi.org/10.1073/pnas.1409853112

Liang J, Crowther TW, Picard N et al (2016) Positive biodiversity-productivity relationship predominant in global forests. Science 354(80-):aaf8957–aaf8957. https://doi.org/10.1126/science.aaf8957

Liu J, Slik JWF (2014) Forest fragment spatial distribution matters for tropical tree conservation. Biol Conserv 171:99–106. https://doi.org/10.1016/j.biocon.2014.01.004

Liu J, Coomes DA, Hu G et al (2018a) Larger fragments have more late-successional species of woody plants than smaller fragments after 50 years of secondary succession. J Ecol in press. https://doi.org/10.1111/1365-2745.13071, https://doi.org/10.1111/1365-2745.13071

Liu J, Wilson M, Hu G et al (2018b) How does habitat fragmentation affect the biodiversity and ecosystem functioning relationship? Landsc Ecol 33:341–352. https://doi.org/10.1007/s10980-018-0620-5

Loreau M, Mouquet N, Gonzalez A (2003) Biodiversity as spatial insurance in heterogeneous landscapes. Proc Natl Acad Sci U S A 100:12765–12770. https://doi.org/10.1073/pnas.2235465100

Lutz JA, Furniss TJ, Johnson DJ et al (2018) Global importance of large-diameter trees. Glob. Ecol, Biogeogr

Magnago LFS, Edwards DP, Edwards FA et al (2014) Functional attributes change but functional richness is unchanged after fragmentation of Brazilian Atlantic forests. J Ecol 102:475–485. https://doi.org/10.1111/1365-2745.12206

Pfeifer M, Lefebvre V, Peres CA et al (2017) Creation of forest edges has a global impact on forest vertebrates. Nature 551:187–191. https://doi.org/10.1038/nature24457

Ratcliffe S, Holzwarth F, Nadrowski K et al (2015) Tree neighbourhood matters—tree species composition drives diversity-productivity patterns in a near-natural beech forest. For Ecol Manage 335:225–234. https://doi.org/10.1016/j.foreco.2014.09.032

Ratcliffe S, Liebergesell M, Ruiz-Benito P et al (2016) Modes of functional biodiversity control on tree productivity across the European continent. Glob Ecol Biogeogr 25:251–262. https://doi.org/10.1111/geb.12406

Ratcliffe S, Wirth C, Jucker T et al (2017) Biodiversity and ecosystem functioning relations in European forests depend on environmental context. Ecol Lett 20:1414–1426

Reich PB, Tilman D, Isbell F et al (2012) Impacts of biodiversity loss escalate through time as redundancy fades. Science 336(80-.):589–592

Roscher C, Temperton VM, Scherer-Lorenzen M et al (2005) Overyielding in experimental grassland communities - irrespective of species pool or spatial scale. Ecol Lett 8:419–429. https://doi.org/10.1111/j.1461-0248.2005.00736.x

Rusterholz HP, Baur B (2010) Delayed response in a plant-pollinator system to experimental grassland fragmentation. Oecologia 163:141–152. https://doi.org/10.1007/s00442-010-1567-7

Sandel B, Svenning J (2013) Human impacts drive a global topographic signature in tree cover. Nat Commun 4:1–7. https://doi.org/10.1038/ncomms3474

Slik JWF, Paoli G, Mcguire K et al (2013) Large trees drive forest aboveground biomass variation in moist lowland forests across the tropics. Glob Ecol Biogeogr 22:1261–1271. https://doi.org/10.1111/geb.12092

Smith IA, Hutyra LR, Reinmann AB et al (2018) Piecing together the fragments: elucidating edge effects on forest carbon dynamics. Front Ecol Environ 16:213–221. https://doi.org/10.1002/fee.1793

Srivastava DS, Vellend M (2005) Biodiversity-ecosystem function research: is it relevant to conservation? Annu Rev Ecol Evol Syst 36:267–294. https://doi.org/10.1146/annurev.ecolsys.36.102003.152636

Staddon P, Lindo Z, Crittenden PD et al (2010) Connectivity, non-random extinction and ecosystem function in experimental metacommunities. Ecol Lett 13:543–552. https://doi.org/10.1111/j.1461-0248.2010.01450.x

Stephenson NL, Das AJ, Condit R et al (2014) Rate of tree carbon accumulation increases continuously with tree size. Nature 507:90–93. https://doi.org/10.1038/nature12914

Steudel B, Hector A, Friedl T et al (2012) Biodiversity effects on ecosystem functioning change along environmental stress gradients. Ecol Lett 15:1397–1405. https://doi.org/10.1111/j.1461-0248.2012.01863.x

Taubert F, Fischer R, Groeneveld J et al (2018) Global patterns of tropical forest fragmentation. Nature 554:519–522. https://doi.org/10.1038/nature25508

Terborgh J, Lopez L, Nuñez P et al (2001) Ecological meltdown in predator-free forest fragments. Science 294:1923–1926. https://doi.org/10.1126/science.1064397

Thompson PL, Rayfield B, Gonzalez A (2017) Loss of habitat and connectivity erodes species diversity, ecosystem functioning, and stability in metacommunity networks. Ecography (Cop) 40:98–108. https://doi.org/10.1111/ecog.02558

Thompson PL, Isbell F, Loreau M et al (2018) The strength of the biodiversity–ecosystem function relationship depends on spatial scale. Proc R Soc B Biol Sci 285:20180038. https://doi.org/10.1098/rspb.2018.0038

Tilman D, Isbell F, Cowles JM (2013) Biodiversity and ecosystem functioning. Annu Rev Ecol Evol Syst 45:141007174654001. https://doi.org/10.1146/annurev-ecolsys-120213-091917

Tobner CM, Paquette A, Gravel D et al (2016) Functional identity is the main driver of diversity effects in young tree communities. Ecol Lett 19:638–647

Vellend M, Verheyen K, Jacquemyn H et al (2006) Extinction debt of forest plants persists for more than a century following habitat fragmentation. Ecology 87:542–548. https://doi.org/10.1890/05-1182

Wardle DA (2016) Do experiments exploring plant diversity-ecosystem functioning relationships inform how biodiversity loss impacts natural ecosystems? J Veg Sci n/a-n/a. https://doi.org/10.1111/jvs.12399

Watson JEM, Evans T, Venter O et al (2018) The exceptional value of intact forest ecosystems. Nat Ecol Evol 2:599–610. https://doi.org/10.1038/s41559-018-0490-x

Weisser WW, Roscher C, Meyer ST et al (2017) Biodiversity effects on ecosystem functioning in a 15-year grassland experiment: Patterns, mechanisms, and open questions. Basic Appl Ecol 23:1–73. https://doi.org/10.1016/J.BAAE.2017.06.002

Wilson MC, Chen X-Y, Corlett RT et al (2016) Habitat fragmentation and biodiversity conservation: key findings and future challenges. Landsc Ecol 31:219–227. https://doi.org/10.1007/s10980-015-0312-3

Wu J, Loucks OL (1995) From balance of nature to hierarchical patch dynamics: a paradigm shift in ecology. Q Rev Biol 70:439–466. https://doi.org/10.1086/419172

Xu W, Viña A, Kong L et al (2017) Reassessing the conservation status of the giant panda using remote sensing. Nat Ecol Evol 1:1635. https://doi.org/10.1038/s41559-017-0317-1

Chapter 17
Basic Principles of Sustainable Land Use Management

Zita Izakovičová, László Miklós and Jana Špulerová

Abstract The paper is aimed at the presentation of the integrative approach assessment of environmental land use conflicts. It presents basic principles of sustainable land use management. The changes in land use are reflected not only in changes of land cover. They are also the cause of changes in landscape structure and are the main initiation of many environmental problems. From this aspect, it is important to engage the land structure studies not only in the studies of land use forms and land cover, but also in the studies of position and relation in landscape (cause–consequence). To find the cause and casualties of these changes is very important for implementation of rationale landscape utilization. The basic tool for suitable landscape utilization is integrated landscape management as a major instrument of sustainable development. It must be based on **understanding landscape systematically as geosystem**. The every point of the landscape is representing the integrating scope, scene in which all natural resources are occurring as layers (geological sources, water and soil sources, climate, biotic sources and morphometric parameters) which are mixing together. It is seen as understanding the space as integration of particular natural sources in given area of landscape. Using one source can negatively affect the quality of other sources. For example, intensive use of soil resources can threaten water resources—negative impacts of chemistry, mechanization, and so on. Therefore, land use needs to be assessed on an integrated basis. From aspect of 'sustainability', the target is to define such landscape management, which would regulate socio-economic development in landscape with its natural, human, cultural and historical potential. It is based on matching the supply that is represented by landscape resources and demand that is represented by community needs and community requirements. The discrepancy

Z. Izakovičová (✉) · L. Miklós · J. Špulerová
Institute of Landscape Ecology, Slovak Academy of Sciences,
Štefánikova 3, 84199 Bratislava, Slovakia
e-mail: zita.izakovicova@savba.sk

L. Miklós
e-mail: miklos@tuzvo.sk

J. Špulerová
e-mail: jana.spulerova@savba.sk

L. Müller and F. Eulenstein (eds.), *Current Trends in Landscape Research*,
Innovations in Landscape Research, https://doi.org/10.1007/978-3-030-30069-2_17

between supply and demand (not respecting the properties of landscape resources) is the determining factor of formation of the environmental problems. The paper presents evaluation of environmental problems resulting from conflict of interests in the land use and methodology for sustainable land use that is based on the integrative approach to the landscape.

Keywords Land use · Land cover · Sustainable development · Integrated landscape management · Landscape as geosystem · Environmental problems

17.1 Introduction: About Landscapes and Land Use

Landscape is a very dynamic system, comprised from two mutually interconnected subsystems (natural and social), which is impacted by constant changes in time and space. Time and space are the two most important universal variables, in which the natural and human forces are together forming and permanently changing the landscape into a cultural and giving it a unique character (Olah and Žigrai 2004). Cultural landscape is an image of the state and development of society, because all social changes appear eventually in change of the structure and utilization of the landscape.

Land use is determined on one side by attributes of natural components, which represent a specific offer for landscape usage by a human. On the other side, the land use is determined by requirements, demands, abilities and overall possibilities of land use by a society (Fig. 17.1). The requirements, demands as well as the overall possibilities are changes in the course of time (change of human conditions, change of priorities, change of ownership, change in technologies, etc.). A value, which was a luxury good in the past, became an essential good. The change in the demands and requirements of human society is then cyclically appearing in changes of forms and intensity of landscape usage. On the other side, the changes can be caused by repercussion of natural environment against a specific land use, if this is becoming unsustainable. For example, improper land use can cause or potentially accelerate the natural risks and hazards, which take part in change of landscape structure and land use to a substantial rate. Mostly, the changes in society, which are influencing the landscape, are faster and variable in comparison with sustained natural process of landscape development. The utilization of the landscape reflects the anthropization and stabilization of the landscape, which are conditioned by the ratio of stabilizing and less stabilizing landscape elements.

In the literature, we can meet two basic terms—land use and land cover. Land cover is according to Earth's surface visible from the space, including natural ecosystem (forest, grassland, water areas, etc.), human-transformed (agroecosystems) or human-created ecosystems (urban areas, roads). Land cover shows the combination of natural, semi-natural (human-changed elements) and artificial elements (human-created elements of landscape structure). The present state of area-anthropization (human-created elements of landscape structure) can be

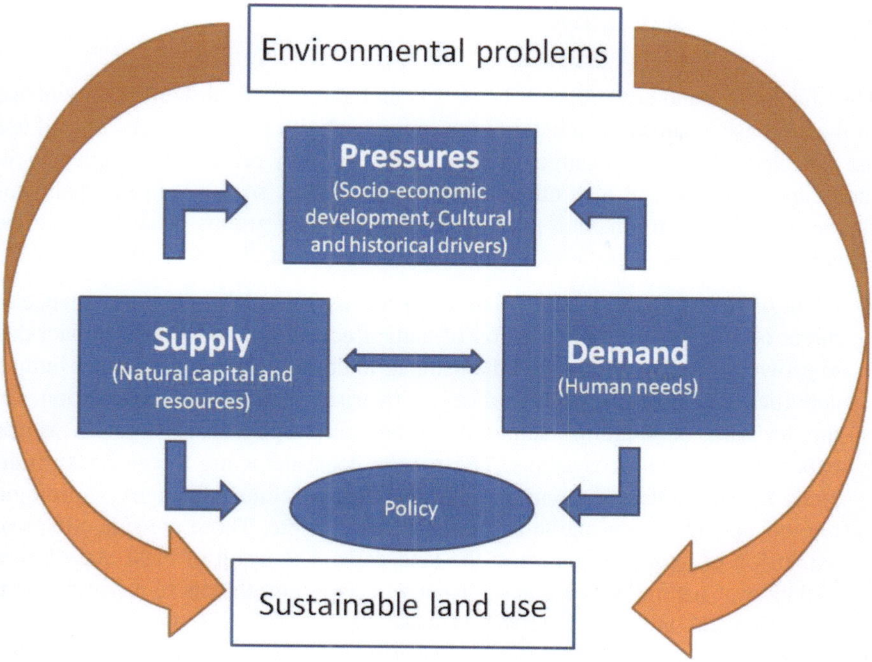

Fig. 17.1 Integrated approach to sustainable land use management (Izakovičová et al. 2018b)

evaluated based on presence and area extent of the particular element of the present landscape structure, if it is a natural area with great landscape-ecological value or vice versa or if its strong anthropicly changed area with low landscape-ecological value. Assessment of land cover applies different statistical and analytical methods, like index of heterogeneity, Shannon index, Index of ecological stability, index of landscape structure, patch density, variability metrics, edge metrics, diversity metrics, etc. (Forman and Godron 1986; Ivanova et al. 2011; Mander and Jongman 1999; Miklos 1996; Miklós et al. 1986; Rehackova et al. 2008; Sklenička 2006; Turner et al. 2013).

Land use is based on set of functions for which the land is used. According to Jensen and Murray (2005), the landscape usage is linked to human activities and its influence in given landscape (e.g. agriculture, forestry, growth of urban area and transport).

Oťahel' et al. (2004) pointed to the difference between land use and land cover, while map of land cover represents objects of Earth's surface via their physiognomic attributes and the map of land use via functional attributes. Land use and land cover are interconnected. Land cover is continually shaped and transformed by land use changes, e.g. when forest is converted to pastures or arable land. Land use change is proximate cause of land cover change (De Sherbinin 2002).

17.2 Land Use Changes

The changes of land cover cannot be understand and explained without knowledge of the land use changes, which condition human activities. The changes in land use are reflected not only in changes of land cover, but they drive also the changes in landscape structure and also other landscape processes, so they are the main initiation of many environmental problems. The following groups of changes can be included to the most significant changes (Izakovičová et al. 1997, 2017):

- ***Change, threat and degradation of natural ecosystem***, firstly in consequence of direct occupation or fragmentation of natural ecosystem and barrier impact due to growth of industrialization, urbanism, agriculture, recreation and other human activities. The outcome of this process is the rise of grade of anthropization, i.e. the increase of share of semi-natural and anthropogenic ecosystem at the expense of natural ecosystems. The natural ecosystems are threaten also indirectly by pressures and threat on their conditions—climate changes, hydrologic regime, spread of non-native and invasive species etc. These are side effects of expansion of human activities in the landscape. The result of these activities is violation of natural development of native ecosystem, threats to environmental stability of the area and biodiversity loss.
- ***Change, threat and degradation of natural resources*** are major threat to the qualitative and quantitative attributes of individual natural resources. As natural resources, we consider individual land-building components, which are capable through its useful attributers to satisfy human needs. These resources as well as biotic resources are threaten directly by land occupation and its depleting in consequence of human activities—e.g. occupation of the best-quality land for building (Fig. 17.2), occupation of forest for agricultural production etc. Indirect threat on natural resources is happening via production of foreign substances, which are a side effect of many form of land use—water pollution as consequence of intensive agriculture, leak of oil from agricultural and forest machines, etc.
- ***Change in quality of environment***—the combination of land cover elements creates various landscape structures, landscape mosaics in the area, which represent basic human living area. The structure and organization of these elements are the basic indicators of quality and aesthetics of the environment. Inappropriate land cover with domination of negative elements can considerably disrupt the aesthetics, landscape image as well as overall quality of environment.
- ***Violation of landscape processes, particular acceleration of natural risks and hazards***—unsuitable land use, e.g. deforestation, vegetation removal, inappropriately intensive use of soil often implicates erosive–accumulative processes, landslide, etc. Vegetation removal appears negatively also on hydrologic regime in the area (accelerated floods, etc.) and can be also interacted factor of climate changes (endangering of microclimate area regulation, lower oxygen production and lower absorption of polluting substances).

Fig. 17.2 The realization of human activities in the landscape causes a significant anthropization of the territory (Photograph: Zita Izakovičová)

17.3 Research Focused on the Evaluation of Landscape Changes

Landscape structure changes represent a wide and very important issue in landscape research. The number of papers in scientific journals that focus on the topic of landscape changes has been increasing significantly during last 10–15 years (Baessler and Klotz 2006; Belcakova 2016; Bürgi et al. 2004; Bürgi et al. 2017; Demková and Lipský 2015; Ivanova et al. 2011; Oťahel' et al. 2004). Among many conferences, workshops and seminars dealing with the topic of landscape changes, the seminar *Landscape change and its ecological consequences in Europe* held in Tilburg in 1995, should be mentioned, from which an important publication was published (Jongman 1996). The international seminar with the characteristic title *Landscape Change: Learning from the past–Visions for the future* was organized in Norwegian Tromso in June 2006. Assessment of the landscape changes represented key topic of IALE word congress, which held in Wageningen in July 2007. The papers are published in the proceeding from the Seventh IALE World Congress (Bunce et al. 2007). The RegioResources conference, held in 2018 in Smolenice, was devoted to changes in landscape use affected by global megatrends (ILE SAS 2018).

Land use as well as general landscape changes are studied in the field both of geography and landscape ecology, apart from other scientific and applied disciplines dealing with landscape issues. The approaches to the assessment of landscape structure changes are very different. The largest number of works is devoted to study territorial organization of the land use forms and land use change in different historical periods. This approach is widely spread at representatives of different landscape-ecological and geographical schools (e.g. Bicik et al. 2001; Bunce et al. 1996; Burgi et al. 2017; Csorba 1996; Dyer 1994; Falťan et al. 2017; Forman and Godron 1986; Kolejka 1987, 2006, 2018; Lieskovsky et al. 2018; Lipský 2000; Lipsky et al. 1999; Muchova et al. 2016; Olah and Žigra 2004; Oťahel' and Feranec 2001; Petrovič 2006; Poudevigne and Alard 1997, Romportl et al. 2013; White and Mladenoff 1994; Zigrai 1996 etc.).

The primary sources for evaluation of landscape changes are historical cartographic and written materials. The most common method is interpretation of air photographs, ortho-photomaps or interpretation of historic maps coming from different mappings. Nowadays, modern geo-information methods (GIS) and accessibility of multidata sources offer great opportunities in retrospective views on the landscape (Bender et al. 2005a, b; Halabuk et al. 2015; Kienast et al. 2015; Mander et al. 2004; Mander and Murka 2003; Palang et al. 2000, Palang et al. 2011; Pauleit et al. 2005; Pauleit 2018; Pauleit and Golding 2005; Quattrochi and Luvall 1999, 2014). From this point of view to the most important can be assigned following projects: Corine Land Cover 2000–2018 (CLC 1990, 2000, 2018), BioScene, BIOPRESS, LANDMAP and LCLUC-NASA (Haddock et al. 2007; Haines-Young et al. 2012; Kumar 2010; Oťahel' et al. 2004; Reschke and Huettich 2014; Roy et al. 2015; Roy and Joshi 2002; Soliva 2007; Štefunková and Hanušin 2016; Pauditšová and Slabeciusova 2014).

Corine Land Cover provides consistent information on land cover and land cover changes across Europe (Pazur et al. 2015). CLC 2018 (CLC 2018) is an update for the reference year 2018 of the first Corine Land Cover database which was finalized in the early 1990s, as part of the European Commission programme to Coordinate Information on the Environment (Oťahel' et al. 2004). It provides information on land cover and land cover changes during the past decade across Europe. The methodology is based on the photograph interpretation of satellite images by the national team of the participating countries. Today, 29 countries and over 100 organizations are involved in the production and dissemination of the CLC 2018 data. CLC 2018 shows the land cover changes in ecosystems such as forests, lakes, pastures and grasslands and the impact of human activities on the land use. The very important information of the projects CLC is information about fragmentation natural areas, too. Forty-four land cover classes map changes over time. These classes are suitable for evaluation of the land cover changes in the small scales. These classes are not satisfactory for mapping at large scales. To assess the land use changes in an urbanized environment, the project Urban Atlas was focused (EEA 2018a).

Similarly, project BIOPRESS was aimed at the evaluation of the land cover changes around NATURA 2000 sites and evaluation of the pressures on biodiversity (BIOPRESS 2003). Five times more changes were calculated from transects than from windows. The main observed land cover changes were different between biogeographical regions.

LANDMAP (Mucher et al. 2003) is a unique system, allowing information about landscape to be gathered, organized and evaluated into a nationally consistent data set. LANDMAP information is collected in a structured and rigorous way that aims to be as objective as possible. Its database contains both relatively objective information such as rock type and historical information, and subjective information as sensory responses and cultural interpretation. LANDMAP information can also be combined with contextual socio-economic information.

BioScene project developed models simulating the biodiversity changes in consensus with land use changes, mostly in connection to agriculture (Tzanopoulos et al. 2005). The individual scenarios were consulted with stakeholder and based on it a sustainable agricultural development strategy in mountain regions was developed.

The project LCLUC (Lieskovsky et al. 2018) was aimed on evaluation of long-term agricultural land use change by reconstructing historic land use change since the mid-1800s for a comprehensive sample of case study regions across the Carpathian Basin.

Systematic attention is paid to monitoring land use changes by European Land use Institute (ELI 2018). The uppermost aim of European Land use Institute (ELI) is to build a sustainable and long-lasting partnership in research, development and capacity development in integrated land use.

One of the most comprehensively developed methods focused on study the cause, change and consequence is DPSIR methodology (Kristensen 2004). DPSIR is a general framework for organizing information about state of the environment. The framework assumes cause–effect relationships between interacting components of social, economic and environmental systems, which are:

- Driving forces of environmental changes—land use changes (e.g. agriculture)
- Pressures on the environment (e.g. land cover changes—deforestation of the landscape)
- State of the environment (increasing of the anthropization of the landscape)
- Impact on the population, economy and ecosystems (e.g. pressures on landscape stability and biodiversity, erosion, flood, creation of mono-functional agricultural and aesthetically unimpressive landscape, etc.)
- Response of the society (e.g. forest and biodiversity protection, revitalization of eroded landscape, etc.).

Variations of the DPSIR framework are included in many national or international strategies or documents like OECD Environmental Indicators (OECD 2001), DSR (UNCSD 1996), US EPA (Bradley and Yee 2015) and it has been applied in an increasing number of research projects. The application of such cause–consequence approach of changes evaluation is essential to ensure sustainable landscape management and use.

Landscape managers and creators, decision-makers and regional planning agencies create an important impulse and a valuable information not only about the history of the area, but also about the optimal, appropriate type of sustainable use of cultural landscape (Antrop 2000; Baránková et al. 2011; Bell et al. 2009; Conway and Lathrop 2005; Nikodemus et al. 2005; Štefunková and Hanušin 2016; Wiggering and Steinhardt 2015).

17.4 Conflict of Interests in the Land Use

Conflict of interests in the land use is based on matching the offer, which is represented by the resources in the region, and demand, which is represented by the community needs of development. The discrepancy between offer and demand (not respecting the properties of landscape resources) is the determining factor of formation various environmental problems. Several sectors claim to use the landscape space. Individual resorts have specific demands for land use. Resort requirements are not always in harmony. The following resorts are mostly involved in the land use (Izakovičová et al. 2018a):

Agriculture: it is a very specific sector. It is, on one hand, obviously responsible for the protection of its own production resource, the **soil resources**, but on other hand, the effort to increase the production causes over-exhaustion of the soils, often degradation by pollution, drought by ameliorations, erosion and compaction with heavy machines. This sector is a specific example for the internal conflict of interest within the same sector (Fig. 17.3).

Forest management and forest resources are crucial for the ecological stability, biodiversity and environment. The Act on forest defines these aspects of the forests as regulation or cultural ecosystem services of forests, which are delineated in practice as forest of special purpose and protective forests. However, on the same

Fig. 17.3 Erosion on agricultural landscape (Photograph: Lothar Mueller)

time, the forest management, which is oriented on timber production, is sometimes considered as main enemy of the nature conservation, causing very diverse problems for biodiversity loss, erosion, accumulation, pollution, etc.

Water management: It is the key sector responsible for good quality and quantity of **water resources**. The Water Act defines activities and duties of water management and in practice the area is projected as protective zones of water resources of a different character. On the other hand, the technical activities of water management, as building dams, regulations of rivers, creating technical measures against floods, are often evaluated as endangering, e.g. when water reservoirs overlap precious habitats, the flow regulation destroys the natural living conditions for animals, when the measures cause considerable lowering or in opposite rise of underground water levels.

Urbanization: The residential areas, including all necessary accessories like parks and other green infrastructure, are considered for **resources** permanently ensuring human health and well-being. The communal activities such as waste management, sewage management, but also the rise of the residential areas and the population itself cause big problems by overlapping the agricultural land, pollution, garbage and non-controlled activities of population in nearby nature.

Recreation: The recreational areas, spas, health resorts and provided recreational activities are considered as natural resource for human health (Fig. 17.4). On the other side, intensive recreational activities often means intensive impact on nature, e.g. the ski resorts in high mountains, dense tourist path in national parks, overpopulated recreation centres in natural parks, noise, garbage, etc. and cause changes and destruction of natural habitats, lowering the ecological stability of these natural resources (Barančok and Barančoková 2013; Špulerová et al. 2016).

Negative influences or pressures of the individual sectors are presented in the Table 17.1.

The assessment of individual sectors based on the character of sectoral activities is given from the **environmental/ecological point of view**. If the same encounter would be assessed by an industrialist, an agriculturist or a forester, they would mark endangered and endangering activities almost fully contrariwise. As a typical example: the ecologists consider the timber production as a big threat to nature conservation. In contrary, the foresters consider the nature conservation as one of the most important barriers of the forestry.

The projected interests **overlap each other** in the space and thus create the encounter of the interests of very diverse character. The activities of sectors for the purposes of the evaluation of conflict of interests can be assessed from environmental aspects (Miklos et al. 2019; Miklós and Izakovičová 1997):

- The positive impact on the landscape and ecosystems has activities aimed at **the nature protection and landscapes protection** (conservation areas of different degrees) and **natural resources protection** (protected zones of geological, soil, water and forest resources). According to the division of ecosystem services, these zones usually provide protection for ecosystems, including supportive, and in some cases even productive services. As an example, the protection of soil

Fig. 17.4 Semi-natural or artificial lake create condition for recreation use—Ružiná water dam, Slovakia (Photograph: Jana Špulerová)

resources focuses primarily on the use of production functions on the high-quality soils. When these activities are encountered with activities of other sectors (mainly with productive sectors), their execution is in most cases **endangered**.

- The negative impact on the landscape and ecosystems causes activities of the stress factors that aggravate the quantity and quality of the landscape as whole as well as individual natural resources and, at the same time, they limit the use of their ecosystem services. These are the activities from the industry, energetics, mining, transport and urbanization sectors. Negative impacts of human activities are manifested by spatial reduction of the areas of natural and semi-natural ecosystems and natural resources, reducing the overall ecological stability of the landscape, creating barriers to natural biota movement, producing contaminants, degrading the environment. As an example, the degraded forest ecosystems and agroecosystems have reduced economic value and limited use due to stress factors. Contaminated soil is not hygienically suitable for growing crops for direct consumption. When these activities encounter with activities of other sectors (mainly with protective sectors), they have character of **endangering** factors.

Table 17.1 Negative influences of economic sectors to the landscape and its resources

Branches	Negative influence
Industry	Taking of the natural ecosystems, occupation of natural resources and areas, barrier effect, production of emissions, noise, smell, negative aesthetic influence and production of the industrial waste
Extraction of the mineral sources	Take of the natural ecosystems, noise, dust, negative aesthetic influence, production of the waste, being of the anthropogenic form of the relief
Transport	Take of the natural ecosystems, fragmentation of habitats, occupation of natural resources and areas, barrier effect, production of the transport emissions, noise, dust, vibration, light effects and negative influence of the sprinkles salts
Urbanization	Occupation of natural resources and areas, barrier effect, take of the natural ecosystems, production of the domestic waste and production of the emissions from local sources
Agriculture	Negative hygienic influences, dust, odour, production of the agricultural waste, negative influences of the agricultural chemicals, take of the natural ecosystems, occupation of natural resources and areas, barrier effect, endangering of the territorial stability of the agricultural landscape and negative influences of the agricultural mechanization—physical degradation of the soil resources
Forestry	Negative influences of the forestry chemicals, occupation of the natural forest ecosystem and negative influences of the inappropriate management of forest ecosystems
Water husbandry	Confusing (eroding) of the water regime of the landscape, barrier effect and occupation of the area
Recreation	Occupation of natural resources and areas, barrier effect, take of the natural ecosystem, production of the domestic waste, production of the emissions from local sources, threatening of the flora and fauna

- Several sectoral activities are of twofold character. On one side, they are important on even crucial for the provision of high-quality environment or natural resources, and on the other side, their execution endangers the optimum use of land, natural resources and ecosystem services.

The evaluation of the encounters of interests is processed in a matrix (Izakovičová et al. 2018a) and they describe main sectoral activities (Fig. 17.5):

- the rows of the matrix incorporate those sectoral activities which aims towards the protection and **conservation of nature and landscape**, **protection of natural resources and human environment**. From the ecological point of view, these activities can be considered as **endangered** by other sectors.
- the columns of the matrix incorporate those sectoral activities which aims towards material production of goods and other technical services, such as the industry, energetic, mining, transport, urbanization, recreation, agriculture, forestry and water management. From the ecological point of view, these activities can be considered as **endangering** the sectors mentioned above.

Stress factors Natural sources	*Industry*	*Urbanization*	*Agriculture*	*Forestry*	*Recreation*	*Type of the environment al problem*
Protected areas	taking of the natural ecosystems, negative influences of the emissions, barrier effect	taking of the natural ecosystems, fragmentation, barrier effect	taking of the natural ecosystems, negative influences of the chemicals	negative consequence s of inappropriate management - holograms, fragmentation, erosion	capture of natural ecosystems destruction of vegetation, noise, creation of wild dumps	**Endangering the landscape ecological stability**
Forest sources	taking of the natural ecosystems, negative influences of the emissions	taking of the natural ecosystems, fragmentation	negative influences of the agricultural chemicals, occupation of forest ecosystems	negative consequence s of inappropriate management - holograms, excessive extraction, excavation of natural ecosystems	taking of the natural ecosystems, negative, ecosystems destruction of vegetation, noise, disturbance of fauna	
Water sources	production of industrial waste water, water consumption	wastewater production, excessive water consumption	endangering of the water sources by chemicals and animal waste	endangering of the water sources by chemicals	wastewater production, excessive water consumption	**Endangering the natural resources**
Soil sources	soil occupation, physical, chemical and biological degradation of the soil	soil occupation, endangering of the soil by emissions from local sources	soil degradation contamination, compactness etc.	plane occupation, negative influence of the forestry chemicals, soil erosion	degradation of the soil – erosion, physical, chemical and biological degradation	
Recreation resources	Air pollutants, noise, dust, radiation	anthroposis of the territory, increasing pressure on recreational areas	noise, smell, dust, negative hygienic influences	degradation of recreational areas due to intensive logging	negative impacts of increased visitors, waste generation, noise	
Human environment	Air pollutants, noise, dust, radiation	anthroposis of the territory, negative aesthetic influences	noise, smell, dust, negative hygienic influences, change of land-use	noise, negative influence of the forestry chemicals	inappropriate localization of recreational objects in an urbanized environment, negative impacts of increased visitors	**Endangering the human resources**

Fig. 17.5 The example of the matrix of endangering activities and endangered landscape resources

- the squares of the matrix represent the **encounters of the interests** of the sectors.

There are two ways for evaluating the encounters, as:

- by evaluation down the columns we ask, how does an endangering activity threatens the endangered factors? Figure 17.5 describes this mode on the

example of the evaluation of how the agriculture threatens the particular endangered factors.
- by evaluation along the rows we ask, how is threatened and endangered factor by endangering activities? Figure 17.5 presents this mode on the example of the evaluation of how are the soil resources threatened by particular endangering activities.

Few other important aspects of the process of evaluation in matrix have to be mentioned. The provided matrix may be considered as a model (even qualitative) of all possible mutual conflicts of all sectors. Of course, the presented graphical forms of the matrix in paper are two-dimensional. However, in a real territory, there are encounters not only of pairs of endangered and endangering activities, there may appear very diverse combinations of several activities. Therefore, the evaluation of encounters and conflicts are much more complex. On the example of study areas, we applied GIS technology, which allows assessing the conflicts of several sectoral activities on the same area in certain time, if they occur together. Therefore, the matrix is actually multidimensional. The GIS technology allows to analyse each conflict of interest quantitatively, according to their area size, including the conflict of the land use with natural conditions.

The conflicts of land use can be documented quantitatively by statistical data, which mirror the changes of areas of land use. Methodically, it is a simply procedure with many example of the studies in the literature (Bezák et al. 2017; Braunisch et al. 2011; de Groot 2005). However, these statistics document the changes in the certain time and certain area only of pairs of land use forms. The list of all binate changes can be considered as integrated picture on the changes, but they are still just a set of binate data.

The evaluation of the conflicts of interests as described above allows a general classification of environmental problems caused by conflicts of interests in the landscape into three groups:

(a) ***Problems of the endangering the ecological stability of the landscape***

These problems arise as the consequence of influence of stress factors to elements with high eco-stabilizing effect like nature conservation areas or other ecologically valuable areas of the landscape (forest, water areas, meadows, pastures, areas of public greenery, etc.). The most frequent problems in this group are the capture of natural ecosystems for realization of investment activities such as residential and recreational or industrial farm buildings and facilities. Significant negative influence on natural ecosystems represents also the exploitation of mineral raw materials. Natural ecosystems are also threatened indirectly, by changing environmental conditions, by production of emissions, by landfilling, by negative impacts of recreational and tourist activities such as vegetation shedding, moulding and other physical damage to vegetation (Piscová 2011). These problems are depicted in first two rows of the matrix (Fig. 17.5). The assessment of the problems of the threat to ecologic stability and biodiversity is given on the example of Bratislava (Fig. 17.6).

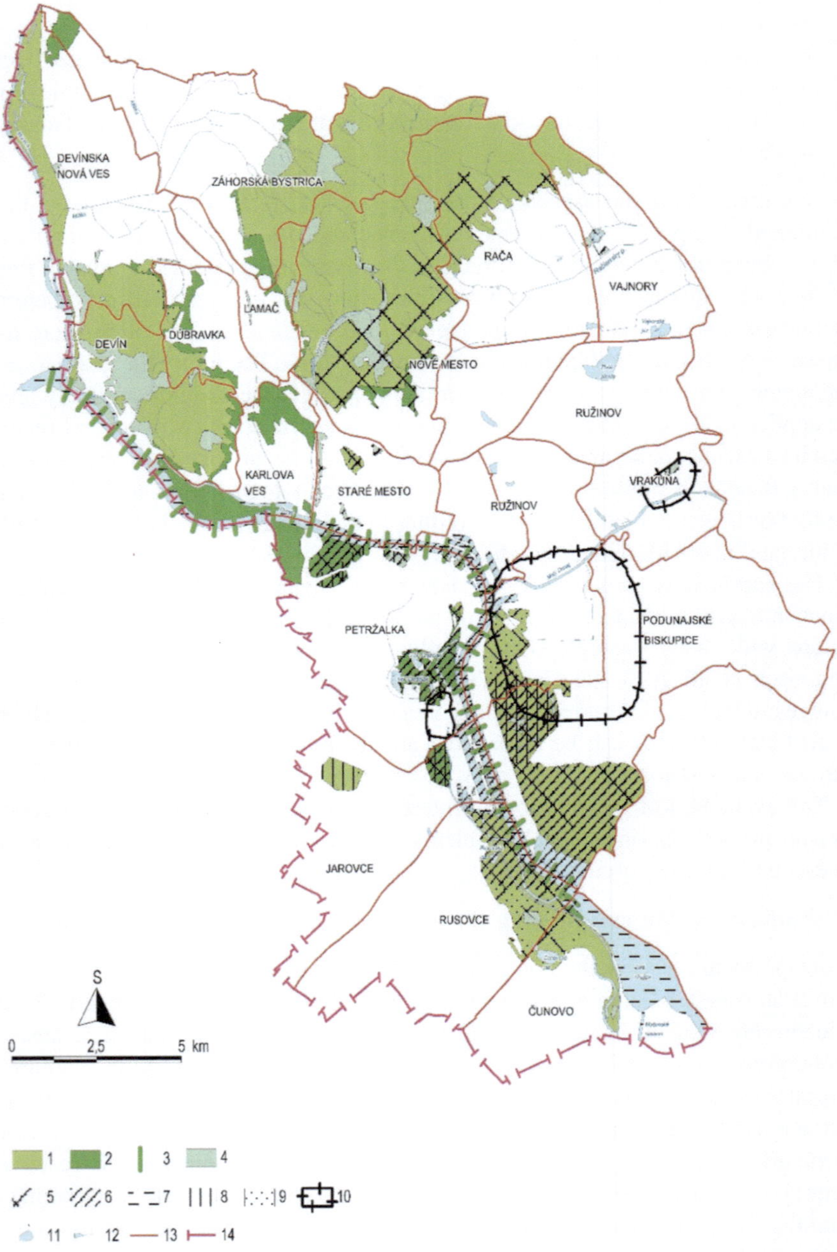

Fig. 17.6 Environmental problems to nature protection, ecological stability and biodiversity. Legend: *Endangered elements:* 1—protected areas in combination with elements of the territorial system of ecological stability and protected forests, 2—elements of the territorial system of ecological stability in combination with protected forests, 3—biocorridors, 4—other forests, *Endangering factors*: 5—damaged vegetation, 6—polluted air, 7—contaminated watercourse, 9—contaminated soils, 10—hygienic zones of anthropogenic objects, *Others elements:* 11—water area, 12—water flow, 13—cadastral border, 14—state border, and environmental problems arising from conflicts of interest are the areas the combination of green and black colours

(b) ***Problems of the endangering the natural resources***

They arise from spatial conflict of stress factors and natural resources. The result of this conflict is deterioration of quality and quantity of natural resources. The most important problems in this category belong to damaging natural resources because of the production of foreign substances like water pollution, soil contamination, forest damage and so on. Frequent is damage of soil resources by physical degradation such as landslides and erosion, compacting. These problems are depicted in third, fourth, fifth and sixth row of the matrix.

(c) ***Problems of the endangering the immediate human environment***

They arise from the impact of stress factors on areas where people permanently live (residential areas) or where they spend the holidays or heal themselves (recreational and health resorts). In this group, the most frequent problems include the load of the residential and recreational environment due to air pollution, dustiness, smell, noise, etc. These problems are depicted in the last two rows of the matrix (Fig. 17.5).

17.5 Sustainable Management of the Landscape

The major international initiatives emerged in the past decade to study land use and land cover changes. They focused on understanding driving forces of land use changes, developing diagnostic models of land cover change and producing regionally and integrated models (Bastian et al. 2012; Brandt et al. 1999; Geist and Lambin 2002; Lambin et al. 2001; Pauditšová and Slabeciusova 2014; Tengberg et al. 2012; Tress and Tress 2001). Integrated approaches study both natural and human aspect of landscape structure. These studies connected with methodical procedures reflecting effects of changes of individual types of land use in various periods on the landscape reveal the rate of sustainability of coexistence of nature and social subsystems in the model area (Antrop 2005; Axelsson et al. 2012; Burgi et al. 2017; Frelichova and Fanta 2015; Schröter et al. 2005). These integrated approaches are needed in the next stage to ensure the sustainable development helps to support and develop land utilization needs.

The need for sustainable landscape management comes from pragmatic needs, as constantly rising environmental and existence problems require it. The sustainable land use is linked to the concept of multifunctionality. The rationale addresses the interdependence of social, economic and environmental effects of land use, taking into account commodities and both negative and positive externalities. Landscape and its ecosystems provides a variety of functions or goods and services, covering production, regulation and cultural services (EEA 2018b; Kumar 2010). Multifunctionality, therefore, is a key feature for implementing sustainable land development (Wiggering and Steinhardt 2015).

The basic tool for sustainable utilization of the landscape is the integrated landscape management what we understand as the environmentally based harmonization of the tools, which regulates the spatial organization and functional utilization of the landscape to eliminate the conflict of interests of the sectors. Sustainable landscape management is based on understanding of the landscape as an integrated natural resource, an interrelated complex of natural resources in certain area. As the area—the space—is representing the integrating scope, scene in which all resources are occurring as layers (geological sources, water and soil sources, climate, biotic sources and morphometric parameters), which are in permanent material-energy connection. This helps to understand the space as integration of particular natural sources in a given area. Every point of the earth surface presents specific homogeneous entity of mutual combination of listed sources (landscape building components, which through its attributes are capable to satisfy human needs and as such in relation to human society act as natural resources) and the relationship between these resources.

The 'integrated' in this case means the systematic assessment and harmonization of the encounters of the interests of all relevant sectors from the environmental point of view. The sustainable landscape management is based on an integrated landscape research in its three basic landscape dimensions—in primary, secondary and tertiary landscape structure, analyses, syntheses and interpretation of the connections and dependencies between these dimensions with the target to define such landscape management, which would harmonize socio-economic development with the natural, human, cultural and historical potential of the landscape. The basic characteristic of three landscape dimensions is as follows (Izakovičová et al. 2018a):

- **Primary landscape structure (PLS)**: It is a set of material elements of the landscape and their relations that constitute the original foundation and condition for other two structures. These are the abiotic elements—the geological base and subsoils, soils, waters, geo-relief and air. The principles of their functioning are not changeable.
- **Secondary landscape structure (SLS)**: It is constituted by **man-influenced, reshaped and created** elements of land cover, which is the result of land use. There belong the elements of real biota, man-made objects and constructions. A very frequent case of the conflict of interests is exactly the disharmony between land use and primary landscape structure.
- **Tertiary landscape structure (TLS)**: It is a set intangible (non-material) socio-economic factors/phenomena displayed to the landscape space as the interests, manifestations and consequences of the activities of individual sectors that are relevant to landscape. These are the protection and other functional zones of nature and natural resources protection, hygienic and safety zones of industrial and infrastructure objects, zones of declared zones of specific environmental measures, administrative boundaries, etc. They are defined in acts, plans and other development documents. Since they are of non-material character, they often **overlap each other**. This fact is one of the main causes of conflict of interests.

17.6 Solving the Conflicts of Interests in the Landscape

The sustainable landscape management is based on an integrated landscape research in its three basic dimensions of landscape, analysing the connections and dependencies between the dimensions with the target to define such landscape management, which would regulate socio-economic development with natural, cultural and historical resources and potentials of landscape. The basic principles of the sustainable landscape management are (Izakovičová et al. 1997):

(a) **preservation of the overall ecological stability of landscape** is the most general and complex condition for conserving gene pool, biological diversity, stability and the natural functioning of ecosystems and through that also for conservation of the natural production capacity of landscape. The preservation of ecological stability is, therefore, primarily achieved by the landscape-ecological optimization of the spatial structure of landscape—through the suitable distribution of landscape elements in space, their proper utilization or protection.

(b) **protection and rational utilization of natural resources**, in particular of air, water, soil, biotic resources and mineral resources. The state of natural resources is determined by their quantity and quality conditions. Protection and rational utilization of natural resources are realized partly through the optimal collocation of objects and activities in the area and by application of the suitable technologies.

(c) **protection of the close human environment** means preserving the quality of air, drinking water and food chain, reducing negative influences like noise, radiation and waste, preservation of aesthetic quality and human environment, etc. The protection of the environment against the unfavourable influences means mainly the optimization of technological processes of production branches and preservation of the aesthetic quality of the environment by the optimal land cover.

(d) **ensuring social and cultural diversity** by respecting the national, religious and culture–historical peculiarities of individual communities that form region. This objective like the preceding one can be ensured through the 'ecologization' and humanization of the above structure, especially by the interaction of economic and legislative tools and by the humanization of social consciousness. The basic goal of this principle is protection and preservation of historical landscape structure—traditional forms of the land use.

The implementation of the sustainable landscape management must consist of following steps.

- To promote the landscape-optimal land use—the landscape-optimal spatial organization and functional utilization of the landscape. This mentioned step amendment is defined as a complex process of reciprocal synchronization of area requirements of human activities with landscape-ecological land conditions, which are resulting from landscape structure.
- The implementation of technological arrangements—to establish effective technologies focused on eliminating production over limit of polluting substances with the goal to minimize their influence on the respective environmental elements with allochthonous substances and other contaminant, also necessary to apply appropriate saving technologies in agriculture and forestry.
- Application of regulative of landscape-ecological optimal land use to sector plans—it is unavoidable to set regulative of usage of the particular resources by production and non-production subjects in order not to prefer the development of one area over another and avoid issues coming from conflicts of interest.
- Implement principles of sustainable development into the awareness of the population, mostly by stakeholders and policy makers—the basis is to create effective system of education on sustainable landscape management and on sustainable development (Fig. 17.7). Only sufficiently educated population is able to promote the principles and criteria of sustainable development in practice.

Fig. 17.7 Nature trails guide the movement of tourists in protected areas (Photograph: Jana Špulerová)

- Assertion of effective tools—ensuring legislative protection, economical tools, etc.—mainly legislative rules and regulations ensuring rational utilization of natural resources, as well as protection of human being, its health and environment. From this aspect, it is needed also the realization of effective economical tools as taxes, duties, and fees for environmental pollution and people's health injury, economical tools eliminating the marginality of regions, social disparities etc.

The consequence of the above means that sustainable landscape management is a method that helps to answer the question **where and how to provide human activities in the landscape** would be in least contradiction to natural conditions (Miklós et al. 1986). This process of the landscape-ecological optimal spatial organization, utilization and protection of landscape results basically to two kinds of proposals: (1) to the proposal of most suitable localization of demanded human activities within the given territory (answer to the question where?—e.g. the most suitable locality for arable land) and successively (2) to the proposal of necessary measurements ensuring proper environmental functioning of those activities on the given locality (answer to the question how?—how to apply the most suitable way of landscape management and ways to reduce natural risks and hazards).

17.7 Landscape-Ecological Optimization of the Land Use

The basic and dominant activity of integrated landscape management is spatial and functional land use optimization. The central conception of this process is based on the understanding of the landscape as geosystem, as complex of natural resources.

The core of the evaluation is the decision-making process based on the confrontation and subsequent proposal of the harmonization way of the:

- complex of the properties of the landscape as a complex natural resources on the one side, with the
- demands and influences of human activities on those properties on the other side.

The realization of the human activities in the landscape has different forms that is manifested by:

- localization of artificial elements into the natural landscape—construction of buildings, lines, etc.
- Large area exploitation of the natural landscape—zones of agriculture, forestry, etc.
- Application of different functional zones in order to protect the natural landscape (protected natural landscapes, protected zones of water resources, etc.), as well as elements of the socio-economic structure (protected zones of buildings and lines, etc.)

- devices, objects, technologies and others, which produce stress factors causing the landscape deterioration–production of heterogeneous polluting substances, soil degradation processes, etc.

Proposal of the landscape-ecological optimum land use is based on the confrontation between the landscape values and the requirements of human activities. The process of evaluation consists of the following steps:

- determination of the functional suitability values (especially environmental limited and restricted values) of individual indexes for each human activity,
- determination of the weighted coefficients of individual indexes for each human activity,
- determination of the total suitability value of each landscape-ecological unit (geosystem) for each human activity.

In order to calculate the total suitability of a geosystem for a human activity, the following formula should be used (Miklós et al. 1986, edited by authors)

$$W(R,G) = \sum_{j=1}^{n} \frac{v_j^{(R)}.s_j^{(R,G)}}{n.100}$$

where

$W(R, G)$—total suitability of geosystem G for activity R
$j = 1$ to n—properties of geosystem from 1 to n
$v_j^{(R)}$ —importance (weighted coefficient) of property j for activity R
$s_j^{(R,\ G)}$—partial suitability of property j in geosystem G for activity R in %
n—number of the properties

Specification of the environmental regulative (limits and restrictions) creates base of the decision-making process. The determined indicators as limits may exclude the realization of a human activity in a given area. This concerns the determination of indicators that do not allow the use of a certain area for a given activity from the environmental principles viewpoint. For example, one of the limits for the development of recreation is the size of protected areas.

We understand the determined values and indicators that significantly limit the intensity of a given activity in an area, although they do not absolutely exclude it, as restrictions. For example, the agricultural production is not excluded in an area of water protection, but its intensity is considerably restricted by the recommended structure of crops, by the restricted use of chemicals and mechanization, etc.

The determination of environmental limits allows an objective and scientific decision-making on the localization of human activities in the landscape. It is necessary to add that this is a very complicated and time-consuming process demanding a multidisciplinary and synergistic approach.

It starts from the existing norms in the society and conditions in nature. If there is no stated norm for a given phenomenon or human activity, the approach starts from a collective evaluation of experts.

Determining the limiting values requires an extensive collection of data on the operation of the landscape system as a whole and on its individual components. It is so because the successful and conflict less implementation of every human activity is to a certain degree dependent upon a wide variety of factors in an environment in which the activities are carried out. Therefore, the properties of landscape as a whole as well as those of its individual components represent certain limits for the localization and realization of individual activities.

The example of the creation of environmental regulative is in Table 17.2.

In general, a set of regulations (limits and restrictions) includes the following categories (Miklós and Izakovičová 1997):

A. ***Abiotic regulations***: Their values are conditioned by the characteristics of the abiotic complex. geomechanical, hydrological, aerodynamic and soil limits belong here. These limits are of a permanent character and cannot be easily changed by technology (e.g. the relative stability of the geological substratum, climate, etc.).
B. ***Biotic regulations***: Their values are derived from the characteristics of the biotic complex. They arise from the needs of living organisms. Up to now, the biotic limits have been respected very little in practice. This situation threatens both gene pool and the biodiversity, ecological stability of the landscape.
C. ***Anthropogenic regulations***: They result from the demands and requirements of human activities that limit the development of other activities through their negative effects or by simply using an area. They comprise, for example, technical, hygienic, protective and other limits. Compared to the preceding types of limits, these indicators can be changed relatively easier. Although the anthropogenic regulations have the character of temporary limits, they represent the group of very serious hygienic and environmental security limits.
D. ***Complex landscape regulations***: They are based on the principles of the functioning landscape as a whole. They include eco-stabilization, localization, carrying capacity, behavioural, aesthetic, cultural–historical and other limits. This is a set of very dynamic limits since they result from the principles of the operation of landscape as a whole and have to respect unavoidably its historical development.

The process of creating regulative is most frequently carried out by means of the decision-making tables. The process of the evaluation is defined by mostly three degrees of the availability for realization of the human activity on the area:

L: Limit—realization of human activities in given area is impossible
0: Restriction—realization of human activity in given area is less suitable, realization is possible under specified conditions
1: Realization of human activity in given area is suitable.

Table 17.2 Example of creation of environmental regulation based on environmental stress factors (Izakovičová et al. 2018b)

	Stress factors							
Land use activities	Air pollution	Noise load area	Soil contamination	Polluted water flows	Damage of vegetation	Radioactivity	Nature reserve	Protected zone of water resources
Forests (F)	1	1	1	1	1	1	1	1
Grassland (G)	1	1	1	1	1	1	1	1
Pastures (P)	L	1	L	–	L	L	L	L
Vineyards (V)	L	1	L	–	L	L	L	0
Forage-crops (C)	L	1	L	–	L	L	L	0
Arable land (A)	0	1	0	–	0	0	L	0
Orchards (O)	0	1	L	–	L	0	L	0
Gardens (GS)	L	1	L	–	L	L	L	0
Recreation (R)	L	L	L	L	L	L	L	0
Sport areas (S)	L	L	L	L	L	L	L	0
Medical areas (M)	L	L	L	–	L	L	L	1
Housing areas (H)	L	L	L	–	L	L	L	L
Farm animals (FA)	L	L	L	–	0	L	L	L
Industrial areas (I)	1	1	1	–	1	1	L	L
Transport areas (T)	1	1	1	–	1	1	L	L
Unlimited activities/ acceptable activities	I, T, F, G/A, O	F, G, P, V, C, A, O, GS, I, T	I, T, F, G/A	F, G	I, TF, G/A, FA	F, G, I, T/O, A	F,G/-	F, G, M/V, C, A, O, GS, R, S

Legend L—environmental limit (limited activities), 0—environmental restriction (restricted activities) and 1—acceptable activities

The limits application is of great importance for sustainable landscape management, where it forms the basis for landscape-ecologically optimized decision-making processes of landscape utilization. They are applied both to the evaluation of the present functional use of the territory and to elaborating the proposals for its further development.

The aim of the landscape-ecological evaluation of the present land use of the territory is to define those landscape-ecologically problematic areas in which the present land use does not correspond with landscape-ecological criteria or to identify the areas where some of the environmental limits restrict the present land use. The aim of the newly proposed, landscape-ecologically optimum land use is to eliminate these problem areas, to foresee possible new problems and thus to create a structure which is in harmony with the natural and socio-economic conditions of the territory. The landscape-ecologically optimum land use is a basic precondition for the sustainable landscape management.

Limits and restriction does not act in the landscape isolated but synergically. The locality of given human activity can be limited or restricted by two or more factors. The determination of limiting and restricting factors for given activity proceeds from evaluation of functional relation between attributes of landscape elements of abiotic, biotic and socio-economic complex and requested human activity. The following principles of limitation are as follows (Izakovičová et al. 2018b):

- abiotic conditions are the determining factors in a given area's diversity. They establish appropriate area utilization. Due to permanent and unique character of these elements, the attributes of abiotic elements become determining factors of human development. Land use management must reduce the risk factors in localities sensitive and predisposed to anthropogenic degradation processes (including erosion, subsidence, landslide, earthquake, etc).
- It is essential to support such development, which does not threaten natural values of landscape units in protected and ecologically valuable areas (localities of the territorial system of ecological stability, localities NATURA 2000). This enables mainly science research, nature protecting or possibly medial, recreational activities, etc.
- Similarly, the development of human activities with negative impact on individual natural resources must be excluded in areas, where natural resources are legally protected and explicit priority must be given to developing activities, which protect individual natural resources.
- All detrimental activities must be excluded from sensitive areas with strong pressure burden. These include areas with air pollution, soil and water contamination and noise pollution.
- Areas without pressure load should be maintained free from activities, which can potentially harm current living quality. This area is suitable mostly for high-quality living development with adequate recreation, ecological and agricultural services.

The outcome of this decision process is:

- Selection of such activities, which cannot be located on the given area from environmental point of view.
- Selection and restriction of activities, which cannot be located on a given area from landscape-ecological point of view.
- Hierarchy of activities which maintains area´s optimal landscape-ecological function.
- Complex of measurements required to protect area´s nature, natural resources and environment.

17.8 Conclusion

Sustainable landscape management is an actual, new-age, but at the same time very complex actual problem which is to be solved by landscape research as integration of the researches of natural, cultural–historical and socio-economical resources in the landscape. The integrated researches allow solving the environmental problems, as well as the existential problems of humankind as negative influences of the floods, climatic changes, health, etc., arising due to the prevailing resorts in land use and protection.

The needs of the sustainable landscape management issue from:

- requirements to improve and ensure the spatial stabilization of the territory. The stated criterion is here formed by the demand to achieve the biological balance in the landscape,
- needs for the nature protection and rational utilization of the natural resources, in particular the protection of the soils, waters, forests and gene pool,
- needs for the protection of cultural and historical resources,
- needs for the regeneration of human resources and for the protection of human health,
- demands on the humanization and aesthetic of the landscape.

These requirements represent the fundamental principles of the sustainable development of the society. Its application in practice contributes not only to elimination of environmental problems, but also to the intensification of socio-economical development of the given areas in harmony with capacity abilities of natural resources. The successful application of sustainable land use management requires multiple social measures at all levels of legislation, economic outcomes, education and teaching. The regulations for optimal land use must be applied to sector plans too.

Acknowledgements This work was supported by the Scientific Grant Agency of Ministry of Education of the Slovak Republic [No. 2/0078/18 "Research of biocultural values of landscape"].

References

Antrop M (2000) Background concepts for integrated landscape analysis. Agr Ecosyst Environ 77:17–28

Antrop M (2005) Why landscapes of the past are important for the future. Landsc Urban Plan, Rural Landsc: Process Futur Strat 70:21–34. https://doi.org/10.1016/j.landurbplan.2003.10.002

Axelsson R, Angelstam P, Elbakidze M, Stryamets N, Johansson KE (2012) Sustainable development and sustainability: landscape approach as a practical interpretation of principles and implementation concepts. J Landsc Ecol 4:5–30. https://doi.org/10.2478/v10285-012-0040-1

Baessler C, Klotz S (2006) Effects of changes in agricultural land-use on landscape structure and arable weed vegetation over the last 50 years. Agr Ecosyst Environ 115:43–50. https://doi.org/10.1016/j.agee.2005.12.007

Barančok P, Barančoková M (2013) Development of sports and recreational activities in the Chopok area (Nízke Tatry Mts.) and protection of important landscape elements. In: Fialová J, Kubíčková H (eds) Public recreation and landscape protection—with man hand in hand: conference proceedings. Mendel Univ, Dept Landscape Management, Brno, pp 27–33

Bastian O, Haase D, Grunewald K (2012) Ecosystem properties, potentials and services—the EPPS conceptual framework and an urban application example. Ecological indicators, challenges of sustaining natural capital and ecosystem services. Quantif, Model Valuat/Account 21:7–16. https://doi.org/10.1016/j.ecolind.2011.03.014

Belcakova I (2016) Approaches to evaluation of landscape scenery—a conceptual contribution. In: Corniello L (ed) World heritage and degradation: smart design, planning and technologies. Scuola Pitagora Editrice, Napoli, pp 480–485

Bell S, Nikodemus O, Penēze Z, Krūze I (2009) Management of cultural landscapes: what does this mean in the former Soviet Union? A case study from Latvia. Landsc Res 34:425–455

Bender O, Boehmer HJ, Jens D, Schumacher KP (2005a) Analysis of land-use change in a sector of Upper Franconia (Bavaria, Germany) since 1850 using land register records. Landsc Ecol 20:149–163. https://doi.org/10.1007/s10980-003-1506-7

Bender O, Boehmer HJ, Jens D, Schumacher KP (2005b) Using GIS to analyse long-term cultural landscape change in Southern Germany. Landsc Urban Plan 70:111–125. https://doi.org/10.1016/j.landurbplan.2003.10.008

Bezák P, Mederly P, Izakovičová Z, Špulerová J, Schleyer C (2017) Divergence and conflicts in landscape planning across spatial scales in Slovakia: an opportunity for an ecosystem services-based approach? Int J Biodivers Sci, Ecosyst Serv Manag 13:119–135. https://doi.org/10.1080/21513732.2017.1305992

Bicik I, Jelecek L, Stepanek V (2001) Land-use changes and their social driving forces in Czechia in the 19th and 20th centuries. Land Use Pol 18:65–73. https://doi.org/10.1016/S0264-8377(00)00047-8

Biopress—Linking pan-European landcover change to pressures on biodiversity [WWW Document] 2003. URL http://www.creaf.uab.es/biopress/index2.htm. Accessed on 12 June 2018

Bradley P, Yee S (2015) Using the DPSIR framework to develop a conceptual model: technical support document

Brandt J, Primdahl J, Reenberg A (1999) Rural land-use and landscape dynamics-analysis of 'driving forces' in space and time. Land-use changes and their environmental impact in rural areas in Europe

Braunisch V, Patthey P, Arlettaz RL (2011) Spatially explicit modelling of conflict zones between wildlife and snow sports: prioritizing areas for winter refuges. Ecol Appl 21:955–967. https://doi.org/10.1890/09-2167.1

Bunce RGH, Barr CJ, Clarke RT, Howard DC, Lane AMJ (1996) Land classification for strategic ecological survey. J Environ Manage 47 37–60. https://doi.org/10.1006/jema.1996.0034

Bunce RGH, Jongman RHG, Hojas L, Weel S (2007) 25 years landscape ecology. Scientific principles in practice. In: book of Abstracts. Proceedings of the 7th IALE world congress Part 1 and 2. international association for landscape, Wageningen, p 524

Bürgi M, Hersperger AM, Schneeberger N (2004) Driving forces of landscape change—current and new directions. Landsc Ecol 19:857–868. https://doi.org/10.1007/s10980-004-0245-8

Bürgi M, Bieling C, von Hackwitz K, Kizos T, Lieskovsky J, Martin MG, McCarthy S, Muller M, Palang H, Plieninger T, Printsmann A (2017) Processes and driving forces in changing cultural landscapes across Europe. Landsc Ecol 32:2097–2112. https://doi.org/10.1007/s10980-017-0513-z

Conway TM, Lathrop RG (2005) Modelling the ecological consequences of land-use policies in an urbanizing region. Environ Manag 35:278–291. https://doi.org/10.1007/s00267-004-4067-x

CORINE Land Cover (CLC). Copernicus global land service providing bio-geophysical products of global land surface [WWW Document], 2018. URL https://land.copernicus.eu/global/. Accessed on 12 Apr 2018

Csorba P (1996) Landscape-ecological change of the land use pattern on the east foothill area of the Tokaj Mountains (Hungary). Ekol Bratisl 15:115–127

de Groot R (2006) Function-analysis and valuation as a tool to assess land use conflicts in planning for sustainable, multi-functional landscapes. Landsc Urban Plan, Landsc Sustain 75:175–186. https://doi.org/10.1016/j.landurbplan.2005.02.016

De Sherbinin A (2002) A CIESIN thematic guide to land-use and land-cover change (LUCC). Center for International Earth Science Information Network (CIESIN) Columbia University Palisades, NY, USA, p 68

Demková K, Lipský Z (2015) Changes in non-forest woody vegetation in the south-western part of the white carpathians (1949–2011). Geografie 120:64–83

Dyer J (1994) Land-use pattern, forest migration, and global warming. Landsc Urban Plan 29: 77–83. https://doi.org/10.1016/0169-2046(94)90019-1

EEA (2018a) Copernicus Land Monitoring Service—Urban Atlas [WWW Document]. European Environment Agency. URL https://www.eea.europa.eu/data-and-maps/data/copernicus-land-monitoring-service-urban-atlas. Accessed on 12 June 2018

EEA (2018b) CICES Version 5.1 [WWW Document]. https://cices.eu/. URL https://cices.eu/. Accessed on 10 June 2018

ELI—The European land-use institute [WWW Document], 2018. URL https://www.eli-web.com/. Accessed on 12 June 2018

Falťan V, Krajcirovičová L, Petrovič F, Khun M (2017) Detailed geoecological research of terroir with the focus on georelief and soil—a case study of Kratke Kesy Vineyards. Ekologia (Bratislava) 36:214–225. https://doi.org/10.1515/eko-2017-0018

Forman RTT, Godron M (1986) Landscape ecology. Wiley, p 640

Frelichova J, Fanta J (2015) Ecosystem service availability in view of long-term land-use changes: a regional case study in the Czech Republic. Ecosyst Health Sustain 1:1–15. https://doi.org/10.1890/EHS15-0024.1

Geist HJ, Lambin EF (2002) Proximate causes and underlying driving forces of tropical deforestation. Bioscience 52:143–150. https://doi.org/10.1641/0006-3568(2002)052%5b0143:PCAUDF%5d2.0.CO,2

Haddock J, Tzanopoulos J, Mitchley J, Fraser R (2007) A method for evaluating alternative landscape management scenarios in relation to the biodiversity conservation of habitats. Ecol Econ 61:277–283. https://doi.org/10.1016/j.ecolecon.2006.02.019

Haines-Young R, Potschin M, Kienast F (2012) Indicators of ecosystem service potential at European scales: mapping marginal changes and trade-offs. Ecol Indic, ChallS Sustain Nat Cap Ecosyst Serv Quantif, Model Valuat/Account 21:39–53. https://doi.org/10.1016/j.ecolind.2011.09.004

Ivanova M, Michaeli E, Boltiziar M, Juhascikova J (2011) Analysis of landscape heterogeneity changes on the example of Hlinne, Vysny Zipov, and Zlatnik village (eastern Slovakia) in the period 1826–2006. Ekologia (Bratislava) 30:269–280. https://doi.org/10.4149/ekol_2011_01_269

Izakovičová Z, Miklos L, Drdoš J (1997) Krajinnoekologické podmienky trvalo udržateľného rozvoja. VEDA publishing of the Slovak Academy of Sciences, Bratislava, p 183 (In Slovak)

Izakovičová Z, Mederly P, Petrovič F (2017) Long-term land use changes driven by urbanisation and their environmental effects (example of Trnava City, Slovakia). Sustainability 9:1553. https://doi.org/10.3390/su9091553

Izakovičová Z, Miklós L, Miklósová V (2018a) Integrative assessment of land use conflicts. Sustainability 10:3270. https://doi.org/10.3390/su10093270

Izakovičová Z, Špulerová J, Petrovič F (2018b) Integrated approach to sustainable land use management. Environments 5:37. https://doi.org/10.3390/environments5030037

Jensen K, Murray F (2005) Intellectual property. Enhanced: intellectual property landscape of the human genome. Science 310:239–240. https://doi.org/10.1126/science.1120014

Jongman RHG (ed) (1996) Ecological and landscape consequences of land use change in Europe. In: Proceedings of the ECNC seminar on land use change and its ecological consequences. European Centre of Nature Conservation, Tilburg

Kienast F, Frick J, van Strien MJ, Hunziker M (2015) The swiss landscape monitoring program—a comprehensive indicator set to measure landscape change. Ecol Model, Use Ecol Indic Model 295:136–150. https://doi.org/10.1016/j.ecolmodel.2014.08.008

Kolejka J (1987) Landscape-historical synthesis materials, methods and results. Ekologia CSFR 6:51–62

Kolejka J (2018) Landscape mapping using GIS and google earth data. Geogr Nat Resour 39: 254–260. https://doi.org/10.1134/S1875372818030095

KolejkaJ (2006) Application of digital landscape model in crisis management. Geogr Tech 1: 101–109

Mucher CA, Bunce RGH, Jongman RHG, Klijn JA, Koomen, E, Metzger, MJ., Wascher, DM (2003) Identification and characterisation of environments and landscapes in Europe (Scientific report). Alterra, Wageningen, p 119

Kristensen P (2004) The DPSIR framework. In: Paper presented at the 27–29 september 2004 workshop on a comprehensive/detailed assessment of the vulnerability of water resources to environmental change in Africa using river basin approach. Presented at the UNEP Headquarters, Nairobi, Kenya, p 10

Kumar P (ed) (2010) The economics of ecosystems and biodiversity: ecological and economic foundations. UNEP/Earthprint, p 454

Lambin EF, Turner BL, Geist HJ, Agbola SB, Angelsen A, Bruce JW, Coomes OT, Dirzo R, Fischer G, Folke C, George PS, Homewood K, Imbernon J, Leemans R, Li X, Moran EF, Mortimore M, Ramakrishnan PS, Richards JF, Skånes H, Steffen W, Stone GD, Svedin U, Veldkamp TA, Vogel C, Xu J (2001) The causes of land-use and land-cover change: moving beyond the myths. Glob Environ Change 11:261–269. https://doi.org/10.1016/S0959-3780(01)00007-3

Lieskovsky J, Kaim D, Balazs P, Boltiziar M, Chmiel M, Grabska E, Kiraly G, Konkoly-Gyuro E, Kozak J, Antalova K, Kuchma T, Mackovcin P, MojsesM Munteanu C, Ostafin K, Ostapowicz K, Shandra O, Stych P, Radeloff VC (2018) Historical land use dataset of the Carpathian region (1819–1980). J Maps 14:644–651. https://doi.org/10.1080/17445647.2018.1502099

Lipský Z (2000) Experience in assessment of landscape character. Ekol (Bratisl)/Ecol (Bratisl) 19:188–198

Lipsky Z, Kopecky M, Kvapil D (1999) Present land use chances in the Czech cultural landscape. Ekol Bratisl 18:31–38

Mander U, Jongman RHG (1999) Ecological and socio-economic consequences of land use changes. Backhuys Publishers, Leiden, pp 269–280

Mander U, Murka M (2003) Landscape coherence: a new criterion for evaluating impacts of land use changes. In: Mander U, Antrop M (eds) Multifunctional landscapes, Vol III: continuity and change. Wit Press, Southampton, pp 15–32

Mander U, Palang H, Ihse M (2004) Development of European landscapes. Landsc Urban Plan 67:1–8. https://doi.org/10.1016/S0169-2046(03)00025-2

Miklós L (1996) Landscape-ecological theory and methodology: a goal oriented application of the traditional scientific theory and methodology to a branch of a new quality. Ekol Bratisl 15: 377–385

Miklós L, Izakovičová Z(1997) Krajina ako geosystém. VEDA publishing of the Slovak Academy of Sciences, p. 153, (In Slovak)

Miklós L, Kočická E, Izakovičova Z, Kočický D, Špinerová A, Diviaková A, Miklósová V (2019) Landscape as a geosystem. Springer International Publishing, p 157

Miklós L, Miklisova D, Reháková Z (1986) Systematization and automatization of decision-making process in LANDEP method. Ekologia ČSFR 5:203–231

Muchova Z, Leitmanova M, Petrovič F (2016) Possibilities of optimal land use as a consequence of lessons learned from land consolidation projects (Slovakia). Ecol Eng 90:294–306. https://doi.org/10.1016/j.ecoleng.2016.01.018

Nikodemus O, Bell S, Grine I, Liepins I (2005) The impact of economic, social and political factors on the landscape structure of the Vidzeme Uplands in Latvia. Landsc Urban Plan 70:57–67. https://doi.org/10.1016/j.landurbplan.2003.10.005

OECD (2001) OECD Environmental indicators. Towards Sustainable Development. Paris, p 155

Olah B, Žigrai F (2004) The meaning of the time-spatial transformation of the landscape for its sustainable use (a case study of the transition zone of the Poľana Biosphere Reserve). Ekol Bratisl 23:231–243

Oťahel' J, Feranec J (2001) Landscape structure of Slovakia in the context of sustainability. Ekol Bratisl 20:88–99

Oťahel' J, Feranec J, Cebecauer T (2004) Landscape changes identified by application of the CORINE land cover databases: case study—Part of the Zahorie lowland. Ekol Bratisl 23: 252–263

Palang H, Alumäe H, Mander Ü (2000) Holistic aspects in landscape development: a scenario approach. Landsc Urban Plan 50:85–94

Palang H, Alumäe H, Printsmann A, Rehema M, Sepp K, Sooväli-Sepping H (2011) Social landscape: ten years of planning 'valuable landscapes' in Estonia. Land Use Policy 28:19–25. https://doi.org/10.1016/j.landusepol.2010.04.004

Pauditšová E, Slabeciusova, B (2014) The modelling as platform for landscape planing. In: Geoconference on informatics, geoinformatics and remote sensing, vol iii SE International multidisciplinary scientific geoconference-SGEM CT 14th International multidisciplinary scientific geoconference (SGEM), 753–760

Pauleit S (2018) The spatial effect of landscape planning. evaluation. Indicators Trends Raumforsch Raumordn 76:87–89. https://doi.org/10.1007/s13147-017-0509-9

Pauleit S, Golding Y (2005) The spatial impact of urban compaction a fine-scale investigation based on Merseyside. Town Plan Rev 76:143–166. https://doi.org/10.3828/tpr.76.2.3

Pauleit S, Ennos R, Golding Y (2005) Modeling the environmental impacts of urban land use and land cover change—a study in Merseyside. UK Landsc Urban Plan 71:295–310. https://doi.org/10.1016/j.landurbplan.2004.03.009

Pazur R, Otahel' J, Maretta M (2015) The distribution of selected CORINE land cover classes in different natural landscapes in Slovakia: methodological framework and applications. Morav Geogr Rep 23:45–56. https://doi.org/10.1515/mgr-2015-0005

Petrovič F (2006) Changes of the landscape with dispersed settlement. Ekologia (Bratislava) 25:201–211

Piscová V (2011) Zmeny vegetácie Tatier na vybraných lokalitách ovplyvnených človekom. VEDA publishing of the Slovak Academy of Sciences, Bratislava, p 227 (In Slovak)

Poudevigne I, Alard D (1997) Landscape and agricultural patterns in rural areas: a case study in the Brionne basin, Normandy. France J Environ Manage 50:335–349. https://doi.org/10.1006/jema.1997.0134

Quattrochi DA, Luvall JC (1999) Thermal infrared remote sensing for analysis of landscape ecological processes: methods and applications. Landsc Ecol 14:577–598. https://doi.org/10.1023/A:1008168910634

Quattrochi DA, Luvall JC (2014) Thermal infrared remote sensing for analysis of landscape ecological processes: current insights and trends. Blackwell Science Publ, Oxford, p 43

RegioResources 21—2018: Global megatrends and landscape, Abstracts.(2018) ILE SAS Bratislava, p 47

Rehackova T, Pauditsova E, Hresko J (2008) The ecological stability and landscape management in Sverzov. Geogr Cassoviensis 2:146–151

Reschke J, Huettich C (2014) Continuous field mapping of Mediterranean wetlands using sub-pixel spectral signatures and multi-temporal Landsat data. Int J Appl Earth Obs Geoinf 28:220–229. https://doi.org/10.1016/j.jag.2013.12.014

Romportl D, Chuman T, Lipsky Z (2013) Landscape typology of Czechia. Geografie 118:16–39

Roy PS, Joshi PK (2002) Forest cover assessment in north-east India—the potential of temporal wide swath satellite sensor data (IRS-1C WiFS). Int J Remote Sens 23:4881–4896. https://doi.org/10.1080/01431160110114475

Roy A, Royer A, Derksen C, Brucker L, Langlois A, Mialon A, Kerr YH (2015) Evaluation of spaceborne L-band radiometer measurements for terrestrial freeze/thaw retrievals in Canada. IEEE J Sel Top Appl Earth Observ Remote Sens 8:4442–4459. https://doi.org/10.1109/JSTARS.2015.2476358

Schröter D, Cramer W, Leemans R, Prentice IC, Araújo M, Arnell NW, Bondeau A, Bugmann H, Carter T et al (2005) Ecosystem service supply and vulnerability to global change in Europe | science. Science 310:1333–1337

Sklenička P (2006) Applying evaluation criteria for the land consolidation effect to three contrasting study areas in the Czech Republic. Land Use Pol 23:502–551. https://doi.org/10.1016/j.landusepol.2005.03.001

Soliva R (2007) Agricultural decline, landscape change, and outmigration: debating the sustainability of three scenarios for a Swiss mountain region. Mt Res Dev 27:124–129. https://doi.org/10.1659/mrd.0907

Špulerová J, Hrnčiarová T, Piscová V, Vlachovičová M, Kalivoda H, Kanka R, Dobrovodská M, Kenderessy P, Miklošová V, Drábová M, Belčáková I (2016) Sustainable tourism development in a selected area of the low Tatras national park—landscape planning versus urban planning. Carpath J Earth Environ Sci 11:485–496

Štefunková D, Hanušin J (2016) Analysis of the spatial and temporal distribution of selected landscape diversity indexes in detailed scale (example of viticultural landscape Svätý Jur). In Landscape and landscape ecology : Proceedings of the 17th International Symposium on Landscape Ecology (proceedings is indexed in WOS). Institute of Landscape Ecology SAS, Bratislava, pp 185–191. ISBN 978-80-89325-28-3

Tengberg A, Fredholm S, Eliasson I, Knez I, Saltzman K, Wetterberg O (2012) Cultural ecosystem services provided by landscapes: assessment of heritage values and identity. Ecosyst Serv 2:14–26. https://doi.org/10.1016/j.ecoser.2012.07.006

Tress B, Tress G (2001) Capitalising on multiplicity: a transdisciplinary systems approach to landscape research. Landsc Urban Plan 57:143–157

Turner MG, Donato DC, Romme WH (2013) Consequences of spatial heterogeneity for ecosystem services in changing forest landscapes: priorities for future research. Landsc Ecol 28:1081–1097. https://doi.org/10.1007/s10980-012-9741-4

Tzanopoulos J, Mitchley J, Pantis J (2005) Modelling the effects of human activity on the vegetation of a northeast Mediterranean island. Appl Veg Sci 8:27–38. https://doi.org/10.1658/1402-2001(2005)008%5b0027:MTEOHA%5d2.0.CO,2

UNCSD (1996) Indicators of sustainable development framework and methodologies. United Nations Commission on Sustainable Development, New York

White M, Mladenoff D (1994) Old-growth forest landscape transitions from pre-european settlement to present. Landsc Ecol 9:191–205. https://doi.org/10.1007/BF00134747

Wiggering H, Steinhardt U (2015) A conceptual model for site-specific agricultural land-use. Ecol Model 295:42–46. https://doi.org/10.1016/j.ecolmodel.2014.08.011

Zigrai F (1996) The relationship between basic and applied landscape-ecological research in Slovakia. Ekol Bratisl 15:387–400

Part III
Tools for Landscape Planning

Chapter 18
Impacts on Landscapes, Biodiversity and the Environment: Methods of Assessment and Compensation

Valery V. Kravchenko

Abstract A review is provided of the methods used to assess the impact of a planned activity on the environment and biodiversity, based on the Russian national environmental impact assessment (EIA) procedure. The review describes methods for identifying influencing factors and their level, and the condition of all components of the environment; for forecasting violations; and for choosing the location of an object. The essence of these methods is based on an assessment of integral and component-specific indicators of environmental risk. This approach allows any available information on the environment to be used, in particular quantitative, semi-quantitative, qualitative, verbal and expert estimates. The forecast is used to develop actions to prevent and decrease environmental violations and reveals other violations for which actions are developed for their natural compensation. This technology helps make the assessment of violations more objective and is used to develop actions which are capable of compensating for violations in an ecologically safe manner.

Keywords Impacts · Landscapes · Biodiversity · Environment · Violations · Methods of assessment and compensation · Russia

18.1 Introduction

The main deficit of the Russian system of environmental impact compensation is the absence of precisely stated requirements for mandatory natural compensation adequate to these impacts. Under existing legislation, only the possibility of such compensation is provided. As a result, the methodological base for the assessment of impacts and the development of measures for their compensation is practically absent in Russia.

V. V. Kravchenko (✉)
V.B. Sochava Institute of Geography of the SB RAS,
1, Ulanbatorskaja St., 664033 Irkutsk, Russia
e-mail: vk@irigs.irk.ru

L. Müller and F. Eulenstein (eds.), *Current Trends in Landscape Research*,
Innovations in Landscape Research, https://doi.org/10.1007/978-3-030-30069-2_18

The system of environmental impact compensation as it exists in Russia is focused mainly on offset payments. These payments, in so-called consolidated form, are paid directly into the federal and regional budgets. Thus, the fees, paid for environmental impacts in accordance with approved fixed rates and methods of environmental impact assessment, are rarely earmarked for environmental measures and almost never for compensation for the concrete impact for which they have been paid.

The existing paradigm of impact compensation, even in monetary terms, is based on an assessment of the impact in relation to one protected asset, although it may affect a few environmental assets, or be of a cumulative type. The problem of integrated assessment of impacts and their compensation is not considered in either legal or methodological documents.

It is thus necessary to create a mechanism to secure environmental impact compensation under these conditions—a natural compensation system supported by legal and methodological principles, excluding monetary compensation. In Russia, there are no principles or methods for real environmental impact compensation, so that the mechanisms for the avoidance of negative impact on the environment work ineffectively.

In this context, a new project was realized with the goal of incorporating German and European experience on avoidance and compensation of environmental impacts into the Russian national system of environmental impact assessment.

For this purpose, an attempt was made to single out major methodological elements of environmental accompaniment of projects from the Russian, German and European experience. A set of these elements forms a body for the assessment of environmental impacts and their compensation, regardless of the country involved or the normative documents used. A clear understanding of the meaning of these elements should allow all participants interested in ecological accompaniment of investment projects to find effective methods of problem solution and compromises for the resolution of conflicts arising during project planning and realization and thus to balance the public interest in further economic development with that of the improvement of the condition of the environment.

The present chapter does not claim to provide unequivocal answers to all questions arising in the context of the ecological accompaniment of investment activity. It is designed to help participants orient themselves more easily in a concrete situation, find required information material and select or develop appropriate methods for problem solution. We hope that this chapter will help readers improve the quality of work on projects, cut the costs of their realization and, ultimately, promote the improvement of the environmental and landscape quality.

18.2 Development of Measures for Impact Compensation as a Final Stage of Environmental Impact Assessment

18.2.1 General Provisions

The major stages of impact assessment (EIA) of a project and their content and methodological approaches have been described in detail in (Kravchenko et al. 2012). This chapter considers only certain methodological aspects of these stages directly related to the development of compensation measures.

The structure of the environmental impact assessment includes the following stages (Fig. 18.1).

Preparatory stage:

- Description of a proposed project, ascertainment of impact factors and their evaluation
- Specification of the scope of the investigation
- Delimitation of the area segments for the investigation: the impact area, the effect area and the compensation area.

Analysis of the territory:

- Specification, description and assessment of affected environmental assets
- Differentiation of areas with different level of conflicts and identification of the main conflicts.

Development of the planned project options:

- Development of options, taking into account conflicts ascertained.

Prediction of impact and comparison of planned project options:

- Forecast of all impact factors on the assets and functions of the environment in the context of their assessment
- Asset-by-asset comparison of options, based on the forecast
- Integrated evaluation of options, choice of preferred option.
- Ascertainment of significant negative impacts.

Development of nature protection measures:

- Development of measures for avoidance, minimization and compensation of significant negative environmental impacts.

The impact assessment documentation and the project documentation include the development of nature protection measures. This chapter provides a package of measures aimed not only at avoidance and minimization, but also at compensation for significant negative environmental impacts which may arise as a result of a proposed project.

Such measures are developed after obtaining forecast of environmental impacts taking into consideration recorded effects. If the impairment is significant, measures are selected for the elimination of the effect that causes this impairment or for

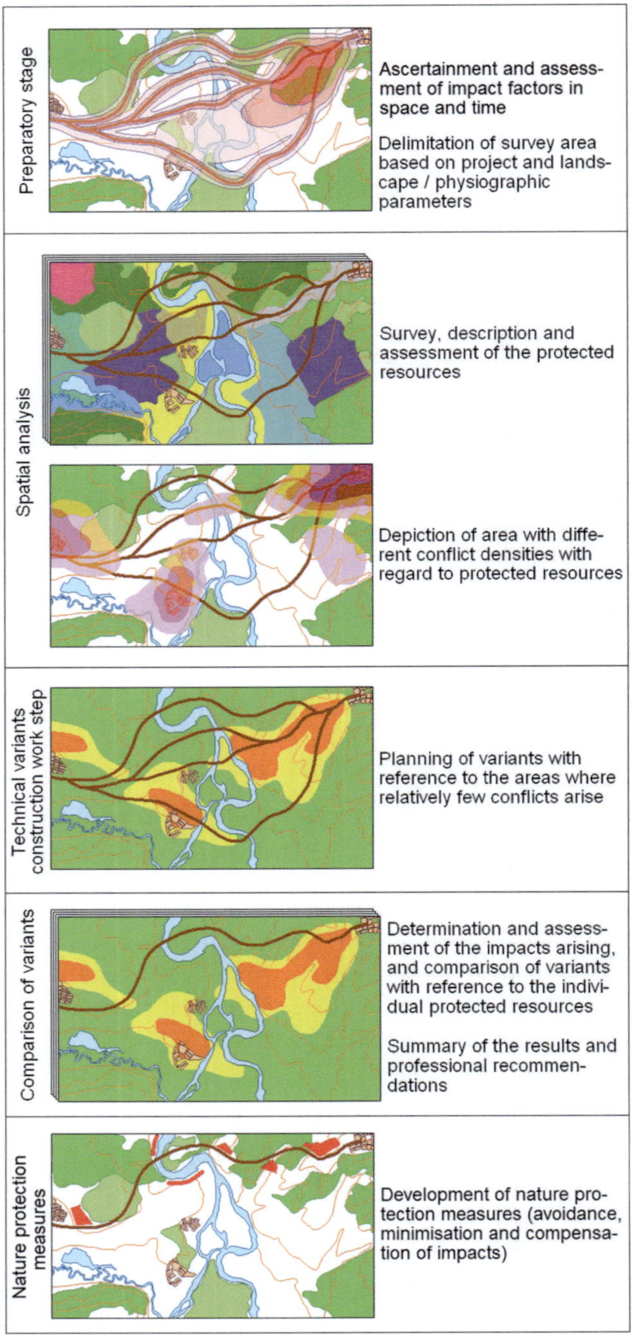

Fig. 18.1 Main stages of environmental impact assessment and the position within the sequence of impact mitigation regulation measures (Kravchenko et al. 2008)

minimization of its level. After the development of such measures and the modification of the original characteristics of the technical project, the impact is re-evaluated. If the remaining impacts still have high level of impairment, measures are developed for compensation of these impacts.

In this chapter, we will therefore call this complex of measures «measures on impact mitigation regulation» or simply «impact mitigation regulation» .

A number of assessments should be made for the development of nature protection measures at each stage of the EIA. A short list of such assessments and their main tasks are given in Table 18.1.

Let us consider the basic methods of realization of the listed stages.

Table 18.1 Substantive methodological task complexes of impact mitigation regulation (Boehme et al. 2005)

Task complexes	Goal and purpose of the evaluation process
Ascertainment of the situation Ascertainment and evaluation of the efficiency and functionality of the natural balance and the landscape quality	Evaluation of the impact area, with the goal of the differentiation of the significance of values and functional characteristics: • As the basis for the evaluation of the impacts (comparison of the situation) • As a reference situation (initial or target situation) • As the basis for the justification of avoidance measures
Estimate of effect Ascertainment and evaluation of the effects expected from the project and from the compensation measures	Evaluation of the changes caused by the impacts of the project, as a component of the forecast/estimation of the impacts • With regard to type and extent/intensity (relevance) • With regard to correspondence with normative goals and principles of the Nature Conservation Act, or local environmental planning Evaluation of site-specific and construction-engineering-related avoidance options Evaluation of the improvements (upgrades) caused by the compensation measures • With regard to type and extent/intensity (relevance) • With regard to correspondence with normative goals and principles of the Nature Conservation Act, or local environmental planning
Compensation determination Evaluation in the context of determination of type (qualitative equivalence) and scope (quantitative equivalence) of negative impacts and compensation	Evaluation of upgrades caused by compensation measures (component of forecast and estimation of effects) • With respect to contribution to upgrade of value/functionality • Evaluation of equivalents of functions according to type and scope (equivalence)

18.2.2 Delimitation of the Area to Be Studied

The area under study is delimited in such a way that all options of expected negative environmental impacts of a project could be fully considered in spatial terms.

The following characteristics are taken into consideration in the delimitation of the area under study:

- Sizes of effect areas with project-related impacts, including significant negative ones
- Spatial relationships of the affected functions of the environmental assets, including those for the planning of compensation measures
- Manners of input and distribution of harmful emissions
- Potentially affected specially protected areas.

The following structural elements are identified on the area under study (Fig. 18.2).

Project site—The project site refers to the area of land to be used directly for the project (site, route, etc.). At the project site, the functions and values of the natural balance and of the quality of the landscape will be directly impacted by the project,

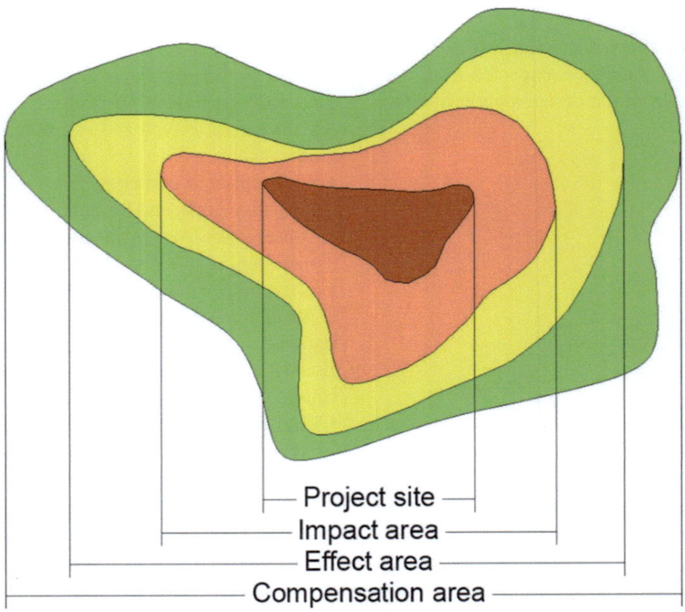

Fig. 18.2 Structural elements of the area for the proposed project

as a result of its construction activities and facilities. It is the point of origin for all negative impacts caused by facilities, construction activities and operations. The project site, with any alternative sites and options, is defined in the project description.

Impact area—The impact area is the area with significant or permanent negative damage due to facilities, construction and operations. It thus encompasses the project site and, depending on the type of project, any adjacent or distant areas which may be affected. Therefore, an analysis of the negative impacts to be caused by a project is essential for delimiting its area of impact. These impacts should be categorized according to the significance or duration of their effects for the efficiency of the natural balance and of the landscape quality.

Effect area—The effect area encompasses the entire area in which project-related negative impacts may have an effect. It generally extends beyond the impact area.

The effect area is usually defined in such a way that it outlines the territory within which negative effects on specific functions of the environmental assets are expected. Depending on the size and specific characteristics of these effects, one or several zones can be determined, e.g. along the road (Fig. 18.3).

Effect areas are delimited on the basis of the significance of effect factors, taking into consideration the sensitivity of environmental assets. Therefore, it is necessary to justify the sensitivity level, if it is not stipulated by industry standards.

Compensation area—The compensation area is the area for which compensation and offset measures are to be provided. It may extend beyond the affected area, particularly in the case of so-called «compensation pool solutions».

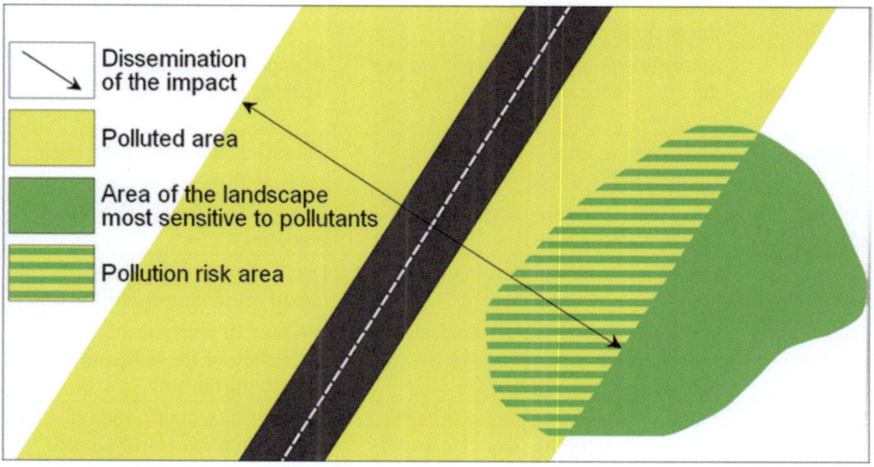

Fig. 18.3 Delimitation of effect and impact areas

18.2.3 Determination and Assessment of Impact Factors

Determination of single impacts. All impact factors are determined at the stages of project realization, in the context of its technical and technological components:

- Impacts caused by construction-related technological processes at construction sites (preparatory measures, construction of sites, installation of technical equipment, etc.)
- Facility-related impacts. Usually, these impacts are irreversible and persistent
- Operationally related impacts (removal of wastes, water and energy supply, discharges and emissions of harmful substances under normal and under emergency conditions, electromagnetic and radiation emission, etc.)
- Impacts caused by the shutdown of facilities, subsequent reclamation and ultimate disposal of industrial wastes (utilization and recycling).

Environmental assets and their functions, which are affected by these factors, are also determined. Impact factors vary depending on the type of activity.

Determination of cumulative impacts. Cumulative impacts are impacts resulted from changes caused by past, present or reasonably predicted future activities which accompany the proposed project and enhance its environmental impact. These include synergic impacts which emerge as a result of the combined intensification of single impacts. The list of major cumulative impacts is given in Table 18.2.

Table 18.2 Major types of cumulative effects (Gassner et al. 2005)

Type of effect	Characteristic	Example
Accumulated effects over time	Permanent or frequently repeated impacts (effects not yet subsided before the next occurs)	Uncontrolled waste disposal on shores of lakes and rivers
Spatial accumulated effects	Spatially concentrated impacts, so that the effect radii overlap	Biotope fragmentation due to frequent road building
Combined effects	Synergism of effects from multiple sources within a medium	Emissions into atmosphere from many sources
Delayed effects	Long delay before effect becomes apparent	Damage to forests, carcinogenic processes
Spatially shifted effects	Effects occur far away from their source	Dams, transfer of emission impurities in atmosphere
Trigger or threshold effects	Sudden deployment of ecological processes that modify the system behaviour in principle	Washout of heavy metals after exhaustion of the buffering capacity of the soil
Structural surprises	Multimedia or multisystemic effects associated with long-term changes in natural systems	Effect of increased carbon dioxide emission on the global climate
Induced effects	Incidental and consequential effects of a primary effect	Road construction causing development of residential areas

As cumulative impacts usually affect several environmental assets, they are described in terms of the assessment of the relationship between these assets.

Assessment of impact factors. Each factor revealed is assessed from

- The scope, frequency and persistence of the impact
- Its radius or effect area
- The intensity of the impact and of the changes it will likely cause within this radius or area.

18.2.4 Description and Assessment of Environmental Assets and Their Functions

Accounting of certain assets and their functions. Only the following environmental assets are considered in the assessment of compensation for significant negative impacts:

- Flora and fauna
- Soils, including the geological environment
- Ground and surface waters
- Air and the climate, and
- Landscape
- Humankind
- Cultural and other material resources.

Another step in the EIA investigation process is the description (inventory) and assessment of these assets and their functions. They are performed in the context of forecast of effects and impacts. All important functions of the enumerated assets are taken into consideration.

Assessment of certain assets and their functions. The significance (value) of a function is usually assessed for a concrete asset. Sometimes, the category sensitivity is assessed for certain functions and impact factors. An evaluation system of criteria is developed for the assessment of these categories.

The following technique is used for the assessment of environmental assets. A level assessment scale is developed for each criterion, and each level is assigned a qualitative or numerical value, e.g. Level III: low value; Level II: medium value; and Level I: high value. The number of such levels depends on the number and credibility of information. The more detailed and reliable this information is, the greater the number of levels which can be assigned. There is no need, however, to strive for extra specification. Generally, three to five levels are enough.

Criteria can be assigned on the basis of either two principles:

- Individual selection of criteria for each level of the assessment scale
- Selection of universal criteria similar for all levels.

Table 18.3 Evaluation framework for the protected asset «Climate/Air» (Kuepfer 2005)

Classification	Evaluation criteria
Level A very high	• Cold air corridors relevant for residential areas • Steep inclines near residential areas (>5°, or 8.5% incline) • Areas which are particularly active in terms of clean air and/or the bio-climate (e.g. forests, large orchard meadow complexes) • Climate protection forest, immission protection forest
Level B high	• Cold air generation areas relevant for human habitation (inclines of 2–5°, or 3.5–8.5% incline; cold air formed there can flow directly into residential areas, or is collected by way of cold air corridors, and thus passed into residential areas) • All other cold air corridors (i.e. those not directly relevant to residential areas); areas active in terms of clean air and/or the bio-climate (e.g. small forests, scattered orchard meadow complexes) • Protective planting
Level C medium	• Cold air generation areas with low incline (cold air generation areas not relevant to residential areas) • Areas in which neither significant cold or fresh air generation is provided, nor in which significant air pollution exists
Level D low	• Areas with little pollution of clean air or the climate, e.g. residential areas with significant greenery
Level E very low	• Areas with major pollution of clean air and/or the climate, which affects adjoining areas, e.g. industrial areas polluting commercial areas

The realization of the first principle means a reference description of the state of the asset to which its particular value is to correspond. The comparison of the real situation with these references at a concrete site allows the assignment of an analysed asset to one or the other evaluation level. Such an approach provides the possibility to assess the situation rapidly, which is very important for preliminary impact assessments. At the same time, it is impossible to investigate in detail all these separate functions of an asset. An example of similar assessment for the asset «Air and climate» using a five-level scale is given in Table 18.3.

The other approach is more universal. It allows a detailed analysis of certain functions of the asset, assigning similar criteria for all value levels. Moreover, the number of levels is variable, which allows for the use of developed criteria for other projects.

The sensitivity of environmental assets and their functions with regard to consequences resulted from a proposed project at various stages of its realization (e.g. noise pollution, emissions of harmful substances, groundwork) is determined within the framework of the analysis of the territory. The data obtained are the basis for delimitation of areas with a high density of conflicts which should be avoided if possible. Pollution of the atmosphere, soil and water, changes in water levels or the microclimate, the emergence of barrier effects, or the fragmentation of territory may have an impact on the functions of one or the other environmental asset. Therefore, it is necessary to determine the sensitivity of each asset to the respective impact factor for a forecast of impacts. The classification of sensitivity results from the type of affected functions with regard to the impact factor under consideration.

For example, determination of the level of sensitivity of a groundwater's aquifer to pollution by harmful substances (an impact factor) is based on the relationship between the occurrence depth of this aquifer to the permeability level of overlying layers. The latter have low sensitivity at deep occurrence of groundwater with surface layers impervious to water. In case of shallow aquifers without surface layers, its sensitivity is high.

A specific sensitivity scale for each environmental asset to the impact factors is determined.

Accounting of relationships between environmental assets. Besides the assessment of certain environmental assets and their functions, it is necessary to take into consideration the relationships of these functions within and between certain environmental assets of their ecosystem. Such an assessment is carried out in three steps:

- Accounting of relationship for each environmental asset
- Asset accounting of relationships
- Accounting of impact transfer from one asset to another.

The majority of certain functions of the environment are realized in combination with several assets. In general, it is possible to distinguish an asset for a concrete function with which it is connected, depends on, to a greater extent, or maximally affects. There is no necessity to estimate this function for all assets in which it is ascertained and with which it is connected. It is enough to assess the function only for a concrete asset. Figure 18.4 shows relationships between assets and their certain functions, and those assets are shown for which it is enough to assess a concrete function.

18.2.5 Forecast of the Impacts of Environmental Assets and Their Functions

The forecast of impacts is based on the results of the analysis and assessment of impact factors and state of certain environmental assets and their functions.

The main input characteristics for the forecast are the following:

- The intensity of the impact on the environmental asset analysed, and its functions and forecast for the assessment of impacts
- The significance (value) of the environmental asset analysed and its functions
- The sensitivity of the environmental asset analysed and its functions
- The characteristic of an effect area or areas, including their assessment in categories of sensitivity and significance (value).

The forecast is based on a comparison of intensity of impact on the analysed asset or its function and its sensitivity or significance (value). These characteristics are expressed in certain grades of estimation described in Sects. 2.3 and 2.4.

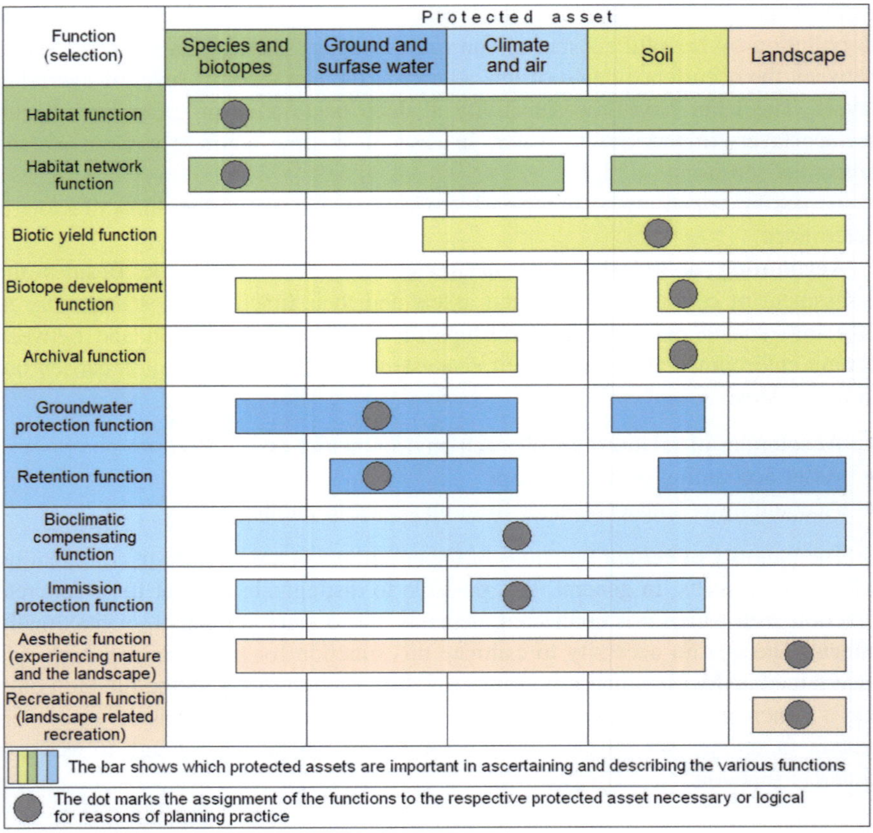

Fig. 18.4 Relationships between protected assets and the functions of nature and the landscape (Kravchenko et al. 2008)

The level of impacts is considered to be higher at those sites in the effect area where the highest intensity of this impact corresponds to the highest sensitivity or significance of the environmental asset analysed or its functions. The accuracy and objectivity of a forecast depend on the estimation scales, i.e. on how precisely and objectively they have been developed.

To compare the intensity of impact on the asset analysed or its function and its sensitivity or significance (value), matrices are used, in which these indices are combined. The principle of construction of such matrices is shown in Fig. 18.5.

These matrices allow to combine maps with information on intensity of impact and significance for individual environment component and for between environment components (Fig. 18.6).

Values of sensitivity or significance (value) range on the basis of evaluation criteria. The same principle is used for ranking intensity of load resulted from impact factors. The intensity of impact will be higher, e.g., at a high level of load

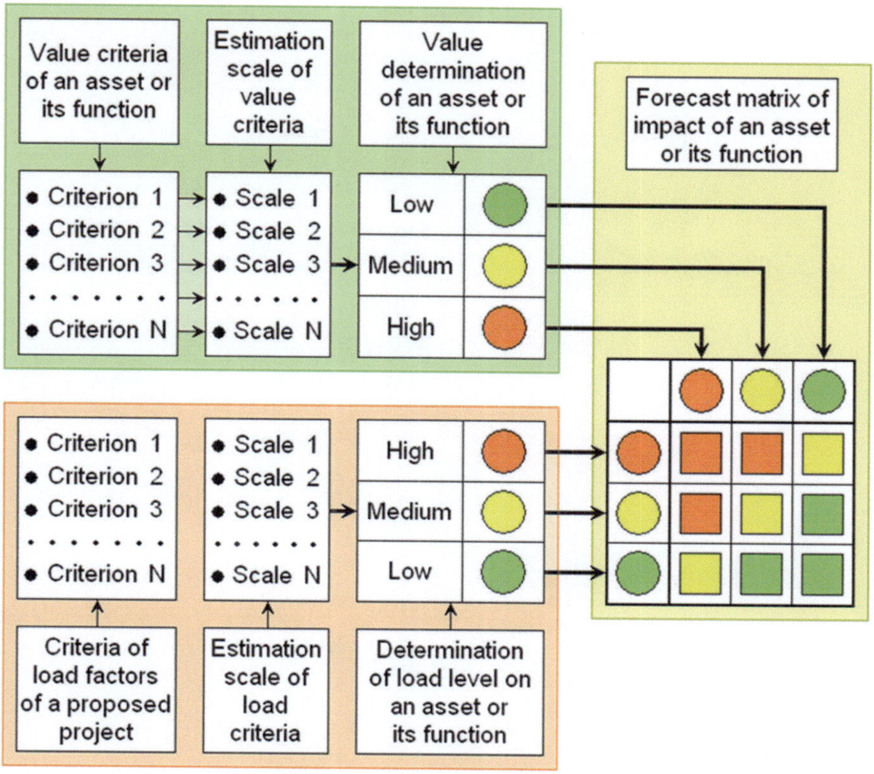

Fig. 18.5 Forecast matrix of the impact on the environmental asset or its function

and a high value of the function. If the load and value are low, insignificant impact is expected. In this case, the matrix is symmetrical. In principle, depending on the specific characteristics of the proposed project, it can be asymmetrical. For example, if the function value is high, the impact will be recognized as significant, regardless of load level. However, the use of such asymmetrical matrices must be justified.

Four different methodological approaches are used to forecast impacts, depending on the types of impact and the conditions under which the particular project is to be realized:

1. Assessment of impacts leading to the direct loss of a natural object or area
2. Assessment of impacts unrelated to sensitivity and significance of the assessed asset or its functions
3. Assessment of impacts related to the value of an asset or its function (simple linkage)
4. Assessment of impacts related to sensitivity and significance (dual linkage).

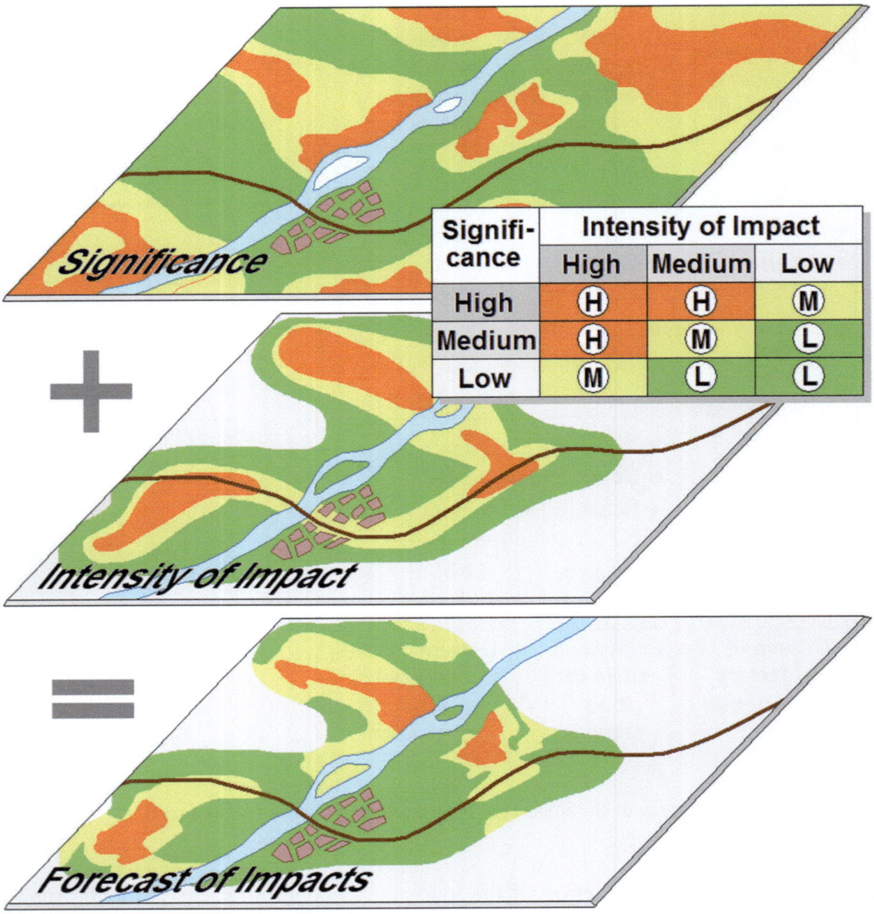

Fig. 18.6 Combining maps based on forecast matrices

Figure 18.7 shows the general scheme of selection of approaches for forecasting. These approaches are described in detail in (Kravchenko et al. 2008, 2014).

18.3 Development of Nature Protection Measures

18.3.1 General Provisions

The following types of measures can be distinguished.

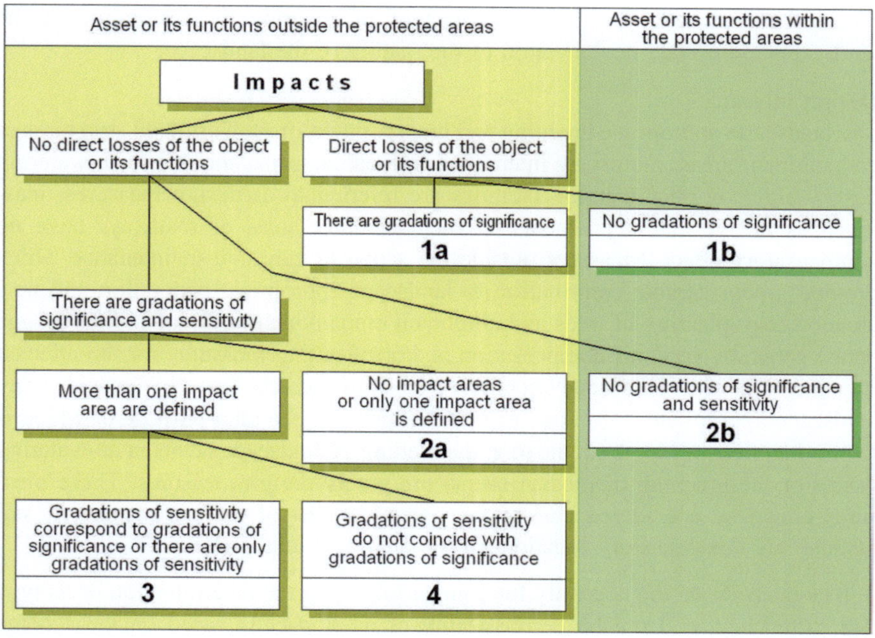

Fig. 18.7 Selection of approaches for impact forecasting. Forecast approaches are marked by numbers: 1—Assessment of loss of object or area; 2—Forecast without linkage; 3—Forecast with simple linkage; 4—Forecast with dual linkage (Kravchenko et al. 2008)

Mitigation measures

Mitigation measures are measures by means of which possible negative impact on nature and the landscape can be wholly (avoidance) or partially (minimization) prevented. These include particularly structural measures (e.g. tunnels, green bridges, passageways, guide facilities) and measures for the protection against temporary endangerment of nature and the landscape (e.g. fencing in, protection of bodies of water and single trees, or protective planting in the context of the construction procedure).

On-site, in-kind compensation measures

This type of compensation refers to conservation and landscape care measures suited to restoring the negatively impacted functions and structures of the natural balance in the same functional manner, or which provide for the restoration or landscape-appropriate new creation of the landscape quality. However, this does not fundamentally mean the identical recreation of the same structures.

Offsetting measures

Offsetting measures are conservation and landscape care measures which become necessary if in-kind compensation is not possible. Offsetting measures should be capable of restoring in an equivalent manner the functions and structures of the natural balance destroyed by the project or of ensuring the restoration or the landscape-appropriate new creation of the quality of the landscape.

Design measures

The landscape-appropriate greening and incorporation of such artificial structures as embankment areas, motorway interchanges, median strips, shoulders, de-watering basins and/or noise protection facilities are referred to as design measures, inasmuch as the vegetation elements of such ancillary areas of roadways have no compensatory effect. Measures outside the scope of care and maintenance which provide a considerable contribution to landscape-appropriate restoration and new creation (e.g. planting of trees and shrubs on embankment areas or in interchange areas) generally serve as compensation and/or offsetting measures for the affected functions of the landscape quality. Measures which do not provide any landscape-appropriate restoration or new creation and the only purpose of which is to provide greening for the route (e.g. the seeding of landscape lawns in de-watering basins or the greening of median strips) are purely design measures. These measures are to be considered part of the operation area of the project and do not assume any compensatory function for the quality of the landscape.

The decision on the necessity for compensation, on the determination of its type and sufficiency, and on the permissibility of an impact is based on the scheme shown in Fig. 18.8. Decision-making is carried out step by step. First, the possibility of impact avoidance is ascertained. If that is not possible because of the technical specific characteristics of the project, an attempt is made to minimize these impacts to the permissible level. If the impacts remain significant after such measures, measures of compensation are developed. If it becomes clear that not all impacts can be compensated completely or partially, chosen choice must be made: what is more important for the public interest—the need for the realization of the project, or the interest in nature conservation. If nature protection takes priority, the proposed project is considered impermissible. Otherwise, offsets are developed. If not all impacts can be offset, approval of the project with compensation payments may be permissible with targeted equivalent environmental improvement at another site. This is the same offsetting, but is organized differently.

It should be noted that monetary compensation may be used only as an exception. In any case, compensation fees are spent on out-of-kind/off-site offsets, when full compensation of impacts has not been achieved within the framework of the proposed project. In Germany, this problem is often solved by using an available offset measure from a «pool» of compensation areas and measures.

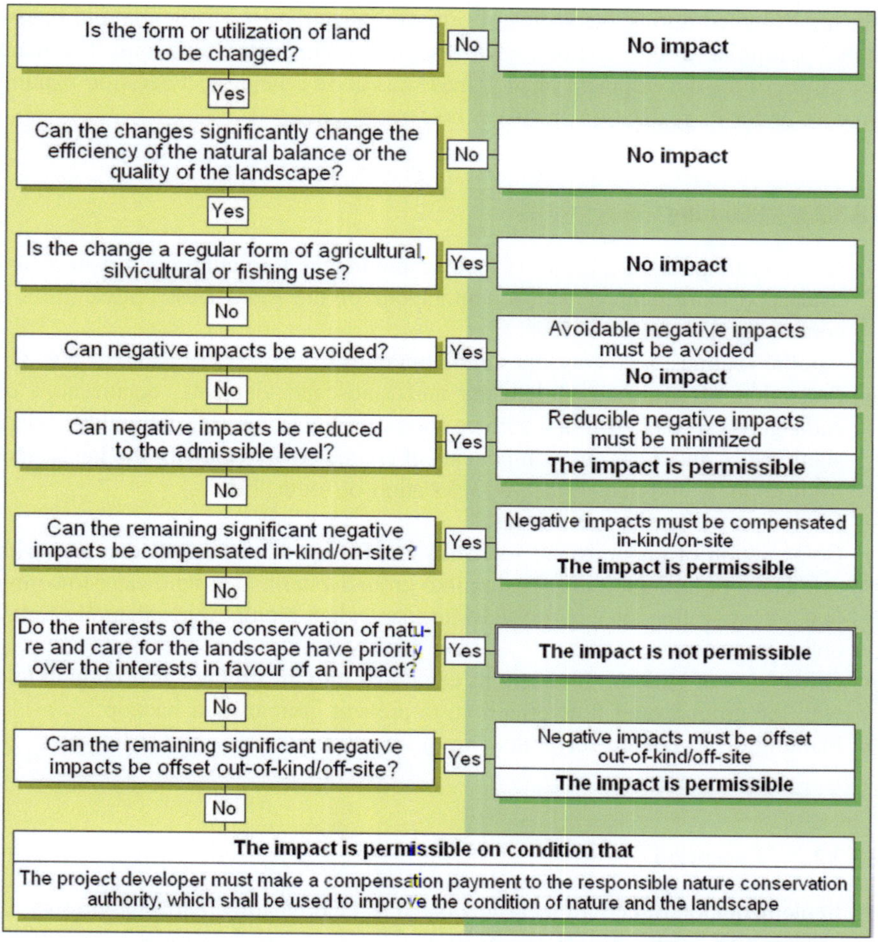

Fig. 18.8 Sequence of assessment and decision-making in the impact mitigation regulation (Kravchenko et al. 2014)

18.3.2 Development of Impact Avoidance and Minimization Measures

The development of such measures is aimed at avoidance of significant negative impacts or their minimization to permissible level. It is necessary to distinguish between avoidance of impacts in the effect and impact areas.

In the effect area, it is necessary to strive to avoid or at least minimize all impacts. Impact avoidance and minimization measures in this area are implemented by modifying the project plans. If it is impossible to change the design of an investment project due to certain specific features, measures are developed which are implemented during the stages of project realization—construction and operation, appropriate to the environmental asset and its impacted functions. The main principles of the development of such measures during the project decision-making process or for its realization are listed below.

18.3.2.1 Planning

- Non-consumption of areas of particular importance for nature conservation and landscape management; decision in favour of the route least critical from a conservationist point of view
- Parallel routing to existing lines and structures
- Protection of near-natural habitats and landscape elements, occurrences of endangered species, soil and surface formations of natural or cultural/historical significance and areas and components of characteristic value for the landscape, from removal and disturbance by relocation or route shift
- Limitation of the construction area
- Construction of underground segments and laying of underground cables with avoidance or limitation of intrusion into groundwater and groundwater lowering (e.g. at excavation sites in groundwater-dependent biotopes, installation of clay drainage system)
- Orientation of building foundations extending into the groundwater body in line with the groundwater flow direction, to prevent groundwater backup
- Planting of vegetation for protection of forests exposed by power line cuts.

18.3.2.2 Execution

- Implementation of certain measures (such as land clearing, earth-moving work, construction work) outside of sensitive periods for certain species
- Restriction of the effects of construction operations (e.g. securing of habitats, trees or sites from vehicular traffic or damage); consideration of the guidelines for the protection of trees and shrubs in the area of construction
- For underground cables, preference for closed construction
- For underground cables in groundwater-dependent habitat types, usually no groundwater lowering during the growing season
- Avoidance of the introduction of non-native topsoil
- Appropriate temporary storage of topsoil and no storage on areas valuable for nature conservation
- Storage and replacement of soils separated by topsoil and subsoil, to permit restoration of the original soil structure
- Loosening up of compacted areas after completion of the construction process

- Careful disposal of wastes, materials, etc., after completion of the construction process
- Transplanting valuable vegetation stands
- Labelling of the overhead lines according to the state of the art, in order to avoid animal losses due to collision.

In the impact area, in the area with project facilities and their infrastructure, avoidance is, first of all, realized mainly by means of the optimization of the site selection for a proposed project. Such optimization is carried out at the EIA site selection stage, seeking the project location with the lowest level of ecological risk. Any remaining significant negative impacts as a result of the direct effect must be compensated.

The most significant difficulties arise when an investor seeks changes in the stipulations made in the initial project in order to balance the interests of nature conservation. These changes require compromise solutions and can be developed in three main directions:

- The principal alternatives of the proposed project from the viewpoint of nature conservation
- Reduction of the proposed project scope
- Selection of other environmentally reasonable sites for the proposed project.

18.3.3 Development of Impact Compensation Measures

The derivation of concrete compensation and offsetting measures for the restoration of the negative impacts upon the balance of nature and the landscape quality is carried out on the basis of:

- The goals and principles of conservation and landscape care, and the precepts of landscape planning derived from it, or from other regional conservationist goals
- The requirements of species protection
- The concept of project-related measures
- The compensability of impact.

Before concrete development of measures, their concepts and the goal of the entire project are stipulated. The measures developed are to correspond to nature conservation goals established for this area within the framework for territorial projects at any level. If there are no such projects for a concrete area, the goal of the measure is to be derived from the general nature protection goals.

As soon as the programme of the measures and its goal are ascertained, certain compensation measures and suitable areas for their implementation are selected according to the scheme shown in Fig. 18.9.

Only areas which will require upgrading and are capable of being upgraded can be considered for compensation and offsetting measures. These preconditions are met if they can be upgraded to a condition which can be categorized as ecologically higher than the baseline condition.

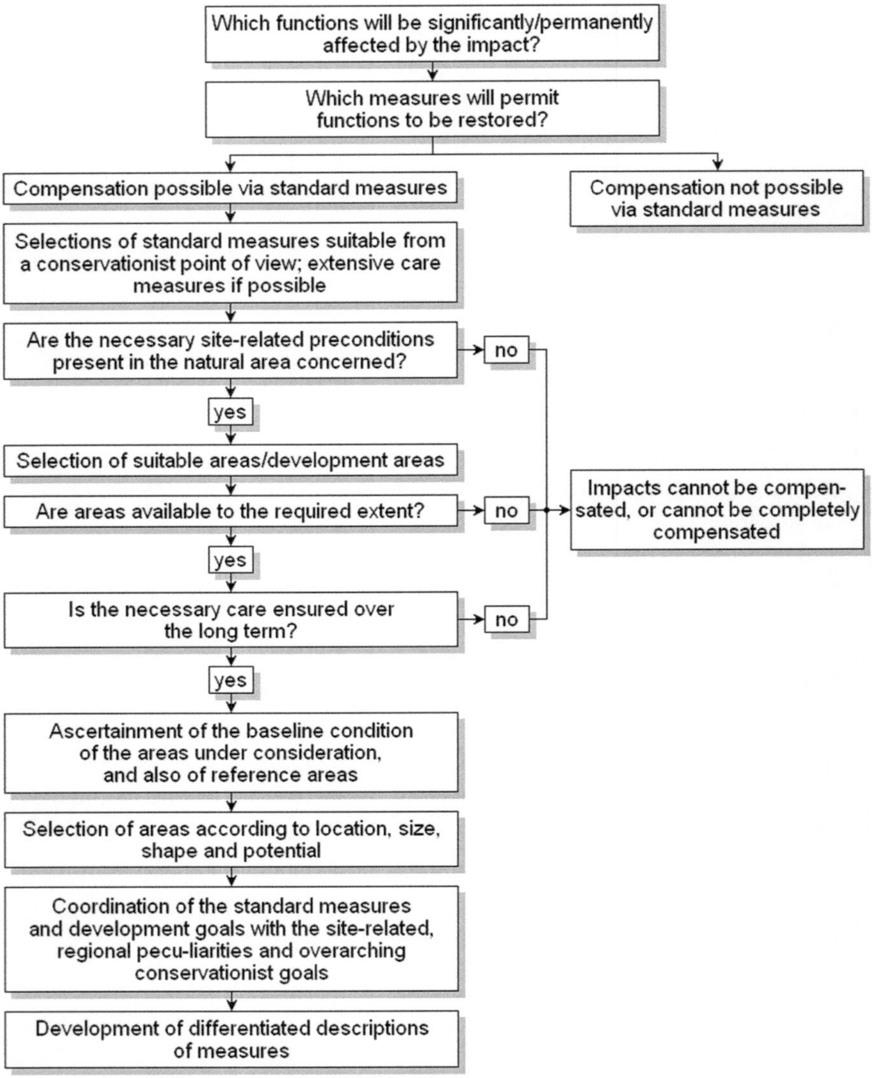

Fig. 18.9 Selection of measures and areas for compensation (Kravchenko et al. 2014)

18.3.4 Standards for the Selection of Compensation and Offsetting Measures

18.3.4.1 Basic Requirements

- High probability of success for actual implementation and permanence
- Permanent effective supervision of the areas must be provided

- Selection of areas on which the natural balance and the quality of the landscape can be upgraded and require upgrading
- Basic suitability of site conditions with regard to the goals of the compensation measures
- No «dual occupancy» of areas which have already been used for compensation measures for other impacts
- No use of areas which could be significantly impacted by planned or foreseeable projects, even if such impacts upon the area would only be indirect
- «Rehabilitation measures» such as ammunition clearance or toxic site rehabilitation are generally excluded
- No crediting of protected area certification (area protection only); land in protected areas can be used if it can be ecologically upgraded, and that would be useful from a conservationist viewpoint.

18.3.4.2 Functional Requirements

Compensation measures require in-kind restoration of the functions and values existing prior to the impact, in a close functional context. That does not mean that the identical restoration is required, but rather that the essential functions which the landscape previously fulfilled must also be able to be fulfilled in the future. This is as a rule possible if the same or similar elements are restored. The more important the lost function is, the closer the relationship of the compensation measures to the affected functions must be in order to be able to be counted as on-site in-kind compensation.

For offsetting measures, the functional relationship is looser. Nonetheless, offsetting measures should have as close as possible an approximation to the criteria of on-site in-kind compensation, and the effectiveness and functionality of nature and the landscape should be restored in a similarly equivalent manner.

18.3.4.3 Spatial Requirements

For recognition as on-site in-kind compensation measures, a close spatial relationship to the impacts to be compensated is necessary. Only those measures can be recognized which affect the space in which the significant negative impacts are to occur. Measures in the direct impact area of operationally caused negative impacts will not be recognized. These are considered landscaping measures which may in exceptional cases be recognized as compensation for negative impacts on the appearance of the landscape. For offsetting measures, the spatial relationship is looser. This compensation can be carried out at a greater distance and in a different manner— «off-site/out-of-kind» . In any case, a spatial connection between the impact and the compensation areas must exist, e.g., within the same natural spatial region.

18.3.4.4 Time-Related Requirements

In order to achieve recognition as on-site/in-kind compensation measures, the functions and values impacted upon must be able to develop effectively to their pre-impact quality within 25 years. Any measures which will require a longer period of time are to be considered offsetting.

The implementation of compensation and offsetting measures should begin simultaneously with the initiation of the impacting project, at the latest, and should be concluded by the time that project is completed, in order to minimize the so-called «time-lag effect». This requirement refers to the technical implementation of compensation measures and follow-up care. Depending on the goal of the particular measure, the development and maintenance care may require a longer period of time.

If negative impacts upon sensitive functions, such as breeding losses for bird species, are expected, compensation and offsetting measures may be necessary even prior to or during the implementation of the impact.

18.3.4.5 Requirements on the Extent of the Area

As a rule, significant negative impacts are to be compensated on an area at least as large as the impacted area.

18.3.5 Compensation or Offsetting Measures?

The main differences between compensation and offsetting measures are in the requirement for the preservation of functional and spatial relationships for compensation measures and the possibility of restoration within a short period of time (usually no more than 25 years for these measures). The preservation of territorial relationships can be visually exemplified.

A territory with two landscape sites was provided for the building project: a meadow with wild fruit trees over 30 years old and a ploughed field. The significance of these sites is assessed at the first step of the development of compensation measures. It is higher for the orchard meadow than for the ploughed field (Fig. 18.10).

In the second step, the level of effect seriousness is assessed. For example, one part of the site has a greater density of buildings than another. To assess this site occupancy index, a two-step scale is used. The threshold value, including impervious coverage, is 0.35 of the total area. A higher level of effect is determined for compact construction planning (Fig. 18.11).

The intensity of impacts is assessed in the third step, by overlapping the plan of the evaluation of the actual condition and the building scheme (Fig. 18.12). Since the plan calls for the use of the entire area for construction, all impacts are

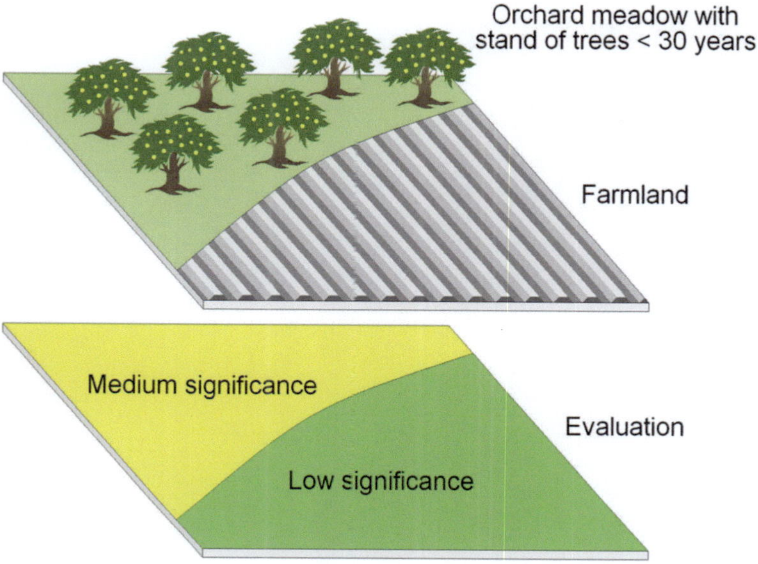

Fig. 18.10 Step 1. Categorization of the planning area prior to construction. Areas with different significance for natural balance and landscape scenery (Kravchenko et al. 2008)

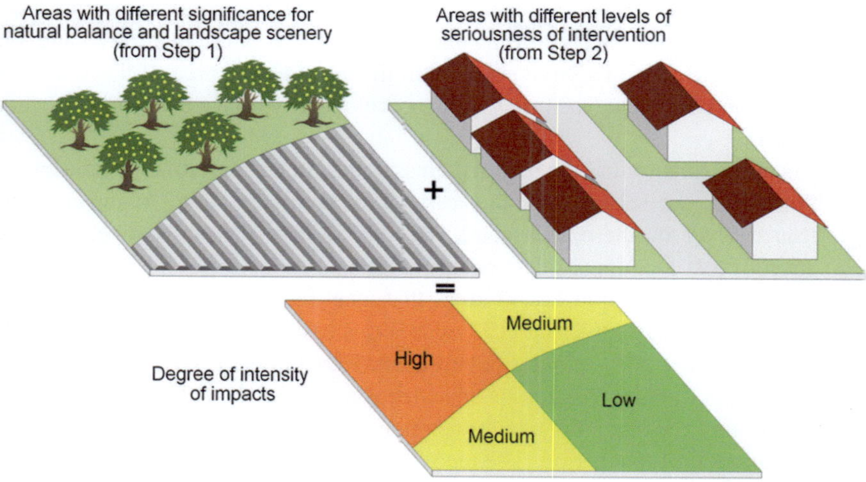

Fig. 18.11 Step 2. Categorization of the planning area according to the planning. Areas with different levels of seriousness of intervention (degree of impervious coverage and degree of use)

considered significant, and it is necessary to compensate the area proportionately to the degree of impact. Methods of scoping of compensation proportional to the level of violations are described in the monograph (Kravchenko et al. 2014).

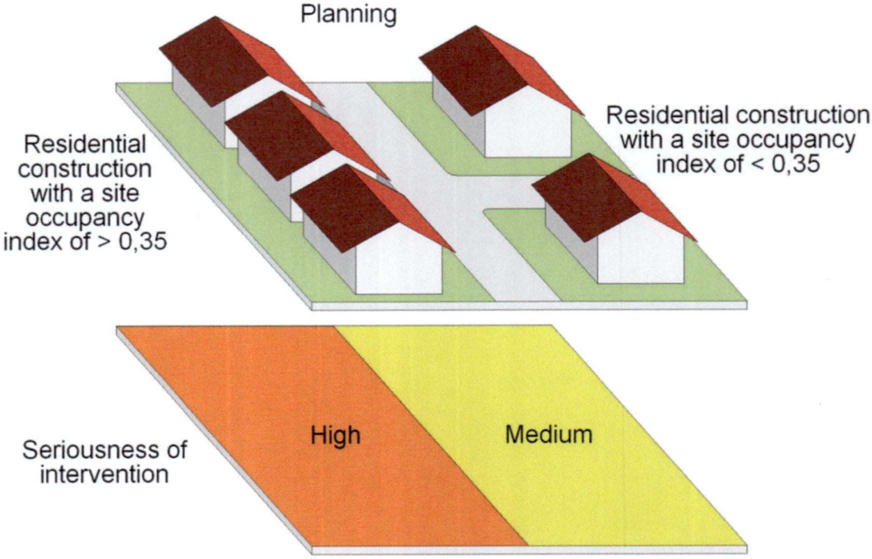

Fig. 18.12 Step 3. Process of derivation of impact intensities

After the ascertainment of the scope of compensation, a type of compensation measures is selected—either a compensation or an offsetting measure. In this case, the selection depends on specific characteristics of the building project and the selection of a site of the required size within the construction area. If this choice is possible, compensation measures are developed within this impact area (Fig. 18.13).

If it is impossible to find the required areas, appropriate areas directly adjacent to the impact area are selected. In this case, it is necessary, first, to increase the significance level of these areas and, second, to take measures to enable the restoration of lost functional relationships (Fig. 18.14).

If it is impossible to find vacant areas that meet the requirements of functional and territorial relationships (Sect. 3.3), offsetting measures are implemented outside the impact area, but under similar environmental conditions (Fig. 18.15).

The selection of compensation or offset measures depends on the possibilities of the restoration of the impaired area for a certain period of time which is determined from the duration of this period. The recovery period depends on how long society can wait for the results of compensation, i.e. improvement of the ecological situation which will have deteriorated due to the project implementation. Approximate terms of recovery of certain ecosystems or biotope types and related possibilities of their compensation or offsetting measures are given in Table 18.4.

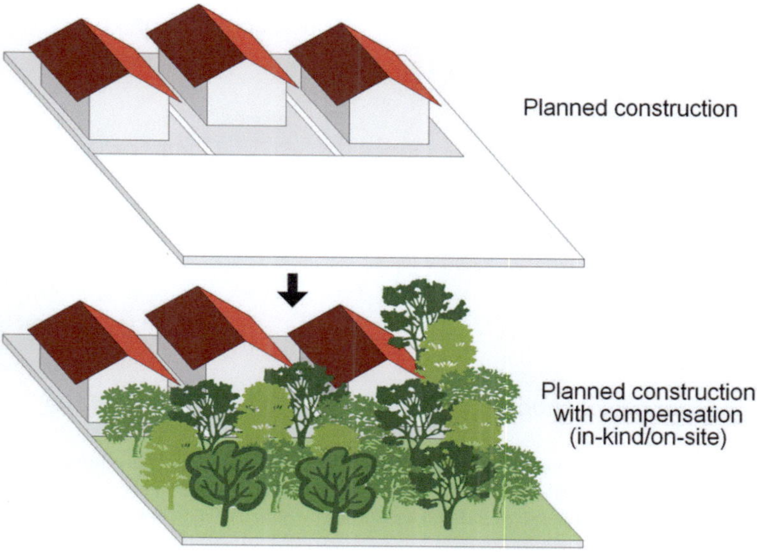

Fig. 18.13 Step 4. Determination of compensation measures. Variant 1. Compensation on the building site

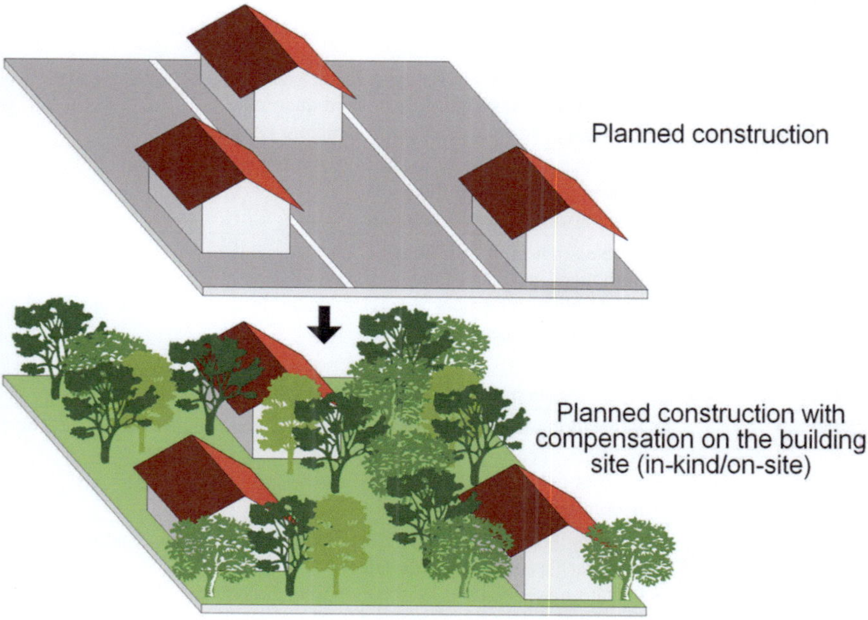

Fig. 18.14 Step 4. Determination of compensation measures. Variant 2. Compensation on other sites covered by the local development plan, e.g., planting a green margin around the estate

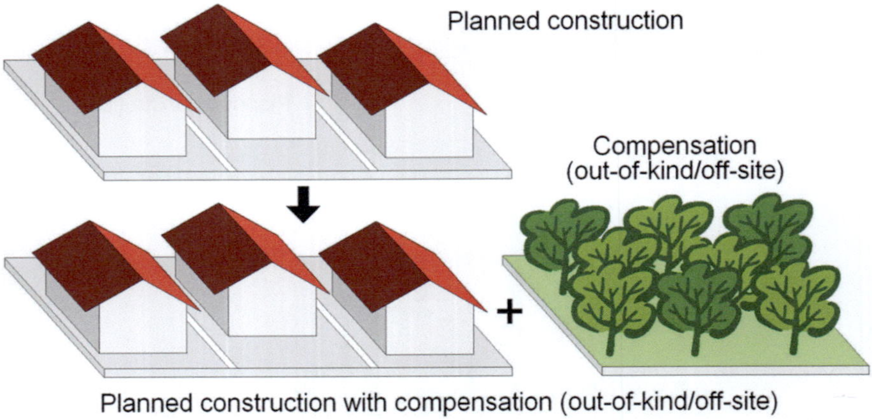

Fig. 18.15 Step 4. Determination of compensation measures. Variant 3. Out-of-kind/off-site offset outside the building zone (Out-of-kind/off-site compensation on another site, e.g. transformation of an intensively used meadow into woodland or into a low-nutrient meadow)

Table 18.4 Development times for ecosystems and biotope types. Possibilities of compensation (on-site and/or offsetting) (Koeppel et al. 1998)

Development time, years	Ecosystems and/or biotope types
Compensable biotope types. Development times <25 years	
<5	• Short-life ruderal vegetation • Pioneer stages of sandy low-nutrient grassland, secondary sandy low-nutrient grassland, e.g. in sand quarries • Eutrophic and/or species-poor, structure-poor ditches
5–15	• Most manifestations of acidic soil shrubbery and hedges • Nitrophilic high forb fields • Long-lasting ruderal fields
15–25	• Species-rich high forb fields • Shrubbery on fallow land • Relatively species-poor secondary sandy low-nutrient grassland and semi-dry grassland • Pioneer forests
Only offsettable biotope types. Development times 25–150 years	
25–50	• Vegetation of secondary oligotrophic and mesotrophic standing bodies of water, species-poor sedge reeds • Species-rich, strongly structured ditches and creeks • Broom heaths on fallow land
50–80 50–150	• Quaking bogs and other land-forming ecosystems on standing bodies of water (species-poor examples) • Certain manifestations of thermophilic shrubbery • Species-rich meadows, mown twice annually • Certain manifestations of open bog woods rich in coniferous trees and birch, on drained bogs, secondary growth

18.4 Conclusions

1. The analysis of the Russian, German and European experiences on the assessment of environmental impacts of the planning project and avoidance and compensation of these impacts has shown that in any country, with the realization of any project, conflicts may emerge between the existing situation and new initiatives and between the facilities under construction and the environment. In the solution of these conflicts, however, priorities are assigned to a number of different criteria, which often have a strongly subjective nuance. Nevertheless, it should be noted that there is a growing tendency to prioritizing long-term environmental values. This tendency is reflected in the development of international and national legislation for the defence of the public interest against unreasonable investment initiatives, in the development of local standards and technical regulations which restrain specific types of activities, and in the development of environmental impact compensation technologies.
2. It should be emphasized that under the conditions of great uncertainty and lack of information, the tasks of assessment, forecast and decision-making in the context of the natural impact compensation are complicated. Moreover, these tasks are often to be solved by people with completely different qualifications. Such a situation requires the use of forms of presentation of project information and methods of data processing, which are available to a wide range of people. From this perspective, it can be very useful for Russia to get to know the range of methods for the solution of impact mitigation regulations applied in Germany. These are generally qualitative, semi-quantitative and expert methods with which not only quantitative, but also extensive verbal information can be managed.
3. Despite the availability of numerous methodological documents on the development of nature protection measures, it is difficult to admit that there is a technology according to which the key issues of the environmental impact assessment can be answered unambiguously and objectively—What directly will be affected and how is this impact to be avoided or compensated? The author of this chapter expresses hope that methodological elements of the measures on avoidance and compensation of impacts offered to the readers can contribute to the creation of such a technology.
4. At the first stages of implementation and efficient application of this technology in Russian practice, it is necessary to test the methods presented on different project types or to develop analogues which can solve the same problems. This is a task for various scientific disciplines, and additional investments and approval by superior authorities are not necessarily required. Such work can be realized within the framework of relatively large planning projects.
5. The implementation of natural environmental impact compensation in Russia is urgent and necessary. This will allow the country to significantly improve the environmental situation and reach the European level in the sphere of environment protection.

References

Boehme C, Bruns E, Bunzel A, Herberg A, Koeppel J (2005) Flaechen- und Massnahmenpools in Deutschland. Ergebnisse aus dem Vorhaben des Bundesamtes fuer Naturschutz. BfN-Schr.-Vertrieb im Landwirtschaftsverl (Naturschutz und Biologische Vielfalt, 6). Muenster

Gassner E, Winkelbrandt A, Bernotat D (2005) UVP. Rechtliche und fachliche Anleitung fuer die Umweltvertraeglichkeitspruefung. 4., völlig neu bearbeitete und erweiterte Auflage, 2005, C. F. Müller Heidelberg. https://d-nb.info/970268459/04. Accessed on 20 Mar 2019

Koeppel J, Feickert U, Spandau L, Strasser H (eds) (1998) Praxis der Eingriffsregelung. Schadenersatz an Natur und Landschaft, UTB Ulmer Stuttgart, p 397

Kravchenko V, May A, Hoppenstedt A, Ignatov A (2008) Assessment of environmental impacts and ecological expertise (Оценкавоздействиянаокружающуюсредуиэкологическа яэкспертиза). Publishing house of the Sochava Institute of Geography of the SB RAS, Irkutsk, Berlin, Bonn, p 199

Kravchenko VV, Hoppenstedt A, Ignatov AV, Wende W (2012) Assessment of environmental impacts and ecological expertise: professional experience of EIA issues in Russia and Germany. Publishing house of the Sochava Institute of Geography of the SB RAS, Irkutsk, Dresden, Berlin, Bonn, Dessau, p 76

Kravchenko VV, May A, Golubeva SG (2014) Assessment, compensation and biodiversity offsets of environmental impacts: the German-Russian compendium. Publishing house of the Sochava Institute of Geography of the SB RAS, Irkutsk, Dresden, Moscow, Bonn, p 196

Kuepfer C (2005) Empfehlungen fuer die Bewertung von Eingriffen in Natur und Landschaft in der Bauleitplanung sowie Ermittlung von Art und Umfang von Kompensationsmaßnahmen sowie deren Umsetzung (Teil A: Bewertungsmodell). Wolfschlugen. http://fachdokumente.lubw.baden-wuerttemberg.de/servlet/is/95966/allgemeine_bewertungsempfehlungen_teil_a.pdf?command=downloadContent&filename=allgemeine_bewertungsempfehlungen_teil_a.pdf. Accessed on 20 Mar 2019

Chapter 19
Assessing the Ecological Status in Regions of Russia and Neighbouring Countries

Viktoria R. Bityukova

Abstract Environmental stress occurs when the level of anthropogenic impact exceeds the resilience of landscapes. A system of indicators of anthropogenic impact and a calculation method for an integral index of the intensity of anthropogenic impact are substantiated in the article. The index reflects the structure of anthropogenic impact (on the air basin, water, land, forest resources, as well as agrarian and radiation effects) and its dynamics for 1990–2016. The study of changes in the structure of environmental landscapes of regions in Russia and Kazakhstan has revealed that considerable contrasts in the placement of industrial complex and population settlement that had formed during the previous development stages are stable. Assessing changes with the help of anthropogenic impact integral index related to natural landscape resilience allows assessing regional changes in Russia and Kazakhstan and revealing main similarities and differences between those changes, as well as examining both industry-specific and spatial patterns forming environmental conditions at the regional level.

Keywords Ecological stress · Ecological situation · Anthropogenic impact · Complex estimation · Landscapes

19.1 Introduction

Ecological status is gradually becoming an increasingly important factor of regional growth in former Soviet countries. Large-scale social and economic, institutional and technological changes since the early 1990s occurring in the course of the economy's adaptation to market conditions and inclusion of Russia and neighbouring countries in the world market have significantly affected the environment, especially at the regional and city level. The past quarter of the century has seen

V. R. Bityukova (✉)
Faculty of Geography, Department of Economic and Social Geography of Russia, Lomonosov Moscow State University, Leninskie Gory, Moscow 119991, Russia
e-mail: bituykova@yandex.ru

L. Müller and F. Eulenstein (eds.), *Current Trends in Landscape Research*, Innovations in Landscape Research, https://doi.org/10.1007/978-3-030-30069-2_19

dramatic changes in the ecological status (ES) in Russia. New conditions have made their appearance, and the importance of ES formation factors has changed: the role of technological factors has gradually decreased, and the role of structural industry shifts has increased since it was relatively high-tech processing industries that experienced maximum output reduction. The greater raw material orientation of economy in Russia and Kazakhstan has amplified anthropogenic impact in mining regions, while the largest cities faced a rapid growth of automobilization being an important factor of anthropogenic impact. *Institutional mechanisms have not succeeded in producing* ES territories; environmental payments do not stimulate a greener economic activity though large private business possesses considerable financial resources for production modernization.

The major role of geographical differentiation of economic activity in the formation of territorial differences in the nature and intensity of anthropogenic impact on the environment should be emphasized. Together with the factor of differentiation of *resilience of various types of natural landscapes* to anthropogenic impact, it is causing territorial heterogeneity of the ecological status. The research of these factors allows constructing not just a stating model, but an explanatory one.

At the same time, the environmental conditions in the industrially developed former Soviet countries rich in natural resources, in particular, Russia and Kazakhstan, still bear the features inherited from the period of the Soviet industrialization. Their role can be revealed through a comparative study of the two countries, the industrial complex of which was formed within a single space.

An integral index of anthropogenic impact has been developed in order to assess changes in environmental conditions in the regions of the Russian Federation and the Republic of Kazakhstan. The scale and method of the research were chosen based on the availability of statistical information for various levels and time periods, comparability of environmental indicators in Russia and Kazakhstan and adequacy of the available statistics. In addition to the analysis of the impact of key factors as well as trends in the ecological status, a comprehensive assessment of the transformation of environmental and economic situation caused by changes in branch and territorial structure of the regional economy is needed. The comprehensive assessment lays the basis for greater efficiency of environmental investment.

19.2 Research Method

Integral indices of environmental conditions started to appear almost simultaneously in the USSR and in the West. But differences in development conditions of this research method and in its purposes, the impossibility of sharing practices predetermined separate ways already at the initial stage of formation of complex indices. Generalizing the available experience of foreign scientists, governmental and international commissions in this field, it is possible to identify a number of features. First, the use of basic indicators (GDP, gross domestic savings, net domestic product) is normalized based on the cost estimate of depletion of natural resources, ecological damage and natural capital (UNEP 2011; Dixon and Xie 2007; OECD 2010). The second is the integral indicators of sustainable development (ecological sustainability, real progress, etc.) including both indicators of quality of life and ecological indicators (Scandinavian Destination 2012, ARCADIS 2015). The third important feature is the initial orientation towards the search of indicators reflecting a contribution of the countries to global problems, i.e. ecological footprint, living planet index (Ewing et al. 2008). And finally, proper environmental indices based on either indicators of anthropogenic impact (air emissions, waste water) or "environmental efficiency" (impact of an ecological factor on human health and ecosystems) (Baeumler et al. 2009; The Green City Index 2012; McGrath 2014).

Large former Soviet countries, such as Kazakhstan and Russia, require a multiscale environmental research: country assessments lacking a regional component are unacceptable for countries as vast as they are. Subregional assessments are even more promising; however, the ecological statistics at the level of districts and municipalities is too scarce for a comprehensive complex assessment of anthropogenic impact over time. Thus, two levels have been chosen for the research: country and regional, allowing assessment of industry-specific and territorial shifts of areas of anthropogenic impact over time.

Key problems which researchers face while elaborating a technique for impact assessment at the regional level are a weak statistical base, frequent change of calculation procedures and incomparability of some indicators within intercountry comparisons due to various approaches to measurements and calculations. For example, a key group of indicators highly relevant to the assessment of anthropogenic impact in the Russian regions connected with the impact on forest resources is less relevant to Kazakhstan. Taking into account a common group for conventional forests and haloxylon desert woodlands in the Kazakhstan statistics, the use of indicators of impact on forest resources demands a careful approach and expert adjustment (Environment Protection in Russia 2017; Environment Protection in Kazakhstan 2015).

Creation of the integral index included four main stages:

Stage 1. Selection of priority components. In this case ecological indicators of almost all significant sources of impact were taken into account: production sector, power industry, motor transport, agriculture and forestry, population as a source of impact (through household waste and indirectly through other indicators), indicators of radiative effects and others.

Stage 2. Selection of indicators for each group was based on the principles of systematicity, reliability, availability of statistics, pronounced territorial differentiation and clearly interpreted dynamics. Each type of anthropogenic impact has a set of absolute and relative (specified in brackets) indicators: A_1—density of air emissions of the production sector and motor transport per the acreage of the cities, production sector and transport; A_2—coefficient of emission toxicity calculated as the ratio of emission volume normalized to a single toxicity to the gross air emission volume; B_1 and B_2—respectively, share of water consumption and wastewater discharge in surface water supply; C_1—solid waste density taking into account their hazard class; C_2—share of disturbed industrial lands; D_1 and D_2—share of cultivated and reclaimed lands, respectively; D_3, D_4 and D_5—mineral, organic fertilizers and pesticides, respectively, per 1 hectare of arable lands; D_6—cattle livestock per the area of pastures; E_1—share of total standing volume of cut timber; E_2—the area of felling per calculated cut; E_3—share of unauthorized felling; F_1—share of inhabitants in radioactive contamination zones; F_2 and F_3—soil pollution density with ^{137}Cs radionuclides of the forests and the area, respectively; F_4—share of lands of the cities, production sector and motor transport polluted by radionuclides; F_5—share of lands of former nuclear test sites and major accidents related to radioactive pollution.

Stage 3. Translation of initial value matrix for assessment indicators into values normalized for addition of different-sized indicators. The following criteria can be used while selecting normalization principles: the nature of asymmetry of separate indicators, their contribution to the final assessment and interpretation of the totals. While searching for an optimum normalization principle, three most widespread methods were applied: ranking, linear scaling and logarithmic scaling. Logarithmic scaling proved to be the most acceptable method for the integral index. For the first time suggested for assessment of territorial differences of anthropogenic impact in the Russian regions, this method not only removes excessive differences of indicators, but also adequately reflects the increasing pollution level parallel to the growing integral index (Bityukova and Kirillov 2011).

$$X = \frac{\log_{X_{\min}} X_i}{\log_{X_{\min}} X_{\max}} \tag{19.1}$$

x_i—index value in i region.

Stage 4. Integral index of anthropogenic impact (IAI) is calculated through the aggregation of group average values for the above-mentioned normalized indicators according to the Formula:

$$\begin{aligned} \mathrm{IAI} = & \left(\frac{A_1 + A_2}{2}\right) + \left(\frac{B_1 + B_2}{2}\right) + \left(\frac{C_1 + C_2}{2}\right) \\ & + \left(\frac{D_1 + D_2 + D_3 + D_4 + D_5 + D_6}{6}\right) \\ & + \left(\frac{E_1 + E_2 + E_3}{3}\right) + \left(\frac{F_1 + F_2 + F_3 + F_4 + F_5}{5}\right) \end{aligned} \tag{19.2}$$

In order to make the assessment more objective, two options of calculation were applied to the integral index: the index of the power of anthropogenic influence (PAI) based on absolute measures and the index of the intensity of anthropogenic influence (IAI) based on relative measures. Use of both absolute and relative measures is especially important taking into account large regions in the studied countries. Application of absolute or relative measures depends on the research objective. If it is focused on the impact on the natural landscape or population health, absolute values are more justified as the whole volume of pollutants caused by a source spreads in the landscape and influences the population.

19.3 Integral Assessment of Anthropogenic Impact

Integral IAIs were calculated for four time periods: 1990—the beginning of the transition period, 1998—the end of crisis, 2003—the end of the first stage of compensatory growth and 2014—the most recent available statistics. In general, the results have demonstrated the adequacy of the developed technique for identification of territorial differentiation of environmental conditions in large countries.

The total index has shown a high, though reducing degree of interregional distinctions in terms of the intensity of environmental pressure on the territory of the two countries (from 3.6 to 1.7 times during 1990–2016). Various changing trends in individual factors have also determined the dynamics of the complex regional index after the collapse of the Soviet Union.

Decreasing AI has become an underlying trend, and internal fluctuations corresponded to the dynamics of economic indicators. At the same time, in the early 2000s, during the maximum increase in oil prices, pollution volumes moved to the extracting regions with low population density. As the result, an export-oriented raw zone Taimyr–Yamal–Urals that concentrates over 55% of industrial production in the country and 70% of air emissions is distinguished based on the environmental pollution (especially, air emissions related to the combustion of associated petroleum gas). Unlike the majority of old developed regions where pollution is localized in the cities, the impact is dispersed across the territory in newly developed extracting regions.

Use of pollution density indices (as compared to absolute values of volumes) moves three types of regions down in rankings, both in Russia, and in Kazakhstan. First, large regions with an intense, but rather localized AI (Krasnoyarsk Krai, Tyumen, Karaganda, Arkhangelsk, Rostov, Vologda, Irkutsk), second, regions with large rivers under serious anthropogenic pressure, but with low density of impact (Novosibirsk, Omsk, Perm) and third, strong agro-industrial regions (Stavropol Krai, Altai Krai, Kostanay) have seen an advancing production decline in the leading agricultural regions with the most intensive agriculture since the 1990s. On the contrary, industrial regions with localized impact (Tula, Lipetsk, Belgorod) and especially federal cities of the Russian Federation and cities of republican status rise in ranking. Agricultural pressure exacerbates ecological stress in the Black Earth and southern regions of the Russian Federation, as well as in traditionally agrarian regions of Kazakhstan—North Kazakhstan and Akmola.

However, centres of ecological stress, unevenly spread across the territory of both Russia and Kazakhstan, are the same for any assessment method. These are the Chelyabinsk, Sverdlovsk, Kemerovo, Moscow and Bryansk regions. All Russian macroregions (Fig. 19.1) have regions with serious anthropogenic impact that distinguishes them from others. Old developed regions are characterized with a more evenly distributed impact density which is more localized beyond the Urals. In general, regional differences are rather levelled.

In Kazakhstan, the centre of ecological stress, Pavlodar-Karagandy-East Kazakhstan area, has actually begun to smear during the considered period due to decreasing impact in the East Kazakhstan region and new centres of ecological stress: oil-extracting West and densely populated South of Kazakhstan.

Despite considerable economic shifts and transition to a new type of economy, the inherited factors still play a crucial role in the ranking of regions. Almost all regions with a very high level of anthropogenic impact have large industries since the USSR that determine specialization of the region. The Chelyabinsk and Sverdlovsk regions and metallurgical and industrial centres of the European part of Russia, i.e. the Lipetsk, Vologda, Tula regions, are the centre of regions with the greatest intensity of anthropogenic impact. Capital regions are an important part of this centre: the Moscow region suffers additional pressure due to an inflow of population and sources of impact, while the Leningrad region has high rates of IAI mostly owing to the industrial development. Radioactive pollution, though having

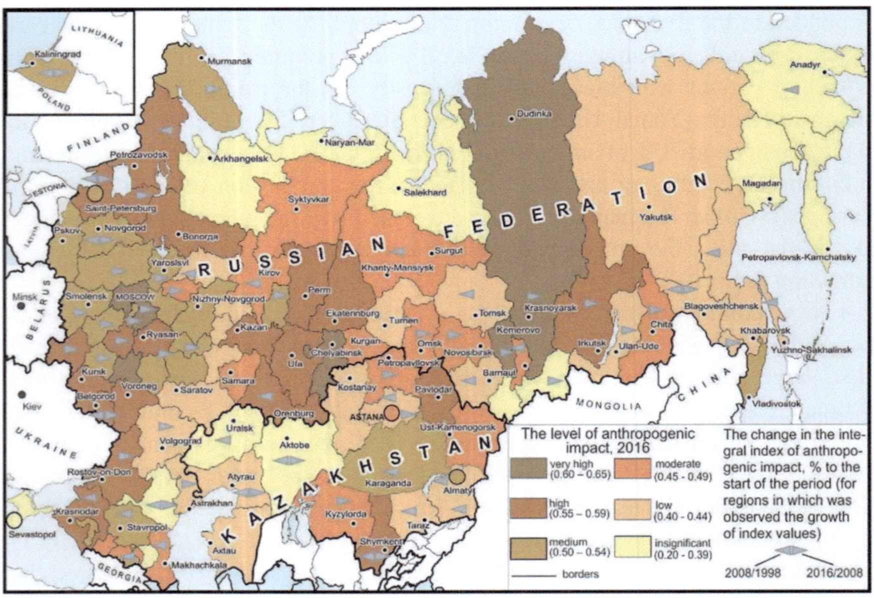

Fig. 19.1 Changes in the integral index of anthropogenic impact in 1998–2016

reduced considerably, is still noticeable in the zones of accidents of the previous period (Bryansk and other regions affected by the Chernobyl accident and the Chelyabinsk region that still bears the footprint of the accident on the Mayak Production Association).

19.4 Stability of Natural Landscapes and Regional Differences in Environmental Stress

Environmental stress occurs when the level of anthropogenic impact exceeds the resilience of landscapes. The ecosystem stability indicator reflects a complex set of environment adaptation responses to anthropogenic impact which allows assimilating pollution, stabilizing climate change, maintaining gas balance in the atmosphere, etc. The integral index used as an integral characteristic of natural ecosystem resilience in this research was calculated as:

- the share of the area of the region with a certain stability score multiplied by the stability index within the area;
- the average stability score was defined based on the maps of the Atlas of natural resources and ecology of Russia: live phytomass (t/ha), annual production of phytomass (t/ha) and structural variety of wildlife sites in different parts of the country (anthropogenic transformation of natural vegetation cover (%)); stability of soil cover (Alyabina and Krechetov 2002; Pikovsky 2002), potential of air pollution (Bezuglaya 1995), surface water pollution (Maslennikova and Skornyakov 2002).

After AI levels and landscape stability levels are summed up according to the matrix, nine aggregated types of regions are formed based on the level of environmental stress. Since the environment stability indicator has a slower response than AI during economic transformations, it is the AI that mainly defines the dynamics of the integral index (Fig. 19.2).

The early 2000s saw an extremely negative tendency when AI was accelerating rapidly (due to growing hydrocarbon extraction) in the most vulnerable natural landscapes: in tundra, including the Arctic tundra. The minimal level of catching and utilization of polluting products in fuel industry posed a serious threat. However, the oil-extracting industry became the leader in the number of investments into crucial nature protection assets, and the situation has changed. In response to air pollution reduction, utilization of associated petroleum gas, decreasing number of oil spills, AI in these territories has taken a downturn in the recent years.

Kazakhstan, in general, possesses a much greater resilience of landscapes than the considerable part of Russia where 63% of the territory is characterized by the lowest pollution resilience. Kazakhstan has 49% of highly stable territories and

Fig. 19.2 Relationship between the level of anthropogenic impact and resilience of natural landscapes in the regions of Russia and Kazakhstan

51% of averagely stable natural landscapes. However, the shifting AI in the Caspian Sea region poses a serious threat.

Environmental rankings of regions based on the complex index are an efficient way of determining the priorities of environmental policy. The integral index provides new insights into the complexity and versatility of the environmental conditions. But a clear understanding of calculation specifics of the complex indicator is necessary to make the ranking a convenient decision-making tool in terms of environmental control and environmental management. The selected AI measurement technique considerably affects the outcome. It is important to use AI density for large countries with isolated intense sources of impact and extensive unaffected territories. Moreover, Kazakhstan that has experienced essential changes of agricultural areas (arable lands, pastures have reduced in the regions by 30–50%) and urban lands during the land reform has the territorial structure of the intensity integral index considerably changed towards increase since 2003, while the indices for 1990 and 1998 proved to be underestimated.

***Key factors maintaining territorial structure of AI areas in Russia and Kazakhstan are*:**

1. Inherited development. The largest AI centres in the territorial structure of Kazakhstan and Russia are still the regions where the industrial base was developed and formed back in the Soviet period: the Urals and industrial regions of the European part, Pavlodar and Karaganda regions. Despite low efficiency

and insignificant economic feasibility related to production and extraction of some minerals, a number of large enterprises, seriously harming the environment, continue to exist. It is all the fault of the existing structure of population settlement (miner's towns) and impossibility to move people from depressive monotowns. This factor is more typical of Russia, than of Kazakhstan.
The influence of the inherited factors is clearly demonstrated in the following areas:

- energy intensity of the leading industries combined with the structure of *fuel balance* in power industry and housing and public utilities in the eastern part of the country. The greatest energy intensity is still typical of large industrial regions with a very high level of environmental stress and of the most environmentally safe regions generally because of the increased coal consumption by municipal services;
- industrial specialization, age and quality of assets. The institutional environment and the policy of companies define the extent of asset modernization, the velocity and sometimes the direction of the process varies. The industrial impact is decreasing, most rapidly in the oil sector, but the industry remains a leading and most dynamic factor influencing the ecological status;
- accumulated radiation pollution is a specific factor in a number of regions, but its role is gradually decreasing.

2. Institutional factor. Despite the stated models of the free market and general privatization, only few foreign companies could establish full control over large industrial enterprises in Russia which has contributed to the stability of the territorial structure of AI areas.

***Factors influencing transformation of AI territorial structure*:**

1. Becoming part of the global resource market. Soaring hydrocarbon prices have caused active development of oil and gas fields in shelf zones, the Arctic zone, the East of Siberia and the Western regions of Kazakhstan. It created new AI centres in the Western Kazakhstan: increase in oil production of the Atyrau region from 2.5 million tons in 1990 up to 31.9 million tons in 2014, development of the Karachaganak oil and gas field in the West Kazakhstan region caused increase in oil production from 4 million tons in 1990 to 13 million tons in 2014.
2. Institutional factor. Emergence of the Western companies in mining and manufacturing industry of Kazakhstan has reduced the level of environmental impact in terms of specific indicators, for example, Kazakhstan has almost twice less emissions than Russia per unit of oil produced. It influences the territorial structure of pollution areas: new industrial centres in Kazakhstan are less significant on a national scale than in Russia (based on relative measures).
3. State environmental policy. A weak state policy in relation to the oil companies has made Khanty-Mansi Autonomous Okrug a "leader" in air pollution for the first time in 2004–2007. Having doubled air emissions, the district "has outrun"

Krasnoyarsk Krai. After signing the resolution "On measures to encourage the reduction in air pollution products flaring gas in flares" in 2009 which sets the target indicator of combustion of associated gas on flares of no more than 5% of the extracted associated gas and provides for an increased payment for its combustion above the limit, large-scale investment has been made. As the result of the utilization coefficient of associated petroleum gas growing from 75.5% in 2011 up to 91.7% in 2013 and 93.2% in 2014, there was a reduction in its combustion in flares and a decrease in pollutant emissions by 21.4%.
4. Reaching the limit of ecological capacity in some regions. Most of regions in Kazakhstan are located within semi-arid and arid zones where ecological limits of water sources, pastures and arable lands are almost reached. When using relative measures, the Mangystau and South Kazakhstan regions demonstrate very high specific rates of anthropogenic impact on limited water sources and land resources. Therefore, any minimum increase of anthropogenic impact in regions with ecological capacity close to its limits leads to a serious degradation of the environment.
5. Demographic changes as a factor transforming the territorial structure of anthropogenic impact are typical, first of all, of Kazakhstan that has AI increasing in the South due to population growth and falling in the North due to depopulation.

Therefore, integral assessment of environmental conditions in the regions of Russia and Kazakhstan has revealed both similarities caused by the consequences of an accelerated industrialization during the Soviet period and differences in the formation of environmental situation caused by different natural and social and economic conditions. The impossibility of blaming decisions made 50 years ago for all problems becomes clear; the reasons for environmental problems are much deeper. The regions of advancing development that possess competitive advantages have high AI level. Rich resource extracting regions see AI strengthening throughout the whole natural complex, faster than the economy develops. The largest city agglomerations experience the effect of the economy of scale that attracts investment and labour, creating greater transport pressure on infrastructure, waste disposal problems, deforestation, environment degradation and depletion of water resources. The state is responsible for the development of the whole territory, and the measures of regional environmental policy gradually improve the situation.

The solution to modern environmental problems lies in a compromise between the demand for an intense economic development and the need for environment protection. Despite considerable positive changes in the national environmental policy, advance economic development regions possessing competitive advantages still tend to have a high AI level.

References

Alyabina YO, Krechetov PP (2002) Self-cleaning potential of soils. Natural resources and ecology of Russia: federal atlas. M.: NIA-Priroda, pp 90–91. (in Russian: Алябина И.О., Кречетов П.П. Потенциальная способность почв к самоочищению. Атлас “Природные ресурсы и экология России”. М.: НИА-Природа, 2002. - С. 90–91)

ARCADIS (2015) Sustainable cities index balancing the economic, social and environmental needs of the world’s leading cities. https://www.arcadis.com/media/E/F/B/%7BEFB74BBB-D788-42EF-A761-4807D69B6F70%7D9185R_Arcadis_whitepaper_2015.pdf. Accessed on 19 Mar 2019

Baeumler A, Chen M, Dastur A, Zhang Y, Filewood R, Al-Jamal KH, Peterson C, Randale M, Pinnoi N (2009) Sino-Singapore Tianjin Eco-City (SSTEC): a case study of an emerging eco-city in China (English). World Bank, Washington, DC. p 168. http://documents.worldbank.org/curated/en/776301468029076278/Sino-Singapore-Tianjin-Eco-City-SSTEC-a-case-study-of-an-emerging-eco-city-in-China, Accessed on 19 Mar 2019

Bezuglaya EY (1995) Air pollution potential. Environment and population health in Russia: federal atlas M., (in Russian: Безуглая Э.Ю. Потенциал загрязнения атмосферы./ В кн. Атлас “Окружающая среда и здоровье населения России” М., 1995)

Bityukova VR, Kirillov PL (2011) Methods of complex estimation of regional differences in ecological stress in Russia. Reg Res 1:56–69 (in Russian: Битюкова В.Р.,Кириллов П.Л. Методы комплексной оценки региональных различий экологиче-ской напряженности в России// Региональные исследования. 2(32):56–69 (2011)

Dixon J, Xie J (2007) Promoting market-oriented ecological compensation mechanisms: payment for ecosystem services in China (Chinese). China addressing water scarcity from analysis to action policy note. World Bank, Washington, DC. http://documents.worldbank.org/curated/en/399131468029072828/Promoting-market-oriented-ecological-compensation-mechanisms-payment-for-ecosystem-services-in-China. Accessed on Mar 19 2019

Environment Protection in Kazakhstan (2015) Environment protection and sustainable development in Kazakhstan. Committee on statistics. Astana/Almaty, 1995–2015 (online source URL:. http://stat.gov.kz. Accessed on 10 Oct 2018) (in Russian: Охрана окружающей среды и устойчивое развитие Казахстана/ Стат. ежегодник/ Комитет по статистике. – Астана/ Алматы, 1995–2015, эл. версия URL: http://stat.gov.kz. Дата обращения 10.10.2018)

Environment Protection in Russia (2017) Federal state statistics service. Statistics of Russia. Moscow, 1995–2017 (online source URL: http://www.gks.ru accessed on 10.10.2018) (in Russian: Охрана окружающей среды в России (2017). Федеральная служба государственной статистики. Москва, 1995-2017 [Электронный ресурс] URL: http://www.gks.ru (дата обращения 10.10.2018).)

Ewing B, Goldfinger S, Wackernagel M, Stechbart M, Rizk S, Reed A, Kitzes J (2008) The ecological footprint atlas 2008. Global Footprint Network, Oakland. https://www.footprintnetwork.org/content/images/uploads/Ecological_Footprint_Atlas_2008.pdf. Accessed on 19 Mar 2019

Maslennikova VV, Skornyakov VA (2002) Self-cleaning of surface water. Ecological atlas of Russia. M., pp 78–79. (in Russian: Масленникова В.В., Скорняков В.А. Самоочищение поверхностных вод, Экологический атлас России М., 2002, с. 78–79.)

McGrath M (ed) (2014) National civic review. In: Special issue: Twenty-Five years of healthy communities: part 1, vol 102. 4, pp 3–78. https://onlinelibrary.wiley.com/toc/15427811/102/4. Accessed on 19 Mar 2019

OECD (2010) Green growth strategy interim report: implementing our commitment for a sustainable future. Paris, 27–28 May 2010. https://www.oecd.org/greengrowth/45312720.pdf. Accessed on 19 Mar 2019

Pikovsky YI (2002) Ability of soils to self-cleaning from hydrocarbons. Natural resources and ecology of Russia: federal atlas. M.: NIA-Priroda, pp 82–83 (in Russian: Пиковский Ю.И.

Способность почв к самоочищению от углеводородов. Атлас “Природные ресурсы и экология России”. М.: НИА-Природа, 2002, pp 82–83)

Scandinavian Destination (2012) Scandinavian destination sustainability index assessing and reporting on the sustainability performance of Scandinavian major meetings destinations. https://www.iccaworld.org/dbs/scandinavian_chapter/files/pictures/2012%20Scandinavian%20Destination%20Sustainability%20Index3.pdf. Accessed on 19 Mar 2019

The Green City Index (2012) A research project conducted by the economist intelligence unit, sponsored by Siemens. Siemens AG corporate communications and government Affairs, Munich, Germany, p 46. https://www.siemens.com/press/pool/de/events/2012/corporate/2012-06-rio20/gci-report-e.pdf. Accessed on 19 Mar 2019

UNEP (2011) Decoupling natural resource use and environmental impacts from economic growth. A report of the working group on decoupling to the international resource panel. In: Fischer-Kowalski M, Swilling M, von Weizsäcker EU, Ren Y, Moriguchi Y, Crane W, Krausmann F, Eisenmenger N, Giljum S, Hennicke P, Romero Lankao P, Siriban Manalang A, UNEP, p 55. https://wedocs.unep.org/rest/bitstreams/15125/retrieve

Chapter 20
Planning Tools for the Protection of the Natural and Cultural Heritage in the Eastern Mediterranean Area

Maria Gabriella Trovato and Dana Ali

Abstract Rapid changes are happening in the Eastern Mediterranean region under the pressure of urban sprawl, rural abandonment and adoption of different agricultural technics, tourism intensification, and movement of displaced populations between the Mediterranean shores. In view of the risks to cultural and natural heritage, the American University of Beirut (AUB) team at the Landscape department developed a Landscape Risk Assessment Model (RA), and Landscape Decision Support System (LDSS) to take into account the protection of landscapes of particular interest, as well as the rational planning of all the landscapes with special emphasis on the use of natural resources. The paper reports the results of the research conducted during the ENPI EU project MEDSCAPES in developing landscape planning tools for the protection of the natural and cultural heritage in the region. The assessment was applied in the study area of each partner country of the ENPI project, allowing for a better understanding of the implications in land use and conservation decision making.

Keywords Landscape Risk Assessment · Landscape Decision Support System · Landscape planning tools

20.1 Introduction

In recent years, the competition of uses for scarce and highly valuable natural resources and the frequency and severity of natural and technological disasters have increased, and this trend is likely to worsen in the years to come. The documents of the European Conference of Ministers Responsible for Regional/Spatial Planning

M. G. Trovato (✉) · D. Ali
Department of Landscape Design and Ecosystem Management, Faculty of Agricultural and Food Sciences, American University of Beirut, Riad El Solh, Beirut 1107 2020, Lebanon
e-mail: mt63@aub.edu.lb

D. Ali
e-mail: dra10@mail.aub.edu

L. Müller and F. Eulenstein (eds.), *Current Trends in Landscape Research*, Innovations in Landscape Research, https://doi.org/10.1007/978-3-030-30069-2_20

(CEMAT) show that sustainable spatial development ensures the coherence of socioeconomic targets with respect to the territory and its ecological and cultural functions, aiming to enhance the quality of life for current and future generations by the creation of sustainable communities able to manage and use resources efficiently (Petrisor 2014), exploiting the ecological and social potential of the economy for innovation, and safeguarding welfare, environmental protection, and social cohesion (Collignon 2009). The MEDSCAPES project, through the development of a methodology to assess and evaluate the landscape of the Eastern Mediterranean and its threats, aimed to support landscape strategies and planning through the introduction of the concept of landscape as a tool to integrate the different features of the region (the ecological and cultural ones).

In the framework of MEDSCAPES and working on the implementation of the European Landscape Convention (ELC 2000), an experience was conducted at the Landscape Design and Ecosystem Management department (LDEM) at the American University of Beirut (AUB) to develop scientific tools designed for the protection, conservation, and management of the natural and cultural heritage of the Eastern Mediterranean region. The objective was to elaborate a Landscape Risk Assessment Model (LRA) to assess, using GIS, the loss of the landscape value, of the chosen areas in the four partner countries, under the pressure of selected and measured hazards, and to develop a Landscape Spatial Decision Support System (LDSS) to support decision making, and assist stakeholders in comparative assessment and selection of options for change.

This paper describes the methodology that was forged and implemented to analyze the landscape hazards, landscape values, and landscape risk. Moreover, it reports on the interactive web-based interface that was created to allow the users to view the results of the Landscape Character Assessment (LCA) and LRA and to alter scenarios in the LRA pertaining to the intensification of urban sprawl patterns hazard and landscape value (LV). This methodology is built upon a process of analysis and investigation of prevailing studies and RA-DSS models developed and implemented in past years. It relied on the results of the LCA conducted in the four Eastern Mediterranean partner countries of the project (Cyprus, Greece, Jordan, and Lebanon) and takes into account two main components: the landscape value, as a resultant of natural and cultural elements and the severity of risks which cause a threat/pressure for change. Four hazard categories were chosen and assessed: desertification, intensification of urban sprawl patterns, erosion, and forest fire.

This study presents some limitations due to the complexity in comparing and interrelating data from four different countries, the difficulty in acquiring the same type of materials in the four partner countries, and the impossibility of monitoring the changes through time due to the lack of data.

20.2 Materials and Methods

20.2.1 Study Area and Available Data

The study area of the LRA and LDSS model covers the level 1 LCA conducted in the four partner countries of the project (Fig. 20.1). The study area in Lebanon represents 45% of the country, chosen to represent its four distinct geomorphological regions, namely the coastal plain, Lebanon Mountain range, the Bekaa Valley, and the Anti-Lebanon Mountain range.

In Jordan, two study areas were chosen, Mujib and Al-Yarmuk. Mujib is located within Madaba and Kerak governorates and extends from the Jordan Rift Valley (JRV) escarpment in the West, at 420 m below sea level, to the central highlands plateau in the East, at elevations greater than 700 m above sea level. Al-Yarmuk is located in the northwestern part of Jordan, includes large urban centers (Irbid and Al Ramtha), as well as extensive agricultural activities, and hosts areas of unique heritage and historical significance.

Two areas were selected in Greece, Epirus and Lesvos. Epirus is located in the Region of Epirus, a predominantly mountainous area characterized by significant landscape diversity, including cultivated settled areas and grazed pastures. Lesvos is the third largest island in the Aegean Sea, with approximately 41% of olive

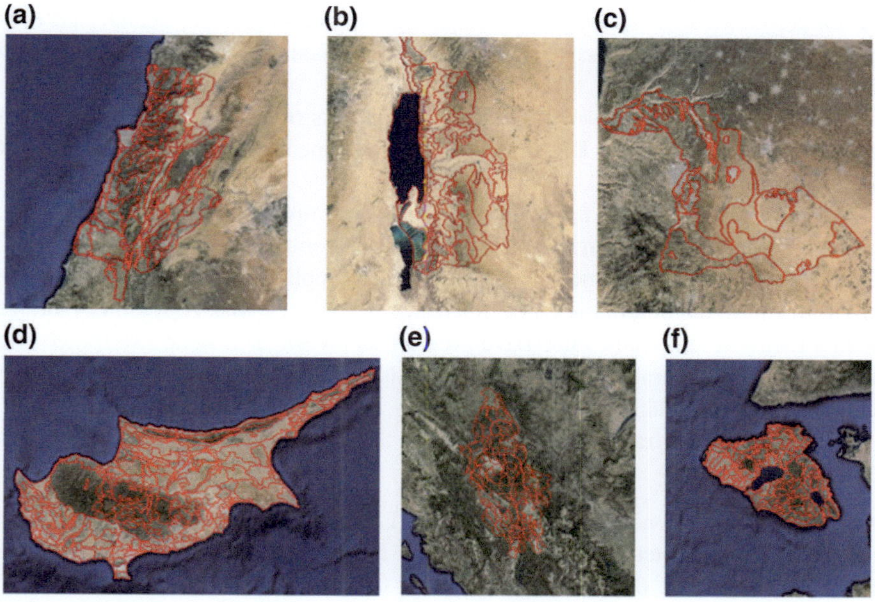

Fig. 20.1 Risk Assessment Model study areas in the four partner countries. **a** Lebanon, Area: 4747 km^2; **b** Jordan-Mujib, Area: 6790 km^2; **c** Jordan-Al-Yarmuk, Area: 3800 km^2; **d** Cyprus, Area: 9251 km^2; **e** Greece-Epirus, Area: 6790 km^2; **f** Greece-Lesvos, Area: 1632 km^2

plantations, 34% with maquis and garrigue, 17% with forest, and 3% with other crops, and finally the remaining land has various uses such as constructions, wetlands.

In Cyprus, the entire island was part of the study. The island is divided into three geomorphological zones: the Trodos Mountain, the Pentadaktylos Range, and the Mesaoria plain.

The four partner countries and six study areas chosen represent a great diversity of geomorphology, land use, settlement patterns, and economy, which has been considered in the development of tools for the study to achieve a comprehensive and representative conceptual model.

20.2.2 Methodology

Disaster management (Bacon 2011) has gained importance recently, and disaster risk reduction has been internationally recognized as a world challenge since the 1992 Earth Summit on Environment and Development in Rio De Janeiro, Brazil, which highlighted the connection between development, environment, and disaster risk reduction (Dransch et al. 2005). In Europe, since the 1980s, the European Commission has been encouraging studies on natural disasters. Under the Seventh Framework Program, a holistic and multidisciplinary approach tackling hazards and the cumulative impacts of multiple hazards, vulnerability, and risk assessment was developed, aimed at reducing and mitigating the environmental, social, and economic effects of natural disasters in an integrated manner (Menoni and Margottini 2011).

The methodology we used is built upon a process of analysis and investigation of prevailing studies and RA-DSS models developed and implemented in past years.

There is much literature today on sustainable risk mitigation involving academic, practitioner, and governmental considerations (Walker et al. 2011). However, while environmental risk assessment has been broadly applied to identify risks on the environment in relation to the types of hazards, Landscape Risk Assessment studies with a focus on landscape value have been a less investigated issue (Martinico and La Rosa 2009). Overall, very few publications connect landscape assessment with land-use decision making.

The LRA-LDSS framework, presented in this paper, incorporates some important key aspects:

- The LRA was conducted at the scale of Landscape Descriptive Units (LDUs) that were delineated by the LCA process. This allowed the tool to visualize the hazard impact on the different Landscape Character Types.
- The LRA was formulated through a risk equation as a result of the relation between the two factors: hazard and landscape value

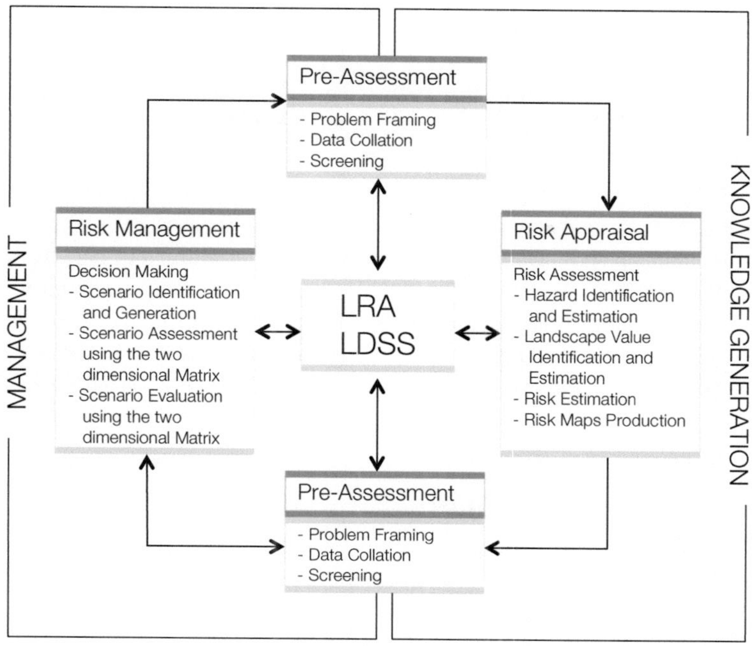

Fig. 20.2 Landscape Risk Assessment conceptual framework

- The framework made use of a two-dimensional matrix as a risk legend, borrowed from the ARMONIA EU project (Armonia 2005) that was reworked and adapted to our study.
- The LRA-LDSS used the scenario-based approach to spatially visualize a series of possible physical landscape transformations due to the increased intensity of a landscape hazard.

Our conceptual framework is the result of the integration of LCA, LRA, and LDSS and describes the operative steps used to derive hazards maps, landscape values maps, risk maps, and scenarios' maps (Fig. 20.2). It is organized as per the following scheme.

20.3 Landscape Risk Assessment (LRA) and Landscape Decision Support System (LDSS)

The Landscape Risk Assessment was conducted in four phases: pre-assessment, risk appraisal, judgment/risk evaluation, and risk management with the Landscape Decision Support System.

20.3.1 Pre-assessment Phase

The pre-assessment phase set out to frame the aims of the model, to verify the availability of data per partner country, to collate LCA results, and to test and assess the model in line with project objectives. The strategy aimed to first build upon the results of the Landscape Character Assessment, where level 1 LCA was incorporated in the hazard analysis, and level 2 LCA assessments were integrated in the landscape value analysis. Data collation followed and continued in parallel with the refinement of the risk assessment model. Discussions on data availability, coherence, and verification took place between the partners and the Lebanese team, in order to achieve a comprehensive and homogenous methodology. The model was further refined due to the limitations of available data among partners.

20.3.2 Risk Appraisal Phase

The risk appraisal phase was rooted on the hazard identification, hazard ranking, and assessment of magnitude per landscape descriptive unit (LDU), as well as landscape value identification and assessment per LDU.

20.3.2.1 Landscape Hazard Identification

In the MEDSCAPES project, landscape hazards were defined as potential threats that are likely to cause damage to the landscape character in the event that they occur (La Rosa and Martinico 2013). They were chosen based on research on the most pressing threats to the Mediterranean region, fieldwork data collection on pressing threats per landscape character type (LCT), discussions with partners, and the ranking of hazards. Four types of quasi-natural and man-made hazards were finally chosen: desertification; erosion; forest fires; and intensification of urban sprawl patterns, reflected in their severity, availability of information, applicability across partner countries, and relevance to the results of LCA.

20.3.2.2 Landscape Hazards Assessment and Ranking

Hazard assessment is an integral component of risk analysis, carried out by processing information from secondary sources (in the forms of maps) and/or through empirical research (direct observation on field). In this study, desertification, erosion, and forest fires were deduced from secondary sources through a qualitative assessment. Intensification of urban sprawl patterns was quantitatively assessed, utilizing both empirical and secondary data sources. Within each partner country,

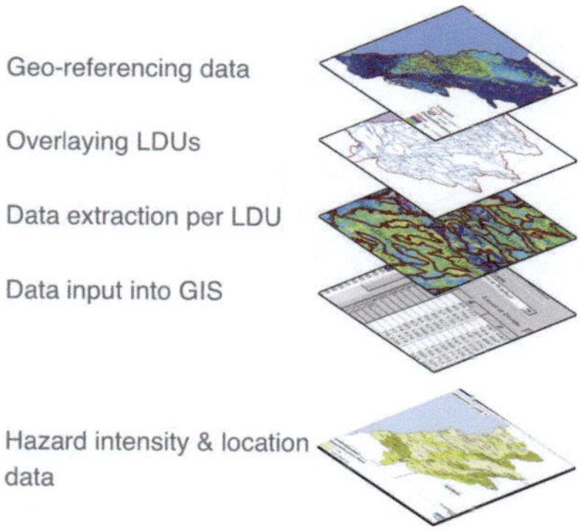

Value ranges used unanimously:

Rank	Value	Value Range
1	Low	0 - 25
2	Low-Medium	25.1 - 50
3	Medium-High	50.1 - 75
4	High	75.1 - 100

Fig. 20.3 Hazard assessment methodology

LDUs formulated the unit of assessment, and the magnitude of risk per each LDU was ranked along four equal intervals (1 as low hazard intensity and 4 as high hazard intensity) Fig. 20.3.

The intensification of urban sprawl patterns was based on a study by La Rosa and Martinico in Sicily, Italy (La Rosa and Martinico 2013) and was assessed using four parameters: existing settlement pattern attribute, landform attribute, geology attribute/slope relation, and the presence of informal settlements. Criteria chosen to weight the four parameters emphasized the importance of existing settlement patterns on future trends (weighted 40%), the relation of geology/slope (weighted 30%), the current trend of settlement of each landform type (weighted 20%), and finally informal settlements (weighted 10%). The assessment, ranking, and indexing methodology are shown in Table 20.1.

20.3.2.3 Landscape Value Identification and Assessment

In the LRA model, landscape value was assessed based on its physical character, adopting LCA level 2 methodology, which accounts for the ecological and cultural quality and integrity of a landscape unit. To reach a comprehensive model that builds upon the LCA, the assessment took into consideration the contemporary methods of evaluating the landscape, the expertise of the involved partner countries (Table 20.2), and the availability of data. The latter were extracted from the LCA process, land cover, and land-use maps, nonetheless satellite imagery, as well as historical/cultural maps, produced by governmental institutions was used. The parameters chosen were ecological integrity, weighted at 80% (naturalness of habitat; habitat continuity; and dominant habitat type) and historical/cultural values, weighted at 20% (archeological sites; heritage/historical entities, trails, centers, and museums), which were both quantitatively assessed.

Table 20.1 Partners' input for hazard ranking

Hazard	Jordan-Yarmuk	Jordan-Mujib	Cyprus	Greece-Lesvos	Greece-Epirus	Lebanon	Total
Intensification of urban sprawl patterns	6	0	5	6	5	6	28
Erosion	5	0	3	5	6	5	24
Desertification	4	4	4	4	4	4	24
Loss of agricultural lands	3	5	2	0	2	3	15
Forest fires	1	0	1	3	1	2	8
Landscape transformation by new planned developments	2	6	6	0	3	1	18
Flood	0	0	0	0	0	0	0

Table 20.2 Partners' input for landscape value assessment criteria

Greece	Cultural values (architecture, traditional settlements, field patterns, terraces)
	Environmental values (vegetation, geomorphology, fauna)
	Historical values (archeological and monumental sites)
	Economic values (agricultural, tourist)
Cyprus	Cultural integrity (change of use, survival of cultural pattern, visual impact of change)
	Ecological integrity (naturalness of LDU, habitat continuity, # of main habitat types, dominant type, intensity of management)
Jordan	Geology
	Natural landforms
	Vegetation
	Fauna environment
	Landscape diversity
	Historical roads + town centers

Ecological Integrity

Ecological integrity assessment was based on the LCA level 2 methodology of the MEDSCAPES project (Fig. 20.4) and took into consideration habitat continuity, dominant habitat type, and the naturalness of the LDU, all weighted equally (33.33%).

Historical/Cultural Values

For the purposes of this study, cultural values were assessed using national maps per partner country and their subsequent ranking of data extracted based on the importance of the site/monument, the dimension of the area/architecture, and on the regional and national significance. Two main components were taken into consideration archeological sites and heritage/historical entities (trails, centers, museums, etc.). The assessment was quantitatively conducted considering the availability of spatial data.

20.3.3 The Judgment/Risk Evaluation Phase

The risk evaluation was conducted by determining the significance of the estimated risks in relation to the loss of the assessed landscape values by applying the MEDSCAPES risk equation:

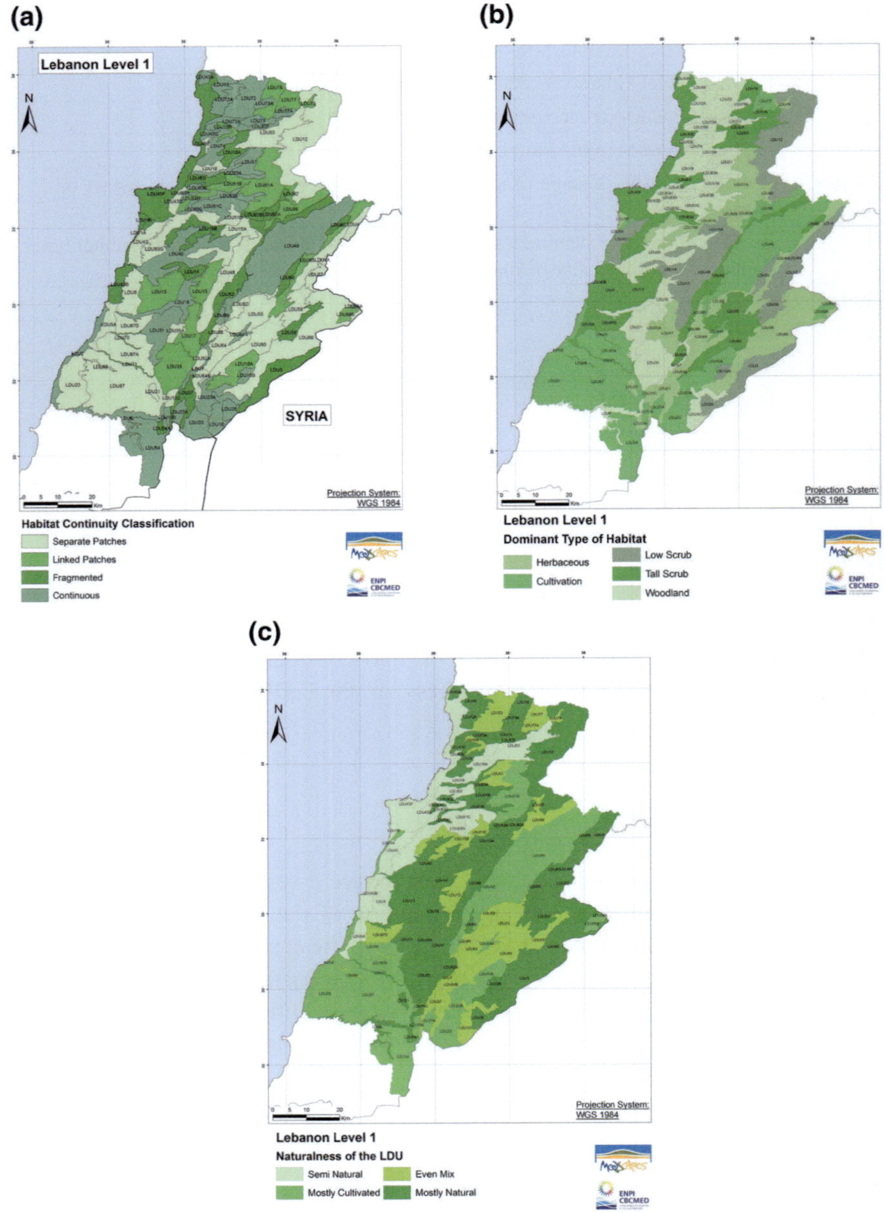

Fig. 20.4 Ecological integrity of the Lebanese study area. **a** Habitat Continuity, **b** Dominant Habitat, **c** Naturalness

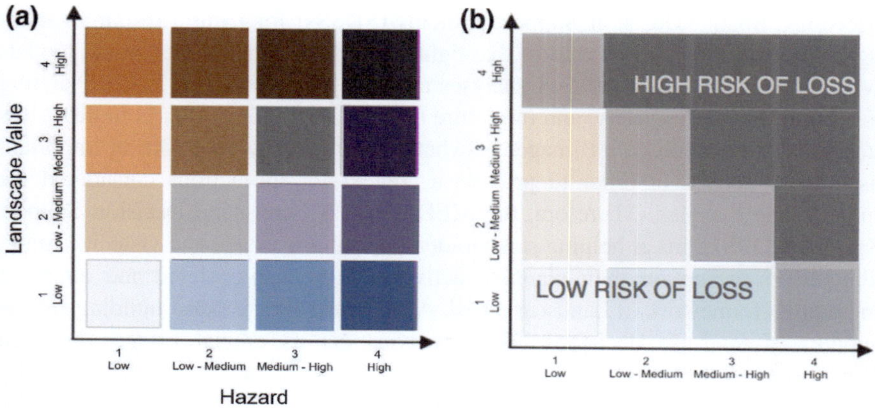

Fig. 20.5 MEDSCAPES Landscape Risk Assessment (LRA) matrix: **a** Color code legend of the LRA; **b** Dark colors = high risk of loss, Light colors = low risk of loss

$$\text{Risk} = \text{Hazard} \times \text{Landscape Value} \tag{1}$$

which was derived from the original equation borrowed from the EU project ARMONIA:

$$\text{Risk} = \text{Hazard} \times \text{Expected losses} \tag{2}$$

where expected losses are understood as vulnerability per value.

In our approach, expressed by the above equation, hazard is a potential threat that is likely to cause damage to the character of the landscape in case it occurs, and landscape value (LV) is the value of the assets in each landscape descriptive unit (LDU), mainly being ecological, natural, and historical/archeological assets. The two components of the equation were visually portrayed using a two-dimensional matrix, adapted from a study by Greiving et al. (2006). In this study, the 4 × 4 matrix with hazard intensity (x-axis) and landscape value (y-axis) yield an integrated risk value. A color-coding method was used for each value in the matrix and serves as legend for the corresponding landscape risk maps generated. Lowest risks in the matrix are closer to the origin, while the highest risks are farthest from the origin (Fig. 20.5).

20.3.4 *The Risk Management Phase: Landscape Decision Support System (LDSS)*

The decision support system is defined as a computer-aided system to support decision making, assisting stakeholders in comparative assessment and selection of options for change. It assists in scenario alteration and generation to support future decisions. Spatial Decision Support Systems (SDSSs) for natural resources management are computer-based tools that tightly integrate decision theory and models with ecological models and GIS analyses and mapping (Rauscher 1999). Planning is a future-oriented activity, and the future being planned for is always more or less uncertain. Nevertheless, an image, of what the future may look like in important aspects, is needed in order to provide a context for the actions considered for inclusion in the plan. Therefore, the MEDSCAPES Landscape Decision Support System (LDSS) aims at helping stakeholders to spatially create and visualize future alternatives to support their planning activities. It is map-centered and offers an integrative framework in Landscape Risk Assessment and scenario building for the support of land-use decisions. The scenarios act as crucial bridges between environmental/landscape science and policy. They influence policymaking by summarizing and synthesizing scientific knowledge in a form that can be used by policymakers to develop policies (2002). The MEDSCAPES Scenario Generator (SG) has been developed with the following capabilities:

- Credible scenarios are translated into changes on the GIS map.
- The SG queries the user for the appropriate information.
- The scenario(/s) are then processed through the model.

Developed as an interactive web-based interface, the LDSS is used for implementing the LRA approach while managing the input data. It allows users to visualize the results of the Landscape Character Assessment, the Landscape Risk Assessment process, with related hazards, landscape values and risk, to create spatially based scenarios for the identification and prioritization of areas and for the definition of adaptation strategies and measures (Fig. 20.6).

The Landscape Decision Support System has the following key features:

- It maps and visualizes information on up to four different hazards and the consequent losses on landscape value at the scale of the LCA level 1 assessment (1:250,000 scale).
- It enables different scenarios to be run which generate information about hazards and landscape loss for all the study areas in the four partner countries and for each of the LDUs, so that different options for mitigating risks or developing land can be compared.
- It enables editing scenarios by altering different indexing weights and LDU ranks.
- It provides a knowledge base on hazards.

20.3.4.1 LDSS Functionality

The LDSS consists of a database, spatial analysis, and bi-dimensional visualization of LCA and risk maps, graphical user interface (GUI), and a simulation model. The database supports cartographic info and permits the users to build spatial relations

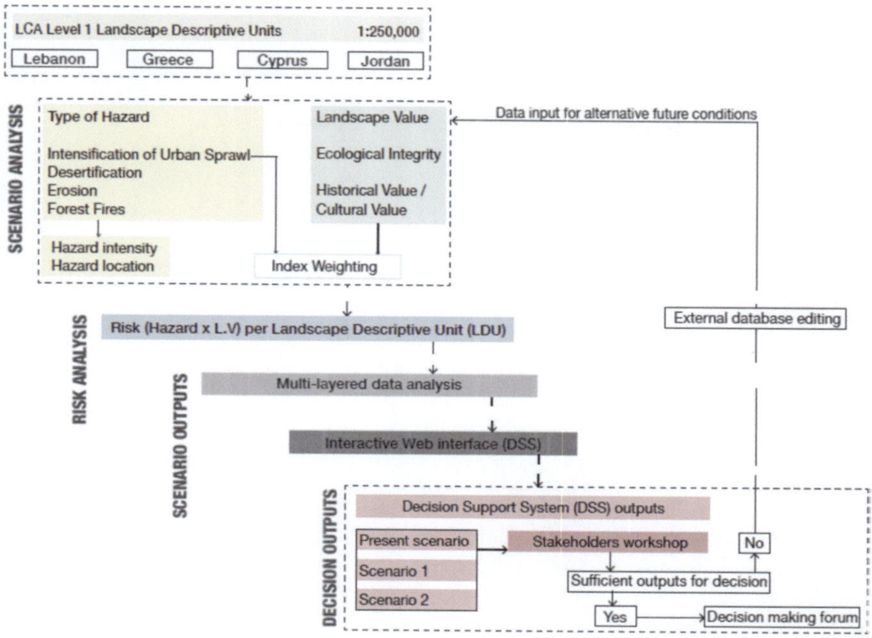

Fig. 20.6 Methodology showing the process of work from LCA to a Landscape Decision Support System passing through the Landscape Risk Assessment

between coordinates and attributes. The spatial analysis and bi-dimensional visualization of risk maps enable the user to interactively explore two-dimensional LCA, hazards, landscape values, and risk maps in the selected areas of the four East Mediterranean partner countries. The GUI facilitates the interaction of the final user with the system. The simulation model permits the user to spatially create scenarios by rank modification (Fig. 20.7) or weight modification. The tool and the GUI closely follow the conceptual model of the LRA, enabling the user to perform the step by step assessment, varying input parameters, testing and comparing different combinations of scenarios. The user's interface is designed in a way to easily allow users to interact, create, and visualize through maps the result(s) of their choice(s).

20.4 Results

The methodology was first tested on Lebanon's study area and was later appropriated unto partner's countries. This allowed for a trial and error period, which was followed by the sharing of the conceptual model with the partners of the project, and a back and forth interaction to reach a satisfactory and representative model. The final outcome reflects the difficulties encountered in the process of work, from unifying data typologies, to ranking and indexing methods when data were either

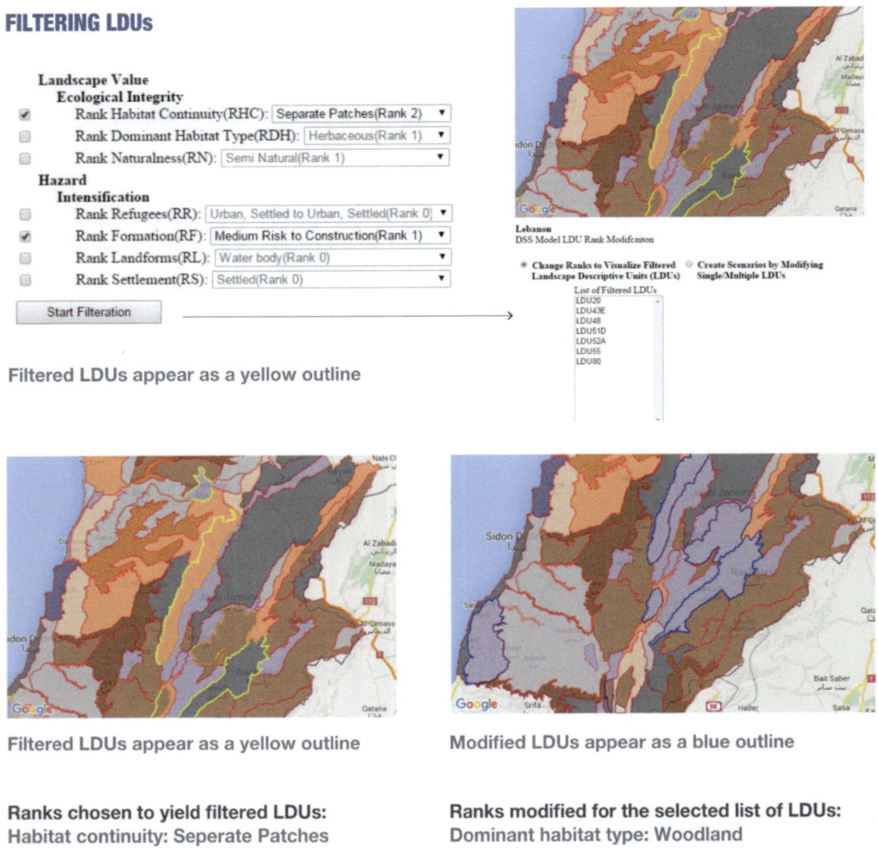

Fig. 20.7 LDSS scenarios create by rank modification

missing or of a different nature (e.g., terminology used, level of detail). This study reveals the necessity to develop systems able to use and process diversified information, in order to produce comparable results. The idea here was to build a model that could be further developed to respond to the differences within each country, for example by taking into account the country's planning rules. In any case, it was a trial to consider the landscape in a model that accounts for hazards, typology, and risk management.

20.4.1 Hazard Maps

Desertification, erosion, forest fire, and intensification of urban sprawl hazard maps were produced for all of Cyprus and for the selected area of Lebanon. In Greece and Jordan, forest fire was not assessed due to impediments in acquisition of the needed information. Furthermore, while the desertification, erosion, and forest fire hazard data are from different sources, and include different parameters of analysis of stressors that were not assessed during the length of the MEDSCAPES project, the final outcome is the result of data collected and compared using the available forms of national information, which were reworked and qualitatively ranked due the uncertainty of the materials (Figs. 20.8 and 20.9).

20.4.2 Landscape Value Maps

At the end of the landscape value assessment, we produced a LV map for all the selected areas for each partner's country, shown in Fig. 20.10 are Lebanon and Cyprus's landscape value maps.

20.4.3 Landscape Risk Maps

The overall landscape risk maps produced within the LRA-LDSS are the result of the MEDSCAPES risk equation that integrates hazard versus landscape value for the LDUs in the four analyzed countries: Cyprus, Jordan, Greece, and Lebanon (Fig. 20.11).

The two-dimensional matrix as a risk legend enabled us to spatially represent the results of our study. Furthermore, it plays an important role in the evaluation process of the effects of the risk in relation to the loss of landscape character of the affected area. For each chosen hazard, it was possible to estimate and judge the risk of a reduction in LV that is directly proportional to the intensity of the hazard. The upper right part of matrix, highlighted as 'high risk of loss,' shows LDUs that are subjected to a greater risk. It means that they are the areas of particular interest for protection/conservation land-use decisions and indicate a more pressing need of attention. In Lebanon, the risk of erosion, for example, is shown in the matrix below, where 57% of LDUs lie within the high risk of loss zone, 31% lie within the medium risk of loss zone, and 9% lie within the low risk of loss zone (Fig. 20.12).

20.4.4 Landscape Decision Support System

The LDSS framework, through weighing and evaluating alternatives, helps decision makers in conceptualizing alternatives, and therefore, in taking the related decisions for future planning. It is conceived in a way to be further upgraded integrating up to date information to produce more credible scenarios. Once an alternative is selected

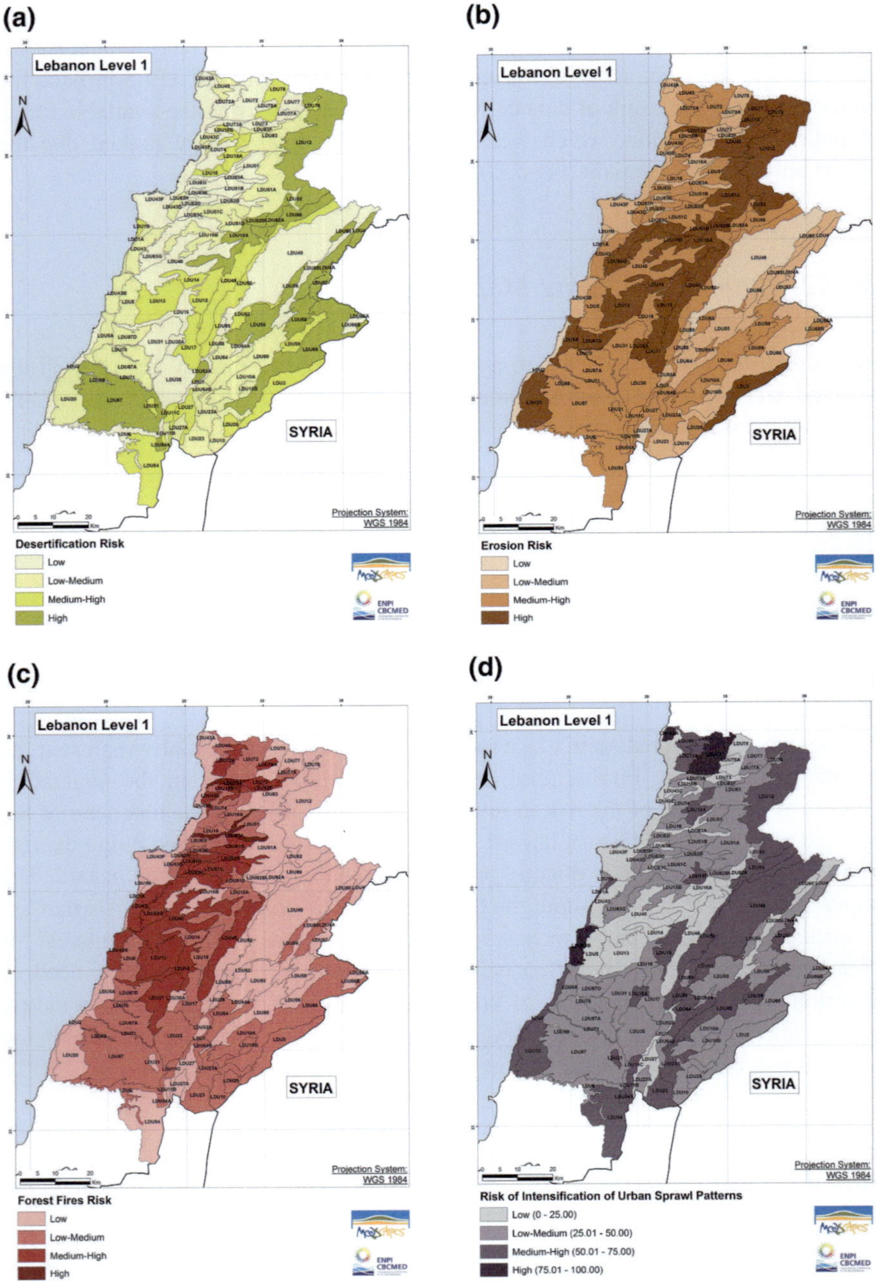

Fig. 20.8 **a** Desertification, **b** Erosion, **c** Forest Fires, and **d** Intensification of urban sprawl hazard maps for the LDUs in Lebanon

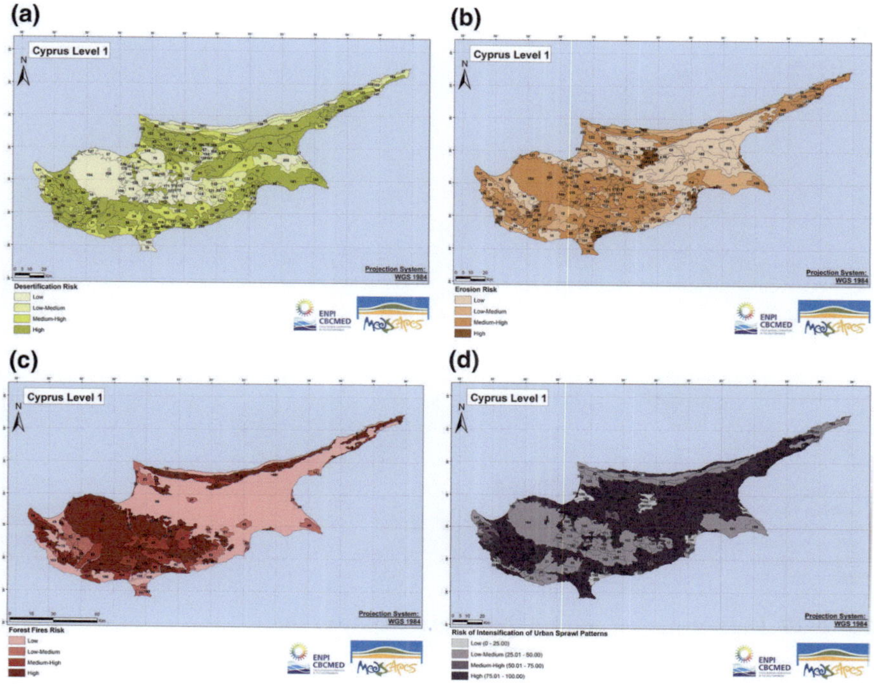

Fig. 20.9 **a** Desertification, **b** Erosion, **c** Forest Fires, and **d** Intensification of urban sprawl hazard maps for the LDUs in Cyprus

and implemented, the results could be compared with the produced scenarios and then a revaluations of criteria has to be done. This tool was built with the aim of spreading awareness on landscape-related topics by visualizing through our user's interface the effects that today's actions could produce in our territories in the East Mediterranean. Since its inception, through workshop and seminars training, we are introducing the audience to the interface and letting them visualize, for example, the consequences that the spread of urbanization could have on the future development and, in particular, on the landscape seen as a human-ecologic system, which can offer a wide range of benefits with positive effects on the quality of life of the local community. On the other hand, the LDSS foresees the user's interaction as manipulation of rank and weight of indicators with particular emphasis on intensification of urban sprawl and ecological integrity. The user-friendly interface allows multiple users to provide input and generate real-time output to support negotiated spatial decisions (Arciniegas and Janssen 2012).

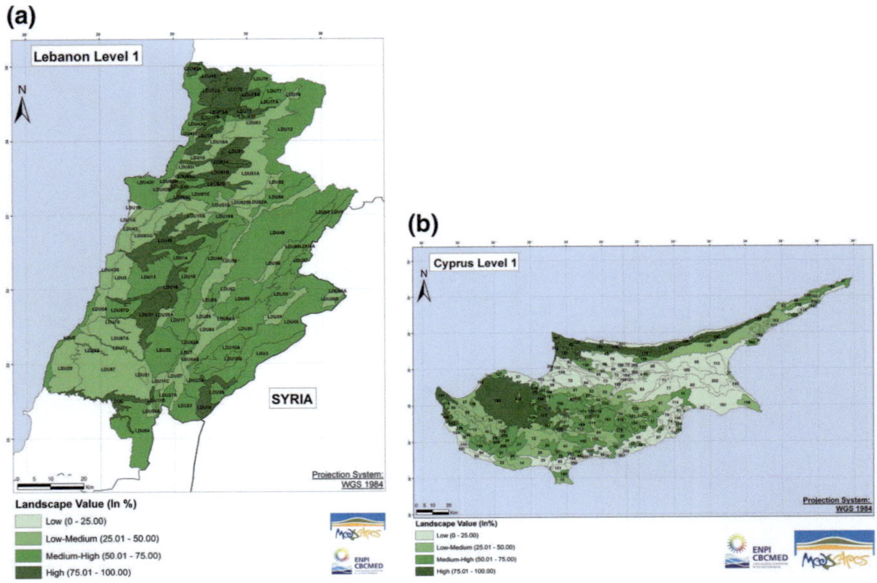

Fig. 20.10 Landscape value map in Lebanon (**a**) and Cyprus (**b**)

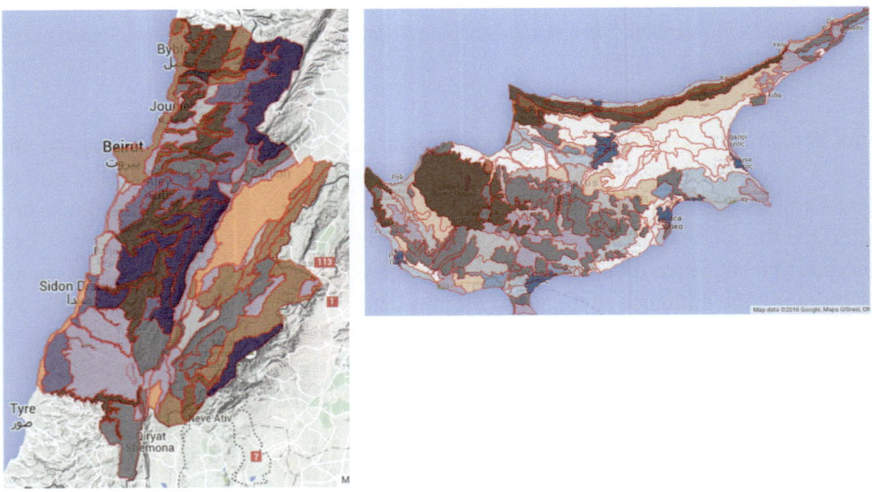

Fig. 20.11 Landscape Risk Assessment (LRA) maps for Lebanon and Cyprus, depicting erosion × landscape value

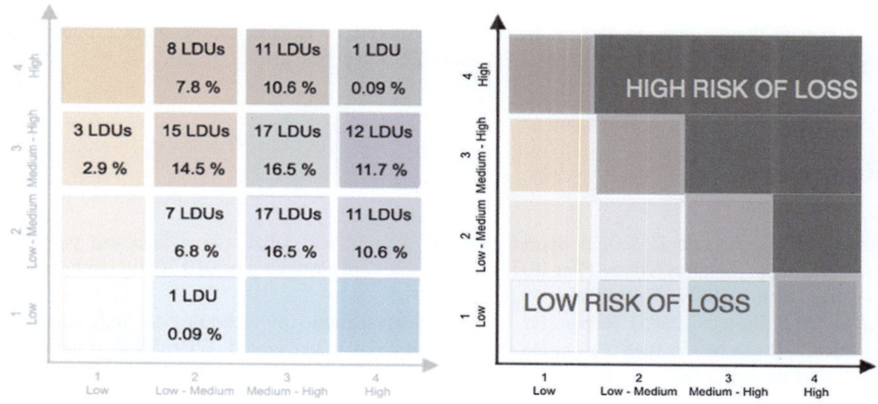

Fig. 20.12 Left: Matrix showing risk of erosion × landscape value, with the number and percentage of LDUs per value of risk of loss in Lebanon. Right: color-coded LRA matrix serving as a map legend

20.5 Conclusions

This study, conducted at the regional scale of the Eastern Mediterranean basin, is an important attempt to develop tools, the LRA, and LDSS to help stakeholders in the process of landscape decision making for the protection of the natural and cultural heritage of the region. The methodology focuses mainly on the physical components of the landscape that were evaluated under the landscape ecology lens with the incorporation of few cultural assets. Though limited, these assets were important in portraying a more thorough picture of the Mediterranean landscape character and of the threats to which it is subjected. Furthermore, the methodology and framework we developed are playing an important and strategic role in advocating the importance of the landscape in the planning process, and in spreading awareness on the protection and management of natural and cultural heritage in the Eastern Mediterranean region.

This was a means of assisting administrators/stakeholders in the planning process through scenarios that were created by modifying weight and rank of the indicators. The generation and selection of scenarios advocate for a bottom-up approach to landscape planning that could complement the conventional practice, and it constitutes a step forward into sustainable land-use protection and management.

Acknowledgments The research was conducted with the financial support of the cross-border cooperation 'Mediterranean Sea Basin Programme—ENPI' of the European Union. My gratitude goes to my University, AUB, and my department in particular, LDEM, for the support and help throughout the duration of the project. A special thanks to all the governmental institutions that directly and indirectly contributed to the smooth development of the research by giving access to resources and data necessary to run the project.

A special thanks goes to Jessica Nicolas and Ammar El Halabi for the work produced in analyzing the data and performing the LRA and LDSS experiments.

References

Arciniegas G, Janssen R (2012) Spatial decision support for collaborative land use planning workshops. Landsc Urban Plan 107:332–342 (2012). https://doi.org/10.1016/j.landurbplan.2012.06.004. ISSN: 0169-2046

ARMONIA Project (2005) Report on the European scenario of technological and scientific standards reached in spatial planning versus natural risk management. WP1: State of the art of spatial planning Deliverable 1.1 Armonia Project. http://tupa.gtk.fi/raportti/arkisto/ut_europe_2007_82.pdf. Accessed Mar 19, 2019

Bacon CM (2011) Disaster risk and sustainable development. In: Wisner B, Gaillard JC, Kelman I (eds) The Routledge handbook of hazards and disaster risk reduction. Routledge London, UK. ISBN: 978-0-415-59065-5 (Print)

Collignon P (2009) Situation et défis du monde rural—Le patrimoine et le paysage au cœur des enjeux du développement territorial durable [Status and challenges of the rural world-heritage and landscape as core goals of sustainable territorial development]. Aménage Territ Eur Paysage 88:99–103

Dransch D, Etter J, Walz, U (2005) Maps for natural risk management. In: Proceedings of the 22nd international cartographic conference, La Coruna, Spain, 9–16 July 2005. https://www.researchgate.net/profile/Ulrich_Walz/publication/241432264_Maps_for_natural_risk_management/links/00463528b0e8545eb1000000/Maps-for-natural-risk-management.pdf. Accessed Mar 19, 2019

ELC (2000) European Landscape Convention (2000). https://rm.coe.int/1680080621, Accessed March 19, 2019

Greiving S, Fleischhauer M, Lückenkötter J (2006) A Methodology for an integrated risk assessment of spatially relevant hazards. J Environ Plan Manage 49(1):1–19. https://doi.org/10.1080/09640560500372800

La Rosa D, Martinico F (2013) Assessment of hazards and risks for landscape protection planning in Sicily. J Environ Manage 127(2013):S156–S167. https://doi.org/10.1016/j.jenvman.2012.05.030

Martinico F, La Rosa D (2009) The use of GIS in landscape protection plan in Sicily. In: Krek A, Rumor M, Zlatanova S, Fendel E (eds) Urban and regional data management (UDMS) Annual 2009. Taylor & Francis: London, UK. ISBN: 978-0-415-55642-2

Menoni S, Margottini C (2011) Foreword. In: Menoni S, Margottini C (eds) Inside risk: a strategy for sustainable risk mitigation. Springer, Milan, Italy. ISBN: 978-88-470-1842-6

Petrisor A-I (2014) GIS-Based methodology for the analysis of regional landscapes and their changes based on land cover and use: a planning perspective aimed at conserving the natural heritage. In: Cracium C, Bostenaru-Dan M (eds) Planning and designing sustainable and resilient landscapes (Chap. 10). Springer, Amsterdam, Netherlans, Earth Sciences & Geography—Geography Series. ISBN 978-94-017-8535-8

Rauscher HM (1999) Ecosystem management decision support for Federal forest in the United States: a review. For Ecol Manag 114(s 2-3):173–197 https://doi.org/10.1016/s0378-1127(98)00350-8

Walker G, Deeming H, Margottini C, Menoni S (2011) Introduction to sustainable risk mitigation for a more resilient Europe. In: Menoni S, Margottini C (eds) Inside risk: a strategy for sustainable risk mitigation. Springer, Milan, Italy. ISBN: 978-88-470-1842-6

Chapter 21
Assessment and Management of Suburbanization Pressure on Landscape in the Munich Region

Johannes Gnaedinger, Cristina Mattos, Tanja Fugiel and Joerg Schaller

Abstract In growing cities and suburban regions, availability of resources and landscape quality is becoming increasingly scarce. We present an analytical exploration of the remaining suitabilities and of the spatial boundaries for further settlement and infrastructure development in the highly dynamic and densely populated region of Munich in southern Germany. We processed existing geodata, land use data and environmental data on the regional scale to determine the remaining potentials and limitations for further settlement development. In doing so, gradients of high to low suitability as well as areas of strict exclusion were distinguished. These generalized and region-wide results were applied and assessed in detail in a subregional development process. There, cross-disciplinary work and inter-municipal coordination were fundamental prerequisites to substantially foster the steering of land use and the avoidance of unstructured growth, resource exploitation and landscape degradation. The political will for landscape preservation strategies and the courage for increasing cooperation in spatial development are the fundamental prerequisites from the municipal to the state level. In the past, several associations for landscape protection were established in the Munich region. Facing the ongoing strong settlement growth, decisive collaboration, enhancement and extension of those activities and initiatives will be required. To support such activities, the above-mentioned spatial analyses and scenarios in inter-municipal processes, projects and regional spatial planning have to be applied and deepened in order to consolidate these processes instead of remaining or falling back into local individualisms. Those are by no means suitable to master the spatial and environmental challenges of today.

Keywords GIS · Inter-municipal cooperation · Landscape quality · Landscape protection · Land use · Regional planning · Spatial development · Spatial planning · Settlement and infrastructure development · Suburbanization · Urban sprawl

J. Gnaedinger (✉) · T. Fugiel · J. Schaller
Prof. Schaller UmweltConsult GmbH, Domagkstrasse 1a, 80807 Munich, Germany
e-mail: j.gnaedinger@psu-schaller.de

C. Mattos · J. Schaller
Esri Deutschland Group GmbH, Ringstrasse 7, 85402 Kranzberg, Germany

L. Müller and F. Eulenstein (eds.), *Current Trends in Landscape Research*,
Innovations in Landscape Research, https://doi.org/10.1007/978-3-030-30069-2_21

21.1 Introduction

The greater Munich Region (or Planning Region 14), comprised of the city of Munich and 186 municipalities in Bavaria, Southern Germany, is one of the fastest growing regions in Germany and in Europe. The stark growth observed, e.g. in the past decade (inhabitants +16%) has resulted in a strong pressure on the region's landscape, natural and cultural resources, intensification of land use and threats to the quality of life. Most of these changes have been driven by population growth and movements, and in consequence settlement and infrastructure developments (Mattos 2007). This tendency will continue at least in the near future, as for the whole region about 350,000 new inhabitants are predicted from 2015 to 2035 (Bayerisches Landesamt fuer Statistik 2019). The scarcity of developable land within the boundaries of the city of Munich has led to extremely high property prices, increased requirements on technical and social infrastructure and restrictions on the ability of low-income populations to settle in Munich (Clark and Moonen 2014). As the city runs out of space for settlement growth, the housing backlog increases, dwelling and settling of companies in the suburban space becomes an alternative, with additional pressure on the regional landscape, increasing travel distances and competition for the most attractive locations.

In this article, we present three approaches related to this complex challenge: (a) the development of GIS suitability models to point out the restrictions as well as to assist in detecting the remaining environmentally compatible settlement potentials, (b) the cross-disciplinary and inter-municipal awareness-rising and steering of further growth and (c) the politically supported establishment and networking of regional landscape initiatives.

21.2 GIS Suitability Models

In face of the Munich Region's planning challenges, intensive spatial data processing is required, where logical sequence of tasks (geoprocessing workflows) must be performed and documented. Geographic information systems (GIS) are essential in this effort, providing the necessary tools for spatial analyses. GIS models facilitate and increase the efficiency of geoprocessing by automating the workflows. *ModelBuilder*, a component of Esri's ArcGIS platform, is a visual programming language for interactively building geoprocessing models on a graphical interface (Allen 2011) (Fig. 21.1). Easy to use, it allows to create and change models by connecting individual geoprocessing tools into a workflow.

Using data created for the Landscape Development Concept of the Munich Region LEK14 (Regierung von Oberbayern 2009; Schaller and Schober 2007), and based on the development goals and principles defined by Bavarian authorities, a framework of GIS models was created to assess settlement suitability and potential future settlement development scenarios in the region, updating and upgrading

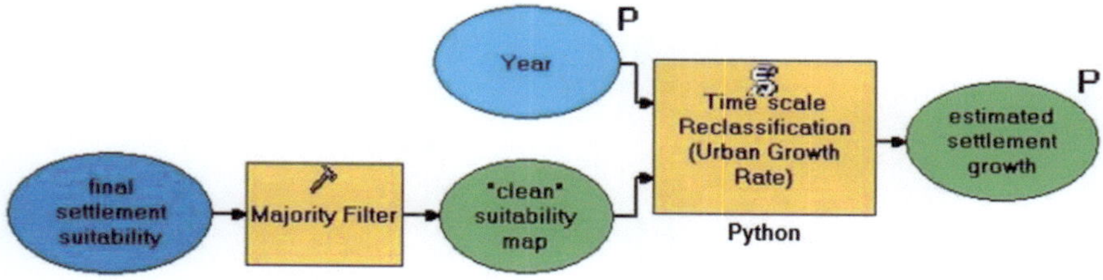

Fig. 21.1 Example of geoprocessing model in *ModelBuilder's* graphical interface

existing preliminary models, testing technologies and methodological approaches that could contribute to the regional planning (Mattos 2007; Schaller and Mattos 2009; Schaller et al. 2009).

Ten settlement development models were designed and implemented in ArcGIS *ModelBuilder*: (1) eight 'individual' suitability models address relevant aspects of settlement development (exclusions, restrictions, physical environment suitability, land cover/land use suitability, socio-economic suitability, proximity to infrastructure, proximity to recreation and scenery beauty) (Fig. 21.2); (2) one main

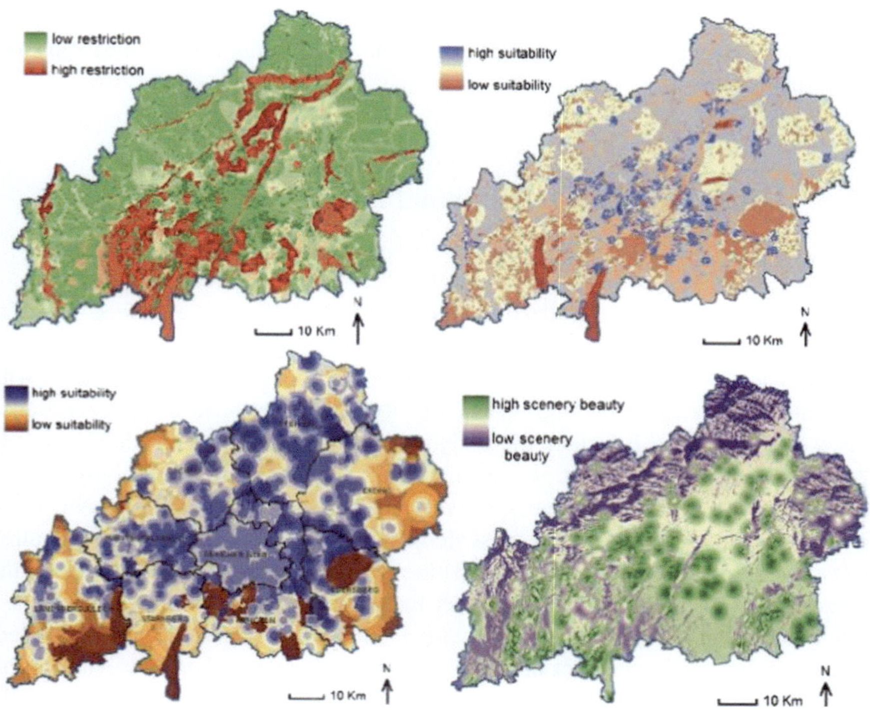

Fig. 21.2 Examples of individual suitability models: restrictions (top, left), land use (top, right), socio-economic (bottom, left) and scenery (bottom, right)

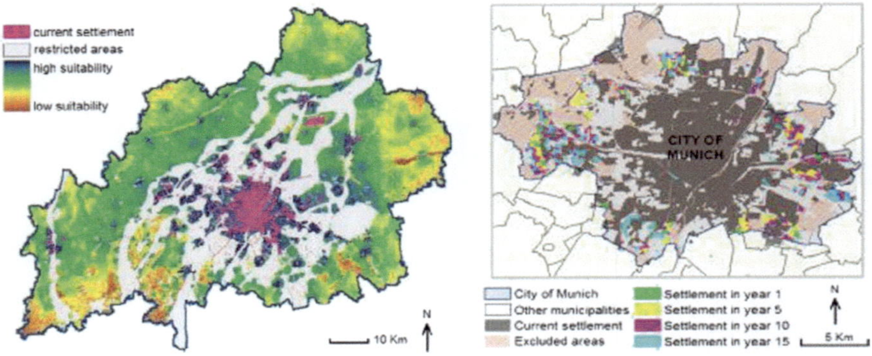

Fig. 21.3 Main suitability model (left) and example of suitability-based spatial allocation of new settlements for a population growth scenario in the city of Munich (right)

suitability model combines and weights the outputs of the eight individual models, adjusts them to each of the four intra-regional groups of municipalities identified with distinct characteristics and development trends and generates a settlement suitability map for the region (Fig. 21.3); (3) a final 'dynamic' model adds a timescale to assess the potential future settlement development in the region based on the suitability mapping and on different prognoses of population growth—'high migration', 'stagnation' and 'airport expansion' (construction of a third runway) scenarios.

21.3 Application in a Subregional Spatial Development Process

These analytical and generalized results were applied and exploited for the north-eastern part of Region 14, the area surrounding the Munich airport (Sweco GmbH et al. 2017). The aim was to determine the effects of the expected growth upon inhabitants, transport and spatial quality and to propose adequate actions. Based on the data of LEK14, on the study presented above, and with additional data sets, e.g. from traffic forecast, we examined (1) how population, economy, workplaces and traffic will develop prospectively until 2030, (2) the resulting challenges for spatial planning in the fields of settlement and transport as well as for nature, landscape and land consumption and (3) the resulting fields of action and possible measures to solve conflicts and to master the current and prospective challenges.

We performed a survey among the participating municipalities, where we asked for their policies in settlement development, for their approaches in solving problems caused by steadily increasing traffic and for their efforts to preserve agricultural land use, nature and landscape.

The analyses of the survey results showed that the municipalities exhibited only few activities to guarantee a long-term sustainable spatial development. Above all, their readiness to collaborate in committed problem-solving was very low in rural parts of the region and somewhat higher in developed and suburbanized areas, probably because of higher pressures as well as the insight that one municipality alone will not master these complex challenges any more.

In a series of feedback workshops and future workshops with mayors, county officials, large companies and superior authorities, a common understanding of the spatial qualities, of the common challenges and main targets for the subregion's future was developed.

As a method for getting a common understanding of the subregion, a set of scenarios (business as usual; rigorous land-saving; development along public transport axes and public transport nodes; Fig. 21.4) was elaborated and discussed. It turned out that there is a common will to find sustainable solutions in the sense of the third scenario, being considered achievable through joint efforts. An urgent task is to overcome the 'egoisms' of the individual municipalities and to establish a stronger common regional planning policy with sound steering functions going beyond the communities' administrative boundaries, rather than a regional plan that predominantly collects the merely individual intentions of the municipalities. Together *fields of action* and finally *recommendations for actions* were prepared by the experts and then discussed, refined and suggested to the political and administrative representatives.

Workshop discussions of the scenarios in Fig. 21.4 showed that scenario 3 was favoured by almost all political representatives. Whilst scenario 1 implies a continuation of planning only on the level of the individual municipalities, and finally badly ordered suburban structures, scenario 2 implies the contrary: any further use of open space for settlement development as well as any extension of roads would immediately be stopped in favour of a concentration of development inside the urban fabric and in favour of a consistent extension of public transport systems instead the motorized individual traffic.

The clear vote for scenario 3 resulted, on the one hand, from safety mindedness (wish for continuation of the high economic prosperity) and, on the other hand, from the insight that current problematic trends in land use and traffic will require changes in political decision processes and consequently in planning. Therefore, scenario 3 was considered a compromise between continued growth (housing, commercial and industry allocation) and a stronger sustainable, resource-saving development, well-ordered spatial structures and landscape preservation.

Which steps are required to put scenario 3 into practice? In the next chapter, we will present how municipalities of the Munich Region tried to preserve landscapes in the past and until today. Many municipalities are engaged in the preservation of exceptional landscapes but at the same time they still consider the 'normal landscapes' just a spatial resource, if not an obstacle for settlement development. Facing the ongoing suburbanization processes and still growing traffic problems, however, we will furthermore argue where the municipalities and the region as a whole

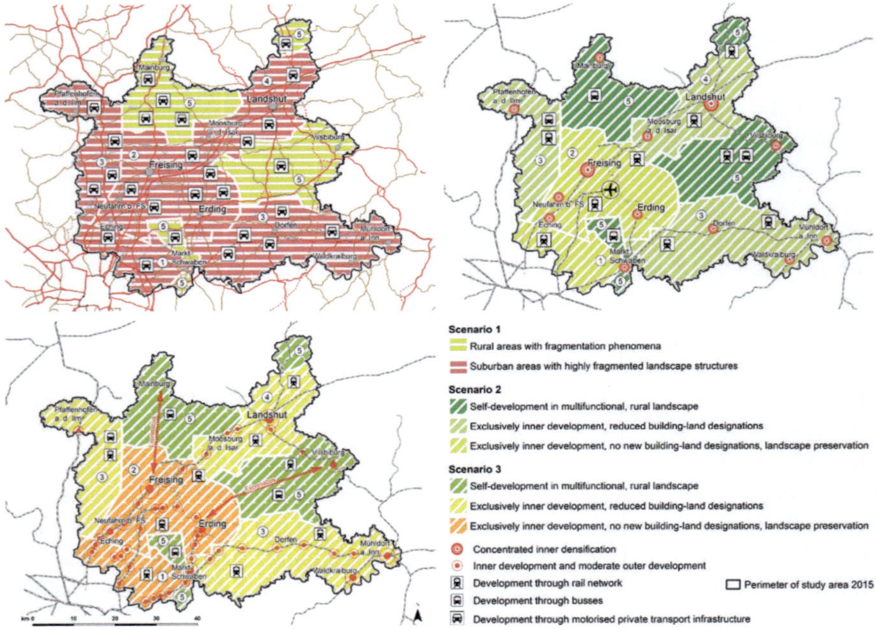

Fig. 21.4 Three future scenarios for a subregion north-east of region 14 Munich. Top left: scenario business as usual (1), with ongoing uncontrolled suburbanization, driven by motorized individual transport. Top right: contrast scenario (2), with rigorous land-saving and change to public transport systems. Bottom left: realistic scenario (3), development along public transport rail axes and rail nodes, with directed and qualified settlement development in subcentres and with preservation of extensive regional open spaces

should go in the long run in order to conduct environmental preservation in the entire remaining open space.

21.4 Regional Strategies for Preservation of Open Spaces and Landscapes

In the past decades, municipalities of different subregions around Munich had founded landscape associations: Isar Valley Association; Heathland Association; Dachau Moss Association; Association for Recreation Areas (Isartalverein 2019; Heideflaechenverein 2019; Erholungsflaechenverein 2019 and Verein Dachauer Moos 2019). The associations' purposes are the protection of outstanding landscapes against pressure from settlement growth, from road construction or from negative effects of recreation, but as well to offer the growing population suitable spaces for recreation. The meaning of those landscape associations is invaluable, as they effectively preserved substantial open spaces from being suburbanized or overexploited in other ways. Doubtlessly the regional tradition of landscape associations is a great success story, a great benefit and a valuable heritage for the region.

However, the pressure in its different manifestations still holds on and even gets stronger. For example, some municipalities feel forced to modify the boundaries of legal nature conservation areas in order to gain more space for the development of residential and business areas. Also, there are long lists of planned new road constructions or road extensions, evidently leading to even more motorized traffic and new settlement allocation in exterior areas. This is why the municipalities as well as the planning region should not just live from their former achievements: the preservation of typical, paradigmatic landscapes through the work of the landscape associations will not be enough anymore. Instead, the municipalities have to go further. They also have to increase the awareness for the 'normal landscapes' or the 'everyday landscapes', because they are particularly exposed to the urban sprawl. The foundation of additional landscape initiatives for normal landscapes however will become as important as those for the exceptional landscapes.

As very promising new examples, two subregions recently made big steps forward as the respective communities started inter-municipal cooperation (RES Regional Development Strategy Fuerstenfeldbruck County; Regional Management Munich South-West). They implemented coordinated land use planning processes by integrating landscape preservation, settlement allocation, infrastructure development and joint social infrastructure. Landscape preservation played a key role when these inter-municipal processes where set up. Now, the processes shall be continued, land use planning shall be done more jointly and greater consideration shall be given to landscape issues.

Together, the areas of the landscape associations and new spatial initiatives form a considerable network around Munich, as Fig. 21.5 shows.

Besides the territories of the landscape associations protected by agreement, there are also formally, respectively, legally protected areas based on nature conservation laws (like Natura 2000/FFH, Nature Conservation Areas/NSG, Landscape Protection Area/LSG) and on the Regional Plan (e.g. Greenways/Gruenzuege), as shown in Fig. 21.6. As mentioned above, the boundaries of at least the LSG do not seem absolutely secure. Similarly, and already more regularly, the Greenways' extensions are getting reduced little by little with each updating, not least because on the official maps they are not delimited by sharp boundaries but by open hatches. We realize that the protective status of restrictive legal categories is not absolutely guaranteed. Even there, open spaces tend to be reduced and landscape qualities as well as ecosystem connections and functions tend to be lost.

The tendencies of disappearance of open spaces depicted so far call for more emphasis especially on the societal and political level. For the realization of the commonly desired scenario 3 (compare Fig. 21.4 and explanations), as well as for effective long-term protection of open spaces, we consider the following political and organizational steps indispensable:

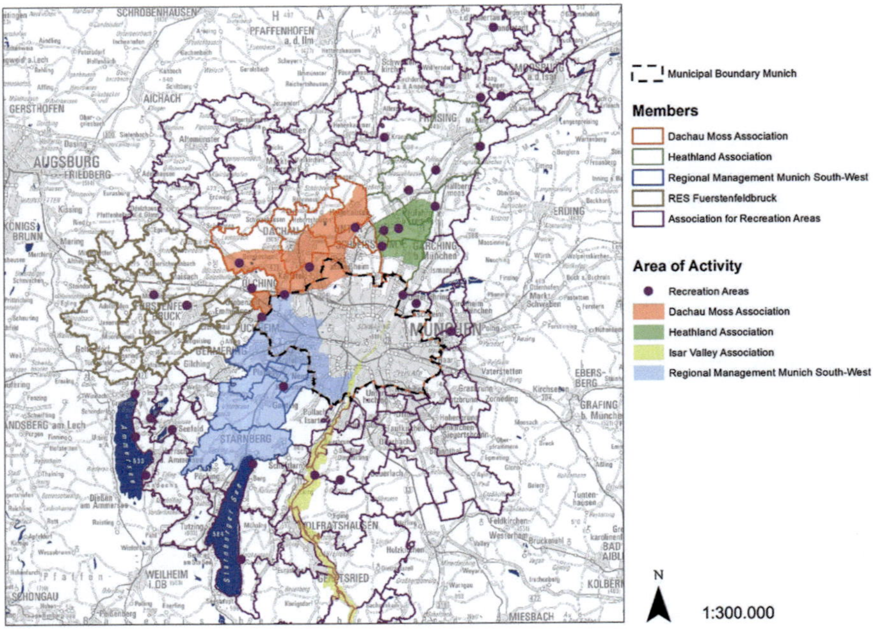

Fig. 21.5 Overview of the territories of the landscape associations as well as of new space-related initiatives (PSU, own cartography)

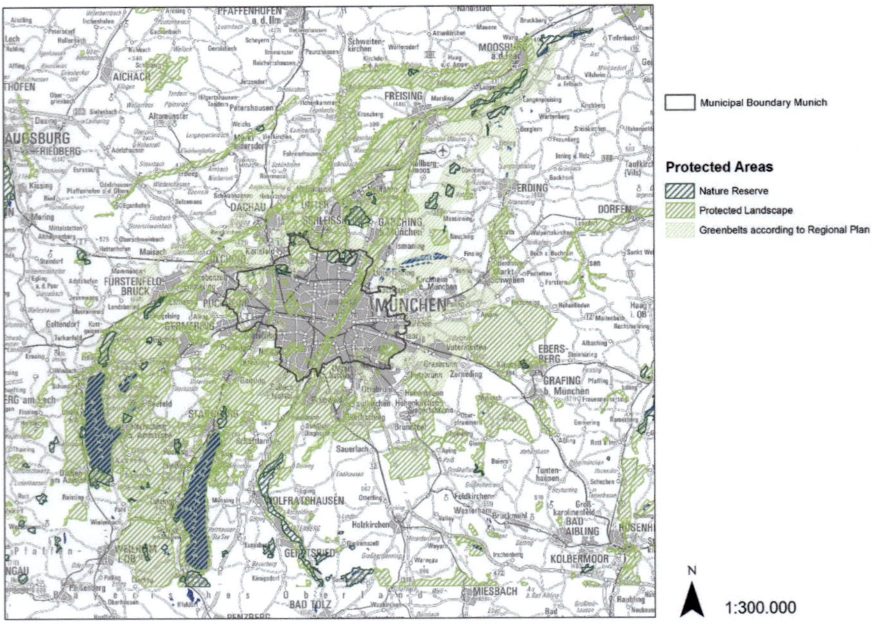

Fig. 21.6 Overview of legally protected areas and regional greenways (PSU, own cartography)

- Instead of just individualistic planning of the communities, inter-municipal planning policies should be established for all relevant types of land use, i.e. integrated spatial planning (like RES Fuerstenfeldbruck and Regional Management South-West). This will minimize competition, strengthen collaboration and synergetic effects among municipalities and will finally lead to better coordinated, more sustainable and clear spatial structures.
- For explicit consideration and awareness of the landscapes, the established landscape associations in the Munich Region should be developed further by connection and extension of their territories, as well as by connection of the associations among each other, and by integrating further regional initiatives and NGOs.
- The landscape associations' contributions, suggestions and demands to land use policy have to have a strong compliance within the planning activities of region and subregions. This will require some organizational and regulatory structures (governance mechanisms).
- In the sense of subsidiarity, governance mechanisms on regional and inter-municipal levels should be channelled and facilitated by the Federal State and the Federal Republic. Furthermore, effective spatial planning instruments along with clearly defined steering competences instead of ongoing deregulation are urgently needed.

To push ahead and to feed the political process, deepening professional studies should be carried out (see also above, GIS Suitability Models), such as spatio-temporal analyses, detection of patterns of land use change, potentials or losses of ecosystem services and capabilities, as well as joint recommendations for actions in landscape networks, land-saving settlement allocation and sustainable mobility systems. All those open spaces worth to be preserved should be clearly delimited in order to finally be saved, managed and enhanced for the next generations.

21.5 Conclusions

1. GIS suitability models help to identify and visualize suitable areas for settlement development, to locate and quantify consequences of alternative developments and to reduce uncertainties about the future.
2. They allow to integrate various information in a systematic way generating quantified, georeferenced, visual outputs.
3. They are invaluable foundations for effective spatial planning, for expert planning and for political decision support.
4. Inter-municipal processes with discussion of alternative development scenarios should be initiated as they can substantially strengthen the municipalities' awareness for joint responsibility and their capacity for action.

5. As a result, the foundation of landscape initiatives, landscape associations or the inter-municipal work on spatial strategies and management structures stresses the public importance and ensures continuous effort in landscape policy.
6. Participatory and integrative processes based on voluntary commitments cannot replace a top-down spatial planning system: in the long run, both approaches (subsidiarity principle) are indispensable for maintaining the quality of space and landscape.

References

Allen DW (2011) Getting to know ArcGIS model Builder. Esri Press, Redlands, p 322. ISBN 9781589482555

Bayerisches Landesamt für Statistik (2019) https://www.statistik.bayern.de. Accessed 05 Jan 2019

Clark G, Moonen T (2014) Munich: a globally fluent metropolitan economy. Brookings Institution & JP Morgan Chase, p 27

Erholungsflaechenverein (2019) Verein zur Sicherstellung ueberoertlicher Erholungsgebiete in den Landkreisen um Muenchen e.V. http://www.erholungsflaechenverein.de. Accessed 05 Jan 2019

Heideflauchenverein (2019) Heideflaechenverein Muenchner Norden e.V. https://heideflaechenverein.de. Accessed 05 Jan 2019

Isartalverein (2019) Isartalverein Muenchen e.V. http://www.isartalverein.de. Accessed 05 Jan 2019

Mattos C (2007) GIS-based modeling of settlement development scenarios in the Munich Region. M.Sc. thesis, Technical University of Munich TUM, Freising, Germany p 146

Regierung von Oberbayern (2009) Landschaftsentwicklungskonzept Region Muenchen (LEK14): Kurzfassung. p 56

Schaller J, Mattos C (2009) ArcGIS modelBuilder applications for landscape development planning in the region of Munich, Bavaria. In: Buhmann E, Pietsch M, Kretzler E (eds) Proceedings of the digital landscape architecture conference DLA 2009, 20–22, Wichmann Verlag, Valletta, Malta, pp 127–136

Schaller J, Schober HM (2007) Landschaftsentwicklungskonzept Region München (LEK14), final project report. Regierung von Oberbayern. Munich, Germany, 612 p

Schaller J, Gehrke T, Strout N (2009) ArcGIS processing models for regional environmental planning in Bavaria. In: Geertman S, Stillwell J (eds). Planning support systems: best practice and new methods. Berlin, Springer, pp 243–264. Print ISBN 978-1-4020-8951-0, Online ISBN 978-1-4020-8952-7

Sweco GmbH, Progonos AG, PSU Prof. Schaller UmweltConsult GmbH, Transver GmbH, ASV Wulfhorst (2017) Fachuebergreifendes Strukturgutachten für den Flughafen Muenchen und sein Umland. Im Auftrag des Freistaat Bayern, vertreten durch das Bayerische Staatsministerium der Finanzen, für Landesentwicklung und Heimat, Oberste Baubehoerde im Bayerischen Staatsministerium des Innern, für Bau und Verkehr, Landkreise Erding und Freising, Flughafen Muenchen GmbH. Preliminary final report, p 322

Verein Dachauer Moos e.V. (2019) https://www.verein-dachauer-moos.de. Accessed 05 Jan 2019

Chapter 22
Gradient Analysis and Surface Metrics for Landscape Ecology

Peter J. Kedron and Amy E. Frazier

Abstract In landscape ecology, spatial pattern analysis is primarily conducted using landscape metrics designed to examine categorical land cover datasets that can be segmented into discrete areas of homogenous coverage. However, this approach to landscape analysis may be inconsistent with many ecological theories which emphasize the continuous, gradient nature of ecological phenomena. To address this challenge, researchers are beginning to develop an alternative gradient surface model, which represents the landscape as a continuous environmental gradient by using ratio rather than categorical data to capture landscape heterogeneity. While the gradient surface model is capable of capturing more landscape heterogeneity, the lack of discrete boundaries between land covers makes it impossible to summarize composition and spatial configuration using conventional landscape metrics. To facilitate spatial pattern analysis of these gradient surfaces, researchers are developing an alternative set of surface metrics to quantify the composition and spatial configuration of features in continuous data formats. Application of surface metrics remains limited in landscape ecology, and it is still unclear how they might be interpreted in many contexts. To facilitate adoption of the gradient surface model and surface metrics, this chapter uses a categorical map of forest land cover in the conterminous USA and a complementary gradient surface model of percent tree canopy cover of the same locations to compute corresponding sets of landscape metrics and surface metrics and identify metric analogues between the two models. Based on this analysis, the ecological meaning of several selected surface metrics is discussed in the context of forest structure. The chapter concludes with key considerations for practitioners and researchers adopting and developing surface metrics.

P. J. Kedron (✉) · A. E. Frazier
School of Geographical Sciences and Urban Planning,
Arizona State University, PO Box 875302, Tempe, AZ 85287-5302, USA
e-mail: peter.kedron@asu.edu

A. E. Frazier
e-mail: amy.frazier@asu.edu

L. Müller and F. Eulenstein (eds.), *Current Trends in Landscape Research*,
Innovations in Landscape Research, https://doi.org/10.1007/978-3-030-30069-2_22

Keywords Landscape ecology · Spatial pattern metrics · Gradient analysis · Surface metrics · Forest

22.1 Introduction

For decades, the "patch-mosaic model" (PMM; Forman 1995), which represents landscapes as mosaics of discrete areas of homogenous land cover, has provided a critical foundation for understanding pattern–process relationships in the landscape. Yet, despite its ubiquity, the PMM neglects the continuous gradient nature of many ecological phenomena and can, therefore, be inconsistent with ecological theory (McGarigal and Cushman 2005; Cushman et al. 2010). As a result, alternative landscape conceptualizations that capture a greater amount of landscape heterogeneity through gradient surface models (GSM) are beginning to emerge (Kupfer 2012; Frazier and Kedron 2017a). However, while these models overcome some of the limitations of patch-based models, the tools for quantifying the two key aspects of landscape structure—composition and configuration—remain underdeveloped.

This chapter is based on a prior study by Kedron et al. (2018) and reviews the state of spatial pattern analysis in both the patch-mosaic and gradient surface paradigms. It also presents an analysis to bridge the methodological divide between the two approaches and facilitate adoption of a gradient perspective in landscape analyses. Specifically, a new brand of metrics for quantifying composition and configuration in landscapes—surface metrics—are evaluated for a set of ecoregions and scales, and their values are compared to traditional landscape metrics to identify analogues between the surface and patch-mosaic paradigms. Based on this analysis, the ecological meaning of several selected surface metrics is discussed in the context of forest structure. The chapter concludes with key considerations for practitioners and researchers adopting surface metrics and continuing their development and use.

22.2 Spatial Pattern Analysis in Landscape Ecology

As a research field, landscape ecology has distinguished itself from other branches of ecology by focusing on the reciprocal relationship between spatial pattern and ecological process (Turner 2005; Kupfer 2011). For nearly 40 years, the spatial pattern analysis portion of landscape ecological research has relied almost exclusively on the development and application of landscape metrics for quantifying patterns in categorical land cover maps. Availability of software platforms such as FRAGSTATS (McGarigal et al. 2012) have made landscape metrics accessible (McGarigal et al. 2009; Uuemaa 2011), allowing them to become so dominant that the current state of knowledge in landscape ecology, including most understanding of pattern–process relationships, rests almost entirely on this approach.

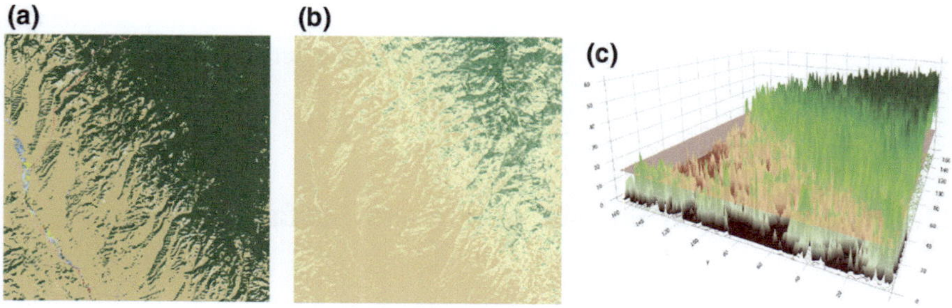

Fig. 22.1 **a** Patch-mosaic model with categorical raster, **b** gradient surface model with continuous raster of percent canopy cover, and **c** 2.5D visualization of continuous raster

However, despite the widespread adoption, the use of patch-based models and landscape metrics is not without critique. Categorical land cover maps collapse intra-pixel heterogeneity, resulting in the loss of potentially relevant ecological information (Fig. 22.1a). This information loss can be exacerbated when data are aggregated to different scales to facilitate analyses (Frazier 2014; Schindler et al. 2015; Bruton et al. 2015). Patch-based models and metrics ignore the continuous gradient nature of the environment (Cushman et al. 2010). Thus, landscape metrics may not measure landscape functionality or provide insight into underlying ecological mechanisms and their relationship to environmental patterns (Kupfer 2012).

To address these issues, new conceptual models, metrics, and software for quantifying landscape patterns are being developed. One promising research stream relies on gradient surface models (GSM) to represent landscape heterogeneity using ratio rather than categorical data (Frazier and Wang 2013). One drawback to these gradient models is that values are continuous, and there are no discrete boundaries between land cover types. Since traditional, patch-based landscape metrics require discrete patches of land cover, they cannot be used to measure composition and configuration in gradient surfaces. Recognizing this limitation, landscape ecologists introduced surface metrics to the field for this purpose (McGarigal et al. 2009).

While surface metrics have been applied in several ecological contexts including flood plains (Scown et al. 2015), habitat connectivity (Moniem and Holland 2013), topography (Zhang et al. 2018), forests (Frazier 2016), and urban areas (Wu et al. 2017), their widespread adoption has been hindered by several factors. First, because many of the surface metrics used in landscape ecology were originally developed for use in mechanical engineering, they lack clear ecological interpretation. Second, like many traditional landscape metrics, surface metrics also suffer from redundancy and correlation because they share common statistical foundations, which can create confusion, misdirected effort, and erroneous conclusions. Finally, limited access to software has slowed metric adoption and corresponding theory development. A critical step in advancing use of the GSM and surface metrics to improve our understanding of pattern process relationships is establishing the ecological meaning of surface metrics across a range of environmental contexts (Kent 2009; Kedron et al. 2018). The study highlighted here provides a step in that direction.

22.3 The Gradient Surface Model and Surface Metrics

Gradient surface models represent the landscape as a continuous environmental gradient by using ratio rather than categorical data to capture a greater degree of landscape heterogeneity (Frazier and Kedron 2017a). The attribute value of each pixel in the continuous raster can be conceptualized and visualized as a 3D surface, where the pixel location is represented by x and y, and the intensity of the value of interest is represented by z. For example, when modeling tree canopy cover using a GSM, the ratio of percent canopy cover for each pixel can be represented in a 2D raster (Fig. 22.1b), or that ratio value can be extruded into 3D to illustrate variations in forest canopy closure across a landscape (Fig. 22.1c). Representing the landscape in these ways clearly demonstrates variation in the intensity of canopy cover. Land cover gradients such as the one shown in Fig. 22.1 can be generated directly from remotely sensed datasets through indices such as the normalized difference vegetation index or procedures such as spectral unmixing, which decompose the spectral signal into its constituent parts and proportions. Gradients can also be generated from categorical maps by statistically combining data through moving windows or other aggregation schemes.

While conceptualizing and modeling the landscape as a gradient better matches the continuous nature of many environmental variables, one major limitation is that these models lack discrete boundaries between land cover classes, making it impossible to summarize composition and spatial configuration using conventional landscape metrics (Frazier 2014; Frazier and Kedron 2017b). Because the majority of landscape metrics cannot be meaningfully applied to GSM landscapes, an alternative set of surface metrics are being adapted and refined to quantify the composition and spatial configuration of features in continuous data formats (McGarigal et al. 2009; Kedron et al. 2018).

Surface metrics have their origins in microscopy and the related field of surface metrology (Barbato et al. 1995) where they were developed to quantify a variety of roughness and texture characteristics, primarily in microscopic surfaces. Dozens of surface metrics have been developed for these purposes, and they generally fall into several categories according to the properties they measure (Table 22.1). For example, amplitude metrics measure variation in surface “height”, where height refers to the magnitude of the value of interest. Several metrics in this category summarize conventional aspects of the statistical distribution of values (e.g., mean, skewness, etc.). These metrics quantify the composition of the landscape variable but do not measure configuration. Applied to an ecological variable such as elevation, amplitude metrics could statistically summarize the distribution of elevation values across the landscape (*Sa*, *Sq*), while also providing insight into local

Table 22.1 Commonly used surface metrics for landscape ecology investigations

Symbol	Name	Description
Amplitude metrics		
Sa	Average roughness	Average absolute deviation of surface heights from mean
Sq	Root mean square roughness	Standard deviation of surface heights
Sp	Maximum peak height	Highest value
Sv	Maximum valley depth	Lowest value
S10z	Ten-point height	Avg. height above mean surface for five highest local maxima plus avg. height below for five lowest local minima
Ssk	Surface skewness	Asymmetry of surface height distribution
Sku	Surface kurtosis	Peakedness of surface distribution
Spatial metrics		
Scl20, *Scl37*	Correlation length 20 and 37%	Horizontal distance of areal autocorrelation function that has fastest decay to 20% and 37%, respectively
Sds	Summit density	Number of local peaks per area
Sfd	Fractal dimension	Calculated for different angles of angular spectrum by analyzing Fourier amplitude spectrum
Srw	Dominant radial wavelength	Dominating wavelength found in radial Fourier spectrum
Srwi	Radial wavelength index	Relative dominance of *Srw* over other radial distances
Std	Dominant texture direction	Angle of dominating texture calculated from Fourier spectrum
Stdi	Texture direction index	Relative dominance of *Std* over other directions of texture
Str20, *Str37*	Texture aspect ratio 20 and 37%	Ratio of fastest to slowest decay to correlation 20% and 37%, respectively, of autocorrelation function
Hybrid metrics		
Sdq	Root mean square slope	Variance in local slope across surface
Sdq6	Area root mean square slope	Similar to *Sdq* but includes more neighbors in slope computation
Sdr	Surface area ratio	Ratio between surface area to area of flat plane with same x,y dimensions
Ssc	Mean summit curvature	Average principal curvature of local maximas on the surface
Abbott curve metrics		
Sbi	Surface bearing index	Ratio of *Sq* to height from top of surface to height at 5% bearing area (z0.05 in Fig. 22.2a)
Sci	Core fluid retention index	Area above Abbott curve in the core zone (refer to Fig. 22.2a)

(continued)

Table 22.1 (continued)

Symbol	Name	Description
SdcX_Y	Height difference of Abbott curve	Height intervals of the Abbott curve: 0–5, 5–10, 10–50, and 50–95
Sk	Core roughness depth	Height difference between the intersection points of the found least mean square line in the Abbott curve
Spk	Reduced peak height	Height of upper left triangle in Abbott curve
Svk	Reduced valley depth	Height of triangle drawn at 100% on Abbott curve
Svi	Valley fluid retention index	Area above Abbott curve in "valley" zone (refer to Fig. 22.2a)

elevation maxima and minima (*S10z*). In this case, "height" refers specifically to topographic height, but in other cases, such as NDVI, "height" would refer to greenness of vegetation.

The metrics in the spatial metrics category measure aspects of spatial configuration of the measured variable. These metrics include summit density (*Sds*), gradient texture and direction (*Std*, *Stdi*), and metrics measuring surface anisotropy (*Scl20*, *Scl37*). For example, when applied to an NDVI surface of vegetation intensity in an agricultural landscape, the texture direction index (*Std*) could provide insight into the directional planting patterns of row crops or tree farms. However, several of these metrics highlight how the microscopy origins of surface metrics complicate ecological interpretation. In particular, *Scl20* and *Scl37* measure the horizontal distance of the areal autocorrelation function that has fastest decay to a 20% and 37% threshold, respectively. The 37% threshold is equal to $1/e$ and is commonly used in microscopy. While e is a universal constant with applications in population ecology, it is not yet clear whether this threshold is the most appropriate for ecological interpretations of surface metrics.

The metrics in the hybrid metrics category measure both amplitude and spatial arrangement and include measures such as the ratio of "vertical" to horizontal slope change (*Sdq* and *Sdq6*) as well as second-derivative measures curvature (*Ssc*) and the surface area ratio (*Sdr*) of an environmental gradient. An ecological application of the hybrid metric *Sdq* (slope) for a forest canopy cover gradient surface could provide an indication of the severity of change in canopy cover in certain areas of the landscape. Areas with abrupt change would have high *Sdq* values while areas with gradual transitions would present low *Sdq* values.

The metrics in the Abbott curve category are all derived from the Abbott curve. The Abbott curve, which is also known as the Abbott-Firestone curve or the bearing area curve, is the cumulative probability density function of the surface profile's height (Fig. 22.2). Abbott curve metrics were originally designed to quantify the bearing properties of industrial surfaces such as how well they would distribute and hold lubricant. These metrics are based on an inversion of the cumulative height distribution histogram and summarize the prevalence of 3D spaces (pits and valleys) across the landscape (*Sci*, *Svi*). For example, in a surface that represents the

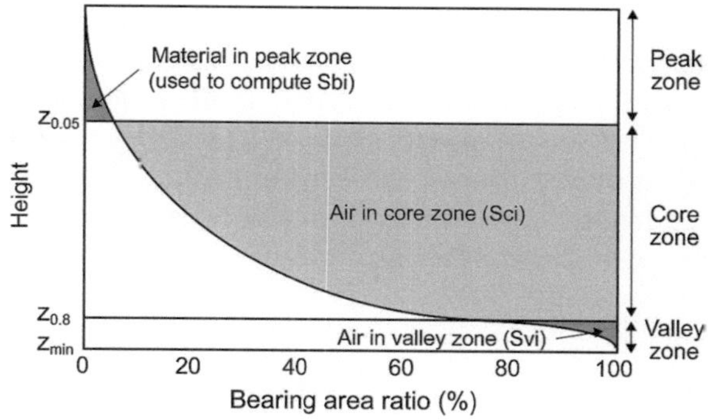

Fig. 22.2 Abbott curve (bearing area curve) illustrating the calculation of surface bearing index, core fluid retention index, and valley fluid retention index. Based on Image Metrology (2017)

percent forest cover in each pixel, the structure of the Abbott curve could indicate whether a landscape has pockets of remnant forest in an otherwise clear agricultural landscape, or, conversely, it could indicate the degree of forest closure after a logging effort.

22.4 Establishing Connections Between Landscape and Surface Metrics

Since surface metrics are only beginning to see widespread application in landscape ecology, there is a limited body of work demonstrating how they might be interpreted in ecological contexts. A small number of studies have established empirical connections between surface metrics and landscape metric through analogues. While identifying analogues is not prerequisite for utilizing surface metrics, it would allow landscape ecologists to draw from the insights developed during 40 years of PMM-based landscape analysis to enrich and expand the emerging GSM paradigm. For example, Fan and Myint (2014) identified relationships between a series of landscape metrics and statistical measures of spatial autocorrelation computed using three vegetation indices. While the spatial autocorrelation measures used in their analysis were not derived from microscopy or surface metrology, they did quantify landscape composition and configuration in a gradient surface, which is the ultimate goal of surface metrics. McGarigal et al. (2009) used a cluster analysis and rank correlation analysis to identify potential analogues between surface metrics and traditional landscape metrics for a range of gradient data types. The authors were able to identify several potential connections, but suggested further research was needed to expand their results. Responding to that call, Kedron et al. (2018) examined potential metric analogues within the forested ecoregions of the USA and expanded the analysis to include the consistency of analogues across spatial scales.

Building on these prior studies, the remainder of this chapter summarizes and expands the analysis performed by Kedron et al. (2018) by highlighting important metric analogues, summarizing lessons from the existing literature, and outlining directions for future research. The research utilizes a patch-mosaic model (categorical map) of forest land cover in the conterminous USA and a complementary gradient surface model of percent tree canopy cover to compute corresponding sets of landscape metrics and surface metrics. The surface metrics are analyzed for clustering across scales. The surface and landscape metrics are then compared through Spearman's rank correlations.

22.5 Spatial Pattern Metric Analogues in the Forested Ecoregions of the Continental United States

22.5.1 Study Region

Four nested levels of ecoregions have been mapped in the USA to facilitate ecosystem management strategies across agencies and organizations (Omernik and Griffith 2014; McMahon et al. 2001). Level I ecoregions are the largest and Level IV are the smallest. This study focuses on Level II ecoregions, specifically eight forested, Level II ecoregions defined by Omernik (1987) (Fig. 22.3). These eight ecoregions were selected because complementary forest data are available in categorical (PMM) and continuous (GSM) formats produced by the Multi-Resolution Land Characteristics Consortium (MRLC). Both datasets are based on Landsat imagery (Coulston et al. 2012, 2013), which facilitates comparison. The categorical land cover raster (hereafter, NLCD) nominally includes 16 classes, which we aggregated into parent classes to create a single forest class. The continuous land cover raster (hereafter, TCC) represents percent tree canopy cover and ranges from 0 to 100%.

EPA Level II Ecoregions	Total no. grid square candidates	Final no. grid squares sampled
5.2 - Mixed Wood Shield	386	99
5.3 – Atlantic Highlands	224	75
6.2 – Western Cordillera	1454	110
8.1 – Mixed Wood Plains	427	113
8.3 – Southeastern US Plains	1662	100
8.4 – Ozark, Ouchita-Appalachian	856	103
8.5 – Mississippi Alluvial & Southeast US Coastal Plains	345	107
13.1 – Upper Gila Mountains	--	--

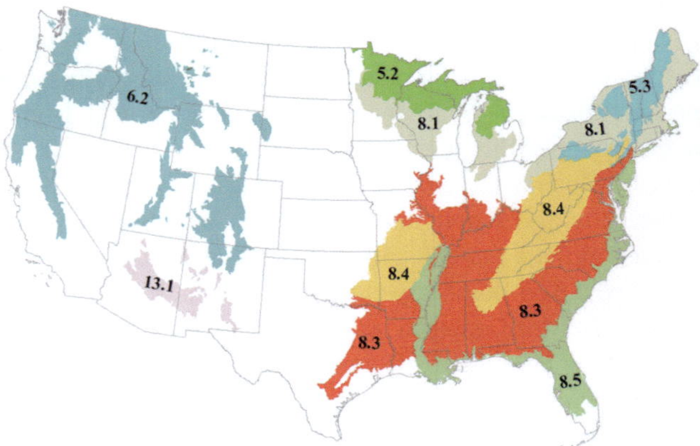

Fig. 22.3 EPA forested Level II ecoregions with the total number of grid candidates and final number of forested landscapes included in analysis

We generated a sampling grid with 20 km^2 sub-regions and randomly selected 125 sub-regions from each of the eight ecoregions for analysis. Prior to sampling, grid squares located within the boundaries of urbanized areas with populations over 500,000 were removed to facilitate a focus on forest cover. In some cases, selected squares did not contain any forest cover, and these were removed prior to analysis (Fig. 22.3). To promote a multi-scale analysis, the NLCD and TCC were aggregated to 60, 120, and 180 m resolutions using majority rules aggregation (with random decision when a tie occurred) for the NLCD and mean statistical aggregation for the TCC. These aggregation operations produced four distinct scales for analysis.

For each sampled square sub-region, at each spatial scale, we computed 28 surface metrics using the TCC data and SPIP™ software (Table 22.1). Following convention, surface metrics were calculated with a correction for surface mean to allow for a relative comparison. We also computed 17 class-level and 7 landscape-level landscape metrics (Table 22.2) using the NLCD data and FRAGSTATS software (McGarigal et al. 2012). Detailed description of the landscape metrics is available at (https://www.umass.edu/landeco/research/fragstats/fragstats.html), while equations and descriptions of the surface metrics are available

Table 22.2 Landscape metrics analyzed as potential analogues of surface metrics

Symbol	Metric	Description
Class-level metrics		
Area/edge metrics		
AREA[a]	Patch area	Area of patch
GYRATE[a]	Radius of gyration	Indicator of patch extent that measures distance between each patch cell and patch centroid
PD	Patch density	Number of patches per unit area
ED	Edge density	Amount of patch edge per unit area
LPI	Largest patch index	Percentage of total landscape comprised by largest patch
LSI	Landscape shape index	Perimeter–area ratio for the landscape
Shape metrics		
PARA[a]	Perimeter–area ratio	Measure of shape complexity that is not standardized to Euclidean geometry
SHAPE[a]	Patch shape complexity	Patch perimeter divided by minimum patch perimeter possible for maximally compact patch of corresponding area
FRAC[a]	Fractal dimension index	Measure of shape complexity across a range of spatial scales
CIRCLE[a]	Circumscribing circle	The smallest circle that can circumscribe a patch, a measure of shape complexity.
CONTIG[a]	Contiguity	Measure of spatial connectedness based on LaGro (1991)

(continued)

Table 22.2 (continued)

Symbol	Metric	Description
Proximity metrics		
PROX[a]	Proximity index	Dimensionless index measuring distance between patches of corresponding patch type considering both size and proximity
ENN[a]	Euclidean nearest neighbor distance	Shortest straight line distance between focal patch and its nearest neighbor of same class
SIMI[a]	Similarity index	Dimensionless index measuring distance between patches of all types and considering both size and proximity
Interspersion metrics		
AI	Aggregation index	Frequency with which different pairs of patch types appear adjacent in the map
Connectivity metrics		
COHESION	Cohesion	Physical connectedness of corresponding patch type
CONNECT	Connectivity index	Number of functional joinings between patches of corresponding patch type using a user-specified distance
Landscape-level metrics		
Interspersion metrics		
CONTAG	Contagion	Measures dispersion and interspersion of patch types
PLADJ	Percent of like adjacencies	Frequency with which different patch types (cells) are located next to each other
Diversity metrics		
PRD	Patch richness density	Number of patch types divided by total landscape area
SHDI	Shannon's diversity index	Measures diversity through the proportion of each land cover type in the landscape
SIDI	Simpson's diversity index	Probability two cells selected at random are different patch types
SHEI	Shannon's evenness index	Measures evenness of patch distribution across the landscape
SIEI	Simpson's evenness index	Measures evenness of patch distribution across the landscape
MSIEI	Modified Simpson's evenness index	Measures evenness of patch distribution across the landscape

Definitions from McGarigal et al. (2012)
[a]Class distribution statistics computed including: mean (MN), area-weighted mean (AM), median (MD) range (RA), standard deviation (SD), and coefficient of variation (CV)

at (http://www.imagemet.com/WebHelp6/Default.htm). Fifteen surface metrics failed to produce valid data and were dropped from subsequent analyses (removal of these metrics is discussed in Sect. 22.5).

22.5.2 Identifying Components of Forest Structure

Components of forested landscape structure were identified using the surface metrics to account for correlation between surface metrics and variation in metric correlation across scales. Component identification followed a two-step process. First, polythetic agglomerative clustering based on a Spearman's rank measure of metric distance was used to identify sets of related metrics in each ecoregion at each scale. Because clustering generated multiple solutions for each ecoregion, scree plots and dendograms were used to identify the optimal number of clusters at each scale. Second, to assess the cohesiveness and universality of surface metrics clustering, the percentage of ecoregions in which metrics clustered was calculated across scales and ecoregions.

Eighteen surface metrics consistently clustered into four or five landscape components across all four scales of analysis. As resolution coarsened, dissimilarity measures between clusters decreased, indicating more overall similarity between metrics following aggregation. At the same time, metric relationships within clusters remained consistent (agglomerative coefficients ranged 0.86–0.93) and closely matched the original rank-based distances observed between surface metrics prior to clustering (cophenetic correlations ranged 0.73–0.89). The set of surface metrics belonging to each cluster also varied minimally across ecoregions and scales (Table 22.3). Collectively, these results suggest that, in the context of forested ecoregions, surface metrics exhibit consistent relationships that are also robust to changes is spatial scale. This finding provides a foundation for the identification of landscape metric analogues and supports the possibility that observed relationships between traditional landscape metrics and surface metrics may be consistent across scales and forested ecoregions.

22.5.3 Establishing Metric Analogues

To aid in the ecological interpretation of surface metrics in the context of forested landscapes, Spearman's rank correlations were calculated for each of the sampled forested landscapes for each spatial scale. When observed, high correlations in different ecoregions and scales indicate potential consistency in the aspects of landscape composition or configuration that surface metrics capture. Consistent correlations also support the transference of ecological theory and interpretation that has been established through PMMs over the past 40 years to GSMs. To

Table 22.3 Surface metric clustering by resolution across the ecoregions

Cluster	Measuring	Surface metrics	Cohesiveness			
			30 m	60 m	120 m	180 m
1	Canopy contiguity	*S10z, Sbi, Sdc0_5*	0.91	0.81	0.81	0.71
2	Canopy distribution and isolation	*Sci, Sp, Sv, Ssk, Sds*	0.77	0.77	0.77	0.71
3a	Canopy roughness, patch complexity and edge	*Sa, Sdr, Sq, Sdq, Sku, Sfd*	0.65	–	–	–
3b		*Sa, Sdr, Sq, Sdq, Sku*	–	0.80	0.80	0.80
4a	Canopy direction and autocorrelation length	*Scl20, Shw, Srwi, Stdi*	0.79	–	–	–
4b		*Scl20, Shw, Srwi, Stdi, Sfd*	–	0.55	–	–
4c		*Scl20, Shw, Srwi, Sfd*	–	–	0.93	0.93

Cohesiveness values indicate how consistently metrics cluster together across ecoregions

simultaneously account for metric redundancy and facilitate interpretation, the correlation structure of metrics is discussed in the context of the five components of landscape structure identified during the analysis of surface metric clustering.

C1: Canopy Contiguity The first component (C1) of forest structure consists of surface metrics measuring the coverage range of areas in the landscape of highest and lowest percent canopy cover. Across ecoregions and scales, this cluster consisted of three metrics (*S10z*, *Sbi*, and *Sdc0-5*), which clustered together 91% of the time at 30 m resolution, 81% of the time at 60 and 120 resolutions, and 71% of the time at the 180 m resolution. The high cohesiveness of this cluster is likely the result of mathematical relationships between the three metrics and the upper limit of percent canopy coverage within the TCC being bound to 100% cover since each metric is computed using pixel measures from the extremes of landscape percent canopy cover. In forested landscapes, these extremes often become homogenous as pixel values in the upper and lower five percent of the percent canopy cover distribution converge. In other ecological contexts, the cohesiveness of these metrics may be reduced. However, observing no relationship is unlikely given their shared mathematical elements.

Regarding analogues with traditional landscape metrics, surface metrics in C1 demonstrated moderate relationships with patch-based measures of landscape shape and contagion with absolute rank correlations ranging from 0.41 to 0.51 (Table 22.4). The highest correlations were observed between *Sdc0-5* and *PLADJ*, *CONTIG, and PARA*, indicating a negative relationship between the range of percent canopy cover in the least forested areas of a landscape and landscape aggregation and like adjacency. Conversely, the surface bearing ratio, *Sbi*, which measures the ratio between variation in percent canopy cover in an entire landscape and the range of percent canopy cover exhibited a positive relationship with the same patch-based measures of landscape aggregation and adjacency. In microscopy, higher values of *Sbi* indicate good bearing properties because they are associated with surfaces with regularized variation in height and fewer extreme outlying peaks or values. In the context of percent forest canopy cover, such surfaces would tend to be characterized by consistently covered areas with fewer areas of minimal coverage, an interpretation reinforced by the observed rank correlations.

Taken together, the cohesiveness and metric analogues of C1 suggest these surface metrics are related to the contiguity and aggregation of a forested landscape and give particular attention to the extremes of the distribution forest canopy cover. However, unlike their patch analogues, the surface metrics of this cluster do not directly measure the spatial configuration of a landscape. Instead, these surface metrics capture characteristics of the distribution of percent canopy cover in a landscape that tend to occur in areas with aggregated and contiguous forest cover.

C2: Canopy Distribution and Isolation The second identified component (C2) of forested landscape structure consists of five surface metrics (*Sci*, *Ssk*, *Sds*, *Sp*, *Sv*) that clustered together over 70% of the time across all four scales of analysis. Two of these five metrics, maximum peak height (*Sp*) and maximum valley depth (*Sv*), did not provide a great amount of information in most forested landscapes. These metrics measure either the maximum or minimum value of pixels in the landscape. In forested ecoregions, it is likely that at there will be at least one 30 m pixel that contains 100% forest cover and at least one that contains 0% forest cover, which makes these metrics largely uninformative.

The three remaining surface metrics in C2 measure different aspects of the distribution of percent canopy cover. Surface skewness (*Ssk*) quantifies the asymmetry of the canopy cover distribution, which provides insight into the percentages of canopy cover most prevalent in the landscape. Core retention index (*Sci*) is a functional parameter that, in microscopy, is typically used to measure the area of a surface acting as a lubricant reserve. In a forested context, this metric could be used as an aggregate indicator of the overall growth potential of areas in the landscape that have some canopy cover but are not as densely covered as those areas in the landscape with high percent canopy cover. Unlike skewness and core retention, surface summit density (*Sds*) is an explicitly spatial measure of the number of local forest canopy cover maxima per unit area. Summit density is a local measure of forest intensity that can be sensitive to variations in percent canopy cover that occur over small distances. For example, in landscapes where forest stands are intermixed with developments or wetland areas that create localized breaks in canopy cover, the resulting pockets of intense canopy coverage would appear as local peaks in the gradient surface raising the summit density value.

Collectively, the surface metrics in C2 characterize distribution of percent canopy cover in forested landscapes. The three metrics of interest—Ssk, Sci, and Sds—all show moderate to strong rank correlations with patch-based measures of

Table 22.4 Spearman's rank correlations for patch-mosaic metric to surface metric pairings for the four components of forested landscape structure

Description	C1: Contiguity			C2: Distribution			C3: Roughness/Complexity					C4: Direction				
	Sbi	Sdc0_5	S10z	Sci	Sds	Ssk	Sa	Sq	Sku	Sdq	Sdr	Scl_20	Srw	Srwi	Sfd	Stdi
Area/edge metrics																
PLAND Forest as percentage of landscape	–	–	0.24	−0.85	0.53	−0.88	–	–	–	–	–	–	–	–	–	–
AREA_MN Mean patch area	–	–	–	−0.77	–	−0.81	–	–	–	–	–	–	–	–	–	–
AREA_CV Coefficient of variation of patch area	–	–	–	–	–	–	0.35	0.29	−0.42	0.30	0.31	–	–	–	–	–
ED Edge Density	–	–	–	–	–	–	0.29	0.22	−0.42	0.58	0.59	−0.31	−0.40	0.24	0.38	0.24
LSI Landscape shape index	–	–	–	–	–	–	–	–	–	0.41	0.42	−0.30	−0.39	0.22	0.38	0.22
Shape metrics																
PARA_AM Area-weighted perimeter area ratio	–	–	–	0.77	–	0.76	–	–	–	–	–	–	–	–	–	–
FRAC_AM Area-weighted mean fractal index	–	–	–	–	–	–	–	–	–	0.38	0.38	–	–	–	–	0.20
CONTIG_AM Area-weighted mean contiguity	0.41	−0.51	–	−0.78	–	−0.76	–	–	–	–	–	–	–	–	–	–

(continued)

Table 22.4 (continued)

Description	C1: Contiguity			C2: Distribution			C3: Roughness/Complexity					C4: Direction				
	Sbi	Sdc0_5	S10z	Sci	Sds	Ssk	Sa	Sq	Sku	Sdq	Sdr	Scl_20	Srw	Srwi	Sfd	Stdi
Interspersion metrics																
PLADJ Percentage of like adjacencies	0.41	−0.51	–	−0.77	–	−0.76	–	–	–	–	–	–	–	–	–	–
CLUMPY Clumpiness index	–	–	–	–	–	–	0.35	0.37	−0.28	–	–	0.52	0.64	−0.41	−0.73	−0.26
Proximity metrics																
ENN_AM Area-weighted mean Euclidean nearest neighbor	–	–	–	0.69	−0.51	0.76	–	–	–	–	–	–	–	–	–	–

patch area, shape, and isolation (Table 22.4). *Sci* and *Ssk* both demonstrated strong negative relationships with forested area measures, *AREA* and *PLAND*, but strong positive relationships with neighbor distance, *ENN_AM*. Increases in the number of pixels in a landscape with low percent canopy cover would correspond to increasing *Ssk* and *Sci*. At the same time, landscapes with more low percent canopy cover pixels will have less forested landscape area, and as those areas split apart greater distances between neighboring forest patches. The splitting of the landscape would increase the perimeter of forest patches while reducing their area, an observation supported by the positive correlation between both metrics and the shape metric *PARA_AM*. This type of landscape would also be likely to be characterized by small *Sds* as local pockets of forest would be farther apart. A conclusion supported by the negative correlation between *Sds* and *ENN_AM*.

C3: Canopy Roughness, Patch Complexity, and Edge The third identified component (C3) of forested landscape structure characterizes variation in percent canopy cover across the landscape. Five surface metrics (*Sa*, *Sq*, *Sku*, *Sdr*, *Sdq*) make up this component. However, only four of those metrics clustered consistently across all four scales of analysis. Kurtosis of the percent canopy cover distribution (*Sku*) was most closely related to *Sa* and *Sq* but only clustered with those metrics in five of the seven ecoregions studied. At the 30 m resolution, C3 also included surface fractal dimension (*Sfd*), which measures the geometric complexity of a surface. However, this metric showed stronger relationships with the surface metrics in the canopy direction and autocorrelation component (C4) at the coarser resolutions. Fractal indices are often scale invariant, which may explain why *Sfd* exhibited inconsistent relationships with other surface metrics more sensitive to changes in scales. Exclusion of *Sfd* from C3 raised cluster cohesiveness from 0.65 to 0.80. For this reason, *Sfd* is excluded from the discussion of C3.

Four of the five surface metrics make up C3 measure variation in the intensity of percent canopy cover across the landscape. Average roughness (*Sa*) and root mean square roughness (*Sq*) both measure mean deviation in percent canopy cover. These two metrics are mathematically related, which suggests that researchers should use discretion when choosing which metrics to include in an analysis and select only one of these to reduce redundancy. Moreover, because *Sa* and *Sq* are incapable of differentiating peaks in percent canopy coverage from valleys in percent canopy coverage, or of differentiating the spacing of differences in canopy cover, we recommend augmenting analysis using these metrics with other surface metrics. Surface area ratio (*Sdr*) is one metric that can be used to further differentiate forested landscapes with similar roughness measures. In microscopy, *Sdr* measures the percentage of additional surface area a surface has when compared to a flat plane. *Sdr* will typically increase with the spatial intricacy of a surface. In the context of forested surfaces, this metric provides insight into the amplitude and spacing of variations in percent canopy cover throughout a landscape. Root mean slope (*Sqd*) is another metric that can be used to differentiate forested landscapes with similar roughness values. Like *Sdr*, *Sdq* is sensitive to amplitude and spacing.

In the context of forested ecoregions, higher values of *Sdq* are likely to be associated with greater differences in percent canopy cover occurring over shorter distances.

The surface metrics in C3 had strong relationships with patch-based measures of edge density, shape, and variation in patch area (Table 22.4). Average roughness (*Sa*) and root mean square roughness (*Sq*) had moderate positive correlations with *AREA_CV* and *CLUMPY*, which suggest that landscapes with greater variation in percent canopy cover tend to also have more aggregated forest patches that vary more widely in size. The surface metrics *Sdq* and *Sdr* correlated highly with patch-based metrics edge density, *ED*, and fractal dimension, *FRAC*. This finding suggests that variation in the slope of percent canopy coverage of a landscape are analogous to edges between patches in the PMM, and that forests with greater roughness tend to have intricately mixed land covers that create complex shapes in the landscape.

C4: Canopy Direction and Autocorrelation Length The final component (C4) of forest structure included surface metrics that measured the direction of the dominant pattern observed in percent canopy coverage (*Stdi*), and the distance between correlated locations in canopy cover landscape (*Scl_20*, *Srwi*). C4 consistently included measures of correlation length, *Scl_20*, and dominant radial wavelength, *Srwi*, but added or dropped fractal dimension, *Sfd*, and dominant texture direction, *Stdi*, depending on scale. At scales where *Stdi* created its own cluster, the remaining surface metrics in C4, including *Sfd*, demonstrated greater cohesiveness (0.93).

Other than *Stdi*, the metrics in C4 showed moderate rank correlations with the patch-based metric *CLUMPY*, which measures class aggregation (Table 22.4). Positive correlation between aggregation and autocorrelation length would be indicative of forests with a small number of forest patches with large areas that also have high percent canopy throughout the entire patch. *Stdi* did not correlate highly with any patch-based metrics, which may be the product of the landscapes under investigation. There may be no clear reason to expect landscapes within forested ecoregion may to exhibit a clear directional pattern in percent canopy coverage. If this is the case, finding correlation with patch-based metrics is unlikely. However, *Stdi* may be useful in the characterization of other landscapes. For example, agricultural landscapes and managed forests may present clear directional characteristics quantifiable using this metric. The surface metrics in C4 appear to have fewer clear PMM analogues than those in the other components of forest structure.

22.6 The Road Ahead

22.6.1 *Removal of Metrics and Scales*

Initially, this investigation included 15 additional surface metrics that were ultimately dropped from the analysis because they failed to return valid results in multiple ecoregions. Considering why these metrics failed to produce usable measures in the context of percent canopy cover of forested areas provides insight into the ecological interpretation of surface metrics and future research directions.

The reasons for surface metric removal varied. First, like *Stdi*, measures of the uniformity of surface texture (*Str20*, *Str37*) may perform poorly in ecological contexts without a dominant directional trend or in contexts where surface variation is minimal. Second, in landscapes characterized by high percent canopy cover in all locations, certain surface metrics (*Sdc5_10*, *Sdc10_50*, *Sdc50_95*) calculated from intervals of the cumulative distribution may not be calculable or practical. In the landscapes examined, *Sdc0_5* (canopy deviation intervals at 0 and 5%) produced acceptable values, but many of the remaining intervals produced zero values. In these types of surfaces, anisotropy may also be minimal or absent, and metrics measuring that characteristic (*Scl37*) may not be reported as well. These findings reinforce the need for practitioners to carefully select the set of surface metrics they will use in a given spatial context.

Similarly, the original analysis of scale conducted for this examination included resolutions up to 480 m. However, analysis was ultimately restricted to resolutions of 180 m and lower because too many surface metrics failed to produce usable values at coarser resolutions. This result is important for researchers in at least three ways. First, it highlights the need for those working with surface metrics to consider whether they can reasonably expect to calculate results at a given resolution. Second, it demonstrates the potential limits to adapting surface metrics for use in ecological contexts. Finally, it suggests that detailed examination of scaling relationships common in the PMM may be difficult to replicate at similar scales in the GSM.

22.6.2 Advice for Practitioners and New Research Directions

Incorporating the GSM and surface metrics into the analysis of landscapes is still maturing, practitioners interested in using or contributing to the development of surface metrics should be mindful of several considerations.

First, not all surface metrics developed for use in industrial applications are appropriate for use in landscape analysis. We highlighted several applications of metrics here for forest investigations, but the usefulness of different metrics will likely vary by application. In addition to some surface metrics failing to compute in certain landscapes, metrics designed to summarize physical features like height, surface area, and void volume may be difficult to translate when the attribute under consideration is non-physical in nature. The attribute examined in this chapter was the percent canopy cover of each individual raster pixel. While this can be visualized and conceptualized as a 3D surface (Fig. 22.1c), for example, spaces between peaks do not have the same interpretations as volumes. For example, in this context, a measure of those spaces, core index *Sci*, could be conceptualized as the closure potential of forest areas. However, this conceptualization makes an assumption about a variety of unobserved factors (e.g., soil type and slope) that may be constraining canopy coverage.

Second, like patch-based landscape metrics, surface metrics demonstrated redundancies that should be accounted for prior to use in analysis. Many surface metrics have very similar mathematical derivations and are designed to quantify related concepts. For example, several of the metrics presented here are derived from the Abbott curve and thus are interdependent. Practitioners should review metrics prior to analysis and attempt to account for potential redundancies to reduce possible biases in subsequent statistical analyses and inferences.

Third, practitioners should give careful consideration to scale, both the resolution (i.e., grain) and spatial extent, of their GSM data. As information about the heterogeneity of the landscape is lost during aggregation, surface metrics become unstable in their characterization of the landscape. In the forested ecoregions examined, many surface metrics could not be computed at resolutions coarser than 180 m. These results suggest that there may be optimal scales at which to use surface metrics in this context.

Fourth, surface metrics should not be considered one-to-one analogues of patch-based landscape metrics. Patch-based landscape metrics quantify either the composition or the configuration of the landscape, with metrics focused on configuration directly considering the spatial arrangement of land cover. Many of the surface metrics analyzed here measure only the composition of the landscape. Because surface metrics were designed to summarize the bearing and fluid retention properties of an entire surface, most metrics focus on the landscape in aggregate. Surface metrics that do directly measure configuration were also the metrics that were most difficult to identify analogues. For example, measures of surface texture direction, slope, and autocorrelation had few robust analogues in the PMM. More work is needed examining the potential meaning of these metrics in different ecological contexts.

Fifth, surface metrics offer a variety of new perspectives that may not be captured by the comparative approach presented here and have yet to be explored in the literature. For example, surface metrics open the possibility of analyzing variation within patches. Because the PMM relies on categorical rasters and patches are defined as contiguous areas of identical categorical classification, this model offer no insight into intra-patch variation. In contrast, in forested landscapes, surface metrics can be used to characterize canopy coverage within patches that were identified by the PMM or alternative threshold approaches. This approach could for example shift how the sub-field understands ecological processes related to functional metrics like core area or edge contrast.

Finally, further work is needed in standardizing and increasing the availability of surface metrics within landscape ecology and related fields. While the mathematical derivations of surface metrics are codified and available through proprietary software packages, few metrics are freely available through programming languages like R or in standalone software programs like FRAGSTATS. While the next planned release of FRAGSTATS plans to incorporate the metrics examined in this paper, until that time, progress remains slowed by the lack of software.

Acknowledgements This chapter is an extension of the analysis conducted by the authors in Kedron et al. (2018). This work was funded, in part, by a grant to the authors from the National Science Foundation (#SBE-1561021). We would also like to thank Harry Connor for his assistance with the development of the figures in this chapter.

References

Barbato G, Carneiro K, Cuppini D, Garnaes J, Gori G, Hughes G, Jensen CP, Jorgensen JF, Jusko O, Livi S, McQuoid H, Nielsen L, Picotto GB, Wilening G (1995) Scanning tunneling microscopy methods for the characterization of roughness and micro hardness measurements. Synthesis report for research contract with the European Union under its programme for applied metrology. European Commission Catalogue number: CD-NA-16145 EN-C Brussels Luxemburg

Bruton MJ, Maron M, Levin N, McAlpine CA (2015) Testing the relevance of binary, mosaic and continuous landscape conceptualisations to reptiles in regenerating dryland landscapes. Landscape Ecol 30(4):715–728

Coulston JW, Jacobs DM, King CR, Elmore IC (2013) The Influence of Multi-season Imagery on Models of Canopy Cover. Photogrammetric Eng Remote Sens 79(5):469–477

Coulston JW, Moisen GG, Wilson BT, Finco MV, Cohen WB, Brewer CK (2012) Modeling Percent Tree Canopy Cover. Photogrammetric Eng Remote Sens 78 (7):715–727

Cushman SA, Gutzweiler K, Evans JS, McGarigal K (2010) The gradient paradigm: a conceptual and analytical framework for landscape ecology. Spatial complexity, informatics, and wildlife conservation. Springer, Japan, pp 83–108

Fan C, Myint S (2014) A comparison of spatial autocorrelation indices and landscape metrics in measuring urban landscape fragmentation. Landscape Urban Plann 121:117–128

Forman RTT (1995) Some general principles of landscape and regional ecology. Landscape Ecol 10(3):133–142

Frazier AE (2014) A new data aggregation technique to improve landscape metric downscaling. Landscape Ecol 29(7):1261–1276

Frazier AE (2016) Surface metrics: scaling relationships and downscaling behavior. Landscape Ecol 31(2):351–363

Frazier AE, Kedron P (2017a) Landscape metrics: past progress and future directions. Curr Land Ecol Rev 2(3):63–72

Frazier AE, Kedron P (2017b) Comparing forest fragmentation in Eastern US forests using patch-mosaic and gradient surface models. Ecol Inform 41:108–115

Frazier AE, Wang L (2013) Modeling landscape structure response across a gradient of land cover intensity. Landscape Ecol 28(2):233–246

Image Metrology (2017) User's guide. http://www.imagemet.com/WebHelp6/Default.htm. Accessed 7 Feb 2019

Kedron P, Frazier A, Ovando-Montejo G, Wang J (2018) Surface metrics for landscape ecology: a comparison of landscape models across ecoregions and scales. Landscape Ecol 33:1489–1504

Kent M (2009) Biogeography and landscape ecology: the way forward—gradients and graph theory. Prog Phys Geog 33(3):424–436

Kupfer J (2011) Theory in landscape ecology and its relevance to biogeography. The SAGE handbook of biogeography. Sage, Thousand Oaks, pp 57–75

Kupfer J (2012) Landscape ecology and biogeography: rethinking landscape metrics in a post-FRAGSTATS landscape. Prog Phys Geogr 36(3):400–420

LaGro J (1991) Assessing patch shape in landscape mosaics. Photogrammetric Eng Remote Sens 57:285–93

McGarigal K, Cushman SA (2005) The gradient concept of landscape structure. In: Wiens J, Moss M (eds) Issues and perspectives in landscape ecology. Cambridge University Press, Cambridge, pp 112–119

McGarigal K, Tagil S, Cushman SA (2009) Surface metrics: an alternative to patch metrics for the quantification of landscape structure. Landscape Ecol 24(3):433–450

McGarigal K, Cushman SA, Ene E (2012) FRAGSTATS v4: spatial pattern analysis program for categorical and continuous maps. University of Massachusetts, Amherst

McMahon G, Gregonis SM, Waltman SW, Omernik JM, Thorson TD, Freeouf JA, Rorick AH, Keys JE (2001) Developing a spatial framework of common ecological regions for the conterminous United States. Environ Manage 28(3):293–316

Moniem HEMA, Holland JD (2013) Habitat connectivity for pollinator beetles using surface metrics. Landscape Ecol 28(7):1251–1267

Omernik JM (1987) Ecoregions of the conterminous United States. Ann Assoc Am Geogr 77 (1):118–125

Omernik JM, Griffith GE (2014) Ecoregions of the conterminous United States: evolution of a hierarchical spatial framework. Environ Manage 54(6):1249–1266

Schindler S, von Wehrden H, Poirazidis K (2015) Performance of methods to select landscape metrics for modelling species richness. Ecol Model 295:107–112

Scown MW, Thoms MC, De Jager NR (2015) Measuring floodplain spatial patterns using continuous surface metrics at multiple scales. Geomorph 245:87–101

Turner MG (2005) Landscape ecology: what is the state of the science? Annu Rev Ecol Evol Syst 36:319–344

Uuemaa E, Roosaare J, Oja T, Mander U (2011) Analysing the spatial structure of the Estonian landscapes: which landscape metrics are the most suitable for comparing different landscapes? Estonian Journ of Ecol 60(1):70

Wu Q, Guo F, Li H, Kang J (2017) Measuring landscape pattern in three dimensional space. Landscape Urban Plan 167:49–59

Zhang Z, Zinda JA, Yang Z, Yin M, Ou Z, Xu Q, Yu Q (2018) Effects of topographic attributes on landscape pattern metrics based on redundancy ordination gradient analysis. Landscape Ecol Eng 14(1):67–77

Part IV
Landscape Characterization: International Case Studies

Chapter 23
Landscape Character and Ecosystem Services Assessment: A Case Study from India

Debbie Bartlett and Sarah Milliken

Abstract This chapter describes a research project in Gujarat, north-west India, which resulted from a request for help with developing a strategy to deal with *Prosopis juliflora*, an invasive thorny shrub also known as 'Mesquite'. After due consideration, we concluded that any attempt at eradication would not only be impractical, it would also be unlikely to succeed. Funding was therefore sought to enable us to apply an integrated approach combining several methods to provide an information base for land use planning, rather than taking a single species approach. Landscape character assessment enabled natural areas in the region to be identified and this was followed by detailed participatory ecosystem services assessment focusing on one specific natural area. This process revealed that, although the spread of *Prosopis juliflora* is a key concern for ecologists, local people value the plant as a source of fuel and other resources for which there was no obvious alternative. All the results were collated into a natural area profile document, which was presented to the focus group participants as well as to other stakeholders and decision-makers as a robust evidence base to inform future land use planning decision-making.

Keywords Landscape · Ecosystem services · Participation · India

---

D. Bartlett (✉)
Faculty of Engineering and Science, University of Greenwich, Central Avenue, Chatham Maritime, Gillingham, Kent ME4 4TB, UK
e-mail: d.bartlett@gre.ac.uk; d.bartlett@greenwich.ac.uk

S. Milliken
Department of Landscape, School of Design, University of Greenwich, Gillingham, UK
e-mail: s.milliken@greenwich.ac.uk

L. Müller and F. Eulenstein (eds.), *Current Trends in Landscape Research*,
Innovations in Landscape Research, https://doi.org/10.1007/978-3-030-30069-2_23

23.1 Background and Objectives

Kachchh district is located in the northern part of Gujarat state, north-west India (Fig. 23.1). It is an arid/semi-arid area and the countryside consists of a mix of productive farmland which is used to grow large-scale crops such as cotton, castor, mango and pomegranate, subsistence farming, salt desert, and extensive grasslands which are of great importance to pastoralists that still practice transhumance. With such a wide range of habitat types, landscape character assessment is a useful tool as a first step towards understanding the area (Tudor 2014). The way in which the different elements provide benefits to the residents and influence land cover and land use can then be investigated by carrying out an assessment of their ecosystem services.

The spread of *P. juliflora* is reducing the availability of grazing land for migratory pastoralists (e.g. Geevan et al. 2003; Koladiya et al. 2016). *P. juliflora* (Fig. 23.2), a non-native shrub originally from South America, was introduced to prevent the salt deserts from encroaching onto the grasslands. While this may have been successful, the unintended consequence has been the rapid spread of this extremely invasive plant which is highly successful at competing with native vegetation, particularly in arid areas with high soil salinity. This plant is considered to be a problem in many countries (Lowe et al. 2004), and it is particularly abundant in the Kachchh region, especially in the Banni, the largest natural grassland in the

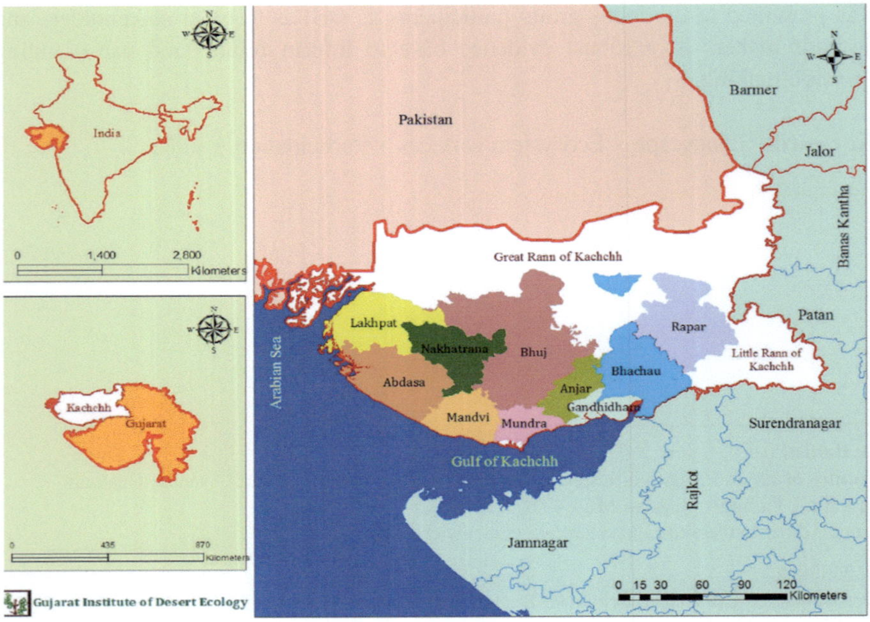

Fig. 23.1 Location of Gujarat state and Kachchh district

Fig. 23.2 *Prosopis juliflora* lining a road in Kachchh where large herds of livestock are a frequent sight

Indian subcontinent, where remote sensing data show that it has spread at a rate of about 25 km^2/year, and it is predicted that by 2020 more than 56% of the grassland will be covered by the shrub (Kathiresan 2006).

Much of the region is rural, with 40–45% of households in Gujarat dependent on natural resource-based livelihoods. The 2001 earthquake, which killed more than 18,000 people in the state, had a significant impact on medium and small-scale industries, including cotton processing and the salt industry, resulting in the loss of employment for thousands of people (Ashok et al. 2001). The state of Gujarat is the second most industrialised state in India, and its rapid development has had a marked impact on the environment, local communities and their livelihoods (Awasthi 2000). The Growth Strategy for Gujarat initiated in 2002–03 has led to a quantum leap in economic growth by providing huge incentives which have successfully encouraged inward investment and industrial development. However, poorly planned growth has exacerbated social inequalities, damaged the environment and compromised the delivery of ecosystem services. This has led to a reduction in funding for education, health and employment and has resulted in environmental damage which has affected rural livelihoods by degrading resources which, along with pollution, has led to reduced productivity and incomes (Hirway 2017).

While Environmental Impact Assessment (EIA) has become an integral part of the decision-making process since it was introduced to India in the 1990s, and while in theory it is required for all developments over a certain size, in practice weak enforcement, including the inability to impose fines, is resulting in many developments taking place without undergoing an EIA, despite their potential to cause negative environmental and socio-economic impacts (Panigrahi and Amirapu 2012). The urgency of encouraging investment to rebuild the regional economy after the 2001 earthquake may have resulted in local social and environmental factors being overlooked (see, for example, Therivel and Wood 2017). Rapid development, especially along the coast, that does not appear to have included a consideration of local social factors in the planning process, combined with environmental pressures (increasing soil salinity, water shortage and the spread of *P. juliflora*), is clearly affecting traditional livelihoods and increasing the vulnerability of communities to the effects of climate change.

23.2 Research Concept

The aim of our research was to introduce an integrated approach to land use planning in India, using Kachchh district as a case study. Landscape character is based on a combination of biophysical and cultural factors rather than administrative boundaries, and in England, this approach has been used to provide guidance for decision-makers that incorporate community views about their environment as well as factual data for all the 'natural character areas' to produce Character Area Profiles (Natural England 2014). These describe natural and cultural features, how the landscape has changed, what has caused this, and identify any drivers likely to cause future change.

This was followed by an assessment of the ecosystem services, an approach that has come to the fore since the Millennium Ecosystem Assessment (United Nations 2005) which examined the state of the environment from the perspective of the services which benefit humans and, in many cases, on which we depend. This enabled the identification of specific environmental opportunities, in order to provide a robust context for decision-making that benefits local communities and informs the development agenda, while at the same time minimising negative impacts on the environment.

We applied this approach to Kachchh district, beginning with the identification of the Natural Character Areas, followed by a participatory ecosystem services assessment exercise, in order to demonstrate an integrated land use planning process.

23.3 Identification of Natural Character Areas in Kachchh

This began with a detailed desk study, gathering together physical, biological and socio-economic data to identify areas within the region with similar characteristics. Topographical, geological and soil maps were used, in combination with remote sensing imagery, to reveal vegetation, bare ground and waterbodies, and how these had changed over time. Books, published papers and reports were reviewed to amass information on livelihoods and socio-economic factors as well natural attributes, such as ecological assemblies/communities, including notable wildlife and designated areas.

The desk study provided the basis for the fieldwork, which adopted a combination of the Landscape Character Assessment method originally developed by the Countryside Commission in the 1990s and updated by Natural England (Tudor 2014). Field sheets were modified to reflect local features such as temples, *P. juliflora* (assessed on the DAFOR scale[1]) and herds of camels, goats and buffalo. The desk study informed the selection of locations for field assessment, which were visited by a team consisting of both Indian and European researchers (Fig. 23.3).

The field sheets were analysed and the results, combined with the desk study, enabled the region to be subdivided into natural areas, shown as different colours in Fig. 23.4. The coastal plain, shown in both red and blue to include the intertidal area, was then selected for in-depth study.

The flat topography of the coastal plain has encouraged the settlement of industries. Development was further encouraged by policies after the 2001 earthquake which introduced tax breaks to encourage investment in the area and the setting up of Special Economic Zones (Ministry of Micro, Small and Medium Enterprises 2015). The rapid industrialisation of the coastal plain has resulted in a proliferation of features such as roads, communication masts and wind turbines (Sharma et al. 2012), and it was for this reason that we chose to focus on the impact of development on the rural communities in this natural character area.

23.4 The Natural Character Area Profile for the Coastal Plain

The Natural Character Area profiling exercise, carried out for the whole region, provided the basis for more detailed research on this particular area which extends for about 406 km, bordering the Kachchh mainland to the north, the Gulf of Kachchh to the south and the Arabian Sea to the west. As one of the three major gulf systems of India, this coast has high biological richness and a variety of habitats, including coral reefs, mangroves, creeks and estuaries. Many village livelihoods depend on the natural habitats and ecosystems, with salt production and fishing having an obvious link to the marine environment. The flat topography is also beneficial for agriculture, horticulture and animal husbandry. The topography and the strategic position, combined with physical and geochemical attributes, have contributed to the recent increase in industrial developments (Thivakaran 2011). The National Highway crosses most of Kachchh district and connects it with its two

[1]**D** = Dominant; **A** = Abundant, **F** = Frequent, **O** = Occasional, **R** = Rare.

Fig. 23.3 Field team in action, comprising both European M.Sc. students and Indian researchers

major and 11 minor ports. Industries include chemical, fertiliser, cement production and mineral mining, among others.

The vast amount of energy required by these industries has led to the establishment of power generation plants such as thermal power stations and wind farms

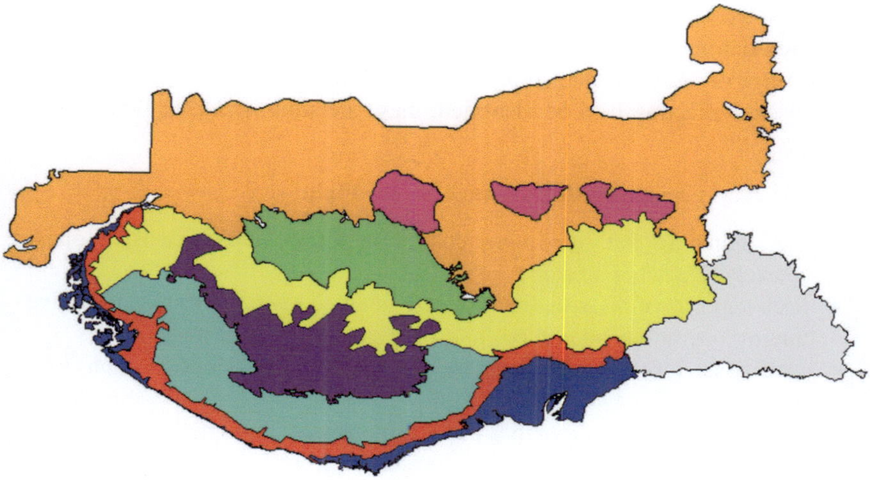

Fig. 23.4 Natural character area map of Kachchh. The coastal plain, the area selected for in-depth study, is shown in red and blue to denote the intertidal range. Other notable areas are the Great Rann (orange) and the Little Rann (grey)

which are especially prevalent along the coast. The location of industrial infrastructure appears to be *ad hoc*, with adverse consequences for landscape, deterioration of natural habitat such as mangroves, and the loss of the traditional livelihoods. This industrial development has encouraged migration towards the coast, which has resulted in the growth of coastal towns and the creation of new settlements, with the conversion of agricultural land into new land uses. Industrial development has also favoured a shift from traditional housing (bungas) made of natural materials, to concrete houses. While the rapid industrialisation of the coastal plain has brought some benefits to the villages, such as an increase in literacy and new job opportunities, these have been limited, and the overall impact has been negative.

Water is a key issue in this arid region. The demand from industry, combined with the removal of key ecosystems such as mangroves, is raising the water table, while water quality is declining due to industrial pollution and the release of salt water into the groundwater. Climate change and rising sea levels exacerbate this as salt water flooding events are more frequent and monsoon rainfall is variable. This combination of factors has resulted in a decrease in drinking water quality and availability and an increase in soil salinity. Traditional livelihoods such as fishing and livestock rearing and agriculture are being severely affected. Fish production is low due to a combination of pollution and reduction of the mangroves which are used as breeding grounds by many species, while the poor-quality soils result in low crop and grassland productivity which in turn affects livestock grazing. In addition, *P. juliflora* is spreading across the coastal area, as it grows better than native plants on saline soils. This shrub is useful for fuel and charcoal production but is generally considered detrimental to both livestock and agriculture.

This stage of research enabled the key characteristics of the coastal plain and the current issues affecting it to be identified; these are summarised in Box 1.

Box 1: Key Characteristics of the Coastal Plain

- Predominantly flat, low-lying coastal landscape with extensive open, sparsely vegetated areas.
- Rich deposits of bauxite, limestone, lignite and bentonite enable intensive mining operations and allied industries such as cement plants.
- Vast areas of salt pans employ 43,000 seasonal workers each year (Fig. 23.5).
- The coastal strip has a higher population density than the hinterland.
- Industrial development has encouraged the growth of coastal towns and the creation of new settlements, reducing the agricultural area.
- There is significant wind farm development evident along the coast (Fig. 23.6).
- Infrastructure such as highways, roads, ports, communication masts and power stations indicate the presence of increasing industrial activity (Fig. 23.7).
- The sandy area between Mandvi and Pingleswar is an important breeding site for sea turtles.
- Extensive mangrove formations and a continental shelf of 164,000 km^2 facilitate rich fishing grounds (Figs. 23.8 and 23.9).
- Agricultural land and herds of livestock indicate significant farming activity (Fig. 23.10).

The basic characterisation (Box 1) gives a comprehensive overview, but it is based on the results of the desk study and the views of the research team working on the project. It would, therefore, have been highly dangerous to draw any conclusions at this stage, as important issues could easily have been overlooked. The next stage, therefore, involved local people in order to determine their views on the changing landscape and to assess the ecosystem services.

23.5 Community Participation in Ecosystem Service Assessment

Eight villages were selected to be involved in the participatory exercise along different parts of the coast (Fig. 23.11).

Fig. 23.5 The salt works at Bachau. Salt production is an important and well-established industry and provides significant seasonal employment

Fig. 23.6 Wind turbines near Lakhapar providing a stark contrast with traditional farming methods

Fig. 23.7 One of many power stations on the coast, demonstrating modern industrialisation among traditional salt pans, where salt has been harvested from sea for centuries

Fig. 23.8 Mangroves which protect the coast from storm damage and soil erosion also provide important nursery grounds for fish to breed

Fig. 23.9 Fisherman in a creek near Mandvi. Fishing is important both commercially and for subsistence in some areas (although most Gujaratis are vegetarian) but catches are reported to have declined significantly over recent years

Fig. 23.10 A large herd of goats. The bare ground with scattered *P. juliflora* is typical of the land around villages

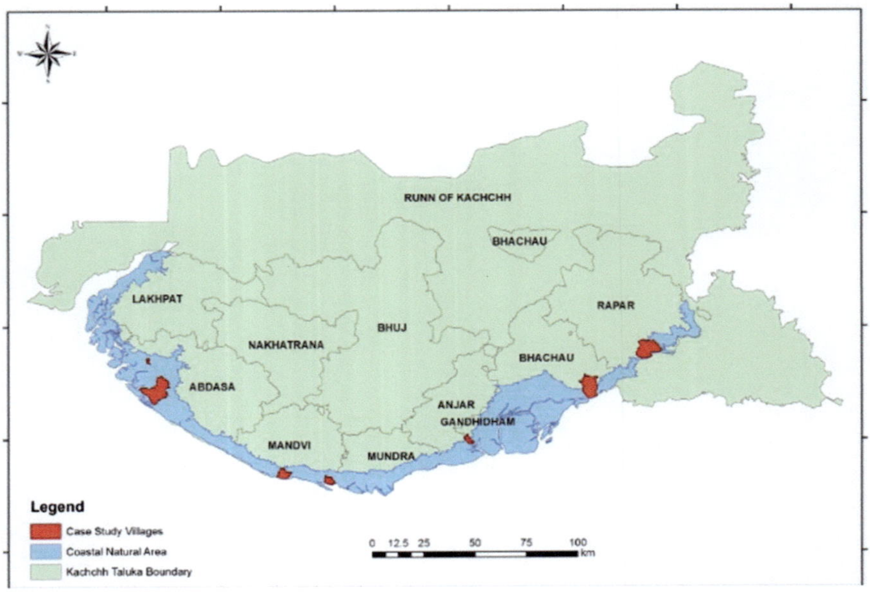

Fig. 23.11 Locations of villages visited in the coastal zone (those close together appear as one)

23.6 Village Profiles

An initial desk study revealed some distinctive features, listed below:

- AshiraVandh contains the Lala Indian Bustard sanctuary
- Kanmer includes hilly land
- Kathada is the only village with forest
- Lakhapar is the only village with significant industry
- Lakhapar and Tragadi have salt pans
- NimdiVandh includes estuary and creek areas.

The villages differ significantly in size, with AshiraVandh being the largest and Tuna and Wandi, combined, being the smallest (Fig. 23.12).

There were cultural differences, with some villages being predominantly Muslim, and others Hindu or a mixture of religions. The population sizes, literacy levels and numbers of individuals identifying as scheduled casts (SC) and scheduled tribes (ST) are given in Table 23.1.

Remote sensing satellite imagery was used to determine land cover in each village and to explore land cover changeover fifteen years, from 1999 to 2014 (Table 23.2). An example, showing interpretation for AshiraVandh in 2014, is given in Fig. 23.13.

The most significant changes are loss of cultivated land, particularly in Lakhapar where over 70 ha have been replaced by industry; increased settlement in Tuna and

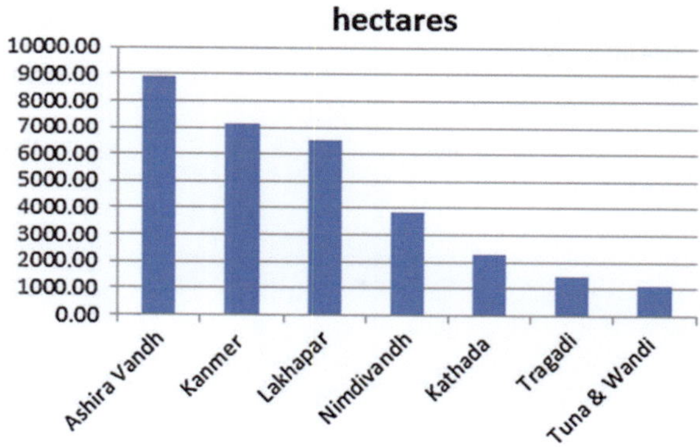

Fig. 23.12 Village areas

Table 23.1 Key to villages: **Wan**, Wandi; **Tun**, Tuna; **Tra**, Tragadi; **Nim**, NimdiVandh; **Lak**, Lakhpar; **Kat**, Kathda; **Kan**, Kanmer; **Ash**, AshiraVandh

Population		Males	Females	% literacy in village (%)	% literacy in males (%)	% literacy in females (%)	SC (%)	ST (%)
Van	388	186	202	71.03	80.23	63.64	31.96	1.03
Tun	5114	2573	2541	50.41	61.68	39.10	13.67	0.02
Tra	1238	636	602	54.52	62.57	46.11	0	0
Nim	–	–	–	–	–	–	–	–
Lak	989	547	442	50.85	67.78	30.23	0	0
Kat	2848	1387	1461	70.95	82.57	60.14	11.48	4.60
Kan	3816	1941	1875	53.58	64.77	41.88	19.42	0.03
Dho	2222	1194	1028	65.94	79.74	49.28	13.95	0.68
Ash	–	–	–	–	–	–	–	–

Table 23.2 Land cover change between 1999 and 2014

Landcover	Ashira Vandh	Kanmer	Kathada	Lakhapar	NimdiVandh	Tragadi	Tuna & Wandi
Settlement	−14.43	0.00	0.00	0.00	+2.60	0.00	+42.42
Cultivated land	−142.24	+12.24	−232.85	−70.68	+4.51	−46.82	−83.12
Industry				+70.67			
Forest			+2.60				
Scrub	+782.27	−5.10	+289.00	−180.48	+20.82	−22.90	+39.59
Mud flat	−662.45	−7.16	+3.11		−28.18	+125.72	
Salt pan				+187.76		+21.19	+4.40
Estuary					−5.01		
Creek					+5.86		
Road	0.00	0.00	0.00	0.00	0.00	0.00	−3.29
River	+51.59	0.00	0.00	0.00	0.00	0.00	0.00
Water bodies	−14.75		−68.27	−7.86		−79.00	

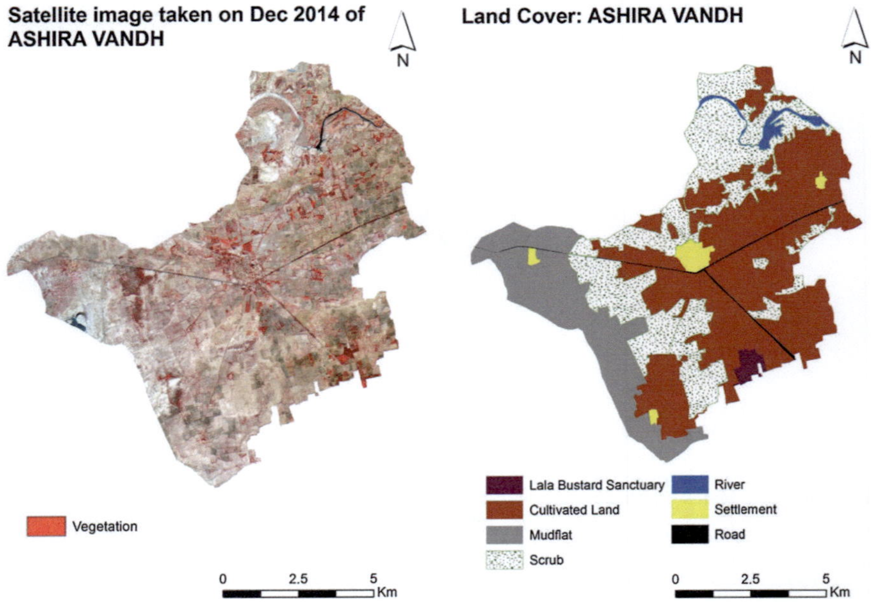

Fig. 23.13 Satellite imagery and interpretation of land cover

Wandi; and replacement of salt pans by *P. juliflora* (classified as scrub in the table) in Lakhapar and Tragadi. Expansion of scrub was also found on former mudflats in AshiraVandh and on formerly cultivated land in Kathada.

23.7 Focus Groups

Focus groups (Fig. 23.14) were held during May and June 2015 using a pre-prepared structured interview schedule to collect information on the topics listed below:

- Livelihoods and alternative livelihoods
- fuel source(s)
- women's views/role
- change/the future
- education
- landscape issues
- topics raised by participants.

Fig. 23.14 Focus group discussion at AshiraVandh

23.8 Analysing the Data

The results from both the landscape character and participatory investigations were analysed to enable an ecosystem services assessment to be carried out. The results for each village were displayed on posters which included photographs taken during the initial visit, and text in both English and Gujarati (Figs. 23.15).

A second visit was made to each village in December 2015, with two copies of the poster along with maps showing land cover change. One poster was given to the villagers, and the other was used to record comments and correct any misunderstandings (Figs. 23.16, 23.17, 23.18 and 23.19), and then retained in order to facilitate the final ecosystem services assessment. The use of the land cover maps

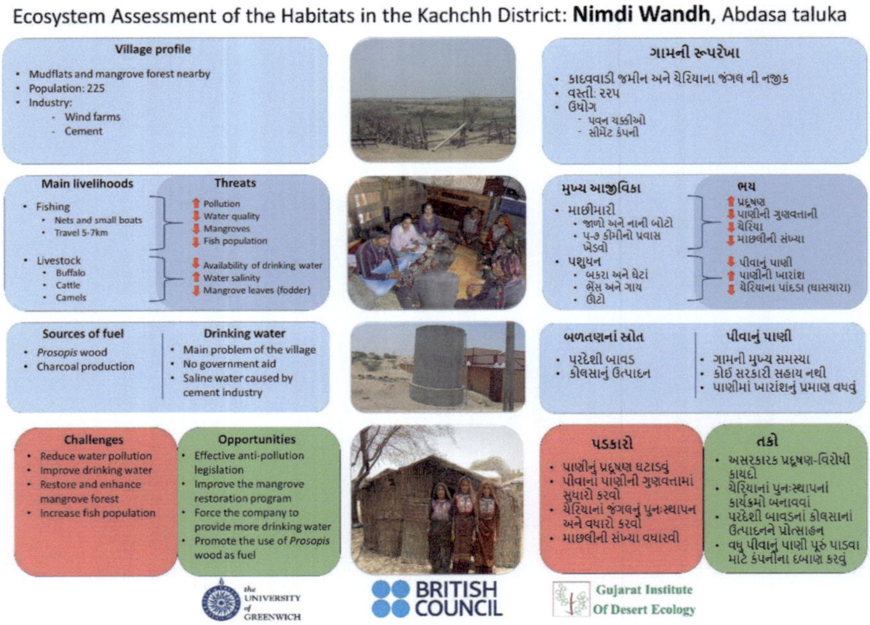

Fig. 23.15 Examples of the posters taken to the villages on the second visit to check that we had correctly represented their views

Fig. 23.16 Villagers in AshiraVandh checking their poster

Fig. 23.17 Ladies, visible in the background in Fig. 23.18, getting their turn to comment

Fig. 23.18 Group discussion about the poster in Wandi

Fig. 23.19 Dayesh Parmar, researcher and translator, explaining the poster to villagers in Tragadi

was not successful as the participants, although very familiar with their locality, were simply not used to the 'bird's eye view'—a learning experience for us.

23.9 Ecosystem Services

These are the benefits provided by ecosystems that contribute to the viability and quality of human life. They can be tangible products such as food and water, flood regulation, control of soil erosion or disease, or non-material benefits such as recreational and spiritual benefits.

The coastal plain of Kachchh district provides a wide range of ecosystem services. Each is derived from the attributes and processes (both natural and cultural features) within the area. The predominant services are summarised below.

Provisioning Services

- **Food provision**: The coastal plain Natural Character Area contains extensive areas of agricultural land with cultivation of different crops such as bajra (pearl millet), jowar (sorghum), cotton, guar (cluster bean) and castor. Large areas of pasture land provide grazing for livestock. The fishing industry is supported by the provision of crustaceans and fishes such as pomfrets, tuna and Bombay duck. Important ecosystems such as mangroves provide fodder for livestock

(leaves and fruit). Salt and edible seaweeds are also provided by marine ecosystems.
- **Medicine**: Native plants are of key importance for traditional practices and medicinal use, for example, gugal (*Commiphora wightii*).
- **Fibre and fuel**: *P. juliflora* and other shrub species provide an important source of wood fuel. The charcoal produced from *P. juliflora* is known to have a high calorific index, thus producing high-quality charcoal. Mangrove leaves are also used as fuel.
- **Minerals**: Bauxite, limestone and bentonite are the main minerals found in the area, leading many industries to settle on the coast.
- **Other**: Despite the negative impact of *P. juliflora,* it provides a number of secondary products, such as honey and gum.

Regulating Services

- **Climate regulation**: Significant carbon storage is provided by extensive areas of mudflats, mangroves and grasslands.
- **Regulating soil quality**: Ecosystems such as mangroves prevent sea water infiltration, thereby reducing the risk of salinisation. Mangroves and their associated habitats also filter and assimilate pollutants from the water, thus reducing their occurrence in the coastal soils. Habitats such as estuaries maintain a constant flux of sediments and nutrients, thereby maintaining soil productivity.
- **Regulating water quality**: Mangroves and their associated habitats preserve water quality and reduce pollution by filtering suspended material and assimilating dissolved nutrients.
- **Regulating coastal flooding and erosion**: The major risk of flooding in the area comes from the sea. The extensive coastal habitats, such as mangroves, creeks, mudflats and sandy dunes, as well as marine ecosystems such as coral reefs, provide an important natural defence against flooding by reducing the impact of wave action on the coastline. In recent years, this service has become highly important due to the rising sea level; coastal habitats provide erosion control as well as protection from cyclones and storms.

Supporting services

- **Species diversity**: Marine ecosystems of the Kachchh coast are very rich in species diversity and abundance. Mangroves, algae (108 species recorded) and seagrass species along with phytoplankton provide primary production which sustains rich faunal life—dolphins, crabs, fish, prawns and shrimps—which are both ecologically and economically important.
- **Breeding sites**: The coast around Mandvi is a breeding site for the endangered green sea turtle, olive ridley turtle and leatherback turtle. Mangroves provide breeding grounds for many marine animals, including commercially important fish, prawns, lobsters and crabs.
- **Shelter**: Mangroves provide shelter for many animals during harsh weather conditions and protection to coastal communities.

- **Habitat maintenance**: Mangroves supply nutrients and oxygen to animals and plants in the ecosystem.

Cultural services

- **Sense of peace/inspiration**: A sense of peace is provided by the coastal landscape.
- **Tranquillity**: Remote areas far from the hubbub of the city can be found on the coast.
- **Recreation**: there are many opportunities for quiet enjoyment of the beautiful beaches; there is horse and camel riding on Mandvi beach, and it is a popular bathing place (Fig. 23.20).
- **Spiritual value**: Many different cultures coexist in the area, with places of significance for different religions. The tranquil and relaxing atmosphere of the area provides an enhanced spiritual experience.
- **Education**: The unique landscape and biodiversity of the area offers opportunities for watching rare species in their natural area. This is especially true for sea turtles and a wide range of both migratory and endemic bird species.

This ecosystem service assessment enabled a number of strategic objectives to be identified with the aim of balancing the priorities of ensuring the livelihoods of local stakeholders while at the same time maintaining important habitats and native wildlife. These were formulated as Statements of Environmental Opportunity (SEO) and are listed below.

Fig. 23.20 Mandvi beach, a popular tourist destination

Fig. 23.21 Front cover of the Natural Character Area profile

- **SEO1**: Maximise the potential of *P. juliflora* for fuel, charcoal and biochar and explore the potential for eradicating it in key areas such as the Naliya grassland.
- **SEO2**: Maintain and enhance the natural and cultural heritage, for example, by promoting sustainable tourism that delivers direct benefit to local people.
- **SEO3**: Formulate a strategy to minimise the negative impact of industry on the environment and local communities and increase the potential for delivering real benefits such as quality jobs for local people.
- **SEO4**: Enhance and maintain mangrove and creek ecosystems and ecosystem services such as flood prevention and fish production while considering the potential for industrial expansion in the area.

The results of the draft Natural Character Area profile (Fig. 23.21) were presented at a workshop held on 29 December 2015 at the Vivekanand Research and Training Institute (VRTI) in Mandvi, to which the villagers and other stakeholders were invited.

23.10 Conclusion

Landscape character and ecosystem services assessment is a well-established technique that has been used by planning authorities in Europe for several decades. Developing countries such as India offer an opportunity to incorporate these approaches before the industrialisation process adversely affects the environment. While preservation of natural ecosystems should neither prevent development nor disadvantage the well-being of local communities, an integrated approach to land use planning, balancing environmental conservation and economic growth, is desirable. The Natural Character Area approach, based on combining Landscape Character Assessment with Participatory Ecosystem Assessment, is proposed as a way to achieve this.

Acknowledgements This project would not have been possible without the support of the British Council UK-India Education and Research Initiative (UKIERI) which provided funding, the M.Sc. students who took part, particularly Eulalia Gomez Martin, and Dayesh Parmar from the Gujarat Institute of Desert Ecology (GUIDE) who, in addition to his role as part of the research team, was the Gujarati interpreter and translator.

References

Ashok KL, Tapas K, Sen R, Kavita RP, Jena R (2001) Economic consequences of the Gujarat earthquake, Gujarat earthquake assessment mission of the Asian Development Bank and the World Bank. Available at www.academia.edu/4617331/Economic_consequences_of_the_Gujarat_earthquake. Accessed on 20 Mar 2019

Awasthi DN (2000) Recent changes in Gujarat industry: issues and evidence. Econ Polit Weekly 35:3183–3192

Geevan CP, Dixit AM, Silori CS (2003) Ecological economic analysis of grassland systems: resource dynamics and management challenges-Kachchh District (Gujarat). Gujarat Institute of Desert Ecology. https://www.academia.edu/26072831/Ecological_Economic_Analysis_of_Grassland_Systems_Resource_Dynamics_and_Management_Challenges-Kachchh_District_Gujarat. Accessed on 20 Mar 2019

Hirway I (2017) The truth behind the Gujarat growth model https://thewire.in/economy/the-truth-behind-the-gujarat-growth-model. Accessed on 20 Mar 2019

Kathiresan RM (2006) Invasion of Prosopis in India. In: Problems posed by the introduction of Prospois spp. in selected countries. FAO, Rome, pp 3–10

Koladiya M, Gajera N, Mahato AKR, Kumar VV, Asari RV (2016) Birds of Banni Grassland. Gujarat Institute of Desert Ecology. Published by the Ravi Sankaran Foundation, 303 pp. ISBN: 978-93-5254-957-3. http://www.ravisankaran.org/wp-content/uploads/2016/06/Birds-of-Banni-Grassland.pdf. Accessed on 20 Mar 2019

Lowe S, Browne M, Boudjelas S, De Poorter M (2004) 100 of the World's worst alien species: a selection form the global invasive species database. Published by the Invasive Species Specialist Group, Species Survival Commission of the IUCN, 12 pp

Ministry of Micro, Small and Medium Enterprises (2015) Brief industrial profile of Kutchdistrict. Government of India. http://dcmsme.gov.in/dips/dip%20kutch.pdf. Accessed on 20 Mar 2019

Natural England (2014) National character areas see https://www.gov.uk/government/publications/national-character-area-profiles-data-for-local-decision-making/national-character-area-profiles . Accessed on 20 Mar 2019

Panigrahi JK, Amirapu S (2012) An assessment of EIA system in India. Environ Impact Assess Rev 35:23–36

Sharma A, Srivastava J, Kar SK, Kumar A (2012) Wind energy status in India: a short review. Renew Sustain Energy Rev 16(2):1157–1164

Therivel R, Wood G (2017) (eds) Methods of environmental and social impact assessment. Taylor and Francis. ISBN 9781138647640, 724 pp

Thivakaran GA (2011) Coastal ecology of Kachchh—status, issues, problems. In: State of the environment in Kachchh. Gujarat Ecological Education and Research (GEER) Foundation and Gujarat Institute of Desert Ecology (GUIDE), pp 59–88

Tudor C (2014) An approach to landscape character assessment. Natural England available at https://assets.publishing.service.gov.uk/government/uploads/system/uploads/attachment_data/file/691184/landscape-character-assessment.pdf. Accessed on 20 Mar 2019

United Nations (2005) Millennium ecosystem assessment. Ecosystems and human well-being: synthesis. Island Press, Washington, DC, 137 p. https://www.millenniumassessment.org/documents/document.356.aspx.pdf. Accessed on 20 Mar 2019

Chapter 24
Landscape-Ecological Conditions and Conflicts of Environmental Management in the Arctic Zone of Russia

Alexander Evseev and Tatiana Krasovskaya

Abstract Landscapes ecological studies nowadays may supply valuable data for advanced environment management. Inadequate recognition of successful social-economic development dependence on regulated exploitation of ecosystem services produced by different landscapes/ecosystems types launches environment management conflicts among different stakeholders, degradation of ecosystems, etc. Management patterns beneficial both to environment and society may be elaborated using ecosystems services assessments based on landscape/ecosystems studies and land-use patterns attributed to different territories. Provisioning ecosystem services use such as food and raw materials, partly water resources production, etc. nowadays are more or less considered in different systems of management practices. But other ecosystem services exploitation pattern, i.e., regulating, supporting and cultural/information services use still reflects "The tragedy of the commons". The main obstacles for elaboration of appropriate ecosystems services exploitation are connected with their inadequate link to landscapes and environment management structure as well as poor quantification of these links. Results of landscape/ecosystems studies revealing ecosystem available services and their monetary assessments necessary for quoting of their exploitation by different stakeholders are presented for the Arctic zone of Russia.

Keywords Landscape · Ecosystems structure · Ecosystems services · Environment management conflicts · Mapping · Russian Arctic

---

A. Evseev · T. Krasovskaya (✉)
Faculty of Geography, Lomonosov Moscow State University,
Leninskie Gory, 119991 Moscow, Russia
e-mail: krasovsktex@yandex.ru

A. Evseev
e-mail: avevseev@yandex.ru

L. Müller and F. Eulenstein (eds.), *Current Trends in Landscape Research*,
Innovations in Landscape Research, https://doi.org/10.1007/978-3-030-30069-2_24

24.1 Landscape Ecological Studies

Landscapes ecological studies nowadays may supply valuable data for advanced environment management corresponding to the goals of sustainable development. Inadequate recognition of successful social-economic development dependence on regulated exploitation of ecosystem services produced by different landscapes/ ecosystems types launches environment management conflicts among different stakeholders, degradation of ecosystems and even loss of nature and culture heritage objects. New data about energy-matter fluxes in anthropogenically modified landscapes draw attention to their management aspects beneficial both to environment and society. The key challenges of management aspects are the following: how to assess existing pools of ecosystem services, how to manage multiple ecosystem services across landscapes to provide their reproduction, what spatial environment limits are necessary to consider for reproduction territories of different ecosystems services. Relevant landscape ecological studies which started at the end of the twentieth century, nowadays became important parts of international sustainable development programs. They demonstrate integrated ecological-economic results necessary for environment management practice (Voinov et al. 2004; Brouwer et al. 2013; The Economics 2015, etc.). Quantitative parameters of landscapes related to ecosystem services may be described in economic terms showing the value of nature. Such integrated approach in landscape studies reflects the principal trend of science development in the twenty-first century, in general: multidisciplinary studies of numerous links in "social–economic–nature" or "man–nature" system.

The Millennium Ecosystem Assessment (2005) presented an overview of ecosystems services exploitation all over the world and clarified importance of ecosystems services proper use for the sake of sustainable development. Ecosystems services international classification was elaborated, reflecting the growing public interest to their assessment for adequate management in economic activities (Haines-Young and Potschin 2017). According to the recently developed global model (GUMBO) linking economic and social welfare with different natural/ ecological factors, the estimated value of global ecosystem services was about 4.5 times higher than that of the gross world product (GWP) in the year 2000 (Boumans et al. 2002).

24.2 Ecosystem Services

Provisioning ecosystem services use such as food and raw materials, partly water resources, etc., nowadays are more or less considered in different systems of management practices. But other ecosystem services exploitation pattern, i.e., regulating, supporting and cultural/information services use still reflects

"The tragedy of the commons" (Hardin 1968), promoting landscapes degradation and environment management conflicts—critical disturbances in "man–nature" system. The main obstacles for elaboration of appropriate ecosystems services exploitation are connected with their inadequate link to landscapes and environment management structure as well as poor quantification of these links.

Quantification process of separate ecosystem services is very rapid today, but as a rule, regulating, supporting and cultural ecosystem services are linked to separate ecosystems pools regardless of accommodating landscape structure supplying their coherence. In spite of great scientific value of creating quantitative database for ecosystem services pools in different landscapes such assessment results appear to be of low importance for territorial planning practice. Economic factors controlling development plans as a rule ignore the value of ecosystems services with the exception for natural (biological) resources production. The key task of ecosystem management is connected with determining multiple ecosystem services value across landscapes (Raudsepp-Hearne et al. 2010). Ecosystem service spatial analysis based on landscape studies helps to identify areas rich or poor in ecosystems services supply for different types of economic activities.

Possibilities of landscape studies and mapping results use for display of spatial patterns of ecosystem services are actively studied nowadays (Burkhard et al. 2014; Semenov and Lysanova 2015; Bastian et al. 2015; Yaneva 2016, etc.). Most often the following approaches are met: (1) "Landscape/ecosystem" based on their regional spatial structure; (2) "River basin" delimiting spatial landscape/ecosystem structure within basins' limits and (3) "Land-use" based on revealing of different types of local ecosystems exploitation. Each of these methods matches different investigation tasks and goals as well as available reference database. In our study, we tried to combine two of these approaches (1 and 3) in order to reveal potential environment management conflicts produced by excessive ecosystem services exploitation in the Arctic zone of Russia.

24.3 Study Region in the Russian Arctic

One of our study areas was situated in Nenets Autonomous okrug (Federal district) and it belongs to eight Arctic regions of Russia which advanced economic development was planned in the nearest future according to the recently adopted State program (2017) (Fig. 24.1).

The coastal Arctic zone of Russia accommodates a great variety of ecosystems patterns. The study area was situated at the northern part of the Russian Plain and belonged mainly to subarctic subzone with an exception of Novaya Zemlya Archipelago and other islands which were in the Arctic zone. Subarctic ecosystems occupied the major part of the territory and were presented by different tundra types: arctic-4.9%, northern-10.3%, southern-57.9%, mountainous-3.5% and

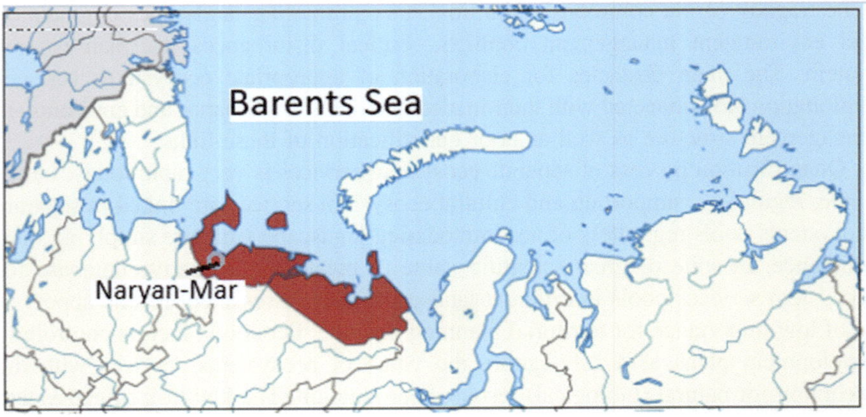

Fig. 24.1 Study area: Nenets Autonomous Okrug

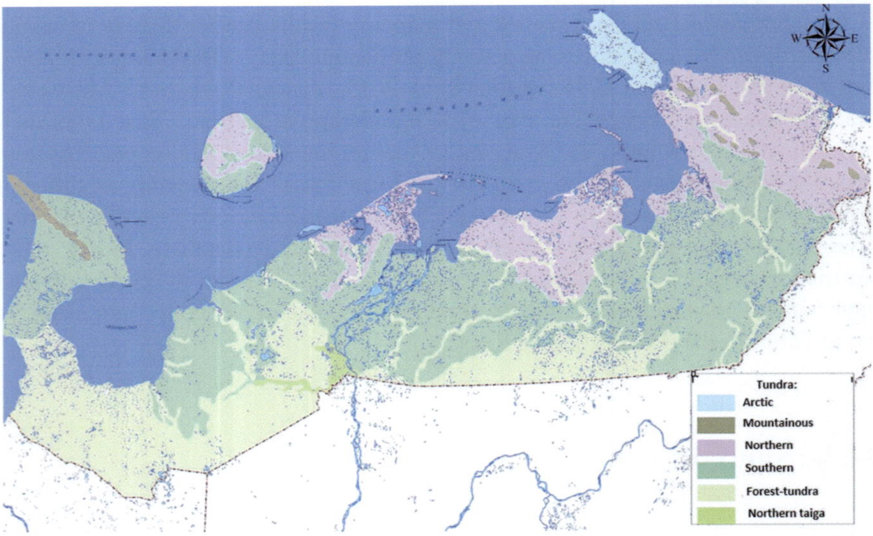

Fig. 24.2 Ecosystems structure of Nenets Autonomous okrug (Geography 2013)

forest-tundra ecosystems (15.4%) which developed at the southern limit of tundra belt and sometime penetrated further to the North via river valleys (15.4% of the territory). The south-western part was occupied by northern taiga forests (8% of the territory) (Figs. 24.2 and 24.3).

Intrazonal ecosystems were presented by bogs and coastal and river meadows. Such ecosystems structure is typical for plains of the whole Arctic zone.

Fig. 24.3 Shrub tundra dominated by dwarf birch (*Betula nana*) and wild rosemary (*Ledum palustre*) near Naryan-Mar. (Photo by Alexandra Bancheva)

24.4 Methodology

The planned economic activities for the study area till 2020–2030 enabled to outline the following groups of stakeholders connected with the major environment management types in the studied region: industrial (mainly extraction of hydrocarbon resources), transport (marine, railway, etc.) and residential, nature conservation (nature protected territories of different types), territories of traditional nature use (TTNU-reindeer breeding, hunting, fishing, etc.), recreation (commercial recreation, ethnic-tourism, etc.). At some territories future development will be based on similar spectrum of ecosystem services exploitation, thus provoking environment management conflicts contradicting sustainable development goals. More than that the different land-use types may overlap each other.

Landscape characteristics necessary for our studies were received from IPY Project MODIL-NAO (2008), Mihalev et al. (2008) and Ecological Atlas of Russia (2018) and completed by our field data. Basic investigation methods in general may be referred to ecological–economic (TEEB 2013) but modified to match the goals of our study: we tried to link ecosystem services pools with landscape/ecosystems spatial structure. Firstly, we examined the landscape structure and determined principle ecosystems services supplied by different ecosystems regarding their future exploitation by different stakeholders. For example, bogs supply water purification and water filtration service important for different stakeholders because it provides water resources supply and partly reduces water purification expenditures. Secondly, the "supply rate" was assessed based on published experimental field data for similar Arctic ecosystems. Knowing future exploitation areas, we received location and size of territories with available spatial pools of ecosystem services necessary for quoting. Finally, monetary assessments were done based on market prices (for water purification, "carbon units" price according to Paris Climate Agreement, etc.) and relevant general assessments (biodiversity preservation) in case of their absence. Limited data on "supply rate" and monetary assessment methods especially for information/cultural ecosystems services forced to regard the results as primary which may be broadened in future. Mapping was done in GIS media. Heterogenous characteristics: regional ecosystems variables presented by ecosystem services, ethnic-cultural data, overlapping modern land-use patterns, etc., were processed with the help of atlas information system (AIS) tools. One of the most interesting AIS characteristics is the possibility of its integration with decision support system (Tikunov 2004).

Landscape-ecological conditions determine available pools for practical use of ecosystems services. Different types of economic activities in this respect may be directed at exploitation of different spectrum of ecosystem services: mineral resources extracting companies and reindeer breeders depend on pools of different provisioning ecosystem services, i.e., reserves of mineral and pasture resources. Ecosystem services of general use by principal stakeholders both for economic and social use are shown in Fig. 24.4. Their joint exploitation may give rise to local environment management conflicts in case quoting absence.

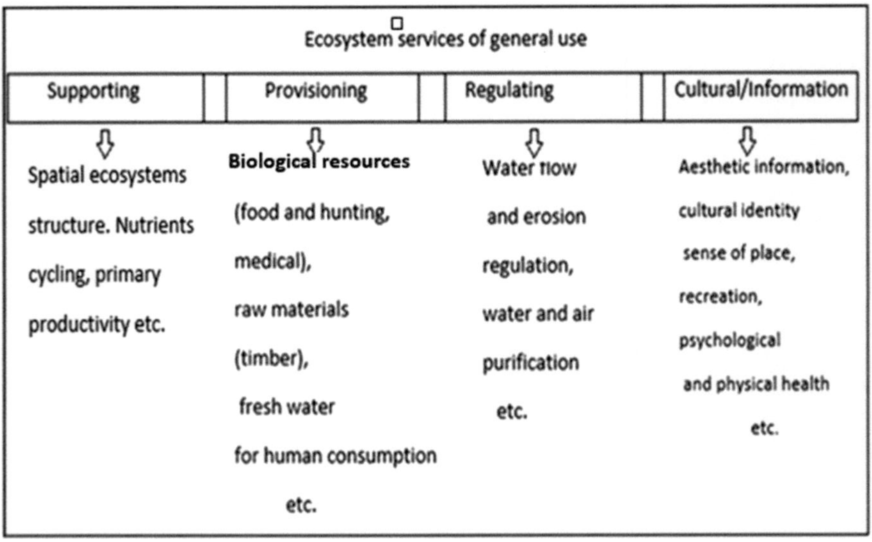

Fig. 24.4 Ecosystems of general use typical for the study area

Table 24.1 Forecast of principle competing patterns of ecosystem services exploitation

Land-use/environment management type	Industrial	Transport	Settlement/residential (nucleus)	TTNU
TTNU	Regulating, supporting	Regulating	Regulating, provisioning, supporting, cultural/information	–
Industrial	–	Regulating	Regulating, supporting	Regulating, supporting
Transport	Regulating	–	Regulating	Regulating
Settlement	Regulating, supporting	Regulating	–	Regulating, provisioning, supporting, cultural/information

Forecast of main of environment management/land-use competing patterns based on ecosystems services use is shown in Table 24.1.

Table 24.1 clearly demonstrates that the most "deficient" ecosystem services for future economic development may be regulating. Landscape/ecosystem studies of their donors enable to find spatial quotas beneficial both to environment stability and economic activities.

24.5 Mapping of Management Conflicts

We tried to apply such approach compiling the map of environment management conflicts at indigenous population territories in the Russian Arctic (Figs. 24.5 and 24.6).

Conflicts types at the map (Fig. 24.6) were presented using temporal and intensity scales elaborated due to processing of field and published data. The existing conflicts in impact regions were shown.

Potential environment management conflicts were also shown for different stakeholders for two regions: Amderma and Indiga the Northern Sea Route ports (small settlements nowadays) (Fig. 24.7). Their revival is connected with the adopted plans of infrastructure development to promote construction/modernization and further exploitation of these sea ports.

Fig. 24.5 Typical environment management conflict—abandoned oil rig on territories of traditional nature use; Nenets autonomous Okrug. (Photo by Ivan Mizin)

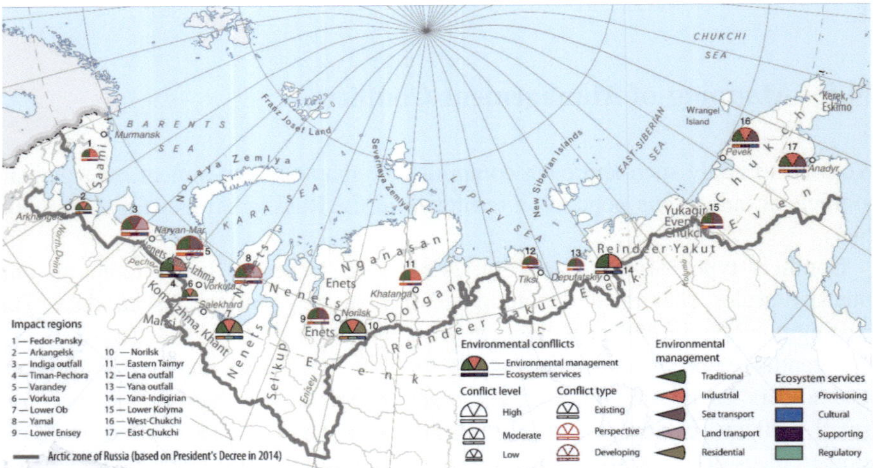

Fig. 24.6 Environment management conflicts at indigenous population territories in the Arctic zone of Russia (Evseev et al. 2018a)

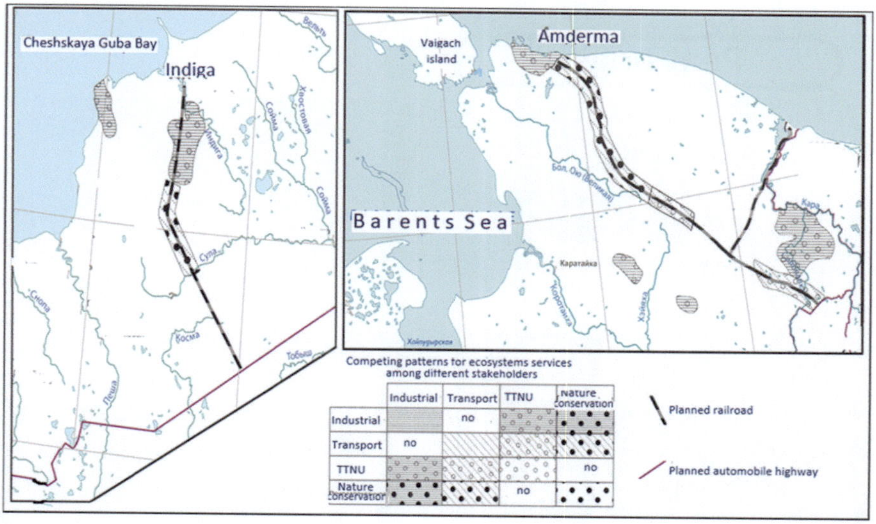

Fig. 24.7 Potential nature management conflicts areas

24.6 Ecosystem Services and Their Valuation in the Numto and Ya'Erv Study Areas

Landscape/ecosystems services structure enabled to determine and assess their available pools and forecast their exploitation in a certain pattern of environment management/land-use. Landscape/ecosystem studies in Numto Nature Park enabled to define ecosystem services structure for forest-tundra and bogs ecosystems typical for this territory. Figure 24.8 demonstrates the structure and percentage from the total value for ecosystem services of peat bogs based on our ecological–economic assessments. These ecosystem services may be considerably reduced by the planned oil extraction. Information/cultural services are missing because of poorly developed assessment methods though their importance is great: they are linked to Nenets TTNU support. Their loss may lead to ethnic-social environment management conflicts.

Our assessments for one of the regions at the study area Ya'Erv TTNU (Fig. 24.9) demonstrated the existing pool for one of the supporting ecosystem services—carbon deposition (important for CO_2 in atmosphere control) (Table 24.2).

Quantification of ecosystem services is a complicated task and its methods still need further development relevant to different environmental conditions. Nevertheless, the primary assessments of ecosystem services are possible nowadays based on local landscapes/ecosystem studies (Voinov et al. 2017; Scholz 2015, etc.). Our relevant recent ecological–economic assessments were based on the analysis of landscape/ecosystem maps (Mihalev et al. 2008) completed by field

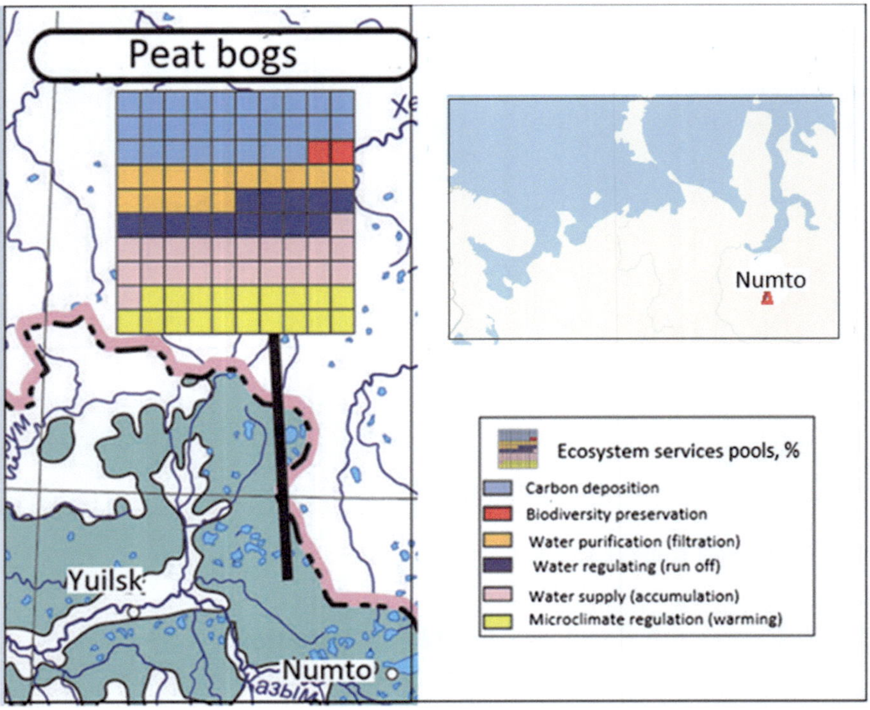

Fig. 24.8 Ecosystem services provided by peat bogs in Numto region

Fig. 24.9 Ya'Erv TTNU in Nenets autonomous Okrug

Table 24.2 Carbon deposition in ecosystems; Ya'Erv study area

	Ecosystems	ha	Annual deposition rate (t/ha)
1	Tundra	331,773	0.44–0.73
2	Forest tundra	7921	0.42
3	Northern taiga	8979	0.9
4	*Bogs*		
	Peat	150,461	0.65
	Grass	14,764	0.18
5	Meadows	35,691	0.23

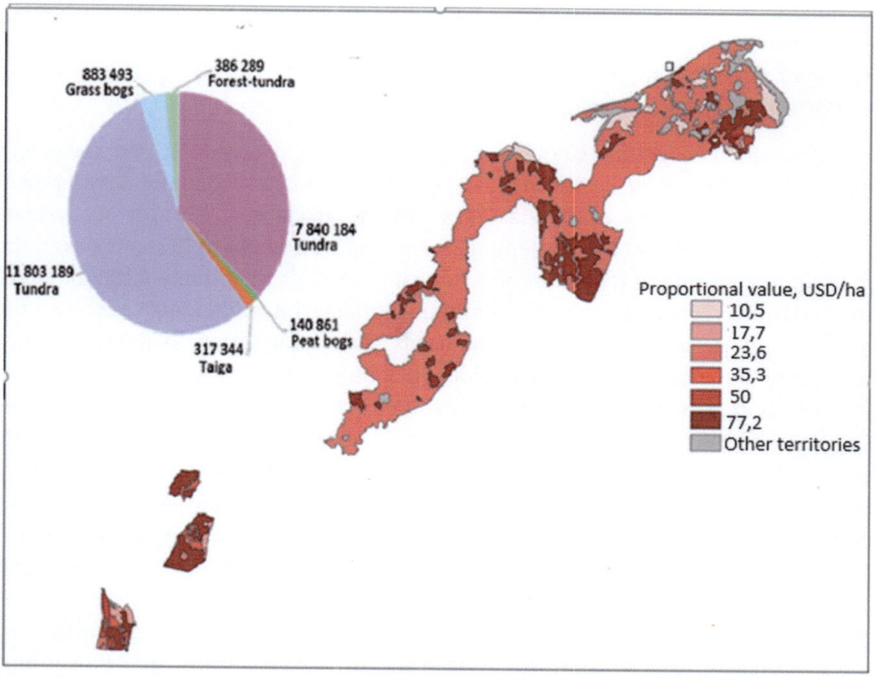

Fig. 24.10 Ecological–economic assessment map for Ya'Erv TTNU based on ecosystems map. Read commas as points (Evseev et al. 2018b)

data. The final assessment results (Fig. 24.10) were presented in economic units (USD) to demonstrate the value of different ecosystem services as productive economic landscape (diagram) determining its market price and possible economic losses in case of conflicts in environmental management as well as costs of compensatory measures.

24.7 Conclusions

Landscape/ecosystem studies form the basis for environment management conflicts forecast and control. Origin and types of land-use/environment management conflicts in the Russian Arctic are common for other regions as well in general, but still have specific regional features. Their specific features include the following (Evseev et al. 2018b):

- Large size of conflicts territories;
- Relatively young origin of the existing conflicts (about half of a century);
- Development in regions with limited resilience to anthropogenic load and unfavorable for economic development on climatic reasons and remoteness;
- Increasing landscapes structure variability because of rapid climate change;
- Direct threat to the global biosphere buffer;
- Possibility of social-ecological destabilization.

Landscapes/ecosystems studies innovative ecological–economic use nowadays appeared to be of great importance for regions of pioneer economic development such as the Arctic zone of Russia. They provide assessment data for ecosystems services pools produced by different landscapes as well as possible responses of the environment leading to reduction of ecosystems services reproduction rate or entire loss. Territorial spatial planning of environment management/land-use matching the principles of sustainable development needs such information to prevent nature management conflicts among stakeholders planning to exploit limited spatial pools of ecosystem services in their activities thus provoking ecological and social-economic conflicts. The Arctic zone is of special concern in this respect. Its ecosystems resilience to anthropogenic impact is low because of climate severity and rehabilitation costs of disturbances are very high if possible. Modern types of environment management (industrial, transport, etc.) may destroy TTNU, which support indigenous population of Northern minorities causing acute ethnic-cultural conflicts. And finally, the Arctic region territory plays an important role in regulation of many global turn-over processes providing nature buffers mitigating negative changes in biosphere. Quoting practice based on landscapes/ecosystems services studies may prevent the development of environmental management conflicts. Each territory of future economic development needs special ecosystem services reproduction areas which size must correspond to the intensity of ecosystem services exploitation. Determination of the appropriate reproduction area size is a very complicated task which needs large scale landscape/ecosystem studies. Nevertheless, the Arctic Council[1] gives the following general environment

---

[1]The Arctic Council is a high-level intergovernmental forum that addresses issues faced by the Arctic governments and the indigenous people of the Arctic. It includes eight permanent members with sovereignty over the lands within the Arctic Circle. Outside these, there are more than a dozen observer states.

management regional recommendations for land-use planning: nature protected territories-15%, "sparing" (adapted to the environment, ex.-TTNU)-30%.

Only three years ago the Arctic Council launched a special ecological–economic studies program for the Arctic region aimed to recognizes the stewardship necessity of its natural capital and reduce the pressures and threats on Arctic ecosystems by mainstreaming values of biodiversity and ecosystem services into policy making. Landscape/ecosystem studies necessity was mentioned in the recent TEEB Arctic documents (The Economics 2015). Several working groups of the Arctic Council perform these studies–SDWG, CAFF, AMAP, etc., nowadays (Arctic Council 2018).

Acknowledgements This research work was funded by the Russian Foundation for Basic Research (grant No. 18-05-00335).

References

Arctic Council (2018) https://arctic-council.org/index.php/en/. Accessed 27 Dec 2018

Bastian O, Grunewald K, Khoroshev AV (2015) The significance of geosystem and landscape concepts for the assessment of ecosystem services: exemplified on a case study in Russia. Landsc Ecol 30(7):1145–1164

Boumans R, Costanza R, Farley J, Wilson M, Portela R, Rotmans J, Villa F, Grasso M (2002) Modeling the dynamics of the integrated earth system and the value of global ecosystem services using the GUMBO model. Ecol Econ 41:529–560

Brouwer R, Brander L, Kuik O, Papyrakis E, Bateman J (2013) A synthesis of approaches to assess and value ecosystem services in the EU in the context of TEEB Amsterdam Institute for Environmental Sciences. VU University of Amsterdam 144 p

Burkhard B, Kandziora M, Hue J, Mueller F (2014) Ecosystems service potentials, flows and demands-concepts for spatial localization, indication and quantification. Landsc Online 34 (1):1–32. https://doi.org/10.3097/LO.201434

Ecological Atlas of Russia (2018) (Экологический атлас России). Feoria Publ. House, Moscow, 510 p

Evseev A, Krasovskaya T, Tikunov V, Tikunova I (2018a) Revealing of land use conflicts at indigenous population territories in the Russian Arctic using atlas information systems methodology. Geoinform Geostat Overv 1:1–6. https://doi.org/10.4172/2327-4581.1000180

Evseev A, Krasovskaya T, Tikunov V, Tikunova I (2018b) New look at territories of traditional nature use traditional nature management lands at the coastal zone of the ice silk road: a case study for the Russian Arctic. Int J Digital Earth 29:1–14. https://doi.org/10.1080/17538947.2017.1423405

Geography (2013) Geography of Nenets autonomous Okrug. General information (География НАО. Общие сведения). http://kraeved-nao.narod.ru/. Accessed 14 Jan 2019

Hardin G (1968) The tragedy of the commons. Science 162(3859):1243–1248

Haines-Young R, Potschin M (2017) Common international classification of ecosystem services (CICES): consultation on version 5.1, 26 p. http://cices.eu. Accessed 19 Nov 2017

IPY Project MODIL-NAO (2008) Monitoring of development of traditional indigenous land use areas in the Nenets Autonomous Okrug. Norwegian Polar Institute/Association of Nenets People "Yasavey", NW Russia, 176 p. https://ipy-nenets.npolar.no/. Accessed 14 Jan 2019

Mihalev G, Murashko O, Kurakin V, Yanina V, Mihaleva L (2008) Assessment of indigenous population of the North, Siberia and the far East of Russia traditional nature management lands quality (Оценка качества земель, являющихся исконной средой обитания коренных малочисленных народов Севера, Сибири и Дальнего Востока Российской Федерации), Moscow, Batany, 13 p

Millennium Ecosystem Assessment (2005) Ecosystems and human well-being: biodiversity synthesis. World Resources Institute, Washington, DC, 86 p

Raudsepp-Hearne C, Peterson G, Bennett E (2010) Ecosystem service bundles for analyzing tradeoffs in diverse landscapes. PNAS 107(11):5242–5247; https://doi.org/10.1073/pnas.0907284107

Scholz M (2015) Wetlands for pollution control. Elsevier, Amsterdam, 556 p

Semenov JU, Lysanova G (2015) The landscape mapping as a base of territorial differentiation of cost for ecosystem services (Ландшафтное картографирование как основа территориальной дифференциации стоимости экосистемных услуг). Geodesy Cartography 1:37–45. https://doi.org/10.22389/0016-7126-2015-895-1-37-45

State Program (2017) Social-economic development of the Arctic zone of the Russian federation up to 2020. New version (2017) (Государственная программа "Социально-экономическое развитие Арктической зоны Российской Федерации на период до 2020 года" Новая редакция) 139 p. http://government.ru/docs/29164/. Accessed 12 Jan 2019

TEEB (2013) The economics of ecosystems and biodiversity (2013): guidance manual for TEEB country studies. Version 1.0, 88 p. http://teebweb.org/. Accessed 20 Mar 2019

The Economics (2015) The economics of ecosystems and biodiversity (TEEB) for the Arctic: a scoping study. Akureyri, Iceland, 22 p

Tikunov VS (2004) Atlas information systems for decision making. In: Basics of geoinformatics (Атласные информационные системы для принятия решений. В кн. Основы геоинформатики), vol 2. Akademia Publication House, Moscow, pp 285–304

Voinov A, Bromley L, Kirk E, Korchak A, Farley J, Moiseenko T, Krasovskaya T, Makarova Z, Megorski V, Selin V, Kharitonova G, Edson R (2004) Understanding human and ecosystem dynamics in the kola Arctic: a participatory integrated study. Arctic 7(4):376–388

Voinov A, Çöltekin A, Chen M, Beydoun G (2017) Virtual geographic environments in socio-environmental modeling: a fancy distraction or a key to communication? Int J Digital Earth 16:1–12

Yaneva R (2016) Mapping ecosystem services supply: challenges and opportunities in geo-spatial analysis. In: Proceedings of 6th international conference on cartography and GIS, 13–17 June 2016. Albena, Bulgaria, pp 232–239. https://doi.org/10.1126/science.162.3859.1243

Chapter 25
Vegetation Monitoring on Quarries in the Russian Far North as a Basis for Creating Models and Analyzing Trends of Landscape Processes

Olga I. Sumina and Elena M. Koptseva

Abstract The area of anthropogenically disturbed lands is constantly increasing; thus, regeneration processes in landscapes, which are substantially determined by vegetation, are being studied all over the world. Industrial expansion to the Russian Far North is causing the appearance of large territories where plant and soil cover are completely destroyed and natural vegetation recovery takes place as primary succession. Direct field monitoring produces the most exact data on vegetation dynamics. This paper presents the results of our long-term study of primary succession on quarries located in 7 areas of forest tundra and southern tundra in North-West Siberia. Vegetation was monitored on 30 stationary key plots (5 * 5 m) and on the whole area of 17 quarries. The biodiversity of vegetation was considered on two levels (species and plant communities), and changes in species composition and plant diversity were analyzed. On two quarries (the New and Old Quarries) located in forest tundra, vegetation mapping allowed us to monitor changes in their vegetation as a dynamic sequence (4, 8, 12, 31, and 35 years after the start of primary succession). A quarry with a complex ground surface relief is a good full-scale model of regeneration processes developing within a heterogeneous landscape. A multivariant model of such processes has been elaborated on the basis of the vegetation maps obtained. The model describes the main trends in natural recovery in heterogeneous landscapes in the North. In diverse quarry habitats, primary succession starts at different times, not simultaneously, and the speed of succession not only differs between different habitats, but also varies with time. The highest difference in speed is observed at the initial stage (the maximum rate is typical of accumulative habitats), while at the next stage the difference between eluvial and transeluvial habitats decreases, and later, the rate of processes differs

---

O. I. Sumina (✉) · E. M. Koptseva
Department of Geobotany and Plant Ecology, Faculty of Biology, St. Petersburg State University, Universitetskaya naberezhnaya, 7/9, 199034 St. Petersburg, Russia
e-mail: o.sumina@spbu.ru

E. M. Koptseva
e-mail: e.koptseva@spbu.ru

L. Müller and F. Eulenstein (eds.), *Current Trends in Landscape Research*, Innovations in Landscape Research, https://doi.org/10.1007/978-3-030-30069-2_25

only between eluvial and accumulative sites. As opposed to the traditional viewpoint that the speed of primary succession becomes slower from stage to stage, our results have also shown that succession could accelerate during the 'middle' stages, concurrently with an increase in mosses' and lichens' participation in communities. Developing vegetation gradually eliminates sharp contrasts in environmental conditions within a quarry, because the plant cover levels out the moisture conditions and nutrient distribution. Thus, the main trend of succession in heterogenic landscapes is an increase in vegetation control over the abiotic environment and the expansion of areas under communities of moderate moistening habitats. Our long-term experiments, knowledge, and data have the potential to be included in international projects aimed at better understanding, monitoring, forecasting, and managing landscape processes in the Arctic.

Keywords Disturbed lands · Quarry · Natural vegetation recovery · Primary succession · Vegetation dynamics · Direct monitoring · Model of landscape processes · Forest tundra · Southern tundra · Russian far north · North-West Siberia

25.1 Introduction

Vegetation is a good indicator not only of environmental conditions but also of the current disturbance level, dynamics, and regenerative potential of terrestrial ecosystems. Ecosystem restoration processes after disturbances are substantially determined by vegetation: (i) plant producers are the basic component in terrestrial ecosystems; (ii) an ecosystem's regenerative potential depends on features of the plant community; and (iii) plant cover stabilizes erosion and other destructive processes in the soil, etc. Thus, vegetation monitoring is the most suitable way to gain information on the condition and dynamics of terrestrial ecosystems, and the most exact data can be obtained by direct field monitoring.

The area of anthropogenically disturbed lands is constantly increasing; therefore, studies on ecosystems' regeneration processes are significant all over the world (Frouz et al. 2008; Rowland et al. 2009; Audet et al. 2015; Marler and del Moral 2018). In the Russian literature, the large disturbed areas formed by severe effects of industry are called 'technogenic landscapes'. We have spent more than 25 years studying the vegetation of disturbed lands in different regions of the Russian Far North, but this publication is focused on data collected in the northern part of West Siberia, the region where oil and gas production is being actively developed, and associated infrastructures are being set up.

Vegetation dynamics have been studied on territories where plant cover and soil cover have been completely destroyed by bulldozing or other mechanical disturbances (quarries, bulldozed sites, etc.). Natural vegetation recovery after such impacts takes place as primary succession, which is defined as 'the establishment and subsequent development of the first assemblage of species on a previously unvegetated surface' (Miles and Walton 1993). Primary successions 'demonstrate the creative role of nature in building up complete and complex ecosystems from simple beginnings' (Bradshaw 1993).

One of the ways to assess the state of ecosystem is biodiversity assessment (Huang et al. 2018; Karadimou et al. 2018). Two levels of biodiversity are considered in our research: species and community diversity. The main goals of our study are to reveal the characteristics of plant cover reestablishment dynamics via direct monitoring and to develop a model of primary succession to analyze trends in landscape processes.

We hypothesized that:

1. The pool of species colonists on disturbed lands is made up of local flora, and alien species from other zones are not significant.
2. Species composition in vegetation on disturbed lands in different regions of the Far North is similar.
3. Biodiversity during primary succession changes gradually and slowly with no sudden jumps.
4. Even if quarries have a similar mineral substrate at the beginning of primary succession, trends differ from one site to another, resulting in the appearance of various types of plant community. Accordingly, the diversity of plant communities in disturbed territory increases.
5. A quarry with a complicated ground surface relief is a good full-scale analogous model to study regeneration processes in ecosystems developing within a heterogeneous landscape.

25.2 Materials and Methods

The results presented in this publication are based on data obtained in 1993–2008 and 2015–2016 in North-West Siberia, in five localities of forest tundra (neighborhood of the towns of Labytnangi, Novy Urengoy, Nadym, Gubkinsky, and the settlement of Pangody) and two localities of southern tundra (the 'Payuta' and '143 km' stations on the railway to Yamal Peninsula) (Fig. 25.1). Detailed descriptions of our field data collection and analysis methods are given in our publications mentioned below.

The vegetation was monitored, including full records of species composition and descriptions of plant communities, on key stationary plots as well as in different quarries.

A quarry without surface leveling after exploitation is a heterogeneous territory with a complicated relief and a large variety of habitats (Fig. 25.2).

Thirty key plots (5 * 5 m) placed within plant communities (characterized by homogeneous stands, constant physiognomy, and dominant species) were set up in different quarry habitats situated in forest tundra and southern tundra. These permanent plots, located in sites with relatively homogeneous environmental

Fig. 25.1 Locations of our study areas in North-West Siberia: 1—Labytnangi town, 2—'143 km' station, 3—'Payuta' station, 4—Nadym town, 5—Pangody settlement, 6—Novy Urengoy town, 7—Gubkinsky town

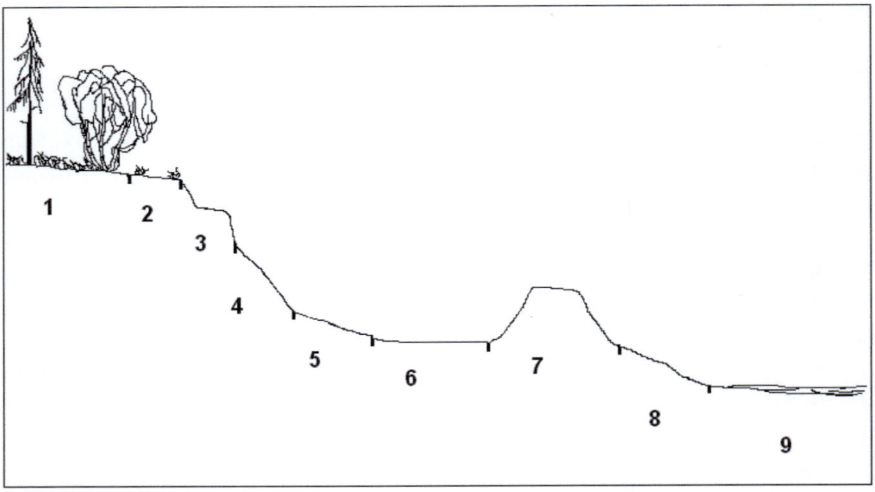

Fig. 25.2 Habitat diversity in a quarry. 1—undisturbed forest tundra community, 2—edge part with fragments of old plant cover, 3—stairs on slopes, 4—steep slopes, 5—alluvium of ground under the slopes, 6—flat surfaces on the quarry floor, 7—heaps of soil at the bottom, 8—gentle slopes, 9—ponds

conditions, were described periodically. Each relevé included a total list of species and estimations of the total vegetation cover, a cover of vascular plants, mosses, and lichens, with cover-abundance estimates for each species.

To assess the communities' diversity in quarries, numerous other plots of 5 * 5 m apart from the key plots were described, so the classification according to the Braun-Blanquet approach made for North-West Siberia was based on 604 relevés (Sumina and Koptseva 2004; Sumina 2012a, 2013). For the nomenclature of vascular plants, we used the system developed by Czerepanov (1995).

During the monitoring, we periodically noted the total list of species in each of 17 quarries. Two of them, situated in forest tundra near the town of Labytnangi, were studied in detail as 'model quarries.' From 1995 to 2003, we mapped their vegetation annually. These sandy-loam quarries both replaced open spruce–larch–birch woodland, were about 1 ha in size, and were situated at a distance of 7 km from each other.

In 1995 (the year mapping started), the first quarry was 4 years old, and we named it the New Quarry (Fig. 25.3a). Another was named the Old Quarry, as it was more than 30 years old when we started mapping there in 1998 (Fig. 25.3b). Analysis of the monitoring data showed that a period of 4 years is optimal to reveal changes in vegetation (Koptseva et al. 2007); thus, the maps of 1995, 1998, 2003

Fig. 25.3 General view of the two model quarries: the New Quarry (a—1994, b—2007) and the Old Quarry (c—1999, d—2007)

(for the New Quarry), and 1999, 2003 (for the Old Quarry) were selected for further processing. Similarities between the two model quarries in terms of environment and size gave us an opportunity to treat the changes in their vegetation as one dynamic sequence (4, 8, 12, 31, and 35 years after the start of primary succession). Maps were created using the computer program MapInfo, which allowed us to estimate the relative area of all selected contours in percentages (Sumina 2010, 2013).

To compare the plant communities formed in quarries with undisturbed vegetation, the 24 relevés of open spruce–larch–birch woodland, birch forest, willows and alder shrubs, as well as tundra with *Betula nana*, were made in plots of 5 * 5 m.

25.3 Results

25.3.1 Biodiversity on the Species Level

In quarries

All our data confirm that the colonization of technogenic habitats is determined by the pool of the local flora. The family of *Poaceae* grasses occupies the first position in species diversity; both this family and the aster family (*Asteraceae)* and willow family (*Salicaceae)* are very typical for disturbed lands of the Russian Far North. The proportion of species with arctic, arcto-alpine, and boreal types of areas does not differ in quarries from that in the local flora. The idea that anthropogenic habitats are mostly colonized by southern boreal species, the process known as 'borealization' in Russian terminology (Gruzdev 1990; Yurtsev 1995; Akul'shina et al. 1996), was not confirmed.

The species composition of vascular plants on technogenically disturbed lands in each study area is very specific. The Jaccard similarity index score was not more than 30%, unlike pioneer mosses and lichens, which are more similar in different areas (Sumina 2013). The difference in the set of vascular plant species is considerable not only between regions with different local floras, but even between quarries within the same region. Every quarry has its own set of species. The widespread idea of the similarity ('unification' in Russian terminology) of floras in anthropogenically disturbed territories (Malyshev 1981; Yurtsev 1995) arises from the comparison of lists of the most widespread and active plants on disturbed lands. However, full lists of species show significant differences and an absence of 'unification.'

According to our findings, the speed and success of disturbed sites' colonization by plants and the species diversity of developing vegetation are affected by such factors as geographical location, which determines the richness of the local flora, heterogeneity of the territory (large variety of habitats), moisture, granulometric composition of the substrate, and time (Sumina 2013).

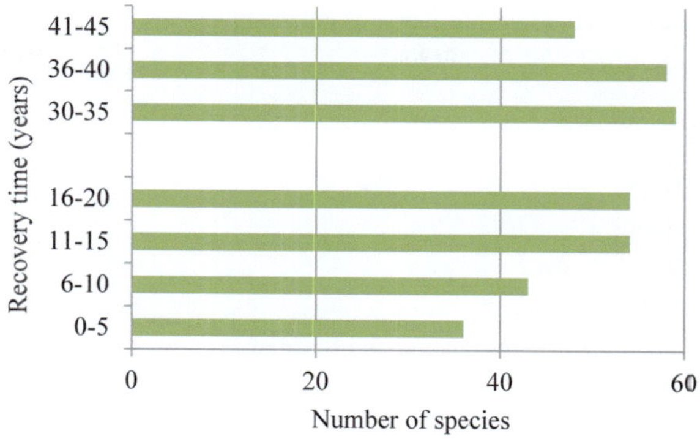

Fig. 25.4 Change in average number of vascular plant species on quarries with different times of natural recovery. No 20–30-year-old quarry was found

The total number of vascular plant species on a quarry increases at the beginning of primary succession (Fig. 25.4). Later, it decreases, because some pioneer species are no longer found in old quarries, but at the same time the mosses and lichens appear which are typical of tundra communities.

In plant communities

Plant communities on disturbed sites differ from undisturbed coenosis not in terms of species richness and total plant cover density, but mostly in respect of dominant species composition and undeveloped spatial structure (Fig. 25.5). The number of vascular plants per 1 m^2 is often comparable with this index in undisturbed tundra.

Typical tundra dominants, mainly shrubs, dwarf shrubs, mosses, and lichens, are not numerous and abundant in the vegetation of quarries, where perennial herbs are active and predominate (Table 25.1).

Fig. 25.5 Undisturbed tundra with herbs, dwarf shrubs, and mosses (**a**) and dense plant community with herbs, young growth of shrubs and trees, and mosses in quarry (**b**). Photographs by I. Ryabtsev (**a**) and N. Glushkovskaya (**b**)

Table 25.1 Main life forms of dominant species on quarries in the study areas

Life forms	Forest tundra		Southern tundra	
	L	NU	km	P
Trees	2	–	–	–
Shrubs	3	2	1	–
Dwarf shrubs	–	–	–	–
Polycarpic herbs	12	9	16	12
Monocarpic herbs, annual and biennial	2	–	1	–
Total number of dominants	19	11	18	12

The number of dominant species of each life form is shown. Study areas: the towns of Labytnangi (L) and Novy Urengoy (NU), the '143 km' (Km) and 'Payuta' (P) stations

25.3.2 Biodiversity on the Community Level

The classification of plant communities developing in quarries of North-West Siberia includes data from four areas (two in forest tundra, the other two in southern tundra). Seven associations, six subassociations, forty-four variants, and eighty-three subvariants of communities were identified. All of them belong to one alliance Chamerio-Matricarion hookeri ISHBIRDIN et al. 1996 (Ishbirdin 2001) of the order Chamerio-Betuletalia nanae KHUSAINOV et al. 1989 (Khusainov et al. 1989; Ishbirdin et al. 1996; Ishbirdin 2001) from the class Matricaria hookeri-Poetea arcticae ISHBIRDIN 2001 (Sumina and Koptseva 2004; Sumina 2012a, 2013).

In each study area, 2 or 3 associations were present (Table 25.2). Some of them were distributed over several areas, for example the association Chamerio-Festucetum ovinae KHUSAINOV and ISHBIRDIN 1989 (Khusainov et al. 1989), combining the communities of the initial stage of succession, has been described in three areas of southern tundra and forest tundra. This association has the highest number of variants (9). Subassociations were always found within only one area. Variants and subvariants display specific characteristics of habitats and/or stage of succession, so their number and diversity directly depend on the heterogeneity of the disturbed site and time of its

Table 25.2 Total number of syntaxa in study areas of North-West Siberia

Study area	Number of			
	Associations	Subassociations	Variants	Subvariants
Southern tundra				
Payuta	3	–	8	11
143 km	2	2	15	25
Forest tundra				
Labytnangi	3	4	13	37
Novy Urengoy	2	–	8	10

recovery. The maximum number of these low-level syntaxa was found near the town of Labytnangi (forest tundra) and near the '143 km' station (southern tundra). These areas both have a large variety of technogenic habitats.

25.3.3 Monitoring of the Key Plots' Vegetation

The data obtained confirm that the recovery process is faster in forest tundra than in southern tundra. The observed stages of primary succession in forest tundra are more numerous (up to 4) than in southern tundra (2, or rarely 3) where each stage takes more time; in other words, natural revegetation is more successful in areas located further south.

In each of the 25 key plots located in forest tundra, there were several constant species that were present in the plot throughout the monitoring period. As a rule, these species were typical native plants (apophytes) such as *Equisetum arvense, Chamaenerion angustifolium, Salix viminalis, Deschampsia obensis, Festuca ovina, Eriophorum scheuchzeri* or young trees and shrubs such as *Betula tortuosa, B. nana, Larix sibirica, Picea obovata, Salix phylicifolia,* and *S. lanata*. Species disappearance was noted in 21 out of 25 plots. Most of these species were pioneers of disturbed land colonization: *Alopecurus aequalis, Polygonum humifusum, Puccinellia hauptiana, Tripleurospermum hookeri, Artemisia tilesii*, others were hydrophytes on sites which started to dry out: *Rorippa palustris, Rumex aquatilis, Eriophorum scheuchzeri, Carex brunnescens,* and *Epilobium palustre*. However, the main change in the species composition noted at all the plots was that it grew richer. During 10 years of monitoring, 53 species settled in communities and remained in them.

The comparison of species composition in key plots in any 2 consecutive years shows a higher level of similarity than a comparison of the first and the last years of the monitoring period (Table 25.3). This means that changes in species sets occur gradually. An overall change in the dominant species, i.e., a change in the community, was noted in only one plot, located in the accumulative type of habitat within the bottom part of a quarry (Fig. 25.6). After 4 years of natural recovery, the open herb community with *Tripleurospermum hookeri* and a total plant cover of 15% was found in this plot. Over a period of 5 years, *Salix viminalis* became dominant, and after 10 years, two species (willow and *Alopecurus aequalis*) prevailed in the community, but later (after 14 years of recovery) only *Salix viminalis* was the most abundant.

Only part of the table is provided, as the data for other plots repeat those presented here.

In most of the plots, the change in the communities proceeded gradually. Once the total plant cover had increased to 50%, it remained nearly constant. Over a period of 10 years, the set of species was enriched by woody plants, mostly trees and shrub species, with dwarf shrubs the latest to appear in communities. Over 30 years, the plant cover density and composition of dominant species alter slightly, but certain changes in the community continue (changes in species composition,

Table 25.3 Comparison of vascular plant species number in key plots in different years (Jaccard index, %)

Plot number	Comparison of 2 consecutive years		Comparison of the first and the last years of observation
	Minimum value	Maximum value	
17	95	100	89
10	88	100	81
19	86	95	80
7	82	93	70
4	76	94	65
21	89	100	55
24	90	98	45
9	63	91	40
11	71	93	40

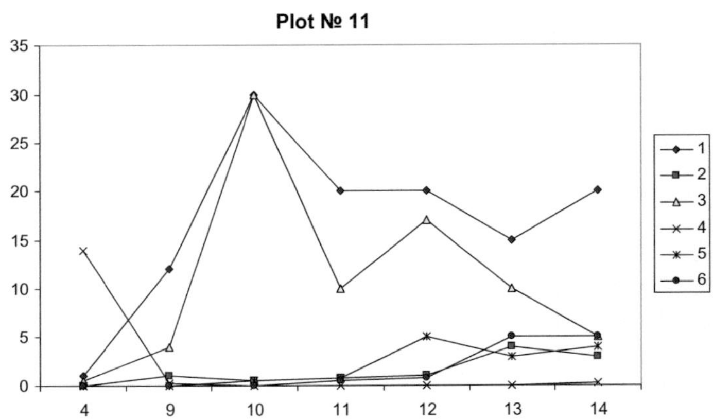

Fig. 25.6 Change in abundance of species in key plot № 11 within 10 years. Axis X—time of recovery (years), axis Y—species abundance (%). 1—*Salix viminalis*, 2—*Eriophorum scheuchzeri*, 3—*Alopecurus aequalis*, 4—*Tripleurospermum hookeri*, 5—*Beckmannia eruciformis*, 6–*Juncus nodosus*

abundance, and dominance, as well as in the set of life forms and spatial structure). Our results demonstrate that even though the vertical and horizontal structures of the community develop simultaneously, stratification by layers takes place faster than pattern formation (Sumina 2012b, 2013, Fig. 25.7).

Fig. 25.7 Plant microcommunities under tree crowns are the element of the community pattern that develop first of all

25.3.4 Monitoring Vegetation in the Model Quarries

As a rule, between 5 and 15% of the quarry area is occupied by vegetation developing on old soil–plant cover which has moved (due to overburdening) to the edges of the quarry. In 3–4 years in such habitats with the most favorable conditions, the plant cover density reaches 75–90%. Taking into account the fact that the recovery process here occurs as secondary succession, we did not consider such habitats for further analysis. The size of the area occupied by vegetation developing in primary succession within the quarry is about 20% after 4 years of recovery and more than 70% after 8 years (Table 25.4). Communities with herbs and communities with shrubs are the most widespread during the initial 8–10 years (about 30–35% and 30–40% of the quarry, respectively). Over a period of 30 years, the area of initial vegetation decreases to 12–15%, while the secondary types retain their area (40%). After 35 years, the territory is mainly (about 50%) colonized by communities with young trees, mosses, and lichens, including many typical tundra species.

In quarries of forest tundra located on the site of a former open spruce–larch–birch woodland, the development of willows and alder shrubs, open woodland, and fens can be predicted (Sumina and Beldiman 2011; Sumina 2013). This forecast is based on the results of a comparison of the most developed communities in the

Table 25.4 Areas (%) of different plant communities in the two model quarries

Parameter	Quarry name				
	New			Old	
Time of natural recovery, years	4	8	12	31	35
Open ground	69.6	13.4	9.4	–	–
Vegetation (in total)	30.4	86.6	90.6	100	100
Communities of secondary succession	9.8	14.8	16.5	4.0	4.8
Communities of primary succession, including:	20.6	71.8	74.1	96.0	95.2
Communities of herbs	6.1	37.6	27.7	14.9	12.2
Communities of herbs and mosses	0.9	0.8	2.5	8.9	5.5
Communities of herbs and shrubs (*Salix, Duschekia*)	13.6	33.4	43.9	40.6	27.0
Communities with young trees (*Larix, Betula*)	–	–	–	0.5	2.0
Communities with young trees, mosses, and lichens	–	–	–	31.1	48.5

(a) (b)

Fig. 25.8 Undisturbed plant communities in the vicinity of the town of Labytnangi: the open spruce–larch–birch woodland (**a**) and birch forest (**b**). Photographs by N. Glushkovskaya

oldest quarries and undisturbed vegetation. The coenosis of open spruce–larch–birch woodland (Fig. 25.8a), birch forest (Fig. 25.8b), willows and alder shrubs, tundra with *Betula nana*, and fens were compared with quarries' plant communities in terms of species composition and plant abundance. Multidimensional scaling by Statistica 6.0 was used for the analysis. The highest similarity with natural alder shrubs was exhibited by their equivalents on quarries, as *Duschekia fruticosa* can change an environment very quickly by forming the dense upper layer (Fig. 25.9), creating leaf litter and increasing the amount of nitrogen in the soil. The tundra with *Betula nana* had no equivalents even on the oldest quarries.

Fig. 25.9 Community with *Duschekia fruticosa* in the New Quarry after 16 years of natural recovery

For the natural recovery of forest vegetation, the exposure of the slope is important. The process is better on western and eastern slopes. On northern slopes, snow melting is too slow and some young growth dies. On the southern slopes, young trees die from spring frost or from root collar burns in the summer (Fig. 25.10).

The angle of the slope also matters. A positive relationship between the tree development and the ground surface inclination angle was revealed. *Pinus sibirica* prefers steeper slopes (up to 20–25°), while birch grows better on slopes less than 15° (Fig. 25.11).

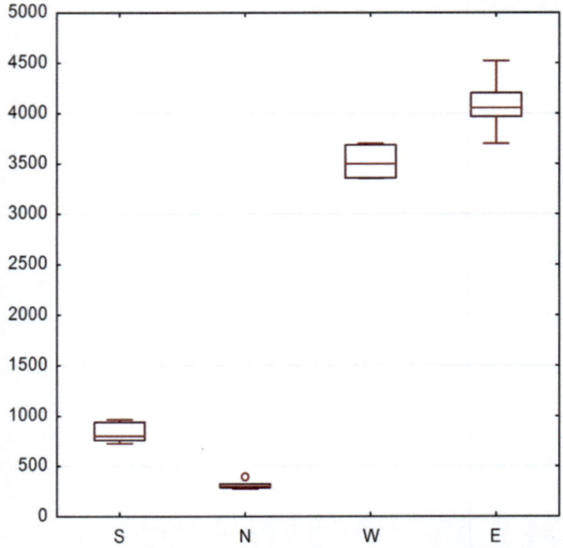

Fig. 25.10 Dependence of woody plant undergrowth density on the slope exposure. Axis X: S—south, N—north, W—west, E—east. Axis Y: number of trees per hectare

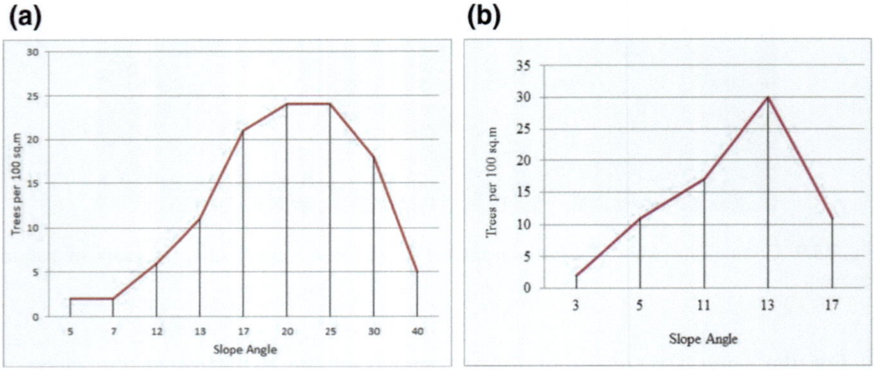

Fig. 25.11 Relationship between the average amount of undergrowth of *Pinus sibirica* (**a**) or *Betula tortuosa* (**b**) and the surface inclination angle

The comparative analysis of five maps of two model quarries revealed the features shared by their recovery processes. In concordance with the obtained results, the primary succession process differs from one habitat to another, depending mainly on substrate moisture. The main trend of succession is an increase in areas with communities of moderate moistening habitats, which occupy more than 50% of a quarry's territory after 35–40 years. The area they occupy in dry, wet, and periodically wet habitats is 20%, 5%, and 15%, respectively (Fig. 25.12). Thus, the

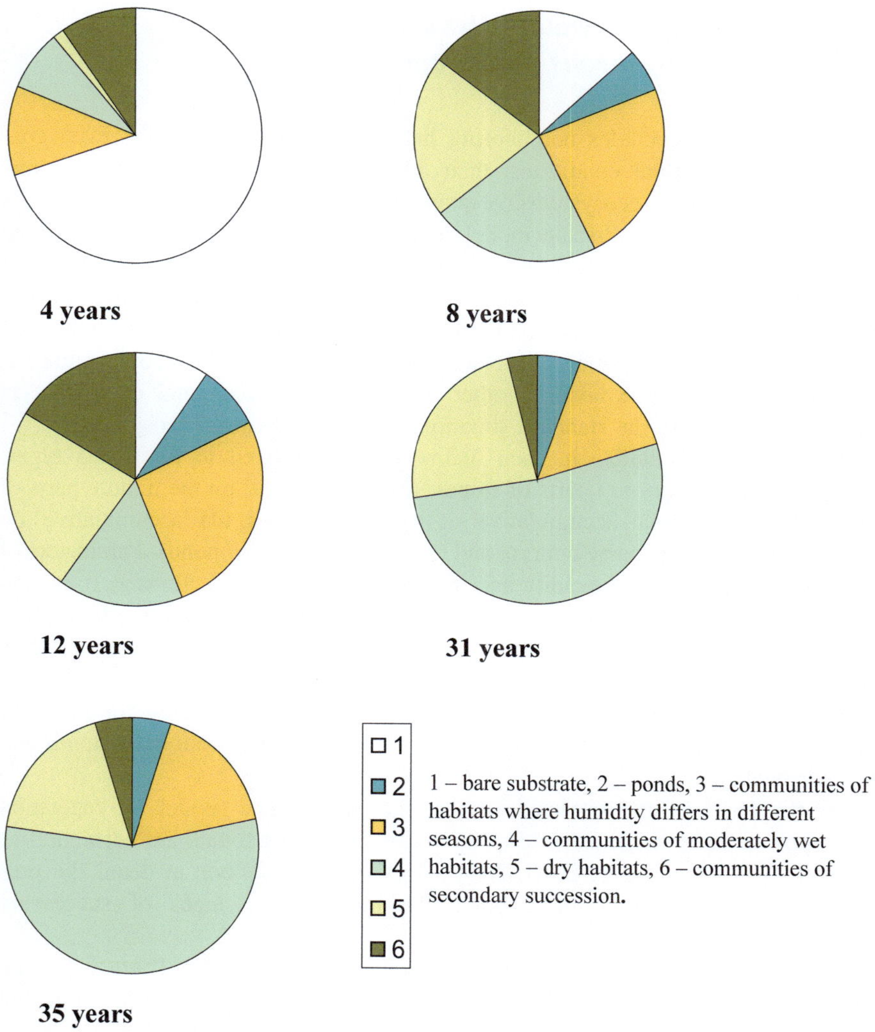

Fig. 25.12 Change during succession in proportion of the areas of communities located in habitats with different humidity. 1—bare substrate, 2—ponds, 3—communities of habitats where humidity differs in different seasons, 4—communities of moderately wet habitats, 5—dry habitats, 6—communities of secondary succession

environmental conditions became balanced (moderate moisturizing prevails) because the development of vegetation (increase in total plant cover and communities' stratification, appearance of woody plants, formation of moss–lichen cover, etc.) promotes environmental stabilization.

25.3.5 Model of Vegetation Dynamics Within a Heterogeneous Landscape

Natural landscape includes neighboring habitats with different, often even contrasting, environmental conditions; their adjacency must influence the dynamic processes going on in ecosystems. A quarry whose ground surface has a complex relief can be a good full-scale model of the regeneration processes developing within a heterogeneous landscape (Sumina 2013, 2014). Compared with other examples of primary succession, quarries used for building material extraction are the best objects of study. There is no influence either from cold water, as in the case of areas near a melting glacier, or substrates with a special chemical composition, as in the case of volcanic lava. In other words, the environmental conditions in quarries are closer to the standard situation in natural terrestrial ecosystems.

The variety of habitats in each quarry can be reduced to five main types: (1) 'eluvial' on the dry top parts of slopes, (2) 'transeluvial' on the middle parts of slopes, (3) 'transeluvial–accumulative' at the feet of slopes, (4) 'accumulative' on the flat surface of the quarry bottom, and (5) 'accumulative' in ponds. Parameters of habitats such as humidity, acidity and the proportion of fine fractions in soils, and soil nutrition increase from Type 1 to Type 5. The difference between habitats is noticeable even at the initial stage of the succession. The speed of succession differs from one habitat to another: Natural recovery is slow in eluvial habitats, and faster in transeluvial–accumulative and accumulative ones, but the earliest formation of sustainable communities is noted in the wet sites of the quarry bottom (in ponds and on their banks).

On the basis of the obtained results, the multivariate model of vegetation development on heterogeneous territory where different habitats are adjacent was elaborated (Sumina 2012c, 2013, 2014). This model describes in detail the processes occurring in different habitats during the five stages of succession (Fig. 25.13).

The start of primary succession is the same in all habitats: There is a bare substrate without plants or soil (zero moment, Fig. 25.13); however, even at that moment there are micromycetes in the ground (Sumina et al. 2010). During primary succession, biological weathering rises, and the influence of mosses and lichens on the mineral substrate is much stronger than that of vascular plants (Sumina and Lessovaia 2016). The substrate acidity gradually increases. Soil under the undisturbed surrounding open woodland is more acidic than on the quarry. The most acidic substrates are in the wet habitats of old quarries (about 35–40 years old and older). Based on our findings, pH values decrease simultaneously with the rise in moisture and plant cover density, which also influences the moisture level (Sumina and Lessovaia 2016). The impact of vegetation on abiotic conditions becomes stronger when woody plants (shrubs and trees) appear. Over time, erosion changes the relief in a quarry: Positive forms are destroyed, and negative ones are filled with moved material. These processes level off moisture conditions, a process which is

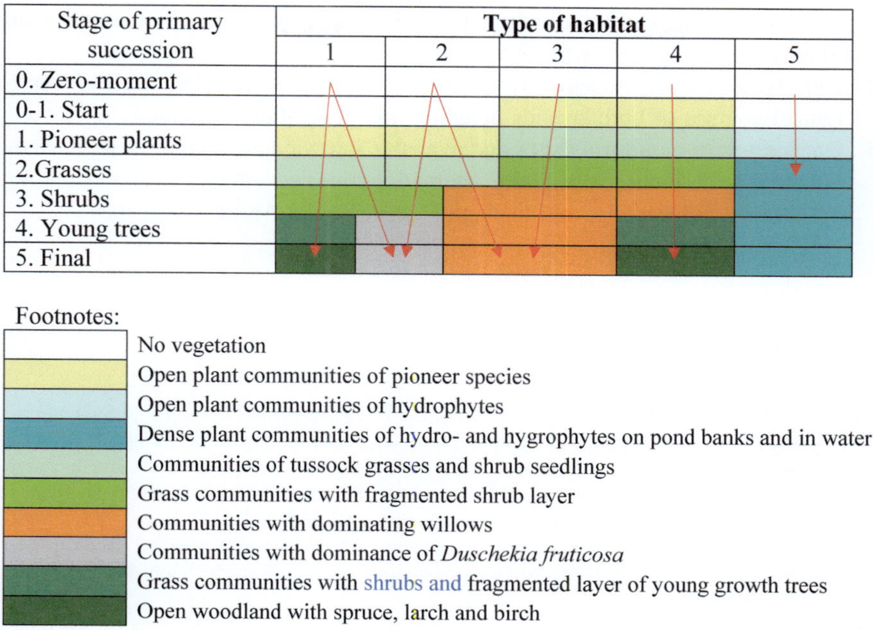

Fig. 25.13 Scheme of the multivariant model of primary succession in heterogeneous landscapes (the example of quarries in forest tundra of North-West Siberia). 1–5 types of habitats—see the text. The arrows indicate the direction of succession

also facilitated by growing vegetation. A detailed description of all successional stages is found in the above-mentioned publications.

25.4 Conclusions

1. Long-term vegetation monitoring is an important tool for understanding landscape processes under the impact of natural factors and human activity.
2. Our results confirmed 4 out of 5 preliminary hypotheses. The 'unification' of plant species composition on disturbed lands was not revealed. The vegetation of different quarries, even those located in the same area, is extremely individual in terms of the composition of species and syntaxa.
3. Natural landscapes are heterogeneous, and quarries used for building material extraction, featuring various habitats, are a good full-scale model to study processes of landscape regeneration. The main trend of succession in heterogenic landscapes is an increase in the vegetation's control over the abiotic environment and leveling out of the main factors, such as moistening and nutrient distribution. A multivariant model of these processes was developed

based on unique field monitoring data. This model should be tested and quantified in other regions of the Arctic.

4. Our studies on technogenically disturbed lands of the Russian Arctic revealed typical courses of both species diversity and vegetation dynamics.
5. Our long-term monitoring sites and available knowledge and data basis have the potential to be included in future international projects aimed at understanding, monitoring, forecasting, and managing processes in Arctic landscapes.

References

Akul'shina NP, Shushpannikova GS, Novakovskaya TV, Poznyanskaya LV (1996) Synanthropic change of flora in anthropogenic habitats in the taiga and tundra of the European Northeast (Синантропное изменение флоры на антропогенных местообитаниях в тайге и тундре Европейского Северо-Востока). In: Vil'chek GE, Sumina OI, Tishkov AA (eds) Flora of anthropogenic habitats of the North, Moscow, pp 31–52 (in Russian)

Audet P, Pinno BD, Thiffaultc E (2015) Reclamation of boreal forest after oil sands mining: anticipating novel challenges in novel environments. Can J For Res 45(3):364–371. https://doi.org/10.1139/cjfr-2014-0330

Bradshaw AD (1993) Introduction: understanding the fundamentals of succession//Primary succession on land. In: Miles J, Walton DWH (eds) Special publication number 12 of the British ecological society. Blackwell Scientific Publications, Oxford, London, pp 1–3

Czerepanov SK (1995) Vascular plants of Russia and neighboring countries (within the limits of former USSR) (Сосудистые растения России и сопредельных государств (в пределах бывшего СССР) Mir i Semia-95. Saint-Petersburg, pp 992 (in Russian)

Frouz J, Prach K, Pizl V, Hanel L, Stary J, Tajovsky K, Materna J, Balik V, Kalcik J, Rehounkova K (2008) Interactions between soil development, vegetation and soil fauna during spontaneous succession in post mining sites. Eur J Soil Biol 44(1):109–121. https://doi.org/10.1016/j.ejsobi.2007.09.002

Gruzdev BI (1990) Synanthropic flora of the coast of the East European tundra (Синантропная флора побережья восточно-европейских тундр). In: Influence of anthropogenic factors on flora and vegetation of the North. Proceedings of the Komi scientific center of the USSR academy of sciences 108. Syktyvkar, pp 28–34. ISSN: 0135-5813 (in Russian)

Huang CB, Zhou ZX, Peng CH, Teng MJ, Wang PC (2018) How is biodiversity changing in response to ecological restoration in terrestrial ecosystems? A meta-analysis in China. Sci Total Environ 650(1):1–9. https://doi.org/10.1016/j.scitotenv.2018.08.320

Ishbirdin AR (2001) Ecological and geographical patterns of synanthropic floras and vegetation formation in residential areas of Russia (Эколого-географические закономерности формирования синантропных флор и растительности селитебных территорий России). Thesis of doctoral dissertation in biology, Moscow, p 342 (in Russian). http://www.dslib.net/botanika/jekologo-geograficheskie-zakonomernosti-formirovanija-sinantropnyh-flor-i.html. Accessed on 22 Mar 2019

Ishbirdin AR, Ishbirdina LM, Khusainov AF (1996) On some regularities in flora and vegetation of populated localities of the North-West Siberia (О некоторых закономерностях флоры и растительности населенных пунктов Севера Западной Сибири). In: Vil'chek GE, Sumina OI, Tishkov AA (eds) Flora of anthropogenic habitats of the North, Moscow, pp 79–101 (in Russian)

Karadimou E, Kallimanis AS, Tsiripidis I, Raus T, Bergmeier E, Dimopoulos P (2018) Functional diversity changes over 100 year of primary succession on a volcanic island: insights into assembly processes. Ecosphere 9(9):1–17. https://doi.org/10.1002/ecs2.2374

Khusainov AF, Ishbirdin AR, Nazirova ZM (1989). Experience in the floristic classification of natural and disturbed tundra vegetation in the area of the "Medvezhye" deposit (Western Siberia) (Опыт флористической классификации естественной и нарушенной растительности тундр района месторождения «Медвежье» (Западная Сибирь). Moscow, VINITI, Biological sciences, p 31 (ВИНИТИ. Биологич. Науки). № 189-B89 (in Russian)

Koptseva EM, Sumina OI, Tyagnereva OS (2007) Monitoring of the biodiversity of vegetation on quarries in forest tundra of Western Siberia through direct stationary observations (Мониторинг биоразнообразия растительности карьеров лесотундры Западной Сибири путем прямых стационарных наблюдений). In: Materials of All-Russian conference "biodiversity of the far North ecosystems: inventory, monitoring, protection" Komi Scientific Centre, Syktyvkar, pp 48–56 (in Russian)

Malyshev LI (1981) Changes in the global floras under anthropogenic impact (Изменение флор земного шара под влиянием антропогенного давления). Biologicheskiye nauki 3:5–20 (in Russian)

Marler TE, del Moral R (2018) Increasing topographic influence on vegetation structure during primary succession. Plant Ecol 219(8):1009–1020. https://doi.org/10.1007/s11258-018-0853-z

Miles J, Walton DWH (eds) (1993) Primary succession on land. Special publication number 12 of the British ecological society. Blackwell Scientific Publications, Oxford, p 309

Rowland SM, Prescott CE, Grayston SJ, Quideau SA, Bradfield GE (2009) Recreating a functioning forest soil in reclaimed oil sands in Northern Alberta: an approach for measuring success in ecological restoration. J Environ Qual 38(4):1580–1590. https://doi.org/10.2134/jeq2008.0317

Sumina OI (2010) Vegetation development on bare grounds: results of long-term study of natural vegetation recovery on two sandy quarries in forest-tundra of the West Siberia (Формирование растительности на свободных субстратах: итоги многолетних наблюдений за зарастанием двух песчаных карьеров в лесотундре Западной Сибири). Botanicheskiy Zh 95(4):562–580 (in Russian)

Sumina OI (2012a) Classification of far North technogenic habitat vegetation: new associations of alliance Chamerio-Matricarion hookeri (Ishbirdin et al. 1996) Ishbirdin 2001 (Классификация растительности техногенных местообитаний Крайнего Севера: новые ассоциации союза Chamerio-Matricarion hookeri (Ishbirdin et al. 1996, Ishbirdin 2001) Rastitel'nost' Rossii 20:67–108 (in Russian)

Sumina OI (2012b) Development of spatial structure of plant communities in the course of primary succession (Формирование пространственной структуры растительных сообществ в ходе первичной сукцессии) Botanicheskiy Zhurnal 97(10):1351–1363 (in Russian)

Sumina OI (2012c) Multivariant model of primary succession of vegetation on heterogeneous territory (on example of quarries in forest-tundra) (Поливариантная модель первичной сукцессии растительности на экотопически гетерогенной территории (на примере карьеров лесотундры). Usp sovremennogo yestestvoznaniya 11(1):112–116 (in Russian)

Sumina OI (2013) Development of vegetation on human-made habitats of the Russian Far North (Формирование растительности на техногенных местообитаниях Крайнего Севера России). Inform-Navigator, Saint-Petersburg, p 340 (in Russian)

Sumina OI (2014) Primary successions on quarries as a full-scale model for study of terrestrial ecosystems development (Первичные сукцессии на карьерах как натурная модель для изучения процессов формирования наземных экосистем). Theor Appl Ecol 1:40–44 (in Russian)

Sumina OI, Beldiman LN (2011) Natural vegetation recovery on quarries in the forest-tundra of Western Siberia: predictions of recovery successions (Зарастание карьеров лесотундры Западной Сибири: прогноз восстановительных сукцессий) Vestnik SPbGU, ser. Biolog 2:13–27 (in Russian)

Sumina OI, Koptzeva EM (2004) Diversity and dynamic of quarries vegetation in forest-tundra zone of Western Siberia (vicinity of town Labytnangi, Yamal-Nenets autonomous District) (Разнообразие и динамика растительности карьеров в лесотундре Западной Сибири (окрестности г. Лабытнанги, Ямало-Ненецкий АО). Rastitel'nost' Rossii 6:83–103 (in Russian)

Sumina OI, Lessovaia SN (2016) Clay minerals in the loose substrate of quarries affected by vegetation in cold environment (Siberia, Russia).In: Frank-Kamenetskaya OV et al (eds) Biogenic-abiogenic interactions in natural and anthropogenic systems. Series lecture notes in earth system sciences. Springer International Publishing Switzerland. https://doi.org/10.1007/978-3-319-24987-2_20, pp 249–259

Sumina OI, Vlasov DYu, Dolgova LL, Safronova EV (2010) Peculiarities of micromycetes communities' formation on the overgrowing sandy quarries of the north of Western Siberia (Особенности формирования сообществ микромицетов в зарастающих песчаных карьерах севера Западной Сибири). Vestnik SPbGU Ser Biol 2:84–90 (in Russian)

Yurtsev BA (ed.) (1995) Anthropogenic dynamics of arctic and subarctic plant cover: principles and methods of study (Антропогенная динамика растительного покрова Арктики и Субарктики: принципы и методы изучения).In: Proceedings of the Komarov Botanical Institute, vol. 15. Russian Academy of Sciences, St. Petersburg, p 185. ISBN 5-201-11088-6

Chapter 26
The Eddy-Covariance Method and Its Application from the Volga Steppe Region to Eastern Siberia

Thomas Foken, Mikhail Strunin and Mathias Göckede

Abstract The eddy-covariance method is a tool that uses the high-frequency turbulent fluctuations within the atmosphere to determine energy and matter fluxes between the atmosphere and the underlying surface. The most important device is the sonic anemometer to measure the fluctuations of the three-component wind vectors, with the first instruments developed in the 1960s in Moscow and tested in the Volga steppe region. This paper presents a general overview on eddy-covariance methodology and instrumentation, including sophisticated updates based on wavelet analysis that help to overcome theoretical limitations with respect to investigating fluxes at high temporal resolution, and above large landscapes. As a first example for these updates, we present high-resolution methane fluxes in the Siberian permafrost area characterized by highly non-steady-state conditions. In a second example, we analyze aircraft measurements in the Lena region, which presents a horizontal distribution of the sensible heat fluxes from the surface to the boundary layer height, demonstrating the effect of the topography on the turbulent fluxes.

Keywords Eddy-covariance method · Sonic anemometer · Methane fluxes · Sensible heat fluxes · Aircraft measurements · Wavelet analysis · Volga step region · Chersky · Lena River · Steppe landscape · Arctic landscape

T. Foken (✉)
Bayreuth Center of Ecology and Environmental Research (BayCEER), University of Bayreuth, 95440 Bayreuth, Germany
e-mail: foken@micrometeorology.de

M. Strunin
Federal State Budgetary Institute "Central Aerological Observatory" of ROSHYDROMET (FSBI "CAO"), Pervomayskaya Str., 3, 141700 Dolgoprudny, Moscow Region, Russia

M. Göckede
Max-Planck-Institute for Biogeochemistry, Hans-Knöll-Straße 10, 07745 Jena, Germany

L. Müller and F. Eulenstein (eds.), *Current Trends in Landscape Research*, Innovations in Landscape Research, https://doi.org/10.1007/978-3-030-30069-2_26

26.1 Introduction

The eddy-covariance method is nowadays a widely used observational approach for determining fluxes of energy and matter between the surface and the atmosphere, with applications not only in meteorology but also in ecology and applied sciences. The first instrumentation devices that facilitated a practical realization of eddy-covariance were developed in Moscow and tested in the Volga steppe region. The presented paper gives an overview of the development of the method from its early stages nearly 70 years ago to the present. Furthermore, through two examples based in Siberia, we demonstrate how eddy-covariance can be applied for landscape-scale investigations. One of the important challenges in climate research is to quantify the potential impact of increasing methane emissions linked to the thawing of permafrost soils in a warming Arctic climate. Here, we present observational studies from Northern Siberia based on the eddy-covariance approach, one focusing on local scale fluxes in the Chersky region and the second on landscape-scale fluxes using aircraft measurements over the Lena region. Both studies aim at improving our understanding of permafrost carbon cycle processes under global climate change, therefore contributing to the reduction of uncertainties in the future climate projections.

26.2 The Eddy-Covariance Method

The eddy-covariance method was originally proposed at about the same time by Montgomery (1948), Swinbank (1951), and Obukhov (1951). Obukhov emphasized the importance of this method for the development of parameterizations of the energy and matter exchange between the atmosphere and the underlying surface, which are today an important part of all weather and climate models. Still, after the presentation of the theoretical foundations, more than ten years of research were required until the first design of today's anemometers was finalized by Bovscheverov and Voronov (1960), and only some years later were similar instruments developed in the USA (Kaimal and Businger 1963) and Japan (Mitsuta 1966). A crucial step toward the general applicability of these new instruments was the organization of sonic anemometer inter-comparison experiments, with one of the first being performed in the Volga steppe region near Tsymlyansk (Tsvang et al. 1973).

The basic equation for the measurements of an energy and matter flux F_x in kinematic units is

$$F_x = \overline{w'x'} \tag{1}$$

with w' giving the fluctuation of the vertical wind component and x' representing the fluctuations of the horizontal wind component in the case of the momentum flux, of the temperature in the case of the sensible heat flux, the water vapor concentration in the case of the latent heat flux (evapotranspiration), or the concentration of a trace gas in the case of any trace gas flux. All of these fluctuations must be measured with a frequency of 10–20 Hz. These high sampling frequencies and the associated storage of the large amount of data were a major obstacle in the early years of application, but due to advances in the development of electronic components even higher measurement frequencies can now be realized.

The most important instrument for the eddy-covariance method is the sonic anemometer. The first sonic anemometers used the phase shift method (Bovscheverov and Voronov 1960; Kaimal and Businger 1963, Fig. 26.1). In this method, the ultrasonic signal emitted by the transmitter is received at several points, and wind velocities can be derived as a function of the phase difference between the transmitted and received signals. Modern sonic anemometers use the travel time principle and a direct time determination (Hanafusa et al. 1982). In this method, a

Fig. 26.1 Russian phase shift anemometer during the experiment ITCE-81(Tsvang et al. 1985) at the Tsymlyansk research area (left) together with the Japanese Kaijo-Denki sonic anemometer PAT 300, working with the travel time principle (Photograph: Foken)

sonic signal is transmitted from both sides of a measurement path and received on the opposite sides. Since the propagation of these signals is influenced by the wind velocity, one signal is faster than the other. The exact travel times of the sonic signals are subsequently used for the determination of the wind velocity:

$$t_{1,2} = \frac{\sqrt{c^2 - u_{\mathrm{n}}^2} \pm u_{\mathrm{d}}}{c^2 - u^2} d \tag{2}$$

where d is the path length, u_{d} is the wind component along the path, u_{n} is the normal component of the wind, and c is the sound velocity.

The difference of the reciprocal travel times gives the wind velocity, and the sum of the reciprocal travel times can be used to derive the sound velocity:

$$\frac{1}{t_1} - \frac{1}{t_2} = \frac{2}{d} u_{\mathrm{d}} \tag{3}$$

$$\frac{1}{t_1} + \frac{1}{t_2} = \frac{2}{d} c \sqrt{1 - \frac{u_{\mathrm{n}}^2}{c^2}} \approx \frac{2}{d} c \tag{4}$$

The sound velocity is a function of temperature and the moisture

$$c^2 = 403T(1 + 0.32e/p) \tag{5}$$

using the partial pressure of water vapor e and the air pressure p. The product of the temperature T and the moisture dependent term is called the sonic temperature and the flux that can be calculated based on the sonic temperature is the buoyancy flux, which can be transformed with the moisture flux into the sensible heat flux.

The moisture and other trace gas fluxes can be measured with trace gas analyzers at a high time resolution according to Lambert–Beer's law

$$I = I_0 \cdot \mathrm{e}^{-k \cdot d \cdot \frac{c}{c_0}} \tag{6}$$

where k is the absorption coefficient, d is the path length, and I_0 the radiation intensity at absorber concentration c_0. Based on this principle, the actual concentration of the trace gas in question can be derived at very high frequency from the fluctuations of the detected radiation intensity at a selected wavelength. Trace gas analyzers can either be installed as an open-path instrument near the sonic anemometer, or alternatively, air from an inlet near the sonic anemometer can be pumped through tubes to the measuring cell of a closed-path analyzer.

Nowadays, the eddy-covariance method is well established with necessary correction methods and quality control tests available in ready-to-use software packages (Aubinet et al 2012). Typically, the averaging time for fluxes is 30 min. However, in certain cases, a higher temporal resolution provides a distinctive advantage, and for both application examples presented in this chapter, flux rates for shorter time intervals and a distribution of the flux to different frequencies (sizes of the turbulent eddies) have been achieved.

The vertical turbulent flux of sensible heat can be calculated with

$$H = c_{p}\rho \overline{w'\theta'}, \tag{7}$$

where c_p is the specific heat at constant pressure, ρ is the air density, w' the fluctuations of the vertical wind component, and θ' are fluctuations of potential temperature. Regarding the sign convention, positive values of H indicate an energy transfer from the surface to the atmosphere. The vertical turbulent flux of latent heat (turbulent water vapor flux converted to thermal units) can be calculated from the expression

$$\lambda E = \lambda \rho \overline{w'q'}, \tag{8}$$

where $E = \overline{w'q'}$ is the turbulent water vapor flux with q' the fluctuations of the specific humidity, and λ is the latent heat of vaporization. From these fluxes, the Obukhov length can be calculated, which characterizes the stability of the atmospheric boundary layer and the dynamics of its development.

26.3 Eddy-Covariance Measurements with Aircrafts

Aircraft measurements allow the application of eddy-covariance techniques, where instead of a sonic anemometer a multi-channel Pitot tube and a hot-wire thermometer are used and instead of a time series, a spatial series is measured. The Russian ILYUSHIN-18D (IL-18D) is an aircraft laboratory (Strunin et al. 2004) carrying special instruments designed to measure atmospheric variables at a very high frequency during field campaigns. A gust-probe system measured the wind speed, wind direction, and air temperature (see Fig. 26.2). Turbulence measurements included horizontal (longitudinal with respect to flight direction) wind speed fluctuations, vertical wind speed fluctuations, air temperature fluctuations, and absolute air humidity fluctuations. A high-response pressure sensor, connected to a Pitot pressure probe and static pressure holes, measured dynamic pressure. A barometer, connected to the static pressure holes of the aircraft pressure system, measured static pressure. A fast-response platinum wire thermometer, specially designed for aircraft conditions, measured the temperature. The angle of attack was based on measured pressure differences in the holes of a spherical probe. Gyros measured variations in pitch angle, and a stable accelerometer measured variations in aircraft vertical acceleration. Doppler radar measured the horizontal components of the aircraft's ground speed, and the vertical component of the aircraft's ground speed fluctuations was calculated using the well-known gas-dynamics and dynamics equations of Lenschow (1972).

Fig. 26.2 Aircraft laboratory Il-18D (left) and aircraft sensors (right) used for eddy-covariance measurements (Strunin 2008)

The ILYUSHIN-18D is perfectly equipped to investigate structure and some properties of the atmospheric boundary layer (ABL) over the non-homogeneous underlying surfaces. For this purpose, a special scheme of flight is needed to facilitate aircraft observations at a high quality. All measurements are made at sampling legs, which are the horizontal flight paths with constant airspeed and altitude of the aircraft. As an example, Fig. 26.3 shows a map of the sampling area and flight paths over non-homogeneous terrain in the vicinity of Yakutsk in Eastern Siberia. Blue and red circles are start and end points of each flight leg. This scheme allows comparison of the data obtained over the same area but under different thermodynamic conditions. Regional flight paths that were about 90 km in the east–west direction were made at five different levels from 100 m up to 1500 m above the highest point of the underlying terrain. Observation sites for the ground-based tower measurements over a larch and pine forest site and over a grass field were located directly under the regional flight paths.

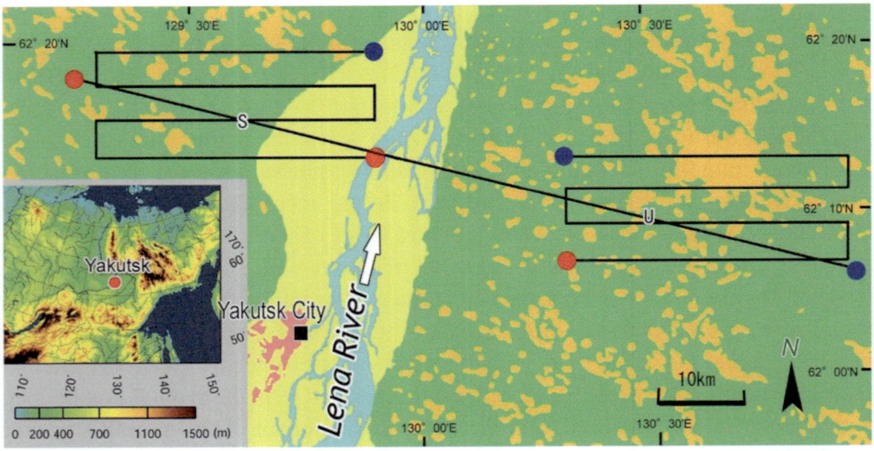

Fig. 26.3 Map of sampling area in the vicinity of Yakutsk city, including the scheme of aircraft observations over the Lena River, a larch and pine forest site (tower **S**), and those over a grass field (tower **U**) (Hiyama et al. 2007)

26.4 Flux Calculations with Wavelet Technology

Over many years, the Fourier transform has been the basic method of spectral analysis of turbulence. However, this method has a number of limitations narrowing the area of its application. The wavelet transform (Grossmann and Morlet 1984) is free of many limitations peculiar to the Fourier method and may be used in studying heterogeneous signals. The wavelet transform for a random function is determined from the expression (Kumar and Foufoula-Georgiou 1994; Torrence and Compo 1998):

$$Wf(a,b) = |a|^{-\frac{1}{2}} \int_{-\infty}^{+\infty} \Psi\left(\frac{t-b}{a}\right) f(t)\mathrm{d}t, \tag{9}$$

where a is the scale parameter and b is the shift parameter (localization). The wavelet function $\Psi(t)$ is called the basis function or basis ("mother") wavelet. The complex Morlet's wavelet is described (for non-dimensional frequency $\omega_0 > 5$) by the formula (Grossmann and Morlet 1984):

$$\Psi(t) = \pi^{-\frac{1}{4}}(\cos\omega_0 t + i\sin\omega_0 t)\mathrm{e}^{-\frac{t^2}{2}}, \tag{10}$$

As a result of application of the wavelet transform to a discrete series, the matrix of the expansion coefficients $Wf(a,b)$ of a random function $f(t)$ is calculated. The energy wavelet spectrum is determined from the expression (Kumar and Foufoula-Georgiou 1994; Torrence and Compo 1998):

$$E_f(a) = C_\Psi^{-1} \int_{-\infty}^{+\infty} |Wf(a,b)|^2 \mathrm{d}b, \tag{11}$$

where C_Ψ is a constant dependant on the basis wavelet. This constant is calculated in the following way:

$$C_\Psi = 2\pi \int_0^{\infty} \frac{|\Psi^*(\omega)|^2}{\omega} \mathrm{d}\omega, \tag{12}$$

and $\Psi^*(\omega)$ is the Fourier image of the "mother" wavelet function. If two random functions $f(t)$ and $g(t)$ are considered, it is possible to define the wavelet cross-spectrum as (Grossmann and Morlet 1984):

$$E_{fg}(a) = C_\Psi^{-1} \int_{-\infty}^{+\infty} Wf(a,b)\overline{Wg(a,b)}\mathrm{d}b, \tag{13}$$

where $\overline{Wg(a,b)}$ is the complex-conjugate wavelet transform of the random function $g(t)$. The function $|Wf(a,b)|^2$, calculated from the real and imaginary matrices of wavelet coefficients, is called the wavelet scalogram of the function $f(t)$, and the product $Wf(a,b)\overline{Wg(a,b)}$ is defined as the cross-calogram of the functions $f(t)$ and $g(t)$. The real $Co_{fg}(a)$ part of the cross-spectrum is the complete analog of the Fourier co-spectra (Lui 1994).

According to the choice of the wavelet function, the result has a better resolution of the frequencies (complex Morlet's wavelet, e.g. used by Strunin and Hiyama 2004) or in the time scale (Mexican hat wavelet, e.g. used Collineau and Brunet 1993). In the following two examples, both versions of wavelet applications were used.

26.5 Application of Eddy-Covariance Techniques and Wavlet Technology

26.5.1 Methane Fluxes in the Siberian Permafrost Area

Permafrost soils in northern high latitudes are characterized by enormous carbon stocks that are estimated to contain about 50% of the global below-ground carbon reservoir. With about two-thirds of the Arctic being classified as wetlands, the prevalence of inundated conditions also implies that permafrost regions hold the potential to become substantial future sources of methane. This very complex region is, therefore, the subject of many scientific investigations. The example presented in this chapter is focused on the Chersky region in North-East Siberia (Kittler et al. 2016; Göckede et al. 2019). In the study area, two eddy-covariance measuring complexes were installed with a sonic anemometer and a closed-path gas analyzer to measure water, methane, and carbon dioxide fluxes (Fig. 26.4).

The area is composed of extensive low-lying wetlands characterized by a patchy structure, but hilly upload tundra with small mountains is also present. Therefore, besides the geochemical situation, the conditions for micrometeorological measurements are also very challenging at times, and the recommendation for eddy-covariance measurements—steady-state conditions within the averaging time of 30 min—cannot always be fulfilled. Intermittent turbulence, katabatic flow, or microfronts are typical conditions observed during the summer months at this observation site, all of which have been shown to particularly influence methane fluxes in the form of high flux emission events (Schaller et al. 2019). To circumvent biases in methane flux calculations that are associated with intermittent turbulence, the flux calculation based on a Mexican hat wavelet that provides a good approach to determine flux rates at a time resolution of about 1 min (Schaller et al. 2017).

Figure 26.5 shows one case study for the application of a wavelet analysis to study a non-steady-state flux situation in a clear night with only a light breeze. Before 23:30, due to the weak turbulent mixing, the surface was decoupled from the

Fig. 26.4 Eddy-covariance tower at the Chersky research site (Photograph: Göckede). The system consists of a heated sonic anemometer (METEK USA1), an open-path gas analyzer for CO_2/H_2O (LiCor 7500, only right picture), and the inlet line for a closed-path gas analyzer for $CH_4/CO_2/H_2O$ (Los Gatos FGGA)

instrumentation level (~5 m a.g.l.), which led to the build-up of a concentrated pool of methane near the ground. After that, increasing turbulence eventually led to the venting of this methane pool toward the instruments at the tower top. Accordingly, the methane concentration increased rapidly by more than 500 nmol mol^{-1} around midnight. At 23:59, the 1 min wavelet flux increased rapidly from 1.9 up to 6.8 nmol mol^{-1} m s^{-1}—this is the beginning of the event, which lasted until 00:07. Exactly, in the time interval 23:30–23:59, the mean flux also increased, but its rate is much lower than the summed 1-min fluxes (Schaller et al. 2017). Based on such findings, a recommendation was derived to include fluxes calculated with a wavelet tool in the case of non-steady-state conditions, in order to avoid systematic biases to long-term flux budgets determined with the standard eddy-covariance data processing approach.

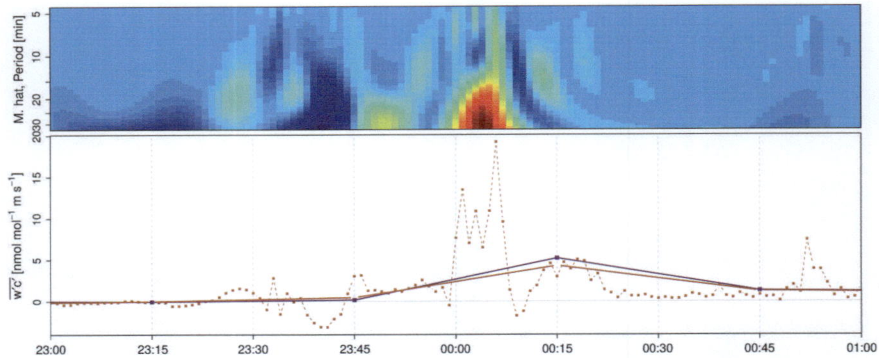

Fig. 26.5 Example of a methane flux at the Chersky research site in the night 2/3 August 2014. The flux determined with the wavelet tool and 1-min time resolution (orange lines) is compared with the conventional calculation with 30 min averaging time (eddy-covariance: violet) in the lower picture, while the upper picture shows the Mexican hat wavelet flux intensity: blue refers to the smallest, green to medium and red to highest methane flux contributions (Schaller et al. 2017)

26.5.2 Definition of the Atmospheric Boundary Layer Height and of the Local Circulations

Vertical profiles of the sensible heat flux H clearly indicate the upper boundary of convective (CBL) or stable boundary layer (SBL), because it is possible to exactly fix the level where $H = 0$. Other parameters do not give such evidence criteria. In case of the SBL, the level where $H= 0$ directly indicates the boundary layer height z_i. In the homogeneous CBL, the sensible heat flux becomes zero at the level of 0.7 z_i. The latter dependence was verified by vertical profiles of potential temperature and specific humidity obtained during numerous aircraft observations and is used now in different scaling models for the convective boundary layer (Stull 1988; Kaimal and Finnigan 1994).

The method was tested based on the aircraft observations of the mesoscale thermal internal boundary layers (MTIBLs) in the Yakutsk area (Strunin et al. 2004) and over the Sea of Okhotsk (Strunin 2008). It was demonstrated that the sensible heat flux was the best indicator of the mixing layer vertical spreading. This method is a good approach for estimating not only the height of homogeneous boundary layer z_i, but also for the fixing of local values of the depth δ_i for the CBL developed over the thermally non-homogeneous terrain. The upper layer of the non-homogeneous CBL is spatially variable, therefore we suggested the method for estimating the CBL local depth based on spatial distributions of turbulent sensible heat fluxes.

The method has the following advantages:

- It has high accuracy due to clear fixing of zero level;
- It allows determination of the local height of the boundary layer over the non-homogeneous terrain;

- It is suitable for use in the aircraft-based data analysis of horizontal distributed parameters.

A few examples of using the method for analyzing non-homogeneous boundary layers are presented below.

Investigations of the CBL over the Lena River and its banks were made in April–June 2000 in the Yakutsk area (Western Siberia) within the framework of the international mission GAME-Siberia (Hiyama et al. 2003). During this observation period, the Lena River—with a width ranging between 5 and 15 km—was substantially colder than the surrounding surface, with temperature differences between banks and water partially exceeding 15°. The sampling scheme employed within the context of this experiment is shown in Fig. 26.3, and examples of the vertical aircraft soundings are presented in Fig. 26.6. The results demonstrate that only during one day of the experiment (on May 1) did vertical profiles allow for an exact determination of the top of the CBL (Fig. 26.6a), while in the other cases, (e.g. on June, 1 and June, 19) the boundary layer heights could not be clearly constrained (Fig. 26.6b and c). Furthermore, vertical sounding could not provide horizontal structures at the CBL top, which is a very important parameter for studying boundary layers over non-homogeneous terrain. Data analysis revealed the development of mesoscale thermal internal boundary layers (MTIBLs) that radically changed the structure of the CBL. MTIBLs were obvious in the structure of virtual potential temperature, specific humidity and, especially, in the structure of vertical turbulent sensible and latent heat fluxes. Figure 26.7 shows the non-monotonous distribution of the turbulent fluxes in vertical cross sections of the sensible heat turbulent fluxes for May 1 and June 1 and 19, 2000, over the Lena River valley, as a more obvious graphical representation of the existence of MTIBLs. Surface temperature is also plotted in the figures to display surface thermal conditions as well as the horizontal scale of surface heterogeneities. Some features of the underlying surface (lowland and the Lena River) are also marked. It was shown in these figures that vertical sensible (both with latent) heat fluxes are sensitive to the presence of large-scale thermal heterogeneity at the surface, and that the fluxes can indicate the existence of MTIBLs. The CBL over homogeneous terrain is typically characterized by strong upward sensible heat fluxes, but these fluxes change their sign at some levels in the CBL. This sign change was used to mark the presence of an MTIBL. The clearest development of an MTIBL occurred on June 19. Downward heat fluxes over the river lowlands denoted stable atmospheric conditions that expanded to the top of the CBL over the cold surface. Vertical cross sections of sensible heat fluxes over the Lena River and surrounding terrain (Fig. 26.7) allow revealing of the distribution of the local values of CBL heights δ_i (Strunin and Hiyama 2005a), which are obtained from the condition that local values of sensible heat fluxes $H_i = 0$ are at the level of 0.7 δ_i.

Another detected phenomenon, besides the arising of MTIBLs described above, was a local breeze circulation (LC), which significantly changed the structure of horizontal advection up to the occurrence of reverse-airflow zones. One example of a LC arising over the Lena River banks is shown in Fig. 26.8 (Hiyama et al. 2007).

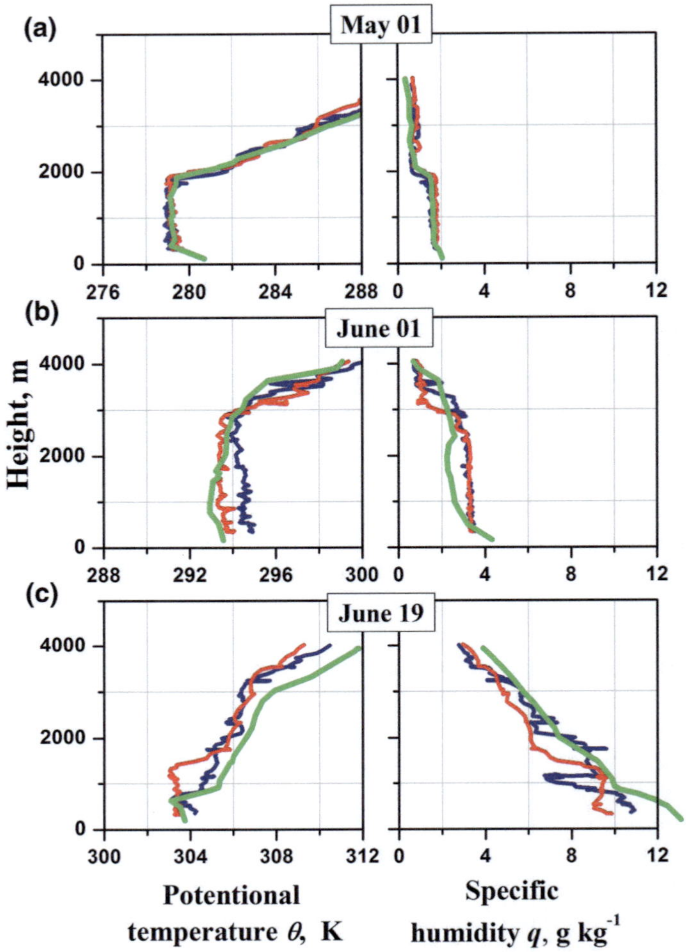

Fig. 26.6 Vertical profiles of potential temperature θ and specific humidity q over the Lena River, based on aircraft observations and radio-soundings (Strunin et al. 2004)

It is important that, unlike the usual land breeze, local vortices were generated by the cold water surface with a relatively small (about 10 km) horizontal extent. The temperature difference between the heated land surface and the cold water body of the river was, by default, considered to be the first cause of the occurrence of a local vortex. However, local vortices are also known to be formed owing to airflow around obstacles on the underlying surface.

Moreover, local vortices occurred only over the windward riverbank, which is characteristic of breeze circulation. At the same time, it is well known that a vortex arising from airflow around an obstacle is located on its leeward side (Skorer 1980). Thus, the height difference between the lowland of the river and its banks did not

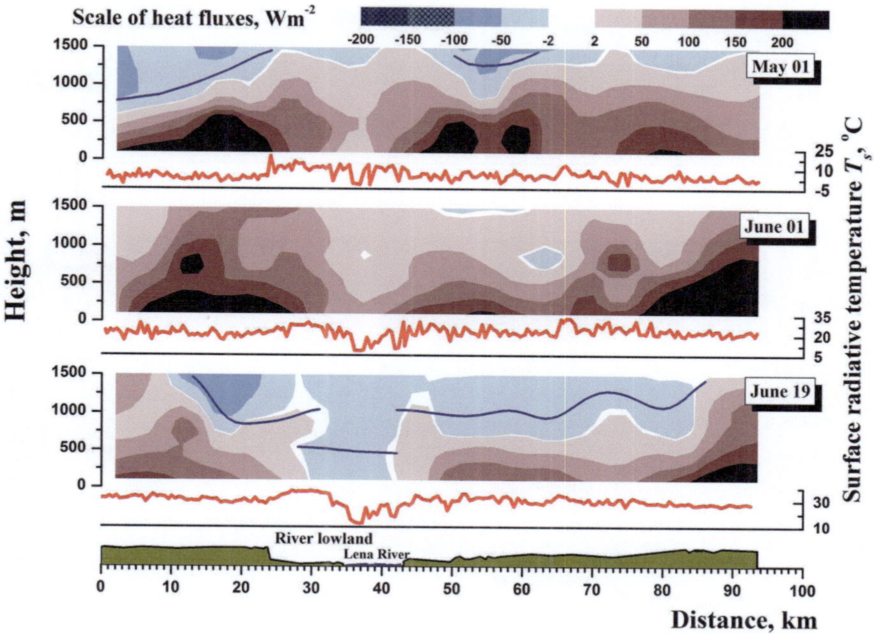

Fig. 26.7 Development of MTIBLs over the Lena River under different thermodynamic conditions (Strunin et al. 2004)

have a noticeable influence on the flow pattern, and the main cause of LC occurrence was the temperature difference between the water and bank surfaces; i.e., the LC had a clearly defined breeze character.

The necessary conditions for the formation of mesoscale layers and local breeze vortices were revealed, and the criteria for the detection of their occurrence were developed. It could furthermore be shown that the occurrence of local vortices could be one of the causes of imbalance in the budgets of heat and water vapor fluxes estimated from data of ground-based measurements.

26.5.3 Wavelet Scalogramms and Cross-Spectra

The application of wavelet techniques to investigate mesoscale thermal internal boundary layers (MTIBL) confirmed that a main feature of the MTIBL was the downward sensible heat fluxes that developed over the relatively cold river under unstable thermodynamic conditions. Wavelet cross-scalograms between vertical wind speed and air temperature fluctuations also reflect this sequence of events (see Fig. 26.9). Cross-scalograms on panels a, b, and c, which correspond to 100-, 150-, and 300-m flight levels, respectively, show strong downward fluxes over the river

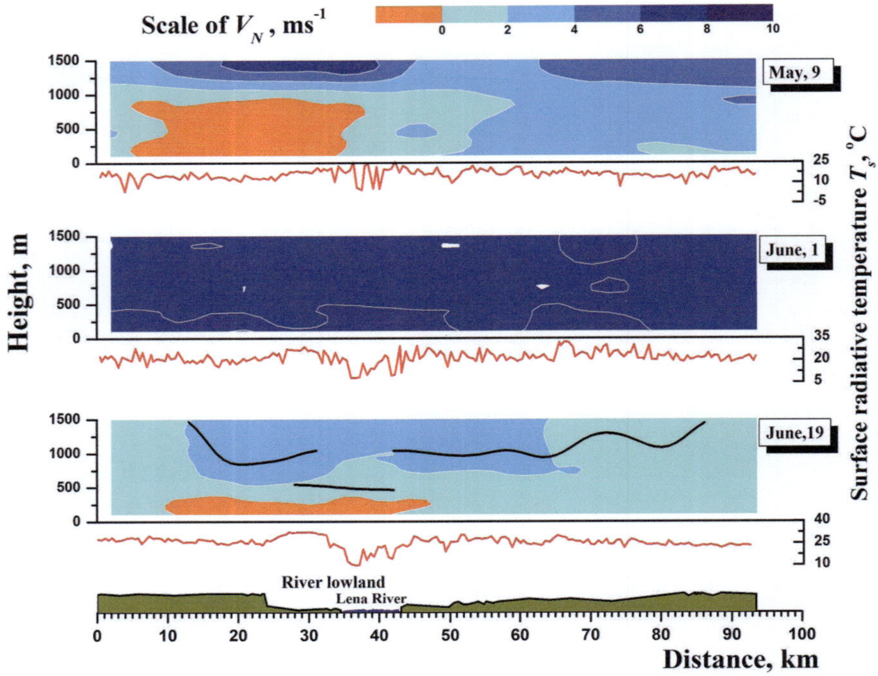

Fig. 26.8 Development of local circulation over the Lena River under different thermodynamic conditions (Hiyama et al. 2007)

lowlands at practically all scales. Cross-scalograms and vertical cross sections of sensible heat flux over the Lena River and its lowland (see Fig. 26.9) show good agreement and provide additional observational support for the MTIBL development. Cross-scalograms were the basis for the calculation of co-spectra between fluctuations of the vertical wind velocity and potential air temperature for different heights and days of the observations of MTIBLs. The co-spectra $\mathrm{Co}_{w\theta}(k)$ (see Eq. 13) were normalized by the values of kinematic heat fluxes using the flux $\left(\overline{w'\theta'}\right)_0$ averaged over the flight path at a height of 100 m (upper level of surface layer). The normalized wavelet co-spectra $\mathrm{Cn}_{w\theta}$, as functions of the normalized wave number *kz*, are given in Fig. 26.10 for different heights based on observations from May 9, 2000. The vertical evolution of the co-spectra over time facilitated the assessment of CBL development under different thermodynamic conditions. It follows from Fig. 26.10 that the behavior of the co-spectra $(\overline{w'\theta'})$ depended on the relative (normalized) height above the underlying surface and the degree of CBL stability.

The pronounced peaks of co-spectra made it possible to separate air motions of different scales within the CBL over the heterogeneous underlying surface for further analysis. It is evident that the development of eddies in the CBL is

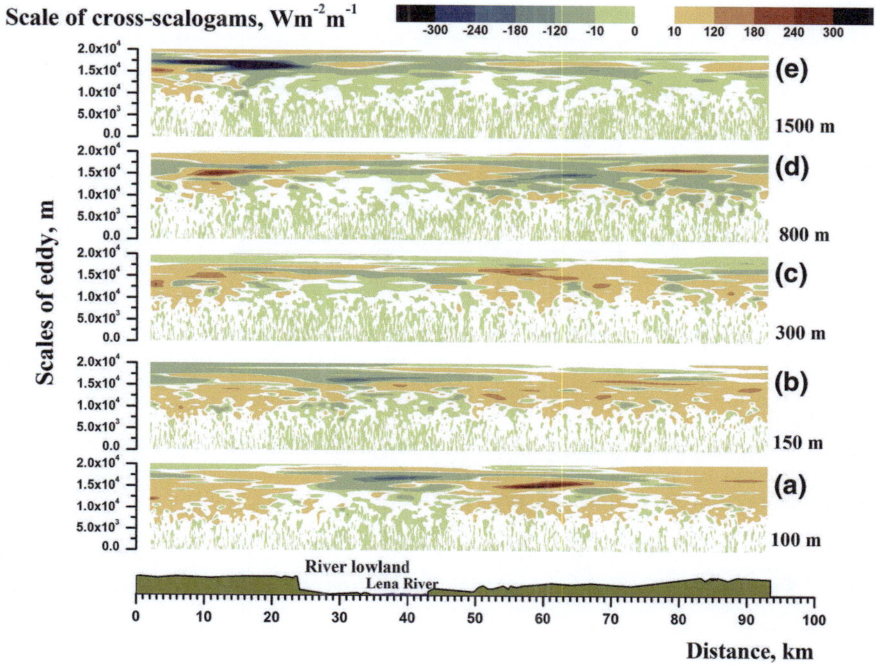

Fig. 26.9 Wavelet cross-scalograms between vertical wind speed and potential air temperature, which is the kinematic heat flux, obtained at five flight heights from 100 m up to 1500 m (panels a–e) during aircraft experiment on June 19, 2000 over the Lena River and its banks (Strunin and Hiyama 2005b)

significantly limited by scales comparable to its thickness (usually 1–2 km). It can, therefore, be assumed that chaotic air motions on scales smaller than the CBL height δ_i must obey the laws of locally isotropic turbulence and may be called purely turbulent. In the CBL, air motions on scales (approximately 2–20 km) that are larger than the CBL thickness must have absolutely different properties and may be regarded as mesoscale motions. The procedure for separating the co-spectra into turbulent and mesoscale portions is explained by the schemes given in Fig. 26.11. One can distinguish four situations typical of the position of the peaks in the co-spectra. In the most evident case, when the turbulent and mesoscale maxima are far from each other, the dimension corresponding to the co-spectrum minimum can be chosen as a boundary between the motions. The area under the co-spectrum curve on the right of the AB line determined the turbulent flux, and the area under the co-spectrum curve on the left of the AB line determined the mesoscale flux.

Based on these findings, we suggested a separation approach for the independent analysis of turbulent and mesoscale motions in the non-homogeneous CBL based on the wavelet analysis. The decomposition of co-spectra made it possible to separately calculate the turbulent and mesoscale fluxes of heat (H_T and H_M) and water vapor (E_T and E_M). The boundaries of the ranges were determined

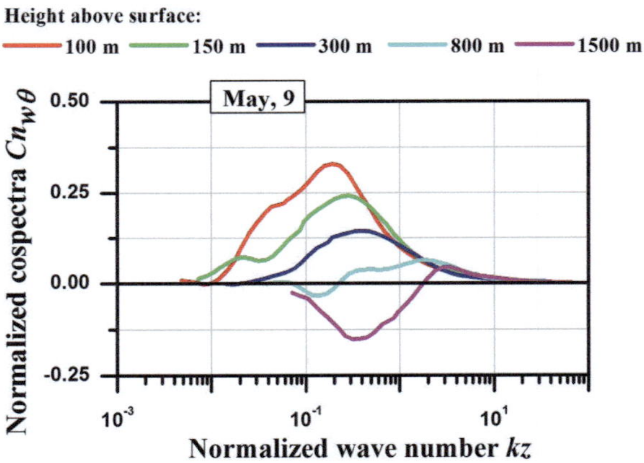

Fig. 26.10 Normalized co-spectra $Cn_{w\theta}$ between the fluctuations in the vertical wind velocity w' and the potential air temperature with respect to the normalized wave number kz for different heights on May, 9, 2000 (Strunin and Hiyama 2005b)

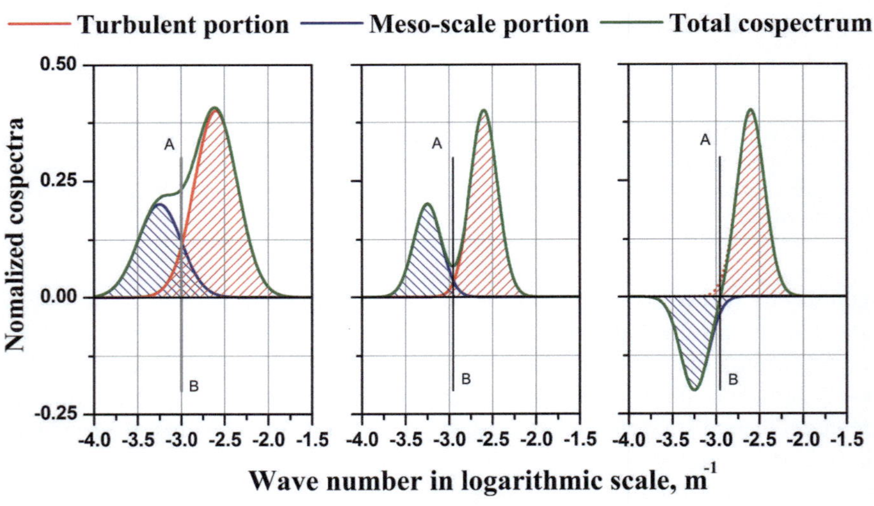

Fig. 26.11 Scheme for separating the co-spectra into turbulent and mesoscale portions. The AB line denotes the boundary scale between the portions of the co-spectra; shaded are the areas under the approximating curves for the turbulent (red curve) and mesoscale (blue curve) portions; the black full curve denotes the entire spectrum (Strunin and Hiyama 2005b)

individually for each sampling leg, but they proved to be almost the same for different fluxes in each leg. The data obtained formed the basis for the construction of empirical similarity models. The vertical similarity profiles for the turbulent and mesoscale fluxes were plotted as functions of the relative height above the land surface. The height was normalized by the reduced CBL thickness. The total flux H_0 defined as the sum of the turbulent and mesoscale portions for a height of 100 m

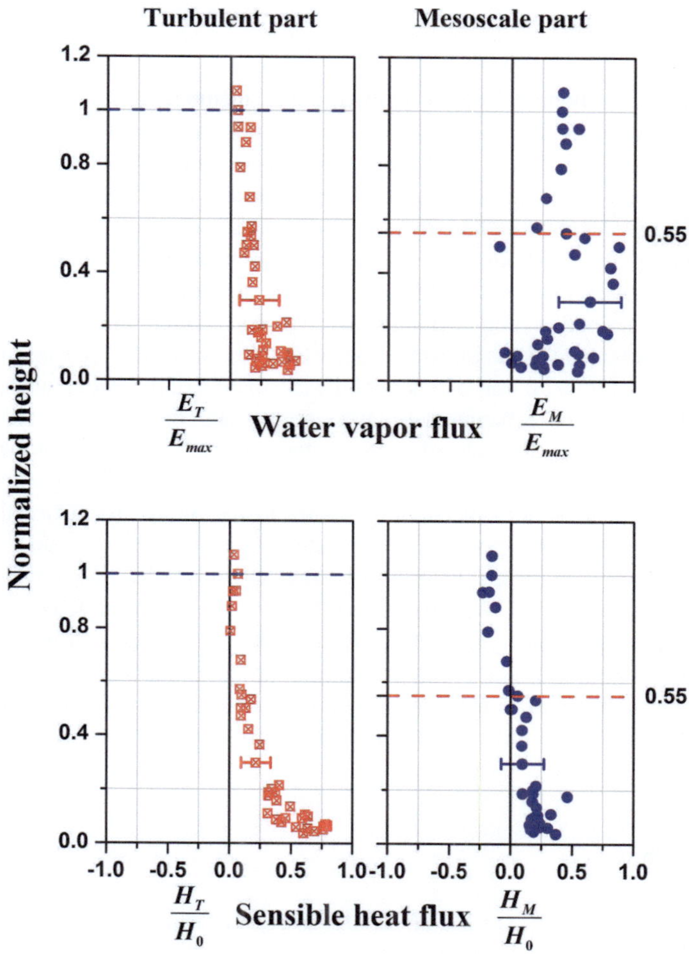

Fig. 26.12 Vertical profiles of the similarity models for the turbulent and mesoscale portions of the heat *H*, water vapor *E* fluxes (Strunin and Hiyama 2005b)

was used as a normalizing quantity for heat fluxes. The water vapor fluxes reached their maxima at a measurement level between 100 and 1500 m. The maximum flux values (E_{max}) averaged over the flight path were used as normalizing quantities. The vertical flux-similarity profiles based on the data obtained from April 24 to June 19, 2000, for all the regional legs are given in Fig. 26.12.

26.6 Conclusion

Following the development of the modern sonic anemometer in Russia, since the 1960s it has facilitated the determination of direct observations of energy and matter exchange processes between surface and atmosphere, being the core instrument for the so-called eddy-covariance method. Since the application of the eddy-covariance technique is restricted by certain theoretical assumptions, in addition to the classical flux calculation for half-hour mean values, modern mathematical methods—like the wavelet technique—have been developed as an extension to also allow flux calculation during periods of non-stationary flow, or for shorter time intervals and for different temporal scales. As shown in our examples, such new developments may significantly contribute to improvement of the calculation of the very non-steady-state methane fluxes in the Siberian permafrost region and therefore the reduction of uncertainties regarding the role of these systems in the context of global climate change. The application for aircraft measurements allows a spatial-frequency analysis of turbulent structures from the smallest scales to the mesoscale. Thus, the use of wavelet analysis makes it possible to separate air-mass motions within the heterogeneous convective boundary layer into turbulent (with scales of 20 m–2 km) and mesoscale (with scales of 2–20 km) motions. Our results demonstrated that empirical profiles of heat and water vapor fluxes for turbulent and mesoscale motions differ significantly from each other and from standard similarity models for the convective boundary layer. It is shown that the well-known similarity models describe the heat fluxes that include both turbulent and partially mesoscale motions.

References

Aubinet M, Vesala T, Papale D (eds) (2012) Eddy covariance: a practical guide to measurement and data analysis. Springer, Dordrecht. https://doi.org/10.1007/978-94-007-2351-1

Bovscheverov VM, Voronov VP (1960) Acoustic rotor. Izv AN SSSR, ser Geofiz. 6: 882–885 (Бовшеверов В.М. и Воронов В.П. (1960) Акустический флюгер. Известия АН СССР, сер. Геофизика 6: 882–885)

Collineau S, Brunet Y (1993) Detection of turbulent coherent motions in a forest canopy. Part I: Wavelet analysis. Bound Layer Meteorol 65:357–379

Göckede M, Kittler F, Schaller C (2019) Quantifying the impact of emission outbursts and non-stationary flow on eddy-covariance CH_4 flux measurements using wavelet techniques. Biogeosciences 16(16):3113–3131

Grossmann A, Morlet J (1984) Decomposition of hardy functions into square integrate wavelets of constant shape. SIAM J Math Anal 15:723–736

Hanafusa T, Fujitana T, Kobori Y, Mitsuta Y (1982) A new type sonic anemometer-thermometer for field operation. Pap Meteorol Geophys 33:1–19

Hiyama T, Strunin MA, Suzuki R, Asanuma J, Mezrin MY, Bezrukova NA, Ohata T (2003) Aircraft observations of the atmospheric boundary layer over a heterogeneous surface in Eastern Siberia. Hydrol Process 17:2885–2911

Hiyama T, Strunin MA, Tanaka H, Ohta T (2007) The development of local circulations around the Lena River and their effect on tower-observed energy imbalance. Hydrol Process 21:2038–2048

Kaimal JC, Businger JA (1963) A continuous wave sonic anemometer-thermometer. J Clim Appl Meteorol 2:156–164

Kaimal JC, Finnigan JJ (1994) Atmospheric boundary layer flows, their structure and measurements. Oxford University Press, New York, Oxford, p 289

Kittler F, Burjack I, Corradi R, Heimann M, Kolle O, Merbold L, Zimov N, Zimov S, Göckede M (2016) Impacts of a decadal drainage disturbance on surface–atmosphere fluxes of carbon dioxide in a permafrost ecosystem. Biogeosciences 13:5315–5332. https://doi.org/10.5194/bg-13-5315-2016

Kumar P, Foufoula-Georgiou E (1994) Wavelet analysis in geophysics: An introduction. Wavelet in geophysics. Academic Press Inc, San Diego, pp 1–43

Lenschow DH (1972) The measurements of air velocity and temperature using the NCAR Buffalo Aircraft Measuring System. Technical Note TN/STR-74, NCAR, Boulder, CO: 39 pp

Lui PC (1994) Wavelet spectrum analysis and ocean wind waves. Wavelet in geophysics. Academic Press Inc, San Diego, pp 151–166

Montgomery RB (1948) Vertical eddy flux of heat in the atmosphere. J Meteorol 5:265–274

Mitsuta Y (1966) Sonic anemometer-thermometer for general use. J Meteor Soc Japan Ser II 44:12–24

Obukhov AM (1951) Characteristics of the micro-structure of the wind in the surface layer of the atmosphere. Izv AN SSSR, ser Geofiz 3:49–68 (Обухов АМ (1951) Характеристики микроструктуры ветра приземном слое атмосферы. Известия АН СССР, сер. Геофизика 3:49–68)

Schaller C, Göckede M, Foken T (2017) Flux calculation of short turbulent events – comparison of three methods. Atmos Meas Tech 10:869–880. https://doi.org/10.5194/amt-10-869-2017

Schaller C, Kittler F, Foken T, Göckede M (2019) Characterisation of short-term extreme methane fluxes related to non-turbulent mixing above an Arctic permafrost ecosystem. Atmos Chem Phys 19(6):4041–4059

Skorer P (1980) Fluid aerodynamics. Izd. Mir, Moscow, 549 pp (Скорер Р. 1980. Аэродинамика окружающей среды, Izd.Мир, Москва: 549 с)

Strunin MA (2008) Development of the mixed convective atmospheric boundary layer above surfaces with changing thermal properties, generating cumulus clouds. Izd. Meteorologija I Gidrologija. Problems of cloud physics, in memorial of SM Shmetera: 362–391 (Струнин МА (2008) Развитие прерывистого конвективного пограничного слоя атмосферы над поверхностью с перемежающимися термическими свойствами, приводящее к образованию облаков кучевых форм. М. Изд. Метеорология и Гидрология. Сб. Вопросы физики облаков памяти СМ Шметера: 362–391)

Strunin MA, Hiyama T (2004) Applying wavelet transforms to analyze aircraft-measured turbulence and turbulent fluxes in the atmospheric boundary layer over eastern Siberia. Hydrol Process 18:3081–3098. https://doi.org/10.1002/hyp.5750

Strunin MA, Hiyama T (2005a) Aircraft observations of the atmospheric boundary layer over the Lena River Lowland: Part 1: Mesoscale structure. Izvest Atmos Ocean 41(2):156–177

Strunin MA, Hiyama T (2005b) Spectral structure of small-scale turbulent and mesoscale fluxes in the atmospheric boundary layer over a thermally inhomogeneous land surface. Bound-Layer Meteorol 117:479–510. https://doi.org/10.1007/s10546-005-2188-2

Strunin MA, Hiyama T, Asanuma J, Ohata T (2004) Aircraft observations of the development of thermal internal boundary layers and scaling of the convective boundary layer over non-homogeneous land surfaces. Bound Layer Meteorol 111:491–522. https://doi.org/10.1023/B:BOUN.0000016542.72958.e9

Stull RB (1988) An introduction to boundary layer meteorology. Kluwer Academic Publishers, Dordrecht, p 666

Swinbank WC (1951) The measurement of vertical transfer of heat and water vapor by eddies in the lower atmosphere. J Meteorol 8:135–145

Torrence C, Compo GP (1998) A practical guide to wavelet analysis. Bull Am Meteor Soc 79:61–78

Tsvang LR, Kaprov BM, Zubkovskij SL, Dyer AJ, Hicks BB, Miyake M, Stewart RW, McDonald JW (1973) Comparison of turbulence measurements by different instruments; Tsimlyansk field experiment 1970. Bound-Layer Meteorol 3:499–521

Tsvang LR, Zubkovskij SL, Kader BA, Kallistratova MA, Foken T, Gerstmann W, Przandka Z, Pretel J, Zelený J, Keder J (1985) International turbulence comparison experiment (ITCE-81). Bound-Layer Meteorol 31:325–348

Chapter 27
In Situ and Remote Water Monitoring in Central Asia—The Central Asian Water (CAWa) Network

Tilo Schöne, Alexander Zubovich, Cornelia Zech, Julia Illigner, Azamat Sharshebaev, Dmitry Mandychev, Abdysamat Shakirov, Nico Stolarczuk, Mahmud H. Haghighi, Lars Gerlitz, Abror Gafurov, Bolot Moldobekov and Jörn Lauterjung

Abstract Water is an important natural resource as it plays a key role for living, agricultural production, and economic development. Central Asia features strongly heterogeneous landscapes in terms of orographic, climatic, and hydrologic conditions, which leads to a spatially non-uniform forming of water resources over the region. The availability and direct access to freshwater often result in disputes between the different countries. Therefore, a reliable trans-boundary water resource management is needed. Within the CAWa project, a network of automatic monitoring stations has been installed especially in remote areas and at higher altitudes providing continuously meteorological and hydrological parameters. In addition to the ground-based monitoring network, water levels and volume changes of selected lakes and reservoirs are provided by satellite-based radar altimetry. All data is stored in an open-access data base to support data sharing between the involved transnational agencies and thus, to allow sustainable decision-making about water management and to contribute to international scientific cooperation.

Keywords Central Asia · CAWa · Water monitoring · Catchments · Landscapes · Hydrometeorological network · ROMPS · SDSS data repository · Radar altimetry · Lake and reservoir water monitoring

T. Schöne (✉) · C. Zech · J. Illigner · N. Stolarczuk · M. H. Haghighi · L. Gerlitz · A. Gafurov · J. Lauterjung
Helmholtz-Zentrum Potsdam Deutsches GeoForschungsZentrum GFZ, Telegrafenberg, Potsdam, Germany
e-mail: tschoene@gfz-potsdam.de

A. Zubovich · A. Sharshebaev · D. Mandychev · A. Shakirov · B. Moldobekov · J. Lauterjung
Central Asian Institute for Applied Geosciences (CAIAG), Bishkek, Kyrgyzstan

L. Müller and F. Eulenstein (eds.), *Current Trends in Landscape Research*, Innovations in Landscape Research, https://doi.org/10.1007/978-3-030-30069-2_27

27.1 Introduction

In the arid and semi-arid region of Central Asia, water is one of the most important natural resource. It is not only the potable water supply for a growing population but also a precondition for the economic development and for the irrigation for agricultural production. The Amu Darya and the Syr Darya, the two major rivers in Central Asia, originate in the high mountains of the Tian Shan and Pamir and are formed from the melt of seasonal snow pack, glacier ice, and precipitation. This important role of the flow formation zone for water management in Central Asia calls for a continuous monitoring of all contributing processes. A sustainable water management requires data for operational runoff forecasts as well as the assessment of changes in the head-water catchments and their water resources (Unger-Shayesteh et al. 2015) resulting from changes in the regional and global climate.

Since 2008, a transnational network of remotely operated multi-parameter stations (ROMPS) has been established in Central Asia (Fig. 27.1) in the frame of the CAWa project (http://www.cawa-project.net) that continuously captures and distributes meteorological and hydrological observations (Schöne et al. 2013). In addition to the ground-based station network, a satellite-based monitoring system has been developed to provide data on water levels and water volume changes of selected lakes and

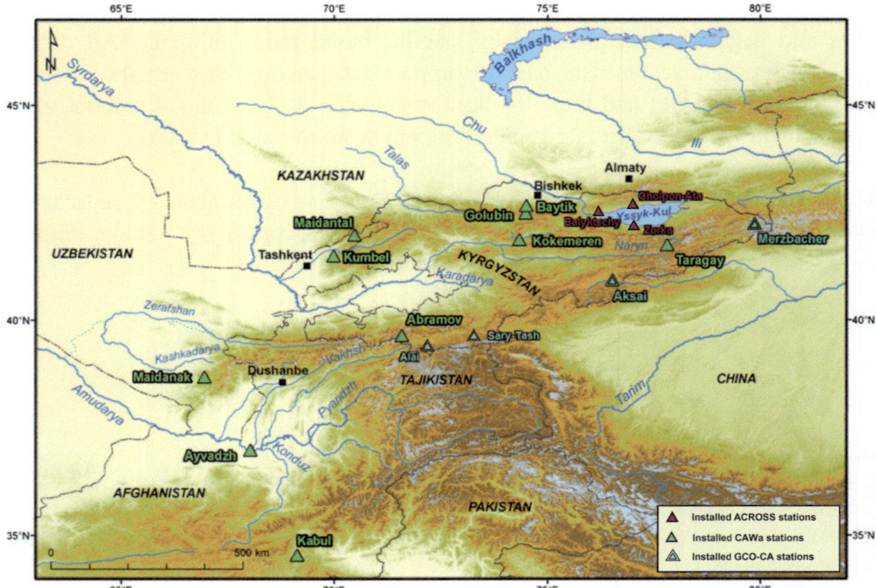

Fig. 27.1 Geographical settings and the ROMPS monitoring network in Central Asia (12/2018). The network consists of stations installed under the CAWa project, GFZ's GCO-CA program, and the ACROSS initiative (Helmholtz Society 2015)

reservoirs in this region. All data is made immediately freely accessible to all national Hydrometeorological Services as well as to the international community through the Sensor Data Storage System (SDSS) of the CAWa project.

27.2 Geographical Settings and Hydrometeorological Monitoring in Central Asia

In the semi-arid to arid regions, water resources are vital for a growing population, food security, and the regional economic development. Knowledge about and monitoring of water resources and the forecasting of seasonal water availability is of utmost importance for water management. In past decades, hydrometeorological monitoring in Central Asia was often based on manually operated posts even in very remote locations and under harsh climatic conditions. Data collection and transmission required human interactions. After the collapse of the Soviet Union in 1991, it became increasingly difficult for the new independent Republics (Kazakhstan, Kyrgyzstan, Tajikistan, Turkmenistan, and Uzbekistan) to continue this monitoring. Also the conflicts in Afghanistan made it difficult to operate monitoring posts in Afghan high mountains. The resulting network thus declined significantly, primarily affecting stations in the flow formation zone at altitudes above 2000 m a.s.l. (e.g., World Bank 2009). Hence, there is a great need to sustain and partly improve the hydrometeorological monitoring in Central Asian's head-water catchments.

Another significant change is that many of the major rivers became of trans-boundary nature after the re-foundation of independent states in 1991 and thus, are now subject to necessary international agreements on water usage. For example, the Aral Sea's main water sources—the Amu Darya and Syr Darya Rivers with their tributaries—now pass six countries: northern Afghanistan, Kazakhstan, Kyrgyzstan, Tajikistan, Turkmenistan, and Uzbekistan. Their water streams are intensively used by all riparian nations, albeit, often with conflicting claims. Countries in the head-water regions are using these resources for generating hydroelectric power in winter, while low-lying counties need the water for irrigation during the summer months (Giese et al. 2004; Sehring 2007; Eschment 2011; Sehring and Diebold 2012). In past decades, overexploitation and sometimes inefficient usage of water resources have led to frequent water shortages in the region (De Martino et al. 2005; UNECE 2011); the desiccation of the Aral Lake being a commonly known example. Long-term effects of, e.g., climate change are likely to aggravate the situation. To mitigate these challenges, information on (regional) water availability is needed for sustainable planning and management (Abdullayev and Rakhmatullaev 2014). To assist such processes, independent and freely available information on water parameters, such as rainfall distribution, soil moisture, glacier, and ice/snow melt, river and groundwater flow, and lake and reservoir volumes, are in high demand (WWAP 2012).

In the 1990s and 2000s, several internationally funded projects already addressed these issues and succeeded to install a number of automatic meteorological and hydrometeorological monitoring stations (World Bank 2018). Beside these positive developments, a number of challenges still remain (Schöne et al. 2013):

- Multi-parameter stations are necessary to consolidate the monitoring network especially at higher altitudes and remote areas. The stations need to be capable of reliable unmanned operation under the prevailing extreme environmental conditions, to have low maintenance requirements and, additionally, to incur very low operational costs.
- Near-real-time data transmission to data centers and users is needed even from remote and uninhabited areas.
- The data management infrastructure should be capable of facilitating data sharing among countries.

These challenges of establishing and operating a regional monitoring network of ROMPS with a special focus on the flow formation zone are addressed in the frame of the CAWa project, the "Global Change Observatory Central Asia (GCO-CA)" of the German Research Center for Geosciences (GFZ), and the long-term strategy of the Central Asian Institute for Applied Geosciences (CAIAG) in Kyrgyzstan. With these programmes, opportunities to support a wide range of operational, societal and scientific tasks, among them weather observations and forecasts, long-term climate monitoring, river discharge monitoring, glacier observations, Earth-System hazard monitoring [e.g., glacier lake outburst floods (Zech et al. 2018), earthquakes (Metzger et al. 2017)] as well as the establishment of water-related early-warning systems are offered. To serve these applications, ROMPS stations combine a multitude of sensors at one station with shared power supply and communication infrastructure (Schöne et al. 2013).

27.3 Hydrometeorological Monitoring

The network which is operated under the CAWa project aims to continuously measure a large number of meteorological and hydrological parameters. The station concept of the ROMPS is described in detail in Schöne et al. (2013) (Fig. 27.2). The mainly acquired meteorological parameters are: wind speed and gust (in 10 m above ground), air temperature (for example, Fig. 27.3), relative humidity, air pressure (all 2 m above ground), precipitation with a tipping bucket, and the incoming and reflected short-wave and long-wave (infrared) radiation by a four-component net radiometer (e.g., Gafurov et al. 2016). With regards to potential application in hydrology up to six soil moisture and soil temperature sensors complement the station setup (typically at 10/20/40/60/80/100 cm below surface). For monitoring the head-water catchments, information about snow height and

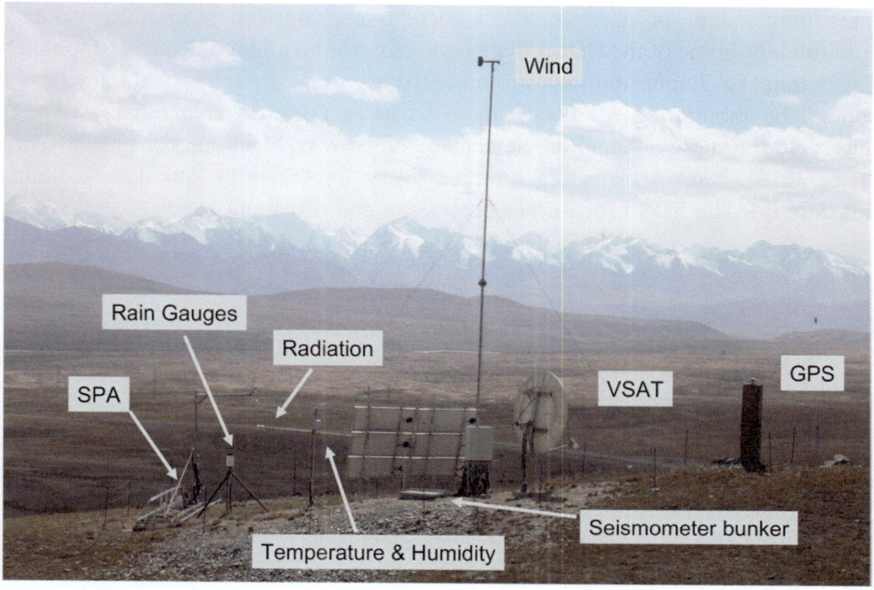

Fig. 27.2 Main components of the ROMPS station at Taragay (Kyrgyzstan). Soil moisture and soil temperature sensors are not visible. This station is additionally equipped with a broadband seismometer from the GEOFON network (https://geofon.gfz-potsdam.de)

snow properties is crucial. Five well-located stations are equipped with a snow pack analysing (SPA) system (Sommer Messtechnik, https://www.sommer.at) to measure snow parameters in combination with a temperature-compensated ultrasonic snow depth/height sensor. River discharge monitoring is continuously performed at four sites using a Doppler-based radar sensor (Sommer Messtechnik) measuring the surface velocity of the river stream. In combination with time-delay measurements (pulse radar), the water level can be determined. The discharge is calculated by

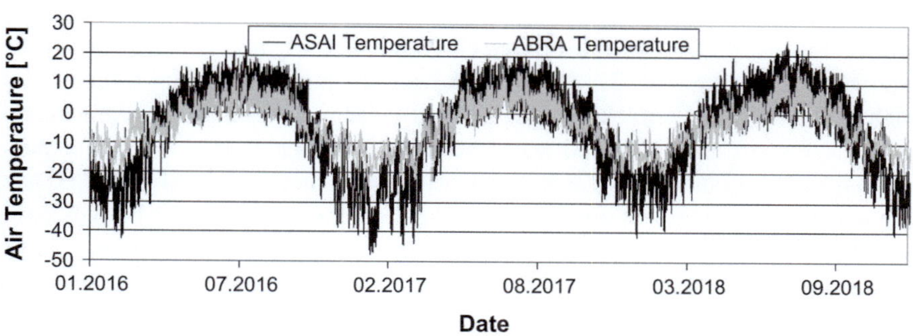

Fig. 27.3 Example of air temperatures at Abramov Glacier ("ABRA", Altitude 4200 m) and Aksai ("ASAI", Altitude 3060 m) (Locations see Fig. 27.1). The coldest recorded temperature in Aksai has been −48 °C (Winter 2017)

integrating values of the river cross section area and the mean flow velocity. Additionally, some of the ROMPS are complemented with broadband seismometers for earthquake monitoring and/or automated high-rate digital optical cameras for continuous monitoring of glacier mass balances (Hoelzle et al. 2017) through observing the snow line and the percentage of snow coverage on the glacier (Huss et al. 2013). Most sensors are sampled every five minutes and data is hourly transferred to the SDSS for further processing and distribution.

27.4 Reservoir, Lake, and River Monitoring in Central Asia with Radar Altimetry

During the past years, remote sensing of water-related resources gained increasingly interest. Different space-based water monitoring missions, such as NASA's Terra and Aqua missions for snow coverage or NASA's SMAP, ESA's SMOS, or EU/ ESA Copernicus SENTINEL's missions for soil moisture retrieval enabled the continuously and spatially complete monitoring of certain water parameters, even in mountainous areas where significant amounts of water is stored in glaciers, lakes, and reservoirs. To get a complete picture of water availability and resources, all sources need to be monitored. This can be done on regional scales by, e.g., satellite gravity observations (e.g., GRACE/GRACE-FO; Gouweleeuw et al. 2018), radar altimetry, or locally by in situ measurements. Public information about the actual stored volume in reservoirs and lakes in a timely and independent manner is available only for a small number of selected water bodies. In Central Asia, a few organizations are dealing with the collection and dissemination of water-related data, but often require front-up registration and/or paid subscriptions. To remotely estimate water body parameters, radar altimetry has become a convenient tool. Since 25 years, space-based radar altimetry continuously measures water surface heights, primarily over oceans. Over the past years, this technology was advanced allowing also precise measurements of inland-water bodies. Radar altimeters deliver ellipsoidal heights by measuring the two-way travel time of a radar pulse in the line of the nadir. Due to the orbital characteristics of the satellites, the repeat interval is fixed (e.g., 10, 27, 35 days). Advanced signal processing technologies, such as the so-called retracking of individual radar pulses, allow the extraction of water levels of smaller inland-water bodies in the nadir of the satellite pass. Although the accuracy of the water levels (water heights) is slightly reduced compared to the traditional open ocean application, a recent study for the Toktogul, Kairakum, and Shardara water chain in Central Asia (Schöne et al. 2018a) demonstrated that the water level of reservoirs can be determined with an accuracy of better than ±30 cm.

Radar altimetry provides water but most hydrological applications require volumes or volume changes. For most reservoirs and lakes, the bathymetry is either not available or unknown preventing the development of an accurate hypsometry or a

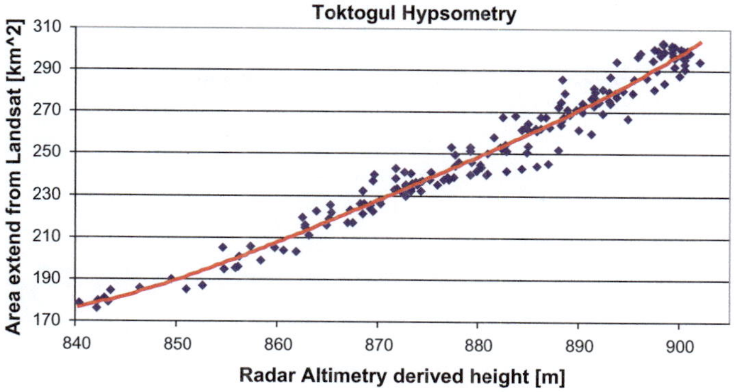

Fig. 27.4 Hypsometric curve of Landsat™ satellite-derived reservoir extends in relation to radar altimetry derived surface heights (heights above geoid)

volumetric curve. In order to provide the complementary information about individual water bodies, remote sensing technology is used. For a large number of lakes and reservoirs in Central Asia, the CAWa project derived time series of water surface area extents using Landsat™ satellite (https://landsat.gsfc.nasa.gov/) images. The surface extents and radar altimetry surface heights are merged and a hypsometric curve which gives volume changes in respect to the lowest observed water level for any height within the definition range. Figure 27.4 shows the resulting hypsometry of the Toktogul Reservoir. Due to errors or inconsistencies in the derived area information and height information, the volumetric curve is estimated only, but still provides sufficient accuracy. Figure 27.5 gives an example for analyzing the water regimes of the reservoirs in the Naryn catchment area, in this case of the Toktogul, Kairakum, and Shardara reservoirs.

Reservoir and lake volumes available in SDSS are not only used in hydrological studies (e.g., Schöne et al. 2018a) but are also of interest in other fields of science, e.g., Earth-system studies (e.g., Neelmeijer et al. 2018; Andersohn et al. 2017).

The monitoring of water table of rivers (stage monitoring) is also of increasing interest for water management. As an example, the Global River Runoff Center (GRDC) database (https://www.bafg.de/GRDC/EN/Home/homepage_node.html) of the World Meteorological Organization (WMO) does not contain recent values from either the Syr Darya or Amu Darya river. The use of radar altimetry for river monitoring is currently technologically limited to larger and well-situated rivers, but the potential of this technology has been demonstrated (Koblinsky et al. 1993). For Central Asia, a few rivers can be monitored with limited accuracy (Fig. 27.6). With the launch of ESA's CryoSat—2 mission in 2010, primarily designed to map the ice-covered oceans, Synthetic Aperture (SAR) altimetry will be able to provide an exciting perspective for inland-water monitoring. The SAR (or Doppler delay) altimetry allows sensing of smaller portions of the radar footprint and thus, to observe water levels of smaller water bodies such as rivers. The planned Surface Water and Ocean Topography mission (SWOT) (launch expected in 2021) with its

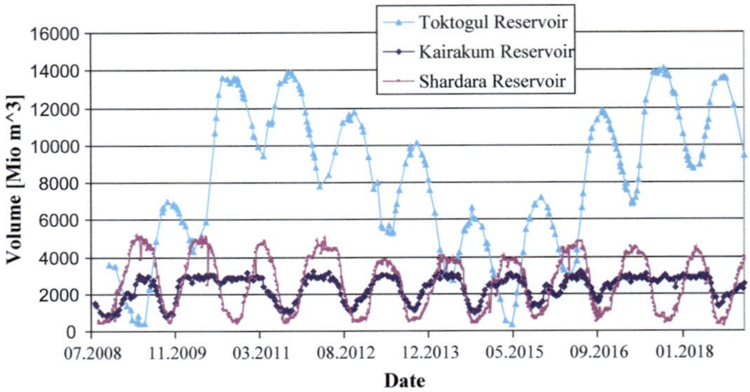

Fig. 27.5 Time series of volume variations since 2008 in the Toktogul, Kairakum, and Shardara reservoirs inferred from satellite radar altimetry (data is available at http://sdss.caiag.kg). The analysis of the time series shows the different aims of water usage, a power scheme for Toktogul reservoir and an irrigation scheme for the Kairakum and Shardara reservoirs

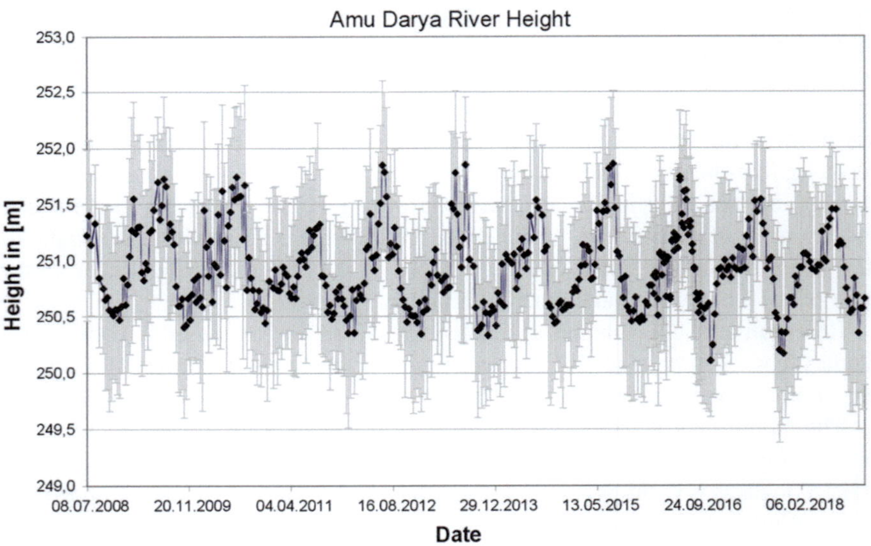

Fig. 27.6 Time series of river heights of the Amu Darya River (65.5586°N, 37.6589°E). The use of such data requires careful analyses due to higher measurement errors

wide-swath altimetry will provide two-dimensional maps of any water body larger than 250 × 250 m every 21 days with an expected accuracy of 20 cm. These new technologies will enable hydrologists to closely monitor water volumes in most lakes, reservoirs, and rivers globally.

27.5 The Sensor Data Storage System (SDSS)

The hydrometeorological observation does not only require the operation of monitoring systems but also the establishment of storage facilities and user interfaces for data download. Thus, one of the aims of the CAWa project is the immediate and unrestricted dissemination of all acquired hydrometeorological parameters to a broad range of users and experts, e.g., national hydrometeorological Services, state and governmental agencies, academia, but also to the general public. The Central Asian Institute for Applied Geosciences (CAIAG) has developed and hosts the sensor data storage system (SDSS) (http://sdss.caiag.kg) allowing easy access and provides long-term availability. The SDSS is the main storage and dissemination system for all Level 0 meteorological and hydrological data acquired by CAWa's ROMPS and other stations operated by GFZ and CAIAG in the frame of the Global Change Observatory in Central Asia (GCO-CA) of GFZ and the ACROSS program (Helmholtz Society 2015). The data is seamlessly integrated into the database of the SDSS immediately after their transmission. Also water level and volume data derived from radar altimetry are automatically processed for selected lakes and reservoirs in Central Asia and continuously added to the SDSS.

A graphical user interface (Fig. 27.7) available in English, Russian, and German language offers the possibility to retrieve data interactively using common Web browsers. The user can select stations as well as parameters of interest, display full or tailored data time series, print charts and download the data as XML or CSV file to be opened by major data analysis tools for further investigation.

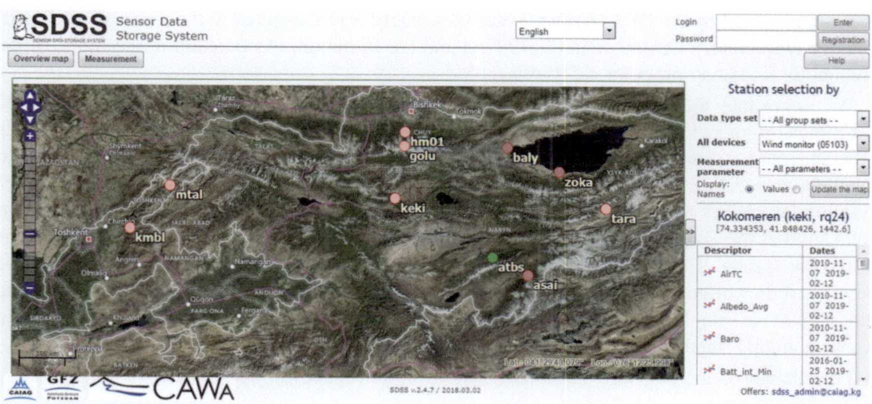

Fig. 27.7 Web interface for the sensor data storage system (http://sdss.caiag.kg) allowing easy and unrestricted data access. No login is required to display and download data

27.6 Conclusions

Over the past decade, the CAWa project contributed in an open and cooperative manner to improve water monitoring in head-water areas of Central Asia. Currently, 14 ROMPS and 7 basic-meteorological stations continuously provide data to national hydrometeorological Services and other users. Five snow-monitoring sites and four river discharge stations complement the operational and scientific basis and, e.g., enable an improvement of water availability forecast. With the use of satellite radar altimetry, water levels and water volume changes for a significant number of lakes and reservoirs are made freely available. In summary:

1. GFZ and CAIAG, in cooperation with national hydrometeorological agencies, research institutes, and universities, are offering unrestricted access to the hydrometeorological data of ROMPS and other stations.
2. Remote sensing technologies, such as radar altimetry, allow assisting the water monitoring in urban as well as remote areas without the help of a local-monitoring infrastructure.
3. With the development of the sensor data storage system (http://sdss.caiag.kg) an easy-to-use interface has been made available for different kinds of users. SDSS is providing a platform for display and download of hydrometeorological data sets.

Acknowledgements The activities presented have been generously funded in the frame of the CAWa project (www.cawa-project.net) by the German Federal Foreign Office as part of the "German Water Initiative for Central Asia" (the so-called "Berlin Process," grant AA7090002). Additional funding for the installation and long-term operation of the monitoring network was and is being provided by the GFZ German Research Center for Geosciences in the frame of the Global Change Observatory Central Asia (GCO-CA) and the Helmholtz Association in the frame of the Helmholtz Roadmap program ACROSS.

We appreciate the support and fruitful cooperation with our colleagues from the national hydrometeorological Services of Kyrgyzstan, Tajikistan, and Uzbekistan without whom, the site selections and station installations would not have been possible. In addition, we would like to thank the researchers of the University of Fribourg for the cooperation in installing optical camera systems for glacier monitoring at Abramov and Golubin glaciers. We extend our thanks to our colleagues at the Ulugh Beg Observatory (Uzbekistan) for the installation and maintenance of the Maidanak station. The installation of the CAWa station in Kabul (Afghanistan) would not have been possible without the excellent work of our colleagues at Kabul Polytechnic University. Also the support of the Norwegian Afghanistan Committee (NAC, http://www.afghanistan.no/) in Kabul is highly appreciated.

This chapter is the extended version of Schöne et al. (2018b).

References

Abdullayev I, Rakhmatullaev S (2014) Data management for integrated water resources management in Central Asia. J Hydroinf 16(6):1425–1440

Andersohn J, Bindrich M, Lang O, Schöne T (2017) Mosul Dam deformation monitoring with TerraSAR-X—an effective complement to terrestrial surveying—*Presentations*. In: ESA Workshop FRINGE (Helsinki 2017). http://gfzpublic.gfz-potsdam.de/pubman/item/escidoc:2650907, accessed 03 Feb 2019

De Martino et al (2005) Environment and security: transforming risks into cooperation—Central Asia—Ferghana/Osh/Khujand area. https://gridarendal-website-live.s3.amazonaws.com/production/documents/:s_document/250/original/ferghana-report-eng_%281%29.pdf, accessed 03 Feb 2019

Eschment B (2011) Wasserverteilung in Zentralasien. Ein unlösbares Problem? Studie der Friedrich Ebert Stiftung. http://library.fes.de/pdf-files/id/08201.pdf, accessed 03 Feb 2019

Gafurov A, Luedtke S, Unger-Shayesteh K, Vorogushyn S, Schöne T, Kalashnikova O, Merz B (2016 July) MODSNOW-Tool—an operational tool for daily snow cover monitoring using MODIS data, EES Special Issue "Water in Central Asia". Environ Earth Sci 75:1078. doi:10/1007/s12665-016-5869-x

Giese E, Sehring J, Trouchine A. (2004) Zwischenstaatliche Wassernutzungskonflikte in Zentralasien, www.uni-giessen.de/fbz/zentren/zeu/forschung/publikationen/publi2/disc18/at_download/file, accessed 03 Feb 2019

Gouweleeuw B, Kvas A, Gruber C, Gain AK, Mayer-Gürr T, Flechtner F, Güntner A (2018) Daily GRACE gravity field solutions track major flood events in the Ganges–Brahmaputra Delta. Hydrol Earth Syst Sci 22:2867—2880. https://doi.org/10.5194/hess-22-2867-2018

HELMHOLTZ Society (2015) Helmholtz-Roadmap für Forschungsinfrastrukturen. https://www.helmholtz.de/fileadmin/user_upload/publikationen/Helmholtz_Roadmap_2015_web_korr_150921.pdf, accessed 03 Feb 2019

Hoelzle M, Azisov E, Barandun M, Huss M, Farinotti D, Gafurov A, Hagg W, Kenzhebaev R, Kronenberg M, Machguth H, Merkushkin A, Moldobekov B, Petrov M, Saks T, Salzmann N, Schöne T, Tarasov Y, Usubaliev R, Vorogushyn S, Yakovlev A, Zemp M (2017) Re-establishing glacier monitoring in Kyrgyzstan and Uzbekistan, Central Asia. Geosci Instrum Method Data Syst 6:397–418. https://doi.org/10.5194/gi-6-397-2017

Huss M, Sold L, Hoelzle M, Stokvis M, Salzmann N, Farinotti D, Zemp M (2013) Towards remote monitoring of sub-seasonal glacier mass balance. Ann Glaciol 54(63):75–83. https://doi.org/10.3189/2013AoG63A427

Koblinsky CJ, Clarke RT, Brenner AC, Frey H (1993) Measurements of river level variations with satellite altimetry. Water Resour Res 29(6):1839–1848

Metzger S, Schurr B, Ratschbacher L, Sudhaus H, Kufner S-K, Schöne T, Zhang Y, Perry M, Bendick R (2017) The 2015 Mw7.2 Sarez strike-slip earthquake in the Pamir interior: response to the underthrusting of India's western promontory. Tectonics 36. https://doi.org/10.1002/2017tc004581

Neelmeijer J, Schöne T, Dill R, Klemann V, Motagh M (2018) Ground deformations around the Toktogul Reservoir, Kyrgyzstan, from Envisat ASAR and Sentinel-1 Data—a case study about the impact of atmospheric corrections on InSAR Time Series. Remote Sens 10:3. https://doi.org/10.3390/rs10030462

Schöne T, Zech C, Unger-Shayesteh K, Rudenko V, Thoss H, Wetzel H-U, Gafurov A, Illigner J, Zubovich A (2013) A new permanent multi-parameter monitoring network in Central Asian high mountains—from measurements to data bases. Geosci Instrum Method Data Syst 2:97–111. https://doi.org/10.5194/gi-2-97-2013

Schöne T, Dusik E, Illigner J, Klein I (2018a) Water in Central Asia: Reservoir monitoring with radar altimetry along the Naryn and Syr Darya Rivers. In: International symposium on earth and environmental sciences for future generations, (International Association of Geodesy

Symposia; 147), Online 2016, pp. 349–357. Springer, Cham. http://doi.org/10.1007/1345_2017_265, accessed 03 Feb 2019

Schöne T, Zubovich A, Zech C, Illigner J, Sharshebaev A, Mandychev D, Shakirov A, Moldobekov B, Lauterjung J (2018b) Water monitoring in Central Asia—the Central Asian water project (CAWA) = Центрально-Азиатский водный мониторинг – проект Вода в Центральной Азии (ЦАВа). In: Sychev VG, Mueller L (eds) Novel methods and results of landscape research in Europe, Central Asia and Siberia, vol I Landscapes in the 21th century: status analyses, basic processes and research concepts, pp. 90–94. Russian Academy of Science, Moscow. https://doi.org/10.25680/2697.2018.71.90.013

Sehring J (2007) Irrigation reform in Kyrgyzstan and Tajikistan. Irrig Drain Syst 21(3–4):277–290. Available at: http://link.springer.com/10.1007/s10795-007-9036-0, accessed 03 Feb 2019

Sehring J, Diebold A (2012) From the Glaciers to the Aral Sea: Water Unites. Trescher Verlag, Berlin. ISBN 978-3897948006

Unger-Shayesteh K, Conrad C, Schöne T, Moldobekov B (2015) Reasonable decision-making on land and water management in Central Asia: how earth observation technologies can contribute (analytical notes from the "German Water Initiative for Central Asia"; 1), GFZ German Center for Geosciences, Potsdam, 6 p., https://doi.org/10.2312/5.4.2015.001r (in Russian)

United Nations Economic Commission for Europe (UNECE) (2011) Second assessment of transboundary rivers, lakes and groundwaters. UNECE, Geneva, http://www.unece.org/fileadmin/DAM/env/water/publications/assessment/English/ECE_Second_Assessment_En.pdf, accessed 03 Feb 2019

World Bank (2009) Improving weather, climate and hydrological services delivery in Central Asia (Kyrgyz Republic, Republic of Tajikistan, and Turkmenistan). Report prepared in the frame of the technical assistance project "An action plan for improving weather and climate service delivery in high-risk, low-income countries in Central Asia", World Bank Russia Country Office, Moscow. https://www.gfdrr.org/sites/default/files/publication/GFDRR_ECA_Hydromet_Study.pdf, accessed 03 Feb 2019

World Bank (2018) Central Asia—hydrometeorology modernization Project: additional financing (English). World Bank Group, Washington, D.C. http://documents.worldbank.org/curated/en/904441533353436310/Central-Asia-Hydrometeorology-Modernization-Project-additional-financing, accessed 03 Feb 2019

WWAP (2012) World water assessment programme. The United Nations World Water Development Report 4: Managing water under uncertainty and risk. UNESCO, Paris. http://unesdoc.unesco.org/images/0021/002156/215644e.pdf, accessed 03 Feb 2019

Zech C, Schöne T, Neelmeijer J, Zubovich A (2018) Continuous kinematic GPS monitoring of a Glacier lake outburst flood. In: International symposium on earth and environmental sciences for future generations, (International Association of Geodesy Symposia; 147), Online 2016, pp. 339–347. Springer, Cham. https://doi.org/10.1007/1345_2016_246

Chapter 28
Peatland Science and Conservation: Contributions of the Greifswald Mire Centre, Germany

Greta Gaudig and Franziska Tanneberger

Abstract Peatlands cover only 3% of the world's land surface, but contain 500 gigatonnes of carbon within their peat—twice as much as all the biomass of the world's forests. Besides their climate impact, peatlands play a significant role for biodiversity, nutrient retention and the regulation of the local climate and the landscape water budget. Peatlands have been the subject of research at the University of Greifswald for about 200 years. In addition to early paleoecological research, Greifswald scientists have also shaped the landscape ecology of mires. In the last 15 years or so, there has been a rapid increase in our understanding of the significance of peatlands for global environmental problems. This applies in particular to greenhouse gas emissions from drained peat soils, which account for about 5% of anthropogenic greenhouse gas emissions worldwide. Since 2015, three Greifswald institutions—the University of Greifswald (research), the Michael Succow Foundation (implementation) and DUENE e.V. (policy advice)—have therefore been working together and are developing the Greifswald Mire Centre (GMC) as an interface between science, practice and politics in relation to peatlands. The work focuses on climate change, biodiversity and sustainable use. The vision of the GMC is a world in which peatlands are perceived as vital and vulnerable systems, natural mires are preserved, degraded peatlands are restored and any use of peatlands is sustainable. The innovative methods developed at the Greifswald Mire Centre for the study of peatlands and their environmentally friendly use by paludiculture have been tested in international projects and are applicable in many regions.

Keywords Peatlands · Landscape conservation · Research projects · Mires · Fens · Bogs · Degradation · Greenhouse gases · Paludiculture

---

G. Gaudig · F. Tanneberger (✉)
Institute of Botany and Landscape Ecology, Greifswald University, Partner in the Greifswald Mire Centre, Soldmannstr. 15, 17487 Greifswald, Germany
e-mail: tanne@uni-greifswald.de

G. Gaudig
Greifswald Mire Centre, c/o Michael Succow Foundation, Ellernholzstr. 1/3, 17489 Greifswald, Germany

L. Müller and F. Eulenstein (eds.), *Current Trends in Landscape Research*, Innovations in Landscape Research, https://doi.org/10.1007/978-3-030-30069-2_28

28.1 Introduction

Peatlands are huge carbon reservoirs. If drained, they turn into hotspots of anthropogenic greenhouse gas emissions: While covering only 0.3% of the world's land surface, drained peatlands cause disproportionally high CO_2 emissions: they are responsible for nearly 5% of the world's anthropogenic CO_2 emissions (2 gigatonnes CO_2 per year, Joosten et al. 2016a). Emission hotspots are Indonesia, the EU, Russia, China and the USA. In a peatland-rich state like Mecklenburg-Vorpommern (where Greifswald is located), drained peatlands constitute the largest single source of greenhouse gas emissions with about 6.2 million t CO_2-eq. per year (MLUV MV 2009). Peatland rewetting is therefore a necessity to achieve the targets of the UN Framework Convention on Climate Change (UNFCCC) and in particular the Paris Agreement, limiting the rise in global average temperature to well below 2 °C above pre-industrial levels.

In addition to their climate impact, natural peatlands are of great importance for biodiversity (Succow and Lange 1984; Succow and Joosten 2001; Wassen et al. 2005; Schipper et al. 2007), which is protected worldwide by the Convention on Biological Diversity (CBD) (United Nations 1992) and the Ramsar Convention (UNESCO 1994). Peatlands also provide numerous other ecosystem services (Bonn et al. 2016), such as the retention of pollutants and the regulation of the local climate and landscape water budget. These functions are also lost as a result of drainage. In coastal areas, the drainage of peatlands also leads to land loss: A considerable proportion of the land area (e.g. in several Asian countries) is threatened with permanent flooding as a result of height losses.

In addition to the preservation of intact, natural peatlands (mires), the rewetting of drained peatlands is urgently needed from an environmental and climate policy point of view (TEEB DE 2014). If they have to be (further) used productively, this should be done as paludiculture, i.e. as wet land use (SRU 2012; Wichtmann et al. 2016). Paludiculture includes traditional ways of peatland use (e.g. reed cutting), but above all includes the development and implementation of new ways for the material and energetic utilisation of biomass from rewetted peatlands. Experiences with peatland rewetting and paludiculture show great potential worldwide for linking many of the 17 UN Sustainable Development Goals (SDGs) and promoting their achievement (Joosten et al. 2016b).

28.2 Development of Peatland Research in Greifswald

Peatlands' special feature of storing history leads again and again to a connection of peatland research with investigation of vegetation, climate and settlement history. Peatland research in and around Greifswald began with Adelbert von Chamisso, who in 1824 provided a description of the mire near the Saline north of Greifswald. Von Chamisso (1824) describes the use of peat and the replenishment of old peat pits with new reed peat. The author points out the discrepancy between the immediate vicinity of the Baltic Sea and the composition of the peat from non-salt tolerant plants. Without specifically discussing a sea level rise or a Baltic Sea transgression, von Chamisso argues that the peat deposits can only be considered and understood embedded in their temporal and spatial development (Michaelis et al. 2016).

In the following period, the peatlands of Vorpommern are mainly described in geological studies such as Koch (1849) and botanical works such as Hornschuch (1837), which deal with peat depths, the state of use and the adapted plant cover as well as sporadic hydrological aspects. In 1820 Christian Friedrich Hornschuch was appointed extraordinary professor for natural history and botany and was co-editor of the botanical journal ‘Flora’ and the ‘Bryologia Germanica’. Hornschuch (1837) describes almost a dozen peatland sites between the Peene valley near Gützkow and the large and small Kieshofer Moor north of Greifswald. The work of Hornschuch shows very clearly the high utilisation pressure that the peatlands in the open landscape of Vorpommern with little forest were exposed to as a result of peat extraction (Michaelis et al. 2016).

Kurd von Bülow was one of the first researchers in Central Europe to use the method of systematic pollen analysis just published by von Post (1916/1918) for his doctorate at the University of Greifswald on the Great Kieshofer Moor (von Bülow 1921, 1928). Von Bülow, who taught at the University of Rostock from 1935 to 1946 and from 1952 after a few intermediate stops, continued to be associated with peatlands and started an epochal, ten-volume, albeit unfinished book project with the ‘Handbuch der Moorkunde’ from 1929. It appeared under his editor-in-chief four, meanwhile classical, works on peatlands: Part I. of Bülow 1929: General peatland geology; Part III. Peus 1932: The wildlife of peatlands; Part IV. Dokturowsky 1938: The peatlands of Eastern Europe and Northern Asia; Part VII: Dachnowski-Stokes & Auer 1933: American peat deposits.

The Great Kieshofer Moor, which was secured as a natural monument on the initiative of Greifswald botanists (Leick 1927a), was also the subject of an early, comprehensive area monograph, published in the ‘Beiträge zur Naturdenkmalpflege’ 1927, i.e. before the passing of a Nature Conservation Act in 1935. Leick (1927b) worked up the reports and representations of the peatland in historical maps. Particularly noteworthy are the investigations of the water balance of raised bogs with regard to the ecological saturation deficit by Leick (1929). Sphagnum peat from the Kieshofer Moor, which was already heathered at that time, served as the object of investigation. Since the time of K. von Bülow, the Kieshofer Moor has repeatedly been the subject of Greifswald peatland research (see also Lange 1994; Jeschke 2003), so that a very good data basis exists for this site, which allows the derivation of time series.

For the period from 1945 to 1990, Werner Rothmaler’s activities must be mentioned as one of the most important foundations for peatland research. From 1953 to 1962, Rothmaler headed the Institute for Agricultural Biology at the University of Greifswald as a botanist. Although Rothmaler himself worked predominantly taxonomically, the establishment of the Department of Taxonomy and

Vegetation Science created the environment for diverse and in-depth work on the floral composition and history of the region (Michaelis et al. 2016). In the following years until 1992, it was mainly Rothmaler's collaborators and students who presented both vegetation studies on peatlands and paleoecological studies with the help of peat deposits (e.g. L. Jeschke, K. Kloss, M. Succow, S. Slobodda, F. Fukarek).

Three aspects of Michael Succow's work on peatlands must be emphasised. First, the separation of ecological and hydrogenetic mire types (Succow 1971, 1981). Closely linked to this distinction is the comprehensive characterisation of the hydrogenetic mire types (Succow 1981, 1988) in the area of the minerotrophic peatlands (fens). Their great diversity is more evident in Central Europe outside the bog districts, but has received little attention to date. In addition, the systematic elaboration of the vegetation form concept for mire vegetation (Succow 1988) must be pointed out (Michaelis et al. 2016).

In addition to botanists, geologists also worked more intensively on the numerous peatlands in the north-east in the second half of the twentieth century. Although geological studies have already dealt intensively with the construction and development of the large peat-filled valleys of today's Mecklenburg-Western Pomerania, much remains unclear. The work of Janke (1978) in particular relates the development of the river valleys to the dynamic development of the Baltic Sea and its precursor stages (Michaelis et al. 2016).

In 1992, Michael Succow was appointed to the Chair of Geobotany and Landscape Ecology at the University of Greifswald as the successor to his doctoral supervisor Franz Fukarek. Thus, the traditional botany was extended by the applied landscape ecological emphasis with focus on peatlands. In order to combine integrating landscape ecology with its practical implementation in nature, environmental and climate protection and to bridge the gap between society and nature, Succow was engaged in 1994 in the establishment of two new professorships, landscape economics and environmental ethics. Thus, in 1996, the degree programme 'Landscape Ecology and Nature Conservation' was established alongside the already established Biology programme.

28.3 The Working Group 'Peatland Science and Paleoecology' and Further Work in Peatland Sciences Since 1996 at the University of Greifswald

28.3.1 Creation and Expansion of the Working Group

In 1996, Succow brought the Dutch peatland scientist and conservationist Hans Joosten to Greifswald to strengthen the team. Hans Joosten received his doctorate here in 1998 and his habilitation in 2002. Joosten established the working group 'Peatland science and paleoecology', which, with some 30 scientific staff, conducts integrative research on peatland at the intersection of paleoecology, ecology, landscape ecology, nature conservation and sustainable use. The research group specialises in the study of lesser-known peat and peatland types regionally and globally. Since 1990, peatland research has been carried out in more than 40 countries on six continents by Greifswald scientists. In 2008, Hans Joosten was appointed Professor of peatland science and paleoecology at the University of Greifswald. As a joint work of Michael Succow and Hans Joosten, the second, fundamentally revised edition of 'Landscape ecological peatland science' was published in 2001 (Succow and Joosten 2001).

When Michael Succow retired in 2006, Stefan Zerbe became his successor before Martin Wilmking, who has been working at the University of Greifswald since 2005 with a focus on 'Ecosystem Dynamics', was appointed Professor of Landscape Ecology in 2010. He expanded his work in peatland science with research projects in the (sub)Arctic and with his own measurements of greenhouse gas fluxes in peatlands. Since 2005 landscape ecology, nowadays 'Environmental Change: Responses and Adaptation', has been one of the five research priorities of the University of Greifswald, in which peatlands also play an important role. In 2006, the Institute of Botany was renamed the Institute of Botany and Landscape Ecology in order to meet the changed requirements.

28.3.2 Final Theses

From 1994 until today more than 150 final theses with peatland topics have been written here, 16 of them dissertations and one habilitation. In 1999, most of the theses were written on peatland topics (17), at a time when Succow and Joosten were teaching at the same time and with the establishment of the new degree course 'Landscape Ecology and Nature Conservation' also work in the fields of landscape economics and environmental ethics took place. More than half of the works deals with the ecology or landscape ecology of peatlands, about one sixth with paleoecological topics (Fig. 28.1). Over the past seven years, more and more work has been done on the climate impact and use of peatlands.

28.3.3 Working Focus: Paleoecology

Three main areas of work are pursued particularly intensively in bog research: paleoecology, greenhouse gas emissions from peatlands and paludiculture.

In addition to geochemical and geophysical investigations, paleoecological research also includes pollen and macrofossil analysis, which is applied to reconstruct the regional landscape and climate history and peatland development in the Late Glacial and Holocene. Research objects are peatlands (and lakes) in Germany (especially north-east Germany) as well as in south-east Asia, Tierra del Fuego and

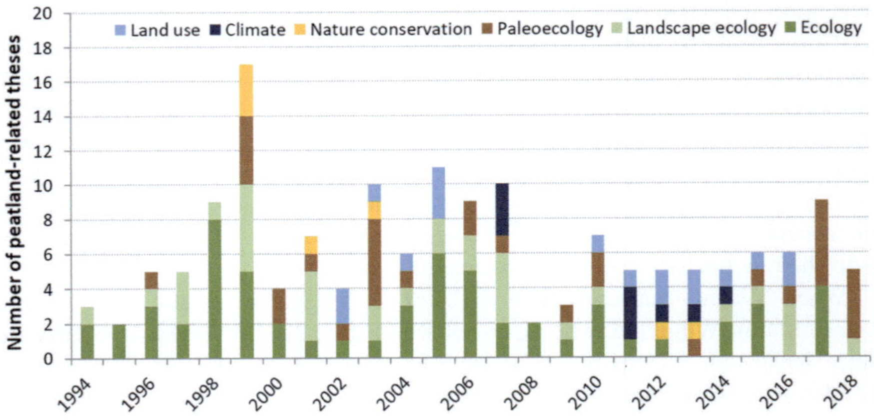

Fig. 28.1 Number of peatland-related theses in the period 1994–2018 at the Institute of Botany and Landscape Ecology of the University of Greifswald, divided by topics (modified after Michaelis et al. 2016). Graduates of Biology and Landscape Ecology and Nature Conservation are included (master, doctoral and 'Habilitation' theses)

in the north-east Siberian Arctic. Special attention is given to increasing the taxonomic, spatial and temporal resolution, in particular the description and identification of non-pollen palynomorphs, the development of innovative methods for the reconstruction of high-resolution vegetation patterns (and related distribution of pollen and spores) and the development of refined sampling techniques.

28.3.4 Working Focus: Method Development for the Estimation of Emissions

Drained peatlands are hotspots of greenhouse gas emissions. Therefore, peatland scientists are investigating the release of greenhouse gases and quantifying the emitted gas quantities on the basis of meta-analyses (Fig. 28.2). Since direct measurements of emissions are very costly and time-consuming, they develop and refine indicators, so-called proxies, to estimate emissions on larger areas. The peatland vegetation in particular proved to be reliable, practicable and extensively applicable in the form of gas emission site types (GESTs; Couwenberg et al. 2011). The methods developed in this way are used in standards, MRV guidelines and methodologies under the Framework Convention on Climate Change (UNFCCC) as well as for certificates of the voluntary carbon market (e.g. VCS, MoorFutures). MoorFutures were the first carbon certificates from peatland rewetting worldwide in 2011. They are now offered in several German federal states and scientifically supervised by the University of Greifswald.

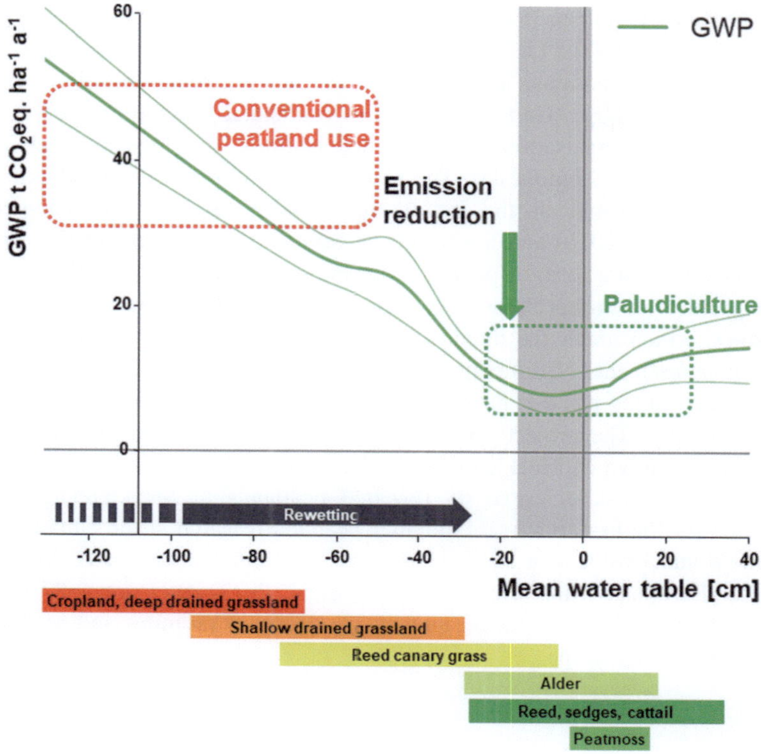

Fig. 28.2 Results of a meta-analysis of greenhouse gas fluxes (CO_2, CH_4) in temperate peatlands and mean water table with typical water table ranges of conventional land use and paludiculture. This figure is based on data of Couwenberg et al. (2011) and Wichtmann et al. (2016). Recent flux measurement results are continuously integrated in the GMC's emission database. Latest analyses show that emissions of drained peatland are even higher. The range for peat moss is slightly smaller than indicated

28.3.5 *Working Focus: Paludiculture*

The type of peatland use influences the climate impact of peatlands. The intensive, drainage-based use leads to serious site changes and environmental problems, including high greenhouse gas emissions. This gave rise to the idea of wet, peat-preserving peatland use in order to reduce environmental damage and at the same time maintain production. The word 'paludiculture' was coined in Greifswald, first mentioned in Joosten (1998) and is now used worldwide. Traditional forms are known from nature conservation management, and 'wet land use' can be crucial for maintaining mire biodiversity (Tanneberger and Kubacka 2018). Within the working group, about 20 employees are currently working on this topic. The fact that paludiculture can make an important contribution to the sustainable land use of peatlands was recognised and rewarded with the award of the European Prize for

Future-Oriented Land Use (Cultura Prize) in 2012, the German Sustainability Prize in 2013 and the 'Land of Ideas' award in 2014/15. The concept of paludiculture is also attracting international attention (e.g. Intergovernmental Panel on Climate Change (IPCC), UNEP (United Nations Environmental Programme), FAO (Food and Agriculture Organisation of the United Nations; see Joosten et al. 2012), European Parliament, Indonesia). The results of the research on paludiculture on fens were summarised in the first comprehensive textbook on paludiculture (Wichtmann et al. 2016) and the results of research and implementation on peat moss cultivation as a form of paludiculture on fens in Gaudig et al. (2018).

The profile was further broadened from 2017 onwards. In addition to landscape economics, which under the direction of Prof. Dr. Volker Beckmann carries out economic assessments of different peatland uses, extensive peatland research projects are now taking place at the Chair of Experimental Plant Ecology of Prof. Dr. Jürgen Kreyling. The project 'WETSCAPES' (2017–2021), funded by the State Excellence Initiative of Mecklenburg-Western Pomerania, carries out comprehensive integrative basic research on rewetted peatlands in Mecklenburg-Western Pomerania for the first time and combines peatland research at the Universities of Greifswald and Rostock. In 2017, H. Joosten, F. Tanneberger and A. Moen finalised the 'European Mires Book' (Joosten et al. 2017) after some 27 years of preparation, providing the first comprehensive overview on mires and peatlands in the biogeographic Europe.

28.4 Safeguarding and Further Development of Research and Teaching from 2021 Onwards

Since 2015, the state government of Mecklenburg-Western Pomerania has agreed that the extraordinary professorship 'Peatland science and pale' makes a significant contribution to raising the profile of Mecklenburg-Western Pomerania as a science location, that its long-term safeguarding is a common interest and that the continuity of the high, internationally leading research quality must be guaranteed. Of particular importance for the state is the development of the links between peatland and climate protection, the possibility of large-scale quantification of greenhouse gas emissions from peatlands, the world's first carbon credits from peatland rewetting developed with the University of Greifswald, and the area of paludicultures and the energetic and material utilisation of biomass from peatlands, for which there have already been very profitable projects for both sides, for example, in cooperation with the Mecklenburg-Western Pomerania state forest. This was emphasised by two state parliament resolutions (2015 and 2016) on peatlands and paludiculture.

After tough negotiations, a target agreement was concluded in December 2018 between the University of Greifswald and the state of Mecklenburg-Western Pomerania represented by the Minister of Education, Science and Culture and the

Minister of Agriculture and Environment. It emphasises that the University of Greifswald contributes to the scientific understanding and preservation of the natural resources of Mecklenburg-Vorpommern in the context of an emerging climate change through its teaching and research activities in the field of peatland research. The state of Mecklenburg-Western Pomerania and the University of Greifswald are committed to placing these activities on a reliable long-term foundation. The nomination of a full professorship for peatland sciences has been announced in March 2019.

This agreement of December 2018 is a historic step for peatland science in Germany: Following a short-term professorship for Grassland and Peatland Science at the Academy of Agricultural Sciences, which Asmus Petersen (1900–1962) held in 1960–1962, and professorships for Agricultural Engineering and Peatland Science at the Humboldt University of Berlin, a permanent professorship for Peatland Science was created for the first time. However, this requires a considerable effort on the part of the state of Mecklenburg-Vorpommern and in particular, the university and is only possible through the involvement of private foundations.

28.5 The Greifswald Mire Centre

28.5.1 Origin and Structure

In Greifswald three non-profit, independent institutions with different expertise are working on mires and peatlands: University of Greifswald, Michael Succow Foundation, and the Institute of Sustainable Development of Landscapes of the Earth (DUENE e.V.). To consolidate achievements and to establish a permanent peatland centre in Greifswald, these three institutions joined forces and founded the Greifswald Mire Centre (GMC) in early 2015 (Fig. 28.3).

The Greifswald Mire Centre is a strategic cooperation between the three institutions, based on a Memorandum of Understanding. The legal form of GMC is a company under civil law, which itself does not participate in legal transactions.

Fig. 28.3 Partners and their focal areas in the Greifswald Mire Centre (from GMC 2019)

The Greifswald Mire Centre has already developed within the last couple of decades, long before it got its name. It builds on:

- 200 years of Greifswald-based peatland expertise,
- Solid and innovative peatland-oriented academic training,
- More than 50 competent and committed peatland scientists and conservationists, as well as
- Four major, well-established peatland databases (GPD, DPPP, PeNCIL, moorwissen.de).

The Greifswald Mire Centre strengthens and consolidates as an integrative umbrella brand all peatland-related activities in Greifswald and develops into a regionally and globally interconnected, influential interface, in which basic and applied research is carried out, know-how is transferred, and inter- and transdisciplinary science-based policy and corporate advice are provided.

The Greifswald Mire Centre has a global orientation with respect to the collection and integration of data and the development of concepts and methodologies. It carries out research and implementation projects mainly in temperate latitudes, especially Europe, and with regional reference to Mecklenburg-Western Pomerania (Fig. 28.4). Figures 28.6, 28.7, 28.8, 28.9, 28.10, and 28.11 (Appendix) provide impressions of research and practical work of the GMC.

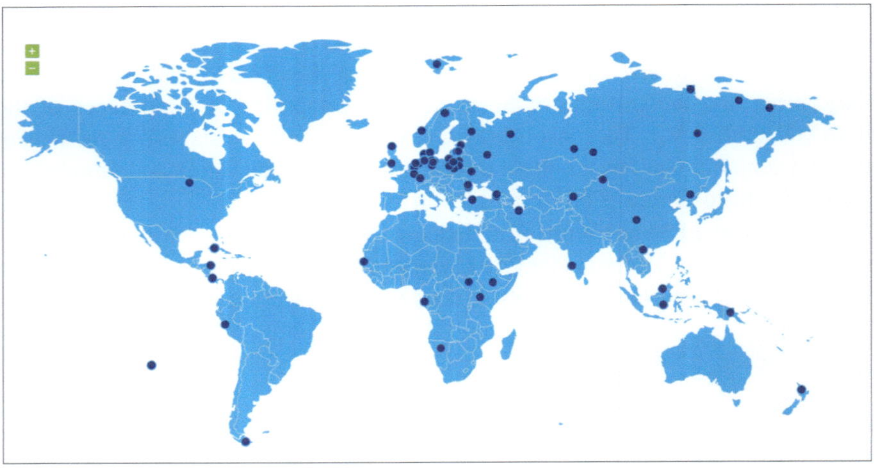

Fig. 28.4 Global distribution of selected ongoing and completed peatland projects of the partners in the GMC (from GMC 2019)

28.5.2 Profile and Niche

GMC is a source of inspiration and ideas at the focal points of peatland research and conservation. It has first-hand knowledge of what is happening at the research front, what is going on in politics, and how implementation proceeds. Special strengths of GMC are:

- Coordination of peatland knowledge: in national and international collaborative research projects (e.g. VIP—Vorpommern Initiative Paludikultur, ERA—Net Projects), organisations (e.g. Secretariat of the International Mire Conservation Group IMCG), book projects (e.g. Paludiculture—Productive use of wet peatlands, Mires and peatlands of Europe) and databases (e.g. Global Peatland Database GPD, Peatland and Nature Conservation International Library PeNCIL)
- Integration: combined ecological, economic, and ethical competence applied in research (e.g. the university research focus for Environmental Change: Responses and Adaptation, ECRA) and teaching (e.g. the international master course Landscape Ecology and Nature Conservation, LENC), meta-analyses of worldwide relevance, and concrete activities, from basic research to global policy consulting
- Communication: target group-oriented preparation and dialogue-oriented mediation of peatland knowledge, both regionally (e.g. project MoorZukunft in MV), nationally (e.g. German Moorschutzdialog) and globally (e.g. policy briefs and side events at environmental conventions, founding partner of the Global Peatlands Initiative, GPI). An extensive peatland knowledge Internet platform can be found in online.

The partners in the GMC are equally entitled. Their institutional and structural diversity brings various peatland expertise and strengths into the GMC:

- University of Greifswald: high-quality (fundamental) research, research networks and projects, and acquisition of research funds, goal-oriented teaching and training, permanent and direct influx of qualified, motivated and creative students
- Michael Succow Foundation: national and international practical implementation (conservation, restoration, sustainable use), policy analysis and communication, trainee and international scholarship programs and acquisition of funds from public and private donors
- DUENE e.V.: (policy) advice and transfer, analysis and development of economic incentives and financing instruments, monetisation, and commodification of ecosystem services and recruitment of public and private consultancy contracts

The Greifswald Mire Centre cooperates with many partners in numerous projects and sees itself as part of a global network of scientists, NGOs and practitioners working in and on peatlands. In order to act as a stimulus, GMC persons participate in the management of important peatland organisations, e.g. the International Mire Conservation Group (IMCG) and the German Peatland Society (DGMT).

GMC was in 2016 one of the founding members of the Global Peatlands Initiative (GPI), in which leading experts, countries and institutions gather to protect peatlands as the world‘s largest terrestrial carbon store, to reduce emissions from peatlands and to progress peatland conservation as a crucial contribution to reach climate protection targets. The main task of GMC in GPI is professional advice. Together with GPI partners GMC co-organises side events at global environmental convention meetings (in particular UNFCCC conferences) and prepares policy briefs and scientific input for delegations.

At present, the work of the GMC is largely financed by third-party funds. Between 2005 and 2015, approximately 15 million euros of peatland-related third-party funds were raised. Since the foundation of the GMC, the GMC management has coordinated the participation in peatland-related application procedures with a jointly financed GMC funding interface. As a result of this and the general growth in staff and capacity, the acquisition of third-party funds was significantly intensified: In the years 2016–2018, a total of approximately 8.8 million euros in peatland-related third-party funds were raised.

28.5.3 Databases

The Peatland and Nature Conservation International Library (PeNCIL) occupies a special position among the databases of the Greifswald Moor Centrum. As a ‘classic’ database, it now contains around 25,000 publications on peatlands in the broadest sense, plus a large number of special prints, maps and photographs. Thanks to the cooperation with the Greifswald University Library, the Michael Succow Foundation and, in particular, the Bernhard and Ursula Plettner Foundation, it has been possible to considerably expand the library and transfer it to the holdings of the University Library as a special library. To date, approximately 5500 book titles have been inventoried and are accessible via the catalogue. A large number of particularly important titles have also been digitised, and the programme library is part of the Digital Library of Mecklenburg-Vorpommern. The ‘Peatland Library’ regularly hosts literature evenings at which Hans Joosten presents books, people and stories about peatlands to a selected audience.

Another important database is the Global Peatland Database (GPD). The GPD is a project of the International Mire Conservation Group (IMCG) and is managed by the Greifswald Mire Centre. It provides an overview of the extent and degradation status of peatlands/organic soils in 268 countries and regions of the world. It mainly contains digital data, but also information from printed sources of the Peatland and Nature Conservation International Library (PeNCIL). A database with thousands of digital photographs of peatlands from all over the world is linked to the GPD. Since 2012, the information on peatlands/organic soils is transferred into a spatially explicit GIS–GPD: a vector-based geographic information system (GIS) on a scale

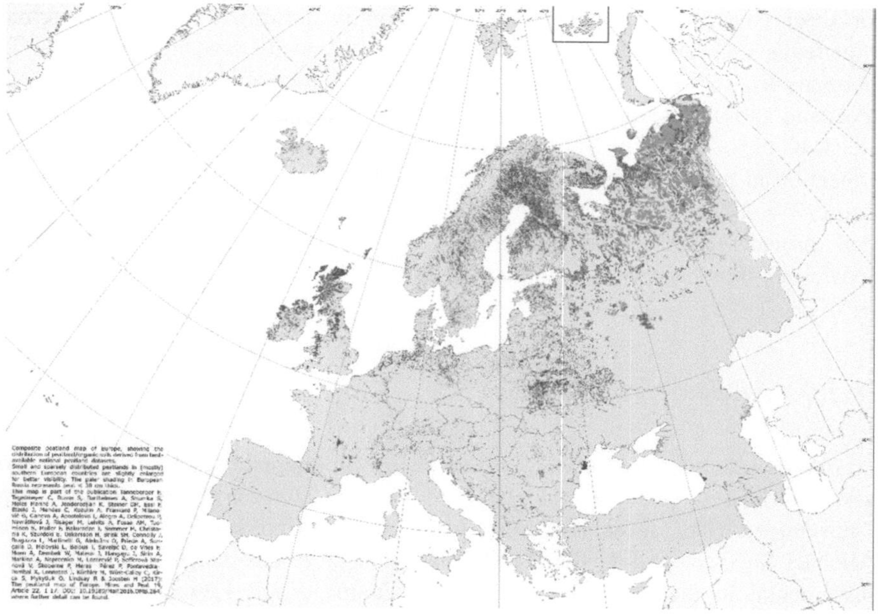

Fig. 28.5 Map of peatlands/organic soils of Europe (from Tanneberger et al. 2017)

of 1:25,000 and a raster-based GIS with 0.01 × 0.01 degree network. The GIS–GPD is continuously updated and improved and can be supplemented with additional information on peat thickness, carbon content, vegetation and mire type. In addition, GIS data sets from other bog mapping projects are evaluated and integrated. A peatland map of Europe was published in 2017 (Tanneberger et al. 2017; Fig. 28.5). Global coverage by 2020 is being sought; a global peatland map has been in progress since 2019. Current projects on data collection and peatland mapping include North-east and West Africa, North-east China, Indonesia, Papua New Guinea, Iran and Cuba.

28.5.4 Application

The concept of paludiculture, i.e. wet peatland cultivation, was developed in Greifswald. International bodies like the Intergovernmental Panel on Climate Change (IPCC), the United Nations Environmental Programme (UNEP), the Food and Agriculture Organisation of the United Nations (FAO) and the European Parliament have recognised the significance of the concept and have made it their own.

In the project 'German Peatland Protection Dialogue—Impulses for Climate Protection, Biodiversity and Sustainable Land Use on Peatlands'—Moor Dialogue for short—the three partners in the GMC jointly networked actors, transferred knowledge and discussed possible solutions to intensify peatland protection in Germany from 2015 to 2019 at federal, state and municipal level. The new project 'Peatland and climate protection—realising practical solutions with land users' at the GMC picks up results from the MoorDialog and intends to implement peatland protection in model regions together with the German Association for Landscape Conservation. The Michael Succow Foundation has implemented numerous peatland projects in Germany, Belarus, Ukraine, Kyrgyzstan and the Russian Federation, thus far rewetting more than 15,000 ha of peatland.

28.5.5 Advice

Greifswald scientists are (or were) active players in international organisations and advisory bodies such as the UNFCCC (Hans Joosten), the IPCC (John Couwenberg, Hans Joosten), FAO (Hans Joosten), UNESCO MAB Committee (Michael Succow) and UNESCO-SCOPE (Hans Joosten). GMC scientists were instrumental in the inclusion of 'Wetland Drainage and Rewetting' as a new land-use activity under the Kyoto Protocol. The UNEP-Report 'Frontiers 2018/19' defines the thawing of Arctic permafrost as one of five pressing but hitherto underestimated environmental threats based on Greifswald peatland expertise. The GMC has strong ties to various administrative levels of environmental protection and nature conservation authorities in Germany and a number of its federal states. Six scientists of the Greifswald Mire Centre have contributed to the TEEB DE report on natural capital and climate policy. Peatland-rich countries and regions are supported by policy briefs, e.g. the Nordic Council of Ministers.

28.6 Conclusions

The importance of peatlands in relation to major societal challenges such as the climate crisis is growing. More than ever, the protection of previously undrained peatlands and the rewetting of drained peatlands and their wet, sustainable use are the order of the day. The Greifswald Mire Centre meets this challenge nationally and internationally in research, implementation and policy consulting.

Appendix: Figures About Research Topics and Practical Work of the GMC

See Figs. 28.6, 28.7, 28.8, 28.9, 28.10, and 28.11.

Fig. 28.6 Field research on biogeochemistry and plant production/decomposition in Peene Valley, NE Germany (Photograph with kind permission of K. Vegelin)

Fig. 28.7 Nursery with *Sphagnum mosses* collected from all over Europe at the University of Greifswald (Photograph G. Gaudig)

Fig. 28.8 Field experiments on Sphagnum growth on a rewetted bog grassland (Photograph G. Gaudig)

Fig. 28.9 Mesocosm compound in the backyard of the Institute of Botany and Landscape Ecology, Greifswald University with our cooperation partners from Warsaw University (Photograph with kind permission of J. Kreyling)

Fig. 28.10 Field work in Western Siberia, where large, virtually undisturbed peatlands are studied (Photograph F. Tanneberger)

Fig. 28.11 The Aquatic Warbler (*Acrocephalus paludicola*), a flagship species for fen mires and key research topic at Greifswald University (Photograph with kind permission of Z. Morkvenas)

References

Bonn A, Allott T, Evans M, Joosten H, Stoneman R (2016) Peatland restoration and ecosystem services: science, practice, policy. Cambridge University Press, Cambridge, 493 p

Couwenberg J, Thiele A, Tanneberger F, Augustin J, Bärisch S, Dubovik D, Liashchynskaya N, Michaelis D, Minke M, Skuratovich A, Joosten H (2011) Assessing greenhouse gas emissions from peatlands using vegetation as a proxy. Hydrobiologia 674:67–89

Gaudig G, Krebs M, Prager A, Wichmann S and 30 others (2018) Sphagnum farming from species selection to the production of growing media: a review. Mires Peat 20(13):1–30

GMC (2019) Greifswald Mire Centre. Strategy 2018–2022. Greifswald. 18 p. https://greifswaldmoor.de/files/dokumente/Jahresberichte%20+%20Strategie/GMC%20strategy%202018-2022_webversion.pdf

Hornschuch CF (1837) Über die Eigenthümlichkeiten der Flora der Torfmoore in der Umgegend von Greifswald. Flora oder Allg Bot Ztg 20:737–747 u. 753–761

Janke W (1978) Untersuchungen zu Aufbau, Genese und Stratigraphie küstennaher Talungen und Niederungen Nordost-Mecklenburgs als Beitrag zu ihrer geoökologischen und landeskulturellen Charakteristik. Diss. B an der Math.-Nat. Fak. der E.-M.-Arndt-Univ. Greifswald

Jeschke L (2003) Das Kieshofer Moor bei Greifswald. Greifsw Geogr Arb 30:35–42

Joosten H (1998) Peat as a renewable resource: the road to paludiculture. In: Malterer T, Johnson K, Stewart J (eds) Peatland restoration & reclamation—techniques and regulatory considerations, 56–63 p. International Peat Society, Jyskä

Joosten H, Sirin A, Couwenberg J, Laine J, Smith P (2016a) The role of peatlands in climate regulation. In: Bonn A, Allott T, Evans M, Joosten H, Stoneman R (eds) Peatland restoration and ecosystem services: science, policy and practice. Cambridge University Press/British Ecological Society, Cambridge, pp 63–76

Joosten H, Gaudig G, Tanneberger F, Wichmann S, Wichtmann W (2016b) Paludicultures: sustainable productive use of wet and rewetted peatlands. In: Bonn A, Allott T, Evans M, Joosten H, Stoneman R (eds) Peatland restoration and ecosystem services: science, policy and practice. Cambridge University Press/British Ecological Society, Cambridge, pp 339–357

Joosten H, Tanneberger F, Moen A (eds) (2017) Mires and peatlands of Europe: status, distribution and conservation. Schweizerbart Science Publishers, Stuttgart, p 730

Koch F (1849) Naturgeschichtliche Bemerkungen über das zwischen Trebel und Recknitztal gelegene Moor. Archiv für Freunde der Naturgeschichte in Mecklenburg 3

Lange M (1994) Das Kieshofer Moor – eine landschaftsökologische Analyse. Dipl.-Arb. Bot. Inst. der E.-M.-Arndt-Univ. Greifswald; 189 p

Leick E (1927a) Das Kieshofer Moor als Naturschutzgebiet. Beitr Naturdenkmalpfl 12:1–10

Leick E (1927b) Das Kieshofer Moor in den kartographischen Darstellungen und Berichten aus früherer Zeit. Beitr Naturdenkmalpfl 12:11–38

Leick E (1929) Zur Frage der Wasserbilanz von Hochmooren. Untersuchungen über das ökologische Sättigungsdefizit. Mitteilungen des naturwissenschaftlichen Vereins für Neuvorpommern und Rügen 56:146–174

Michaelis D, Abel S, Gaudig G (2016) 200 Jahre Moorforschung in Greifswald – Ein Über- und Ausblick. TELMA - Berichte der Deutschen Gesellschaft für Moor- und Torfkunde 46:195–212

MLUV MV (2009) Konzept zum Schutz und zur Nutzung der Moore. Fortschreibung des Konzeptes zur Bestandssicherung und zur Entwicklung der Moore. Schwerin

Schipper AM, Zeefat R, Tanneberger F, Van Zuidam JP, Hahne W, Schep S, Loos S, Bleuten W, Joosten H, Lapshina ED, Wassen MJ (2007) Vegetation characteristics and eco-hydrological processes in a pristine mire in the Ob River valley (Western Siberia). Plant Ecol 193:131–145

SRU (2012) Umweltgutachten 2012. Verantwortung in einer begrenzten Welt. Erich Schmidt Verlag. 420 p

Succow M, Joosten H (eds) (2001) Landschaftsökologische Moorkunde. Schweizerbart, Stuttgart, p 622

Succow M (1971) Die Talmoore des nordostdeutschen Flachlandes, ein Beitrag zur Charakterisierung des Moortyps „Niedermoor". Archiv für Naturschutz und Landschaftsforschung 11:133–168
Succow M (1981) Landschaftsökologische Kennzeichnung und Typisierung der Moore der DDR. 256 S. u. Anlagenbd.; Diss. B, AdL der DDR
Succow M, Lange E (1984) The mire types of the German Democratic Republic. In: Moore PD (ed) European mires. Academic Press, London, pp 149–175
Succow M (1988) Landschaftsökologische Moorkunde. Fischer, Jena, p 340
Tanneberger F, Tegetmeyer C, Busse S, Barthelmes A, Shumka S, Moles Mariné A, Jenderedjian K, Steiner GM, Essl F, Etzold J, Mendes C, Kozulin A, Frankard P, Milanović Đ, Ganeva A, Apostolova I, Alegro A, Delipetrou P, Navrátilová J, Risager M, Leivits A, Fosaa AM, Tuominen S, Muller F, Bakuradze T, Sommer M, Christanis K, Szurdoki E, Oskarsson H, Brink SH, Connolly J, Bragazza L, Martinelli G, Aleksāns O, Priede A, Sungaila D, Melovski L, Belous T, Saveljić D, de Vries F, Moen A, Dembek W, Mateus J, Hanganu J, Sirin A, Markina A, Napreenko M, Lazarević P, Šefferová Stanová V, Skoberne P, Heras Pérez P, Pontevedra-Pombal X, Lonnstad J, Küchler M, Wüst-Galley C, Kirca S, Mykytiuk O, Lindsay R, Joosten H (2017) The peatland map of Europe. Mires and Peat 19 (Article 22):1–17
Tanneberger F, Kubacka J (2018) The aquatic warbler conservation handbook. Brandenburg State Office for Environment (LfU), Potsdam, p 260
TEEB DE (2014) Naturkapital Deutschland. Naturkapital und Klimapolitik – Synergien und Konflikte. Kurzbericht für Entscheidungsträger. Technische Universität Berlin, Helmholtz-Zentrum für Umweltforschung – UFZ, Leipzig
UNESCO (1994) Convention on Wetlands of International Importance especially as Waterfowl Habitats. Ramsar 2.2.1971. https://www.ramsar.org/sites/default/files/documents/library/scan_certified_e.pdf, accessed on Mar 30, 2019
United Nations (1992) Convention on Biological Diversity. https://www.cbd.int/doc/legal/cbd-en.pdf, accessed on Mar 30, 2019
von Bülow K (1921) Das Kieshofer Moor bei Greifswald. Diss. an der Math.-Nat. Fak. der Univ. Greifswald: 38 p.; Greifswald
von Bülow K (1928) Zur Geologie des Kieshofer Moores. Mitteilungen des naturwissenschaftlichen Vereins für Neuvorpommern und Rügen 52/56
von Chamisso A (1824) Untersuchung eines Torfmoors bei Greifswald und ein Blick auf die Insel Rügen. - Karstens Archiv für Bergbau und Hüttenwesen 8(1):129–139
von Post L (1916) Skogträdpollen i sydsvenska torfmosselagerföljder (föredragsreferat). Geol. Fören, Stockholm Förhandl, p 38
von Post L (1918) Skogträdpollen i sydsvenska torfmosselagerföljder. Forh. Ved. 16. skand. Naturforskermøte 1916, Kristiana, 433–465
Wassen MJ, Olde Venterink H, Lapshina ED, Tanneberger F (2005) Endangered plants persist under phosphorus limitation. Nature 437(7058):547–551
Wichtmann W, Schröder C, Joosten H (2016) Paludikultur – Bewirtschaftung nasser Moore. Kimaschutz – Biodiversität – regionale Wertschöpfung. Schweizerbart Science Publishers, Stuttgart. 272 p

Chapter 29
Agricultural Land Conversion and Food Policy in Indonesia: Historical Linkages, Current Challenges, and Future Directions

Mohammad Rondhi, Pravitasari Anjar Pratiwi, Vivi Trisna Handini, Aryo Fajar Sunartomo and Subhan Arif Budiman

Abstract The problem of agricultural land conversion (ALC) is inseparable to the food security issue in Indonesia. The government has set ALC as the primary threat to the existence of food security. The monoculture food policy in Indonesia has developed an unbalanced food system which relies heavily on rice and shaped the rice-eating habit among Indonesians. The government has formed an act about sustainable agricultural land (LP2B) to protect productive agricultural land and to mitigate the risk of decreasing rice production. The government also encourages the diversification of food to lessen the dependence on rice. However, it is difficult to implement both policies. The formation of LP2B will promote the conflict of interests between stakeholders involved (farmers, real estate developer, and commercial business). On the other hand, food diversification will not be sufficient unless followed by massive effort to stimulate the establishment of supporting institutions (the market for the agricultural product, processing industries, and the market for the consumer). Thus, it is required to limit the rate of ALC with a different and innovative approach. In this chapter, we explain how increasing agricultural land rent encourages farmers to retain farmland. Furthermore, we also identify what significant factors that affect the rent of agricultural land.

M. Rondhi (✉) · P. A. Pratiwi · V. T. Handini
Department of Agribusiness, University of Jember, Jember 68121, Indonesia
e-mail: rondhi.faperta@unej.ac.id

P. A. Pratiwi
e-mail: pravita.anjar@unej.ac.id

V. T. Handini
e-mail: vivi.trisna@unej.ac.id

A. F. Sunartomo
Department of Agricultural Extension, University of Jember, Jember 68121, Indonesia
e-mail: aryo.faperta@unej.ac.id

S. A. Budiman
Department of Soil Science, University of Jember, Jember 68121, Indonesia
e-mail: sabudiman@unej.ac.id

L. Müller and F. Eulenstein (eds.), *Current Trends in Landscape Research*, Innovations in Landscape Research, https://doi.org/10.1007/978-3-030-30069-2_29

Keywords Agriculture · Land conversion · Landscape · Indonesia · Food policy · LP2B · Food diversification · Land rent

29.1 Introduction: Intersection Between Land and Food Policy in Indonesia

The conversion of agricultural land is a significant threat to the food security of a nation (Agus and Irawan 2006; FAO 2011). Especially in Indonesia, where rice as the the most important staple food crop, is mostly grown in convertible agricultural land (Irawan 2011). Historically, the self-sufficiency of rice has always been the top priority of food policy in each Indonesian government regime (Hafsah and Sudaryanto 2004). It shows that rice has a critical political value.

Consequently, it is not only crucial to food security but more importantly also vital to national security. Thus, the logical framework to achieve rice self-sufficiency (RSS) is by fostering the factors which increase rice production and suppressing the factors which decrease it. Diagrammatically, this framework is provided in Fig. 29.1.

One of the significant suppressing factors to achieve RSS is agricultural land conversion (ALC) (Pasandaran et al. 2006). As the Indonesian economy is experiencing rapid industrialization, more land is needed to support this process (Azadi et al. 2011). The conversion of productive farmland mostly supplied this demand since the cost of development in that area is much cheaper than in an established neighborhood (Bhatta 2010; Omrani et al. 2015). Commonly, ALC started to occur in the area near the urban core and formed an area termed as peri-urban (Pribadi and Pauleit 2016). Peri-urban area is the result of quick urbanization and economic development in the urban core which requires more spaces for either housing, commercial, or public uses (Liu et al. 2014; Magsi et al. 2017). Previously, urbanization occurred only in the metropolitan area such as the national and the provincial capital city. Currently, there are many new urban cores in the district and sub-district level which further increase the rate of ALC and the number of peri-urban areas (Hudalah and Firman 2012a).[1] Most of these districts have an agricultural-based economic structure and play a vital role in supplying food for the population. Thus, the ALC in these areas has more crucial implications for food security than those in the metropolitan area.

Moreover, urbanization is the major cause of the loss of valued agricultural landscapes (Schulp et al. 2019). Urbanization is often accompanied by the over-exploitation of natural resources which changes agricultural landscapes completely

[1]Indonesia has two levels of the regional administrative goverment. The first regional administrative goverment is province (Provinsi) and the second regional administrative goverment is district (Kabupaten). Each district consists of sub-districts (Kecamatan) which in turn consisted of several villages (Desa).

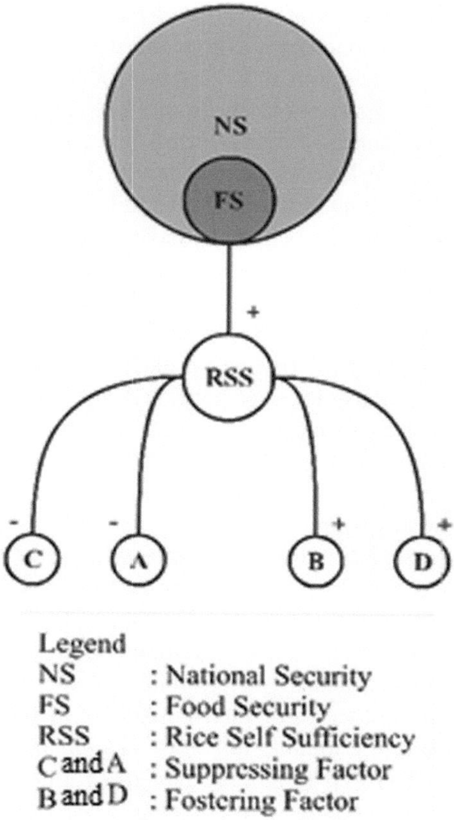

Fig. 29.1 Framework to achieve rice self-sufficiency

(Plieninger et al. 2016). Also, it makes agricultural landscapes less dominant, aggregated but more unstable, and fragmented but irregular (Su et al. 2014). Furthermore, as cultural landscapes are defined as a result of the interaction between the human affair and the natural aspect of landscapes, the higher economic value of peri-urban land compared to those in the rural area has influenced the shapes and forms of the houses and generally in the form of the landscape in both areas. Thus, understanding how land economic value contributes to the development of landscape in both areas is important for the planning of development policy in the area.

ALC poses a significant threat to food security because of Indonesians' high dependence on rice (Ariani 2004). Indonesia is a mega-biodiversity nation with over 800 varieties of food crops (Fajar 2014). However, the monoculture food policy of the Indonesian government which focuses only on rice has shaped the habit of rice eating among Indonesian people. Also, the monoculture nature of food policy has caused an imbalanced development between supporting institutions for rice and another food commodity (Sumarno 2004). Thus, the food diversification policy is difficult to achieve. Currently, Indonesians are the major rice consumer in the world with over 100 kg/capita/year of rice consumption (OECD/FAO 2013). The Indonesian government encourages food diversification policy to lessen dependency on rice. However, it is worthy to note that it requires 70 years and many resources to shape this rice-eating habit. It is evident that the same or even more substantial amount of time and resources are required to change this habit. Thus, at least in the next few decades, rice is still the main food in Indonesia.

Consequently, ALC will still be the main threat to food security. Moreover, numerous investigations have expressed the significance of farming in peri-urban zones. Peri-urban farming can enhance the state of nature by alleviating the urban high-temperature impact and decreasing flood hazard (Malaque and Yokohari 2007; Yokohari et al. 2000), making green spaces for the urban populace (Ives and Kendal 2013; Tassinari et al. 2013; Zasada 2011), what's more, a wellspring of pay particularly for poor people (Bryld 2003; Zezza and Tasciotti 2010). Additionally, in the development setting, the indispensable role of peri-urban agriculture has frequently been expressed (McGee 2010). Peri-urban agriculture plays a vital role in supporting food security in the urban region, with its agricultural resource and vicinity to the urban center. Peri-urban area additionally assumed a primary job in the sustainability of urban environment (Allen 2003), providing job opportunities, and alters the course of environmental degradation in the urban region (Douglass 1998). In this manner, it is essential to control the issue of ALC in Indonesia.

The Indonesian government created a regulation to control the rate of ALC which contains two essential systems (Government of Indonesia 2009). The first system focuses on the formation of inconvertible farmland zone under the name of *Lahan Pertanian Pangan Berkelanjutan* (LP2B, in English: sustainable agricultural land). LP2B cannot be converted for non-agricultural use for 20 years. The second system focuses on motivating farmers to retain farmland through various forms of compensation. The type of compensations includes diminishing tax for the land, enhancing farm infrastructure, and compensating the accomplishments of farmers. The former system focuses on limiting ALC by avoiding the conversion of farmland. The latter system focuses on limiting ALC through increasing the motivations of landowners to retain their land in agricultural use.

Some researchers have demonstrated that the use of legal power to limit ALC has given rise to illegal farmland conversion (Chen et al. 2015; Wang and Scott 2008; Zhong et al. 2014). Economically, the regulation that limits ALC contradicts with the motive of the sellers (farmers who own farmland) and the buyers (parties who intend to convert farmland for other uses). As rapid economic growth increases the land rent for non-agricultural purposes, the economic benefits accrued to farmers by retaining their farmland are relatively low. Thus, it drives farmers to sell their land (Ettema et al. 2012; van Vliet et al. 2015). On the other hand, the buyers (whether firm willing to expand their business, a real estate developer, or government willing to build physical infrastructure) need more land with low buying and development costs. Thus, because there is an agreement in economic motives between buyers and sellers, the occurrence of ALC is unavoidable.

Conversely, the policy focus on incentivizing farmers to retain their farmland is proven to be effective in limiting ALC. For example in Europe, decoupled payments and environmental schemes increase land value because those payments are included in land value (Ciaian et al. 2013; Feichtinger and Salhofer 2013; Kilian et al. 2012; Latruffe and Le Mouël 2009). The payment makes farmers reluctant to convert the land for other uses since the payment improves land value. Moreover, conflict among farmers and between farmers and non-farmers has emerged as the result of increasing land value. Increasing land value limits ALC in two ways: first, by decreasing farmer's economic motive to sell the land, since it is more profitable for them to retain their land, and second, increasing the buying costs of agricultural land, in the form of conflict costs. The conflict costs arise because farmers are interested in obtaining the economic incentive of retaining the farmland.

Based on that background, this chapter focuses on measuring the agricultural land rent (as the proxy for land economic value) relative to non-agricultural land and identifying the determinants of rent for each type of land. In explaining those purposes, this chapter summarized the result of our previous study, which is conducted in two different areas, the first one is in peri-urban area where rapid ALC occurred and the second one is in rural area which has never experienced ALC (Rondhi et al. 2018). Besides, before explaining those results, in the next section, we first describe the historical course of food policy in Indonesia. The purpose of this section is to appreciate the impact of continuous rice-based food policy on high Indonesian dependent on rice and indirectly on the existence of agricultural land. The second section is followed by the explanation of the current challenge of agricultural land and food policy in Indonesia. This is the challenge of the establishment of LP2B to limit ALC and food diversification to reduce the dependency of Indonesians to rice. It is in the fourth section that we explained the result of our previous studies regarding agricultural and non-agricultural land rent. In the fifth section, we showed the cultural landscapes in the peri-urban and rural area. In the sixth section, we proposed how to use the finding of this study as a future direction for policy aimed to control ALC in Indonesia. Finally, we conclude this chapter in the last section.

29.2 Historical Course of Food Policy in Indonesia

After gained independence in 1945, Indonesia has been led by seven presidents for 74 years. Consequently, it has experienced numerous political interests and various policy packages. However, in case of food policy, RSS has always been the top priority, whether for shortest-reigned president (B. J. Habibie) or the longest-reigned president (Soeharto). In general, there are four groups of Indonesian presidential. President Soekarno led the first era and was named *Orde Lama* (Old Era), while President Soeharto led the second era and was named Orde Baru (New Era). The third was named the transition era which was led by President B. J. Habibie. After the transition, four presidents have led Indonesia, and this era is named *Era Reformasi* (Reform Era). Figure 29.2 shows the overall course of food policy in Indonesia.

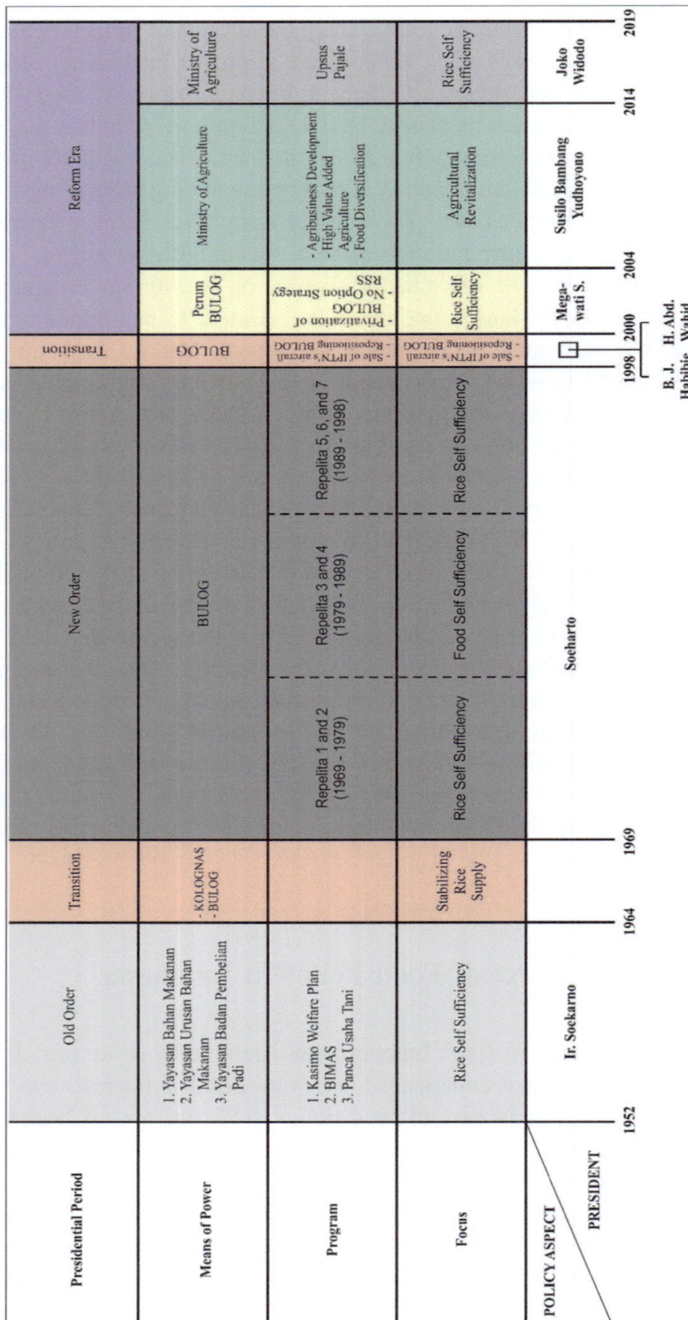

Fig. 29.2 Historical course of food policy in Indonesia

29.2.1 Food Policy in the Old Order (1952–1964)

Soekarno led the Old Era and lasted for 22 years. The food policy in this regime mainly focused on rice and had strong political interests. The first policy was aimed to achieve RSS through *Program Kesejahteraan Kasimo* (Kasimo Welfare Plan) (Nusantoro and Cramb 1990). *Yayasan Bahan Makanan* (the Foundation for Food) was established to carry this plan and did its job from 1950 to 1952. In 1953, the government changed the foundation's name to *Yayasan Urusan Bahan Makanan* (the Foundation for Food Supply). However, the core function of this body remained the same. The government issued the second policy in 1956 under the name *Swasembada Beras Melalui Program Sentra Padi* (Rice Self-Sufficiency through Rice Center Program). The second policy remained to focus on achieving RSS; however, in the second program the government started to focus on rice production. In 1956, the government established *Yayasan Badan Pembelian Padi* (the Foundation of Rice Purchasing Agency). The main job of this agency is to buy rice from farmers and also act as the price-regulating authority for rice. In 1963, the government introduced maize as the substitution of rice, since typically Indonesian farmers planted maize after rice. In 1964, the government wanted to further support the rice farming by issuing farm intensification program, *BIMAS* (Mass Extension Program) and *Panca Usaha Tani* (Five Farming's Principles) (Soen 1968).

Soekarno's regimes collapsed in 1965 and followed by the formation of a transitional government in 1965–1967. In this transition period, the government formed KOLOGNAS (National Logistic Command). It acted like the YBPP in the Soekarno era. However, in 1967, the KOLOGNAS was halted, and the government formed BULOG (National Logistics Agency), which mainly acted as a sole buyer of rice. This agency was the initial period of the most extended Indonesian government regime led by Soeharto (Nawiyanto 2017).

29.2.2 Food Policy in the New Order 1969–1998

It was in the New Order where agriculture received considerable attention from the government, especially rice-based agriculture. The policy was started with a ten-year development program under *Repelita 1 and 2* (*Rencana Pembangunan Lima Tahun*/five-year development plan) in the period of 1969-1979. The main aim of the policy remained the same, achieving self-sufficiency in rice production. The policy started with the addition of BULOG task as National Food Buffer Stock, alongside this policy, is the use of National Food Balance (*Neraca Pangan Nasional*) as the standard for food security. However, the BULOG in this era did not focus only on rice but also on sugar, and wheat import (1971), meat procurement for the capital city (Jakarta) started in 1974, and soybean import control (1977). The first ten years of the New Order also marked with the rise of Indonesian Farmers Union (Serikat Petani Indonesia) in 1973 and the introduction of the Green

Revolution to achieve RSS in 1974. The end of this period showed the introduction of the floor price for maize, soybean, green bean, and peanuts (Soen 1968).

The second ten-year period of the New Order showed remarkable growth for rice production. Although the national focus changed to food self-sufficiency, which was marked by the expanding tasks of BULOG as a price control authority for rice, wheat flour, sugar, meat, and other food commodities, Indonesia achieved its first and only rice self-sufficiency in this period. It was in 1984 where Indonesian rice production exceeds its consumption. The RSS achieved in 1984 affected the direction of food policy in the third development period. The national focus changed back to rice self-sufficiency. In 1995, the entire BULOG personnel were officially appointed as government officials (PNS, *Pegawai Negeri Sipil*) along with its privileges and incentives. The subsequent major policy showed the narrowing tasks of BULOG. In 1997, BULOG was appointed to control only the price for rice and sugar. In 1998, the narrowing of BULOG tasks continued by controlling only the price for rice. The third period was planned for Repelita 5, 6, and 7 and started in 1989. However, the New Order regime cannot retain its power, collapsed in 1998, and marked the beginning of the Reform Era.

29.2.3 Food Policy in the Transition Period 1998–2000

This period was the age of turmoil in the history of Indonesian politics. The economy was collapsed, coupled with unstable national security. It was more challenging to sustain food production in this period. However, rice was still the primary focus of the government. The first president in this period (B. J. Habibie) tried to stabilize national food supply (rice). Habibie sold the IPTN's aircraft for Thai's rice in 1998/1999. As the national stability increased, the shift in focus of national policy occurred during the second president (Abdurrahman Wahid). In 2000, the government reaffirmed the task of BULOG as the rice logistic management authority which focused on stabilizing supply, distribution, and the price of rice. Abdurrahman Wahid was projected to be in the office from 1999 to 2004; however, due to political circumstances, the office was assumed by Megawati in 2000.

29.2.4 Food Policy in the Reform Era 2000–Now

The course of Indonesian politics has shown the significant role of BULOG in food policy. Back then, BULOG was a not-for-profit government agency and acted mostly to stabilize the national rice logistics. The previous government, either in Old Order, New Order, or in the transition period was only modifying the role of BULOG by expanding or narrowing its tasks, it was only in the era of Megawati that BULOG experienced significant changes. In 2003, the government privatized

BULOG into a general corporation form. By the privatization of BULOG, it focused more on business activity rather than public services. However, although the government reduced its role in national rice logistics (through reducing the role of BULOG), in 2004 the government declared RSS as the sole priority for national food security. It was more challenging to achieve RSS than ever before, given the reduced priority of government in rice policy.

The era of Susilo Bambang Yudhoyono (SBY) realized the difficulty of achieving RSS. Consequently, he focused on agricultural revitalization rather than on individual rice policy. In the ten years of SBY, the food policy not only focused on rice but also on food diversification and strengthening the agribusiness sector that adds value to agricultural products. However, with only ten years, the dependence of Indonesians on rice does not significantly decrease. It was exacerbated by the current Indonesian government (Joko Widodo) which focuses on food policy on achieving the self-sufficiency in rice, maize, and soybean by increasing the investment in agricultural irrigation infrastructure.

The previously described course of Indonesian politics showed that rice was essential to Indonesians, and politically, it is more critical for the government. With this condition, it is difficult or even impossible to completely change the rice-eating habit of Indonesians in the next few years. Thus, the government focuses on minimizing the suppressing factors which limit rice production, such as the conversion of fertile farmland by forming LP2B as the sustainable agricultural land which is inconvertible for other uses. Also, the promotion of "eat local food" is also encouraged today. However, both of these policies have critical challenges. The next section will discuss the challenges of these policies.

29.3 Current Challenges of Food Policy in Indonesia

The importance of preserving rice-based policy in general and protecting farmland from conversion and diversifying food sources, in particular, has been discussed in the previous sections. In this section, we will focus on describing why these policies are relevant to Indonesian conditions and the challenges that arise in the implementation stages. The protection of farmland has been regulated under *UU NO. 41 Tahun 2009*. There are two significant policies contained in this regulation, the formation of legally protected farmland (LP2B) and the effort to improve the economic value of agricultural activities to incentive farmers to retain their land. Meanwhile, food diversification has been regulated under *Perpres No. 22 Tahun 2009* and its main aim was to foster the consumption of locally produced food.

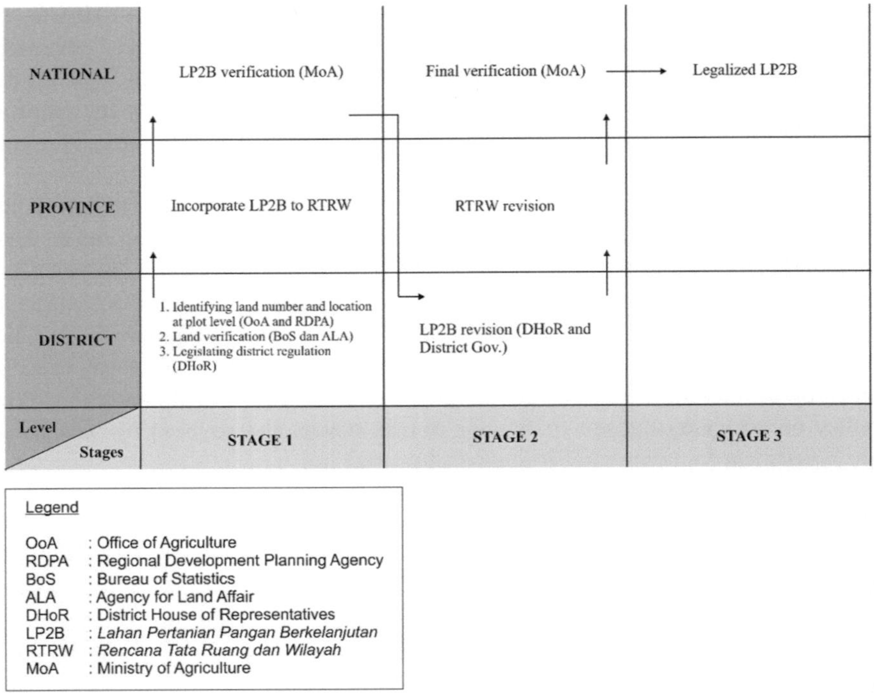

Fig. 29.3 Mechanism of LP2B formation

29.3.1 Farmland Protection Under UU No. 41 Tahun 2009

Forming legally protected farmland is conceptually sound in restricting its conversion to other uses. However, instinctively, it contradicts with the economic motives of the related stakeholders as has been found that there were many problems arisen in the formation of LP2B, whether in the district, provincial, or national level. There are three stages consisting of three administrative levels involved in the formation of LP2B. The three stages are submission, verification, and implementation stages. Diagrammatically, Fig. 29.3 represents the process.

The formation of LP2B initiated at the district level. The district government, through its executive body, identifies the number (area) and location of farmland in the district. The executive body involved is the Office of Agriculture (OoA) and Regional Development Planning Agency (RDPA).[2] Technically, the field extension

[2]Officially, OoA is *Dinas Pertanian Kabupaten* while RDPA is *Badan Perencanaan Pembangunan Daerah (BAPPEDA).*

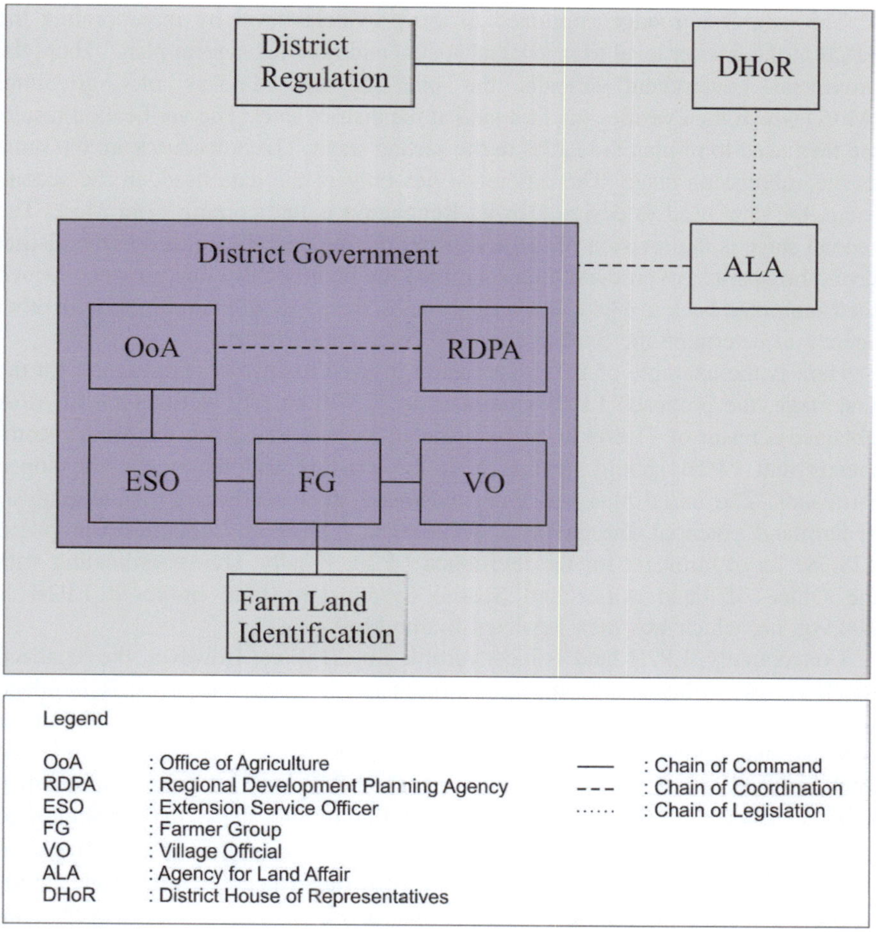

Fig. 29.4 Mechanism of LP2B formation at the district level

officials of OoA cooperate with farmer group and village officials to enumerate the field at the plot level. The obtained data is used as the basis to plan District Regulation which is legalized by the District House of Representatives (DHoR).[3] The data obtained is also confirmed with the data gathered by the Bureau of Statistics (BoS) and the Agency for Land Affair (ALA).[4] Diagrammatically, Fig. 29.4 shows the mechanism in the district level.

[3]DHoR is Dewan Perwakilan Rakyat Daerah (DPRD) in the official term.

[4]BoS: *Badan Pusat Statistik*; ALA: *Badan Pertanahan Nasional* (BPN).

The submission stage continued to the provincial level by incorporating the LP2B at the district level to provincial spatial and regional system plan.[5] Then, the provincial government submits the plan to the Ministry of Agriculture (MoA) which then verifies the farmland at the district level. The verification results are then used to re-plan the LP2B in the second stage. The procedures are the same as the submission stage. The difference lies only in the data used. In the second stage, the data used to design District Regulation is that coming from MoA. The second stage is the revision of the first. After the regulation legalized in the district level, the district government then submitted the final LP2B to the province, which then submitted back to MoA. In the final stages, the MoA will coordinate with other agency to determine the final figure of LP2B in each district.

Here is the example of LP2B formation in the Province of Yogyakarta. In the first stage, the proposed LP2B amounted to 51,770 ha. The verification of MoA obtained a result of 71,868 ha of farmland in Yogyakarta. Also, the MoA recommends that LP2B should include both the existing and the newly developed farmlands. The initial plan was then revisioned, after conducting spatial analysis of farmland potential, the province government (Office of Agriculture) proposed 128,381 ha of farmland for the revisioned LP2B. Finally, after coordinating with the Office of Land Affair and Spatial System, the final proposed LP2B is 104.905 ha, which was then legalized by the MoA.

Conceptually, LP2B land is inconvertible for 20 years. However, the legalized LP2B is always lower than the actual farmland. It means that there is farmland which is intended for conversion. Most of these lands are located near the urban core and play a vital role in the urban environment. In East Java, for example, the total farmland in 2017 is 1.086.486 ha, and the LP2B is 952.285 ha. It means that there are 134.201 ha of convertible farmland. Figure 29.5 shows the distribution of farmland in the heavily urbanized and less urbanized region in East Java. Based on the level of urbanization, the Province of East Java is categorized into four groups in the order from the highest to the lowest level of urbanization, megapolitan, metropolitan, intermediate city, and small city.

In the megapolitan region, LP2B comprises 49% of productive farmland. In the metropolitan region, LP2B comprises 94% of productive farmland. However, it is only 4% of the total farmland in the province. In the intermediate city, the LP2B is 51 percent of the total land. However, the rate of ALC in the period of 2012–2017 is 39% which is the highest compared to another region.

Furthermore, the intermediate region has 59% of the total farmland in the province. Thus, the high rate of ALC in the intermediate region causes a significant loss of farmland. In the small city where the urbanization level is low, the LP2B land comprises 94% of the total land and also the rate of ALC is the lowest compared to the other region. This figure shows that the ALC will continue to occur and with the highest rate will be likely to occur in the intermediate city which currently experiences rapid

[5]The provincial government has a spatial system which contained in *Rencana Tata Ruang dan Wilayah* (RTRW).

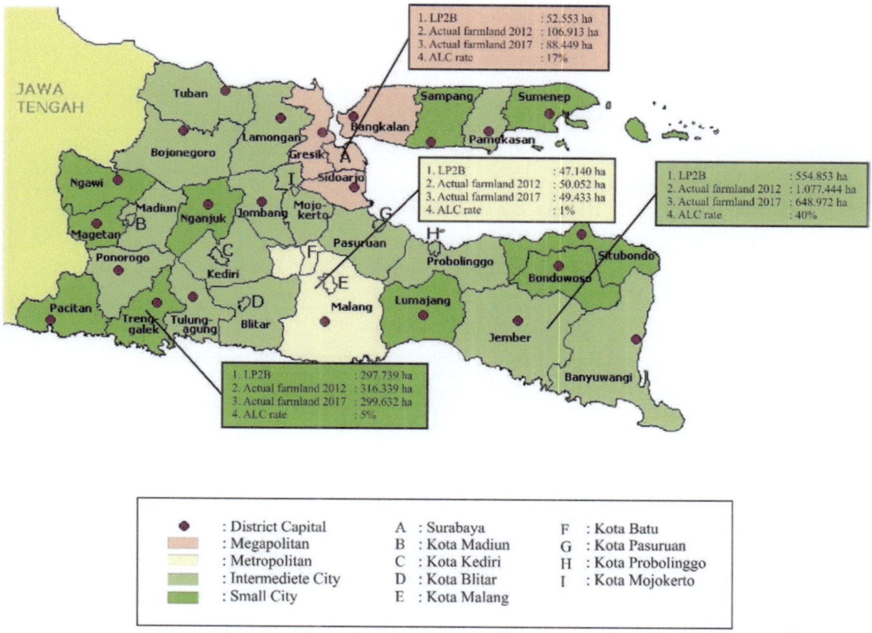

Fig. 29.5 LP2B and actual farmland (2012 and 2017) in East Java

growth of urbanization. The agricultural land conversion poses a severe risk, given the importance of farmland to the urban environment.

In general, LP2B attempts to reduce the rate of conversion through state power by prohibiting the conversion of particular farmland. However, if we look at the root causes of ALC, farmers (the seller) decide not to retain their land because the economic incentive they receive is high. On the other hand, the buyer receives higher economic incentive by transforming agricultural land for other uses, such as real estate, commercial site, and industrial plant. Consequently, the government faces parties with conflicting interests, and this will reduce the effectiveness of LP2B policy. The government task would be easier by aligning their interest with the farmers to retain the farmland. One way to motivate farmer to retain their land is by increasing the economic incentive of agricultural activity.

29.3.2 Food Diversification and Its Challenges

Indonesia has a rich variety of food crops. It is not less than 800 food crops grown in Indonesia. Currently, there are four crops available as a substitute for rice as the source of carbohydrate, e.g., cassava, maize, sweet potato, and sago. These food crops grow in the majority of the Indonesian region. Table 29.1 shows the distribution of the growing region of these crops.

Table 29.1 Growing region (a) for cassava, (b) for maize (c) for sweet potato (d) for sago in Indonesia

Food crop	Growing region	
	Province	District
Cassava	North Sumatera	Serdang Bedagai
	Bangka Belitung	Bangka Barat; Bangka; Belitung; Belitung Timur
	Lampung	Lampung Timur; Lampung Utara; Lampung Tengah; Tulang Bawang
	West Java	Cimahi; Bandung; Ciamis
	Central Java	Banjarnegara; Boyolali; Kebumen; Wonogiri
	Yogyakarta	Bantul; Kulon Progo; Gunung Kidul
	East Java	Trenggalek; Malang; Pacitan
	West Nusa Tenggara	Lombok Barat; Lombok Tengah; Lombok Timur; Dompu; Lombok Utara; West Sumbawa
	Southeast Sulawesi	Muna
	Maluku	Maluku Tenggara Barat; Maluku
	East Kalimantan	Nunukan
	Southeast Kalimantan	Sukamara
Maize	Central Java	Magelang; Temanggung; Boyolali; Semarang; Batang
	East Java	Kediri; Bangkalan; Tulungagung; Lumajang; Ponorogo
	East Nusa Tenggara	Lembata; Flores Timur; TTS; TTU; Alor; Ende
	North Sulawesi	Minahasa Utara
	Gorontalo	
Sweet potato	Aceh	
	North Sumatera	Medan
	Bengkulu	Bengkulu Utara
	Riau Islands	
	Banten	Serang
	West Java	Kuningan
	West Papua	Manokwari; Sorong Selatan
	Papua	Wamena

(continued)

Table 29.1 (continued)

Food crop	Growing region	
	Province	District
Sago	Riau	Pekanbaru
	Riau Islands	Karimun; Natuna
	Central Kalimantan	Sampit
	West Kalimantan	Pontianak
	Central Sulawesi	Parigi Moutong; Poso
	Southeast Sulawesi	Kendari
	North Sulawesi	Sangihe
	Maluku	Ambon; Seram; Maluku Tengah
	Papua	Keerom; Jayapura
	West Papua	Sorong Selatan

Source Hariyanto (2017)

The lack of supporting infrastructure and policy to encourage the consumption of this food is the main reason behind the low acceptability of this food to Indonesian. Although the government issued a regulation to encourage food diversification, it was not followed by the establishment of a market for these crops, the processing plant, and the mass commercialization of the food made with these crops. Consequently, few Indonesian know this food and still fewer are willing to consume it.

There must be a considerable effort from the government if food diversification is to be achieved. However, the government should focus on strengthening their role as the stimulator in the on and off the farm. In the on-farm, the government should stimulate farmer to grow these alternative food crops by stimulating the establishment of a market for these crops. In the off-farm, the government should stimulate the private sectors to develop new food product using these alternative crops as the raw materials. The involvement of private sectors will foster the development and innovation of food product based on this alternative food crops, but more importantly, commercialization and heavy promotion of this food will follow.

29.4 Land Economic Value as the Tool to Control Agricultural Land Conversion

Both of the challenges previously described show that rice is still the primary food in Indonesia. Consequently, the preservation of farmland is crucial to food security. To preserve farmland, it is crucial to control ALC. In this section, we present the result of our study which shows that the main driving factors of ALC are the relative land value between agricultural and non-agricultural uses.[6]

29.4.1 *Study Background*

According to traditional location theory, urban land creates higher rent compared to agricultural land and causes the occurrence of ALC (Irwin and Bockstael 2002, 2007; North 1955). The characteristics of ALC in developing countries such as Indonesia are rapid and unplanned. Rapid urbanization causes a growing number of urban areas in Indonesia. Unfortunately, this area is surrounded by productive agricultural land (Imhoff et al. 1997). Consequently, as the urban core grows, the subsequent agricultural land was converted to support this growth. The rate of ALC grows faster since both the seller (farmer) and the buyer (real estate developer, a firm, or government) will receive higher rent by converting those lands. It means that there is a significantly higher rent for urban land compared to agricultural land.

Conversely, zero ALC occurred in an area with higher rent for agricultural land. Many factors affect the rent for agricultural land, and identifying these factors can help in controlling ALC since the leading causes of ALC are the lower rent created from agricultural land related to urban land. Increasing rent for agricultural land will slow down the rate of ALC. The rent for agricultural land can be increased by manipulating factors which significantly increase it.

The general purpose of this study is to examine the rent for agricultural and urban land. The study was conducted in the peri-urban area with high rate of ALC and in a rural area where the ALC has not yet been recorded. The point of conducting this study in these areas was to compare the role of land rent in slowing or accelerating the rate of ALC. Also, this study also identifies factors that affect the rent for agricultural and urban land in both areas.

The primary purpose of this investigation is that it exhibits that land rent is the primary driver of ALC as has been previously mentioned in several empirical studies. However, few policy implications to control ALC are based on these findings. In the case of Indonesia, the study of urbanization has primarily focused on big megapolitan cities (Firman 1999, 2000; Firman and Dharmapatni 1994; Hudalah and Firman 2012b). Few studies are conducted in the smaller region which

[6]The full paper can be found in Rondhi et al. (2018).

experience the initial phase of urbanization. Studying these cities is essential. Furthermore, controlling ALC in the early urbanization process will be less costly than in the advanced urban area.

29.4.2 Methods for Measuring Land Rent

Two villages in Jember district in the Province of East Java Indonesia were selected for this study. The first village (Kepanjen) represents the rural area which has never experienced ALC during 2009–2016. The second village (Antirogo) represents the peri-urban area with 8.6% ALC rate during 2009–2016 (BPS-Statistics of Jember Regency 2017). The selection of these locations was to compare the rent for agricultural and non-agricultural land in the area with a rapid ALC and in the area with no ALC. Figure 29.6 shows the relative location of the studied area to Indonesia.

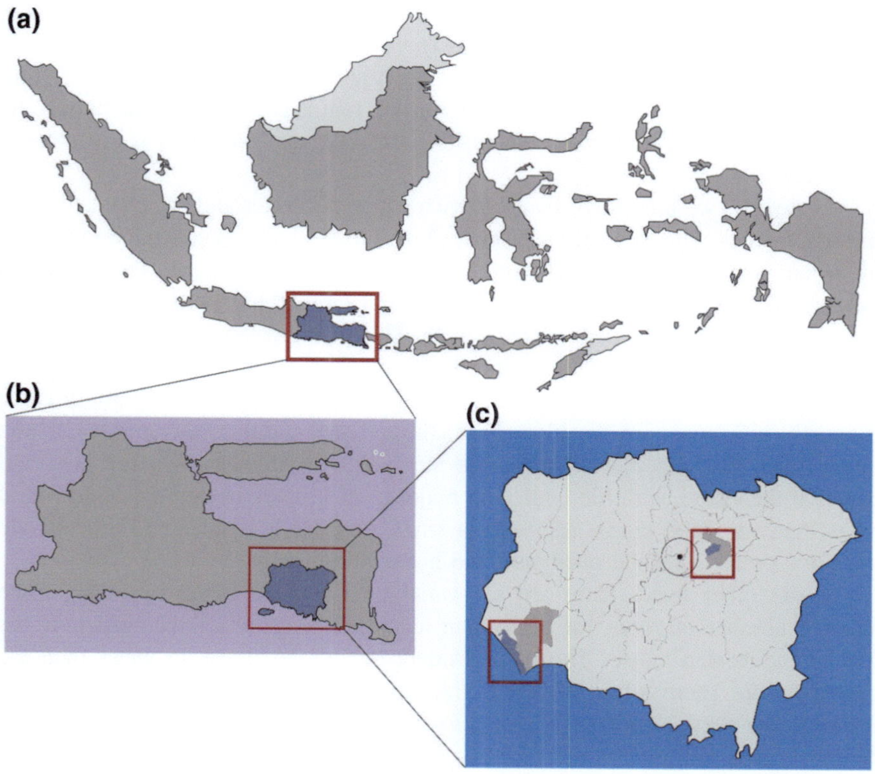

Fig. 29.6 Research site, **a** East Java, **b** Jember, and **c** Jember district map; the upper box shows the peri-urban area (Antirogo); the lower box shows the rural area (Kepanjen); the circle indicates the urban core

Agricultural land in both villages consisted of land planted with a variety of crops. However, the rural area has more various crops with 11 distinct cropping patterns. Meanwhile, the peri-urban area has fewer crops' variety with only two cropping patterns. The non-agricultural land identified in this study was land for housing. The development of housing is the primary purpose of converting farmland in the peri-urban area in Jember district. The new housing in Antirogo was developed in the formerly converted farmland. However, there is an interesting pattern in the farmland conversion. The productive rice field is changed first to moorland before housing construction begins. Moorland is a rice field which is abandoned and left uncultivated for a few years. Rice field abandonment is the pattern in which the housing developer converts agricultural land since they cannot directly convert productive rice field.

The primary data was collected from 200 respondents consisted of 100 farmers and 100 homeowners, the data collected through a survey questionnaire. The questionnaire consisted of two sections. The first section focuses on calculating the rent of farmland and housing, while the second section focuses on exploring the socioeconomic characteristics of landowners. We measure land rent as the annual profit obtained from each type of land. The agricultural land rent was calculated by deducting the total farming cost from the total revenue. The total cost consisted of both the implicit and the explicit costs. The implicit cost is accrued from unpurchased farm input such as seed from the previous harvest, organic fertilizer from livestock manure, and family labor. The housing rent is calculated as the annual net rental fee obtained by the homeowners. The rental fee was estimated as the imputed rental fee.[7]

Meanwhile, we identified seven characteristics of farmer and six characteristics of homeowners. The characteristics cover both physical and socioeconomic aspects of farmers and homeowners. The rural farmers have more land than the peri-urban farmers do. In case of irrigation access, the rural farmers have closer access to the irrigation canal than the peri-urban farmers. The average distance of rural farmer's land to irrigation is only 42 m, while the peri-urban farmers have land which is located 2053 m away on average from an irrigation canal. The sale of farm produce is mostly conducted on the plot. Thus, the average distance of farmland to the market and the road in both areas are similar. Most farmers own the land and perceive their land as fertile land. The majority of farmers in peri-urban and rural areas cultivate food crops, and only a small portion of farmers (33%) cultivate non-food crops (seasonal plantation and horticultural crops).

On the other hand, there are six variables identified for homeowners. The average number of room for peri-urban and rural houses is three, which is the typical Indonesian house. The rural resident has easier access to water than the peri-urban do. There is four percent of the peri-urban resident sampled in this study reported that they do not have access to sufficient water. It shows that there is a degradation in the quality and quantity of water resources in the peri-urban area.

[7]Imputed rental fee is the rental fee obtained by the homeowners had the house been rented.

Peri-urban area has easier access to road and downtown area. The downtown is the sub-district center (kecamatan) of each village which is vital to village life since the significant purchase of the villagers occurred in this area. Thus, it plays an important economic role.

29.4.3 Results and Discussion

29.4.3.1 The Dynamic of Land Rent in the Peri-urban and Rural Area

The result demonstrates that peri-urban area has a higher value of housing land rent compared to agricultural land. Conversely, in the rural area, the agricultural land has higher rent compared to housing. The critical distinguishing characteristic between the two is that the agricultural-housing rent gap in the peri-urban area is significantly higher compared to those in the rural area. In average, the agricultural land rent in peri-urban has a value of Rp 4447/m^2/year, and the average land rent for housing is Rp 39,904/m^2/year. Meanwhile, in the rural area, the average agricultural land rent is Rp 6047/m^2/year, and the rent for housing is Rp 5059/m^2/year on average.

Housing creates 700% higher value compared to agricultural land in the peri-urban area. The wide agricultural-housing land rent gap is the main reason for the rapid conversion of farmland in this area. Moreover, the converted agricultural land creates higher rent when used for housing. The converted farmland creates a rent of Rp 7917/m^2/year which is two times higher than when it used for agricultural purposes. In the rural area, although agricultural land has a higher land rent, it was only 19% higher compared to housing rent. It shows that currently, agriculture is still the main economic activity in the rural area. However, the narrow gap between agricultural and housing land rents indicates that the land can be converted easily to non-agricultural use.

Many studies demonstrated that urbanization and economic growth in the urban region cause the conversion of farmland (Wasilewski and Krukowski 2004; Kontgis et al. 2014; Phuc et al. 2014; Peerzado et al. 2018). Urbanization causes people to live in the urban area and increases the demand for housing. The development of new housing is often conducted in flat agricultural land. Similar conditions occurred in the peri-urban area of our study. The economy that overgrows and is moving away from agriculture causes rapid urbanization. The urbanization started to impact the nearby agricultural region which followed by the conversion of productive farmland and the degradation of environmental quality. The farmers choose to sell their land to capitalize on the rising land price. Also, they regard that works in agriculture are much less enticing than in the other sectors. Meanwhile, although currently there is no ALC in the rural area, the narrow gap in land rent indicates that ALC can happen any time in the future since farmers tend to sell their land when there is a rising price for the land.

The systematic effort is required to control the rate of ALC in the peri-urban area and to mitigate the risk of ALC in the rural area. The effort should focus on increasing the rent for agricultural land both in the peri-urban and in the rural areas. Thus, it is crucial to identify the factors that affect rent. The next results of our study found that land area, accessibility (to irrigation, market, and road), and cropping pattern significantly affect agricultural land rent. Land area has a negative impact on land rent since the cost of farming in larger land requires more input than in smaller land. The farm yield potential is not maximized since the farmer tends to use constrained family labor.

The accessibility of farmland to irrigation, market, and road increases the land rent (Brown and Barrows 1985; Joshi et al. 2017; Maddison 2009). It is following the compensation scheme of *UU 41 Tahun 2009.* The incentive set in the regulation is in the form of establishment of agricultural infrastructures such as irrigation canal and road. The accessibility of farmland to the road enhances accessibility to the market. However, this kind of incentive is useful only in the rural area where no ALC occurred. The irrigation canal and road are built as a network, and the network will cease to function when some part of it was damaged. In the peri-urban area, the ALC occurred partially and damaged some part of irrigation and road network. The partly damaged irrigation and road network will impact the overall network. Thus, it is essential to identify the kind of incentive should be given to a farmer in the peri-urban area.

29.4.3.2 Urban Farmer and the New Era of Peri-urban Agriculture

The agricultural land rent in the peri-urban area shows an exceptional value. The range value of the rent is between Rp −416/m^2/year and Rp 10,975/m^2/year. It means there are farmers in the peri-urban area who choose to remain in farming even with the expense of profit. Yagi and Garrod (2018) identify the growing number of the farmers in the peri-urban area who choose not to participate in land speculation and motivated to remain in farming as a hobby or to retain family tradition, and they termed this type of farmer as an urban farmer. The critical point of the urban farmer is that they do not want to sell their land even if it costs them the farm profit.

The existence of an urban farmer will limit the rate of ALC in the peri-urban area. Systematic identification of the urban farmer will help to target the incentive contained in *UU NO. 41 Tahun 2009.* Decreasing land tax, supporting research into high yield variety, and reward for farm achievement will likely be adequate for urban farmer since it will improve their motivation to remain in farming. In our data, the farmer who has negative rent is relatively younger compared to other farmers. However, our data consists only one farmer of this type. Further research is required to identify the characters of the urban farmer.

A carefully planned incentive mechanism accurately targeted to the urban farmer is the most promising policy to reduce the rate of ALC in the peri-urban area. Since agricultural land is privately owned, the sale of farmland is at the control of farmer, and if the monetary incentive they receive by retaining the farmland is high, it is likely that they will retain the farmland. The form of incentive should be based on the incentive contained in the *UU No. 41 Tahun 2009*. The compensation scheme in the UU No.41 Tahun 2009 is adequate to motivate urban farmer. Furthermore, promoting the implementation of this regulation would have another advantage in the form of government support and resources.

29.4.3.3 Commercial Farming in the Rural Area and the Degradation of Land Quality

One of the consequences of economic development in the agricultural sector is the shift of subsistence farming toward commercial farming (Nafziger 2012). One of the variables that significantly affect agricultural land rent is the cropping pattern. Non-food cropping pattern is becoming more popular in the rural area. Farmer started to plant non-food crops, especially horticulture because they produce higher rent than the food crops.

Most farmers still cultivate food crops, especially rice and maize. However, the number of farmers who cultivated horticultural crops is growing. The majority of farmers cultivated food crops. Also, some farmers cultivated non-food crops. The rural farmers mix food and horticultural crops, while farmers in the peri-urban area mix food crops with seasonal plantation crops. The result of this study reveals that the non-food crop pattern yielded higher rent compared to food crop pattern. There are 26% of rural farmers and 40% of peri-urban farmers who cultivated non-food crops. Horticulture and seasonal plantation crops need high farming cost and impose a high risk; thus, only a small number of farmers are able to provide the cost and bear the risk.

There is a serious problem caused by the increasing number of farmers in the rural area who planted horticultural crops. The application of pesticides in horticulture farming in the rural area is high, especially in the form of fungicides and herbicides. The average pesticide application frequency is three times a week. It is significantly higher than food crops which require only one to two times application every two weeks. As a consequence, the farmland in the rural area is being polluted by pesticide residue.

Also, the absorption of pesticide residue by the soil in Kepanjen is accelerated with its sandy structure. The water consumed by the rural resident will also be polluted by the residues since the farmland is close to the residential area. Watermelon, one of the leading horticultural crops, is cultivated three to four times annually. The high planting frequency increases the amount of pesticide residue in the soil. Also, the crop residue is left on the field. Consequently, since the pesticide residue also contained in the crop remnants, the amount of residue in the soil increases accordingly. Horticulture generates high rent, but it should be appropriately managed to limit the risk of soil and water quality degradation.

29.5 The Cultural Landscapes in the Peri-urban and Rural Area

29.5.1 The Landscapes in the Urban Core

Cultural landscapes are defined as a result of the interaction between human and landscapes which is evolving. One of the major causes of the changes in landscapes is urbanization and rapid economic growth. Urbanization started with high intensity of economic activity in the urban core, and then the following area near the urban is affected accordingly. The urban core is the center of economic, government, and cultural activities. The activity in the urban core can be represented by the existence of the building in which these activities take place.

There are two facilities in the urban core which represent the economic, governmental, and religious activities. First, the economic activity is represented by the largest traditional market in Jember (Fig. 29.7). In Indonesia, most of the retail transaction is conducted in the traditional market. The retail purchase in the traditional market accounts for 80% of the total retail sale. Thus, it plays a central role in the Indonesian economy. The largest traditional market in Jember is called Pasar Tanjung, in which Pasar means market. Second, the town square represents the city center and consisted of city park (Fig. 29.8), government office (Fig. 29.9), and religious building (Fig. 29.10). All of these buildings are surrounding the central plaza.

Fig. 29.7 Front side of the market *Pasar Tanjung*

Fig. 29.8 Central park of Jember

Fig. 29.9 Office of Jember district government

Fig. 29.10 Great Mosque of Jember

29.5.2 The Landscapes in the Peri-urban Area

The existence of these facilities influences the nearby area. The formerly rural area near the urban core is affected by the increasing rate of population growth. As the urban core grows, the housing area for the urban population will be pushed away from the urban core. Consequently, the rural area near the urban core will be transformed into a peri-urban area where the characteristics of rural life are being pushed away by the urban population. It is evident in the shapes of landscapes both in agricultural and in housing landscapes.

The expanding housing area threatens agricultural land in the peri-urban area. In the peri-urban area, the agricultural land was very close to housing. Consequently, the expansion of housing will consume the following farmland. As shown in Fig. 29.11, although there is still vast farmland, the housing was located at the edge of the farmland. As the population growth in this area, the conversion of the farmland is inevitable. Coupled with the existence of the permanent road, the development of housing is certain in the future.

In the case of housing, the peri-urban area consists of two types of neighborhood. The first is the housing of the native resident. The first group was characterized by pertinent rural characteristics, both in the physical aspects and in the social aspects. Physically, the first neighborhood is relatively poor, and most of the economic activity is still in agriculture. The physical form of the house is quite large with a wide yard around the house (Fig. 29.12) since the price of land when the house was built was low. In terms of style, the house is a traditional house.

Fig. 29.11 Farmland in the peri-urban area and the housing in the edge of the farmland area and concrete road to access the housing

Fig. 29.12 Traditional peri-urban house which possessed by the native peri-urban resident

Fig. 29.13 Contemporary peri-urban house which is quite small relative to the traditional peri-urban house

On the other hand, the second neighborhood occupies the newly developed housing which was the result of the growing urban population and has different characteristics. The house was built in the formerly farmland. Hence, it was the result of farmland conversion. Most of these lands are high priced, and it affects the type and size of the house. In terms of size, the house is quite small; typically, it was built on land, not more than 100 m^2. Consequently, it does not have an extensive yard like those in the traditional peri-urban house. Furthermore, each home is attached to the other house (Fig. 29.13). However, the house building is contemporary in style.

The building of the new housing requires a lot of building material, such as sand and rock. Most of this material was obtained from the extraction of the hill in the peri-urban farmland (Label D). The mining of these materials harms the ecosystem in the peri-urban area since the hill plays a role in preserving water resources. Figure 29.14 shows the overexploited hill for the extraction of sand and rock for building material. The mining activity changes the agricultural landscape entirely in the peri-urban area since the exploited hill is mined until it has the same level as the farmland.

Fig. 29.14 Sand and rock mining in the peri-urban area

29.5.3 The Landscapes in the Rural Area

The different conditions occurred in the rural area, where the landscape is dominated by farmland. Agriculture is a significant economic activity in the rural area. It is shown in the extensive farmland and various crops planted in the rural area. There can be more than one crop planted in a farmland area (Fig. 29.15), while the housing in a rural area is relatively small than the agricultural area. The house in the rural area has similar characteristics as the traditional house in the peri-urban area. However, the rural house is relatively better than the traditional peri-urban house in terms of quality of the building (Fig. 29.16). Generally, the house is the proxy for wealth in the farmer community; the better the house, the wealthier is the farmer. Based on this measure, the farmer in the rural area is better off than peri-urban farmer. It is also shown in the land rent measurement of farmland in both areas.

29.6 Policy Implications

Indonesia is a big nation, with over 250 million population, and the Indonesian food system receives a massive burden. The monoculture food policy in Indonesia has developed supporting institutions only for rice and shaped rice-eating habit among Indonesian people. Currently, the primary threat to food security in

Fig. 29.15 Farmland in a rural area with three crops planted in the area (peanut, rice, and watermelon)

Fig. 29.16 Rural house which has a wide yard and contemporary style

Indonesia is the loss of productive farmland due to ALC since it reduces rice production. Also, the difficulty of implementing food diversification policy exacerbated the problem. Thus, it is crucial to limit ALC and to strengthen the food diversification policy. Based on the discussion in the previous sections, we propose two policy implications for food security in Indonesia.

First, we propose the limiting of ALC as a short-term strategy to maintain food security in Indonesia. The result of our previous study presented in Sect. 4 demonstrates that significant difference between agricultural and non-agricultural land rents is the primary driver of ALC. The increasing housing demand from the urban resident will increase the rate of ALC in the peri-urban area. Also, the narrow gap between agricultural and non-agricultural land rents in the rural area imposes a possibility of ALC in this area. The narrow gap of land rent indicates that the resistance of farmers to sell the farmland is not quite strong. The increasing land price will be likely to move farmers to sell the land.

It is vital to identify the peri-urban farmer to limit the rate of ALC in the peri-urban area. A peri-urban farmer tends to retain their farmland. The farming motivation of peri-urban farmer is entirely different from the typical farmer. The farming motivation of peri-urban farmer is to satisfy their hobby or to preserve the family farming tradition. Since they tend to retain the farmland, systematic identification of peri-urban farmer and a targeted farming compensation to them will strengthen their motivation in farming and increase the possibility of them retaining the farmland. Also, in rural areas, the identification of commercial farmers and the way they behave are vital to mitigate the risk of ALC in the rural area and the risk of environmental degradation due to the overuse of pesticides.

There are three options to limit the rate of ALC in the peri-urban area and to mitigate the risk of ALC and environmental degradation in the rural area. First, the current compensation scheme in UU No. 41 Tahun 2009 should be focused on the peri-urban farmer. Since peri-urban farmer has a stronger non-economic motive than the typical farmer, this compensation will strengthen their motivation in farming. Second, there should be an effort to mitigate the risk of pesticide residue from commercial farming in the rural area. The cultivation of non-food crops increases the rent of agricultural land. However, pesticide residue causes environmental degradation. Leaving the problem of commercial farming unsolved will sacrifice the sustainability of agriculture itself. Third, commercial farming should be granted access to timely information regarding market and farm innovation. Commercial farmer is responsive to new information, and the improved channel of information to this farmer will improve their productivity and farming practice.

Second, we encourage strengthening food diversification policy as a long-term strategy to improve food security in Indonesia. There should be a keen national interest to make food diversification effort effective. Specifically, the government should focus on stimulating the development of supporting institutions for alternative foods. The kind of institutions that must be stimulated is the market for the primary product of alternative food crops, the processing industry of the food made from these crops, and the consumer market for food products made from these crops. In stimulating the second and the third institutions, the government should put the focus on stimulating the private sectors.

Currently, the effort to induce food innovation to alternative food crops is aimed at academic and research institutions primarily. Consequently, the products cannot pass the commercialization stages since both of these institutions are focused on research and development alone. There should be a collaboration between academic and research institutions with the private sectors. However, the implementation of this collaboration is difficult. It will be more beneficial if the private sectors are included from the beginning. The inclusion of private sectors will improve the chance of alternative food products to be commercially distributed. Furthermore, the food policy in Indonesia mainly focused on food production. In the longer term, the food policy should also focus on mainstreaming the balance consumption of healthy food and improve the affordability and stability of this food.

29.7 Conclusions and Future Research Directions

This chapter describes the dynamic of agricultural land conversion in the rural and peri-urban area in East Java, Indonesia. It explains the critical position played by ALC in the effort to achieve food security in Indonesia. The monoculture nature of food policy in Indonesia which focused on rice has developed a food system which relies heavily on rice and shaped the rice-eating habit among Indonesians. ALC becomes a primary threat to food security in Indonesia since most of the converted land is located in the agricultural region. Figure 29.5 shows that the LP2B which is the primary tool to limit ALC only protects half of the productive farmland. The remaining farmland was converted at a rapid rate with the highest occurred in the intermediate city where the urbanization has only just begun. Thus, it is vital to protect the non-protected farmland in this region. One of the feasible solutions to limit the rate of ALC is by increasing the rent of agricultural land in this area.

The analysis of land rent for agricultural and non-agricultural land demonstrated that it is highly correlated with the rate of ALC. In the peri-urban area where rapid ALC takes place, non-agricultural land rent is 700% higher than agricultural land. Conversely, the rent for agricultural land in the rural area where no ALC occurred is higher compared to non-agricultural land. Thus, increasing agricultural land rent will be likely to limit the rate of ALC. Consequently, it is logical that the effort to reduce the rate of ALC must focus on the effort to increase the rent for agricultural land. Below, we pointed out the direction for future research in the effort to reduce ALC:

1. The study of ALC should be focused on the intermediate city as shown in Fig. 29.5. Currently, the majority of farmland are located in this region. Moreover, also, this region is experiencing the initial stage of urbanization. Thus, it is better to mitigate the negative impact of urbanization earlier in this

region. The focused study location should be in the peri-urban area and the area where new urban core has emerged. This location experienced the highest rate of ALC.

2. It is vital to identify the characteristics of peri-urban farmer systematically. Peri-urban farmer potentially has the vital role in limiting the ALC in the peri-urban area since their motive of farming is entirely different from the typical farmer in which they tend to retain their farming even at the expense of profit. Targeting the compensation scheme contained in UU NO. 41 Tahun 2009 to these farmers will improve their motivation in farming and will encourage them to retain the land.
3. It is also essential to mitigate the risk of pesticide overuse by a commercial farmer in rural areas. The high level of pesticide is applied to horticulture farming in the rural area. Although horticulture farming increases agricultural land rent and limits the ALC, the pesticide overuse will degrade environmental quality in the long term.
4. Research into the development and evolution of cultural landscapes in a peri-urban and rural area will provide extensive insight into how the landscapes evolved as the consequence of urbanization.

References

Agus F, Irawan (2006) Agricultural land conversion as a threat to food security and environmental quality. J Litbang Pertan 25(3):90–98

Allen A (2003) Environmental planning and management of the peri-urban interface: perspectives on an emerging field. Environ Urban 15(1):135–148

Ariani M (2004) Dinamika Konsumsi Beras Rumah Tangga dan Kaitannya dengan Diversifikasi Konsumsi Pangan. In: Kasryono F, Pasandaran E, Fagi AM (eds) Ekon. Padi dan Beras di Indones. (Badan Penelitian dan Pengembangan Pertanian, Bogor), 541–558

Azadi H, Ho P, Hasfiati L (2011) Agricultural land conversion drivers: a comparison between less developed, developing and developed countries. L Degrad Dev 22(6):596–604

Bhatta B (2010) Analysis of urban growth and sprawl from remote sensing data. Springer, Berlin

BPS-Statistics of Jember Regency (2017) Jember Regency in Figures. BPS Kabupaten Jember, Jember

Brown K, Barrows R (1985) The impact of soil conservation investments on land prices. Am J Agric Econ 67:943–947

Bryld E (2003) Potentials, problems, and policy implications for urban agriculture in developing countries. Agric Hum Values 20:79–86

Chen Z, Wang Q, Chen Y, Huang X (2015) Is illegal farmland conversion ineffective in China? Study on the impact of illegal farmland conversion on economic growth. Habitat Int 49:294–302

Ciaian P, Kancs D, Swinnen J (2013) The impact of decoupled payments on land prices in the EU. In: Swinnen J, Knops L (eds) Land, labour cap. Mark. Eur. Agric. Divers. Under A Common Policy (Centre For European Policy Studies (CEPS), Brussels), 28–42

Douglass M (1998) A regional network strategy for reciprocal rural-urban linkages: an agenda for policy research with reference to Indonesia. Third World Plann Rev 20(1):1

Ettema D, Jong K de, Timmermans H, Bakema A (2012) Puma: multi-agent modelling of urban systems. Model. Land-use Chang. Springer Netherlands, Dordrecht, 237–258

Fajar J (2014) Sia-Siakan Keanekaragaman Sumber Pangan, Indonesia Terjebak Kebijakan Pangan Monokultur Beras. https://www.mongabay.co.id/2014/10/24/sia-siakan-keanekaragaman-sumber-pangan-indonesia-terjebak-kebijakan-pangan-monokultur-beras/. Retrieved Jan 5 2019

FAO (2011) The state of the world's land and water resources for food and agriculture - managing system at risk (Rome, Italy).http://www.fao.org/3/i1688e/i1688e.pdf. Retrieved Jan 5, 2019

Feichtinger P, Salhofer K (2013) What do we know about the influence of agricultural support on agricultural land prices? In: Swinnen J, Knops L (eds) Ger J Agric Econ (Centre For European Policy Studies (CEPS), Brussels), 71–85

Firman T (1999) From 'global city' to 'city of crisis': Jakarta metropolitan region under economic turmoil. Habitat Int 23(4):447–466

Firman T (2000) Rural to urban land conversion in Indonesia during boom and bust periods. Land Use Policy 17(1):13–20

Firman T, Dharmapatni IAI (1994) The challenges to sustainable development in Jakarta metropolitan region. Habitat Int 18(3):79–94

Government of Indonesia (2009) Perlindungan Lahan Pertanian Berkelanjutan (Government of Indonesia, Indonesia)

Hafsah MJ, Sudaryanto T (2004) Sejarah Intensifikasi Padi dan Prospek Pengembangannya. In: Kasryono F, Pasandaran E, Fagi AM (eds) Ekon. Padi dan Beras di Indones. (Badan Penelitian dan Pengembangan Pertanian, Bogor), 17–29

Hariyanto B (2017) Pangan Lokal. http://adelyadesi.lecture.ub.ac.id/files/2017/05/Pangan-Lokal-di-Gizi-Unibraw.pdf

Hudalah D, Firman T (2012) Beyond property: industrial estates and post-suburban transformation in Jakarta Metropolitan Region. Cities 29(1):40–48

Imhoff ML, Lawrence WT, Stutzer DC, Elvidge CD (1997) A technique for using composite DMSP/OLS "city lights" satellite data to map urban area. Remote Sens Environ 61(3):361–370

Irawan B (2011) Konversi Lahan Sawah di Jawa Barat: Kecenderungan dan Pengaruhnya terhadap Produksi Padi Sawah. Kasryno, FaisalPasaribu SM, Saliem HP, Soeparno H, Pasandaran E, eds. Konversi dan Fragm. Lahan Ancaman Terhadap Kemandirian Pangan. (IPB Press, Bogor), 125–148

Irwin EG, Bockstael NE (2002) Interacting agents, spatial externalities and the evolution of residential land use patterns. J Econ Geogr 2(1):31–54

Irwin EG, Bockstael NE (2007) The evolution of urban sprawl: evidence of spatial heterogeneity and increasing land fragmentation. Proc Natl Acad Sci 104(52):33–46

Ives CD, Kendal D (2013) Values and attitudes of the urban public towards peri-urban agricultural land. Land Use Policy 34:80–90

Joshi J, Ali M, Berrens RP (2017) Valuing farm access to irrigation in Nepal: a hedonic pricing model. Agric Water Manag 181:35–46

Kilian S, Antón J, Salhofer K, Röder N (2012) Impacts of 2003 CAP reform on land rental prices and capitalzation. Land Use Policy 29(4):789–797

Kontgis C, Schneider A, Fox J, Saksena S, Spencer JH, Castrence M (2014) Monitoring peri-urbanization in the greater Ho Chi Minh City metropolitan area. Appl Geogr 53:377–388

Latruffe L, Le Mouël C (2009) Capitalization of government support in agricultural land prices: what do we know? J Econ Surv 23(4):659–691

Liu Y, Fang F, Li Y (2014) Key issues of land use in China and implications for policy making. Land Use Policy 40:6–12

Maddison DJ (2009) A spatio-temporal model of farmland values. J Agric Econ 60(1):171–189

Magsi H, Torre A, Liu Y, Javed Sheikh M (2017) Land use conflicts in the developing countries: proximate driving forces and preventive measures. Pak Dev Rev 56(1):19–30

Malaque IR, Yokohari M (2007) Urbanization process and the changing agricultural landscape pattern in the urban fringe of Metro Manila, Philippines. Environ Urban 19(1):191–206

McGee TG (2010) Building liveable cities in Asia in the twenty-first century research and policy challenges for the urban future of Asia. Malays J Environ Manag 11(1):14–28
Nafziger EW (2012) Economic development. Cambridge University Press, Cambridge, 838 p
Nawiyanto N (2017) The politics of food and food security during Indonesia's old order (1945–1965). Lembaran Sej 10(1):65–73
North DC (1955) Location theory and regional economic growth. J Polit Econ 63(3):243–258
Nusantoro A, Cramb RA (1990) Land-use targets in Javanese agriculture: an institutional analysis. In: 34th annual conference of the Australian agricultural economics society, University of Queensland, Brisbane, 1–23
OECD/FAO (2013) OECD-FAO agricultural outlook 2012–2021. OECD Publishing. http://dx.doi.org/10.1788/agr_outlook-2013-en
Omrani H, Abdallah F, Charif O, Longford NT (2015) Multi-label class assignment in land-use modelling. Int J Geogr Inf Sci 29(6):1023–1041
Pasandaran E, Syam M, Las I (2006) Degradasi sumber daya alam: ancaman bagi kemandirian pangan nasional. In: Kasryno, FaisalPasaribu SM, Saliem HP, Soeparno H, Pasandaran E (eds) Konversi dan Fragm. Lahan Ancaman Terhadap Kemandirian Pangan. IPB Press, Bogor, 35–53
Peerzado MB, Magsi H, Sheikh MJ (2018) Land use conflicts and urban sprawl: conversion of agriculture lands into urbanization in Hyderabad, Pakistan. J Saudi Soc Agric Sci
Phuc NQ, Van Westen ACM, Zoomers A (2014) Agricultural land for urban development: the process of land conversion in Central Vietnam. Habitat Int 41:1–7
Plieninger T, Draux H, Fagerholm N, Bieling C, Bürgi M, Kizos T, Kuemmerle T, Primdahl J, Verburg PH (2016) The driving forces of landscape change in Europe: a systematic review of the evidence. Land Use Policy 57:204–214
Pribadi DO, Pauleit S (2016) Peri-urban agriculture in Jabodetabek Metropolitan Area and its relationship with the urban socioeconomic system. Land Use Policy 55:265–274
Rondhi M, Pratiwi PA, Handini VT, Sunartomo AF, Budiman SA (2018) Agricultural land conversion, land economic value, and sustainable agriculture: a case study in East Java, Indonesia. Land 7(4)
Schulp CJE, Levers C, Kuemmerle T, Tieskens KF, Verburg PH (2019) Mapping and modelling past and future land use change in Europe's cultural landscapes. Land Use Policy 80 (April):332–344
Soen SK (1968) Prospects for agricultural development in Indonesia. Centre for Agricultural Publications and Documentation, Wageningen
Su S, Ma X, Xiao R (2014) Agricultural landscape pattern changes in response to urbanization at ecoregional scale. Ecol Indic 40:10–18
Sumarno (2004) Kemandirian Pangan Nasional Mengapa Sulit Dicapai dan Apa Yang Harus Dilakukan. Memperkuat Kemamp. Swasembada Pangan. IAARD Press, Bogor, 283–310
Tassinari P, Torreggiani D, Benni S (2013) Dealing with agriculture, environment and landscape in spatial planning: a discussion about the Italian case study. Land Use Policy 30(1):739–747
van Vliet J, de Groot HLF, Rietveld P, Verburg PH (2015) Manifestations and underlying drivers of agricultural land use change in Europe. Landsc Urban Plan 133:24–36
Wang Y, Scott S (2008) Illegal farmland conversion in China's urban periphery: local regime and national transitions. Urban Geogr 29(4):327–347
Wasilewski A, Krukowski K (2004) Land conversion for suburban housing: a study of urbanization around Warsaw and Olsztyn, Poland. Environ Manage 34(2):291–303
Yagi H, Garrod G (2018) The future of agriculture in the shrinking suburbs: the impact of real estate income and housing costs. Land Use Policy 76(March):812–822
Yokohari M, Takeuchi K, Watanabe T, Yokota S (2000) Beyond greenbelts and zoning—a new planning concept for the environment of Asian mega-cities. Landsc Urban Plan 47:159–171

Zasada I (2011) Multifunctional peri-urban agriculture—a review of societal demands and the provision of goods and services by farming. Land Use Policy 28(4):639–648

Zezza A, Tasciotti L (2010) Urban agriculture, poverty, and food security: empirical evidence from a sample of developing countries. Food Policy 35(4):265–273

Zhong T, Huang X, Ye L, Scott S (2014) The impacts on illegal farmland conversion of adopting remote sensing technology for land inspection in China. Sustainability 6(7):4426–4451

Chapter 30
The Role of Natura 2000 at Maintaining Dynamic Landscapes in Europe Over the Last Two Decades: Implications for Conservation

Virgilio Hermoso, Alejandra Morán-Ordóñez and Lluís Brotons

Abstract The Natura 2000 network aims to promote the long-term persistence of both biodiversity and traditional uses. European landscapes have undergone large transformations in the past decades, mainly associated with the abandonment of less productive lands, mostly in mountains, and concentration of intensive agriculture in the most productive areas. These changes could pose important management challenges and offer new opportunities to the achievement of the network's goals. Here, we evaluate the change in land cover within Natura 2000 in the last two decades, explore the role of different drivers in observed changes and assess the impacts of these changes in the structure of landscape. We found that within Natura 2000, landscape has been highly dynamic in the last two decades with more than 20% of the area under protection undergoing land cover changes. However, this change was smaller within than outside Natura 2000. The most systematic transitions involved both, succession processes towards naturalization (e.g. transitional woodland—TRW—to forest—FOR) and anthropization (e.g. mosaic—MOS—to arable—ARA—and pastures—PAS). Changes across land cover categories had also significant effects on the landscape configuration towards a higher homogenization, mainly driven by the increase of patch size and clumpiness of FOR and ARA, while a higher fragmentation of TRW and open areas (OPEN). Given these changes, two different strategies would be needed to enhance the role of Natura 2000, (i) tighter control to ensure anthropization, mainly intensive agriculture, does not compromise conservation goals within protected areas (PAs) and (ii) tackle more effectively the ecological and socio-economic effects of abandonment in less

V. Hermoso (✉) · A. Morán-Ordóñez · L. Brotons
Centre de Ciència i Tecnològia Forestal de Catalunya (CEMFOR-CTFC), Crta. Sant Llorenç de Morunys km 2, Solsona 25280, Spain
e-mail: virgilio.hermoso@gmail.com

V. Hermoso · A. Morán-Ordóñez · L. Brotons
CREAF, Cerdanyola del Vallés 08193, Spain

L. Brotons
CSIC, Cerdanyola del Vallés 08193, Spain

L. Müller and F. Eulenstein (eds.), *Current Trends in Landscape Research*, Innovations in Landscape Research, https://doi.org/10.1007/978-3-030-30069-2_30

productive areas to halt loss of key habitats. On the other hand, changes in composition and structure of landscape open new conservation opportunities derived from enhanced connectivity, at least within PAs.

Keywords Corine Land Cover · Landscape · Habitats · Land abandonment · Transition matrix · Clumpiness · Common Agriculture Policy

30.1 Introduction

European landscapes have undergone large transformations in the past decades, mainly associated with the abandonment of the less productive agricultural lands (Young et al. 2007) and their consequent re-naturalization through a process of vegetation regeneration and forest expansion (Gellrich et al. 2007; Ameztegui et al. 2010; Hatna and Bakker 2011). Most of these changes have been promoted by the Common Agricultural Policy (CAP). The implementation of the CAP has led to a polarization of production systems, by concentrating intensive agriculture in the most productive lands while driving a decline of traditional agricultural practices and extensive livestock farming in marginal areas such as mountains (Mitchley et al. 2007, Ameztegui et al. 2010; Álvarez-Martínez et al. 2014). The concentration of agricultural production implies drastic changes not only in the intensity of land use but also in the configuration of the landscape. In order to increase productivity and allow the use of large machinery, there has been a trend towards unifying small agricultural fields into expanded monocultures (Dobrev et al. 2014) resulting in more homogenous landscape diversity (Brotons et al. 2004).

The designation of protected areas (PAs) is a common mechanism for addressing the global decline of biodiversity. The Natura 2000 network is the centrepiece of the EU's Biodiversity Strategy, aiming at halting and reversing the loss of biodiversity in Europe by 2020 (EC 2011a). This network comprises more than 27,000 Natura 2000 sites, which cover >1 million km^2 (18.3% of the land surface and about 6% of marine area), representing the world's largest network of PAs (EEA 2012). This network includes sites designated for protecting species and/or habitats listed as a priority in the Habitats (92/43/EEC) and the Birds Directives (2009/147/EC). However, Natura 2000 is not a system of strict nature reserves where all human activities are limited and strictly controlled to ensure protection of conservation values; on the contrary, it seeks to harmonize, mostly on private land, biodiversity conservation and other legitimate land uses. Member States must, however, ensure that the sites are managed in a sustainable manner, both ecologically and economically. For this reason, the designation of an area as a Natura 2000 site aims at promoting the long-term persistence of the biodiversity for which it was declared, as well as traditional uses that sustain biodiversity and local socio-economic activities. It would then be expected that the Natura 2000 fostered landscape stability and maintenance of traditional land uses in the long term.

The process of generalized land abandonment in marginal productive areas mentioned above could threaten the achievement of some of the fundamental goals of Natura 2000, such as the maintenance of traditional land uses and habitats, but also offer opportunities for habitat restoration, rewilding and the recovery of threatened species (Fernández et al. 2017). For instance, 41 out of the 231 priority habitat types listed in the European Habitats Directive are associated with low-intensity agricultural management, including semi-natural grasslands and hay meadows (Halada et al. 2011). Loss of traditional practices, together with landscape restructuring, has negative effects on the populations of many species, like farmland birds (Donald et al. 2006; Baldi and Batary 2011; Santana et al. 2013), mammals (Delibes-Mateos et al. 2009) or plant species (Luoto et al. 2003). To address this threat, the CAP funds the designation of Less Favored Areas (LFAs-Regulation 1257/1999) to prevent rural abandonment and maintain cultural landscapes (Dax 2005; Stoate et al. 2009). The application of this policy has actually resulted in a significant increase in the area of LFAs going from a third of the European Utilized Agricultural Area in 1975 to more than half in 2005 (Dax 2005; MacDonald et al. 2000).

Despite the effort made by the EU in the last decades to address this land cover change process, recent studies keep reporting the same pattern of land abandonment, loss of agricultural surface and expansion of forest within some specific Natura 2000 sites (e.g. Butsic et al. 2016; Flores Ribeiro et al. 2018). However, there is little known about long-term land cover changes and their drivers at the whole extent of the Natura 2000 network (but see Kallimanis et al. 2014 for the period 2000–2006). This makes it difficult to extrapolate to large-scale and/or long-term changes (Hatna and Bakker 2011) and assess the potential consequences within PAs.

Here, we analyse the changes in land cover composition and spatial configuration undergone within the whole extent of Natura 2000 over the last two decades, grouped in three main classes (sensu Martínez-Fernández et al. 2015): naturalization, anthropization and internal changes in natural areas. This classification was used to strengthen the link between the changes observed and their implications for conservation. For example, anthropization processes are usually associated with increases in human pressure and higher risks to the achievement of conservation goals. We also assess how the land cover changes observed within the boundaries of the Natura 2000 sites compared to those of neighbouring areas outside, over the evaluated time period. Given the role of Natura 2000 and the efforts made by the EU at maintaining traditional agricultural landscapes, we would expect the designation of a site as part of the network to foster stability of the landscape. Our results should help assess the effectiveness of these policies and guide their future application to ensure the achievement of the EU's conservation commitments as well as assessing the risks and opportunities for conservation derived from land cover changes within the Natura 2000 network.

30.2 Mainland Cover Changes Over the Last Two Decades

We evaluated the change in land cover across all terrestrial PAs included in the Natura 2000 network across Europe. We use land cover stability (understood as the maintenance of land cover class over time in a given spatial unit) as a proxy of the stability of habitats of community interest (those listed in the Habitat's Directive), given the lack of spatial data coverage of the latter across Europe. The location and extent of PAs were sourced from European Environmental Agency (version as of March 2017; http://www.eea.europa.eu/data-and-maps/data/natura-1#tab-gis-data). This data set contains spatial information on all Natura 2000 sites that have been declared to date (n = 27,308 PAs covering over 1.3 million km^2) as well as descriptive data, such as date of designation or the area covered by the PA.

To evaluate land cover changes over time, we used a temporal series of Corine Land Cover (CLC) maps sourced from the Copernicus Land service (http://land.copernicus.eu/global/). These maps cover the whole of Europe at 100 m resolution and are available for four different time periods: 1990, 2000, 2006 and 2012. Using these time series of land cover maps, we produced an initial (t_0) and final (t_1) land cover map for each Natura 2000 site. The initial state was derived from the closest CLC map available to the date of the PA designation. To be conservative and ensure that the observed changes within each PA might result from its protection status, we used as our t_0 map the next available Corine to the date when the PA was designated. As an example, for Natura 2000 sites designated in 2004, we used the Corine 2006 as an initial stage. For this reason, sites designated after 2006 were excluded, leaving the final database of PAs in 25,270 sites (>90% of all sites). Moreover, given that there were no data available for Sweden and Finland for CLC 1990, all sites designated in these countries before 2000 were assigned the CLC map of 2000 as their initial stage. We then used CLC 2012 to characterize land cover at the final stage for the analyses (t_1).

To make the comparisons between the initial (t_0) and final (t_1) land cover maps easier and more robust, and in order to minimize potential mismatches of the thematic resolution of the four CLC, we grouped the original 44 CLC classes into the eight categories following the LEAC classification (Land and Ecosystems Accounts; EEA 2006a; Table 30.1). This simplified the number of potential land cover comparisons to 64 (8×8), compared to the 1936 possible combinations between the 44 original categories. These eight categories have also been recommended for global and ecological analysis (EEA 2006a; Gómez and Páramo 2005) and previously used in similar assessments (e.g. Martínez-Fernández et al. 2015).

To assess changes in land cover within Natura 2000 between t_0 and t_1, we constructed a transition matrix (Pontius et al. 2004) based on these eight land cover categories for the full extent of the Natura 2000 as considered for this study. This transition matrix is a rectangular matrix that represents the number of grid cells that underwent change from land cover i to j for each of the potential combinations of land cover categories. The values on the diagonal show the persistence within each

Table 30.1 Correspondence between Land and Ecosystem Accounts classification—LEAC (EEA 2006a) used in this study and land cover categories in CLC level 3

LEAC land cover category	CLC_CODE
Artificial surfaces (ARTF)	111 Continuous urban fabric; 112 Discontinuous urban fabric; 121 Industrial or commercial units; 122 Road and rain networks and associated land; 123 Port areas; 124 Airports; 131 Mineral extraction sites; 132 Dump sites; 133 Construction sites; 141 Green urban areas; 142 Sport and leisure facilities
Arable Land (ARA)	211 Non-irrigated arable land; 212 Permanently irrigated land; 213 Rice field; 221 Vineyards; 222 Fruit trees and berry plantations; 223 Olive groves; 241 Annual crops associated with permanent crops
Pastures (PAS)	231 Pastures
Mosaic Farming (MOS)	242 Complex cultivation patterns; 243 Land principally occupied by agriculture, with significant areas of natural vegetation; 244 Agroforestry areas
Standing Forests (FOR)	311 Broad-leaved forest; 312 Coniferous forest; 313 Mixed forest
Natural Grassland, Heathland and Shrubland (GRSH)	321 Natural grasslands; 322 Moors and heathland; 323 Sclerophyllous vegetation
Transitional Woodland and Shrubs (TRW)	324 Transitional woodland–shrub
Open area (OPEN)	331 Beaches, dunes, sands; 332 Bare rocks; 333 Sparsely vegetated areas; 334 Burnt areas

category over the evaluated time frame, while values off the diagonal indicate transitions between land cover categories. We used the cross-tabulation tools in ArcGIS (ESRI Inc.) to build the transition matrix for the whole study area (all Natura 2000 sites altogether) between t_0 and t_1 at 100 m resolution. We then calculated the proportion of the landscape that remained stable under each land cover category and the proportion that underwent transition from land cover i to j. These proportions were calculated as the ratio between the total number of grid cells under each transition i to j and the total area of land cover i at t_0.

We also compared the observed patterns of change within the Natura 2000 and outside. With this aim, we constructed a transition matrix as explained above for a 5 km buffer around all Natura 2000 sites considered in the analyses. We constrained this second transition matrix to a buffer to focus on areas close to Natura 2000 sites, rather than comparing them against very distant areas where landscape composition and land use might be very different. Given that we used three different starting points to build t_0 (1990, 2000 and 2006 depending on when each Natura 2000 site had been designated), we computed a different transition matrix for each starting point (1991–2012, 2000–2012, 2006–2012) for the areas outside N2000 sites. The sum of the values of the diagonal of these transition matrices was used to calculate the proportion of the landscape that remained stable in the 5 km buffer areas around PAs, and to compare them against the values within the Natura 2000 sites.

Table 30.2 Summary statistics for the different land cover categories as considered in this study (see correspondence with CLC and codes on Table 30.1)

	Proportion Natura 2000 at t_0	Gain	Loss	Total change	Swap	Absolute value of net change
ARTF	1.1	0.3	0.3	0.6	0.6	0.1
ARA	12.8	2.0	2.2	4.2	4.1	0.2
PAS	7.2	1.7	1.3	3.0	2.6	0.4
MOS	7.5	2.0	2.6	4.5	4.0	0.6
FOR	43.5	6.0	4.9	11.0	10.0	1.1
TRW	7.0	2.2	4.4	6.6	4.5	2.2
GRSH	16.3	5.1	3.7	8.7	7.3	1.4
OPEN	4.6	1.2	1.3	2.6	2.5	0.1
Total	100.0	20.6	20.6	20.6	17.6	3.0

Numbers show the proportion of each land cover categories at the time of PA designation and the land cover changes observed between t_0 and t_1 (gains, losses, net changes and swaps) expressed in terms of per cent of the overall landscape
Source Hermoso et al. (2018)

We found that 20.6% of land cover within Natura 2000 sites changed in the period t_0–t_1 (Table 30.2). This change was, however, lower than the change observed in the 5 km buffer outside PAs, where there was a 51.1, 40.3 and 32.9% change for the periods 1990–2012, 2000–2012 and 2006–2012, respectively. Changes within Natura 2000 corresponded mainly to the changes in forest (FOR), transitional woodland and shrubs (TRW) and grassland, heathland and shrubs (GRSH) categories that together accounted for near 2/3 of the total changes observed (Table 30.2). The role of TRW in this share was significant given the smaller proportion of land cover it represented within Natura 2000 at t_0 (Table 30.2).

30.3 Processes Leading to Changes in Land Cover

The lack of net change as calculated above at the whole Natura 2000 scale does not necessarily mean lack of change across the landscape between two given land cover categories. This is possible because changes from land cover *i* to *j* at a given location could be compensated by the changes from *j* to *i* elsewhere. These 'swaps' between land cover categories could then lead to underestimation of landscape dynamics (Pontius et al. 2004). For this reason, we used the information on the transition matrix to disaggregate the values of total change underwent by each land cover category into different indicators of change as suggested by Pontius et al. (2004). We computed the net gain and loss of each land cover category and the contribution of swap versus the absolute value of change (gain or loss) to the observed values change by using the macros provided in Pontius et al. (2004). Net gain and loss represent the total area gained and lost by each category from/to other categories; absolute net change represents the area that was either gained or lost by each category as a result from the balance between net gain and loss and measured as the maximum of the gain and loss minus the min of the gain and loss (Pontius et al. 2004); finally, swap represents the amount of area that underwent a change but did not contribute to absolute net change and was measured as two times the minimum of the gain and losses (Pontius et al. 2004).

We found that the changes identified within the Natura 2000 sites were mainly driven by swap processes, or changes in the spatial location of land cover categories, while only 3.0% of the land cover underwent a net change in the period t_0–t_1 (Table 30.2). This net change was mainly driven by gains in pasturelands (PAS) and FOR and losses in mosaic farming (MOS).

In order to evaluate to what extent the transitions mentioned above were significant or not, we further compared the observed values of net gain and losses to what it would be expected by a random process of land cover change for all pairwise combinations of land cover categories off the diagonal in the transition matrix. This calculation takes into consideration the area occupied by each land cover category at t_0, since larger changes would be expected in common/widespread land cover categories. The expected gains and losses were computed by distributing the observed gain/loss of each land cover j across all other categories proportionally to their relative area (total number of grid cells of each category j at t_0) as recommended by Pontius et al. (2004). We then calculated a change score as the difference between the observed and expected values relative to the magnitude of the expected change [(observed–expected)/expected], which is similar to the ratio used in Chi-squared tests. This score represents the proportional difference between the observed changes against what it would be expected by chance, so values over 0 indicate that the transition from land cover i to j was larger than what it would be expected by chance. Negative values indicate that the transition from land cover i to j was smaller than expected by chance (Pontius et al. 2004), and then uncertainty remains about what land cover category assumed the observed loss in i. We were interested in positive scores that indicated substantial changes from one land cover category to another and considered a transition *systematic* whenever it appeared with positive scores in both gains and losses following Martínez-Fernández et al. (2015).

After comparing the observed values of change against the expected by chance, the transitions that posed the largest changes in the landscape (*systematic* transitions) were TRW to FOR, followed by TRW to GRSH (Fig. 30.1). However, when we accounted for the initial area of each land cover category at t_0 and compared the observed changes against the expected changes by random processes, we also found systematic transitions among other less prevalent land cover categories (see Fig. 30.2 for systematic transitions). The most important systematic transitions include changes from MOS to arable land (ARA) and MOS to PAS (anthropization) with a prevalence in the observed data that was twofold the expected by chance;

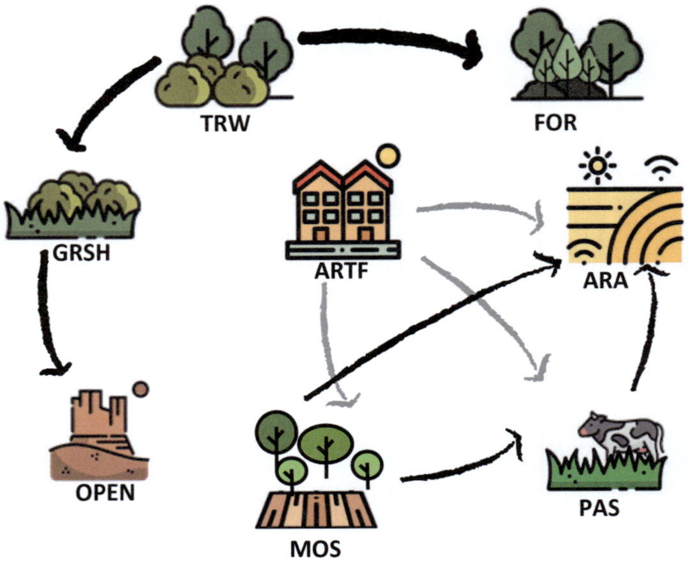

Fig. 30.1 Summary of the most systematic land cover transitions between t_0 and t_1 within the boundaries of the Natura 2000 sites (land cover class codes in Table 30.1). Arrow's thickness represents the value of the ratio (observed–expected)/expected, being observed the proportion of Natura 2000 that underwent the transition, while the expected was computed by distributing the observed gain/loss for each land cover *j* across all other categories proportional to their relative area (total number of grid cells of each category *j* at t_0) as recommended by Pontius et al. (2004). Therefore, the thicker and the darker the arrow, the larger the transition change between two land cover categories

these imply a landscape homogenization through conversion to agricultural areas. Systematic transitions towards naturalization mainly consisted of changes from artificial surfaces (ARTF) to PAS and ARA that occurred in the observed data almost three times than the expected by chance. The transitions between GRSH and open areas (OPEN) (in both directions) were the most systematic regarding succession processes. It was also remarkable the transition from TRW to FOR (natural succession towards recovery; Fig. 30.1) that occurred over 1.7 times expected by chance (Fig. 30.1).

30.4 Changes in Landscape Structure and Connectivity

To explore if the transitions between t_0 and t_1 also implied a change in the landscape structure across the Natura 2000 network, we evaluated the changes over time of three different landscape metrics: (1) number of patches, being a patch an aggregation of two or more contiguous cells of the same land cover category; (2) average size of patches and (3) a clumpiness index that measures the degree of

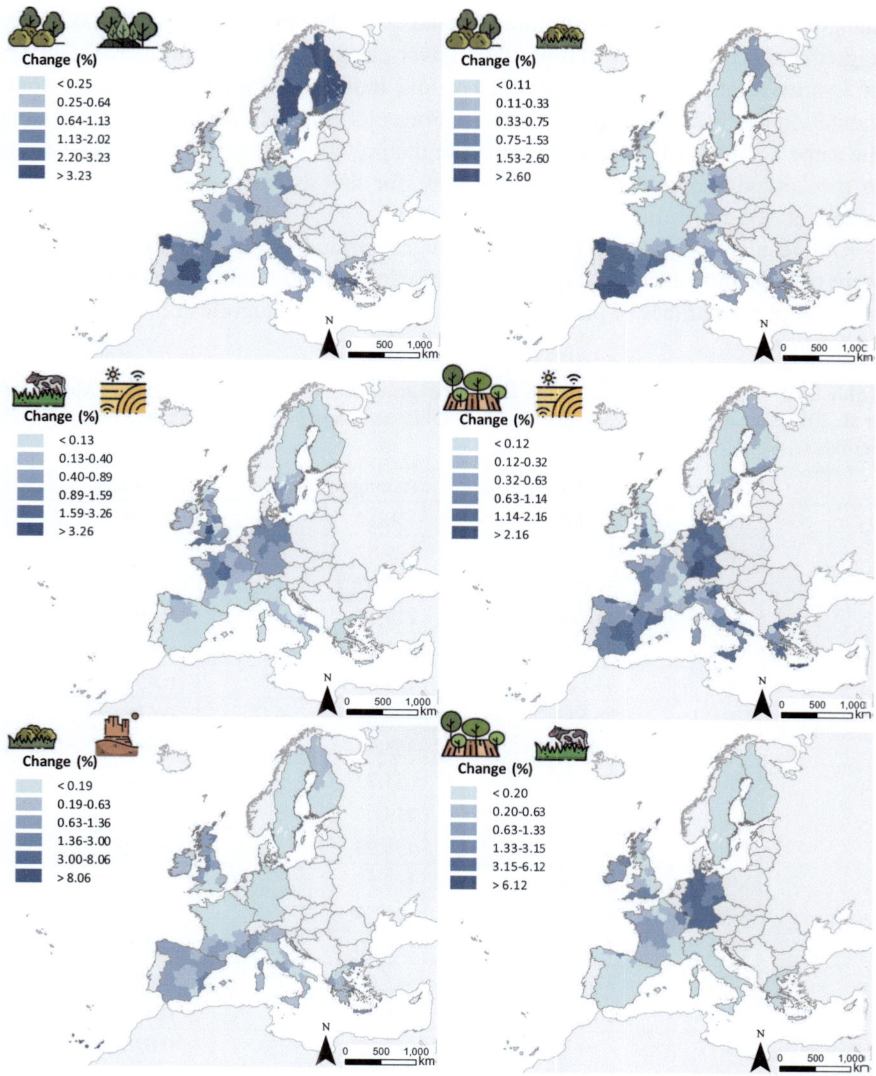

Fig. 30.2 Maps of the six most systematic transitions—shown in Fig. 30.1—across NUTS regions. Values show the average change across all Natura 2000 sites within each region. The boundaries of the NUTS regions were sourced from Eurostat (http://ec.europa.eu/eurostat/). We used level 2 of this regional classification for the whole of Europe, except for Germany, The Netherlands, Belgium, Denmark and England. In these latter countries, the level 2 territorial division resulted in extremely small spatial units, and we used level 1 in order to homogenize the size of regions for further analyses

spatial aggregation of cells within land cover categories and compares it to a random distribution of cells among categories. This index ranges between −1 and 1, being −1 indicative of maximum disaggregation, 0 an aggregation of cells within categories not different from random and 1 when the category is maximally aggregated (i.e. a single patch per land cover category). Larger number of patches of less average size and less clumped would indicate a dominant landscape fragmentation process. We used the software Fragstats v4.2 (McGarigal et al. 2012) and the same t_0 and t_1 CLC maps described in the previous sections. Due to limitations in the computational capacity of Fragstats, for this analysis, we resampled these rasters at 500 m resolution.

Our results suggest that observed changes in land cover over the period t_0–t_1 were also accompanied by changes in landscape configuration. Despite the overall increase in the number of patches across all land cover categories (29% increase on

Table 30.3 Summary statistics of the landscape structure obtained from Fragstats (McGarigal et al. 2012), for the two different periods used in this study and proportional change between both periods (t_1–t_0/t_0)

	Category	Number patches	Average patch area	Clumpiness index
t_0	ARTF	15089	50.52	0.230
	ARA	43,240	199.62	0.544
	PAS	31,839	152.07	0.456
	MOS	43,849	115.15	0.398
	FOR	54,224	542.64	0.498
	TRW	33,124	143.41	0.457
	GRSH	25,533	431.64	0.616
	OPEN	8621	363.98	0.690
t_1	ARTF	22,042	51.72	0.234
	ARA	55,155	213.05	0.549
	PAS	45,706	158.98	0.452
	MOS	51,654	119.58	0.402
	FOR	65,978	606.59	0.507
	TRW	40,415	110.60	0.400
	GRSH	32,196	415.01	0.618
	OPEN	10,846	310.59	0.667
Change	ARTF	0.46	0.02	0.02
	ARA	0.28	0.07	0.01
	PAS	0.44	0.05	−0.01
	MOS	0.18	0.04	0.01
	FOR	0.22	0.12	0.02
	TRW	0.22	−0.23	−0.13
	GRSH	0.26	−0.04	0.00
	OPEN	0.26	−0.15	−0.03
	Average	0.29	−0.02	−0.01

The clumpiness index measures the degree of spatial aggregation of cells within land cover categories and compares it to a random distribution of cells among categories land cover codes which are shown in Table 30.1

Source Hermoso et al. (2018)

average; Table 30.3) that would point towards a higher landscape fragmentation overall across the Natura 2000 network, the change in patch size and clumpiness indicated that Natura 2000 underwent an overall homogenization. FOR and ARA land covers showed the largest increase in average patch size and spatial aggregation (clumpiness). On the other hand, patches of TRW and OPEN became smaller and more fragmented (Table 30.3).

30.5 Implications of Landscape Change for Conservation

Our results show that the landscape within Natura 2000 sites has been quite dynamic in the last two decades or since their designation as part of the network, with more than 20% of the area under protection undergoing changes, especially in small protected areas. However, the pace at which landscape changed was smaller within that outside Natura 2000 where the percentage of change was over 50% indicating that the designation of an area as part of the network may foster landscape stability. These changes were mainly driven by swap processes rather than by net changes, as also observed by Kallimanis et al. (2014) and Feranec et al. (2010). These spatial swaps are relevant since not all of them occurred within the same Natura 2000 sites, which in the end translates into significant net gains or losses locally. Changes across land cover categories had a significant effect on the landscape configuration towards a higher homogenization, mainly driven by the increase of patch size and clumpiness of FOR and ARA, while a higher fragmentation of TRW and OPEN. These results should not be taken as an indication of the effectiveness of Natura 2000 (or lack of it) at protecting species and habitats, which is beyond the scope of this study. However, the changes in landscape composition and spatial configuration have implications for the conservation of key habitats and the species they sustain. Therefore, the changes reported here should be considered when designing/applying policies to halt processes such as landscape homogenization reported here or account for new conservation opportunities associated with the new composition and configuration of landscape (e.g. Fernández et al. 2017).

Overall, the main processes of land cover change that we identified are the response to a polarization of land uses mainly driven by either the abandonment of the less productive areas, or the intensification of agricultural and pastoral activities in the most productive ones. These patterns largely reported elsewhere outside protected areas in developed countries (Hill et al. 2008) and were also observed within Natura 2000 boundaries. These changes could be driven by socio-economic factors such as loss of population and traditional activities (e.g. grazing or low-intensity logging) in rural areas and the effects of EU policies such as the CAP (Navarro and Pereira 2012; Kallimanis et al. 2014). There are direct implications for conservation of these changes, as the net loss in some land cover categories such as MOS and TRW observed within Natura 2000 could threaten the persistence of priority habitats associated with them. For example, MOS and TRW hold priority habitats listed in the Habitats Directive such as Natural and Semi-Natural Grassland Formations (e.g. Dehesas with evergreen *Quercus* spp.), Temperate Heath and Scrub (e.g. Bushes with *Pinus mugo* and *Rhododendron hirsutum*) or Sclerophyllous Scrub (e.g. Arborescent matorral with *Juniperus* spp.). The transitions behind the loss of these land cover categories respond, however, to different processes so they should be addressed in different ways for conservation purposes. On the other hand, not all changes pose risks to conservation, such as the increase in clumpiness of FOR patches, that might improve connectivity within Natura 2000 sites for forest-dependent species (although the same process could also compromise connectivity among populations of open habitat-dependent species). Moreover, naturalization processes, such as the loss of ARTF surfaces could also reduce human pressure on natural habitats.

The loss of TRW was mainly driven by a natural vegetation succession process leading to the expansion of FOR, an internal change in natural areas towards higher naturalization (Martínez-Fernández et al. 2015; Forest Europe 2015). However, this transition could partly be reflecting afforestation policies on TRW areas with fast growing species (e.g. pine and eucalyptus) for commercial purposes, although the area under this type of land use is small and has overall experienced a reduction over the last decades (Forest Europe 2015). This contributed to the overall homogenization of the landscape. The important increase in FOR within the Natura 2000 could be driven by the high prevalence of mountain landscapes within protected areas, systems prone to land abandonment (e.g. Gellrich and Zimmerman 2007; Lasanta et al. 2009). This over representation of mountain systems responds to the fact that PA has been traditionally designated by using ad hoc criteria such as the relative lack of value for major commercial land uses, high value for their scenery, recreation, tourist potential and historical protection for uses such as hunting or water supply (Pressey 1994). This transition has two sides to conservation although it poses a risk of loss of priority TRW habitats, the increase in FOR, associated with the increase in clumpiness might pose opportunities to conservation derived from enhanced connectivity. An additional systematic transition involving loss of TRW was towards GRSH, indicating an internal change in natural areas towards degradation (Martínez-Fernández et al. 2015), occurred mostly in the Mediterranean region and especially in the Iberian Peninsula (Fig. 30.2). The mechanisms behind this transition are, however, not as easy to explain as the previous. It could be related to a natural succession processes after fire, which have a higher prevalence in the Mediterranean region (Tedim et al. 2015), that occurred on TRW areas. However, given the similarity in composition and structure between both categories (both of them include shrub formations, for example), this transition could also be driven by subjectivity in photointerpretation when defining TRW and GRSH across different CLC maps (EEA 2006b).

The loss of MOS was driven by a land use intensification process and its transformation mainly into ARA and PAS secondarily. Both transitions indicate anthropization processes and could pose higher pressure to the maintenance of biodiversity. The transition from MOS to ARA might be a consequence of the agricultural intensification process, probably related to the consolidation of agricultural land on monocultures in more productive areas promoted by the CAP policy. Our analyses of landscape structure showed that ARA suffered the second largest increase in average patch size within Natura 2000 after FOR. The consequences of this change go beyond the loss of priority habitats associated with MOS, as mentioned above, but also the simplification of the structure of the landscape (Brotons et al. 2004) and its consequences on biodiversity loss (e.g. Luoto et al. 2003; Delibes-Mateos et al. 2009; Santana et al. 2013). We were not able to assess whether this land consolidation process was also associated with an increase in intensity of use due to limitations of the thematic resolution of the CLC data set. However, the land consolidation process usually comes with more intensive agricultural practices (e.g. use of large machinery, pesticides and fertilizers) that do not align with the objectives of Natura 2000.

Our results show that most of the changes in land cover at the Natura 2000 scale occurred through swap processes that led to the changes in the composition and structure of the landscape. One of the strongest transitions, although not systematic according to the criteria used here, was the swap between FOR and GRSH. This swap was more intense towards southern Europe and co-occurred in space so the loss in FOR to GRSH was replaced by gain in FOR from GRSH in nearby areas (at least when assessed at the regional scale). Although these transitions might have had a small effect on land cover composition due to mutual replacement, they could pose important structural modifications in the configuration of the landscape associated with changes in the spatial allocation of these land cover categories (e.g. increasing or decreasing clumpiness and connectivity). There were, however, other swap processes, such as that observed between PAS and MOS, that did not co-occur at regional scale (MOS → PAS mainly occurred in Germany that was not accompanied by changes PAS → MOS; Fig. 30.2), leading to significant changes in land cover composition at that scale. Accounting for these swaps is important from a conservation point of view to avoid underestimating the dynamism of the landscape at local and regional scales.

We acknowledge that our results and conclusions are subjected to potential classification errors in CLC (Büttner et al. 2002) and then should be taken cautiously. However, the reliability of the different CLC databases has been assessed against alternative data sources resulting in >85% accuracy (Büttner et al. 2002, EEA 2006b). Although the thematic accuracy varies across land cover categories [e.g. 22 out of the 44 categories could not be validated for CLC2006 mostly artificial surfaces and wetlands (EEA 2006b)], the reliability of the largest CLC categories, such as arable land and coniferous forest have been reported to be between 90 and 95%, being also high for agroforestry and permanently irrigated land (EEA 2006b). We also acknowledge that some of the areas identified as stable over the time period evaluated could have been subjected to the management of different intensities, potentially leading to changes in ecological conditions (e.g. forest might have been thinned, inducing changes in forest structure and composition). However, these changes could not be detected with the information available for the whole of Europe, and therefore, could not be reported here.

30.6 Conclusions

We found that the landscape within the Natura 2000 network is dynamic, although land cover changes occur at different rates than in the immediate surroundings beyond its boundaries (here a 5 km buffer). Net changes in land cover categories at the whole Natura 2000 network scale were dominated by swap processes that defined most of the 20% land cover change observed. This uncovers a dynamic landscape at the local level within protected areas in Europe. Some of the changes (i.e. anthropization) could compromise the goals of this network and deserve attention from a policy and management perspective. On the other hand, naturalization transitions or internal changes within natural or semi-natural habitats call for a stronger recognition of dynamic processes in ensuring the conservation potential of the N2000 network in the long term. Our results point towards two different processes that require attention from the policy perspective: (1) more control of land use change within PAs to ensure intensification does not compromise conservation goals and (2) tackle more effectively the ecological and socio-economic effects of abandonment in less productive areas.

Acknowledgements V. H and A. M-O were supported by the Spanish Government through the 'Ramón y Cajal' and 'Juan de la Cierva' fellowship programs (RYC-2013-13979 and IJCI-2016-30349, respectively), and the FORESTCAST project (CGL2014-59742-C2-2-R).

References

Álvarez-Martínez JM, Suárez-Seoane S, Stoorvogel JJ, Calabuig EL (2014) Influence of land use and climate on recent forest expansion: a case study in the Eurosiberian-Mediterranean limit of north-west Spain. J Ecol 102:905–919. https://doi.org/10.1111/1365-2745.12257

Ameztegui A, Brotons L, Coll L (2010) Land-use changes as major drivers of mountain pine (*Pinus uncinata* Ram.) expansion in the Pyrenees. Global Ecol Biogeogr 19:632–641. https://doi.org/10.1111/j.1466-8238.2010.00550.x

Baldi A, Batary P (2011) Spatial heterogeneity and farmland birds: different perspectives in Western and Eastern Europe. Ibis 153:875–876. https://doi.org/10.1111/j.1474-919X.2011.01169.x

Brotons L, Manosa S, Estrada J (2004) Modeling the effects of irrigation schemes on the distribution of steppe birds in Mediterranean farmland. Biodivers Conserv 13:1039–1058. https://doi.org/10.1023/B:BIOC.0000

Büttner G, Feranec J, Jaffrain G (2002) CORINE land cover update 2000. Technical Report no 89. European Environmental Agency, Copenhagen, Denmark, 56 pp. http://www.eea.europa.eu/publications/technical_report_2002_89. Last access: 20 Feb 2019

Butsic V, Munteanu C, Griffiths P, Knorn J, Radeloff VC, Lieskovský J, Mueller D, Kuemmerle T (2016) The effect of protected areas on forest disturbance in the Carpathian Mountains 1985–2010. Conserv Biol 3:570–580

Dax T (2005) The redefinition of Europe's less favoured areas. In: Rural development in Europe—3rd annual conference—Funding European Rural Development in 2007–2013. MPRA paper no. 711

Delibes-Mateos M, Farfan MA, Olivero J, Marquez AL, Vargas JM (2009) Long-term changes in game species over a long period of transformation in the Iberian Mediterranean landscape. Environ Manage 43:1256–1268. https://doi.org/10.1007/s00267-009-9297-5

Dobrev V, Popgeorgiev G, Plachiyski D (2014) Effects of the Common Agricultural Policy on the coverage of grassland habitats in Besaparski Ridove Special Protection Area (Natura 2000), Southern Bulgaria. Acta Zool Bulg 5:147–155

Donald PF, Sanderson FJ, Burfield IJ, van Bommel FPJ (2006) Further evidence of continent-wide impacts of agricultural intensification on European farmland birds, 1990–2000. Agr Ecosyst Environ 116:189–196. https://doi.org/10.1016/j.agee.2006.02.007

EEA (2006a) Land accounts for Europe 1990–2000: towards integrated land and ecosystem accounting. In: European Environment Agency Report No.11. OPOCE, Luxembourg, 107 pp. http://www.eea.europa.eu/publications/eeareport200611. Accessed August 2017

EEA (2006b) The thematic accuracy of CORINE Land Cover 2000. Assessment using LUCAS (Land use/cover area frame statistical survey). Technical report No. 7. ISBN 1725-2237. European Environment Agency

EEA (2012) Protected areas in Europe: an overview. EEA Report No 5/2012. EU, Luxembourg

Feranec J, Jaffrain G, Soukup T, Hazeu G (2010) Determining changes and flows in European landscapes 1990–2000 using CORINE land cover data. Appl Geogr 30:19–35. https://doi.org/10.1016/j.apgeog.2009.07.003

Fernández N, Navarro LM, Pereira HM (2017) Rewilding: a call for boosting ecological complexity in conservation. Conserv Lett 10:276–278. https://doi.org/10.1111/conl.12374

Flores Ribeiro P, Catela Nunes L, Beja P, Reino L, Santana J, Moreira F, Lima Santos J (2018) A spatially explicit choice model to assess the impact of conservation policy on high nature value farming systems. Ecol Econ 145:331–338. https://doi.org/10.1016/j.ecolecon.2017.11.011

Forest Europe (2015) State of Europe's Forests. Available at: http://www.foresteurope.org/docs/fullsoef2015.pdf. Last accessed June 2018

Gellrich M, Zimmermann NE (2007) Investigating the regional-scale pattern of agricultural land abandonment in the Swiss mountains: a spatial statistical modelling approach. Landscape Urban Plan 79:65–76. https://doi.org/10.1016/j.landurbplan.2006.03.004

Gellrich M, Baur P, Koch B, Zimmermann NE (2007) Agricultural land abandonment and natural forest re-growth in the Swiss mountains: a spatially explicit economic analysis. Agr Ecosyst Environ 118:93–108. https://doi.org/10.1016/j.agee.2006.05.001

Gómez O, Páramo F (2005) Environmental accounting. Methodological guidebook. In: Data processing of land cover flows. Internal Report of the European Topic Centre on Terrestrial Environment, with the support of the EEA, Barcelona. http://www.eea.europa.eu/data-and-maps/data/land-cover-accounts-leac-based-on-corine-land-cover-changes-database-1990-2000/. Accessed 20 Feb 2019

Halada L, Evans D, Romão C, Petersen JE (2011) Which Habitats of European Importance Depend on Agricultural Practices? Biodivers Conserv 20:2365–2378

Hatna E, Bakker MM (2011) Abandonment and expansion of Arable Land in Europe. Ecosystems 14:720–731. https://doi.org/10.1007/s10021-011-9441-y

Hermoso V, Morán-Ordóñez A, Ll Brotons (2018) Assessing the role of Natura 2000 at maintaining dynamic landscapes in Europe over the last two decades: implications for conservation. Landsc Ecol 33:1447–1460. https://doi.org/10.1007/s10980-018-0683-3

Hill J, Stellmes M, Udelhoven Th, Röder A., Sommer S (2008) Mediterranean desertification and land degradation. Mapping related land use change syndromes based on satellite observations. Glob Planet Chang 64:146–157

Kallimanis AS, Touloumis K, Tzanopoulos J, Mazaris A, Apostolopoulou E, Stefanidou S, Scott A.V, Potts SG, Pantis JD (2014) Vegetation coverage change in the EU: patterns inside and outside Natura 2000 protected areas. Biodivers Conserv 24:579–591

Lasanta T, Arnaéz J, Errea MP, Ortigosa L, Ruiz-Flaño P (2009) Mountain pastures, environmental degradation, and landscape remediation: the example of a Mediterranean policy initiative. App Geogr 29:308–319. https://doi.org/10.1016/j.apgeog.2008.09.006

Luoto M, Rekolainen S, Aakkula J, Pykala J (2003) Loss of plant species richness and habitat connectivity in grasslands associated with agricultural change in Finland. Ambio 32:447–452. https://doi.org/10.1579/0044-7447-32.7.447

MacDonald D, Crabtree JR, Wiesinger G, Dax T, Stamou N, Fleury P, Gutierrez Lazpita J, Gibon A (2000) Agricultural abandonment in mountain areas of Europe: environmental consequences and policy response. J Environ Manag 59:47–69. https://doi.org/10.1006/jema.1999.0335

Martínez-Fernández J, Ruiz-Benito P, Zavala MA (2015) Recent land cover changes in Spain across biogeographical regions and protection levels: implications for conservation policies. Land Use Policy 44:62–75. https://doi.org/10.1016/j.landusepol.2014.11.021

McGarigal K, Cushman SA, Ene E (2012) FRAGSTATS v4: Spatial Pattern Analysis Program for Categorical and Continuous Maps. Computer software program produced by the authors at the University of Massachusetts, Amherst. Available at the following web site: http://www.umass.edu/landeco/research/fragstats/fragstats.html

Mitchley J, Tzanopoulos J, Cooper T (2007) Reconciling the conservation of biodiversity with declining agricultural use in the Mountains of Europe: the challenge of interdisciplinary research. In: Price MF (ed) Mountain research and management: integrated approaches. Earthscan Publication, London, pp 164–177

Navarro LM, Pereira HM (2012) Rewilding abandoned landscapes in Europe. Ecosystems 15:900–912. https://doi.org/10.1007/s10021-012-9558-7

Pontius RG, Shusas E, McEachern M (2004) Detecting important categorical land changes while accounting for persistence. Agr Ecosyst Environ 101:251–268. https://doi.org/10.1016/j.agee.2003.09.008

Pressey RL (1994) Ad hoc reservations: forward or backward steps in developing representative reserve systems? Conserv Biol 8:662–668. https://doi.org/10.1046/j.1523-1739.1994.08030662.x

Santana J, Reino L, Stoate C, Borralho R, Carvalho CR, Schindler S, Moreira F, Bugalho MN, Ribeiro PF, Santos JL, Vaz A, Morgado R, Porto M, Beja P (2013) Mixed effects of long-term conservation investment in Natura 2000 farmland. Conserv Lett 7:1–10. https://doi.org/10.1111/conl.12077

Stoate C, Báldi A, Beja P, Boatman ND, Herzon I, Van Doorn A, De Snoo GR, Rakosy L, Ramwell C (2009) Ecological impacts of early 21st century agricultural change in Europe—a review. J Environ Manage 91:22–46. https://doi.org/10.1016/j.jenvman.2009.07.005

Tedim F, Xanthopoulos G, Leone V (2015) Forest fires in Europe: facts and challenges. In: Paton D (ed) Wildfire hazards, risks and disasters, pp. 77–99. Elsevier, Amsterdam

Young J, Richards C, Fischer A, Halada L, Kull T, Kuzniar A, Tartes U, Uzunov Y, Watt A (2007) Conflicts between biodiversity conservation and human activities in the central and eastern European countries. Ambio 36:545–550. https://doi.org/10.1579/0044-7447(2007)36%5B545%3ACBBCAH%5d2%2E0%2ECO%3B2

MIX
Papier aus verantwortungsvollen Quellen
Paper from responsible sources
FSC® C105338

If you have any concerns about our products,
you can contact us on
ProductSafety@springernature.com

In case Publisher is established outside the EU,
the EU authorized representative is:
Springer Nature Customer Service Center GmbH
Europaplatz 3, 69115 Heidelberg, Germany

Printed by Libri Plureos GmbH
in Hamburg, Germany